Format for Spreadsheet Functions on Excel®

Present worth:	**Contents of ( )**
$= \mathbf{PV}(i\%,n,A,F)$	for constant A series; single F value
$= \mathbf{NPV}(i\%,\mathbf{second\_cell:last\_cell}) + \mathbf{first\_cell}$	for varying cash flow series
Future worth:	
$= \mathbf{FV}(i\%,n,A,P)$	for constant A series; single P value
Annual worth:	
$= \mathbf{PMT}(i\%,n,P,F)$	for single amounts with no A series
$= \mathbf{PMT}(i\%,n,\mathbf{NPV})$	to find AW from NPV; embed NPV function
Number of periods (years):	
$= \mathbf{NPER}(i\%,A,P,F)$	for constant A series; single P and F

(Note: The PV, FV, and PMT functions change the sense of the sign. Place a minus in front of the function to retain the same sign.)

Rate of return:	
$= \mathbf{RATE}(n,A,P,F)$	for constant A series; single P and F
$= \mathbf{IRR}(\mathbf{first\_cell:last\_cell})$	for varying cash flow series
Interest rate:	
$= \mathbf{EFFECT}(r\%,m)$	for nominal r, compounded m times per period
$= \mathbf{NOMINAL}(i\%,m)$	for effective annual i, compounded m times per year
Depreciation:	
$= \mathbf{SLN}(P,S,n)$	straight line depreciation for each period
$= \mathbf{DDB}(P,S,n,t,d)$	double declining balance depreciation for period t at rate d (optional)
$= \mathbf{DB}(P,S,n,t)$	declining balance, rate determined by the function
$= \mathbf{VBD}(P,0,n,\mathbf{MAX}(0,t-1.5),$ $\mathbf{MIN}(n,t-0.5),d)$	MACRS depreciation for year t at rate d for DDB or DB method
Logical IF function:	
$= \mathbf{IF}(\mathbf{logical\_test,value\_if\_true,value\_if\_false})$	for logical two-branch operations

Relations for Discrete Cash Flows with End-of-Period Compounding

Type	Find/Given	Factor Notation and Formula	Relation	Sample Cash Flow Diagram
Single Amount	F/P Compound amount	$(F/P,i,n) = (1+i)^n$	$F = P(F/P,i,n)$	
	P/F Present worth	$(P/F,i,n) = \dfrac{1}{(1+i)^n}$	$P = F(P/F,i,n)$ (Sec. 2.1)	
Uniform Series	P/A Present worth	$(P/A,i,n) = \dfrac{(1+i)^n - 1}{i(1+i)^n}$	$P = A(P/A,i,n)$	
	A/P Capital recovery	$(A/P,i,n) = \dfrac{i(1+i)^n}{(1+i)^n - 1}$	$A = P(A/P,i,n)$ (Sec. 2.2)	
	F/A Compound amount	$(F/A,i,n) = \dfrac{(1+i)^n - 1}{i}$	$F = A(F/A,i,n)$	
	A/F Sinking fund	$(A/F,i,n) = \dfrac{i}{(1+i)^n - 1}$	$A = F(A/F,i,n)$ (Sec. 2.3)	
Arithmetic Gradient	P_G/G Present worth	$(P/G,i,n) = \dfrac{(1+i)^n - in - 1}{i^2(1+i)^n}$	$P_G = G(P/G,i,n)$	
	A_G/G Uniform series	$(A/G,i,n) = \dfrac{1}{i} - \dfrac{n}{(1+i)^n - 1}$ (Gradient only)	$A_G = G(A/G,i,n)$ (Sec. 2.5)	
Geometric Gradient	P_g/A_1 and g Present worth	$P_g = \begin{cases} \dfrac{A_1\left[1 - \left(\dfrac{1+g}{1+i}\right)^n\right]}{i - g} & g \neq i \\[4mm] A_1\dfrac{n}{1+i} & g = i \end{cases}$ (Gradient and base A_1)	$g \neq i$ $g = i$ (Sec. 2.6)	

Eighth Edition

ENGINEERING ECONOMY

Leland Blank, P. E.
Texas A & M University
American University of Sharjah, United Arab Emirates

Anthony Tarquin, P. E.
University of Texas at El Paso

McGraw Hill Education

ENGINEERING ECONOMY, EIGHTH EDITION

Published by McGraw-Hill Education, 2 Penn Plaza, New York, NY 10121. Copyright © 2018 by McGraw-Hill Education. All rights reserved. Printed in the United States of America. Previous editions © 2012, 2005, and 2002. No part of this publication may be reproduced or distributed in any form or by any means, or stored in a database or retrieval system, without the prior written consent of McGraw-Hill Education, including, but not limited to, in any network or other electronic storage or transmission, or broadcast for distance learning.

Some ancillaries, including electronic and print components, may not be available to customers outside the United States.

This book is printed on acid-free paper.

1 2 3 4 5 6 7 8 9 LWI 21 20 19 18 17

ISBN 978-0-07-352343-9
MHID 0-07-352343-7

Chief Product Officer, SVP Products & Markets: *G. Scott Virkler*
Vice President, General Manager, Products & Markets: *Marty Lange*
Vice President, Content Design & Delivery: *Betsy Whalen*
Managing Director: *Thomas Timp*
Global Brand Manager: *Raghothaman Srinivasan*
Director, Product Development: *Rose Koos*
Product Developer: *Jolynn Kilburg*
Marketing Manager: *Nick McFadden*
Director of Digital Content: *Chelsea Haupt, Ph.D.*
Director, Content Design & Delivery: *Linda Avenarius*
Program Manager: *Lora Neyens*
Content Project Managers: *Jane Mohr, Emily Windelborn, and Sandra Schnee*
Buyer: *Jennifer Pickel*
Design: *Studio Montage, St. Louis, MO*
Content Licensing Specialist: *Lorraine Buczek*
Cover Image: *pocket watch: Andrew Unangst/Getty Images; currency: © Frank van den Bergh/Getty Images*
Compositor: *MPS Limited*
Printer: *LSC Communications*

All credits appearing on page or at the end of the book are considered to be an extension of the copyright page.

Library of Congress Cataloging-in-Publication Data

Blank, Leland T. | Tarquin, Anthony J.
   Engineering economy / Leland Blank, P.E., Texas A & M University,
   American University of Sharjah, United Arab Emirates, Anthony Tarquin,
   P.E., University of Texas at El Paso.
   Eighth edition. | New York : McGraw-Hill Education, [2017] | Includes index.
   LCCN 2016044149| ISBN 9780073523439 (acid-free paper) | ISBN 0073523437
   (acid-free paper)
   LCSH: Engineering economy—Textbooks. | Economics—Textbooks.
   LCC TA177.4 .B58 2017 | DDC 658.15—dc23
   LC record available at https://lccn.loc.gov/2016044149

The Internet addresses listed in the text were accurate at the time of publication. The inclusion of a website does not indicate an endorsement by the authors or McGraw-Hill Education, and McGraw-Hill Education does not guarantee the accuracy of the information presented at these sites.

Required=Results

©Getty Images/iStockphoto

McGraw-Hill Connect®
Learn Without Limits

Connect is a teaching and learning platform that is proven to deliver better results for students and instructors.

Connect empowers students by continually adapting to deliver precisely what they need, when they need it, and how they need it, so your class time is more engaging and effective.

73% of instructors who use **Connect** require it; instructor satisfaction **increases** by 28% when **Connect** is required.

Connect's Impact on Retention Rates, Pass Rates, and Average Exam Scores

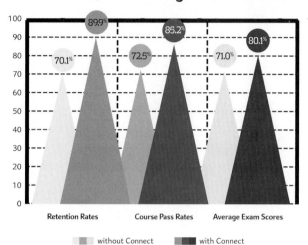

Using **Connect** improves retention rates by **19.8%**, passing rates by **12.7%**, and exam scores by **9.1%**.

Analytics

Connect Insight®

Connect Insight is Connect's new one-of-a-kind visual analytics dashboard—now available for both instructors and students—that provides at-a-glance information regarding student performance, which is immediately actionable. By presenting assignment, assessment, and topical performance results together with a time metric that is easily visible for aggregate or individual results, Connect Insight gives the user the ability to take a just-in-time approach to teaching and learning, which was never before available. Connect Insight presents data that empowers students and helps instructors improve class performance in a way that is efficient and effective.

Impact on Final Course Grade Distribution

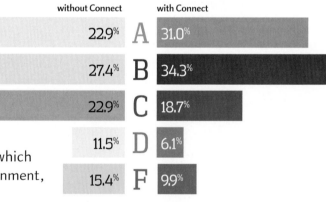

without Connect		with Connect
22.9%	A	31.0%
27.4%	B	34.3%
22.9%	C	18.7%
11.5%	D	6.1%
15.4%	F	9.9%

Students can view their results for any **Connect** course.

Mobile

Connect's new, intuitive mobile interface gives students and instructors flexible and convenient, anytime–anywhere access to all components of the Connect platform.

Adaptive

THE **ADAPTIVE** **READING EXPERIENCE**
DESIGNED TO TRANSFORM THE WAY STUDENTS READ

More students earn **A's** and **B's** when they use McGraw-Hill Education **Adaptive** products.

SmartBook®

Proven to help students improve grades and study more efficiently, SmartBook contains the same content within the print book, but actively tailors that content to the needs of the individual. SmartBook's adaptive technology provides precise, personalized instruction on what the student should do next, guiding the student to master and remember key concepts, targeting gaps in knowledge and offering customized feedback, and driving the student toward comprehension and retention of the subject matter. Available on tablets, SmartBook puts learning at the student's fingertips—anywhere, anytime.

Over **8 billion questions** have been answered, making McGraw-Hill Education products more intelligent, reliable, and precise.

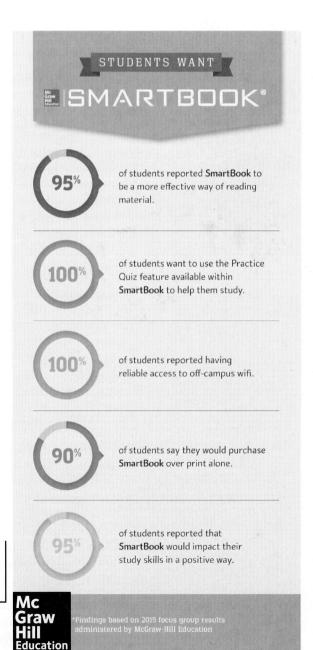

STUDENTS WANT

SMARTBOOK®

95% of students reported **SmartBook** to be a more effective way of reading material.

100% of students want to use the Practice Quiz feature available within **SmartBook** to help them study.

100% of students reported having reliable access to off-campus wifi.

90% of students say they would purchase **SmartBook** over print alone.

95% of students reported that **SmartBook** would impact their study skills in a positive way.

McGraw Hill Education

*Findings based on 2015 focus group results administered by McGraw-Hill Education

www.mheducation.com

CONTENTS

LEARNING STAGE 2

BASIC ANALYSIS TOOLS

LEARNING STAGE 2

EPILOGUE: SELECTING THE BASIC ANALYSIS TOOL

LEARNING STAGE 3

MAKING BETTER DECISIONS

PREFACE TO EIGHTH EDITION

This new edition includes the time-tested approach and topics of previous editions and introduces significantly new print and electronic features useful to learning about and successfully applying the exciting field of engineering economics. Money makes a huge difference in the life of a corporation, an individual, and a government. Learning to understand, analyze, and manage the money side of any project is vital to its success. To be professionally successful, every engineer must be able to deal with the time value of money, economic facts, inflation, cost estimation, tax considerations, as well as spreadsheet and calculator use. This book is a great help to the learner and the instructor in accomplishing these goals by using easy-to-understand language, simple graphics, and online features.

What's New and What's Best ● ● ●

This eighth edition has new digital features and retains the time-tested features that make the book reliable and easy to use. Plus the supporting online materials are updated to enhance the teaching and learning experience.

Exciting new features in print:
- All new end-of-chapter problems
- Expanded questions for either review or preparation for the Fundamentals of Engineering (FE) Exam

Valuable new features in digital content:
- McGraw-Hill Connect
  - Online video presentations with closed captioning to serve as learning support tools
  - Algorithmic end-of chapter problems that present a new set of parameters and estimates every time the problem is opened
  - SmartBook, an adaptive reading experience

Familiar features retained in this edition:
- Easy-to-read language
- End-of-chapter case studies
- Ethical considerations in economic analyses
- Progressive examples for improved understanding of concepts
- Hand and spreadsheet example solutions
- Spreadsheet solutions with on-image comments and Excel® functions
- Vital concepts and guidelines located in margins and appendix
- Flexible chapter ordering

How to Use This Text ● ● ●

This textbook is best suited for a one-semester or one-quarter undergraduate course. Students should be at the sophomore level or above with a basic understanding of engineering concepts and terminology. A course in calculus is not necessary; however, knowledge of the concepts in advanced mathematics and elementary probability will make the topics more meaningful.

Practitioners and professional engineers who need a refresher in economic analysis and cost estimation will find this book very useful as a reference document as well as a learning medium.

Chapter Organization and Coverage Options ● ● ●

The textbook contains 19 chapters arranged into four learning stages, *as indicated in the flowchart* on the next page, and five appendices. Each chapter starts with a statement of purpose and specific learning outcomes for each section. Chapters include a summary, numerous end-of-chapter

CHAPTERS IN EACH LEARNING STAGE

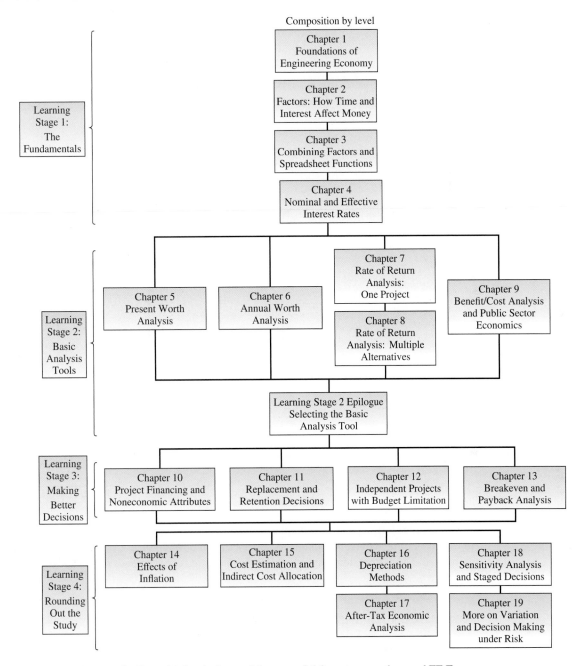

problems (essay and numerical), multiple-choice problems useful for course review and FE Exam preparation, and a case study.

The appendices are important elements of learning for this text:

Appendix A Using Spreadsheets and Microsoft Excel®

Appendix B Basics of Accounting reports and business ratios

Appendix C Code of Ethics for Engineers (from NSPE)

Appendix D Alternate methods for equivalence calculations

Appendix E Glossary of concepts and terms

There is considerable flexibility in the sequencing of topics and chapters once the first six chapters are covered, *as shown in the progression graphic* on the next page. If the course is designed to emphasize sensitivity and risk analysis, Chapters 18 and 19 can be covered immediately after

CHAPTER AND TOPIC PROGRESSION OPTIONS

Topics may be introduced at the point indicated or any point thereafter
(Alternative entry points are indicated by ←– – –)

Numerical progression through chapters		Inflation	Cost Estimation	Taxes and Depreciation	Sensitivity, Staged Decisions, and Risk

1. Foundations
2. Factors
3. More Factors
4. Nominal and Effective *i*
5. Present Worth
6. Annual Worth

7. Rate of Return
8. More ROR
9. Benefit/Cost

10. Financing and Noneconomic Attributes
11. Replacement
12. Capital Budgeting
13. Breakeven and Payback

14. Inflation

15. Estimation

16. Depreciation
17. After-Tax

18. Sensitivity, Decision Trees, and Real Options
19. Risk and Simulation

Learning Stage 2 (Chapter 9) is completed. If depreciation and tax emphasis are vitally important to the goals of the course, Chapters 16 and 17 can be covered once Chapter 6 (annual worth) is completed. The progression graphic can help in the design of the course content and topic ordering.

Resources for Instructors and Students ● ● ●

LEARNING OUTCOMES:

- Each chapter begins with a purpose, list of topics, and learning outcomes (ABET style) for each corresponding section. This behavioral-based approach sensitizes the reader to what is ahead, leading to improved understanding and learning.

☑ CONCEPTS AND GUIDELINES:

- To highlight the fundamental building blocks of the course, a checkmark and title in the margin call attention to particularly important concepts and decision-making guidelines. Appendix E includes a brief description of each fundamental concept.

IN-CHAPTER EXAMPLES:

- Numerous in-chapter examples throughout the book reinforce the basic concepts and make understanding easier. In many cases, the example is solved using separately marked hand and spreadsheet solutions.

PE *PROGRESSIVE EXAMPLES:*

- Several chapters include a progressive example—a more detailed problem statement introduced at the beginning of the chapter and expanded upon throughout the chapter in specially marked examples. This approach illustrates different techniques and some increasingly complex aspects of a real-world problem.

 ONLINE PRESENTATIONS:

- An icon in the margin indicates the availability of an animated voice-over slide presentation that summarizes the material in the section and provides a brief example for learners who need a review or prefer video-based materials. Presentations are keyed to the sections of the text.

SPREADSHEETS:

- The text integrates spreadsheets to show how easy they are to use in solving virtually any type of engineering economic analysis problem. Cell tags or full cells detail built-in functions and relations developed to solve a specific problem.

FE EXAM AND COURSE REVIEWS:

- Each chapter concludes with several multiple-choice, FE Exam–style problems that provide a simplified review of chapter material. Additionally, these problems cover topics for test reviews and homework assignments.

Digital Resources ● ● ●

ALGORITHMIC END-OF-CHAPTER PROBLEMS:

- Available through the online homework platform Connect, algorithmic end-of-chapter problems can be assigned for homework, practice, exams, or quizzes. Problems include algorithmically generated values so that each student receives different numbers, while responses are auto-graded to provide immediate feedback to the student.

SMARTBOOK:

- Also available through Connect is SmartBook which contains the same content within the print book, but actively tailors that content to the needs of the individual. SmartBook's adaptive technology provides precise, personalized instruction on what the student should do next, guiding the student to master and remember key concepts, targeting gaps in knowledge and offering customized feedback to drive the student toward comprehension and retention of the subject matter.

ACKNOWLEDGMENT OF CONTRIBUTORS

It takes the input and efforts of many individuals to make significant improvements in a textbook. We wish to give special thanks to the following persons for their contributions to this edition.

Jack Beltran, Beltran and Associates
Neal McCollom, University of Texas at Arlington
Sallie Sheppard, Texas A&M University

If you discover errors that require correction in the next printing of the textbook or in updates of the online resources, please contact us. We hope you find the contents of this edition helpful in your academic and professional activities.

Leland Blank   lelandblank@yahoo.com
Anthony Tarquin   atarquin@utep.edu

LEARNING STAGE 1

The Fundamentals

The fundamentals of engineering economy are introduced in these chapters. When you have completed stage 1, you will be able to understand and work problems that account for the **time value of money, cash flows** occurring at different times with different amounts, and **equivalence** at different interest rates. The techniques you master here form the basis of how an engineer in any discipline can take **economic value** into account in virtually any project environment.

The factors commonly used in all engineering economy computations are introduced and applied here. Combinations of these factors assist in moving monetary values forward and backward through time and at different interest rates. Also, after these chapters, you should be comfortable using many of the spreadsheet functions.

Many of the terms common to economic decision making are introduced in learning stage 1 and used in later chapters. A checkmark icon in the margin indicates that a new **concept or guideline** is introduced at this point.

CHAPTER 1

Foundations of Engineering Economy

Malcolm Fife/age fotostock

LEARNING OUTCOMES

Purpose: Understand and apply fundamental concepts and use the terminology of engineering economics.

SECTION	TOPIC	LEARNING OUTCOME
1.1	Description and role	• Define engineering economics and the time value of money; identify areas of application.
1.2	Engineering economy study approach	• Understand and identify the steps in an engineering economy study.
1.3	Ethics and economics	• Identify areas in which economic decisions can present questionable ethics.
1.4	Interest rate	• Perform calculations for interest rates and rates of return.
1.5	Terms and symbols	• Identify and use engineering economic terminology and symbols.
1.6	Cash flows	• Understand cash flows and how to graphically represent them.
1.7	Economic equivalence	• Describe and calculate economic equivalence.
1.8	Simple and compound interest	• Calculate simple and compound interest amounts for one or more time periods.
1.9	MARR and opportunity cost	• State the meaning and role of Minimum Attractive Rate of Return (MARR) and opportunity costs.
1.10	Spreadsheet functions	• Identify and use some Excel© functions commonly applied in engineering economics.

The need for engineering economy is primarily motivated by the work that engineers do in performing analyses, synthesizing, and coming to a conclusion as they work on projects of all sizes. In other words, engineering economy is at the heart of **making decisions**. These decisions involve the fundamental elements of **cash flows of money, time, and interest rates.** This chapter introduces the basic concepts and terminology necessary for an engineer to combine these three essential elements in organized, mathematically correct ways to solve problems that will lead to better decisions.

1.1 Why Engineering Economy and the Time Value of Money are Important ● ● ●

Decisions are made routinely to choose one alternative over another by engineers on the job; by managers who supervise the activities of others; by corporate presidents who operate a business; and by government officials who work for the public good. Most decisions involve money, called **capital** or **capital funds,** which is usually limited in amount. The decision of where and how to invest this limited capital is motivated by a primary goal of **adding value** as future, anticipated results of the selected alternative are realized. Engineers play a vital role in capital investment decisions based upon their ability and experience to design, analyze, and synthesize. The factors upon which a decision is based are commonly a combination of economic and noneconomic elements. Engineering economy deals with the economic factors. By definition,

Engineering economy involves formulating, estimating, and evaluating the expected economic outcomes of alternatives designed to accomplish a defined purpose. Mathematical techniques simplify the economic evaluation of alternatives.

Because the formulas and techniques used in engineering economics are applicable to all types of money matters, they are equally useful in business and government, as well as for individuals. Therefore, besides applications to projects in your future jobs, what you learn from this book and in this course may well offer you an economic analysis tool for making personal decisions such as car purchases, house purchases, and purchases on credit for enjoyment, e.g., electronics, games, drones, vacations, etc.

To be financially literate is very important as an engineer and as a person. Unfortunately many people do not have the fundamental understanding of concepts such as financial risk and diversification, inflation, numeracy, and compound interest. You will learn and apply these basic concepts, and more, through the study of engineering economy. A comprehensive Standard & Poors[1] study reported in 2015 evaluated the financial literacy of people worldwide using the survey results of more than 150,000 people interviewed in 148 countries. Results indicated that worldwide only one out of three adults are able to answer correctly three out of four simple questions that indicate an understanding in the areas mentioned above—inflation, compound interest, risk, and diversification. Scandinavian countries (e.g., Denmark, Norway, and Sweden), plus Germany, Canada, and the United Kingdom have acceptably good scores (67% to 71% financial literacy), the United States is mediocre at 57% (14th worldwide), while countries such as Cambodia, Armenia, and Haiti are low (15% to 18%). An obvious conclusion is that a college graduate in engineering anywhere in the world must be financially literate to responsibly and successfully function in his or her professional and personal activities.

Other terms that mean the same as *engineering economy* are *engineering economic analysis, capital allocation study, economic analysis,* and similar descriptors.

People make decisions; computers, mathematics, concepts, and guidelines assist people in their decision-making process. Since most decisions affect what will be done, the time frame of engineering economy is primarily the **future.** Therefore, the numbers used in engineering economy are **best estimates of what is expected to occur.** The estimates and the decision usually involve four essential elements:

Cash flows

Times of occurrence of cash flows

Interest rates for time value of money

Measure of worth for selecting an alternative

[1]Klapper, L., Lusardi, A., and van Oudheusden, P. "Financial Literacy around the World: Insights from the Standard & Poor's Ratings Services Global Financial Literacy Survey", 2015, Standards & Poors, and Gallup World Poll. (accessed December 2015). http://www.finlit.mhfi.com.

Since the estimates of cash flow amounts and timing are about the future, they will be somewhat different than what is actually observed, due to changing circumstances and unplanned events. In short, the variation between an amount or time estimated now and that observed in the future is caused by the stochastic (random) nature of all economic events. **Sensitivity analysis** is utilized to determine how a decision might change according to varying estimates, especially those expected to vary widely.

The criterion used to select an alternative in engineering economy for a specific set of estimates is called a **measure of worth.** The measures developed and used in this text are

Present worth (PW)	Future worth (FW)	Annual worth (AW)
Rate of return (ROR)	Benefit/cost (B/C)	Capitalized cost (CC)
Payback period	Profitability index	Economic value added (EVA)

All these measures of worth account for the fact that money makes money over time. This is the concept of the **time value of money.**

Time value of money

It is a well-known fact that money **makes** money. The time value of money explains the change in the amount of money **over time** for funds that are owned (invested) or owed (borrowed). **This is the most important concept in engineering economy.**

The time value of money is very obvious in the world of economics. If we decide to invest capital (money) in a project today, we inherently expect to have more money in the future than we invested. If we borrow money today, in one form or another, we expect to return the original amount plus some additional amount of money. An engineering economic analysis can be performed on future estimated amounts or on past cash flows to determine if a specific measure of worth, e.g., rate of return, was achieved.

Engineering economics is applied in an extremely wide variety of situations. Samples are:

- Equipment purchases and leases
- Chemical processes
- Cyber security
- Construction projects
- Airport design and operations
- Sales and marketing projects
- Transportation systems of all types
- Product design
- Wireless and remote communication and control
- Manufacturing processes
- Safety systems
- Hospital and healthcare operations
- Quality assurance
- Government services for residents and businesses

In short, any activity that has money associated with it—which is just about everything—is a reasonable topic for an engineering economy study.

EXAMPLE 1.1

Cyber security is an increasingly costly dimension of doing business for many retailers and their customers who use credit and debit cards. A 2014 data breach of U.S.-based Home Depot involved some 56 million cardholders. Just to investigate and cover the immediate direct costs of this identity theft amounted to an estimated $62,000,000, of which $27,000,000 was recovered by insurance company payments. This does not include indirect costs, such as, lost future business, costs to banks, and cost to replace cards. If a cyber security vendor had proposed in 2006 that a $10,000,000 investment in a malware detection system could guard the company's computer and payment systems from such a breach, would it have kept up with the rate of inflation estimated at 4% per year?

Solution

As a result of this data breach, Home Depot experienced a direct out-of-pocket cost of $35,000,000 after insurance payments. In this chapter and in Chapter 2, you will learn how to determine the future equivalent of money at a specific rate. In this case, the estimate of $10,000,000 after 8 years (from 2006 to 2014) at an inflation rate of 4% is equivalent to $13,686,000.

The 2014 equivalent cost of $13.686 million is significantly less than the out-of-pocket loss of $35 million. The conclusion is that the company should have spent $10 million in 2006. Besides, there may be future breaches that the installed system will detect and eliminate.

This is an extremely simple analysis; yet, it demonstrates that at a very elementary level, it is possible to determine whether an expenditure at one point in time is economically worthwhile at some time in the future. In this situation, we validated that a previous expenditure (malware detection system) should have been made to overcome an unexpected expenditure (cost of the data breach) at a current time, 2014 here.

1.2 Performing an Engineering Economy Study ● ● ●

An engineering economy study involves many elements: problem identification, definition of the objective, cash flow estimation, financial analysis, and decision making. Implementing a structured procedure is the best approach to select the best solution to the problem.

The steps in an engineering economy study are as follows:

1. Identify and understand the problem; identify the objective of the project.
2. Collect relevant, available data and define viable solution alternatives.
3. Make realistic cash flow estimates.
4. Identify an economic measure of worth criterion for decision making.
5. Evaluate each alternative; consider noneconomic factors; use sensitivity analysis as needed.
6. Select the best alternative.
7. Implement the solution and monitor the results.

Technically, the last step is not part of the economy study, but it is, of course, a step needed to meet the project objective. There may be occasions when the best economic alternative requires more capital funds than are available, or significant noneconomic factors preclude the most economic alternative from being chosen. Accordingly, steps 5 and 6 may result in selection of an alternative different from the economically best one. Also, sometimes more than one project may be selected and implemented. This occurs when projects are independent of one another. In this case, steps 5 through 7 vary from those above. Figure 1–1 illustrates the steps above for one alternative. Descriptions of several of the elements in the steps are important to understand.

Problem Description and Objective Statement   A succinct statement of the problem and primary objective(s) is very important to the formation of an alternative solution. As an illustration, assume the problem is that a coal-fueled power plant must be shut down by 2025 due to the production of excessive sulfur dioxide. The objectives may be to generate the forecasted electricity needed for 2025 and beyond, plus to not exceed all the projected emission allowances in these future years.

Alternatives   These are stand-alone descriptions of viable solutions to problems that can meet the objectives. Words, pictures, graphs, equipment and service descriptions, simulations, etc. define each alternative. The best estimates for parameters are also part of the alternative. Some parameters include equipment first cost, expected life, salvage value (estimated trade-in, resale, or market value), and annual operating cost (AOC), which can also be termed maintenance and operating (M&O) cost, and subcontract cost for specific services. If changes in income (revenue) may occur, this parameter must be estimated.

Detailing all viable alternatives at this stage is crucial. For example, if two alternatives are described and analyzed, one will likely be selected and implementation initiated. If a third, more attractive method that was available is later recognized, a wrong decision was made.

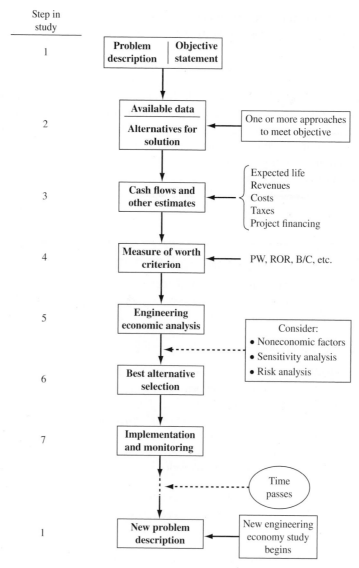

Figure 1–1

Steps in an engineering economy study.

Cash Flows   All cash flows are estimated for each alternative. Since these are future expenditures and revenues, the results of step 3 usually prove to be inaccurate when an alternative is actually in place and operating. When cash flow estimates for specific parameters are expected to vary significantly from a *point estimate* made now, risk and sensitivity analyses (step 5) are needed to improve the chances of selecting the best alternative. Sizable variation is usually expected in estimates of revenues, AOC, salvage values, and subcontractor costs. Estimation of costs is discussed in Chapter 15, and the elements of variation (risk) and sensitivity analysis are included throughout the text.

Engineering Economy Analysis   The techniques and computations that you will learn and use throughout this text utilize the cash flow estimates, time value of money, and a selected measure of worth. The result of the analysis will be one or more numerical values; this can be in one of several terms, such as money, an interest rate, number of years, or a probability. In the end, a specific measure of worth mentioned in the previous section will be used to select the best alternative.

Before an economic analysis technique is applied to the cash flows, some decisions about what to include in the analysis must be made. Two important possibilities are taxes and inflation. Federal, state or provincial, county, and city taxes will impact the costs of every alternative. An after-tax analysis includes some additional estimates and methods compared to a before-tax analysis. If taxes and inflation are expected to impact all alternatives equally, they may be disregarded in the analysis. However, if the size of these projected costs is important, taxes and inflation should be considered. Also, if the impact of inflation over time is important to the decision, an additional set of computations must be added to the analysis; Chapter 14 covers the details.

Selection of the Best Alternative   The measure of worth is a primary basis for selecting the best economic alternative. For example, if alternative A has a rate of return (ROR) of 15.2% per year and alternative B will result in an ROR of 16.9% per year, B is better economically. However, there can always be **noneconomic** or **intangible factors** that must be considered and that may alter the decision. There are many possible noneconomic factors; some typical ones are:

- Market pressures, such as need for an increased international presence
- Availability of certain resources, e.g., skilled labor force, water, power, tax incentives
- Government laws that dictate safety, environmental, legal, or other aspects
- Corporate management's or the board of director's interest in a particular alternative
- Goodwill offered by an alternative toward a group, for example, employees, union, county, etc.

As indicated in Figure 1–1, once all the economic, noneconomic, and risk factors have been evaluated, a final decision of the "best" alternative is made.

At times, only one viable alternative is identified. In this case, the **do-nothing (DN) alternative** must be included in the evaluation and may be chosen provided the measure of worth and other factors result in the alternative being a poor choice. The do-nothing alternative maintains the status quo.

Whether we are aware of it or not, we use criteria every day to choose between alternatives. For example, when you drive to campus or work, you decide to take the "best" route. But how did you define *best?* Was the best route the safest, shortest, fastest, cheapest, most scenic, or what? Obviously, depending upon which criterion or combination of criteria is used to identify the best, a different route might be selected each time. In economic analysis, **financial units (dollars or other currency)** are generally used as the tangible basis for evaluation. Thus, when there are several ways of accomplishing a stated objective, the alternative with the lowest overall cost or highest overall net income is selected.

1.3 Professional Ethics and Economic Decisions ● ● ●

Many of the fundamentals of engineering ethics are intertwined with the roles of money and economic-based decisions in the making of professionally ethical judgments. Some of these integral connections are discussed here, plus sections in later chapters discuss additional aspects of ethics and economics. For example, Chapter 9, Benefit/Cost Analysis and Public Sector Economics, includes material on the ethics of public project contracts and public policy. Although it is very limited in scope and space, it is anticipated that this coverage of the important role of economics in engineering ethics will prompt further interest on the part of students and instructors of engineering economy.

The terms **morals** and **ethics** are commonly used interchangeably, yet they have slightly different interpretations. Morals usually relate to the underlying tenets that form the character and conduct of a person in judging right and wrong. Ethical practices can be evaluated by using a code of morals or **code of ethics** that forms the standards to guide decisions and actions of individuals and organizations in a profession, for example, electrical, chemical, mechanical, industrial, or civil engineering. There are several different levels and types of morals and ethics.

Universal or common morals   These are fundamental moral beliefs held by virtually all people. Most people agree that to steal, murder, lie, or physically harm someone is wrong.

It is possible for **actions** and **intentions** to come into conflict concerning a common moral. Consider the World Trade Center buildings in New York City. After their collapse on September 11, 2001, it was apparent that the design was not sufficient to withstand the heat generated by the firestorm caused by the impact of an aircraft. The structural engineers who worked on the design surely did not have the intent to harm or kill occupants in the buildings. However, their design actions did not foresee this outcome as a measurable possibility. Did they violate the common moral belief of not doing harm to others or murdering?

Individual or personal morals   These are the moral beliefs that a person has and maintains over time. These usually parallel the common morals in that stealing, lying, murdering, etc. are immoral acts.

It is quite possible that an individual strongly supports the common morals and has excellent personal morals, but these may conflict from time to time when decisions must be made. Consider the engineering student who genuinely believes that cheating is wrong. If he or she does not know how to work some test problems, but must make a certain minimum grade on the final exam to graduate, the decision to cheat or not on the final exam is an exercise in following or violating a personal moral.

Professional or engineering ethics   Professionals in a specific discipline are guided in their decision making and performance of work activities by a formal standard or code. The code states the commonly accepted standards of honesty and integrity that each individual is expected to demonstrate in her or his practice. There are codes of ethics for medical doctors, attorneys, and, of course, engineers.

Although each engineering profession has its own code of ethics, the **Code of Ethics for Engineers** published by the National Society of Professional Engineers (NSPE) is very commonly used and quoted. This code, reprinted in its entirety in Appendix C, includes numerous sections that have direct or indirect economic and financial impact upon the designs, actions, and decisions that engineers make in their professional dealings. Here are three examples from the Code:

"Engineers, in the fulfillment of their duties, shall hold paramount the *safety, health, and welfare of the public*." (section I.1)

"Engineers shall *not accept financial or other considerations*, including free engineering designs, from material or equipment suppliers for specifying their product." (section III.5.a)

"Engineers using designs supplied by a client recognize that the *designs remain the property of the client* and may not be duplicated by the engineer for others without express permission." (section III.9.b)

As with common and personal morals, conflicts can easily rise in the mind of an engineer between his or her own ethics and that of the employing corporation. Consider a manufacturing engineer who has recently come to firmly disagree morally with war and its negative effects on human beings. Suppose the engineer has worked for years in a military defense contractor's facility and does the detailed cost estimations and economic evaluations of producing fighter jets for the Air Force. The Code of Ethics for Engineers is silent on the ethics of producing and using war materiel. Although the employer and the engineer are not violating any ethics code, the engineer, as an individual, is stressed in this position. Like many people during a declining national economy, retention of this job is of paramount importance to the family and the engineer. Conflicts such as this can place individuals in real dilemmas with no or mostly unsatisfactory alternatives.

At first thought, it may not be apparent how activities related to engineering economics may present an ethical challenge to an individual, a company, or a public servant in government service. Many money-related situations, such as those that follow, can have ethical dimensions.

In the design stage:

- Safety factors are compromised to ensure that a price bid comes in as low as possible.
- Family or personal connections with individuals in a company offer unfair or insider information that allows costs to be cut in strategic areas of a project.
- A potential vendor offers specifications for company-specific equipment, and the design engineer does not have sufficient time to determine if this equipment will meet the needs of the project being designed and costed.

During the construction or implementation phase:

- Price materials and equipment from a high-quality, reputable vendor, but actually purchase the items from an inferior, cheaper supplier.

- Construct with below-standard or sub-code materials when detection is difficult. (For example, in locations where water is sourced from desalination, use mixing water that is not completely desalinated for structural concrete, which will accelerate corrosion of imbedded steel.)

While the system is operating:

- Delayed or below-standard maintenance can be performed to save money when cost overruns exist in other segments of a project.
- Opportunities to purchase cheaper repair parts can save money for a subcontractor working on a fixed-price contract.
- Safety margins are compromised because of cost, personal inconvenience to workers, tight time schedules, etc.

A good example of the last item—safety is compromised while operating the system—is the situation that arose in 1984 in Bhopal, India (Martin and Schinzinger 2005, pp. 245–8). A Union Carbide plant manufacturing the highly toxic pesticide chemical methyl isocyanate (MIC) experienced a large gas leak from high-pressure tanks. Some 500,000 persons were exposed to inhalation of this deadly gas that burns moist parts of the body. There were 2500 to 3000 deaths within days, and over the following 10-year period, some 12,000 death claims and 870,000 personal injury claims were recorded. Although Union Carbide owned the facility, the Indian government had only Indian workers in the plant. Safety practices clearly eroded due to cost-cutting measures, insufficient repair parts, and reduction in personnel to save salary money. However, one of the surprising practices that caused unnecessary harm to workers was the fact that masks, gloves, and other protective gear were not worn by workers in close proximity to the tanks containing MIC. Why? Unlike plants in the United States and other countries, there was no air conditioning in the Indian plant, resulting in high ambient temperatures in the facility.

Many ethical questions arise when corporations operate in international settings where the corporate rules, worker incentives, cultural practices, and costs in the home country differ from those in the host country. Often these ethical dilemmas are fundamentally based in the economics that provide cheaper labor, reduced raw material costs, less government oversight, and a host of other cost-reducing factors. When an engineering economy study is performed, it is important for the engineer performing the study to consider all ethically related matters to ensure that the cost and revenue estimates reflect what is likely to happen once the project or system is operating.

It is important to understand that the translation from universal morals to personal morals and professional ethics does vary from one culture and country to another. As an example, consider the common belief (universal moral) that the awarding of contracts and financial arrangements for services to be performed (for government or business) should be accomplished in a fair and transparent fashion. In some societies and cultures, corruption in the process of contract making is common and often "overlooked" by the local authorities, who may also be involved in the affairs. Are these immoral or unethical practices? Most would say, "Yes, this should not be allowed. Find and punish the individuals involved." Yet, such practices do continue, thus indicating the differences in interpretation of common morals as they are translated into the ethics of individuals and professionals.

EXAMPLE 1.2

Jamie is an engineer employed by Burris, a United States–based company that develops subway and surface transportation systems for medium-sized municipalities in the United States and Canada. He has been a registered professional engineer (PE) for the last 15 years. Last year, Carol, an engineer friend from university days who works as an individual consultant, asked Jamie to help her with some cost estimates on a metro train job. Carol offered to pay for his time and talent, but Jamie saw no reason to take money for helping with data commonly used by him in performing his job at Burris. The estimates took one weekend to complete, and once Jamie delivered them to Carol, he did not hear from her again; nor did he learn the identity of the company for which Carol was preparing the estimates.

Yesterday, Jamie was called into his supervisor's office and told that Burris had not received the contract award in Sharpstown, where a metro system is to be installed. The project estimates were prepared by Jamie and others at Burris over the past several months. This job was greatly needed by Burris, as the country and most municipalities were in a real economic slump, so much so that Burris was considering furloughing several engineers if the Sharpstown bid was not accepted. Jamie was told he was to be laid off immediately, not because the bid was rejected, but because he had been secretly working without management approval for a prime consultant of Burris' main competitor. Jamie was astounded and angry. He knew he had done nothing to warrant firing, but the evidence was clearly there. The numbers used by the competitor to win the Sharpstown award were the same numbers that Jamie had prepared for Burris on this bid, and they closely matched the values that he gave Carol when he helped her.

Jamie was told he was fortunate, because Burris' president had decided to not legally charge Jamie with unethical behavior and to not request that his PE license be rescinded. As a result, Jamie was escorted out of his office and the building within one hour and told to not ask anyone at Burris for a reference letter if he attempted to get another engineering job.

Discuss the ethical dimensions of this situation for Jamie, Carol, and Burris' management. Refer to the NSPE Code of Ethics for Engineers (Appendix C) for specific points of concern.

Solution

There are several obvious errors and omissions present in the actions of Jamie, Carol, and Burris' management in this situation. Some of these mistakes, oversights, and possible code violations are summarized here.

Jamie

- Did not learn identity of company Carol was working for and whether the company was to be a bidder on the Sharpstown project
- Helped a friend with confidential data, probably innocently, without the knowledge or approval of his employer
- Assisted a competitor, probably unknowingly, without the knowledge or approval of his employer
- Likely violated, at least, Code of Ethics for Engineers section II.1.c, which reads, "Engineers shall not reveal facts, data, or information without the prior consent of the client or employer except as authorized or required by law or this Code."

Carol

- Did not share the intended use of Jamie's work
- Did not seek information from Jamie concerning his employer's intention to bid on the same project as her client
- Misled Jamie in that she did not seek approval from Jamie to use and quote his information and assistance
- Did not inform her client that portions of her work originated from a source employed by a possible bid competitor
- Likely violated, at least, Code of Ethics for Engineers section III.9.a, which reads, "Engineers shall, whenever possible, name the person or persons who may be individually responsible for designs, inventions, writings, or other accomplishments."

Burris' management

- Acted too fast in dismissing Jamie; they should have listened to Jamie and conducted an investigation
- Did not put him on administrative leave during a review
- Possibly did not take Jamie's previous good work record into account

These are not all ethical considerations; some are just plain good business practices for Jamie, Carol, and Burris.

1.4 Interest Rate and Rate of Return ● ● ●

Interest is the manifestation of the time value of money. Computationally, interest is the difference between an ending amount of money and the beginning amount. If the difference is zero or negative, there is no interest. There are always two perspectives to an amount of interest—interest paid and interest earned. These are illustrated in Figure 1–2. Interest is **paid** when a person or organization

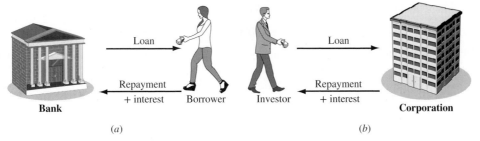

Figure 1–2
(*a*) Interest paid over time to lender. (*b*) Interest earned over time by investor.

borrowed money (obtained a loan) and repays a larger amount over time. Interest is **earned** when a person or organization saved, invested, or lent money and obtains a return of a larger amount over time. The numerical values and formulas used are the same for both perspectives, but the interpretations are different.

    Interest paid on borrowed funds (a loan) is determined using the original amount, also called the *principal,*

$$\text{Interest} = \text{amount owed now} - \text{principal} \qquad [1.1]$$

When interest paid over a *specific time unit* is expressed as a percentage of the principal, the result is called the **interest rate.**

$$\text{Interest rate (\%)} = \frac{\text{interest accrued per time unit}}{\text{principal}} \times 100\% \qquad [1.2]$$

The time unit of the rate is called the **interest period.** By far the most common interest period used to state an interest rate is 1 year. Shorter time periods can be used, such as 1% per month. Thus, the interest period of the interest rate should always be included. If only the rate is stated, for example, 8.5%, a 1-year interest period is assumed.

EXAMPLE 1.3

An employee at LaserKinetics.com borrows $10,000 on May 1 and must repay a total of $10,700 exactly 1 year later. Determine the interest amount and the interest rate paid.

Solution

The perspective here is that of the borrower since $10,700 repays a loan. Apply Equation [1.1] to determine the interest paid.

$$\text{Interest paid} = \$10,700 - 10,000 = \$700$$

Equation [1.2] determines the interest rate paid for 1 year.

$$\text{Percent interest rate} = \frac{\$700}{\$10,000} \times 100\% = 7\% \text{ per year}$$

EXAMPLE 1.4

Stereophonics, Inc. plans to borrow $20,000 from a bank for 1 year at 9% interest for new recording equipment. (*a*) Compute the interest and the total amount due after 1 year. (*b*) Construct a column graph that shows the original loan amount and total amount due after 1 year used to compute the loan interest rate of 9% per year.

Solution

(*a*) Compute the total interest accrued by solving Equation [1.2] for interest accrued.

$$\text{Interest} = \$20,000(0.09) = \$1800$$

The total amount due is the sum of principal and interest.

$$\text{Total due} = \$20{,}000 + 1800 = \$21{,}800$$

(*b*) Figure 1–3 shows the values used in Equation [1.2]: $1800 interest, $20,000 original loan principal, 1-year interest period.

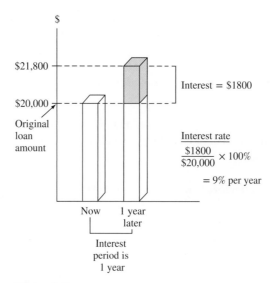

Figure 1–3
Values used to compute an interest rate of 9% per year. Example 1.4.

Comment

Note that in part (*a*), the total amount due may also be computed as

$$\text{Total due} = \text{principal}(1 + \text{interest rate}) = \$20{,}000(1.09) = \$21{,}800$$

Later we will use this method to determine future amounts for times longer than one interest period.

From the perspective of a saver, a lender, or an investor, **interest earned** (Figure 1–2*b*) is the final amount minus the initial amount, or principal.

$$\text{Interest earned} = \text{total amount now} - \text{principal} \qquad [1.3]$$

Interest earned over a specific period of time is expressed as a percentage of the original amount and is called **rate of return (ROR).**

$$\text{Rate of return (\%)} = \frac{\text{interest accrued per time unit}}{\text{principal}} \times 100\% \qquad [1.4]$$

The time unit for rate of return is called the **interest period,** just as for the borrower's perspective. Again, the most common period is 1 year.

The term **return on investment (ROI)** is used equivalently with ROR in different industries and settings, especially where large capital funds are committed to engineering-oriented programs.

The numerical values in Equations [1.2] and [1.4] are the same, but the term *interest rate paid* is more appropriate for the borrower's perspective, while the *term rate of return earned* applies for the investor's perspective.

EXAMPLE 1.5

(*a*) Calculate the amount deposited 1 year ago to have $1000 now at an interest rate of 5% per year.

(*b*) Calculate the amount of interest earned during this time period.

Solution

(*a*) The total amount accrued ($1000) is the sum of the original deposit and the earned interest. If X is the original deposit,

$$\text{Total accrued} = \text{deposit} + \text{deposit(interest rate)}$$

$$\$1000 = X + X(0.05) = X(1 + 0.05) = 1.05X$$

The original deposit is

$$X = \frac{1000}{1.05} = \$952.38$$

(*b*) Apply Equation [1.3] to determine the interest earned.

$$\text{Interest} = \$1000 - 952.38 = \$47.62$$

In Examples 1.3 to 1.5 the interest period was 1 year, and the interest amount was calculated at the end of one period. When more than one interest period is involved, e.g., the amount of interest after 3 years, it is necessary to state whether the interest is accrued on a *simple* or *compound* basis from one period to the next. This topic is covered later in this chapter.

Since **inflation** can significantly increase an interest rate, some comments about the fundamentals of inflation are warranted at this early stage. By definition, inflation represents a decrease in the value of a given currency. That is, $10 now will not purchase the same amount of gasoline for your car (or most other things) as $10 did 10 years ago. The changing value of the currency affects market interest rates.

In simple terms, interest rates reflect two things: a so-called real rate of return *plus* the expected inflation rate. The real rate of return allows the investor to purchase more than he or she could have purchased before the investment, while inflation raises the real rate to the market rate that we use on a daily basis.

Inflation

The safest investments (such as government bonds) typically have a 3% to 4% real rate of return built into their overall interest rates. Thus, a market interest rate of, say, 8% per year on a bond means that investors expect the inflation rate to be in the range of 4% to 5% per year. Clearly, inflation causes interest rates to rise.

From the borrower's perspective, the rate of inflation is another interest rate *tacked on to the real interest rate*. And from the vantage point of the saver or investor in a fixed-interest account, inflation *reduces the real rate of return* on the investment. Inflation means that cost and revenue cash flow estimates increase over time. This increase is due to the changing value of money that is forced upon a country's currency by inflation, thus making a unit of currency (such as the dollar) worth less relative to its value at a previous time. We see the effect of inflation in that money purchases less now than it did at a previous time. Inflation contributes to

- A reduction in purchasing power of the currency
- An increase in the CPI (consumer price index)
- An increase in the cost of equipment and its maintenance
- An increase in the cost of salaried professionals and hourly employees
- A reduction in the real rate of return on personal savings and certain corporate investments

In other words, inflation can materially contribute to changes in corporate and personal economic analysis.

Commonly, engineering economy studies assume that inflation affects all estimated values equally. Accordingly, an interest rate or rate of return, such as 8% per year, is applied throughout the analysis without accounting for an additional inflation rate. However, if inflation were explicitly taken into account, and it was reducing the value of money at, say, an average of 4% per year, then it would be necessary to perform the economic analysis using an inflated interest rate. (The rate is 12.32% per year using the relations derived in Chapter 14.)

1.5 Terminology and Symbols ● ● ●

The equations and procedures of engineering economy utilize the following terms and symbols. Sample units are indicated.

P = value or amount of money at a time designated as the present or time 0. Also P is referred to as present worth (PW), present value (PV), net present value (NPV), discounted cash flow (DCF), and capitalized cost (CC); monetary units, such as dollars

F = value or amount of money at some future time. Also F is called future worth (FW) and future value (FV); dollars

A = series of consecutive, equal, end-of-period amounts of money. Also A is called the annual worth (AW) and equivalent uniform annual worth (EUAW); dollars per year, euros per month

n = number of interest periods; years, months, days

i = interest rate per time period; percent per year, percent per month

t = time, stated in periods; years, months, days

The symbols P and F represent one-time occurrences: A occurs with the same value in each interest period for a specified number of periods. It should be clear that a present value P represents a single sum of money at some time prior to a future value F or prior to the first occurrence of an equivalent series amount A.

It is important to note that the symbol A always represents a uniform amount (i.e., the same amount each period) that extends through *consecutive* interest periods. Both conditions must exist before the series can be represented by A.

The interest rate i is expressed in percent per interest period, for example, 12% per year. Unless stated otherwise, assume that the rate applies throughout the entire n years or interest periods. The decimal equivalent for i is always used in formulas and equations in engineering economy computations.

All engineering economy problems involve the element of time expressed as n and interest rate i. In general, every problem will involve at least four of the symbols P, F, A, n, and i, with at least three of them estimated or known.

Additional symbols used in engineering economy are defined in Appendix E.

EXAMPLE 1.6

Today, Julie borrowed $5000 to purchase furniture for her new house. She can repay the loan in either of the two ways described below. Determine the engineering economy symbols and their value for each option.

(*a*) Five equal annual installments with interest determined at 5% per year.
(*b*) One payment 3 years from now with interest determined at 7% per year.

Solution
(*a*) The repayment schedule requires an equivalent annual amount A, which is unknown.

$$P = \$5000 \qquad i = 5\% \text{ per year} \qquad n = 5 \text{ years} \qquad A = ?$$

(*b*) Repayment requires a single future amount F, which is unknown.

$$P = \$5000 \qquad i = 7\% \text{ per year} \qquad n = 3 \text{ years} \qquad F = ?$$

EXAMPLE 1.7

You plan to make a lump-sum deposit of $5000 now into an investment account that pays 6% per year, and you plan to withdraw an equal end-of-year amount of $1000 for 5 years, starting next year. At the end of the sixth year, you plan to close your account by withdrawing the remaining money. Define the engineering economy symbols involved.

Solution

All five symbols are present, but the future value in year 6 is the unknown.

$$P = \$5000$$

$$A = \$1000 \text{ per year for 5 years}$$

$$F = ? \text{ at end of year 6}$$

$$i = 6\% \text{ per year}$$

$$n = 5 \text{ years for the } A \text{ series and 6 for the } F \text{ value}$$

EXAMPLE 1.8

Last year Jane's grandmother offered to put enough money into a savings account to generate $5000 in interest this year to help pay Jane's expenses at college. (*a*) Identify the symbols, and (*b*) calculate the amount that had to be deposited exactly 1 year ago to earn $5000 in interest now, if the rate of return is 6% per year.

Solution

(*a*) Symbols P (last year is -1) and F (this year) are needed.

$$P = ?$$

$$i = 6\% \text{ per year}$$

$$n = 1 \text{ year}$$

$$F = P + \text{interest} = ? + \$5000$$

(*b*) Let F = total amount now and P = original amount. We know that $F - P = \$5000$ is accrued interest. Now we can determine P. Refer to Equations [1.1] through [1.4].

$$F = P + Pi$$

The $5000 interest can be expressed as

$$\text{Interest} = F - P = (P + Pi) - P$$

$$= Pi$$

$$\$5000 = P(0.06)$$

$$P = \frac{\$5000}{0.06} = \$83,333.33$$

1.6 Cash Flows: Estimation and Diagramming ●●●

As mentioned in earlier sections, cash flows are the amounts of money estimated for future projects or observed for project events that have taken place. All cash flows occur during specific time periods, such as 1 month, every 6 months, or 1 year. Annual is the most common time period. For example, a payment of $10,000 once every year in December for 5 years is a series of 5 outgoing cash flows. And an estimated receipt of $500 every month for 2 years is a series of 24 incoming cash flows. Engineering economy bases its computations on the timing, size, and direction of cash flows.

Cash inflows are all types of receipts, including sales, revenues, incomes, money from a loan when received from the lender, and savings generated by project and business activity. A **plus sign** indicates a cash inflow.

Cash flow

Cash outflows are all types of costs, including disbursements, expenses, deposits into retirement or savings accounts, loan repayments, and taxes caused by projects and business activity. A **negative or minus sign** indicates a cash outflow. When a project involves only costs, the minus sign may be omitted for some techniques, such as benefit/cost analysis.

Of all the steps in Figure 1–1 that outline the engineering economy study, estimating cash flows (step 3) is the most difficult, primarily because it is an attempt to predict the future. Some examples of cash flow estimates are shown here. As you scan these, consider how the cash inflow or outflow may be estimated most accurately.

Cash Inflow Estimates

Income: +$150,000 per year from sales of solar-powered watches

Savings: +$24,500 tax savings from capital loss by equipment salvage value

Receipt: +$750,000 received on large business loan plus accrued interest

Savings: +$150,000 per year saved by installing more efficient air conditioning

Revenue: +$50,000 to +$75,000 per month in sales for extended battery life iPhones

Cash Outflow Estimates

Operating costs: −$230,000 per year annual operating costs for software services

First cost: −$800,000 next year to purchase replacement earthmoving equipment

Expense: −$20,000 per year for loan interest payment to bank

Initial cost: −$1 to −$1.2 million in capital expenditures for a water recycling unit

All of these are **point estimates,** that is, *single-value estimates* for cash flow elements of an alternative, except for the last revenue and cost estimates listed above. They provide a **range estimate,** because the persons estimating the revenue and cost do not have enough knowledge or experience with the systems to be more accurate. For the initial chapters, we will utilize point estimates. The use of risk and sensitivity analysis for range estimates is covered in the later chapters of this book.

Once all cash inflows and outflows are estimated (or determined for a completed project), the **net cash flow** for each time period is calculated.

$$\text{Net cash flow} = \text{cash inflows} - \text{cash outflows} \qquad [1.5]$$
$$NCF = R - D \qquad [1.6]$$

where NCF is net cash flow, R is receipts, and D is disbursements.

At the beginning of this section, the *timing, size, and direction of cash flows* were mentioned as important. Because cash flows may take place at any time during an interest period, as a matter of convention, all cash flows are assumed to occur at the end of an interest period.

End-of-period convention

> The end-of-period convention means that all cash inflows and all cash outflows are assumed to take place at the **end of the interest period** in which they actually occur. When several inflows and outflows occur within the same period, the *net* cash flow is assumed to occur at the *end* of the period.

In assuming end-of-period cash flows, it is important to understand that future (F) and uniform annual (A) amounts are located at the end of the interest period, which is not necessarily December 31. If in Example 1.7 the lump-sum deposit took place on July 1, 2017, the withdrawals will take place on July 1 of each succeeding year for 6 years. Remember, end of the period means end of interest period, not end of calendar year.

The **cash flow diagram** is a very important tool in an economic analysis, especially when the cash flow series is complex. It is a graphical representation of cash flows drawn on the *y* axis with a time scale on the *x* axis. The diagram includes what is known, what is estimated, and what is needed. That is, once the cash flow diagram is complete, another person should be able to work the problem by looking at the diagram.

Cash flow diagram time $t = 0$ is the present, and $t = 1$ is the end of time period 1. We assume that the periods are in years for now. The time scale of Figure 1–4 is set up for 5 years. Since the end-of-year convention places cash flows at the ends of years, the "1" marks the end of year 1.

While it is not necessary to use an exact scale on the cash flow diagram, you will probably avoid errors if you make a neat diagram to approximate scale for both time and relative cash flow magnitudes.

The direction of the arrows on the diagram is important to differentiate income from outgo. A vertical arrow pointing up indicates a positive cash flow. Conversely, a down-pointing arrow

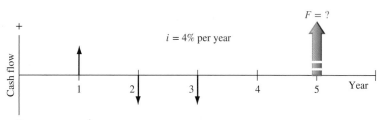

Figure 1–4
A typical cash flow time scale for 5 years.

Figure 1–5
Example of positive and negative cash flows.

indicates a negative cash flow. **We will use a bold, colored arrow to indicate what is unknown and to be determined.** For example, if a future value F is to be determined in year 5, a wide, colored arrow with $F = ?$ is shown in year 5. The interest rate is also indicated on the diagram. Figure 1–5 illustrates a cash inflow at the end of year 1, equal cash outflows at the end of years 2 and 3, an interest rate of 4% per year, and the unknown future value F after 5 years. The arrow for the unknown value is generally drawn in the opposite direction from the other cash flows; however, the engineering economy computations will determine the actual sign on the F value.

Before the diagramming of cash flows, a perspective or vantage point must be determined so that + or − signs can be assigned and the economic analysis performed correctly. Assume you borrow $8500 from a bank today to purchase an $8000 used car for cash next week, and you plan to spend the remaining $500 on a new paint job for the car 2 weeks from now. There are several perspectives possible when developing the cash flow diagram—those of the borrower (that's you), the banker, the car dealer, or the paint shop owner. The cash flow signs and amounts for these perspectives are as follows.

Perspective	Activity	Cash Flow with Sign, $	Time, week
You	Borrow	+8500	0
	Buy car	−8000	1
	Paint job	−500	2
Banker	Lender	−8500	0
Car dealer	Car sale	+8000	1
Painter	Paint job	+500	2

One, and only one, of the perspectives is selected to develop the diagram. For your perspective, all three cash flows are involved and the diagram appears as shown in Figure 1–6 with a time scale of weeks. Applying the end-of-period convention, you have a receipt of +$8500 now (time 0) and cash outflows of −$8000 at the end of week 1, followed by −$500 at the end of week 2.

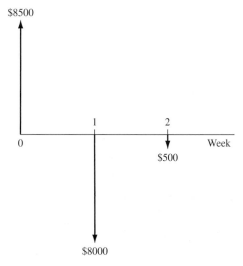

Figure 1–6
Cash flows from perspective of borrower for loan and purchases.

EXAMPLE 1.9

Each year Exxon-Mobil expends large amounts of funds for mechanical safety features throughout its worldwide operations. Carla Ramos, a lead engineer for Mexico and Central American operations, plans expenditures of $1 million *now* and each of the next 4 years just for the improvement of field-based pressure-release valves. Construct the cash flow diagram to find the equivalent value of these expenditures at the end of year 4, using a cost of capital estimate for safety-related funds of 12% per year.

Solution

Figure 1–7 indicates the uniform and negative cash flow series (expenditures) for five periods, and the unknown F value (positive cash flow equivalent) at exactly the same time as the fifth expenditure. Since the expenditures start immediately, the first $1 million is shown at time 0, not time 1. Therefore, the last negative cash flow occurs at the end of the fourth year, when F also occurs. To make this diagram have a full 5 years on the time scale, the addition of the year −1 completes the diagram. This addition demonstrates that year 0 is the end-of-period point for the year −1.

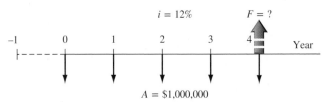

Figure 1–7
Cash flow diagram, Example 1.9.

EXAMPLE 1.10

An electrical engineer wants to deposit an amount P now such that she can withdraw an equal annual amount of $A_1 = \$2000$ per year for the first 5 years, starting 1 year after the deposit, and a different annual withdrawal of $A_2 = \$3000$ per year for the following 3 years. How would the cash flow diagram appear if $i = 8.5\%$ per year?

Solution

The cash flows are shown in Figure 1–8. The negative cash outflow P occurs now. The withdrawals (positive cash inflow) for the A_1 series occur at the end of years 1 through 5, and A_2 occurs in years 6 through 8.

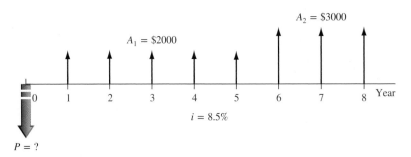

Figure 1–8
Cash flow diagram with two different A series, Example 1.10.

EXAMPLE 1.11

A rental company spent $2500 on a new air compressor 7 years ago. The annual rental income from the compressor has been $750. The $100 spent on maintenance the first year has increased each year by $25. The company plans to sell the compressor at the end of next year for

$150. Construct the cash flow diagram from the company's perspective and indicate where the present worth now is located.

Solution

Let now be time $t = 0$. The incomes and costs for years -7 through 1 (next year) are tabulated below with net cash flow computed using Equation [1.5]. The net cash flows (one negative, eight positive) are diagrammed in Figure 1–9. Present worth P is located at year 0.

End of Year	Income	Cost	Net Cash Flow
−7	$ 0	$2500	$−2500
−6	750	100	650
−5	750	125	625
−4	750	150	600
−3	750	175	575
−2	750	200	550
−1	750	225	525
0	750	250	500
1	750 + 150	275	625

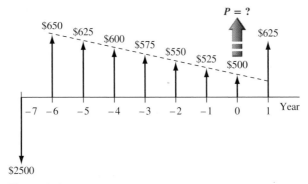

Figure 1–9
Cash flow diagram, Example 1.11.

1.7 Economic Equivalence ●●●

Economic equivalence is a fundamental concept upon which engineering economy computations are based. Before we delve into the economic aspects, think of the many types of equivalency we may utilize daily by transferring from one scale to another. Some example transfers between scales are as follows:

Length:
12 inches = 1 foot   3 feet = 1 yard   39.370 inches = 1 meter
100 centimeters = 1 meter   1000 meters = 1 kilometer   1 kilometer = 0.621 mile

Pressure:
1 atmosphere = 1 newton/meter^2 = 10^3 pascal = 1 kilopascal

Often equivalency involves two or more scales. Consider the equivalency of a *speed* of 110 kilometers per hour (kph) into miles per minute using conversions between distance and time scales with three-decimal accuracy.

Speed:
1 mile = 1.609 kilometers   1 hour = 60 minutes
110 kph = 68.365 miles per hour (mph)   68.365 mph = 1.139 miles per minute

Four scales—time in minutes, time in hours, length in miles, and length in kilometers—are combined to develop these equivalent statements on speed. Note that throughout these statements, the fundamental relations of 1 mile = 1.609 kilometers and 1 hour = 60 minutes are applied. If a fundamental relation changes, the entire equivalency is in error.

Now we consider economic equivalency.

Economic equivalence

> **Economic equivalence** is a combination of **interest rate** and **time value of money** to determine the different amounts of money at different points in time that are equal in economic value.

As an illustration, if the interest rate is 6% per year, $100 today (present time) is equivalent to $106 one year from today.

$$\text{Amount accrued} = 100 + 100(0.06) = 100(1 + 0.06) = \$106$$

If someone offered you a gift of $100 today or $106 one year from today, it would make no difference which offer you accepted from an economic perspective. In either case you have $106 one year from today. However, the two sums of money are equivalent to each other *only* when the interest rate is 6% per year. At a higher or lower interest rate, $100 today is not equivalent to $106 one year from today.

In addition to future equivalence, we can apply the same logic to determine equivalence for previous years. A total of $100 now is equivalent to $100/1.06 = $94.34 one year ago at an interest rate of 6% per year. From these illustrations, we can state the following: $94.34 last year, $100 now, and $106 one year from now are equivalent at an interest rate of 6% per year. The fact that these sums are equivalent can be verified by computing the two interest rates for 1-year interest periods.

$$\frac{\$6}{\$100} \times 100\% = 6\% \text{ per year}$$

and

$$\frac{\$5.66}{\$94.34} \times 100\% = 6\% \text{ per year}$$

The cash flow diagram in Figure 1–10 indicates the amount of interest needed each year to make these three different amounts equivalent at 6% per year.

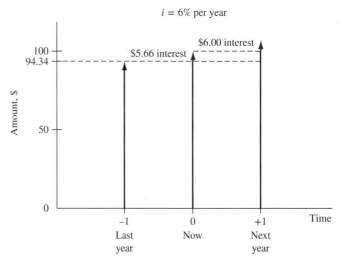

Figure 1–10
Equivalence of money at 6% per year interest.

EXAMPLE 1.12

Manufacturers make backup batteries for computer systems available to Batteries+Bulbs dealers through privately owned distributorships. In general, batteries are stored throughout the year, and a 5% cost increase is added each year to cover the inventory carrying charge for the distributorship owner. Assume you own the City Center Batteries+Bulbs outlet. Make the calculations necessary to show which of the following statements are true and which are false about battery costs.

(a) The amount of $98 now is equivalent to a cost of $105.60 one year from now.
(b) A truck battery cost of $200 one year ago is equivalent to $205 now.
(c) A $38 cost now is equivalent to $39.90 one year from now.
(d) A $3000 cost now is equivalent to $2887.14 one year earlier.
(e) The carrying charge accumulated in 1 year on an investment of $20,000 worth of batteries is $1000.

Solution

(a) Total amount accrued = 98(1.05) = $102.90 ≠ $105.60; therefore, it is false. Another way to solve this is as follows: Required original cost is 105.60/1.05 = $100.57 ≠ $98.
(b) Equivalent cost 1 year ago is 205.00/1.05 = $195.24 ≠ $200; therefore, it is false.
(c) The cost 1 year from now is $38(1.05) = $39.90; true.
(d) Cost now is 2887.14(1.05) = $3031.50 ≠ $3000; false.
(e) The charge is 5% per year interest, or $20,000(0.05) = $1000; true.

Comparison of alternative cash flow series requires the use of equivalence to determine when the series are economically equal or if one is economically preferable to another. The keys to the analysis are the interest rate and the timing of the cash flows. Example 1.13 demonstrates how easy it is to be misled by the size and timing of cash flows.

EXAMPLE 1.13

Howard owns a small electronics repair shop. He wants to borrow $10,000 now and repay it over the next 1 or 2 years. He believes that new diagnostic test equipment will allow him to work on a wider variety of electronic items and increase his annual revenue. Howard received 2-year repayment options from banks A and B.

Year	Amount to Pay, $ per year	
	Bank A	Bank B
1	−5,378.05	−5,000.00
2	−5,378.05	−5,775.00
Total paid	−10,756.10	−10,775.00

After reviewing these plans, Howard decided that he wants to repay the $10,000 after only 1 year based on the expected increased revenue. During a family conversation, Howard's brother-in-law offered to lend him the $10,000 now and take $10,600 after exactly 1 year. Now Howard has three options and wonders which one to take. Which one is economically the best?

Solution

The repayment plans for both banks are economically equivalent at the interest rate of 5% per year. (This is determined by using computations that you will learn in Chapter 2.) Therefore, Howard can choose either plan even though the bank B plan requires a slightly larger sum of money over the 2 years.

The brother-in-law repayment plan requires a total of $600 in interest 1 year later plus the principal of $10,000, which makes the interest rate 6% per year. Given the two 5% per year options from the banks, this 6% plan should not be chosen as it is not economically better than the other two. Even though the sum of money repaid is smaller, the timing of the cash flows and the interest rate make it less desirable. The point here is that cash flows themselves, or their sums, cannot be relied upon as the primary basis for an economic decision. The interest rate, timing, and economic equivalence must be considered.

1.8 Simple and Compound Interest ● ● ●

The terms *interest, interest period,* and *interest rate* (introduced in Section 1.4) are useful in calculating equivalent sums of money for one interest period in the past and one period in the future. However, for more than one interest period, the terms *simple interest* and *compound interest* become important.

   Simple interest is calculated using the principal only, ignoring any interest accrued in preceding interest periods. The total simple interest over several periods is computed as

> **Simple interest = (principal)(number of periods)(interest rate)**   [1.7]
> $$I = Pni$$

where *I* is the amount of interest earned or paid and the interest rate *i* is expressed in decimal form.

EXAMPLE 1.14

GreenTree Financing lent an engineering company $100,000 to retrofit an environmentally unfriendly building. The loan is for 3 years at 10% per year simple interest. How much money will the firm repay at the end of 3 years?

Solution
The interest for each of the 3 years is

$$\text{Interest per year} = \$100,000(0.10) = \$10,000$$

Total interest for 3 years from Equation [1.7] is

$$\text{Total interest} = \$100,000(3)(0.10) = \$30,000$$

The amount due after 3 years is

$$\text{Total due} = \$100,000 + 30,000 = \$130,000$$

   The interest accrued in the first year and in the second year does not earn interest. The interest due each year is $10,000 calculated only on the $100,000 loan principal.

In most financial and economic analyses, we use **compound interest** calculations.

For *compound interest,* the interest accrued for each interest period is calculated on the **principal plus the total amount of interest accumulated in all previous periods.** Thus, compound interest means interest on top of interest.

   Compound interest reflects the effect of the time value of money on the interest also. Now the interest for one period is calculated as

> **Compound interest = (principal + all accrued interest)(interest rate)**   [1.8]

In mathematical terms, the interest I_t for time period t may be calculated using the relation.

$$I_t = \left(P + \sum_{j=1}^{j=t-1} I_j \right)(i)$$ [1.9]

EXAMPLE 1.15

Assume an engineering company borrows $100,000 at 10% per year compound interest and will pay the principal and all the interest after 3 years. Compute the annual interest and total amount due after 3 years. Graph the interest and total owed for each year, and compare with the previous example that involved simple interest.

Solution

To include compounding of interest, the annual interest and total owed each year are calculated by Equation [1.8].

Interest, year 1:	$100,000(0.10) = \$10,000$
Total due, year 1:	$100,000 + 10,000 = \$110,000$
Interest, year 2:	$110,000(0.10) = \$11,000$
Total due, year 2:	$110,000 + 11,000 = \$121,000$
Interest, year 3:	$121,000(0.10) = \$12,100$
Total due, year 3:	$121,000 + 12,100 = \$133,100$

The repayment plan requires no payment until year 3 when all interest and the principal, a total of $133,100, are due. Figure 1–11 uses a cash flow diagram format to compare end-of-year (a) simple and (b) compound interest and total amounts owed. The differences due to compounding are clear. An extra $133,100 – 130,000 = $3100 in interest is due for the compounded interest loan.

Note that while simple interest due each year is constant, the compounded interest due grows geometrically. Due to this geometric growth of compound interest, the difference between simple and compound interest accumulation increases rapidly as the time frame increases. For example, if the loan is for 10 years, not 3, the extra paid for compounding interest may be calculated to be $59,374.

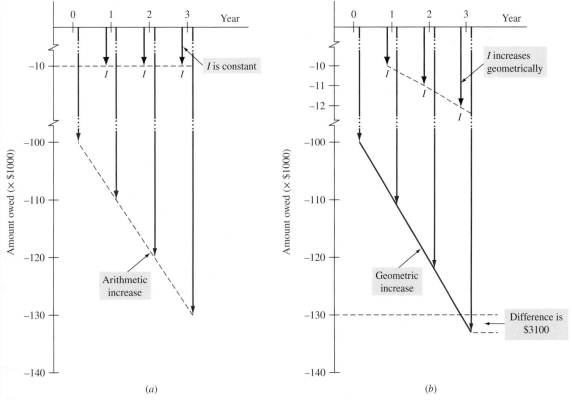

Figure 1–11

Interest I owed and total amount owed for (a) simple interest (Example 1.14) and (b) compound interest (Example 1.15).

A more efficient way to calculate the total amount due after a number of years in Example 1.15 is to utilize the fact that compound interest increases geometrically. This allows us to skip the year-by-year computation of interest. In this case, the **total amount due at the end of each year** is

Year 1: $\$100{,}000(1.10)^1 = \$110{,}000$

Year 2: $\$100{,}000(1.10)^2 = \$121{,}000$

Year 3: $\$100{,}000(1.10)^3 = \$133{,}100$

This allows future totals owed to be calculated directly without intermediate steps. The general form of the equation is

$$\textbf{Total due after } \textit{n} \textbf{ years} = \textbf{principal}(1 + \textbf{interest rate})^{n \text{ years}} \qquad \textbf{[1.10]}$$
$$= P(1 + i)^n$$

where i is expressed in decimal form. The total due after n years is the same as the future worth F, defined in Section 1.5. Equation [1.10] was applied above to obtain the $\$133{,}100$ due after 3 years. This fundamental relation will be used many times in the upcoming chapters.

We can combine the concepts of interest rate, compound interest, and equivalence to demonstrate that different loan repayment plans may be equivalent, but differ substantially in amounts paid from one year to another and in the total repayment amount. This also shows that there are many ways to take into account the time value of money.

EXAMPLE 1.16

Table 1–1 details four different loan repayment plans described below. Each plan repays a $\$5000$ loan in 5 years at 8% per year compound interest.

- **Plan 1: Pay all at end.** No interest or principal is paid until the end of year 5. Interest accumulates each year on the total of principal and all accrued interest.
- **Plan 2: Pay interest annually, principal repaid at end.** The accrued interest is paid each year, and the entire principal is repaid at the end of year 5.
- **Plan 3: Pay interest and portion of principal annually.** The accrued interest and one-fifth of the principal (or $\$1000$) are repaid each year. The outstanding loan balance decreases each year, so the interest (column 2) for each year decreases.
- **Plan 4: Pay equal amount of interest and principal.** Equal payments are made each year with a portion going toward principal repayment and the remainder covering the accrued interest. Since the loan balance decreases at a rate slower than that in plan 3 due to the equal end-of-year payments, the interest decreases, but at a slower rate.

(*a*) Make a statement about the *equivalence* of each plan at 8% compound interest.

(*b*) Develop an 8% per year *simple* interest repayment plan for this loan using the same approach as plan 2. Comment on the total amounts repaid for the two plans.

Solution

(*a*) The amounts of the annual payments are different for each repayment schedule, and the total amounts repaid for most plans are different, even though each repayment plan requires exactly 5 years. The difference in the total amounts repaid can be explained by the time value of money and by the partial repayment of principal prior to year 5.

A loan of $\$5000$ at time 0 made at 8% per year compound interest is equivalent to each of the following:

Plan 1	$\$7346.64$ at the end of year 5
Plan 2	$\$400$ per year for 4 years and $\$5400$ at the end of year 5
Plan 3	Decreasing payments of interest and partial principal in years 1 ($\$1400$) through 5 ($\$1080$)
Plan 4	$\$1252.28$ per year for 5 years

An engineering economy study typically uses plan 4; interest is compounded, and a constant amount is paid each period. This amount covers accrued interest and a partial amount of principal repayment.

TABLE 1-1	Different Repayment Schedules Over 5 Years for $5000 at 8% Per Year Compound Interest			
(1) End of Year	(2) Interest Owed for Year	(3) Total Owed at End of Year	(4) End-of-Year Payment	(5) Total Owed After Payment
Plan 1: Pay All at End				
0				$5000.00
1	$400.00	$5400.00	—	5400.00
2	432.00	5832.00	—	5832.00
3	466.56	6298.56	—	6298.56
4	503.88	6802.44	—	6802.44
5	544.20	7346.64	$-7346.64	
Total			$-7346.64	
Plan 2: Pay Interest Annually; Principal Repaid at End				
0				$5000.00
1	$400.00	$5400.00	$-400.00	5000.00
2	400.00	5400.00	-400.00	5000.00
3	400.00	5400.00	-400.00	5000.00
4	400.00	5400.00	-400.00	5000.00
5	400.00	5400.00	-5400.00	
Total			$-7000.00	
Plan 3: Pay Interest and Portion of Principal Annually				
0				$5000.00
1	$400.00	$5400.00	$-1400.00	4000.00
2	320.00	4320.00	-1320.00	3000.00
3	240.00	3240.00	-1240.00	2000.00
4	160.00	2160.00	-1160.00	1000.00
5	80.00	1080.00	-1080.00	
Total			$-6200.00	
Plan 4: Pay Equal Annual Amount of Interest and Principal				
0				$5000.00
1	$400.00	$5400.00	$-1252.28	4147.72
2	331.82	4479.54	-1252.28	3227.25
3	258.18	3485.43	-1252.28	2233.15
4	178.65	2411.80	-1252.28	1159.52
5	92.76	1252.28	-1252.28	
Total			$-6261.40	

(*b*) The repayment schedule for 8% per year simple interest is detailed in Table 1–2. Since the annual accrued interest of $400 is paid each year and the principal of $5000 is repaid in year 5, the schedule is exactly the same as that for 8% per year compound interest, and the total amount repaid is the same at $7000. In this unusual case, simple and compound interest result in the same total repayment amount. Any deviation from this schedule will cause the two plans and amounts to differ.

TABLE 1-2	A 5-Year Repayment Schedule of $5000 at 8% per Year Simple Interest			
End of Year	Interest Owed for Year	Total Owed at End of Year	End-of-Year Payment	Total Owed After Payment
0				$5000
1	$400	$5400	$-400	5000
2	400	5400	-400	5000
3	400	5400	-400	5000
4	400	5400	-400	5000
5	400	5400	-5400	0
Total			$-7000	

1.9 Minimum Attractive Rate of Return ● ● ●

For any investment to be profitable, the investor (corporate or individual) expects to receive more money than the amount of capital invested. In other words, a fair *rate of return,* or *return on investment,* must be realizable. The definition of ROR in Equation [1.4] is used in this discussion, that is, amount earned divided by the principal.

Engineering alternatives are evaluated upon the prognosis that a reasonable ROR can be expected. Therefore, some reasonable rate must be established for the selection criteria (step 4) of the engineering economy study (Figure 1–1).

Minimum Attractive Rate
of Return (MARR)

> The Minimum Attractive Rate of Return (MARR) is a reasonable rate of return established for the evaluation and selection of alternatives. A project is not economically viable unless it is **expected to return at least the MARR.** MARR is also referred to as the *hurdle rate, cutoff rate, benchmark rate,* and *minimum acceptable rate of return.*

Figure 1–12 indicates the relations between different rate of return values. In the United States, the current U.S. Treasury Bill return is sometimes used as the benchmark safe rate. The MARR will always be higher than this, or a similar, safe rate. The MARR is not a rate that is calculated as a ROR. The MARR is established by (financial) managers and is used as a criterion against which an alternative's ROR is measured, when making the accept/reject investment decision.

To develop a foundation-level understanding of how a MARR value is established and used to make investment decisions, we return to the term **capital** introduced in Section 1.1. Although the MARR is used as a criterion to decide on investing in a project, the size of MARR is fundamentally connected to how much it costs to obtain the needed capital funds. It always costs money in the form of interest to raise capital. The interest, expressed as a percentage rate per year, is called the **cost of capital.** As an example on a personal level, if you want to purchase a new widescreen smart TV, but do not have sufficient money (capital), you could obtain a bank loan for, say, a cost of capital of 9% per year and pay for the TV in cash now. Alternatively, you might choose to use

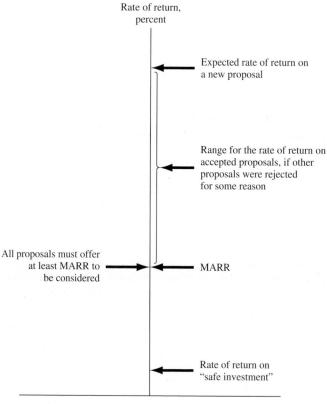

Figure 1–12
Size of MAAR relative to other rate of return values.

your credit card and pay off the balance on a monthly basis. This approach will probably cost you at least 15% per year. Or, you could use funds from your investment account that earns 5% per year and pay cash. This approach means that you also forgo future returns from these funds. The 9%, 15%, and 5% rates are your cost of capital estimates to raise the capital for the system by different methods of capital financing. In analogous ways, corporations estimate the **cost of capital** from different sources to raise funds for engineering projects and other types of projects.

Cost of capital

In general, capital is developed in two ways—equity financing and debt financing. A combination of these two is very common for most projects. Chapter 10 covers these in greater detail, but a snapshot description follows.

Equity financing   The corporation uses its own funds from cash on hand, stock sales, or retained earnings. Individuals can use their own cash, savings, or investments. In the example above, using money from the 5% investment account is equity financing.

Debt financing   The corporation borrows from outside sources and repays the principal and interest according to some schedule, much like the plans in Table 1–1. Sources of debt capital may be bonds, loans, mortgages, venture capital pools, and many others. Individuals, too, can utilize debt sources, such as the credit card (15% rate) and bank options (9% rate) described above.

Combinations of debt-equity financing mean that a **weighted average cost of capital (WACC)** results. If the smart TV is purchased with 40% credit card money at 15% per year and 60% savings account funds earning 5% per year, the weighted average cost of capital is 0.4(15%) + 0.6(5%) = 9% per year.

For a corporation, the *established MARR* used as a criterion to accept or reject an investment alternative will usually be *equal to or higher than the WACC* that the corporation must bear to obtain the necessary capital funds. So the inequality

$$ROR \geq MARR > WACC \qquad [1.11]$$

must be correct for an accepted project. Exceptions may be government-regulated requirements (safety, security, environmental, legal, etc.), economically lucrative ventures expected to lead to other opportunities, etc.

Often there are many alternatives that are expected to yield a ROR that exceeds the MARR as indicated in Figure 1–12, but there may not be sufficient capital available for all, or the project's risk may be estimated as too high to take the investment chance. Therefore, new projects that are undertaken usually have an expected return at least as great as the return on another alternative that is not funded. The expected rate of return on the unfunded project is called the **opportunity cost.**

The opportunity cost is the rate of return of a forgone opportunity caused by the inability to pursue a project. Numerically, it is the **largest rate of return of all the projects not accepted (forgone) due to the lack of capital funds or other resources.** When no specific MARR is established, the de facto MARR is the opportunity cost, that is, the ROR of the first project not undertaken due to unavailability of capital funds.

Opportunity cost

As an illustration of opportunity cost, refer to Figure 1–12 and assume a MARR of 12% per year. Further, assume that a proposal, call it A, with an expected ROR = 13% is not funded due to a lack of capital. Meanwhile, proposal B has a ROR = 14.5% and is funded from available capital. Since proposal A is not undertaken due to the lack of capital, its estimated ROR of 13% is the *opportunity cost;* that is, the opportunity to make an additional 13% return is forgone.

An opportunity cost can also be expressed in monetary terms. For example, if you lent money to a friend and he did not repay you, you have lost the opportunity to utilize these funds for other purposes of your own.

1.10 Introduction to Spreadsheet Use ● ● ●

The functions on a computer spreadsheet can greatly reduce the amount of hand work for equivalency computations involving *compound interest* and the terms P, F, A, i, and n. The use of a calculator to solve most simple problems is preferred by many students and professors as described in Appendix D. However, as cash flow series become more complex, the spreadsheet offers a good alternative. Microsoft Excel© is used throughout this book because it is readily available and easy to use. Appendix A is a primer on using spreadsheets and Excel©. The functions used in engineering economy are described there in detail, with explanations of all the parameters. Appendix A also includes a section on spreadsheet layout that is useful when the economic analysis is presented to someone else—a coworker, a boss, or a professor.

The following example demonstrates the use of a spreadsheet to develop relations that calculate interest and cash flows. Once set up, the spreadsheet can be used to perform sensitivity analysis for estimates that are subject to change. We will illustrate the use of spreadsheets throughout the chapters. There are problems at the end of many chapters entitled "Exercises for Spreadsheets" that are specifically written for spreadsheet solution. (*Note:* The spreadsheet examples may be omitted, if spreadsheets are not used in the course. A solution by hand is included in virtually all examples.)

EXAMPLE 1.17

A Japan-based architectural firm has asked a United States-based software engineering group to infuse GPS sensing capability via satellite into monitoring software for high-rise structures in order to detect greater than expected horizontal movements. This software could be very beneficial as an advance warning of serious tremors in earthquake-prone areas in Japan and the United States. The inclusion of accurate GPS data is estimated to increase annual revenue over that for the current software system by $200,000 for each of the next 2 years, and by $300,000 for each of years 3 and 4. The planning horizon is only 4 years due to the rapid advances made internationally in building-monitoring software. Develop spreadsheets to answer the questions below.

(*a*) Determine the total interest and total revenue after 4 years, using a compound rate of return of 8% per year.
(*b*) Repeat part (*a*) if estimated revenue increases from $300,000 to $600,000 in years 3 and 4.
(*c*) Repeat part (*a*) if inflation is estimated to be 4% per year. This will decrease the *real rate of return* from 8% to 3.85% per year (Chapter 14 shows why).

Solution by Spreadsheet

Refer to Figure 1–13*a* to *d* for the solutions. All the spreadsheets contain the same information, but some cell values are altered as required by the question. (Actually, all the questions can be answered on one spreadsheet by changing the numbers. Separate spreadsheets are shown here for explanation purposes only.)

The spreadsheet functions are constructed with reference to the cells, not the values themselves, so that sensitivity analysis can be performed without function changes. This approach treats the value in a cell as a *global variable* for the spreadsheet. For example, the 8% rate in cell B2 will be referenced in all functions as B2, not 8%. Thus, a change in the rate requires only one alteration in the cell B2 entry, not in every relation where 8% is used. See Appendix A for additional information about using cell referencing and building spreadsheet relations.

(*a*) Figure 1–13*a* shows the results, and Figure 1–13*b* presents all spreadsheet relations for estimated interest and revenue (yearly in columns C and E, cumulative in columns D and F). As an illustration, for year 3 the interest I_3 and revenue plus interest R_3 are

$$I_3 = \text{(cumulative revenue through year 2)(rate of return)}$$
$$= \$416,000(0.08)$$
$$= \$33,280$$

$$R_3 = \text{revenue in year 3} + I_3$$
$$= \$300,000 + 33,280$$
$$= \$333,280$$

The detailed relations shown in Figure 1–13*b* calculate these values in cells C8 and E8.

Cell C8 relation for I_3: = F7*B2
Cell E8 relation for CF_3: = B8 + C8

	A	B	C	D	E	F
1			Part (a) - Find totals in year 4			
2	i =	8.0%				
3/4	End of Year	Revenue at end of year, $	Interest earned during year, $	Cumulative interest, $	Revenue during year with interest, $	Cumulative revenue with interest, $
5	0					
6	1	200,000	0	0	200,000	200,000
7	2	200,000	16,000	16,000	216,000	416,000
8	3	300,000	33,280	49,280	333,280	749,280
9	4	300,000	59,942	109,222	359,942	1,109,222
10			109,222			1,109,222

(a) Total interest and revenue for base case, year 4

	A	B	C	D	E	F
1			Part (a) - Find totals in year 4			
2	i =	0.08				
3/4	End of Year	Revenue at end of year, $	Interest earned during year, $	Cumulative interest, $	Revenue during year with interest, $	Cumulative revenue with interest, $
5	0					
6	1	200000	0	=C6	=B6 + C6	=E6
7	2	200000	=F6*$B$2	=C7 + D6	=B7 + C7	=E7 + F6
8	3	300000	=F7*$B$2	=C8 + D7	=B8 + C8	=E8 + F7
9	4	300000	=F8*$B$2	=C9 + D8	=B9 + C9	=E9 + F8
10			=SUM(C6:C9)		=SUM(E6:E9)	

(b) Spreadsheet relations for base case

	A	B	C	D	E	F
1			Part (b) - Find totals in year 4 with increased revenues			
2	i =	8.0%				
3/4	End of Year	Revenue at end of year, $	Interest earned during year, $	Cumulative interest, $	Revenue during year with interest, $	Cumulative revenue with interest, $
5	0					
6	1	200,000	0	0	200,000	200,000
7	2	200,000	16,000	16,000	216,000	416,000
8	3	600,000	33,280	49,280	633,280	1,049,280
9	4	600,000	83,942	133,222	683,942	1,733,222
10			133,222		1,733,222	
11						
12		Revenue changed				
13						

(c) Totals with increased revenue in years 3 and 4

	A	B	C	D	E	F	
1			Part (c) - Find totals in year 4 considering 4% inflation				Rate of return changed
2	i =	3.85%					
3/4	End of Year	Revenue at end of year, $	Interest earned during year, $	Cumulative interest, $	Revenue during year with interest, $	Cumulative revenue with interest, $	
5	0						
6	1	200,000	0	0	200,000	200,000	
7	2	200,000	7,700	7,700	207,700	407,700	
8	3	300,000	15,696	23,396	315,696	723,396	
9	4	300,000	27,851	51,247	327,851	1,051,247	
10			51,247		1,051,247		

(d) Totals with inflation of 4% per year considered

Figure 1–13
Spreadsheet solutions with sensitivity analysis, Example 1.17a to c.

The equivalent amount after 4 years is $1,109,022, which is comprised of $1,000,000 in total revenue and $109,022 in interest compounded at 8% per year. The shaded cells in Figure 1–13a and b indicate that the sum of the annual values and the last entry in the cumulative columns must be equal.

(b) To determine the effect of increasing estimated revenue for years 3 and 4 to $600,000, use the same spreadsheet and change the entries in cells B8 and B9 as shown in Figure 1–13c. Total interest increases 22%, or $24,000, from $109,222 to $133,222.

(c) Figure 1–13d shows the effect of changing the original i value from 8% to an inflation-adjusted rate of 3.85% in cell B2 on the first spreadsheet. [Remember to return to the

$300,000 revenue estimates for years 3 and 4 after working part (b).] Inflation has now reduced total interest by 53% from $109,222 to $51,247, as shown in cell C10.

Comment

When you are working with an Excel© spreadsheet, it is possible to display all of the entries and functions on the screen as shown in Figure 1–13b by simultaneously touching the <Ctrl> and <`> keys, which may be in the upper left of the keyboard on the key with <~>.

A total of seven spreadsheet functions can perform most of the fundamental engineering economy calculations. The functions are great supplemental tools, but they do not replace the understanding of engineering economy relations, assumptions, and techniques. Using the symbols P, F, A, i, and n defined in the previous section, the functions most used in engineering economic analysis are formulated as follows.

To find the present value P: $= \mathbf{PV}(i\%,n,A,F)$

To find the future value F: $= \mathbf{FV}(i\%,n,A,P)$

To find the equal, periodic value A: $= \mathbf{PMT}(i\%,n,P,F)$

To find the number of periods n: $= \mathbf{NPER}(i\%,A,P,F)$

To find the compound interest rate i: $= \mathbf{RATE}(n,A,P,F)$

To find the compound interest rate i of any series: $= \mathbf{IRR}(\mathbf{first\_cell{:}last\_cell})$

To find the present value P of any series: $= \mathbf{NPV}(i\%, \mathbf{second\_cell{:}last\_cell}) + \mathbf{first\_cell}$

If some of the parameters don't apply to a particular problem, they can be omitted and zero is assumed. For readability, spaces can be inserted between parameters within parentheses. If the parameter omitted is an interior one, the comma must be entered. The last two functions require that a series of numbers be entered into contiguous spreadsheet cells, but the first five can be used with no supporting data. In all cases, the function must be preceded by an equals sign (=) in the cell where the answer is to be displayed.

One of the characteristics of an Excel© spreadsheet to remember is that the first three functions listed above (PV, FV, and PMT) will always display the answer with the opposite sign of that entered for the A, F, or P values in parenthesis. This indicates that the cash flow resultant is in the opposite direction of the cash flow amounts entered. In order to display the same sign, simply enter a minus sign prior to the function name.

To understand how the spreadsheet functions work, look back at Example 1.6a, where the equivalent annual amount A is unknown, as indicated by $A = ?$. (In Chapter 2, we learn how engineering economy factors calculate A, given P, i, and n.) To find A using a spreadsheet function, simply enter the PMT function $= \text{PMT}(5\%,5,5000)$. Figure 1–14 is a screen image of a spreadsheet with this PMT function entered into cell B4. The answer ($1154.87) is displayed. The answer may appear in red and in parentheses, or with a minus sign on your screen to indicate a negative amount from the perspective of a reduction in the account balance. The right side of Figure 1–14 presents the solution to Example 1.6b. The future value F is determined by using the FV function. The FV function appears in the formula bar. Many examples throughout this text include cell tags, as shown here, to indicate the format of important entries.

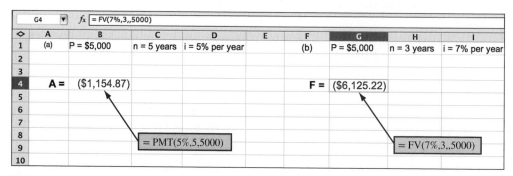

Figure 1–14
Use of spreadsheet functions PMT and FV, Example 1.6.

CHAPTER SUMMARY

Engineering economy is the application of economic factors and criteria to evaluate alternatives, considering the time value of money. The engineering economy study involves computing a specific economic measure of worth for estimated cash flows over a specific period of time.

The concept of *equivalence* helps in understanding how different sums of money at different times are equal in economic terms. The differences between simple interest (based on principal only) and compound interest (based on principal and interest upon interest) have been described in formulas, tables, and graphs. This power of compounding is very noticeable, especially over extended periods of time, and for larger sums of money.

The MARR is a reasonable rate of return established as a hurdle rate to determine if an alternative is economically viable. The MARR is always higher than the return from a safe investment and the cost to acquire needed capital.

Also, we learned a lot about cash flows:

End-of-year convention for cash flow location

Net cash flow computation

Different perspectives in determining the cash flow sign

Construction of a cash flow diagram

Difficulties with estimating future cash flows accurately

PROBLEMS

Basic Concepts

1.1 When faced with different alternatives, there are many criteria that can be used to identify the best one. What is the tangible criterion that is usually used in an engineering economic analysis?

1.2 Identify at least three noneconomic attributes that may be used as evaluation criteria in the decision-making process.

1.3 Define the term measure of worth. Identify three different commonly-applied measures.

1.4 List three evaluation criteria besides the economic ones for selecting the best automobile to purchase.

1.5 Identify the following factors as either tangible or intangible: sustainability, installation cost, transportation cost, simplicity, taxes, resale value, morale, rate of return, dependability, inflation, acceptance by others, and ethics.

1.6 Identify at least five areas of personal finances in which engineering economic analysis can be used by you in the future.

Ethics

1.7 After starting the wall and roof construction on a custom-designed home, FD, the prime contractor, realized he was not going to make the big profit that he had anticipated due to the many unique features of the home for which he had not correctly estimated the cost to the owners. The contract was fixed-price with

no more than 10% override on the total cost. As time proceeded, FD took the choices of the owners on appliances, finishing work on the floors and walls, and many other features and purchased look-alike substitutes from questionable-quality, internationally-based manufacturers and vendors.

After living in the house for only 3 years, the owners were so disappointed with the quality of work that they decided to bring a law suit against FD and his company for breach of contract. However, they needed sound reasons upon which to base their legal claims against FD, were the case to go to court. One of the owners, being an engineer, decided to consult the Code of Ethics for Engineers to gain insight into what may be substantial violations performed during the construction phase.

(a) Help the owners by suggesting Code violations and provide a brief logic as to why each one sited from the Code may be a sound basis for legal action.

(b) Discuss the applicability to this situation of the Engineer's Code of Ethics.

(c) Suggest other resources that may be of use to the owners as they ponder legal action.

1.8 Stefanie is a design engineer with an international railroad locomotive manufacturing company in the state of Illinois. Management wants to return some of the engineering design work to the United States rather than export all of it to India, where their primary design work have been accomplished for the last decade. This transfer will employ more people locally and could improve the economic conditions for families in and around Illinois.

Stefanie and her design team were selected as a test case to determine the quality and speed of the design work they could demonstrate on a more fuel-efficient diesel locomotive. None of her team members or she has done such a significant design job themselves, because their jobs had previously entailed only the interface with the subcontracted engineers in India. One of her team members had a great design idea on a key element that will improve fuel efficiency by approximately 15%. She told Stefanie it came from one of the Indian-generated documents, but that it would probably be okay for the team to use it and remain silent as to its origin, since it was quite clear the U.S. management was about to cancel the foreign contract. Though reluctant at first, Stefanie did go forward with a design that included the efficiency improvement, and no mention of the origin of the idea was made at the time of the oral presentation or documentation delivery. As a result, the Indian contract was cancelled and full design responsibility was transferred to Stefanie's group.

Consult the NSPE Code of Ethics for Engineers (Appendix C) and identify sections that are points of concern about Stefanie's decisions and actions.

1.9   Consider the common moral that stealing is wrong. Hector is with a group of friends in a local supermarket. One of Hector's buddies takes a high-energy drink from a 6-pack on the shelf, opens it, drinks it, and returns the empty can to the package, with no intention of paying for it. He then invites the others to do the same saying "It's only one drink. Others do it all the time." All the others, except Hector, have now consumed a drink of their choice. Personally, Hector believes this is a form of stealing. State three actions that Hector can take and evaluate them from the personal moral perspective.

1.10   Claude is a fourth-year engineering university student who has just been informed by his instructor that he made a very low grade on his Spanish language final test for the year. Though he had a passing score prior to the final, his final grade was so low that he has now flunked the entire year and will likely have to extend his graduation another semester or two.

Throughout the year, Claude, who hated the course and his instructor, has copied homework, cheated on tests, and never seriously studied for anything in the course. He did realize during the semester that he was doing something that even he considered wrong morally and ethically. He knew he had done badly on the final. The classroom was reconfigured for the final exam in a way that he could not get any answers from classmates, and, cell phones were collected prior to the exam, thus removing texting and using WhatsApp possibilities to friends outside the classroom who may help him on the final exam. Claude is now face-to-face with the instructor in her office. The question to Claude is "What have you been doing throughout this year to make passing scores repeatedly, but demonstrate such a poor command of Spanish on the final exam?"

From an ethical viewpoint, what options does Claude have in his answer to this question? Also, discuss some of the possible effects that this experience may have upon Claude's future actions and moral dilemmas.

Interest Rate and Rate of Return

1.11   Abby purchased 100 shares of her dad's favorite stock for $25.80 per share exactly 1 year ago, commission free. She sold it today for a total amount of $2865. She plans to invest the entire amount in a different corporation's stock today, but must now pay a $50 commission fee. If she plans to sell this new stock exactly 1 year from now and realize the same return as she has just made, what must be the total amount she receives next year? Include the commission fee as a part of the purchase price, but neglect any tax effects.

1.12   In order to build a new warehouse facility, the regional distributor for Valco Multi-position Valves borrowed $1.6 million at 10% per year interest. If the company repaid the loan in a lump sum amount after 2 years, what was (a) the amount of the payment, and (b) the amount of interest?

1.13   RKI Instruments borrowed $4,800,000 from a private equity firm for expansion of its facility for manufacturing carbon monoxide monitors. The company repaid the loan after 1 year with a single payment of $5,184,000. What was the interest rate on the loan?

1.14   Callahan Construction borrowed $2.6 million to finance the construction of an entertainment complex in a smart community development project. The company made "interest only" payments of $312,000 each year for 3 years and then repaid the principal in a single lump sum payment of $2.6 million. What was the interest rate on the loan?

1.15   Which of the following 1-year investments has the highest rate of return: $12,500 that yields $1125 in interest, $56,000 that yields $6160 in interest, or $95,000 that yields $7600 in interest?

1.16 A new engineering graduate who started a consulting business borrowed money for 1 year to furnish the office. The amount of the loan was $45,800, and it had an interest rate of 10% per year. However, because the new graduate had not built up a credit history, the bank made him buy loan-default insurance that costs $900. In addition, the bank charged a loan set-up fee of 1% of the loan principal. What was the effective interest rate the engineer paid for the loan?

Terminology and Symbols

In problems 1.17 to 1.21, identify the four engineering economy symbols and their values. Use a question mark with the symbol whose value is to be determined.

1.17 Determine the amount of money FrostBank might loan a housing developer who will repay the loan 2 years from now by selling eight water-view lots at $240,000 each. Assume the bank's interest rate is 10% per year.

1.18 Bodine Electric, based in Des Moines, Iowa, USA, makes gear motors with a three-stage, selectively-hardened gearing cluster that is permanently lubricated. If the company borrows $20 million for a new distribution facility in Europe, how much must it pay back each year to repay the loan in six equal annual payments at an interest rate of 10% per year?

1.19 DubaiWorks manufactures angular contact ball bearings for pumps that operate in harsh environments. The company invested $2.4 million in a process that resulted in net profits of $760,000 per year for five consecutive years. What rate of return did the company make?

1.20 A capital investment firm placed $1.5 million 2 years ago to acquire part-ownership in an innovative chip making company. How long will it take from the date of their initial investment for their share of the chip company to be worth $3 million, if the company is realizing a 20% per year return?

1.21 Southwestern Moving and Storage wants to have enough money to purchase a new tractor-trailer in 3 years. If the unit will cost $250,000, how much should the company set aside each year provided the account earns 9% per year?

Cash Flows

1.22 Identify the following as cash inflows or outflows to a privately-owned water company: well drilling, maintenance, water sales, accounting, government grants, issuance of bonds, energy cost, pension plan contributions, heavy equipment purchases, used equipment sales, stormwater fees, and discharge permit revenues.

1.23 The annual cash flows (in $1000 units) for Browning Brothers Glass Works are summarized.
 (a) Determine the total net cash flow over the 5 years.
 (b) Calculate the percentage of revenues represented by expenses for each year.

Year	1	2	3	4	5
Revenue, $	521	685	650	804	929
Expenses, $	610	623	599	815	789

1.24 Bucknell, Inc. uses the calendar year as its fiscal year. Determine the total net cash flow recorded at the end of the fiscal year.

Month	Receipts, $1000	Disbursements, $1000
Jan	300	500
Feb	950	500
Mar	200	400
Apr	120	400
May	600	500
June	900	600
July	800	300
Aug	900	300
Sept	900	200
Oct	500	400
Nov	400	400
Dec	1800	700

1.25 To attract new customers, EP Employees Credit Union advertised that they will begin paying 3% interest every quarter on all savings accounts. (Their competitors pay interest every 6 months.) The credit union uses March 31st, June 30th, September 30th, and December 31st as quarterly interest periods. Determine (a) the end-of period totals in the account, and (b) the interest paid each quarter on the total. Assume there are no withdrawals and that quarterly interest is not redeposited.

Month	Deposit, $
Jan	50
Feb	70
Mar	0
Apr	120
May	20
June	0
July	150
Aug	90
Sept	0
Oct	40
Nov	110
Dec	0

1.26   Construct a cash flow diagram that represents the amount of money that will be accumulated in 7 years from an initial investment of $20,000 now and $3,500 per year for 7 years at an interest rate of 8% per year.

1.27   Construct a cash flow diagram to find the present worth in year 0 of a $400 expenditure in year 3, a $900 receipt in year 4, and $100 expenses in each of years 5 and 6 at an interest rate of 15% per year.

Equivalence

1.28   Use economic equivalence to determine the amount of money or value of i that makes the following statements correct.
   (a) $5000 today is equivalent to $4275 exactly 1 year ago at $i = $____% per year.
   (b) A car that costs $28,000 today will cost $____ a year from now at $i = 4$% per year.
   (c) At $i = 4$% per year, a car that costs $28,000 now, would have cost $____ one year ago.
   (d) Last year, Jackson borrowed $20,000 to buy a preowned boat. He repaid the principal of the loan plus $2750 interest after only 1 year. This year, his brother Henri borrowed $15,000 to buy a car and expects to pay it off in only 1 year plus interest of $2295. The rate that each brother paid for his loan is ____% for Jackson and ____% per year for Henri.
   (e) Last year, Sheila turned down a job that paid $75,000 per year. This year, she accepted one that pays $81,000 per year. The salaries are equivalent at $i = $____% per year.

1.29   Vebco Water and Gas received a contract for a seawater desalination plant wherein the company expected to make a 28% rate of return on its investment. (a) If Vebco invested $8 million the first year, what was the amount of its profit in that year? (b) What amount would have to be invested to realize the same monetary amount of return if the rate decreases to 15% per year?

1.30   A publicly traded construction company reported that it just paid off a loan that it received 1 year earlier. If the total amount of money the company paid was $1.6 million and the interest rate on the loan was 10% per year, how much money had the company borrowed 1 year ago?

1.31   During a recession, the price of goods and services goes down because of low demand. A company that makes Ethernet adapters is planning to expand its production facility at a cost of $1,000,000 one year from now. However, a contractor who needs work has offered to do the job for $790,000 provided the company will do the expansion now

instead of 1 year from now. If the interest rate is 10% per year, how much of a discount is the company receiving?

1.32   Bull Built, a design/build engineering company that has always given year-end bonuses in the amount of $4,000 to each of its engineers, is having cash flow problems. The president said that although Bull Built couldn't give bonuses this year, next year's bonus would be large enough to make up for this year's omission. If the interest rate is 10% per year, what is the equivalent bonus that engineers should receive next year?

1.33   State University tuition and fees can be paid using one of two plans.

Early-bird: Pay total amount due *one year in advance* and get a 10% discount.

On-time: Pay total amount due when classes start.

   If the cost of tuition and fees is $20,000 per year, (a) how much is paid in the early-bird plan, and (b) at an interest rate of 6% per year, what is the equivalent amount of the savings compared to paying when classes start, that is, 1 year later than the early-bird plan?

Simple and Compound Interest

1.34   Durco Automotive needs a $1 million balance in its contingency fund 3 years from now. The CFO (chief financial officer) wants to know how much to deposit now into Durco's high-yield investment account. Determine the amount if it grows at a rate of 20% per year (a) simple interest, and (b) compound interest.

1.35   TMI Systems, a company that customizes software for construction cost estimates, repaid a loan obtained 3 years ago at 7% per year simple interest. If the amount that TMI repaid was $120,000, calculate the principal of the loan.

1.36   The Nicor family is planning to purchase a new home 3 years from now. If they have $240,000 now, how much will be available 3 years from now? The fund grows at a compound rate of 12% per year.

1.37   (a) At a simple interest rate of 12% per year, determine how long it will take $5000 to increase to twice as much. (b) Compare the time it will take to double if the rate is 20% per year simple interest.

1.38   Valley Rendering, Inc. is considering purchasing a new flotation system for grease recovery. The company can finance a $150,000 system at 5% per year compound interest or 5.5% per year simple interest. If the total amount owed is due in a single payment at the end of 3 years, (a) which interest rate should the company select, and (*b*) how much is the difference in interest between the two schemes?

1.39 Iselt Welding has extra funds to invest for future capital expansion. If the selected investment pays simple interest, what annual interest rate would be required for the amount to grow from $60,000 to $90,000 in 5 years?

1.40 In order to make CDs look more attractive as an investment than they really are, some banks advertise that their rates are higher than their competitors' rates; however, the fine print says that the rate is based on simple interest. If you were to deposit $10,000 at 10% per year simple interest in a CD, what compound interest rate would yield the same amount of money in 3 years? Solve by formula and write the spreadsheet function to display the *i* value.

1.41 Fill in the missing values (A through D) for a loan of $10,000 if the interest rate is compounded at 10% per year.

End of Year	Interest for Year	Amount Owed After Interest	End of Year Payment	Amount Owed After Payment
0	—	—	—	10,000
1	1000	11,000	2000	9,000
2	900	9,900	2000	A
3	B	C	2000	D

1.42 A company that manufactures general-purpose transducers invested $2 million 4 years ago in high-yield junk bonds. If the bonds are now worth $2.8 million, what rate of return per year did the company make on the basis of (*a*) simple interest, and (*b*) compound interest? (*c*) What is the spreadsheet function to find the answer for compound interest?

MARR and Opportunity Cost

1.43 Identify the following as either equity or debt financing: bonds, stock sales, retained earnings, venture capital, short term loan, capital advance from friend, cash on hand, credit card, home equity loan.

1.44 What is the weighted average cost of capital for a corporation that finances an expansion project using 40% retained earnings and 60% venture

capital? Assume the interest rates are 10% for equity financing and 16% for debt financing.

1.45 Managers from different departments in Bensen Systems, a large multinational corporation, have offered seven projects for consideration by the corporate office. A staff member for the chief financial officer used key words to identify the projects and then listed them in order of projected rate of return as shown below. If the company wants to grow rapidly through high leverage and uses only 5% equity financing that has a cost of equity capital of 10% and 95% debt financing with a cost of debt capital of 19%, which projects should the company undertake?

Project ID	Projected ROR, % per year
Inventory	30.0
Technology	28.4
Warehouse	21.9
Maintenance	19.5
Products	13.1
Energy	9.6
Shipping	8.2

1.46 Your boss, whose background is in financial planning, is concerned about the company's high-weighted average cost of capital of 21%. He has asked you to determine what combination of debt-equity financing would lower the company's WACC to 13%. If the cost of the company's equity capital is 6% and the cost of debt financing is 28%, what debt-equity mix would you recommend?

1.47 Last month you lent a work colleague $5000 to cover some overdue bills. He agreed to pay you in 1 month with interest at 2% for the month, thus owing you $5100. Today, when the repayment is due, he asked you to extend the loan for another month and he would pay you the $5100 next month. In the meantime, you have had the offer to invest as much as you wish in an oil-well venture that is expected to pay 25% per year and a hot new IT stock that is estimated to return 30% the first year. If you let your colleague have another month, what is the opportunity cost of your decision? (*Note*: Express your answer in dollar and percentage amounts.)

EXERCISES FOR SPREADSHEETS

1.48 Write the engineering economy symbol that corresponds to each of the following spreadsheet functions.

(*a*) PV
(*b*) PMT
(*c*) NPER
(*d*) IRR
(*e*) FV
(*f*) RATE

1.49 State the purpose for each of the following six built-in spreadsheet functions.
(a) PV($i\%,n,A,F$)
(b) FV($i\%,n,A,P$)
(c) RATE(n,A,P,F)
(d) IRR(first_cell:last_cell)
(e) PMT($i\%,n,P,F$)
(f) NPER($i\%,A,P,F$)

1.50 For the following five spreadsheet functions, (a) write the values of the engineering economy symbols P, F, A, i, and n, using a ? for the symbol that is to be determined, and (b) state whether the displayed answer will have a positive sign, a negative sign, or it can't be determined from the entries.

(1) = FV(8%,10,3000,8000)
(2) = PMT(12%,20,−16000)
(3) = PV(9%,15,1000,600)
(4) = NPER(10%,−290,,12000)
(5) = FV(5%,5,500,−2000)

1.51 Emily and Madison both invest $1000 at 10% per year for 4 years. Emily receives simple interest and Madison gets compound interest. Use a spreadsheet and cell reference formats to develop relations that show a total of $64 more interest for Madison at the end of the 4 years. Assume no withdrawals or further deposits are made during the 4 years.

ADDITIONAL PROBLEMS AND FE EXAM REVIEW QUESTIONS

1.52 All of the following are examples of noneconomic factors except:
(a) Availability of resources
(b) Goodwill
(c) Customer acceptance
(d) Profit

1.53 Another name for noneconomic attributes is:
(a) Sustainability
(b) Intangible factors
(c) Equivalence
(d) Evaluation criteria

1.54 All of the following are noneconomic attributes except:
(a) Sustainability
(b) Morale
(c) Taxes
(d) Environmental acceptability

1.55 All of the following mean the same as Minimum Attractive Rate of Return except:
(a) Hurdle rate
(b) Inflation rate
(c) Benchmark rate
(d) Cutoff rate

1.56 The time it would take for money to double at a simple interest rate of 5% per year is closest to:
(a) 10 years
(b) 12 years
(c) 15 years
(d) 20 years

1.57 At a compound interest rate of 10% per year, the amount that $10,000 one year ago is equivalent to now is closest to:
(a) $8264
(b) $9091
(c) $11,000
(d) $12,100

1.58 The compound interest rate per year that amounts of $1000 one year ago and $1345.60 one year hence are equivalent to is closest to:
(a) 8.5% per year
(b) 10.8% per year
(c) 20.2% per year
(d) None of the above

1.59 The simple interest rate per year that will accumulate the same amount of money in 2 years as a compound interest rate of 20% per year is closest to:
(a) 20.5%
(b) 21%
(c) 22%
(d) 23%

1.60 In order to finance a new project costing $30 million, a company borrowed $21 million at 16% per year interest and used retained earnings valued at 12% per year for the remainder of the financing. The company's weighted average cost of capital for the project was closest to:
(a) 12.5%
(b) 13.6%
(c) 14.8%
(d) 15.6%

1.61 A company that utilizes carbon fiber 3-D printing wants to have money available 2 years from now to add new equipment. The company currently has $650,000 in a capital account and it plans to deposit $200,000 now and another $200,000 one year from now. The total amount available in 2 years, provided it returns a compounded rate of 15% per year, is closest to:
(a) $1,354,100
(b) $1,324,100
(c) $1,125,125
(d) $1,050,000

CASE STUDY

COST OF ELECTRICITY WITH RENEWABLE SOURCES ADDED

Background

Pedernales Electric Cooperative (PEC) is the largest member-owned electric co-op in the United States with over 232,000 meters in 12 Central Texas counties. PEC has a capacity of approximately 1300 MW (megawatts) of power, of which 277 MW, or about 21%, is from renewable sources. The latest addition is 60 MW of power from a wind farm in south Texas close to the city of Corpus Christi. A constant question is how much of PEC's generation capacity should be from renewable sources, especially given the environmental issues with coal-generated electricity and the rising costs of hydrocarbon fuels.

Wind and nuclear sources are the current consideration for the PEC leadership as Texas is increasing its generation by nuclear power and the state is the national leader in wind farm–produced electricity.

Consider yourself a member of the board of directors of PEC. You are an engineer who has been newly elected by the PEC membership to serve a 3-year term as a director-at-large. As such, you do not represent a specific district within the entire service area; all other directors do represent a specific district. You have many questions about the operations of PEC, plus you are interested in the economic and societal benefits of pursuing more renewable source generation capacity.

Information

Here are some data that you have obtained. The information is sketchy, as this point, and the numbers are very approximate. Electricity generation cost estimates are national in scope, not PEC-specific, and are provided in cents per kilowatt-hour (¢/kWh).

	Generation Cost, ¢/kWh	
Fuel Source	Likely Range	Reasonable Average
Coal	4 to 9	7.4
Natural gas	4 to 10.5	8.6
Wind	4.8 to 9.1	8.2
Solar	4.5 to 15.5	8.8

National average cost of electricity to residential customers: 11¢/kWh

PEC average cost to residential customers: 10.27 ¢/kWh (from primary sources) and 10.92 ¢/kWh (renewable sources)

Expected life of a generation facility: 20 to 40 years (it is likely closer to 20 than 40)

Time to construct a facility: 2 to 5 years

Capital cost to build a generation facility: $900 to $1500 per kW

You have also learned that the PEC staff uses the well-recognized *levelized energy cost* (LEC) method to determine the price of electricity that must be charged to customers to break even. The formula takes into account the capital cost of the generation facilities, the cost of capital of borrowed money, annual maintenance and operation (M&O) costs, and the expected life of the facility. The LCOE (levelized cost of electricity) formula, expressed in dollars per kWh for ($t = 1, 2, \ldots, n$), is

$$\text{LCOE} = \frac{\displaystyle\sum_{t=1}^{t=n} \frac{I_t + M_t + F_t}{(1 + i)^t}}{\displaystyle\sum_{t=1}^{t=n} \frac{E_t}{(1 + i)^t}}$$

where I_t = capital investments made in year t
M_t = annual maintenance and operating (M&O) costs for year t
F_t = fuel costs for year t
E_t = amount of electricity generated in year t
n = expected life of facility
i = discount rate (cost of capital)

Case Study Exercises

1. If you wanted to know more about the new arrangement with the wind farm in south Texas for the additional 60 MW per year, what types of questions would you ask of a staff member in your first meeting with him or her?

2. Much of the current generation capacity of PEC facilities utilizes coal and natural gas as the primary fuel source. What about the ethical aspects of the government's allowance for these plants to continue polluting the atmosphere with the emissions that may cause health problems for citizens and further the effects of global warming? What types of regulations, if any, should be developed for PEC (and other generators) to follow in the future?

3. You developed an interest in the LCOE relation and the publicized cost of electricity of 10.27¢/kWh for this year. You wonder if the addition of 60 MW of wind-sourced electricity will make any difference in the LCOE value for this next year. You did learn the following:

   This is year $t = 11$ for LCOE computation purposes

   $n = 25$ years

   $i = 5\%$ per year

   $E_{11} = 5.052$ billion kWh

   LCOE last year was 10.22 ¢/kWh (last year's breakeven cost to customers)

   From these sketchy data, can you determine the value of unknowns in the LCOE relation for this year? Is it possible to determine if the wind farm addition of 60 MW makes any difference in the electricity rate charged to customers? If not, what additional information is necessary to determine the LCOE with the wind source included?

CHAPTER 2

Factors: How Time and Interest Affect Money

© Noel Hendrickson/Blend Images LLC

LEARNING OUTCOMES

Purpose: Derive and use the engineering economy factors to account for the time value of money.

SECTION	TOPIC	LEARNING OUTCOME
2.1	*F/P* and *P/F* factors	• Derive and use factors for single amounts—compound amount (*F/P*) and present worth (*P/F*) factors.
2.2	*P/A* and *A/P* factors	• Derive and use factors for uniform series—present worth (*P/A*) and capital recovery (*A/P*) factors.
2.3	*F/A* and *A/F* factors	• Derive and use factors for uniform series—compound amount (*F/A*) and sinking fund (*A/F*) factors.
2.4	Factor values	• Use linear interpolation in factor tables or spreadsheet functions to determine factor values.
2.5	Arithmetic gradient	• Use the present worth (*P/G*) and uniform annual series (*A/G*) factors for arithmetic gradients.
2.6	Geometric gradient	• Use the geometric gradient series factor (*P/A,g*) to find present worth.
2.7	Find *i* or *n*	• Use equivalence relations to determine *i* (interest rate or rate of return) or *n* for a cash flow series.

he cash flow is fundamental to every economic study. Cash flows occur in many configurations and amounts—isolated single values, series that are uniform, and series that increase or decrease by constant amounts or constant percentages. This chapter develops derivations for all the commonly used **engineering economy factors** that take the time value of money into account.

The application of factors is illustrated using their mathematical forms and a standard notation format. Spreadsheet functions are used in order to rapidly work with cash flow series and to perform sensitivity analysis.

If the derivation and use of factors are not covered in the course, alternate ways to perform time value of money calculations are summarized in Appendices A and D.

2.1 Single-Amount Factors (F/P and P/F) ● ● ●

The **most fundamental factor in engineering economy** is the one that determines the amount of money F accumulated after n years (or periods) from a *single* present worth P, with interest compounded one time per year (or period). Recall that compound interest refers to interest paid on top of interest. Therefore, if an amount P is invested at time $t = 0$, the amount F_1 accumulated 1 year hence at an interest rate of i percent per year will be

$$F_1 = P + Pi$$
$$= P(1 + i)$$

where the interest rate is expressed in decimal form. At the end of the second year, the amount accumulated F_2 is the amount after year 1 plus the interest from the end of year 1 to the end of year 2 on the entire F_1.

$$F_2 = F_1 + F_1 i$$
$$= P(1 + i) + P(1 + i)i \qquad [2.1]$$

The amount F_2 can be expressed as

$$F_2 = P(1 + i + i + i^2)$$
$$= P(1 + 2i + i^2)$$
$$= P(1 + i)^2$$

Similarly, the amount of money accumulated at the end of year 3, using Equation [2.1], will be

$$F_3 = F_2 + F_2 i$$

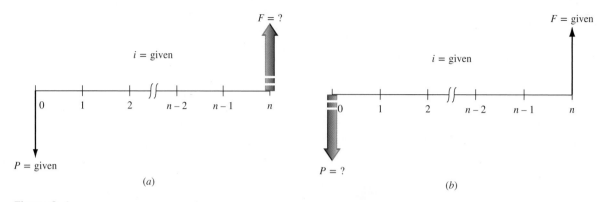

Figure 2–1
Cash flow diagrams for single-payment factors: (*a*) find *F*, given *P*, and (*b*) find *P*, given *F*.

Substituting $P(1 + i)^2$ for F_2 and simplifying, we get

$$F_3 = P(1 + i)^3$$

From the preceding values, it is evident by mathematical induction that the formula can be generalized for *n* years. To find *F*, given *P*,

$$F = P(1 + i)^n \qquad\qquad [2.2]$$

The factor $(1 + i)^n$ is called the *single-payment compound amount factor* (SPCAF), but it is usually referred to as the **F/P factor.** This is the conversion factor that, when multiplied by *P*, yields the future amount *F* of an initial amount *P* after *n* years at interest rate *i*. The cash flow diagram is seen in Figure 2–1*a*.

Reverse the situation to **determine the *P* value for a stated amount *F*** that occurs *n* periods in the future. Simply solve Equation [2.2] for *P*.

$$P = F\left[\frac{1}{(1 + i)^n}\right] = F(1 + i)^{-n} \qquad\qquad [2.3]$$

The expression $(1 + i)^{-n}$ is known as the *single-payment present worth factor* (SPPWF), or the **P/F factor.** This expression determines the present worth *P* of a given future amount *F* after *n* years at interest rate *i*. The cash flow diagram is shown in Figure 2–1*b*.

Note that the two factors derived here are for *single payments;* that is, they are used to find the present or future amount when only one payment or receipt is involved.

> A standard notation has been adopted for all factors. The notation includes two cash flow symbols, the interest rate, and the number of periods. It is always in the general form (*X/Y,i,n*). The letter *X* represents what is sought, while the letter *Y* represents what is given. For example, *F/P* means *find F when given P.* The *i* is the interest rate in percent, and *n* represents the number of periods involved.

Using this notation, (*F/P*,6%,20) represents the factor that is used to calculate the future amount *F* accumulated in 20 periods if the interest rate is 6% per period. The *P* is given. The standard notation, simpler to use than formulas and factor names, will be used hereafter.

Table 2–1 summarizes the standard notation and equations for the *F/P* and *P/F* factors. This information is also included on the first pages of this text.

TABLE 2–1	*F/P* and *P/F* Factors: Notation and Equations				
Factor			**Standard Notation**	**Equation**	**Excel**
Notation	**Name**	**Find/Given**	**Equation**	**with Factor Formula**	**Function**
(*F/P,i,n*)	Single-payment compound amount	*F/P*	$F = P(F/P,i,n)$	$F = P(1 + i)^n$	$= FV(i\%,n,,P)$
(*P/F,i,n*)	Single-payment present worth	*P/F*	$P = F(P/F,i,n)$	$P = F(1 + i)^{-n}$	$= PV(i\%,n,,F)$

To simplify routine engineering economy calculations, tables of factor values have been prepared for interest rates from 0.25% to 50% and time periods from 1 to large *n* values, depending on the *i* value. These tables, found at the rear of the book, have a colored edge for easy identification. They are arranged with factors across the top and the number of periods *n* down the left side. The word *discrete* in the title of each table emphasizes that these tables utilize the end-of-period convention and that interest is compounded once each interest period. For a given factor, interest rate, and time, the correct factor value is found at the intersection of the factor name and *n*. For example, the value of the factor (*P*/*F*,5%,10) is found in the *P*/*F* column of Table 10 at period 10 as 0.6139. This value is determined by using Equation [2.3].

$$(P/F,5\%,10) = \frac{1}{(1+i)^n}$$

$$= \frac{1}{(1.05)^{10}}$$

$$= \frac{1}{1.6289} = 0.6139$$

For **spreadsheets,** a future value *F* is calculated by the FV function using the format

$$= FV(i\%,n,,P) \tag{2.4}$$

A present amount *P* is determined using the PV function with the format

$$= PV(i\%,n,,F) \tag{2.5}$$

These functions are included in Table 2–1. Refer to Appendix A or Excel online help for more information on the use of FV and PV functions.

EXAMPLE 2.1

Sandy, a manufacturing engineer, just received a year-end bonus of $10,000 that will be invested immediately. With the expectation of earning at the rate of 8% per year, Sandy hopes to take the entire amount out in exactly 20 years to pay for a family vacation when the oldest daughter is due to graduate from college. Find the amount of funds that will be available in 20 years by using (*a*) hand solution by applying the factor formula and tabulated value, and (*b*) a spreadsheet function.

Solution

The cash flow diagram is the same as Figure 2–1*a*. The symbols and values are

$$P = \$10,000 \qquad F = ? \qquad i = 8\% \text{ per year} \qquad n = 20 \text{ years}$$

(*a*) *Factor formula:* Apply Equation [2.2] to find the future value *F*. Rounding to four decimals, we have

$$F = P(1 + i)^n = 10,000(1.08)^{20} = 10,000(4.6610)$$
$$= \$46,610$$

Standard notation and tabulated value: Notation for the *F*/*P* factor is (*F*/*P*,*i*%,*n*).

$$F = P(F/P,8\%,20) = 10,000(4.6610)$$
$$= \$46,610$$

Table 13 provides the tabulated value. Round-off errors can sometimes cause a slight difference in the final answer between these two methods.

(*b*) *Spreadsheet:* Use the FV function to find the amount 20 years in the future. The format is that shown in Equation [2.4]; the numerical entry is = FV(8%,20,,10000). The spreadsheet will appear similar to that in the right side of Figure 1–14, with the answer ($46,609.57) displayed. (You should try it on your own computer now.) The FV function has performed the computation in part (*a*) and displayed the result.

The equivalency statement is: If Sandy invests $10,000 now and earns 8% per year every year for 20 years, $46,610 will be available for the family vacation.

EXAMPLE 2.2 The Steel Plant Case

As discussed in the introduction to this chapter, the HBNA plant will require an investment of $200 million to construct. Delays beyond the anticipated implementation year of 2020 will require additional money to construct the plant. Assuming that the cost of money is 10% per year compound interest, use both **tabulated factor values** and **spreadsheet functions** to determine the following for the board of directors of the Italian company that plans to develop the plant.

(a) The equivalent investment needed in 2023 if the plant is delayed for 3 years.

(b) The equivalent investment needed in 2023 if the plant is constructed sooner than originally planned.

Solution

Figure 2–2 is a cash flow diagram showing the expected investment of $200 million ($200 M) in 2020, which we will identify as time $t = 0$. The required investments 3 years in the future and 4 years in the past are indicated by $F_3 = ?$ and $P_{-4} = ?$, respectively.

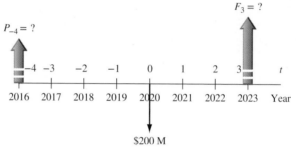

Figure 2–2
Cash flow diagram for Example 2.2a and b.

(a) To find the equivalent investment required in 3 years, apply the F/P factor. Use $1 million units and the tabulated value for 10% interest (Table 15).

$$F_3 = P(F/P,i,n) = 200(F/P,10\%,3) = 200(1.3310)$$
$$= \$266.2 \quad (\$266,200,000)$$

Now, use the FV function on a spreadsheet to find the same answer, $F_3 = \$266.20$ million. (Refer to Figure 2–3, left side.)

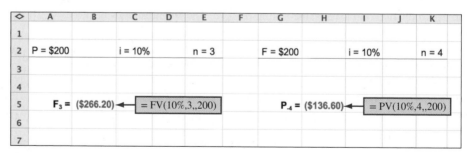

Figure 2–3
Spreadsheet functions for Example 2.2.

(b) The year 2016 is 4 years prior to the planned construction date of 2020. To determine the equivalent cost 4 years earlier, consider the $200 M in 2020 ($t = 0$) as the future value F and apply the P/F factor for $n = 4$ to find P_{-4}. (Refer to Figure 2–2.) Table 15 supplies the tabulated value.

$$P_{-4} = F(P/F,i,n) = 200(P/F,10\%,4) = 200(0.6830)$$
$$= \$136.6 \quad (\$136,600,000)$$

The PV function = PV(10%,4,,200) will display the same amount as shown in Figure 2–3, right side.

This equivalence analysis indicates that at $136.6 M in 2016, the plant would have cost about 68% as much as in 2020, and that waiting until 2023 will cause the price tag to increase about 33% to $266 M.

2.2 Uniform Series Present Worth Factor and Capital Recovery Factor (*P/A* and *A/P*) ●●●

The equivalent present worth *P* of a uniform series *A* of end-of-period cash flows (investments) is shown in Figure 2–4*a*. An expression for the present worth can be determined by considering each *A* value as a future worth *F*, calculating its present worth with the *P/F* factor, Equation [2.3], and summing the results.

$$P = A\left[\frac{1}{(1+i)^1}\right] + A\left[\frac{1}{(1+i)^2}\right] + A\left[\frac{1}{(1+i)^3}\right] + \cdots$$
$$+ A\left[\frac{1}{(1+i)^{n-1}}\right] + A\left[\frac{1}{(1+i)^n}\right]$$

The terms in brackets are the *P/F* factors for years 1 through *n*, respectively. Factor out *A*.

$$P = A\left[\frac{1}{(1+i)^1} + \frac{1}{(1+i)^2} + \frac{1}{(1+i)^3} + \cdots + \frac{1}{(1+i)^{n-1}} + \frac{1}{(1+i)^n}\right] \qquad [2.6]$$

To simplify Equation [2.6] and obtain the *P/A* factor, multiply the *n*-term geometric progression in brackets by the (*P/F*,*i*%,1) factor, which is $1/(1 + i)$. This results in Equation [2.7]. Now subtract the two equations, [2.6] from [2.7], and simplify to obtain the expression for *P* when $i \neq 0$ (Equation [2.8]).

$$\frac{P}{1+i} = A\left[\frac{1}{(1+i)^2} + \frac{1}{(1+i)^3} + \frac{1}{(1+i)^4} + \cdots + \frac{1}{(1+i)^n} + \frac{1}{(1+i)^{n+1}}\right] \qquad [2.7]$$

$$\frac{1}{1+i}P = A\left[\frac{1}{(1+i)^2} + \frac{1}{(1+i)^3} + \cdots + \frac{1}{(1+i)^n} + \frac{1}{(1+i)^{n+1}}\right]$$
$$- \qquad P = A\left[\frac{1}{(1+i)^1} + \frac{1}{(1+i)^2} + \cdots + \frac{1}{(1+i)^{n-1}} + \frac{1}{(1+i)^n}\right]$$

$$\frac{-i}{1+i}P = A\left[\frac{1}{(1+i)^{n+1}} - \frac{1}{(1+i)^1}\right]$$

$$P = \frac{A}{-i}\left[\frac{1}{(1+i)^n} - 1\right]$$

$$P = A\left[\frac{(1+i)^n - 1}{i(1+i)^n}\right] \qquad i \neq 0 \qquad [2.8]$$

The term in brackets in Equation [2.8] is the conversion factor referred to as the *uniform series present worth factor* (USPWF). It is the **P/A factor** used to calculate the **equivalent P value in year 0** for a uniform end-of-period series of *A* values beginning at the end of period 1 and extending for *n* periods. The cash flow diagram is shown in Figure 2–4*a*.

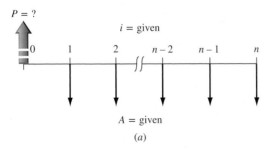

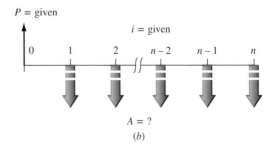

Figure 2–4
Cash flow diagrams used to determine (*a*) *P*, given a uniform series *A*, and (*b*) *A*, given a present worth *P*.

TABLE 2–2		P/A and A/P Factors: Notation and Equations			
	Factor		**Find/Given**	**Factor**	**Standard**	**Excel**
Notation	**Name**			**Formula**	**Notation Equation**	**Function**
$(P/A,i,n)$	Uniform series present worth	P/A	$\dfrac{(1+i)^n - 1}{i(1+i)^n}$	$P = A(P/A,i,n)$	$= \mathrm{PV}(i\%,n,A)$
$(A/P,i,n)$	Capital recovery	A/P	$\dfrac{i(1+i)^n}{(1+i)^n - 1}$	$A = P(A/P,i,n)$	$= \mathrm{PMT}(i\%,n,P)$

To reverse the situation, the present worth P is known and the equivalent uniform series amount A is sought (Figure 2–4b). The first A value occurs at the end of period 1, that is, one period after P occurs. Solve Equation [2.8] for A to obtain

$$A = P\left[\frac{i(1+i)^n}{(1+i)^n - 1}\right] \qquad [2.9]$$

The term in brackets is called the *capital recovery factor* (CRF), or **A/P factor.** It calculates the **equivalent uniform annual worth A** over n years for a given P in year 0, when the interest rate is i.

Placement of P

> The P/A and A/P factors are derived with the present worth P and the first uniform annual amount A **one year (period) apart.** That is, the present worth P *must always* be located **one period prior to the first A.**

The factors and their use to find P and A are summarized in Table 2–2 and on the first pages of this text. The standard notations for these two factors are $(P/A,i\%,n)$ and $(A/P,i\%,n)$. Tables at the end of the text include the factor values. As an example, if $i = 15\%$ and $n = 25$ years, the P/A factor value from Table 19 is $(P/A,15\%,25) = 6.4641$. This will find the equivalent present worth at 15% per year for any amount A that occurs uniformly from years 1 through 25.

Spreadsheet functions can determine both P and A values in lieu of applying the P/A and A/P factors. The PV function calculates the P value for a given A over n years and a separate F value in year n, if it is given. The format is

$$= \mathrm{PV}(i\%,n,A,F) \qquad [2.10]$$

Similarly, the A value is determined by using the PMT function for a given P value in year 0 and a separate F, if given. The format is

$$= \mathrm{PMT}(i\%,n,P,F) \qquad [2.11]$$

Table 2–2 includes the PV and PMT functions.

EXAMPLE 2.3

How much money should you be willing to pay now for a guaranteed $600 per year for 9 years starting next year, at a rate of return of 16% per year?

Solution
The cash flows follow the pattern of Figure 2–4a, with $A = \$600$, $i = 16\%$, and $n = 9$. The present worth is

$$P = 600(P/A,16\%,9) = 600(4.6065) = \$2763.90$$

The PV function $= \mathrm{PV}(16\%,9,600)$ entered into a single spreadsheet cell will display the answer $P = (\$2763.93)$.

EXAMPLE 2.4 The Steel Plant Case

As mentioned in the chapter introduction of this case, the HBNA plant may generate a revenue base of $50 million per year. The president of the Italian parent company Baleez may have reason to be quite pleased with this projection for the simple reason that over the 5-year planning horizon, the expected revenue would total $250 million, which is $50 million more than the initial investment. With money worth 10% per year, address the following question from the president: Will the initial investment be recovered over the 5-year horizon with the time value of money considered? If so, by how much extra in present worth funds? If not, what is the equivalent annual revenue base required for the recovery plus the 10% return on money? Use both tabulated factor values and spreadsheet functions.

Solution

Tabulated value: Use the P/A factor to determine whether A = $50 million per year for $n = 5$ years starting 1 year after the plant's completion ($t = 0$) at $i = 10\%$ per year is equivalently less or greater than $200 M. The cash flow diagram is similar to Figure 2–4*a*, where the first A value occurs 1 year after P. Using $1 million units and Table 15 values,

$$P = 50(P/A,10\%,5) = 50(3.7908)$$

$$= \$189.54 \quad (\$189,540,000)$$

The present worth value is less than the investment plus a 10% per year return, so the president should not be satisfied with the projected annual revenue.

To determine the minimum required to realize a 10% per year return, use the A/P factor. The cash flow diagram is the same as Figure 2–4*b*, where A starts 1 year after P at $t = 0$ and $n = 5$.

$$A = 200(A/P,10\%,5) = 200(0.26380)$$

$$= \$52.76 \text{ per year}$$

The plant needs to generate $52,760,000 per year to realize a 10% per year return over 5 years.

Spreadsheet: Apply the PV and PMT functions to answer the question. Figure 2–5 shows the use of = PV($i\%,n,A,F$) on the left side to find the present worth and the use of = PMT($i\%,n,P,F$) on the right side to determine the minimum A of $52,760,000 per year. Because there are no F values, it is omitted from the functions. The minus sign placed before each function name forces the answer to be positive, since these two functions always display the answer with the opposite sign entered on the estimated cash flows.

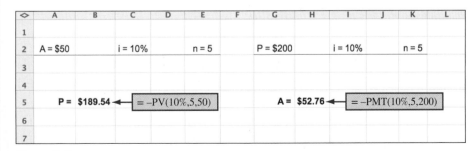

Figure 2–5
Spreadsheet functions to find *P* and *A* for the steel plant case, Example 2.4.

2.3 Sinking Fund Factor and Uniform Series Compound Amount Factor (*A*/*F* and *F*/*A*) ● ● ●

The simplest way to derive the A/F factor is to substitute into factors already developed. If P from Equation [2.3] is substituted into Equation [2.9], the following formula results.

$$A = F\left[\frac{1}{(1+i)^n}\right]\left[\frac{i(1+i)^n}{(1+i)^n - 1}\right]$$

$$A = F\left[\frac{i}{(1+i)^n - 1}\right] \qquad [2.12]$$

The expression in brackets in Equation [2.12] is the **A/F or sinking fund factor.** It determines the **uniform annual series A** that is equivalent to a given future amount F. This is shown graphically in Figure 2–6*a*, where A is a uniform annual investment.

Placement of *F*

> The uniform series A begins at the **end of year (period) 1** and continues **through the year of the given F.** The last A value and F occur at the same time.

Equation [2.12] can be rearranged to find F for a stated A series in periods 1 through n (Figure 2–6*b*).

$$F = A\left[\frac{(1+i)^n - 1}{i}\right] \qquad [2.13]$$

The term in brackets is called the *uniform series compound amount factor* (USCAF), or **F/A factor.** When multiplied by the given uniform annual amount A, it yields the **future worth of the uniform series.** It is important to remember that the future amount F occurs in the same period as the last A.

Standard notation follows the same form as that of other factors. They are $(F/A,i,n)$ and $(A/F,i,n)$. Table 2–3 summarizes the notations and equations. They are also included on the first pages of the book.

As a matter of interest, the uniform series factors can be symbolically determined by using an abbreviated factor form. For example, $F/A = (F/P)(P/A)$, where cancellation of the P is correct. Using the factor formulas, we have

$$(F/A,i,n) = [(1+i)^n]\left[\frac{(1+i)^n - 1}{i(1+i)^n}\right] = \frac{(1+i)^n - 1}{i}$$

For solution by spreadsheet, the FV function calculates F for a stated A series over n years. The format is

$$= \textbf{FV}(i\%,n,A,P) \qquad [2.14]$$

The P may be omitted when no separate present worth value is given. The PMT function determines the A value for n years, given F in year n and possibly a separate P value in year 0. The format is

$$= \textbf{PMT}(i\%,n,P,F) \qquad [2.15]$$

If P is omitted, the comma must be entered so the function knows the last entry is an F value.

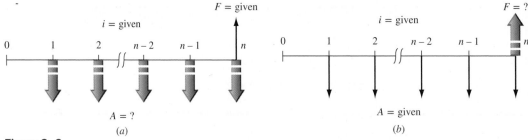

Figure 2–6
Cash flow diagrams to (*a*) find A, given F, and (*b*) find F, given A.

TABLE 2–3		*F/A* and *A/F* Factors: Notation and Equations			
Notation	**Factor Name**	**Find/Given**	**Factor Formula**	**Standard Notation Equation**	**Excel Functions**
$(F/A,i,n)$	Uniform series compound amount	F/A	$\dfrac{(1 + i)^n - 1}{i}$	$F = A(F/A,i,n)$	$= \mathrm{FV}(i\%,n,A)$
$(A/F,i,n)$	Sinking fund	A/F	$\dfrac{i}{(1 + i)^n - 1}$	$A = F(A/F,i,n)$	$= \mathrm{PMT}(i\%,n,F)$

EXAMPLE 2.5

The president of Ford Motor Company wants to know the equivalent future worth of a $1 million capital investment each year for 8 years, starting 1 year from now. Ford capital earns at a rate of 14% per year.

Solution

The cash flow diagram (Figure 2–7) shows the annual investments starting at the end of year 1 and ending in the year the future worth is desired. In $1000 units, the *F* value in year 8 is found by using the *F/A* factor.

$$F = 1000(F/A,14\%,8) = 1000(13.2328) = \$13{,}232.80$$

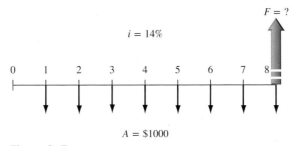

Figure 2–7
Diagram to find *F* for a uniform series, Example 2.5.

EXAMPLE 2.6 The Steel Plant Case

Once again, consider the HBNA case presented at the outset of this chapter, in which a projected $200 million investment can generate $50 million per year in revenue for 5 years starting 1 year after start-up. A 10% per year time value of money has been used previously to determine *P*, *F*, and *A* values. Now the president would like the answers to a couple of new questions about the estimated annual revenues. Use tabulated values, factor formulas, or spreadsheet functions to provide the answers.

(*a*) What is the equivalent future worth of the estimated revenues after 5 years at 10% per year?

(*b*) Assume that, due to the economic downturn, the president predicts that the corporation will earn only 4.5% per year on its money, not the previously anticipated 10% per year. What is the required amount of the annual revenue series over the 5-year period to be economically equivalent to the amount calculated in (*a*)?

Solution

(*a*) Figure 2–6*b* is the cash flow diagram with *A* = $50 million. Note that the last *A* value and *F* = ? both occur at the end of year *n* = 5. We use tabulated values and the spreadsheet function to find *F* in year 5.

Tabulated value: Use the *F/A* factor and 10% interest factor table. In $1 million units, the future worth of the revenue series is

$$F = 50(F/A,10\%,5) = 50(6.1051)$$
$$= \$305.255 \quad (\$305{,}255{,}000)$$

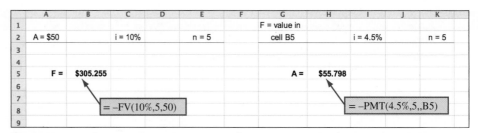

Figure 2–8
Spreadsheet functions to find *F* and *A* at *i* = 4.5% for the steel plant case,
Example 2.6.

If the rate of return on the annual revenues were 0%, the total amount after 5 years would be $250,000,000. The 10% per year return is projected to grow this value by 22%.

Spreadsheet: Apply the FV factor in the format $= -\text{FV}(10\%,5,50)$ to determine $F = \$305.255$ million. Because there is no present amount in this computation, P is omitted from the factor. See Figure 2–8, left side. (As before, the minus sign forces the FV function to result in a positive value.)

(*b*) The president of the Italian company planning to develop the steel plant is getting worried about the international economy. He wants the revenue stream to generate the equivalent that it would at a 10% per year return, that is, $305.255 million, but thinks that only a 4.5% per year return is achievable.

Factor formula: The A/F factor will determine the required A for 5 years. Since the factor tables do not include 4.5%, use the formula to answer the question. In $1 million units,

$$A = 305.255(A/F,4.5\%,5) = 305.255\left[\frac{0.045}{(1.045)^5 - 1}\right] = 305.255(0.18279)$$
$$= \$55.798$$

The annual revenue requirement grows from $50 million to nearly $55,800,000. This is a significant increase of 11.6% each year.

Spreadsheet: It is easy to answer this question by using the $= \text{PMT}(i\%,n,,F)$ function with $i = 4.5\%$ and $F = \$305.255$ found in part (*a*). We can use the cell reference method (described in Appendix A) for the future amount F. Figure 2–8, right side, displays the required A of $55.798 per year (in $1 million units).

2.4 Factor Values for Untabulated *i* or *n* Values ●●●

Often it is necessary to know the correct numerical value of a factor with an *i* or *n* value that is not listed in the compound interest tables in the rear of the book. Given specific values of *i* and *n*, there are several ways to obtain any factor value.

- Use the formula listed in this chapter or the front cover of the book.
- Use an Excel function with the corresponding *P*, *F*, or *A* value set to 1.
- Use linear interpolation in the interest tables.

When the **formula** is applied, the factor value is accurate since the specific *i* and *n* values are input. However, it is possible to make mistakes since the formulas are similar to each other, especially when uniform series are involved. Additionally, the formulas become more complex when gradients are introduced, as you will see in the following sections.

A **spreadsheet function** determines the factor value if the corresponding *P*, *A*, or *F* argument in the function is set to 1 and the other parameters are omitted or set to zero. For example, the *P/F* factor is determined using the PV function with *A* omitted (or set to 0) and $F = 1$, that is, PV(*i*%,*n*,,1) or PV(*i*%,*n*,0,1). A minus sign preceding the function identifier causes the factor to have a positive value. Functions to determine the six common factors are as follows.

Factor	To Do This	Excel Function
P/F	Find *P*, given *F*.	$= -PV(i\%,n,,1)$
F/P	Find *F*, given *P*.	$= -FV(i\%,n,,1)$
P/A	Find *P*, given *A*.	$= -PV(i\%,n,1)$
A/P	Find *A*, given *P*.	$= -PMT(i\%,n,1)$
F/A	Find *F*, given *A*.	$= -FV(i\%,n,1)$
A/F	Find *A*, given *F*.	$= -PMT(i\%,n,,1)$

Figure 2–9 shows a spreadsheet developed explicitly to determine these factor values. When it is made live in Excel, entering any combination of *i* and *n* displays the exact value for all six factors. The values for $i = 3.25\%$ and $n = 25$ years are shown here. As we already know, these same functions will determine a final *P*, *A*, or *F* value when actual or estimated cash flow amounts are entered.

Linear interpolation for an untabulated interest rate *i* or number of years *n* takes more time to complete than using the formula or spreadsheet function. Also, interpolation introduces some level of inaccuracy, depending upon the distance between the two boundary values selected for *i* or *n*, as the formulas themselves are nonlinear functions. Interpolation is included here for individuals who wish to utilize it in solving problems. Refer to Figure 2–10 for a graphical description of the following explanation. First, select two tabulated values (x_1 and x_2) of the parameter for which the factor is requested, that is, *i* or *n*, ensuring that the two values surround and are not too distant from the required value *x*. Second, find the corresponding tabulated factor values (f_1 and f_2). Third, solve for the unknown, linearly interpolated value *f* using the formulas below, where the differences in parentheses are indicated in Figure 2–10 as *a* through *c*.

	A	B	C	D
1	i =	n =		
2	**3.25%**	**25.00**	Enter requested i and n	
3			**Value obtained**	
4	**Factor**	**Value**	**with this function**	
5	P/F	0.44952	`= -PV($A$2,$B$2,,1)`	
6	P/A	16.93786	`= -PV($A$2,$B$2,1)`	
7				
8	F/P	2.22460	`= -FV($A$2,$B$2,,1)`	
9	F/A	37.67993	`= -FV($A$2,$B$2,1)`	
10				
11	A/F	0.02654	`= -PMT($A$2,$B$2,,1)`	
12	A/P	0.05904	`= -PMT($A$2,$B$2,1)`	
13				

Figure 2–9
Use of Excel functions to display factor values for any *i* and *n* values.

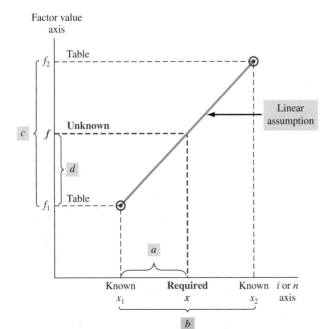

Figure 2–10
Linear interpolation in factor value tables.

$$f = f_1 + \frac{(x - x_1)}{(x_2 - x_1)}(f_2 - f_1) \qquad [2.16]$$

$$f = f_1 + \frac{a}{b}c = f_1 + d \qquad \textbf{[2.17]}$$

The value of d will be positive or negative if the factor is increasing or decreasing, respectively, in value between x_1 and x_2.

EXAMPLE 2.7

Determine the P/A factor value for $i = 7.75\%$ and $n = 10$ years, using the three methods described previously.

Solution

Factor formula: Apply the P/A factor relation in Equation [2.8] or from the summary page at the front of the text. Showing 5-decimal accuracy,

$$(P/A,7.75\%,10) = \frac{(1 + i)^n - 1}{i(1 + i)^n} = \frac{(1.0775)^{10} - 1}{0.0775(1.0775)^{10}} = \frac{1.10947}{0.16348}$$

$$= 6.78641$$

Spreadsheet: Utilize the spreadsheet function in Figure 2–9, that is, $= -PV(7.75\%,10,1)$, to display 6.78641.

Linear interpolation: Use Figure 2–10 as a reference for this solution. Apply the Equation [2.16] and [2.17] sequence, where x is the interest rate i, the bounding interest rates are $i_1 = 7\%$ and $i_2 = 8\%$, and the corresponding P/A factor values are $f_1 = (P/A,7\%,10) = 7.0236$ and $f_2 = (P/A,8\%,10) = 6.7101$. With 4-place accuracy,

$$f = f_1 + \frac{(i - i_1)}{(i_2 - i_1)}(f_2 - f_1) = 7.0236 + \frac{(7.75 - 7)}{(8 - 7)}(6.7101 - 7.0236)$$

$$= 7.0236 + (0.75)(-0.3135) = 7.0236 - 0.2351$$

$$= 6.7885$$

Comment

Note that since the P/A factor value decreases as i increases, the linear adjustment is negative at -0.2351. As is apparent, linear interpolation provides an approximation to the correct factor value for 7.75% and 10 years, plus it takes more calculations than using the formula or spreadsheet function. It is possible to perform two-way linear interpolation for untabulated i and n values; however, the use of a spreadsheet or factor formula is recommended.

2.5 Arithmetic Gradient Factors (*P/G* and *A/G*) ●●●

Assume a manufacturing engineer predicts that the cost of maintaining a robot will increase by $5000 per year until the machine is retired. The cash flow series of maintenance costs involves a constant gradient, which is $5000 per year.

An **arithmetic gradient** series is a cash flow series that either increases or decreases by a **constant amount** each period. The amount of change is called the **gradient.**

Formulas previously developed for an A series have year-end amounts of equal value. In the case of a gradient, each year-end cash flow is different, so new formulas must be derived. First, assume that the cash flow at the end of year 1 is the **base amount** of the cash flow series and, therefore, not part of the gradient series. This is convenient because in actual applications, the base amount is usually significantly different in size compared to the gradient. For example, if you purchase a used car with a 1-year warranty, you might expect to pay the gasoline and insurance costs during the first year of operation. Assume these cost $2500; that is, $2500 is the base amount. After the first year, you absorb the cost of repairs, which can be expected to increase

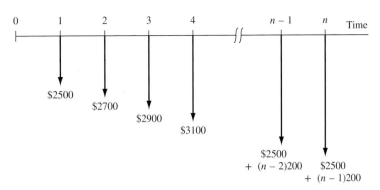

Figure 2–11
Cash flow diagram of an arithmetic gradient series.

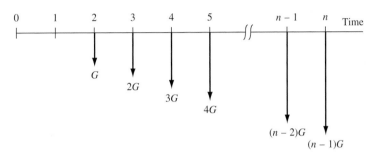

Figure 2–12
Conventional arithmetic gradient series without the base amount.

each year. If you estimate that total costs will increase by $200 each year, the amount the second year is $2700, the third $2900, and so on to year n, when the total cost is $2500 + (n − 1)200$. The cash flow diagram is shown in Figure 2–11. Note that the gradient ($200) is first observed between year 1 and year 2, and the base amount ($2500 in year 1) is not equal to the gradient.

Define the symbols G for gradient and CF_n for cash flow in year n as follows.

G = constant arithmetic change in cash flows from one time period to the next; G may be positive or negative.

$$\text{CF}_n = \text{base amount} + (n − 1)G \qquad [2.18]$$

It is important to realize that the base amount defines a uniform cash flow series of the size A that occurs each time period. We will use this fact when calculating equivalent amounts that involve arithmetic gradients. If the base amount is ignored, a generalized arithmetic (increasing) gradient cash flow diagram is as shown in Figure 2–12. Note that the gradient begins between years 1 and 2. This is called a **conventional gradient.**

EXAMPLE 2.8

A local university has initiated a logo-licensing program with the clothier Holister, Inc. Estimated fees (revenues) are $80,000 for the first year with uniform increases to a total of $200,000 by the end of year 9. Determine the gradient and construct a cash flow diagram that identifies the base amount and the gradient series.

Solution

The year 1 base amount is $CF_1 = \$80,000$, and the total increase over 9 years is

$$CF_9 − CF_1 = 200,000 − 80,000 = \$120,000$$

Equation [2.18], solved for G, determines the arithmetic gradient.

$$G = \frac{(CF_9 − CF_1)}{n − 1} = \frac{120,000}{9 − 1}$$
$$= \$15,000 \text{ per year}$$

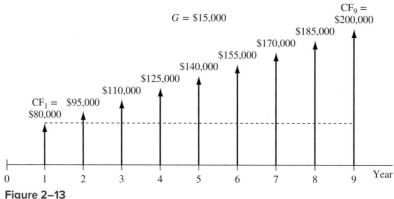

Figure 2–13
Diagram for gradient series, Example 2.8.

The cash flow diagram (Figure 2–13) shows the base amount of $80,000 in years 1 through 9 and the $15,000 gradient starting in year 2 and continuing through year 9.

The **total present worth** P_T for a series that includes a base amount A and conventional arithmetic gradient must consider the present worth of both the uniform series defined by A and the arithmetic gradient series. The addition of the two results in P_T.

$$P_T = P_A \pm P_G \qquad [2.19]$$

where P_A is the present worth of the uniform series only, P_G is the present worth of the gradient series only, and the + or − sign is used for an increasing (+G) or decreasing (−G) gradient, respectively.

The corresponding equivalent annual worth A_T is the sum of the base amount series annual worth A_A and gradient series annual worth A_G, that is,

$$A_T = A_A \pm A_G \qquad [2.20]$$

Three factors are derived for arithmetic gradients: the P/G factor for present worth, the A/G factor for annual series, and the F/G factor for future worth. There are several ways to derive them. We use the single-payment present worth factor ($P/F,i,n$), but the same result can be obtained by using the F/P, F/A, or P/A factor.

In Figure 2–12, the present worth at year 0 of only the gradient is equal to the sum of the present worths of the individual cash flows, where each value is considered a future amount.

$$P = G(P/F,i,2) + 2G(P/F,i,3) + 3G(P/F,i,4) + \cdots$$
$$+ [(n-2)G](P/F,i,n-1) + [(n-1)G](P/F,i,n)$$

Factor out G and use the P/F formula.

$$P = G\left[\frac{1}{(1+i)^2} + \frac{2}{(1+i)^3} + \frac{3}{(1+i)^4} + \cdots + \frac{n-2}{(1+i)^{n-1}} + \frac{n-1}{(1+i)^n}\right] \qquad [2.21]$$

Multiplying both sides of Equation [2.21] by $(1+i)^1$ yields

$$P(1+i)^1 = G\left[\frac{1}{(1+i)^1} + \frac{2}{(1+i)^2} + \frac{3}{(1+i)^3} + \cdots + \frac{n-2}{(1+i)^{n-2}} + \frac{n-1}{(1+i)^{n-1}}\right] \qquad [2.22]$$

Subtract Equation [2.21] from Equation [2.22] and simplify.

$$iP = G\left[\frac{1}{(1+i)^1} + \frac{1}{(1+i)^2} + \cdots + \frac{1}{(1+i)^{n-1}} + \frac{1}{(1+i)^n}\right] - G\left[\frac{n}{(1+i)^n}\right] \qquad [2.23]$$

The left bracketed expression is the same as that contained in Equation [2.6], where the P/A factor was derived. Substitute the closed-end form of the P/A factor from Equation [2.8]

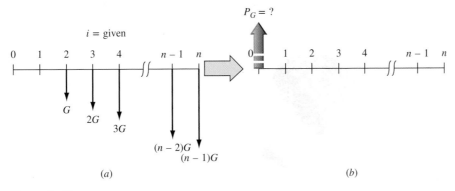

Figure 2–14
Conversion diagram from an arithmetic gradient to a present worth.

into Equation [2.23] and simplify to solve for P_G, the present worth of the gradient series only.

$$P_G = \frac{G}{i}\left[\frac{(1+i)^n - 1}{i(1+i)^n} - \frac{n}{(1+i)^n}\right] \qquad [2.24]$$

Equation [2.24] is the general relation to **convert an arithmetic gradient G (not including the base amount) for n years into a present worth at year 0.** Figure 2–14*a* is converted into the equivalent cash flow in Figure 2–14*b*. The *arithmetic gradient present worth factor,* or ***P/G* factor,** may be expressed in two forms:

$$(P/G,i,n) = \frac{1}{i}\left[\frac{(1+i)^n - 1}{i(1+i)^n} - \frac{n}{(1+i)^n}\right]$$

or $\qquad\qquad (P/G,i,n) = \dfrac{(1+i)^n - in - 1}{i^2(1+i)^n} \qquad\qquad\qquad [2.25]$

> Remember: The conventional arithmetic gradient starts in year 2, and P is located in year 0.

☑
Placement of gradient P_G

Equation [2.24] expressed as an engineering economy relation is

$$P_G = G(P/G,i,n) \qquad [2.26]$$

which is the rightmost term in Equation [2.19] to calculate total present worth. The G carries a minus sign for decreasing gradients.

The equivalent uniform annual series A_G for an arithmetic gradient G is found by multiplying the present worth in Equation [2.26] by the $(A/P,i,n)$ formula. In standard notation form, the equivalent of algebraic cancellation of P can be used.

$$A_G = G(P/G,i,n)(A/P,i,n)$$
$$= G(A/G,i,n)$$

In equation form,

$$A_G = \frac{G}{i}\left[\frac{(1+i)^n - 1}{i(1+i)^n} - \frac{n}{(1+i)^n}\right]\left[\frac{i(1+i)^n}{(1+i)^n - 1}\right]$$

$$A_G = G\left[\frac{1}{i} - \frac{n}{(1+i)^n - 1}\right] \qquad [2.27]$$

which is the rightmost term in Equation [2.20]. The expression in brackets in Equation [2.27] is called the *arithmetic gradient uniform series factor* and is identified by **(*A/G,i,n*).** This factor converts Figure 2–15*a* into Figure 2–15*b*.

The *P/G* and *A/G* factors and relations are summarized on the first pages of the text. Factor values are tabulated in the two rightmost columns of factor values at the rear of this text.

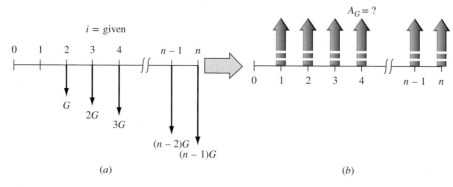

Figure 2–15
Conversion diagram of an arithmetic gradient series to an equivalent uniform annual series.

There is no direct, single-cell spreadsheet function to calculate P_G or A_G for an arithmetic gradient. Use the NPV function to display P_G and the PMT function to display A_G after entering all cash flows (base and gradient amounts) into contiguous cells. General formats for these functions are

$$= \text{NPV}(i\%, second\_cell:last\_cell) + first\_cell \qquad [2.28]$$
$$= \text{PMT}(i\%, n, cell\_with\_P_G) \qquad [2.29]$$

The word entries in italic are cell references, not the actual numerical values. (See Appendix A, Section A.2, for a description of cell reference formatting.) These functions are demonstrated in Example 2.10.

An **F/G factor** (*arithmetic gradient future worth factor*) to calculate the future worth F_G of a gradient series can be derived by multiplying the P/G and F/P factors. The resulting factor, $(F/G,i,n)$, in brackets, and engineering economy relation is

$$F_G = G \left(\frac{1}{i}\right) \left[\left(\frac{(1+i)^n - 1}{i} \right) - n \right]$$

EXAMPLE 2.9

Neighboring parishes in Louisiana have agreed to pool road tax resources already designated for bridge refurbishment. At a recent meeting, the engineers estimated that a total of $500,000 will be deposited at the end of next year into an account for the repair of old and safety-questionable bridges throughout the area. Further, they estimated that the deposits will increase by $100,000 per year for only 9 years thereafter, then cease. Determine the equivalent (*a*) present worth, and (*b*) annual series amounts, if public funds earn at a rate of 5% per year.

Solution

(*a*) The cash flow diagram of this conventional arithmetic gradient series from the perspective of the parishes is shown in Figure 2–16. According to Equation [2.19], two computations must be made and added: the first for the present worth of the base amount P_A and the second for the present worth of the gradient P_G. The total present worth P_T occurs in year 0. This is illustrated by the partitioned cash flow diagram in Figure 2–17. In $1000 units, the total present worth is

$$P_T = 500(P/A,5\%,10) + 100(P/G,5\%,10)$$
$$= 500(7.7217) + 100(31.6520)$$
$$= \$7026.05 \quad (\$7,026,050)$$

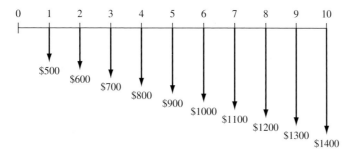

Figure 2–16
Cash flow series with a conventional arithmetic gradient (in $1000 units),
Example 2.9.

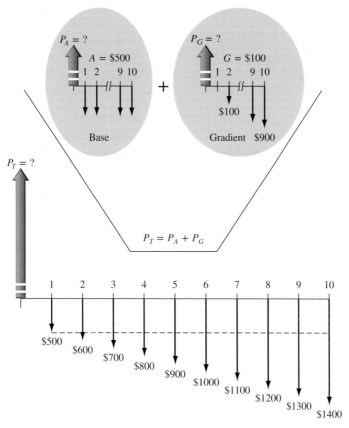

Figure 2–17
Partitioned cash flow diagram (in $1000 units), Example 2.9.

(*b*) Here, too, it is necessary to consider the gradient and the base amount separately. The
total annual series A_T is found by Equation [2.20] and occurs in years 1 through 10.

$$A_T = 500 + 100(A/G,5\%,10) = 500 + 100(4.0991)$$

$$= \$909.91 \text{ per year} \quad (\$909,910)$$

Comment

Remember: The *P/G* and *A/G* factors determine the present worth and annual series of the
gradient only. Any other cash flows must be considered separately.

    If the present worth is already calculated [as in part (*a*)], P_T can be multiplied by an *A/P*
factor to get A_T. In this case, considering round-off error,

$$A_T = P_T(A/P,5\%,10) = 7026.05(0.12950)$$

$$= \$909.873 \quad (\$909,873)$$

EXAMPLE 2.10 The Steel Plant Case   PE

The announcement of the HBNA steel plant states that the $200 million (M) investment is planned for 2020. Most large investment commitments are actually spread out over several years as the plant is constructed and production is initiated. Further investigation may determine, for example, that the $200 M is a present worth in the year 2020 of anticipated investments during the next 4 years (2021 through 2024). Assume the amount planned for 2021 is $100 M with constant decreases of $25 M each year thereafter. As before, assume the time value of money for investment capital is 10% per year to answer the following questions using tabulated factors and spreadsheet functions, as requested below.

(a) In equivalent present worth values, does the planned decreasing investment series equal the announced $200 M in 2020? Use both tabulated factors and spreadsheet functions.
(b) Given the planned investment series, what is the equivalent annual amount that will be invested from 2021 to 2024? Use both tabulated factors and spreadsheet functions.
(c) (This optional question introduces **Excel's Goal Seek tool.**) What must be the amount of yearly constant decrease through 2024 to have a present worth of exactly $200 M in 2020, provided $100 M is expended in 2021? Use a spreadsheet.

Solution

(a) The investment series is a decreasing arithmetic gradient with a base amount of $100 M in year 1 (2021) and $G = \$-25$ M through year 4 (2024). Figure 2–18 diagrams the cash flows with the shaded area showing the constantly declining investment each year. The P_T value at time 0 at 10% per year is determined by using tables and a spreadsheet.

Tabulated factors: Equation [2.19] with the minus sign for negative gradients determines the total present worth P_T. Money is expressed in $1 million units.

$$P_T = P_A - P_G = 100(P/A,10\%,4) - 25(P/G,10\%,4) \qquad [2.30]$$

$$= 100(3.1699) - 25(4.3781)$$

$$= \$207.537 \quad (\$207,537,000)$$

In present worth terms, the planned series will exceed the equivalent of $200 M in 2020 by approximately $7.5 M.

Spreadsheet: Since there is no spreadsheet function to directly display present worth for a gradient series, enter the cash flows in a sequence of cells (rows or columns) and use the NPV function to find present worth. Figure 2–19 shows the entries and function NPV(i%,second_cell:last_cell). There is no first_cell entry here because there is no investment per se in year 0. The result displayed in cell C9, $207.534, is the total P_T for the planned series. (Note that the NPV function does not consider two separate series of cash flows as is necessary when using tabulated factors.)

The interpretation is the same as in part (a); the planned investment series exceeds the $200 M in present worth terms by approximately $7.5 M.

(b) *Tabulated factors:* There are two equally correct ways to find A_T. First, apply Equation [2.20] that utilizes the A/G factor, and second, use the P_T value obtained above and the A/P factor. Both relations are illustrated here, in $1 million units.

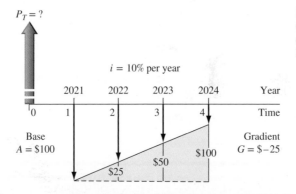

Figure 2–18
Cash flow diagram for decreasing gradient in $1 million units, Example 2.10.

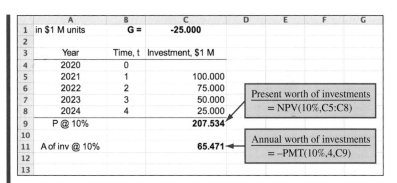

Figure 2–19
Spreadsheet solution for
Example 2.10*a* and *b*.

Use Equation [2.20]:

$$A_T = 100 - 25(A/G,10\%,4) = 100 - 25(1.3812)$$

$$= \$65.471 \text{ per year} \quad (\$65,471,000)$$

Use P_T:

$$A_T = 207.537(A/P,10\%,4) = 207.537(0.31547)$$

$$= \$65.471 \text{ per year}$$

Spreadsheet: Apply the PMT function in Equation [2.29] to obtain the same $A_T =$ \$65.471 per year (Figure 2–19).

(*c*) (Optional) The **Goal Seek tool** is described in Appendix A. It is an excellent tool to apply when one cell entry must equal a specific value and only one other cell can change. This is the case here; the NPV function (cell C9 in Figure 2–19) must equal \$200, and the gradient *G* (cell C1) is unknown. This is the same as stating $P_T = 200$ in Equation [2.30] and solving for *G*. All other parameters retain their current value.

Figure 2–20 (top) pictures the same spreadsheet used previously with the Goal Seek template added and loaded. When OK is clicked, the solution is displayed; $G = \$-26.721$. Refer to Figure 2–20 again. This means that if the investment is decreased by a constant annual amount of \$26.721 M, the equivalent total present worth invested over the 4 years will be exactly \$200 M.

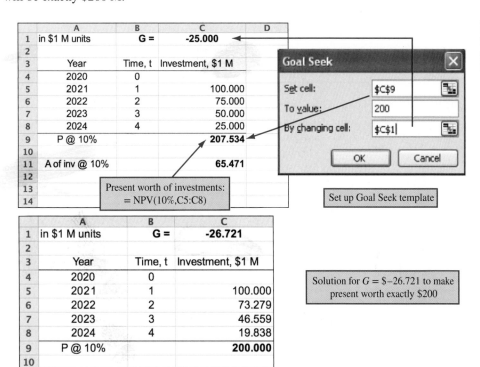

Figure 2–20
Solution for arithmetic gradient using Goal Seek, Example 2.10*c*.

2.6 Geometric Gradient Series Factors ●●●

It is common for annual revenues and annual costs such as maintenance, operations, and labor to go up or down by a constant percentage, for example, +5% or −3% per year. This change occurs every year on top of a starting amount in the first year of the project. A definition and description of new terms follow.

> A **geometric gradient** series is a cash flow series that either increases or decreases by a **constant percentage** each period. The uniform change is called the **rate of change.**
>
> g = **constant rate of change,** in decimal form, by which cash flow values increase or decrease from one period to the next. The gradient g can be + or −.
>
> A_1 = **initial cash flow in period 1** of the geometric series
>
> P_g = **present worth** of the entire geometric gradient series, including the initial amount A_1

Note that the initial cash flow A_1 **is not considered separately** when working with geometric gradients.

Figure 2–21 shows increasing and decreasing geometric gradients starting at an amount A_1 in time period 1 with present worth P_g located at time 0. The relation to determine the total present worth P_g **for the entire cash flow series** may be derived by multiplying each cash flow in Figure 2–21a by the P/F factor $1/(1 + i)^n$.

$$P_g = \frac{A_1}{(1+i)^1} + \frac{A_1(1+g)}{(1+i)^2} + \frac{A_1(1+g)^2}{(1+i)^3} + \cdots + \frac{A_1(1+g)^{n-1}}{(1+i)^n}$$

$$= A_1\left[\frac{1}{1+i} + \frac{1+g}{(1+i)^2} + \frac{(1+g)^2}{(1+i)^3} + \cdots + \frac{(1+g)^{n-1}}{(1+i)^n}\right] \qquad [2.31]$$

Multiply both sides by $(1 + g)/(1 + i)$, subtract Equation [2.31] from the result, factor out P_g, and obtain

$$P_g\left(\frac{1+g}{1+i} - 1\right) = A_1\left[\frac{(1+g)^n}{(1+i)^{n+1}} - \frac{1}{1+i}\right]$$

Solve for P_g and simplify.

$$P_g = A_1\left[\frac{1 - \left(\frac{1+g}{1+i}\right)^n}{i-g}\right] \qquad g \neq i \qquad [2.32]$$

The term in brackets in Equation [2.32] is the $(P/A,g,i,n)$ or *geometric gradient series present worth factor* for values of g not equal to the interest rate i. When $g = i$, substitute i for g in Equation [2.31] and observe that the term $1/(1 + i)$ appears n times.

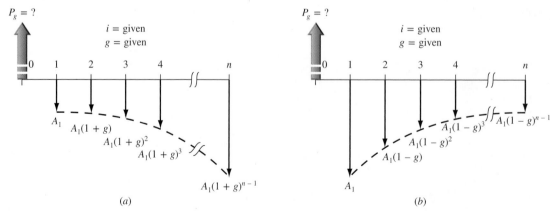

Figure 2–21
Cash flow diagram of (*a*) increasing and (*b*) decreasing geometric gradient series and present worth P_g.

$$P_g = A_1\left(\frac{1}{(1+i)} + \frac{1}{(1+i)} + \frac{1}{(1+i)} + \cdots + \frac{1}{(1+i)}\right)$$

$$P_g = \frac{nA_1}{(1+i)} \qquad\qquad\qquad\qquad [2.33]$$

Placement of gradient P_g

The $(P/A,g,i,n)$ factor calculates P_g in period $t = 0$ for a geometric gradient series **starting in period 1** in the amount A_1 and increasing by a constant rate of g each period.

The equation for P_g and the $(P/A,g,i,n)$ factor formula are

$$P_g = A_1(P/A,g,i,n) \qquad\qquad [2.34]$$

$$(P/A,g,i,n) = \begin{cases} \dfrac{1 - \left(\dfrac{1+g}{1+i}\right)^n}{i - g} & g \neq i \\[4mm] \dfrac{n}{1+i} & g = i \end{cases} \qquad [2.35]$$

It is possible to derive factors for the equivalent A and F values; however, it is easier to determine the P_g amount and then multiply by the A/P or F/P factor.

As with the arithmetic gradient series, there are no direct spreadsheet functions for geometric gradient series. Once the cash flows are entered, P and A are determined using the NPV and PMT functions, respectively.

EXAMPLE 2.11

A coal-fired power plant has upgraded an emission control valve. The modification costs only $8000 and is expected to last 6 years with a $200 salvage value. The maintenance cost is expected to be high at $1700 the first year, increasing by 11% per year thereafter. Determine the equivalent present worth of the modification and maintenance cost by hand and by spreadsheet at 8% per year.

Solution by Hand

The cash flow diagram (Figure 2–22) shows the salvage value as a positive cash flow and all costs as negative. Use Equation [2.35] for $g \neq i$ to calculate P_g. Total P_T is the sum of three present worth components.

$$P_T = -8000 - P_g + 200(P/F,8\%,6)$$

$$= -8000 - 1700\left[\frac{1 - (1.11/1.08)^6}{0.08 - 0.11}\right] + 200(P/F,8\%,6)$$

$$= -8000 - 1700(5.9559) + 126 = \$-17{,}999$$

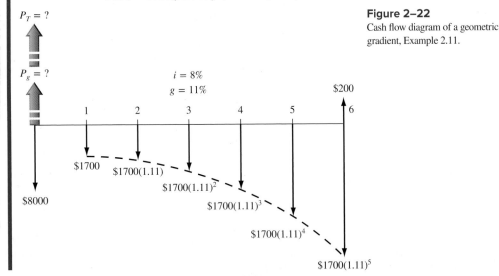

Figure 2–22
Cash flow diagram of a geometric gradient, Example 2.11.

Solution by Spreadsheet

Figure 2–23 details the spreadsheet operations to find the geometric gradient present worth P_g and total present worth P_T. To obtain $P_T = \$-17,999$, three components are summed—first cost, present worth of estimated salvage in year 6, and P_g. Cell tags detail the relations for the second and third components; the first cost occurs at time 0.

Comment

The relation that calculates the $(P/A,g,i\%,n)$ factor is rather complex, as shown in the cell tag and formula bar for C9. If this factor is used repeatedly, it is worthwhile using cell reference formatting so that A_1, i, g, and n values can be changed and the correct value is always obtained. Try to write the relation for cell C9 in this format.

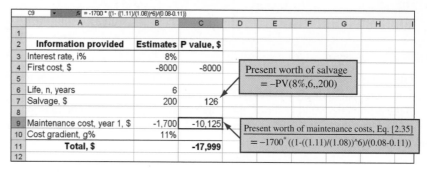

Figure 2–23
Geometric gradient and total present worth calculated via spreadsheet, Example 2.11.

EXAMPLE 2.12 The Steel Plant Case

Now let's go back to the proposed HBNA steel plant. The revenue series estimate of $50 million annually is quite optimistic, especially since there are other steel fabricators operating in the area. Therefore, it is important to be sensitive in our analysis to possibly declining and increasing revenue series, depending upon the longer-term success of the plant's marketing, quality, and reputation. Assume that revenue may start at $50 million by the end of the first year, but then decreases geometrically by 12% per year through year 5. Determine the **present worth** and **future worth** equivalents of all revenues during this 5-year time frame at the same rate used previously, that is, 10% per year.

Solution

The cash flow diagram appears much like Figure 2–21b, except that the arrows go up for revenues. In year 1, $A_1 = \$50$ M and revenues decrease in year 5 to

$$A_1(1-g)^{n-1} = 50\ M(1-0.12)^{5-1} = 50\ M(0.88)^4 = \$29.98\ M$$

First, we determine P_g in year 0 using Equation [2.35] with $i = 0.10$ and $g = -0.12$, then we calculate F in year 5. In $1 million units,

$$P_g = 50\left[\frac{1-\left(\frac{0.88}{1.10}\right)^5}{0.10-(-0.12)}\right] = 50[3.0560]$$

$$= \$152.80$$

$$F = 152.80(F/P,10\%,5) = 152.80(1.6105)$$

$$= \$246.08$$

This means that the decreasing revenue stream has a 5-year future equivalent worth of $246.080 M. If you look back to Example 2.6, we determined that the F in year 5 for the uniform revenue series of $50 M annually is $305.255 M. In conclusion, the 12% declining geometric gradient has lowered the future worth of revenue by $59.175 M, which is a sizable amount from the perspective of the owners.

2.7 Determining *i* or *n* for Known Cash Flow Values ● ● ◐

When all the cash flow values are known or have been estimated, the *i* value (interest rate or rate of return) or *n* value (number of years) is often the unknown. An example for which *i* is sought may be stated as follows: A company invested money to develop a new product. After the net annual income series is known following several years on the market, determine the rate of return *i* on the investment. There are several ways to find an unknown *i* or *n* value, depending upon the nature of the cash flow series and the method chosen to find the unknown. The simplest case involves only single amounts (*P* and *F*) and solution utilizing a spreadsheet function. The most difficult and complex involves finding *i* or *n* for irregular cash flows mixed with uniform and gradient series utilizing solution by hand and calculator. The solution approaches are summarized below, followed by examples.

Single Amounts—*P* and *F* Only

Hand or Calculator Solution   Set up the equivalence relation and (1) solve for the variable using the factor formula, or (2) find the factor value and interpolate in the tables.

Spreadsheet Solution   Use the IRR or RATE function to find *i* or the NPER function to find *n*. (See below and Appendix A for details.)

Uniform Series—*A* Series

Hand or Calculator Solution   Set up the equivalence relation using the appropriate factor (*P/A*, *A/P*, *F/A*, or *A/F*), and use the second method mentioned above.

Spreadsheet Solution   Use the IRR or RATE function to find *i* or the NPER function to find *n*.

Mixed *A* Series, Gradients, and/or Isolated Values

Hand or Calculator Solution   Set up the equivalence relation and use (1) trial and error or (2) the calculator functions.

Spreadsheet Solution   Use the IRR or RATE function to find *i* or the NPER function to find *n*. (This is the recommended approach.)

Besides the PV, FV, and NPV functions, other spreadsheet functions useful in determining *i* are IRR (internal rate of return) and RATE, and NPER (number of periods) to find *n*. The formats are shown here and on the first pages of the text, with a detailed explanation in Appendix A. In all three of these functions, at least one cash flow entry must have a sign opposite that of others in order to find a solution.

$$= \textbf{IRR(first\_cell:last\_cell)} \qquad\qquad \textbf{[2.36]}$$

To use IRR to find *i*, enter all cash flows into contiguous cells, including zero values.

$$= \textbf{RATE}(n,A,P,F) \qquad\qquad \textbf{[2.37]}$$

The single-cell RATE function finds *i* when an *A* series and single *P* and/or *F* values are involved.

$$= \textbf{NPER}(i\%,A,P,F) \qquad\qquad \textbf{[2.38]}$$

NPER is a single-cell function to find *n* for single *P* and *F* values, or with an *A* series.

EXAMPLE 2.13

If Laurel made a $30,000 investment in a friend's business and received $50,000 five years later, determine the rate of return.

Solution

Since only single amounts are involved, i can be determined directly from the P/F factor.

$$P = F(P/F,i,n) = F\frac{1}{(1+i)^n}$$

$$30,000 = 50,000\frac{1}{(1+i)^5}$$

$$0.600 = \frac{1}{(1+i)^5}$$

$$i = \left(\frac{1}{0.6}\right)^{0.2} - 1 = 0.1076 \quad (10.76\%)$$

Alternatively, the interest rate can be found by setting up the standard P/F relation, solving for the factor value, and interpolating in the tables.

$$P = F(P/F,i,n)$$

$$30,000 = 50,000(P/F,i,5)$$

$$(P/F,i,5) = 0.60$$

From the interest tables, a P/F factor of 0.6000 for $n = 5$ lies between 10% and 11%. Interpolate between these two values to obtain $i = 10.76\%$.

EXAMPLE 2.14

Pyramid Energy requires that for each of its offshore wind power generators $5000 per year be placed into a capital reserve fund to cover unexpected major rework on field equipment. In one case, $5000 was deposited for 15 years and covered a rework costing $100,000 in year 15. What rate of return did this practice provide to the company? Solve by hand and spreadsheet.

Solution by Hand

The cash flow diagram is shown in Figure 2–24. Either the A/F or F/A factor can be used. Using A/F,

$$A = F(A/F,i,n)$$

$$5000 = 100,000(A/F,i,15)$$

$$(A/F,i,15) = 0.0500$$

From the A/F interest tables for 15 years, the value 0.0500 lies between 3% and 4%. By interpolation, $i = 3.98\%$.

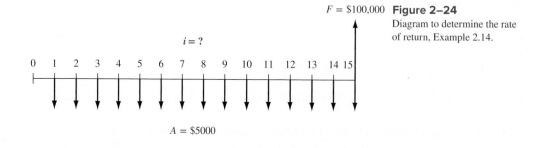

$F = \$100,000$ **Figure 2–24**
Diagram to determine the rate of return, Example 2.14.

$i = ?$

0 1 2 3 4 5 6 7 8 9 10 11 12 13 14 15

$A = \$5000$

Solution by Spreadsheet

Refer to the cash flow diagram (Figure 2–24) while completing the spreadsheet (Figure 2–25). A single-cell solution using the RATE function can be applied since $A = \$-5000$ occurs each year and $F = \$100,000$ takes place in the last year of the series. The function = RATE(15,−5000,,100000) displays the value $i = 3.98\%$. This function is fast, but it allows only limited sensitivity analysis because all the A values have to change by the same amount. The IRR function is much better for answering "what if " questions.

To apply the IRR function, enter the value 0 in a cell (for year 0), followed by –5000 for 14 years and in year 15 enter +95,000 (Figure 2–25). In any cell enter the IRR function. The answer $i = 3.98\%$ is displayed. It is advisable to enter the year numbers 0 through n (15 in this example) in the column immediately to the left of the cash flow entries. The IRR function does not need these numbers, but it makes the cash flow entry activity easier and more accurate. Now any cash flow can be changed, and a new rate will be displayed immediately via IRR.

◇	A	B	C	D	E
1				Year	Cash flow, $
2	**Using RATE**			0	0
3	3.98%			1	-5000
4				2	-5000
5	*i* using RATE function			3	-5000
6	= **RATE(15,-5000,,100000)**			4	-5000
7				5	-5000
8				6	-5000
9	**Using IRR**			7	-5000
10	3.98%			8	-5000
11				9	-5000
12	*i* using IRR function			10	-5000
13	= **IRR(E2:E17)**			11	-5000
14				12	-5000
15				13	-5000
16				14	-5000
17				15	95000
18					

Figure 2–25
Use of RATE and IRR functions to determine *i* value for a uniform series, Example 2.14.

EXAMPLE 2.15   The Steel Plant Case

From the introductory comments about the HBNA plant, the annual revenue is estimated to be $50 million. All analysis thus far has taken place at 10% per year; however, the parent company has made it clear that its other international plants are able to show a 20% per year return on the initial investment. Determine the number of years required to generate 10%, 15%, and 20% per year returns on the $200 million investment at the new site.

Solution

If hand solution is utilized, the present worth relation can be established and the n values interpolated in the tables for each of the three rate of return values. In $1 million units, the relation is

$$P = -200 + 50(P/A,i\%,n) \qquad (i = 10\%, 15\%, 20\%)$$

$$(P/A,i\%,n) = 4.00$$

This is a good opportunity to utilize a spreadsheet and repeated NPER functions from Equation [2.38], since several i values are involved. Figure 2–26 shows the single-cell = NPER($i\%$,50,−200) function for each rate of return. The number of years (rounded up) to produce at least the required returns are

Return, $i\%$	Years
10	6
15	7
20	9

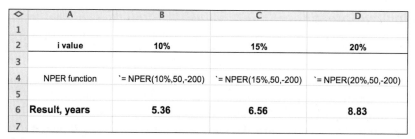

◇	A	B	C	D
1				
2	i value	10%	15%	20%
3				
4	NPER function	`= NPER(10%,50,-200)	`= NPER(15%,50,-200)	`= NPER(20%,50,-200)
5				
6	Result, years	5.36	6.56	8.83
7				

Figure 2–26
Use of NPER function to find *n* values for various rate of return requirements, Example 2.15.

CHAPTER SUMMARY

Formulas and factors derived and applied in this chapter perform equivalence calculations for present, future, annual, and gradient cash flows. Capability in using these formulas and their standard notation manually and with spreadsheets is critical to complete an engineering economy study. Using these formulas and spreadsheet functions, you can convert single cash flows into uniform cash flows, gradients into present worths, and much more. Additionally, you can solve for rate of return *i* or time *n*.

PROBLEMS

Use of Interest Tables

2.1 Look up the numerical value for the following factors from the compound interest factor tables.
1. $(F/P,10\%,7)$
2. $(A/P,12\%,10)$
3. $(P/G,15\%,20)$
4. $(F/A,2\%,50)$
5. $(P/G,35\%,15)$

Determination of *F*, *P*, and *A*

2.2 The Department of Traffic Security of a city is considering the purchase of a new drone for aerial surveillance of traffic on its most congested streets. A similar purchase 4 years ago cost $1,200,000. At an interest rate of 7% per year, what is the equivalent value today of the previous $1,200,000 expenditure?

2.3 Pressure Systems, Inc. manufactures high-accuracy liquid level transducers. It is investigating whether it should update in-place equipment now or wait and do it later. If the cost now is $200,000, what will be the equivalent amount 3 years from now at an interest rate of 10% per year?

2.4 How much money should a bank be willing to loan a real estate developer who will repay the loan by selling seven lakefront lots at $120,000 *each* 2 years from now? Assume the bank's interest rate is 10% per year.

2.5 A rotary engine is the essence of a vertical takeoff and landing (VTOL) personal aircraft known as the Moller Skycar M400. It is a flying car known as a personal air vehicle (PAV) and it is expected to make its first untethered flight in 2020. The PAV has been under development for 30 years at a total cost of $100 million. Assuming the $100 million was spent in an equal amount each year, determine the future worth at the end of the 30-year period at an interest rate of 10% per year.

2.6 What is the present worth of an expenditure of $25,000 in year 8 if the interest rate is 10% per year?

2.7 Calculate the present worth of 10 uniform payments of $8000 that begin 1 year from now at an interest rate of 10% per year.

2.8 Atlas Long-Haul Transportation is considering installing Valutemp temperature loggers in all of its refrigerated trucks for monitoring temperatures during transit. If the systems will reduce insurance claims by $100,000 in each of the next 2 years, how much should the company be willing to spend now if it uses an interest rate of 12% per year?

2.9 Determine the size of your investment account 30 years from now (when you plan to retire) if you deposit $12,000 each year, beginning 1 year from now, and the account earns interest at a rate of 10% per year.

2.10 How much could BTU Oil & Gas Fracking afford to spend on new equipment each year for the next 3 years if it expects a profit of $50 million 3 years from now? Assume the company's MARR is 20% per year.

2.11 Thompson Mechanical Products is planning to set aside $150,000 now for possible replacement of large synchronous refiner motors when it becomes necessary. If the replacement isn't needed for 5 years, how much will the company have in its investment set-aside account? Assume a rate of return of 18% per year.

2.12 Electric car maker Gentech signed a $75 million contract with Power Systems, Inc. to automate a major part of its assembly line system. If Power Systems will be paid 2 years from now, when the systems are ready, determine the present worth of the contract at 18% per year interest.

2.13 Labco Scientific sells high-purity chemicals to universities, research laboratories, and pharmaceutical companies. The company wants to invest in new equipment that will reduce shipping costs by better matching the size of the completed products with the size of the shipping container. The new equipment is estimated to cost $450,000 to purchase and install. How much must Labco save each year for 3 years in order to justify the investment at an interest rate of 10% per year?

2.14 Loadstar Sensors is a company that makes load/force sensors based on capacitive sensing technology. For a major plant expansion project, the company wants to have $30 million 5 years from now. If the company already has $15 million in an investment account for the expansion, how much more must the company add to the account now so that it will have the $30 million 5 years from now? The funds earn interest at the rate of 10% per year.

2.15 Meggitt Systems, a company that specializes in extreme-high-temperature accelerometers, is investigating whether it should update certain equipment now or wait and do it later. If the cost now is $280,000, what is the equivalent amount 2 years from now at an interest rate of 12% per year?

2.16 Henry Mueller Supply Co. sells vibration control equipment for wind turbines exposed to harsh environmental factors. Annual cash flows for an 8-year period are shown in the table. Determine the future worth of the cash flows at an interest rate of 10% per year.

Year	1	2	3	4	5	6	7	8
Revenue, $1000	200	200	200	200	200	200	200	200
Expenses, $1000	90	90	90	90	90	90	90	90

2.17 Stanley, Inc. makes self-clinching fasteners for stainless steel applications. It expects to acquire new time-saving punching equipment 4 years from now. If the company sets aside $125,000 each year, determine the amount available in 4 years at an earning rate of 10% per year.

2.18 China spends an estimated $100,000 per year on cloud seeding efforts which include using antiaircraft guns and rocket launchers to fill the sky with silver iodide. In the United States, utilities that run hydroelectric dams are among the most active cloud seeders because they believe it is a cost-effective way to increase limited water supplies by 10% or more. If the yields of cash crops will increase by 4% per year for the next 3 years because of extra irrigation water captured behind dams during cloud seeding, what is the future worth (in year 3) of the *extra value* of the cash crops? Assume the interest rate is 10% per year and the value of the cash crops without the extra irrigation water would be $600,000 per year.

2.19 American Gas Products manufactures a device called a Can-Emitor that empties the contents of old aerosol cans in 2 to 3 seconds. This eliminates the need to dispose of the cans as hazardous waste. If a paint manufacturing company can save $90,000 per year in waste disposal costs, how much could the company afford to spend now on the Can-Emitor if it wants to recover its investment in 3 years at an interest rate of 20% per year?

2.20 Durban Moving and Storage wants to have enough money available 5 years from now to purchase a new tractor-trailer. If the estimated cost is $250,000, how much should the company set aside each year if the funds earn 9% per year?

2.21 The Public Service Board (PSB) awarded two contracts worth a combined $3.07 million to increase the depth of a retention basin and reconstruct a spillway that was severely damaged in a flood 2 years ago. The PSB president stated that, surprisingly, the bids were $1,150,000 lower than PSB engineers estimated. If the projects are assumed to have a 20-year life, what is the annual worth of the savings at an interest rate of 5% per year?

2.22 Syringe pumps oftentimes fail because reagents adhere to the ceramic piston and deteriorate the seal. Trident Chemical developed an integrated polymer dynamic seal that provides a higher sealing force on the sealing lip, resulting in extended seal life. One of Trident's customers expects to reduce down time by 30% as a result of the new seal design. If lost production would have cost the company $110,000 per year for each of the next 4 years, how much could the company afford to spend now on the new seals? Use a MARR of 12% per year.

2.23 The cost of a fence that can detect poacher intrusion into a National Wildlife Preserve is $3 million per mile. If the effective life of the fence is 10 years, determine the equivalent annual cost of a 10-mile long fence at an interest rate of 8% per year.

2.24 A small oil company wants to replace its Micro Motion Coriolis flowmeters with nickel-based steel alloy flowmeters from the Emerson F-Series. The replacement process will cost the company $50,000 three years from now. How much money must the company set aside each year beginning 1 year from now in order to have the total amount in 3 years? Assume the company earns a generous 20% per year on investment funds.

Factor Values

2.25 Determine the numerical value of the following factors using (a) linear interpolation, (b) the formula, and (c) the spreadsheet function from Figure 2-9.
1. $(P/F,8.4\%,15)$
2. $(A/F,17\%,10)$

2.26 Find the numerical value of the following factors by (a) linear interpolation, (b) using the appropriate formula, and (c) a spreadsheet function.
1. $(F/A,19\%,20)$
2. $(P/A,26\%,15)$

2.27 Find the numerical value of the following factors using (a) linear interpolation and (b) the appropriate formula.
1. $(F/P,18\%,33)$
2. $(A/G,12\%,54)$

2.28 For the factor $(F/P,10\%,43)$, find the percent difference between the interpolated and formula-calculated values, assuming the formula-calculated value is the correct one.

Arithmetic Gradient

2.29 A cash flow sequence starts in year 1 at $4000 and decreases by $300 each year through year 9. Determine (a) the value of the gradient G; (b) the amount of cash flow in year 5; and (c) the value of n for the $(P/G,i\%,n)$ factor.

2.30 An arithmetic cash flow gradient series equals $500 in year 1, $600 in year 2, and amounts increasing by $100 per year through year 9. At $i = 10\%$ per year, determine the present worth of the cash flow series in year 0.

2.31 NMTeX Oil owns several gas wells in Carlsbad, NM. Revenue from the wells has been increasing according

to an arithmetic gradient for the past 5 years. The revenue in year 1 from well no. 24 was $390,000 and it increased by $15,000 each year thereafter. Determine (a) the revenue in year 3, and (b) the equivalent annual worth of the revenue through year 5. Assume an interest rate of 10% per year.

2.32 Solar Hydro manufactures a revolutionary aeration system that combines coarse and fine bubble aeration components. This year (year 1) the cost for check valve components is $9,000. Based on closure of a new contract with a distributor in China and volume discounts, the company expects this cost to decrease. If the cost in year 2 and each year thereafter decreases by $560, what is the equivalent annual cost for a 5-year period at an interest rate of 10% per year?

2.33 For the cash flow revenues shown below, find the value of G that makes the equivalent annual worth in *years 1 through 7* equal to $500. The interest rate is 10% per year.

Year	Cash Flow, $	Year	Cash Flow, $
0		4	200 + 3G
1	200	5	200 + 4G
2	200 + G	6	200 + 5G
3	200 + 2G	7	200 + 6G

2.34 A low-cost noncontact temperature measuring tool may be able to identify railroad car wheels that are in need of repair long before a costly structural failure occurs. If the BNF railroad saves $100,000 in year 1, $110,000 in year 2, and amounts increasing by $10,000 each year for 5 years, what is the future worth of the savings in year 5 at an interest rate of 10% per year?

2.35 For the cash flows below determine the amount in year 1, if the annual worth in years 1 through 9 is $3500 and the interest rate is 10% per year.

Year	1	2	3	4	5	6	7	8	9
Cost, $1000	A	A+40	A+80	A+120	A+160	A+200	A+240	A+280	A+320

2.36 Apple Computer wants to have $2.1 billion available 5 years from now in order to finance initial production of a device that, based on your behavior, will learn how to monitor and control nearly all of the electronic devices in your home, such as thermostat, coffee pot, TV, sprinkler system, etc. using Internet of Things (IOT) technology. The company expects to set aside uniformly increasing amounts of money each year to meet its goal. If the amount set aside at the end of year 1 is $100 million, how much will the uniform increase, G, have to be each year? Assume the investment funds grow at a rate of 18% per year.

2.37 Tacozza Electric, which manufactures brush dc servo motors, budgeted $95,000 per year to pay for specific components over the next 5 years. If the company expects to spend $55,000 in year 1, how much of a uniform (arithmetic) increase each year is the company expecting in the cost of this part? Assume the company uses an interest rate of 10% per year.

2.38 The future worth in year 10 of an arithmetic gradient cash flow series for years 1 through 10 is $500,000. If the gradient increase each year, G, is $3,000, determine the cash flow in year 1 at an interest rate of 10% per year.

Geometric Gradient

2.39 Assume you were asked to prepare a table of compound interest factor values (like those in the back of this book) used in calculating the present worth of a geometric gradient series. Determine the two values for $n = 1$ and 2 for an interest rate of 10% per year and a rate of change g of 5% per year.

2.40 A company that manufactures purgable hydrogen sulfide monitors will make deposits such that each one is 7% larger than the preceding one. How large must the first deposit at the end of year 1 be if the deposits extend through year 10 and the fourth deposit is $5550? Use an interest rate of 10% per year.

2.41 Calculate the present worth of a geometric gradient series with a cash flow of $35,000 in year 1 and increases of 5% each year through year 6. The interest rate is 10% per year.

2.42 To improve crack detection in military aircraft, the Air Force combined ultrasonic inspection procedures with laser heating to identify fatigue cracks. Early detection of cracks may reduce repair costs by as much as $200,000 per year. If the savings start at the end of year 1 and increase by 3% each year through year 5, calculate the present worth of these savings at an interest rate of 10% per year.

2.43 A civil engineer planning for her retirement places 10% of her salary each year into a high-technology stock fund. If her salary this year (end of year 1) is $160,000 and she expects her salary to increase by 3% each year, what will be the future worth of her retirement fund after 15 years provided it earns 7% per year?

2.44 El Paso Water (EPW) purchases surface water for treatment and distribution to EPW customers from the County Water Improvement District during the irrigation season. A new contract between the two entities resulted in a reduction in future price increases in the cost of the water from 8% per year to 4% per year for the next 20 years. The cost of water next year (which is year 1 of the new contract) will be $260 per acre-ft. Using an interest rate of 6% per year,

(a) Determine the present worth of the savings (in terms of $/acre-ft) to EPW between the old and the new contracts.

(b) Determine the total present worth of the savings over the life of the contract if EPW uses 51,000 acre-ft per year.

2.45 Toselli Animation plans to offer its employees a salary enhancement package that has revenue sharing as its main component. Specifically, the company will set aside 1% of total sales revenue for year-end bonuses. The sales are expected to be $5 million the first year, $5.5 million the second year, and amounts increasing by 10% each year for the next 5 years. At an interest rate of 8% per year, what is the equivalent annual worth in years 1 through 5 of the bonus package?

2.46 A northern California consulting firm wants to start saving money for replacement of network servers. If the company invests $5000 at the end of year 1 but decreases the amount invested by 5% each year, how much will be available 5 years from now at an earning rate of 8% per year?

Interest Rate and Rate of Return

2.47 A start-up company that makes robotic hardware for CIM (computer integrated manufacturing) systems borrowed $1 million to expand its packaging and shipping facility. The contract required the company to repay the lender through an innovative mechanism called "faux dividends," a series of uniform annual payments over a fixed period of time. If the company paid $290,000 per year for 5 years, what was the interest rate on the loan?

2.48 Your grandmother deposited $10,000 in an investment account on the day you were born to help pay the tuition when you go to college. If the account was worth $50,000 seventeen years after she made the deposit, what was the rate of return on the account?

2.49 An A&E firm planning for a future expansion deposited $40,000 each year for 5 years into a sinking (investment) fund that was to pay an unknown rate of return. If the account had a total of $451,000 immediately after the fifth deposit, what rate of return did the company make on these deposits?

2.50 Parkhill, Smith, and Cooper, a consulting engineering firm, pays a bonus to each engineer at the end of the year based on the company's profit for that year. If the company's initial investment was

$1.2 million, what rate of return has it made if each engineer's bonus has been $3000 per year for the past 10 years? Assume the company has six engineers and that the bonus money represents 5% of the company's profit.

2.51 For a 5-year period, determine the compound interest rate per year that is equivalent to a simple interest rate of 15% per year.

2.52 A person's credit score is important in determining the interest rate they have to pay on a home mortgage. According to Consumer Credit Counseling Service, a homeowner with a $100,000 mortgage and a 580 credit score will pay $90,325 more in interest charges over the life of a 30-year loan than a homeowner with the same mortgage and a credit score of 720. How much higher would the interest rate per year have to be in order to account for this much difference in interest charges, if the $100,000 loan is repaid in a single lump sum payment at the end of 30 years?

2.53 The business plan for KnowIt, LLC, a start-up company that manufactures portable multi-gas detectors, showed equivalent annual cash flows of $400,000 for the first 5 years. If the cash flow in year 1 was $320,000 and the constant increase thereafter was $50,000 per year, what interest rate was used in the calculation? (Solve using factors or a spreadsheet as requested by your instructor.)

Number of Years

2.54 RKE & Associates is considering the purchase of a building it currently leases for $30,000 per year. The owner of the building put it up for sale at a price of $170,000, but because the firm has been a good tenant, the owner offered to sell it to RKE for a cash price of $160,000 now. If purchased now, how long will it be before the company recovers its investment at an interest rate of 15% per year? Solve by spreadsheet function or factor.

2.55 A systems engineer who invested wisely can retire now because she has $2,000,000 in her self-directed retirement account. Determine how many years she can withdraw (a) $100,000 per year, or (b) $150,000 per year (beginning 1 year from now) provided her account earns at a rate of 5% per year. (c) Explain why the increased annual withdrawal from $100,000 to $150,000 per year is important.

2.56 How many years will it take for a uniform annual deposit of size A to accumulate to 10 times the size of a single deposit at a rate of return of 10% per year?

2.57 Demco Products, a company that manufactures stainless steel control valves, has a fund for equipment replacement that contains $500,000. The company plans to spend $85,000 each year on new equipment. (a) Estimate the number of years it will take to reduce the fund to no more than $85,000 at an interest rate of 10% per year. (b) Use the NPER function to determine the exact number of years.

2.58 A company that manufactures ultrasonic wind sensors invested $1.5 million 2 years ago to acquire part-ownership in an innovative chip-making company. How long will it take from the date of their initial investment for their share of the chip-making company to be worth $6 million, if that company is growing at a rate of 25% per year?

2.59 A trusted friend told you that a cash flow sequence that started at $3000 in year 1 and increased by $2000 each year would be worth $15,000 in 12 years at a rate of return of 10% per year. Is she correct?

2.60 You are a well-paid engineer with a well-established international corporation. In planning for your retirement, you are optimistic and expect to make an investment of $10,000 in year 1 and increase this amount by 10% each year. How long will it take for your account to have a future worth of $2,000,000 at a rate of return of 7% per year?

EXERCISES FOR SPREADSHEETS

These are introductory spreadsheet problems to become familiar with using Excel functions to solve single-factor problems. The development of correct functions is the primary goal of working them. The statements and questions may be similar to or an extension of previous problems.

2.61 A solar-powered personal aircraft with VTOL capability has been under development for the past 30 years by a group of engineers and physicists. SPPAV, as the plane will be termed, is expected to be available for its final test flight in exactly 3 years from now. Over the previous 30 years, a total of $100 million has been spent in its development. Assuming the $100 million was spent in an equal amount each year, and assuming an interest rate of 10%, determine the following:
(a) The value of the total investment now, after the 30 years.
(b) The value of the total investment at the expected time of the final test flight in three more years, assuming the same amount is spent for each of the next three years.

(c)   The value of the total investment at the expected time of the final test flight in three more years, assuming that *twice* the amount is spent for each of the *next* 3 *years* than that for the previous 30 years.

2.62   You expect to contribute to an investment fund for your retirement over the next 30 years with an annual deposit of a yet-to-be determined amount. Assume your goal is to have $2 million available when you stop the annual desposits and that the fund is able to return 10% per year every year.

(a)   Determine if you will reach your goal for either of the following two deposit scenarios: (1) $12,000 each and every year; (2) $8000 at the end of next year for 15 years, followed by $15,000 deposits in each of years 16 through 30.

(b)   Determine the exact number of years necessary to accumulate the $2 million if $12,000 is deposited each year until the goal is achieved.

(c)   For a little more of a challenge, use only the FV function to determine the number of years necessary to attain the $2 million goal for the second deposit scenario, assuming the $15,000 is deposited annually until the goal is achieved.

2.63   (This is a restatement of Problem 2.36.) Apple Computer wants to have $2.1 billion available 5 years from now in order to finance initial production of a device that applies IOT technology for home use. The company expects to set aside

uniformly increasing amounts of money each year to meet its goal, starting with $100 million at the end of year 1. How much will the constant increase, G, have to be each year at a rate of return of 18% per year? Try your skill by using the Goal Seek tool to find the required gradient. Start the evaluation with $G = 50 million per year.

2.64   For the data in Problem 2.60, find the year that your retirement account first exceeds (a) $2 million, and (b) then $3 million. In setting up the spreadsheet you wish to know the amount that must be deposited each year. Use any spreadsheet functions that you choose.

2.65   The SEWA (Southwestern Electricity and Water Authority) authorized construction projects totaling $1.07 million to improve desalinization plant efficiency and salinity-reduction technology for reject chemicals. Three bids from potential vendors were received in the amounts of $1.06, $1.053, and $1.045 million. Assume the savings will be realized immediately for each bid, were it accepted. Use an expected life of 10 years and $i = 6\%$ per year to do the following *for the savings* anticipated from the lower-than authorized bids:

(a)   Determine the equivalent present worth of the savings.

(b)   Determine the equivalent annual worth of the savings.

(c)   Develop a column chart for the equivalent annual worth for the savings for each bid.

ADDITIONAL PROBLEMS AND FE EXAM REVIEW QUESTIONS

2.66   A manufacturer of prototyping equipment wants to have $3,000,000 available 10 years from now so that a new product line can be initiated. If the company plans to deposit money each year, starting 1 year from now, the equation that represents how much the company is required to deposit each year at 10% per year interest to have the $3,000,000 immediately after the last deposit is:

(a)   3,000,000(A/F,10%,10)

(b)   3,000,000(A/F,10%,11)

(c)   3,000,000 + 3,000,000(A/F,10%,10)

(d)   3,000,000(A/P,10%,10)

2.67   The amount of money the Teachers Credit Union should be willing to loan a developer who will

repay the loan in a lump sum amount of $840,000 two years from now at the bank's interest rate of 10% per year is:

(a) $694,180

(b) $99,170

(c) $1,106,400

(d) $763,650

2.68   The cost of updating an outdated production process is expected to be $81,000 four years from now. The equivalent present worth of the update at 6% per year interest is closest to:

(a) $51,230

(b) $55,160

(c) $60,320

(d) $64,160

2.69   A single deposit of $25,000 was made by your grandparents on the day you were born 25 years ago. The balance in the account today if it grew at 10% per year is closest to:
(a) $201,667   (b) $241,224
(c) $270,870   (d) $296,454

2.70   A chip manufacturing company wants to have $10 million available 5 years from now in order to build new warehouse and shipping facilities. If the company can invest money at 10% per year, the amount that it must deposit *each year* in years 1 through 5 to accumulate the $10 million is closest to:
(a) $1,638,000   (b) $2,000,000
(c) $2,638,000   (d) $2,938,000

2.71   An engineer who believes in "save now; play later" wants to retire in 30 years with $2.0 million. At 8% per year interest, the amount the engineer will have to invest each year (starting in year 1) to reach the $2 million goal is closest to:
(a) $17,660   (b) $28,190
(c) $49,350   (d) $89,680

2.72   The cost of tuition at a large public university was $390 per credit hour 5 years ago. The cost today (exactly 5 years later) is $585. The annual rate of increase is closest to:
(a)   5%   (b)   7%   (c)   9%   (d)   11%

2.73   The Gap has some of its jeans stone-washed under a contract with Vietnam Garment Corporation (VGC). If VGC's estimated operating cost per machine is $26,000 for year 1 and it increases by $1500 per year through year 5, the equivalent uniform annual cost per machine over years 1 to 5, at an interest rate of 8% per year, is closest to:
(a) $30,850   (b) $28,770
(c) $26,930   (d) $23,670

2.74   Adams Manufacturing spent $30,000 on a new sterilization conveyor belt, which resulted in a cost savings of $4202 per year. The length of time it should take to recover the investment at 8% per year is closest to:
(a)   Less than 6 years (b)   7 years
(c)   9 years   (d)   11 years

2.75   The present worth of an increasing geometric gradient is $23,632. The interest rate is 6% per year and the rate of change series is 4% per year. If the cash flow in year 1 is $3,000, the year in which the gradient ends is:
(a)   7   (b)   9   (c)   11   (d)   12

2.76   At $i = 8\%$ per year, the annual worth for years 1 through 6 of the cash flows shown is closest to:
(a) $302   (b) $421
(c) $572   (d) $824

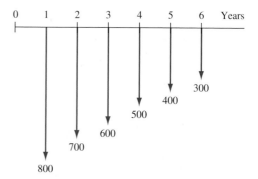

2.77   Chemical costs associated with a packed-bed flue gas incinerator for odor control have been decreasing uniformly for 5 years because of increases in efficiency. If the cost in year 1 was $100,000 and it decreased by $5,000 per year through year 5, the present worth of the costs at 10% per year is closest to:
(a) $344,771   (b) $402,200
(c) $515,400   (d) $590,700

2.78   The winner of a multistate mega millions lottery jackpot worth $175 million was given the option of taking payments of $7 million per year for 25 years, beginning 1 year from now, or taking $109.355 million now. The interest rate that renders the two options equivalent to each other is closest to:
(a)   4%   (b)   5%   (c)   6%   (d)   7%

2.79   Maintenance costs for a regenerative thermal oxidizer increased according to an arithmetic gradient for 5 years. The cost in year 1 was $7000. If the interest rate is 10% per year and the present worth of the costs for a 5-year period was $28,800, the amount of the yearly increase, G, was closest to:
(a) $1670   (b) $945
(c) $620   (d) $330

2.80   Aero Serve, Inc. manufactures cleaning nozzles for reverse pulse jet dust collectors. The company spent $40,000 on a production control system that will increase profits by $11,096 per year for 5 years. The rate of return per year on the investment is closest to:
(a)   20%
(b)   16%
(c)   12%
(d)   Less than 11%

CASE STUDY

THE AMAZING IMPACT OF COMPOUND INTEREST

Background on Five Situations

1. The first mass produced automobile was the Ford Model T, initially manufactured and sold in 1909 for $825. The rate of inflation in the United States over the period 1909 to 2015 has averaged 3.10% per year. You just purchased a new car for $28,000. You wonder what the cost of this same car might be 50 years from now when your son may be purchasing a similar car for his daughter (your granddaughter) so she can attend college. Also, you wonder what the value of the Model T will be 50 years in the future, that is, in the year 2065.

2. The purchase price of Manhattan Island, where much of the city of New York is concentrated, in the year 1626 was $24. After 391 years in 2017, you wonder what the value of the land might be, provided it has appreciated in value at a rate of 6% per year, every year.

3. Last week, your friend Jeremy borrowed $200 from a pawn shop operator because he was totally broke. He was to pay the operator $230 after 1 week, but missed the payment. At first, you thought this was "no big deal," but then started to realize that the interest was $30 the first week alone and would increase, compounded at the same rate until the total debt was repaid. When your friend told you he would pay off the loan in a year (provided the operator didn't come after him), you gave him the results of your analysis. He was shocked and paid the total amount immediately.

4. In 1939, two people teamed up to manufacture and market electronic test equipment. By 1957, the initial capital investment from themselves and a few friends that amounted to only $80,805.12 in 1939 had increased in value to an equivalent of $1 million. After this, the company skyrocketed to become a world leader in electronic equipment, computers, and a wide range of other products. If the net cash flow averaged $150,000 per year from 1957 to 2017 (60 years) at the same rate, these two individuals would be quite wealthy.

5. Assume that when your great-grandmother was 25 years old, she received an engagement ring from her husband-to-be. He paid $50 for the ring containing a single, high-quality diamond. When she passed away at the age of 90, the ring went to your grandmother, who kept it for 60 years and then gave it to your mother. After 30 years of keeping the ring in a safe place, she gave it to you on your 24th birthday. Today is your 48th birthday and you have just discovered the ring in a desk drawer, forgotten for all these years. If this high-grade diamond has been now appraised as a collector's grade stone, which has appreciated in value at an average rate of 4% per year, every year, since it was first purchased, you wonder what the ring might be valued at today.

Team Exercises

For each situation described, do the following using a five-person team.

(a) Determine the annual compound interest or inflation rate and discuss the differences in these rates from one situation to another.

(b) First, each member of the team should make an estimate (a guess) of the beginning and ending amounts of money involved *for each situation*. Each team member should now calculate the two amounts for one selected situation. Between team members, discuss the accuracy of their first estimates compared to the actual amounts calculated.

CHAPTER 3

Combining Factors and Spreadsheet Functions

Steve Allen/Brand X Pictures

LEARNING OUTCOMES

Purpose: Use multiple factors and spreadsheet functions to find equivalent amounts for cash flows that have nonstandard placement.

SECTION	TOPIC	LEARNING OUTCOME
3.1	Shifted series	• Determine the P, F, or A values of a series starting at a time other than period 1.
3.2	Shifted series and single cash flows	• Determine the P, F, or A values of a shifted series and randomly placed single cash flows.
3.3	Shifted gradients	• Make equivalence calculations for shifted arithmetic or geometric gradient series that increase or decrease in size of cash flows.

ost estimated cash flow series do not fit exactly the series for which the factors, equations, and spreadsheet functions in Chapter 2 were developed. For a given sequence of cash flows, there are usually several correct ways to determine the equivalent present worth P, future worth F, or annual worth A. This chapter explains how to combine engineering economy factors and spreadsheet functions to address more complex situations involving shifted uniform series, gradient series, and single cash flows.

3.1 Calculations for Uniform Series That Are Shifted ● ● ●

When a uniform series begins at a time other than at the end of period 1, it is called a **shifted series**. In this case several methods can be used to find the equivalent present worth P. For example, P of the uniform series shown in Figure 3–1 could be determined by any of the following methods:

- Use the P/F factor to find the present worth of each disbursement at year 0 and add them.
- Use the F/P factor to find the future worth of each disbursement in year 13, add them, and then find the present worth of the total, using $P = F(P/F,i,13)$.
- Use the F/A factor to find the future amount $F = A(F/A,i,10)$, and then compute the present worth, using $P = F(P/F,i,13)$.
- Use the P/A factor to compute the "present worth" $P_3 = A(P/A,i,10)$ (which will be located in year 3, not year 0), and then find the present worth in year 0 by using the $(P/F,i,3)$ factor.

Typically the last method is used for calculating the present worth of a uniform series that does not begin at the end of period 1. For Figure 3–1, the "present worth" obtained using the P/A factor is located in year 3. This is shown as P_3 in Figure 3–2. Note that a P value is always located *1 year or period prior* to the beginning of the first series amount. Why? Because the P/A factor was derived with P in time period 0 and A beginning at the end of period 1. The most common mistake made in working problems of this type is improper placement of P. Therefore, it is extremely important to remember:

> The present worth is always located **one period prior** to the first uniform series amount when using the P/A factor.

Placement of P

To determine a future worth or F value, recall that the F/A factor derived in Section 2.3 had the F located in the *same* period as the last uniform series amount. Figure 3–3 shows the location of the future worth when F/A is used for Figure 3–1 cash flows.

> The future worth is always located in the **same period as the last** uniform series amount when using the F/A factor.

Placement of F

It is also important to remember that the number of periods n in the P/A or F/A factor is equal to the number of uniform series values. It may be helpful to *renumber* the cash flow diagram to avoid errors in counting. Figures 3–2 and 3–3 show Figure 3–1 renumbered to determine $n = 10$.

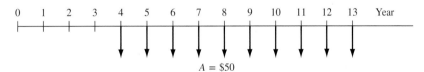

Figure 3–1
A uniform series that is shifted.

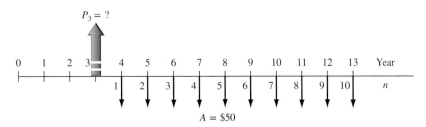

Figure 3–2
Location of present worth and renumbering for n for the shifted uniform series in Figure 3–1.

Figure 3–3
Placement of F and renumbering for n for the shifted uniform series of Figure 3–1.

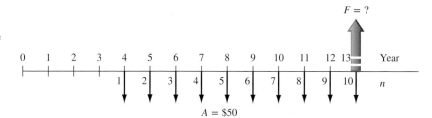

As stated above, several methods can be used to solve problems containing a uniform series that is shifted. However, it is generally more convenient to use the uniform series factors than the single-amount factors. Specific steps should be followed to avoid errors:

1. Draw a diagram of the positive and negative cash flows.
2. Locate the present worth or future worth of each series on the cash flow diagram.
3. Determine n for each series by renumbering the cash flow diagram.
4. Draw another cash flow diagram representing the desired equivalent cash flow.
5. Set up and solve the equations.

These steps are illustrated below.

EXAMPLE 3.1

The offshore design group at Bechtel just purchased upgraded CAD software for $5000 now and annual payments of $500 per year for 6 years starting 3 years from now for annual upgrades. What is the present worth in year 0 of the payments if the interest rate is 8% per year?

Solution

The cash flow diagram is shown in Figure 3–4. The symbol P_A is used throughout this chapter to represent the present worth of a uniform annual series A, and P_A' represents the present worth at a time other than period 0. Similarly, P_T represents the total present worth at time 0. The correct placement of P_A' and the diagram renumbering to obtain n are also indicated. *Note that P_A' is located in actual year 2, not year 3. Also, $n = 6$, not 8, for the P/A factor.* First find the value of P_A' of the shifted series.

$$P_A' = \$500(P/A,8\%,6)$$

Since P_A' is located in year 2, now find P_A in year 0.

$$P_A = P_A'(P/F,8\%,2)$$

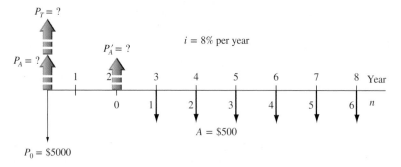

Figure 3–4
Cash flow diagram with placement of P values, Example 3.1.

The total present worth is determined by adding P_A and the initial payment P_0 in year 0.

$$P_T = P_0 + P_A$$

$$= 5000 + 500(P/A,8\%,6)(P/F,8\%,2)$$

$$= 5000 + 500(4.6229)(0.8573)$$

$$= \$6981.60$$

The more complex that cash flow series become, the more useful are the spreadsheet functions. When the uniform series A is shifted, the NPV function is used to determine P, and the PMT function finds the equivalent A value. The NPV function, like the PV function, determines the P values, but NPV can handle any combination of cash flows directly from the cells. As we learned in Chapter 2, enter the net cash flows in contiguous cells (column or row), making sure to enter "0" for all zero cash flows. Use the format

$$\textbf{NPV}(\textbf{\textit{i}\%,second\_cell:last\_cell}) + \textbf{first\_cell}$$

First_cell contains the cash flow for year 0 and must be listed separately for NPV to correctly account for the time value of money. The cash flow in year 0 may be 0.

The easiest way to find an equivalent A over n years for a shifted series is with the PMT function, where the P value is from the NPV function above. The format is the same as we learned earlier; the entry for P is a cell reference, not a number.

$$\textbf{PMT}(\textbf{\textit{i}\%,\textit{n},cell\_with\_\textit{P},\textit{F}})$$

Alternatively, the same technique can be used when an F value was obtained using the FV function. Now the last entry in PMT is "cell_with_F."

It is very fortunate that any parameter in a spreadsheet function can itself be a function. Thus, it is possible to write the PMT function in a single cell by embedding the NPV function (and FV function, if needed). The format is

$$\textbf{PMT}(\textbf{\textit{i}\%,\textit{n},NPV}(\textbf{\textit{i}\%,second\_cell:last\_cell}) + \textbf{first\_cell,\textit{F}})\qquad\text{[3.1]}$$

Of course, the answer for A is the same for the two-cell operation or a single-cell, embedded function. All three of these functions are illustrated in Example 3.2.

EXAMPLE 3.2

Recalibration of sensitive measuring devices costs \$8000 per year. If the machine will be recalibrated for each of 6 years starting 3 years after purchase, calculate the 8-year equivalent uniform series at 16% per year. Show hand and spreadsheet solutions.

Solution by Hand

Figure 3–5a and b shows the original cash flows and the desired equivalent diagram. To convert the \$8000 shifted series to an equivalent uniform series over all periods, first convert the uniform series into a present worth or future worth amount. Then either the A/P factor or the A/F factor can be used. Both methods are illustrated here.

Present worth method. (Refer to Figure 3–5a.) Calculate P'_A for the shifted series in year 2, followed by P_T in year 0. There are 6 years in the A series.

$$P'_A = 8000(P/A,16\%,6)$$

$$P_T = P'_A(P/F,16\%,2) = 8000(P/A,16\%,6)(P/F,16\%,2)$$

$$= 8000(3.6847)(0.7432) = \$21{,}907.75$$

The equivalent series A' *for 8 years* can now be determined via the A/P factor.

$$A' = P_T(A/P,16\%,8) = \$5043.60$$

Future worth method. (Refer to Figure 3–5a.) First calculate the future worth F in year 8.

$$F = 8000(F/A,16\%,6) = \$71{,}820$$

The A/F factor is now used to obtain A' over all 8 years.

$$A' = F(A/F,16\%,8) = \$5043.20$$

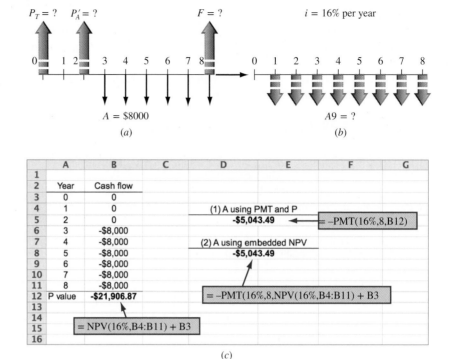

Figure 3–5

(*a*) Original, and (*b*) equivalent cash flow diagrams, and (*c*) spreadsheet functions to determine *P* and *A*, Example 3.2.

Solution by Spreadsheet

(Refer to Figure 3–5*c*.) Enter the cash flows in B3 through B11 with entries of "0" in the first three cells. Use the NPV function to display $P = \$21,906.87$.

There are two ways to obtain the equivalent *A* over 8 years. Of course, only one of these PMT functions needs to be entered. (1) Enter the PMT function making direct reference to the *P* value (see cell tag for D/E5) or (2) use Equation [3.1] to embed the NPV function into the PMT function (see cell tag for D/E8).

3.2 Calculations Involving Uniform Series and Randomly Placed Single Amounts ●●●

When a cash flow includes both a uniform series and randomly placed single amounts, the procedures of Section 3.1 are applied to the uniform series and the single-amount formulas are applied to the one-time cash flows. This approach, illustrated in Examples 3.3 and 3.4, is merely a combination of previous ones. For spreadsheet solutions, it is necessary to enter the net cash flows before using the NPV and other functions.

EXAMPLE 3.3

An engineering company in Wyoming that owns 50 hectares of valuable land has decided to lease the mineral rights to a mining company. The primary objective is to obtain long-term income to finance ongoing projects 6 and 16 years from the present time. The engineering company makes a proposal to the mining company that it pay \$20,000 per year for 20 years beginning 1 year from now, plus \$10,000 six years from now and \$15,000 sixteen years from now. If the mining company wants to pay off its lease immediately, how much should it pay now if the investment is to make 16% per year?

Solution

The cash flow diagram is shown in Figure 3–6 from the owner's perspective. Find the present worth of the 20-year uniform series and add it to the present worth of the two one-time amounts to determine P_T.

$$P_T = 20,000(P/A,16\%,20) + 10,000(P/F,16\%,6) + 15,000(P/F,16\%,16)$$

$$= \$124,075$$

Note that the $20,000 uniform series starts at the end of year 1, so the P/A factor determines the present worth at year 0.

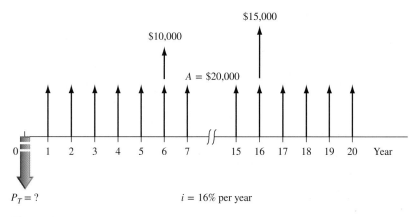

Figure 3–6
Diagram including a uniform series and single amounts, Example 3.3.

When you calculate the A value for a cash flow series that includes randomly placed single amounts and uniform series, **first convert everything to a present worth or a future worth.** Then you obtain the A value by multiplying P or F by the appropriate A/P or A/F factor. Example 3.4 illustrates this procedure.

EXAMPLE 3.4

A design-build-operate engineering company in Texas that owns a sizable amount of land plans to lease the drilling rights (oil and gas only) to a mining and exploration company. The contract calls for the mining company to pay $20,000 per year for 20 years beginning 3 years from now (i.e., beginning at the end of year 3 and continuing through year 22) plus $10,000 six years from now and $15,000 sixteen years from now. Utilize engineering economy relations by hand and by spreadsheet to determine the five *equivalent values* listed below at 16% per year.

1. Total present worth P_T in year 0
2. Future worth F in year 22
3. Annual series over all 22 years
4. Annual series over the first 10 years
5. Annual series over the last 12 years

Solution by Hand

Figure 3–7 presents the cash flows with equivalent P and F values indicated in the correct years for the P/A, P/F, and F/A factors.

1. **P_T in year 0:** First determine P_A' of the series in year 2. Then P_T is the sum of three P values: the series present worth value moved back to $t = 0$ with the P/F factor, and the two P values at $t = 0$ for the two single amounts in years 6 and 16.

$$P_A' = 20,000(P/A,16\%,20)$$

$$P_T = P_A'(P/F,16\%,2) + 10,000(P/F,16\%,6) + 15,000(P/F,16\%,16)$$

$$= 20,000(P/A,16\%,20)(P/F,16\%,2) + 10,000(P/F,16\%,6)$$
$$+ 15,000(P/F,16\%,16)$$

$$= \$93,625 \tag{3.2}$$

2. **F in year 22:** To determine F in year 22 from the original cash flows (Figure 3–7), find F for the 20-year series and add the two F values for the two single amounts. Be sure to carefully determine the n values for the single amounts: $n = 22 - 6 = 16$ for the \$10,000 amount and $n = 22 - 16 = 6$ for the \$15,000 amount.

$$F = 20,000(F/A,16\%,20) + 10,000(F/P,16\%,16) + 15,000(F/P,16\%,6) \tag{3.3}$$

$$= \$2,451,626$$

3. **A over 22 years:** Multiply $P_T = \$93,625$ from (1) on the previous page by the A/P factor to determine an equivalent 22-year A series, referred to as A_{1-22} here.

$$A_{1-22} = P_T(A/P,16\%,22) = 93,625(0.16635) = \$15,575 \tag{3.4}$$

An alternate way to determine the 22-year series uses the F value from (2) above. In this case, the computation is $A_{1-22} = F(A/F,16\%,22) = \$15,575$.

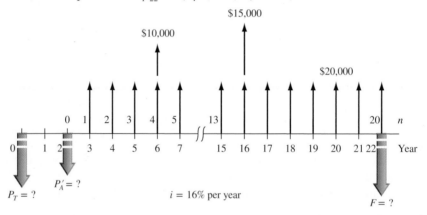

Figure 3–7
Diagram for Example 3.4.

4. **A over years 1 to 10:** This and (5), which follows, are special cases that often occur in engineering economy studies. The equivalent A series is calculated for a number of years different from that covered by the original cash flows. This occurs when a defined *study period* or *planning horizon* is preset for the analysis. (More is mentioned about study periods later.) To determine the equivalent A series for years 1 through 10 only (call it A_{1-10}), the P_T value *must be used* with the A/P factor for $n = 10$. This computation transforms Figure 3–7 into the equivalent series A_{1-10} in Figure 3–8a.

$$A_{1-10} = P_T(A/P,16\%,10) = 93,625(0.20690) = \$19,371 \tag{3.5}$$

5. **A over years 11 to 22:** For the equivalent 12-year series for years 11 through 22 (call it A_{11-22}), the F value *must be used* with the A/F factor for 12 years. This transforms Figure 3–7 into the 12-year series A_{11-22} in Figure 3–8b.

$$A_{11-22} = F(A/F,16\%,12) = 2,451,626(0.03241) = \$79,457 \tag{3.6}$$

Notice the huge difference of more than \$60,000 in equivalent annual amounts that occurs when the present worth of \$93,625 is allowed to compound at 16% per year for the first 10 years. This is another demonstration of the time value of money.

Solution by Spreadsheet

Figure 3–9 is a spreadsheet image with answers for all five questions. The \$20,000 series and the two single amounts have been entered into separate columns, B and C. The zero cash flow

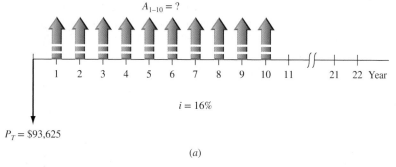

$$A_{1-10} = ?$$

$$i = 16\%$$

$$P_T = \$93,625$$

(a)

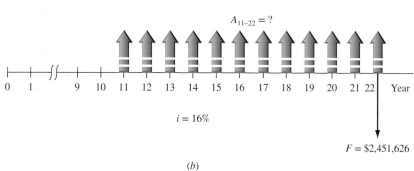

$$A_{11-22} = ?$$

$$i = 16\%$$

$$F = \$2,451,626$$

(b)

Figure 3–8

Cash flows of Figure 3–7 converted to equivalent uniform series for (a) years 1 to 10 and (b) years 11 to 22.

	A	B	C	D	E	F
1		**Interest rate**		**16.00%**		
2						
3		Cash flows		Value	Spreadsheet function used	
4	Year	Series	Single	calculated	to obtain result in column F	Result
5	0	0	0	Present worth		
6	1	0	0	of series	`= NPV(D1,B6:B27) + B5`	$88,122
7	2	0	0			
8	3	$ 20,000	0			
9	4	$ 20,000	0	Present worth		
10	5	$ 20,000	0	of singles	`= NPV(D1,C6:C27) + C5`	$5,500
11	6	$ 20,000	$ 10,000			
12	7	$ 20,000	0			
13	8	$ 20,000	0			
14	9	$ 20,000	0	1. Present worth	`= F6 + F10` or	$93,622
15	10	$ 20,000	0	total	`= NPV(D1,B6:B27) + NPV(D1,C6:C27)`	$93,622
16	11	$ 20,000	0			
17	12	$ 20,000	0	2. Future worth		
18	13	$ 20,000	0	total	`= FV(D1,22,0,-F14)`	$2,451,621
19	14	$ 20,000	0			
20	15	$ 20,000	0	3. Annual series		
21	16	$ 20,000	$ 15,000	over 22 years	`= PMT(D1,22,-F14)`	$15,574
22	17	$ 20,000	0			
23	18	$ 20,000	0	4. Annual series		
24	19	$ 20,000	0	for first 10 years	`= PMT(D1,10,-F14)`	$19,370
25	20	$ 20,000	0			
26	21	$ 20,000	0	5. Annual series		
27	22	$ 20,000	0	for last 12 years	`= PMT(D1,12,0,-F18)`	$79,469

Figure 3–9

Spreadsheet using cell reference format. Example 3.4.

values are all entered so that the functions will work correctly. This is an excellent example demonstrating the versatility of the NPV, FV, and PMT functions. To prepare for sensitivity analysis, the functions are developed using cell reference format or global variables, as indicated in the column E function. This means that virtually any number—the interest rate, any cash flow estimate in the series or the single amounts, and the timing within the 22-year time frame—can be changed and the new answers will be immediately displayed.

1. Present worth values for the series and single amounts are determined in cells F6 and F10, respectively, using the NPV function. The sum of these in F14 is $P_T = \$93,622$, which corresponds to the value in Equation [3.2]. Alternatively, P_T can be determined directly via the sum of two NPV functions, shown in row 15.
2. The FV function in row 18 uses the P value in F14 (preceded by a minus sign) to determine F 22 years later. This is significantly easier than Equation [3.3].
3. To find the 22-year A series starting in year 1, the PMT function in row 21 references the P value in cell F14. This is effectively the same procedure used in Equation [3.4] to obtain A_{1-22}.

For the spreadsheet enthusiast, it is possible to find the 22-year A series value directly by using the PMT function with embedded NPV functions. The cell reference format is = PMT(D1,22,−(NPV(D1,B6:B27)+B5 + NPV(D1,C6:C27)+C5)).

4. and 5. It is quite simple to determine an equivalent uniform series over any number of periods using a spreadsheet, provided the series starts one period after the P value is located or ends in the same period that the F value is located. These are both true for the series requested here. The results in F24 and F27 are the same as A_{1-10} and A_{11-22} in Equations [3.5] and [3.6], respectively.

Comment

Remember that **round-off error** will always be present when comparing hand and spreadsheet results. The spreadsheet functions carry more decimal places than the tables during calculations. Also, be very careful when constructing spreadsheet functions. It is easy to miss a value, such as the P or F in PMT and FV functions, or a minus sign between entries. Always check your function entries carefully before touching <Enter>.

3.3 Calculations for Shifted Gradients ● ● ●

In Section 2.5, we derived the relation $P = G(P/G,i,n)$ to determine the present worth of the **arithmetic gradient** series. The P/G factor, Equation [2.25], was derived for a present worth in year 0 with the gradient first appearing in year 2.

Placement of gradient P

> The present worth of an **arithmetic gradient** will always be located **two periods before the gradient starts.**

Refer to Figure 2–14 as a refresher for the cash flow diagrams.

The relation $A = G(A/G,i,n)$ was also derived in Section 2.5. The A/G factor in Equation [2.27] performs the equivalence transformation of a **gradient only** into an A series from years 1 through n (Figure 2–15). Recall that the base amount must be treated separately. Then the equivalent P or A values can be summed to obtain the equivalent total present worth P_T and total annual series A_T.

A conventional gradient series starts between periods 1 and 2 of the cash flow sequence. A gradient starting at any other time is called a **shifted gradient.** The n value in the P/G and A/G factors for a shifted gradient is determined by renumbering the time scale. The period in which the *gradient first appears is labeled period 2.* The n value for the gradient factor is determined by the renumbered period where the last gradient increase occurs.

Partitioning a cash flow series into the arithmetic gradient series and the remainder of the cash flows can make very clear what the gradient n value should be. Example 3.5 illustrates this partitioning.

EXAMPLE 3.5

Fujitsu, Inc. has tracked the average inspection cost on a robotics manufacturing line for 8 years. Cost averages were steady at $100 per completed unit for the first 4 years, but have increased consistently by $50 per unit for each of the last 4 years. Analyze the gradient increase, using the P/G factor. Where is the present worth located for the gradient? What is the general relation used to calculate total present worth in year 0?

Solution

The cash flow diagram in Figure 3–10a shows the base amount $A = \$100$ and the arithmetic gradient $G = \$50$ starting between periods 4 and 5. Figure 3–10b and c partitions these two series. Gradient year 2 is placed in actual year 5 of the entire sequence in Figure 3–10c. It is clear that $n = 5$ for the P/G factor. The $P_G = ?$ arrow is correctly placed in gradient year 0, which is year 3 in the cash flow series.

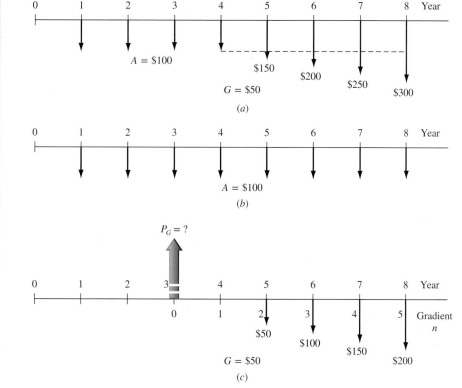

Figure 3–10
Partitioned cash flow, $(a) = (b) + (c)$, Example 3.5.

The general relation for P_T is taken from Equation [2.19]. The uniform series $A = \$100$ occurs for all 8 years, and the $G = \$50$ gradient present worth P_G appears in year 3.

$$P_T = P_A + P_G = 100(P/A,i,8) + 50(P/G,i,5)(P/F,i,3)$$

It is important to note that the A/G factor *cannot* be used to find an equivalent A value in periods 1 through n for cash flows involving a shifted gradient. Consider the cash flow diagram of Figure 3–11. To find the equivalent annual series in years 1 through 10 for the gradient series only, first find the present worth P_G of the gradient in actual year 5, take this present worth back to year 0, and annualize the present worth for 10 years with the A/P factor. If you apply the annual series gradient factor $(A/G,i,5)$ directly, the gradient is converted into an equivalent annual series over years 6 through 10 only, not years 1 through 10, as requested.

> **To find the equivalent A series of a shifted gradient through all the n periods, first find the present worth of the gradient at actual time 0, then apply the $(A/P,i,n)$ factor.**

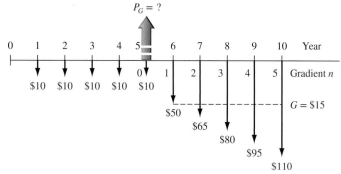

Figure 3–11
Determination of G and n values used in factors for a shifted gradient.

EXAMPLE 3.6

Set up the engineering economy relations to compute the equivalent annual series in years 1 through 7 for the cash flow estimates in Figure 3–12.

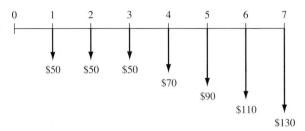

Figure 3–12
Diagram of a shifted gradient, Example 3.6.

Solution

The base amount annual series is $A_B = \$50$ for all 7 years (Figure 3–13). Find the present worth P_G in actual year 2 of the $20 gradient that starts in actual year 4. The gradient n is 5.

$$P_G = 20(P/G,i,5)$$

Bring the gradient present worth back to actual year 0.

$$P_0 = P_G(P/F,i,2) = 20(P/G,i,5)(P/F,i,2)$$

Annualize the gradient present worth from year 1 through year 7 to obtain A_G.

$$A_G = P_0(A/P,i,7)$$

Finally, add the base amount to the gradient annual series.

$$A_T = 20(P/G,i,5)(P/F,i,2)(A/P,i,7) + 50$$

For a spreadsheet solution, enter the original cash flows into adjacent cells, say, B3 through B10, and use an embedded NPV function in the PMT function. The single-cell function is = PMT(i%,7,−NPV(i%, B3:B10)).

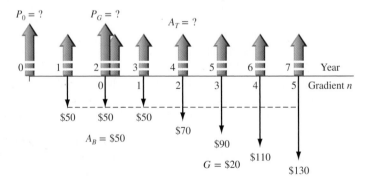

Figure 3–13
Diagram used to determine A for a shifted gradient, Example 3.6.

In Section 2.6, we derived the relation $P_g = A_1(P/A,g,i,n)$ to determine the present worth of a **geometric gradient** series, including the initial amount A_1. The factor was derived to find the present worth in year 0, with A_1 in year 1 and the first gradient appearing in year 2.

Placement of gradient P

The present worth of a **geometric gradient series** will always be located **two periods before the gradient starts** and the **initial amount is included** in the resulting present worth. Refer to Figure 2–21 as a refresher for the cash flows.

Equation [2.35] is the formula used for the factor. It is not tabulated.

EXAMPLE 3.7

Chemical engineers at a Coleman Industries plant in the Midwest have determined that a small amount of a newly available chemical additive will increase the water repellency of Coleman's tent fabric by 20%. The plant superintendent has arranged to purchase the additive through a 5-year contract at $7000 per year, starting 1 year from now. He expects the annual price to increase by 12% per year thereafter for the next 8 years. Additionally, an initial investment of $35,000 was made now to prepare a site suitable for the contractor to deliver the additive. Use $i = 15\%$ per year to determine the equivalent total present worth for all these cash flows.

Solution

Figure 3–14 presents the cash flows. The total present worth P_T is found using $g = 0.12$ and $i = 0.15$. Equations [2.34] and [2.35] are used to determine the present worth P_g for the entire geometric series at actual year 4, which is moved to year 0 using $(P/F,15\%,4)$.

$$P_T = 35{,}000 + A(P/A,15\%,4) + A_1(P/A,12\%,15\%,9)(P/F,15\%,4)$$

$$= 35{,}000 + 7000(2.8550) + \left[7000 \frac{1 - (1.12/1.15)^9}{0.15 - 0.12} \right](0.5718)$$

$$= 35{,}000 + 19{,}985 + 28{,}247$$

$$= \$83{,}232$$

Note that $n = 4$ in the $(P/A,15\%,4)$ factor because the $7000 in year 5 is the base cash flow of the gradient A_1.

For solution by spreadsheet, enter the cash flows of Figure 3–14. If cells B1 through B14 are used, the function to find $P = \$83{,}230$ is

$$\text{NPV}(15\%,\text{B2:B14})+\text{B1}$$

The fastest way to enter the geometric series is to enter $7000 for year 5 (into cell B6) and set up each succeeding cell multiplied by 1.12 for the 12% increase.

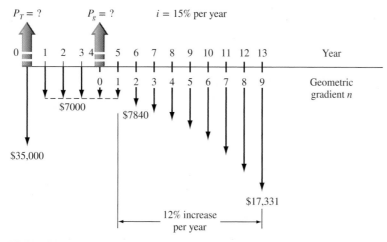

Figure 3–14
Cash flow diagram including a geometric gradient with $g = 12\%$, Example 3.7.

Decreasing arithmetic and geometric gradients are common, and they are often **shifted gradient series**. That is, the constant gradient is $-G$ or the percentage change is $-g$ from one period to the next, and the first appearance of the gradient is at some time period (year) other than year 2 of the series. Equivalence computations for present worth P and annual worth A are basically the same as discussed in Chapter 2, except for the following.

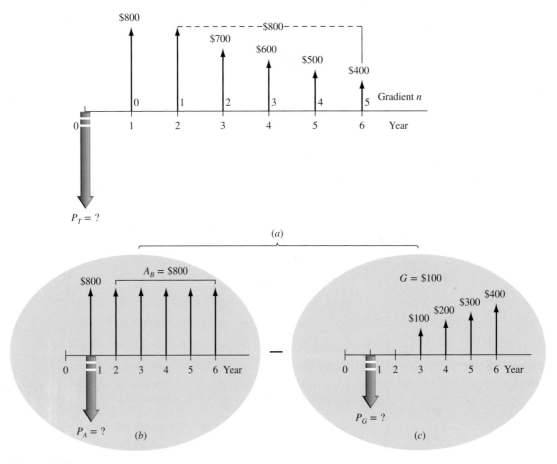

Figure 3–15
Partitioned cash flow of a shifted arithmetic gradient, $(a) = (b) - (c)$.

For shifted, decreasing gradients:

- The base amount A (arithmetic) or initial amount A_1 (geometric) is the *largest amount* in the first year of the series.
- The gradient amount is *subtracted from* the previous year's amount, not added to it.
- The amount used in the factors is $-G$ for arithmetic and $-g$ for geometric gradient series.
- The present worth P_G or P_g is located 2 years prior to the appearance of the first gradient; however, a P/F factor is necessary to find the present worth in year 0.

Figure 3–15 partitions a decreasing arithmetic gradient series with $G = \$-100$ that is shifted 1 year forward in time. P_G occurs in actual year 1, and P_T is the sum of three components.

$$P_T = \$800(P/F,i,1) + 800(P/A,i,5)(P/F,i,1) - 100(P/G,i,5)(P/F,i,1)$$

EXAMPLE 3.8

Morris Glass Company has decided to invest funds for the next 5 years so that development of "smart" glass is well funded in the future. This type of new-technology glass uses electro-chrome coating to allow rapid adjustment to sun and dark in building glass, as well as assisting with internal heating and cooling cost reduction. The financial plan is to invest first, allow appreciation to occur, and then use the available funds in the future. All cash flow estimates are in $1000 units, and the interest rate expectation is 8% per year.

Years 1 through 5: Invest $7000 in year 1, decreasing by $1000 per year through year 5.

Years 6 through 10: No new investment and no withdrawals.

Years 11 through 15: Withdraw $20,000 in year 11, decreasing 20% per year through year 15.

Determine if the anticipated withdrawals will be covered by the investment and appreciation plans. If the withdrawal series is over- or underfunded, what is the exact amount available in year 11 (the beginning of the withdrawal series), provided all other estimates remain the same?

Solution by Hand

Figure 3–16 presents the cash flow diagram and the placement of the equivalent P values used in the solution. Calculate the present worth of both series in actual year 0 and add them to determine if the investment series is adequate to fund the anticipated withdrawals.

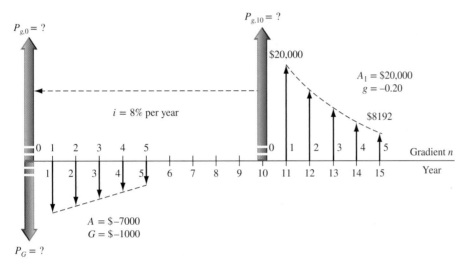

Figure 3–16
Investment and withdrawal series, Example 3.8.

Investment series: Decreasing, conventional arithmetic series starting in year 2 with $A = \$-7000$, $G = \$-1000$, and gradient $n = 5$ years. The present worth in year 0 is

$$P_G = -[7000(P/A,8\%,5) - 1000(P/G,8\%,5)]$$
$$= \$-20{,}577$$

Withdrawal series: Decreasing, shifted geometric series starting in year 12 with $A_1 = \$20{,}000$, $g = -0.20$, and gradient $n = 5$ years. If the present worth in year 10 is identified as $P_{g,10}$, the present worth in year 0 is $P_{g,0}$. Use Equation [2.35] for the $(P/A,-20\%,8\%,5)$ factor.

$$P_{g,0} = P_{g,10}(P/F,i,n) = A_1(P/A,g,i,n)(P/F,i,n) \qquad [3.7]$$
$$= 20{,}000 \left\{ \dfrac{1 - \left[\dfrac{1 + (-0.20)}{1 + 0.08} \right]^5}{0.08 - (-0.20)} \right\} (0.4632)$$
$$= 20{,}000(2.7750)(0.4632)$$
$$= \$25{,}707$$

The net total present worth is

$$P_T = -20{,}577 + 25{,}707 = \$+5130$$

This means that more is withdrawn than the investment series earns. Either additional funds must be invested or less must be withdrawn to make the series equivalent at 8% per year.

To find the exact amount of the initial withdrawal series to result in $P_T = 0$, let A_1 be an unknown in Equation [3.7] and set $P_{g,0} = -P_G = 20{,}577$.

$$20{,}577 = A_1(2.7750)(0.4632)$$
$$A_1 = \$16{,}009 \text{ in year } 11$$

The geometric series withdrawal would be 20% less each year.

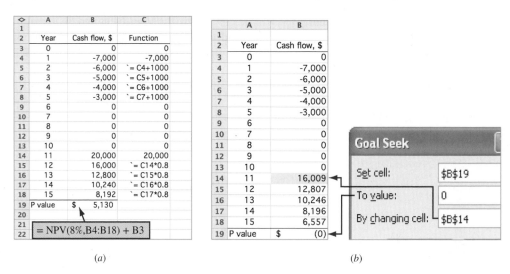

Figure 3–17
Spreadsheet solution, Example 3.8. (*a*) Cash flows and NPV function and (*b*) Goal Seek to determine initial withdrawal amount in year 11.

Solution by Spreadsheet

See Figure 3–17*a*. To determine if the investment series will cover the withdrawal series, enter the cash flows (in column B using the functions shown in column C) and apply the NPV function shown in the cell tag to display $P_T = \$+5130$ directly. As above, the $+$ sign indicates that, from a time value of money perspective, there is more withdrawn than invested and earned.

The Goal Seek tool is very handy in determining the initial withdrawal amount that results in $P_T = 0$ (cell B19). Figure 3–17*b* shows the template and result $A_1 = \$16,009$ in year 11. Each succeeding withdrawal is 80% of the previous one.

Comment

If the withdrawal series is fixed as estimated initially and the investment series base amount A can be increased, the Goal Seek tool can be used to, again, set $P_T = 0$ (cell B19). However, now establish the entry in B4 as the changing cell. The response is $A = \$-8285$, and, as before, succeeding investments are $1000 less.

CHAPTER SUMMARY

In Chapter 2, we derived the equations to calculate the present, future, or annual worth of specific cash flow series. In this chapter, we learned how to use these equations on cash flow series that are shifted in time from those for which the basic relations are derived. For example, when a uniform series does not begin in period 1, we still use the P/A factor to find the "present worth" of the series, except the P value is located one period ahead of the first A value, not at time 0. For arithmetic and geometric gradients, the P value is two periods ahead of where the gradient starts. With this information, it is possible to solve for P, A, or F for any conceivable cash flow series.

We have used the power of spreadsheet functions in determining P, F, and A values by single-cell entries and using cash flow estimates entered into a series of spreadsheet cells. Though spreadsheet solutions are fast, they do remove some of the understanding of how the time value of money and the factors change the equivalent value of money.

PROBLEMS

Present Worth Calculations

3.1   A commercial real estate developer plans to borrow money to finance an upscale mall in an exclusive area of the city. The developer plans to get a loan that will be repaid with uniform payments of $400,000 over a 15-year period beginning in year 2 and ending in year 15. How much will a bank be willing to loan at an interest rate of 10% per year?

3.2 A maker of micromechanical systems can reduce product recalls by 10% with the installation of new packaging equipment. If the cost of the new equipment is expected to be $40,000 four years from now, how much could the company afford to spend now (instead of 4 years from now) at a minimum attractive rate of return of 12% per year?

3.3 A proximity sensor attached to the tip of an endoscope could reduce risks during eye surgery by alerting surgeons to the location of critical retinal tissue. If an eye surgeon expects that by using this technology he may avoid lawsuits of $0.6 and $1.35 million 2 and 5 years from now, respectively, how much could he afford to spend now provided his out-of-pocket costs for the lawsuits would be only 10% of the total amount of each suit? Use an interest rate of 10% per year.

3.4 Industrial Electric Services has a contract with an Embassy in Mexico to provide maintenance for scanners and other devices in the building. What is the present worth of the contract (in year 0) if the company will receive a total of nine $12,000 payments beginning in year 2 and ending in year 10 and the interest rate is 10% per year.

3.5 Civil Engineering consulting firms that provide services to outlying communities are vulnerable to a number of factors that affect the financial condition of the communities, such as bond issues, real estate developments, etc. A small consulting firm entered into a fixed-price contract with a spec home builder, resulting in a stable income of $320,000 per year in years 1 through 4. At the end of that time, a mild recession slowed the development, so the parties signed another contract for $150,000 per year for two more years. Determine the present worth of the two contracts at an interest rate of 10% per year.

3.6 Because unintended lane changes by distracted drivers are responsible for 43% of all highway fatalities, Ford Motor Co. and Volvo launched a program to jointly develop a technology to prevent accidents by sleepy or distracted drivers. A device costing $260 tracks lane markings and sounds an alert during lane change. If these devices are included in 100,000 new cars per year beginning 3 years from now, determine the present worth of the cost over a 10-year period at an interest rate of 10% per year.

3.7 Pittsburgh Custom Products (PCP) purchased a new machine for ram cambering large I beams. PCP expects to bend 80 beams at $2000 per beam in each of the first 3 years, after which it expects to bend 100 beams per year at $2500 per beam through year 8. If the company's minimum attractive rate of return is 18% per year, what is the present worth of the expected revenue?

3.8 Centrum Water & Gas provides standby power to pumping stations using diesel-powered generators. An alternative is the use of natural gas to power the generators, but it will be a few years before the gas is available at remote sites. Centrum estimates that by switching to gas, it will save $15,000 per year, starting 3 years from now through the end of year 20. At an interest rate of 8% per year, determine the present worth of the projected savings.

3.9 A large water utility is planning to upgrade its SCADA system for controlling well pumps, booster pumps, and disinfection equipment for centralized monitoring and control. Phase I will reduce labor and travel costs by $28,000 per year. Phase II will reduce costs by an additional $20,000 per year, that is $48,000. If phase I savings occur in years 0, 1, 2, and 3 and phase II savings occur in years 4 through 10, what is the present worth of the upgraded system in years 1 to 10 at an interest rate of 8% per year?

3.10 Costs associated with the manufacture of miniature high-sensitivity piezoresistive pressure transducers is, $73,000 per year. A clever industrial engineer found that by spending $16,000 now to reconfigure the production line and reprogram two of the robotic arms, the cost will go down to $58,000 next year and $52,000 in years 2 through 5. Using an interest rate of 10% per year, determine the present worth of the savings due to the reconfiguration. (Hint: Include the reconfiguration cost.)

Annual Worth Calculations

3.11 Determine the equivalent annual worth for years 1 through 10 of a uniform series of payments of $20,000 that begins in year 3 and ends in year 10. Use an interest rate of 10% per year. Also, write the single-cell spreadsheet function to find A.

3.12 AutomationDirect, which makes 6-inch TFT color touch screen HMI panels, is examining its cash flow requirements for the next 5 years. The company expects to replace office machines and computer equipment at various times over the 5-year planning period. Specifically, the company expects to spend $7000 two years from now, $9000 four years from now, and $15,000 five years from now. What is the annual worth over years 1 through 5 of the planned expenditures at an interest rate of 10% per year? Also, write the single-cell spreadsheet function to find A.

3.13 What is the equivalent annual cost in years 1 through 7 of a contract that has a first cost of $70,000 in year 0 and annual costs of $15,000 in years 3 through 7? Use an interest rate of 10% per year.

3.14   The net cash flows associated with development and sale of a new product are shown. Determine the *beginning of period* annual worth (i.e., for years 0 through 5) at an interest rate of 12% per year. The cash flows are in $1000 units.

Year	1	2	3	4	5	6
Cash Flow, $1000	−120	−50	+90	+90	+90	+90

3.15   Improvised explosive devices (IEDs) are responsible for many deaths in times of strife and war. Unmanned ground vehicles (robots) can be used to disarm the IEDs and perform other tasks as well. If the robots cost $140,000 each and the Military Arms Unit signs a contract to purchase 4000 of them now and another 6000 one year from now, what is the equivalent annual cost of the contract over a 4-year period at 10% per year interest?

3.16   How much will Kingston Technologies have to pay each year in eight equal payments, starting 2 years from now, to repay a $900,000 loan? The interest rate is 8% per year.

3.17   The operating cost for a pulverized coal cyclone furnace is expected to be $80,000 per year. The steam produced will be needed for only 6 years beginning now (i.e., years 0 through 5). What is the equivalent annual worth in years 1 through 5 of the operating cost at an interest rate of 10% per year?

3.18   An entrepreneurial electrical engineer has approached a large utility with a proposal that promises to reduce the utility's power bill by at least 15% per year for the next 5 years through installation of patented surge protectors. The proposal states that the engineer will get $20,000 now and annual payments that are equivalent to 75% of the power savings achieved from the devices. Assuming the savings are the same every year at 15% and that the utility's power bill is $1 million per year, what would be the equivalent uniform amount for years 1 to 5 of the payments to the engineer at an interest rate of 6% per year?

3.19   For the cash flows shown, find the value of *x* that makes the equivalent annual worth in years 1 through 7 equal to $300 per year. Use an interest rate of 10% per year.

Year	Cash Flow, $	Year	Cash Flow, $
0	x	4	300
1	300	5	300
2	300	6	300
3	300	7	x

3.20   Kenworth Imaging got a $700,000 loan that came with a choice of two different repayment schedules. In Plan 1, the company would have to repay the loan in 4 years with four equal payments at an interest rate of 10% per year. In Plan 2, the company would repay the loan in 3 years, with each payment twice as large as the preceding one. How much larger in dollars was the final payment in Plan 2 than the final payment in Plan 1?

Future Worth Calculations

3.21   Lifetime Savings Accounts, known as LSAs, allow people to invest after-tax money without being taxed on any of the gains. If an engineer invests $10,000 now and $10,000 each year for the next 20 years, how much will be in the account immediately after the last deposit, provided the account grows by 10% per year?

3.22   Calculate the equivalent future cost in year 10 for the following series of disbursements at an interest rate of 10% per year. Also, write a single-cell spreadsheet function that will determine the future worth.

Year	Disbursement, $	Year	Disbursement, $
0		5	4000
1	4000	6	5000
2	4000	7	5000
3	4000	8	5000
4	4000	9	5000

3.23   How much money would be accumulated 18 years from now from deposits of $15,000 per year for five consecutive years, starting 3 years from now, if the interest rate is 8% per year?

3.24   A manufacturer of industrial grade gas handling equipment wants to have $500,000 in an equipment replacement contingency fund 10 years from now. If the company plans to deposit a uniform amount of money each year *beginning now* and continuing through year 10 (total of 11 deposits), what must be the size of each deposit? Assume the account grows at a rate of 10% per year.

3.25   For the cash flows shown, find the future worth in (a) year 5, and (b) year 4. Assume an *i* of 10% per year.

Year	0	1	2	3	4	5
Cash Flow, $	0	0	3000	3000	3000	3000

3.26   For the cash flows shown, calculate the future worth in year 9 using $i = 10\%$ per year.

Year	0	1	2	3	4	5	6
Cash Flow, $	200	200	200	200	300	300	300

3.27   For the cash flows shown, calculate the future worth in year 8 at $i = 10\%$ per year.

Year	0	1	2	3	4	5	6	7
Cash Flow, $	500	500	500	0	800	800	800	800

3.28 For the cash flows shown in the diagram, determine the value of x and $2x$ that will make the future worth in year 8 equal to $100,000.

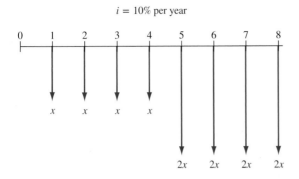

$i = 10\%$ per year

Random Single Amounts and Uniform Series

3.29 Find the present worth for the following cash flow series (in $1000 units).

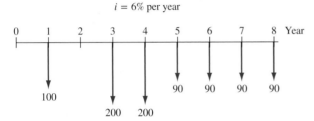

$i = 6\%$ per year

3.30 A company that manufactures air-operated drain valve assemblies currently has $85,000 available to pay for plastic components over a 5-year period. If the company spent only $42,000 in year 1, what uniform annual amount can the company spend in each of the next 4 years to deplete the entire budget? Let $i = 10\%$ per year.

3.31 Marcia observed the following cash flow series (in $1000 units) in an accounting report at work. The actual amounts in years 1 and 4 are missing; however, the report states that the present worth in year 0 was $300,000 at an interest rate of 10% per year. Calculate the value of x.

Year	0	1	2	3	4	5	6
Cash Flow, $1000	40	x	40	40	x	40	40

3.32 El Paso Water is planning to install wind turbines to provide enhanced evaporation of reverse osmosis concentrate from its inland desalting plant. The company will spend $1.5 million in year 1 and $2 million in year 2. Annual maintenance is expected to cost $65,000 per year through year 10. Determine the equivalent annual cost of the project in years 1 through 10 at an interest rate of 6% per year. Also, develop a single-cell spreadsheet function to display the total A value.

3.33 A 5-year plan to raise extra funds for public schools involves an "enrichment tax" that will raise $56 for every student the first year, increasing by $1 per student per year thereafter. There are 50,000 students in the district in year one, 51,000 in year 2, with increases of 1000 students per year thereafter. Calculate the future worth in year 5 of the enrichment plan over a 5-year planning period at an interest rate of 8% per year. Solve this problem using (*a*) tabulated factors, and (*b*) a spreadsheet.

3.34 A recently hired CEO (chief executive officer) wants to reduce future production costs to improve the company's earnings, thereby increasing the value of the company's stock. The plan is to invest $70,000 now and $50,000 in each of the next 2 years to improve productivity. By how much must annual costs decrease in years 3 through 10 to recover the investment plus a return of 15% per year?

3.35 A construction management company is examining its cash flow requirements for the next 7 years. The company expects to replace software and infield computing equipment at various times over a 7-year planning period. Specifically, the company expects to spend $6000 one year from now, $9000 three years from now, and $10,000 each year in years 6 through 10. What is the future worth in year 10 of the planned expenditures, at an interest rate of 12% per year?

3.36 The following series of revenues and expenses in $1000 units were recorded by a mobile phone business. At a return rate of 10% per year, (*a*) determine the net future worth in year 10. (*b*) Did the owners make their expected 10% return? How do you know?

Year	Revenues, $	Expenses, $
0	0	−2500
1–4	700	−200
5–10	2000	−300

Shifted Gradients

3.37 Find the present worth in year 0 for the cash flows shown. Let $i = 10\%$ per year.

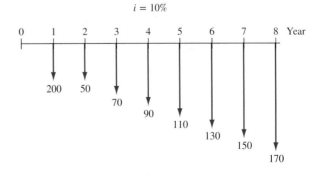

$i = 10\%$

3.38 Silastic-LC-50 is a liquid silicon rubber designed to provide excellent clarity, superior mechanical properties, and short cycle time for high speed manufacturers. One high-volume manufacturer used it to achieve smooth release from molds. The company's projected growth will result in silicon costs of $26,000 in years 1 and 2, with costs increasing by $2000 per year in years 3 through 5. At an interest rate of 10% per year, what is the present worth of these costs?

3.39 For the cash flows shown, calculate the equivalent annual worth in years 1 through 4 at an interest rate of 10% per year.

Year	0	1	2	3	4
Cash Flow, $	250,000	275,000	300,000	325,000	375,000

3.40 A build-to-operate (BTO) company signed a contract to operate Alamosa County industrial wastewater treatment plants for the next 20 years. The contract will pay the company $2.5 million now and amounts increasing by $200,000 each year through year 20. At an interest rate of 10% per year, what is the present worth *now*? Solve using (*a*) tabulated factors, and (*b*) a spreadsheet.

3.41 Nippon Steel's expenses for heating and cooling a large manufacturing facility are expected to increase according to an arithmetic gradient beginning in year 2. If the cost is $550,000 this year (year 0) and will be $550,000 again in year 1, but then are estimated to increase by $40,000 each year through year 12, what is the equivalent annual worth in years 1 to 12 of these energy costs at an interest rate of 10% per year?

3.42 The Pedernales Electric Cooperative estimates that the present worth *now* of increased revenue from an investment in renewable energy sources is $12,475,000. There will be no new revenue in years 1 or 2, but in year 3 revenue will be $250,000, and thereafter it will increase according to an arithmetic gradient through year 15. What is the required gradient, if the expected rate of return is 15% per year? Solve (*a*) by tabulated factors, and (*b*) using a spreadsheet.

3.43 A software company that installs systems for inventory control using RFID technology spent $600,000 per year for the past 3 years in developing their latest product. The company optimistically hopes to recover its investment in 5 years on a single contract beginning immediately (year 0). The company is negotiating a

contract that will pay $250,000 now and a to-be-agreed-upon annual increase of a constant amount each year through year 5. How much must the income increase (an arithmetic gradient) each year, if the company wants to realize a return of 15% per year?

3.44 The future worth in year 8 for the cash flows shown is $20,000. At an interest rate of 10% per year, what is the value, *x*, of the cash flow in year 4?

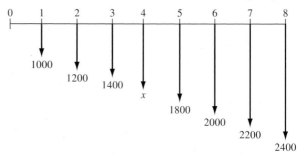

3.45 The annual worth for years 1 through 8 of the cash flows shown is $30,000. What is the amount of *x*, the cash flow in year 3, at *i* = 10% per year? Solve using (*a*) tabulated factors, and (*b*) a spreadsheet.

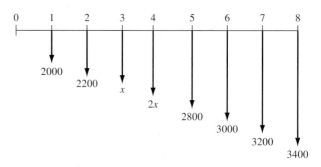

3.46 United Fruits now offers volume discounts by the 10-kilogram (10-kilo) pack to small, as well as large, supermarket chains for organically grown in-season fruit shipped directly from the orchard. Over the last 8 years, the food market chain CoopForAll has purchased the number of 10-kilo packs shown below when United Fruits has California free-stone peaches in season. Through a special arrangement, the price per 10-kilo pack has remained constant at $15.00 per pack. United Fruits will now charge CoopForAll according to the volume discount schedule shown. The no-discount price this year is $19.95 per 10-kilo pack.

Years ago	8	7	6	5	4	3	2	1
10-kilo packs	100	150	500	800	1100	1400	1700	2000

10-kilo packs purchased	0–100	101–250	251–1000	1001–10,000	10,001–50,000	50,001–100,000	> 100,000
Discount, %	None	10% off	Added 10%	Added 10%	Added 20%	Added 20%	Added 20%

Based on the progressive discounts of 10% more for increased-size orders, if CoopForAll buys the same amount as last year, the total cost with discounts will be: 2000 packs × $19.95[1 − (3 × 0.10)] = 2000 × 13.97 = $27,940. This is less than $30,000, the cost for 2000 packs at the flat rate of $15.00 per pack.

Calculate the equivalent worth *now* of the annual costs for the 8 years of prior purchases at the flat rate of (*a*) $15.00 per pack, and (*b*) the cost had they been purchased using the discount schedule. Let $i = 8\%$ per year. (Hint: This is a future worth value now at the end of year 1.)

3.47 Calculate the present worth of all costs for a newly acquired machine with an initial cost of $29,000, no trade-in value, a life of 10 years, and an annual operating cost of $13,000 for the first 4 years, increasing by 10% per year thereafter. Use an interest rate of 10% per year.

3.48 Dakota Hi-C Steel signed a contract that will generate revenue of $210,000 now, $226,800 in year 1, and amounts increasing by 8% per year through year 5. Calculate the future worth of the contract at an interest rate of 8% per year. Also, develop the FV function that will correctly determine F for revenues placed in cells B2 through B7.

3.49 Wrangler Western has some of its jeans stonewashed under a contract with an independent contractor, Almos Garment Corp. If Almos' operating cost is $22,000 per year for years 1 and 2 and then it increases by 8% per year through year 10, what is the year-zero present worth of the machine operating cost at an interest rate of 10% per year?

3.50 McCarthy Construction is trying to bring the company-funded portion of its employee retirement fund into compliance with HB-301. The company has already deposited $500,000 in each of the last 5 years. If the company increases its deposits beginning in year 6 by 15% per year each year through year 20, how much will be in the fund immediately after the last deposit, provided the fund grows at a rate of 12% per year? Solve using (*a*) tabulated factors, and (*b*) a spreadsheet.

Shifted Decreasing Gradients

3.51 Find the present worth at time 0 of the chrome plating costs shown in the cash flow diagram. Assume $i = 10\%$ per year.

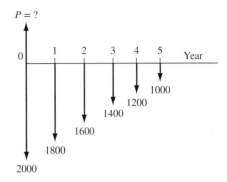

3.52 For the cash flows shown in the diagram, determine the equivalent uniform annual worth over years 1 through 5 at an interest rate of 10% per year.

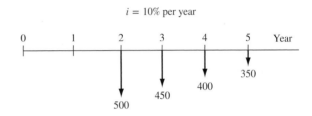

3.53 The pumping cost for delivering water from the Ohio River to Wheeling Steel for cooling hot rolled steel was $1.8 million for the first 4 years. An effective energy conservation program resulted in a reduced cost of $1.77 million in year 5, $1.74 million in year 6, and amounts decreasing by $30,000 each year through year 10. What is the present worth of the pumping costs in year 0 at an interest rate of 12% per year?

3.54 Prudential Realty has an escrow account for one of its property management clients that contains $20,000. How long will it take to deplete the account if the client withdraws $5000 now, $4500 one year from now, and amounts decreasing by $500 each year? Assume the account earns interest at a rate of 8% per year.

3.55 For the cash flow diagram shown, determine the value of *G* that will make the present worth in year 0 equal to $2500 at an interest rate of 10% per year.

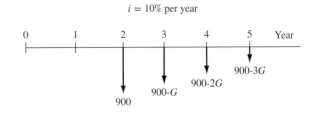

3.56   The City of San Antonio is considering various options for providing water in its 50-year plan, including desalting. One brackish aquifer is expected to yield desalted water that will generate revenue of $4.1 million per year for the first 4 years, after which less production will decrease revenue by 10% per year each year. If the aquifer will be totally depleted in 20 years, what is the present worth of the desalting option revenue at an interest rate of 6% per year?

3.57   Revenue from gas wells that have been in production for at least 5 years tends to follow a decreasing geometric gradient. One particular rights holder received royalties of $4000 per year for years 1 through 6; however, beginning in year 7, income decreased by 15% per year each year through year 14. Calculate the future value in year 14 of the royalty income from the wells provided all of it was invested at 10% per year.

EXERCISES FOR SPREADSHEETS

3.58   A large water utility is planning to upgrade its SCADA system for controlling well pumps, booster pumps, and disinfection equipment so that everything can be centrally controlled. Phase I will reduce labor and travel costs by $28,000 per year. Phase II will reduce costs by an additional $20,000 per year, that is, $48,000. Phase I savings should occur in years 0, 1, 2, and 3 and phase II savings should occur in years 4 through 10. Let $i = 8\%$ per year.
   (a)   Determine the present worth of the upgraded system for years 1 to 10.
   (b)   The utility General Manager had hoped for a present worth of at least $400,000. Determine the interest rate at which this would be correct.

3.59   The operating cost of a pulverized coal cyclone furnace is expected to be $80,000 per year. The steam produced will be needed for 6 years—now through year 5. Determine the equivalent annual worth, A, in years 1 through 5 of the operating costs at an interest rate of 10% per year.

3.60   Hansus Enterprises is a start-up small business specializing in software systems for machine-based high-school student tutoring in science and math. Sheryl Hansus, the owner, has spoken with lenders to obtain a loan of $50,000 now (year 0), and the same amount in years 3 and 6. Two different repayment schedules are available. They are:

   Schedule A: Pay a uniform amount of $19,500 in years 3 through 12.

   Schedule B: Pay a uniform amount of $20,000 in years 1 through 6, with "balloon" payments of an additional $20,000 in year 2 and a final amount of $40,000 in year 7.

To assist in the decision of which schedule to select, determine the following:
   (a)   Total amount repaid for each schedule. Which is smaller?
   (b)   The equivalent annual worth of the loan amounts and of the repayments at $i = 5\%$ per year over a 12-year evaluation period. Which schedule has the smaller loss per year? Why is this the case?

3.61   (This is Problem 3.19 repeated for spreadsheet solution.) For the cash flows shown, find the value of x that makes the equivalent annual worth in years 1 through 7 equal to $300 per year. Use an interest rate of 10% per year. Start the analysis at $x = \$300$ and use Goal Seek to solve.

Year	Cash Flow, $	Year	Cash Flow, $
0	x	4	300
1	300	5	300
2	300	6	300
3	300	7	x

3.62   For the cash flows shown, find the future worth in (a) year 5, and (b) year 4. Assume an i of 10% per year. (This is Problem 3.25 above.)

Year	0	1	2	3	4	5
Cash Flow, $	0	0	3000	3000	3000	3000

3.63   For the cash flows shown (in $1000 units), calculate the value of x that makes the present worth in year 0 equal to $300,000 at an interest rate of 10% per year.

Year	0	1	2	3	4	5	6
Cash Flow, $1000	40	x	40	40	x	40	40

3.64 Jowel Smithers and Sonda Richards merged all of their financial resources and opened a mobile phone sales and repair business in the upscale Domain Mall in Los Angeles 10 years ago with an initial investment of $2.5 million. As revenues increased significantly over the first 4 years, they took out salaries that totaled $500,000 each year for the two of them. With increased success, they decided to pay themselves a total of $1.7 million each year from years 5 through 10. They hoped at the initiation of the joint business venture to make 10% per year on the investment in terms of salaries.

(a) Determine the equivalent future worth after 10 years of the salaries.

(b) If they had been satisfied with a $5 million future worth, calculate the size that the initial investment could have been instead of the $2.5 million, provided they had the funds available.

3.65 The negative arithmetic gradient series in Problem 3.44 is missing the value in year 4. Solve for this value on a spreadsheet. In doing so, display both the full amount of the missing payment and the extra amount over the expected amount of $-1600.

ADDITIONAL PROBLEMS AND FE EXAM REVIEW QUESTIONS

3.66 A uniform series of payments begins in year 4 and ends in year 11. If you use the P/A factor with $n = 8$, the P value you get will be located in year:

(a) 0   (b) 3   (c) 4   (d) 5

3.67 Detrich Products is planning to upgrade an aging manufacturing operation 5 years from now at a cost of $100,000. If the company plans to deposit money into an account each year for 4 years beginning 2 years from now (first deposit is in year 2) to pay for the expansion, the amount of the deposit at 10% per year interest is closest to:

(a) $30,211
(b) $21,547
(c) $16,380
(d) $14,392

3.68 An arithmetic gradient has cash flows of $1000 in year 4, $1200 in year 5, and amounts increasing by $200 per year through year 10. If you use the factor $200(P/G,10\%,n)$ to find P_G in year 3, the value of n you have to use in the P/G factor is:

(a) 10   (b) 9   (c) 8   (d) 7

3.69 For the diagram shown below, the respective values of n to calculate the present worth in year 0 by the equation $P_0 = 100(P/A,10\%,n_1)(P/F,10\%,n_2)$ are:

(a) $n_1 = 6, n_2 = 1$
(b) $n_1 = 5, n_2 = 2$
(c) $n_1 = 7, n_2 = 1$
(d) $n_1 = 7, n_2 = 2$

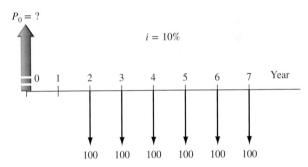

3.70 You plan to pay $38,000 cash for the new truck you want to buy 5 years from now. You are a very astute investor; all your money earns at 20% per year. If you have already saved $9500, the amount your rich aunt has to give you 2 years from now (as a graduation present) in order for you to have the total amount of $38,000 is closest to:

(a) <$7500
(b) $7654
(c) $8,310
(d) $9,880

3.71 At a return rate of 20% per year, the amount of money you must deposit for five consecutive years starting 3 years from now for the account to contain $50,000 fifteen years from now is closest to:

(a) $1565
(b) $1759
(c) $1893
(d) $2093

3.72 You found a report stating that the equivalent annual worth of chemical costs associated with a water treatment process were $125,000 per year for a 5-year period. The report stated that the cost in year 1 was $190,000 and the cost decreased by a uniform amount each year over that 5-year period. However, it did not say how much the decrease was each year. If the interest rate was 20% per year, the amount of the annual decrease, G, is closest to:

(a) $27,358   (b) $31,136
(c) $33,093   (d) $39,622

3.73 If you borrow $24,000 now at an interest rate of 10% per year and promise to repay the loan with payments of $3695 per year starting 1 year from now, the number of payments that you will have to make is closest to:

(a) 7   (b) 8   (c) 11   (d) 14

3.74 The maker of a motion-sensing towel dispenser is considering adding new products to enhance offerings in the area of touchless technology. If the company does not expand its product line now, it will definitely do so in 2 years. Assume the interest rate is 10% per year. The amount the company can afford to spend *now* if the cost 2 years from now is estimated to be $100,000 is closest to:

(a) $75,130
(b) $82,640
(c) $91,000
(d) $93,280

3.75 Assume you borrow $10,000 today and promise to repay the loan in two payments, one in year 2 and the other in year 4, with the one in year 4 being only half as large as the one in year 2. At an interest rate of 10% per year, the size of the payment in year 4 is closest to:

(a) $4280   (b) $3975
(c) $3850   (d) $3690

3.76 For the cash flows shown, you have been asked to calculate the present worth (in year 0) using $i = 10\%$ per year. Which of the following solutions is *not correct*?

Year	0	1	2	3	4	5	6
Cash Flow, $	200	200	200	200	300	300	300

(a) $P = 200 + 200(P/A,10\%,3) + 300(P/A,10\%,3)(P/F,10\%,3)$
(b) $P = 200(P/A,10\%,4) + 300(P/A,10\%,3)(P/F,10\%,3)$
(c) $P = [200(F/A,10\%,7) + 100(F/A,10\%,3)](P/F,10\%,6)$
(d) $P = [200(P/A,10\%,7) + 100(P/A,10\%,3)(P/F,10\%,4)](F/P,10\%,1)$

3.77 In order to establish a contingency fund to replace equipment after unexpected breakdowns, a manufacturer of thin-wall plastic bottles plans to deposit $100,000 now and $150,000 two years from now into an investment account. Assuming the account grows at 15% per year, the equation that does *not* represent the future value of the account in year 5 is:

(a) $F = 100,000(F/P,15\%,5) + 150,000(F/P,15\%,3)$
(b) $F = [100,000(F/P,15\%,2) + 150,000](F/P,15\%,3)$
(c) $F = [100,000 + 150,000(P/F,15\%,2)](F/P,15\%,5)$
(d) $F = 100,000(F/P,15\%,5) + 150,000(F/P,15\%,2)$

CASE STUDY

PRESERVING LAND FOR PUBLIC USE

Background and Information

The Trust for Public Land (TPL) is a national organization that purchases and oversees the improvement of large land sites for government agencies at all levels. Its mission is to ensure the preservation of the natural resources, while providing necessary, but minimal, development for recreational use by the public. All TPL projects are evaluated at 7% per year, and TPL reserve funds earn 7% per year.

A southern U.S. state, which has long-term groundwater problems, has asked the TPL to manage the purchase of 10,000 acres of aquifer recharge land and the development of three parks of different use types on the land. The 10,000 acres will be acquired in increments over the next 5 years with $4 million expended immediately on purchases. Total annual purchase amounts are expected to decrease 25% each year through year 5 and then cease for this particular project.

A city with 1.5 million citizens immediately to the southeast of this acreage relies heavily on the aquifer's water. Its citizens passed a bond issue last year, and the city government now has available $3 million for the purchase of land. The bond interest rate is an effective 7% per year.

The engineers working on the park plan intend to complete all the development over a 3-year period starting in year 4, when the amount budgeted is $550,000. Increases in construction costs are expected to be $100,000 each year through year 6.

At a recent meeting, the following agreements were made:

- Purchase the initial land increment now. Use the bond issue funds to assist with this purchase. Take the remaining amount from TPL reserves.
- Raise the remaining project funds over the next 2 years in equal annual amounts.
- Evaluate a financing alternative (suggested informally by one individual at the meeting) in which the TPL provides all funds, except the $3 million available now, until the park's development is initiated in year 4.

Case Study Exercises

1. For each of the 2 years, what is the equivalent annual amount necessary to supply the remaining project funds?

2. If the TPL did agree to fund all costs except the $3 million bond proceeds now available, determine the equivalent annual amount that must be raised in years 4 through 6 to supply all remaining project funds. Assume the TPL will not charge any extra interest over the 7% to the state or city on the borrowed funds.

3. Review the TPL website (www.tpl.org). Identify some economic and noneconomic factors that you believe must be considered when the TPL is deciding to purchase land to protect it from real estate development.

CHAPTER 4

Nominal and Effective Interest Rates

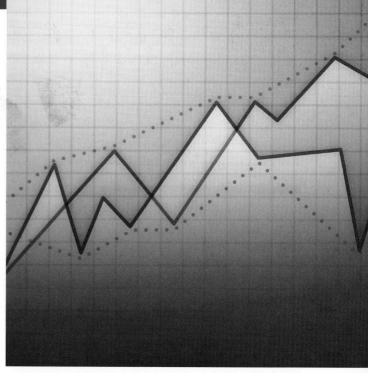

Chad Baker/Getty Images

LEARNING OUTCOMES

Purpose: Make computations for interest rates and cash flows that are on a time basis other than a year.

SECTION	TOPIC	LEARNING OUTCOME
4.1	Statements	• Understand interest rate statements that include nominal and effective rates.
4.2	Effective annual rate	• Derive and use the formula for an effective annual interest rate.
4.3	Effective rate	• Determine the effective interest rate for any stated time period.
4.4	Payment period and compounding period	• Determine the payment period (PP) and compounding period (CP) for equivalence computations.
4.5	Single cash flows with PP ≥ CP	• Perform equivalence calculations for single-amount cash flows and PP ≥ CP.
4.6	Series cash flows with PP ≥ CP	• Perform equivalence calculations for series and gradient cash flows and PP ≥ CP.
4.7	Single amounts and series with PP < CP	• Perform equivalence calculations for cash flows with PP < CP.
4.8	Continuous compounding	• Derive and use the effective interest rate formula for interest rates that are compounded continuously.
4.9	Varying rates	• Perform equivalency calculations for interest rates that vary from one time period to another.

I n all engineering economy relations developed thus far, the interest rate has been a constant, annual value. For a substantial percentage of the projects evaluated by professional engineers in practice, the interest rate is compounded more frequently than once a year; frequencies such as semiannually, quarterly, and monthly are common. In fact, weekly, daily, and even continuous compounding may be experienced in some project evaluations. Also, in our own personal lives, many of our financial considerations—loans of all types (home mortgages, credit cards, automobiles, boats), checking and savings accounts, investments, stock option plans, etc.— have interest rates compounded for a time period shorter than 1 year. This requires the introduction of two new terms—**nominal and effective interest rates.**

This chapter explains how to understand and use nominal and effective interest rates in engineering practice and in daily life situations. Equivalence calculations for any compounding frequency in combination with any cash flow frequency are presented.

PE

The Credit Card Offer Case: Today, Dave received a special offer of a new credit card from Chase Bank linked with the major airline that he flies frequently. It offers a generous bonus package for signing up by a specific date about 60 days from now. The bonus package includes extra airline points (once the first purchase is made), priority airport check-in services (for 1 year), several free checked-bag allowances (for up to 10 check-ins), extra frequent-flyer points on the airline, access to airline lounges (provided he uses the card on a set time basis), plus several other rewards (rental car discounts, cruise trip amenities, and floral order discounts).

The annual fee of $85 for membership does not start until the second year, and balance transfers from other credit cards have a low transfer fee, provided they are made at the time of initial membership.

Dave has a credit card currently with a bank that he is planning to leave due to its poor customer service and high monthly fees. If he enrolls, he will transfer the $1000 balance on the current card to the new Chase Bank card.

In the page that accompanies the offer letter, "pricing information" is included. This includes interest rates, interest charges, and fees. A summary of several of these rates and fees follows.

APR (annual percentage rate) for purchases and balance transfers*	**14.24% per year** (sum of the current U.S. Government prime rate of 3.25% and 10.99%, which is the APR added to determine the balance transfer APR for Chase Bank)
APR for cash and overdraft advances*	**19.24% per year**
Penalty APR for late minimum payment, exceeding credit limit, and returned unpaid payments*†	**29.99% per year (maximum penalty APR)**

Fees are listed as follows:

Annual membership	$85; free the first year
Balance transfers	$5 or 3% of each transfer, whichever is greater
Cash advances	$10 or 3% of each advance, whichever is greater
Late payment	$39 each occurrence, if balance exceeds $250
Over the credit limit	$39 each occurrence
Returned check or payment	$39 each occurrence

*All APR rates are variable, based on a current 3.25% prime rate with 10.99% added to determine purchase/balance transfer APR; with 15.99% added to determine cash/overdraft APR; and with 26.99% added to determine penalty APR.
†The penalty APR applies indefinitely to future transactions. If no minimum payment is received within 60 days, the penalty APR applies to all outstanding balances and all future transactions on the account.

(Continued)

This case is used in the following topics (and sections) of this chapter:

Nominal and effective interest rate statements (4.1)

Effective annual interest rates (4.2)

Equivalence relations: Series with PP ≥ CP (4.6)

4.1 Nominal and Effective Interest Rate Statements ● ● ●

In Chapter 1, we learned that the primary difference between simple interest and compound interest is that compound interest includes interest on the interest earned in the previous period, while simple interest does not. Here we discuss *nominal and effective interest rates*, which have the same basic relationship. The difference here is that the concepts of nominal and effective must be used when interest is compounded more than once each year. For example, if an interest rate is expressed as 1% per month, the terms *nominal* and *effective* interest rates must be considered.

Understanding and correctly handling effective interest rates are important in engineering practice as well as for individual finances. The interest amounts for loans, mortgages, bonds, and stocks are commonly based upon interest rates compounded more frequently than annually. The engineering economy study must account for these effects. In our own personal finances, we manage most cash disbursements and receipts on a nonannual time basis. Again, the effect of compounding more frequently than once per year is present. First, consider a **nominal interest rate**.

Nominal interest rate *r*

A nominal interest rate *r* is an interest rate that **does not account** for compounding. By definition,

r = interest rate per time period × number of periods   [4.1]

A nominal rate may be calculated for *any time period longer than the time period stated* by using Equation [4.1]. For example, the interest rate of 1.5% per month is the same as each of the following nominal rates.

Time Period	Nominal Rate by Equation [4.1]	What This Is
24 months	1.5 × 24 = 36%	Nominal rate per 2 years
12 months	1.5 × 12 = 18%	Nominal rate per 1 year
6 months	1.5 × 6 = 9%	Nominal rate per 6 months
3 months	1.5 × 3 = 4.5%	Nominal rate per 3 months

Note that none of these rates mention anything about compounding of interest; they are all of the form "*r*% per time period." These nominal rates are calculated in the same way that simple rates are calculated using Equation [1.7], that is, interest rate *times* number of periods.

After the nominal rate has been calculated, the **compounding period (CP)** must be included in the interest rate statement. As an illustration, again consider the nominal rate of 1.5% per month. If we define the CP as 1 month, the nominal rate statement is 18% per year, *compounded monthly*, or 4.5% per quarter, *compounded monthly*. Now we can consider an **effective interest rate**.

Effective interest rate *i*

An effective interest rate *i* is a rate wherein the **compounding of interest is taken into account**. Effective rates are commonly expressed on an annual basis as an effective annual rate; however, any time basis may be used.

The most common form of interest rate statement when compounding occurs over time periods shorter than 1 year is "% per time period, compounded CP-ly," for example, 10% per year, compounded monthly, or 12% per year, compounded weekly. An effective rate may not always include the compounding period in the statement. If the CP is not mentioned, it is understood to be the same as the time period mentioned with the interest rate. For example, an interest rate of "1.5% per month" means that interest is compounded each month; that is, CP is 1 month. An equivalent effective rate statement, therefore, is 1.5% per month, compounded monthly.

All of the following are effective interest rate statements because either **they state they are effective** or the **compounding period is not mentioned**. In the latter case, the CP is the same as the time period of the interest rate.

Statement	CP	What This Is
$i = 10\%$ per year	CP not stated; CP = year	Effective rate per year
$i =$ effective 10% per year, compounded monthly	CP stated; CP = month	Effective rate per year
$i = 1\frac{1}{2}\%$ per month	CP not stated; CP = month	Effective rate per month
$i =$ effective $1\frac{1}{2}\%$ per month, compounded monthly	CP stated; CP = month	Effective rate per month; terms *effective* and *compounded monthly* are redundant
$i =$ effective 3% per quarter, compounded daily	CP stated; CP = day	Effective rate per quarter

All nominal interest rates can be converted to effective rates. The formula to do this is discussed in the next section.

> All interest formulas, factors, tabulated values, and spreadsheet functions must use an effective interest rate to properly account for the time value of money.

The term **APR (Annual Percentage Rate)** is often stated as the annual interest rate for credit cards, loans, and house mortgages. This is the same as the **nominal rate**. An APR of 15% is the same as a nominal 15% per year or a nominal 1.25% on a monthly basis. Also the term **APY (Annual Percentage Yield)** is a commonly stated annual rate of return for investments, certificates of deposit, and saving accounts. This is the same as an **effective rate**. The names are different, but the interpretations are identical. As we will learn in the following sections, the effective rate is always greater than or equal to the nominal rate, and similarly APY ≥ APR.

Based on these descriptions, there are always three time-based units associated with an interest rate statement.

Interest period (t)—The period of time over which the interest is expressed. This is the t in the statement of $r\%$ per time period t, for example, 1% *per month*. The time unit of 1 year is by far the most common. It is assumed when not stated otherwise.

Compounding period (CP)—The shortest time unit over which interest is charged or earned. This is defined by the compounding term in the interest rate statement, for example, 8% per year, *compounded monthly*. If CP is not stated, it is assumed to be the same as the interest period.

Compounding frequency (m)—The number of times that compounding occurs within the interest period t. If the compounding period CP and the time period t are the same, the compounding frequency is 1, for example, 1% *per month, compounded monthly*.

Consider the (nominal) rate of 8% per year, compounded monthly. It has an interest period t of 1 year, a compounding period CP of 1 month, and a compounding frequency m of 12 times per year. A rate of 6% per year, compounded weekly, has $t = 1$ year, CP = 1 week, and $m = 52$, based on the standard of 52 weeks per year.

In previous chapters, all interest rates had t and CP values of 1 year, so the compounding frequency was always $m = 1$. This made them all effective rates because the interest period and compounding period were the same. Now, it will be necessary to express a nominal rate as an effective rate on the same time base as the compounding period.

An effective rate can be determined from a nominal rate by using the relation

$$\text{Effective rate per CP} = \frac{r\% \text{ per time period } t}{m \text{ compounding periods per } t} = \frac{r}{m} \qquad [4.2]$$

As an illustration, assume $r = 9\%$ per year, compounded monthly; then $m = 12$. Equation [4.2] is used to obtain the effective rate of $9\%/12 = 0.75\%$ per month, compounded monthly. Note that changing the interest period t does not alter the compounding period, which is 1 month in this illustration. Therefore, $r = 9\%$ per year, compounded monthly, and $r = 4.5\%$ per 6 months, compounded monthly, are two expressions of the same interest rate.

EXAMPLE 4.1

Three different bank loan rates for electric generation equipment are listed below. Determine the effective rate on the basis of the compounding period for each rate.

(*a*) 9% per year, compounded quarterly.
(*b*) 9% per year, compounded monthly.
(*c*) 4.5% per 6 months, compounded weekly.

Solution

Apply Equation [4.2] to determine the effective rate per CP for different compounding periods. The graphic in Figure 4–1 indicates the effective rate per CP and how the interest rate is distributed over time.

Nominal r% per t	Compounding Period (CP)	Compounding Frequency (m)	Effective Rate per CP $\left(\frac{r}{m}\right)$	Distribution over Time Period t
(*a*) 9% per year	Quarter	4	2.25%	
(*b*) 9% per year	Month	12	0.75%	
(*c*) 4.5% per 6 months	Week	26	0.173%	

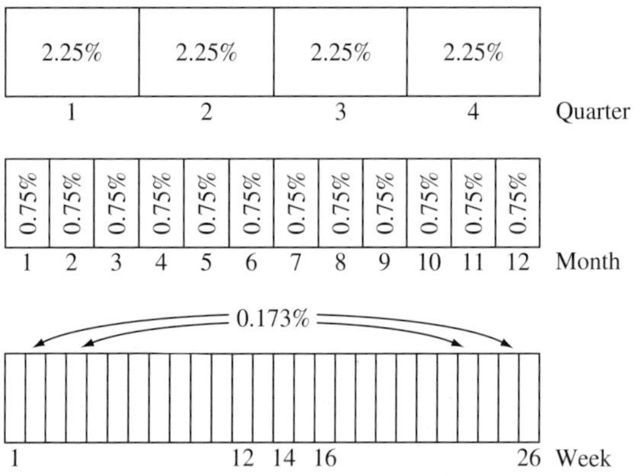

Figure 4–1
Relations between interest period t, compounding period CP, and effective interest rate per CP.

Sometimes it is not obvious whether a stated rate is a nominal or an effective rate. Basically there are three ways to express interest rates, as detailed in Table 4–1. The right column includes a statement about the effective rate. For the first format, a nominal interest rate is given and the compounding period is stated. The effective rate must be calculated (discussed in the next sections). In the second format, the stated rate is identified as effective (or APY could also be used), so the rate is used directly in computations.

In the third format, no compounding period is identified, for example, 8% per year. This rate is effective over a compounding period equal to the stated interest period of 1 year in this case. The effective rate for any other time period must be calculated.

TABLE 4–1 Various Ways to Express Nominal and Effective Interest Rates

Format of Rate Statement	Example of Statement	What about the Effective Rate?
Nominal rate stated, compounding period stated	8% per year, compounded quarterly	Find effective rate for any time period (next two sections)
Effective rate stated	Effective 8.243% per year, compounded quarterly	Use effective rate of 8.243% per year directly for annual cash flows
Interest rate stated, no compounding period stated	8% per year	Rate is effective for CP equal to stated interest period of 1 year; find effective rate for all other time periods

EXAMPLE 4.2 The Credit Card Offer Case

As described in the introduction to this case, Dave has been offered what is described as a credit card deal that should not be refused—at least that is what the Chase Bank offer letter implies. The balance transfer APR interest rate of 14.24% is an annual rate, with no compounding period mentioned. Therefore, it follows the format of the third entry in Table 4–1, that is, interest rate stated, no CP stated. Therefore, we should conclude that the CP is 1 year, the same as the annual interest period of the APR. However, as Dave and we all know, credit card payments are required monthly.

(*a*) First, determine the effective interest rates for *compounding periods of 1 year and 1 month* so Dave knows some effective rates he might be paying when he transfers the $1000 balance from his current card.

(*b*) Second, assume that immediately after he accepts the card and completes the $1000 transfer, Dave gets a bill that is due 1 month later. What is the amount of the total balance he owes?

Now, Dave looks a little closer at the fine print of the "pricing information" sheet and discovers a small-print statement that Chase Bank uses the daily balance method (including new transactions) to determine the balance used to calculate the interest due at payment time.

(*c*) We will reserve the implication of this new finding until later, but for now help Dave by determining the *effective daily interest rate* that may be used to calculate interest due at the end of 1 month, provided the CP is 1 day.

Solution

(*a*) The interest period is 1 year. Apply Equation [4.2] for both CP values of 1 year ($m = 1$ compounding period per year) and 1 month ($m = 12$ compounding periods per year).

CP of year: Effective rate per year = $14.24/1 = 14.24\%$

CP of month: Effective rate per month = $14.24/12 = 1.187\%$

(*b*) The interest will be at the monthly effective rate, plus the balance transfer fee of 3%.

$$\text{Amount owed after 1 month} = 1000 + 1000(0.01187) + 0.03(1000)$$
$$= 1000 + 11.87 + 30$$
$$= \$1041.87$$

Including the $30 fee, this represents an interest rate of $(41.87/1000)(100\%) = 4.187\%$ for only the 1-month period.

(*c*) Again apply Equation [4.2], now with $m = 365$ compounding periods per year.

CP of day: Effective rate per day = $14.24/365 = 0.039\%$

4.2 Effective Annual Interest Rates ● ● ○

In this section, effective **annual interest rates** are calculated. Therefore, the year is used as the interest period t, and the compounding period CP can be any time unit less than 1 year. For example, we will learn that a *nominal* 18% per year, compounded quarterly, is the same as an *effective* rate of 19.252% per year.

The symbols used for nominal and effective interest rates are

$\qquad r$ = nominal interest rate per year

$\quad$CP = time period for each compounding

$\quad m$ = number of compounding periods per year

$\qquad i$ = effective interest rate per compounding period = r/m

$\quad i_a$ = effective interest rate per year

The relation $i = r/m$ is exactly the same as Equation [4.2].

Figure 4–2
Future worth calculation
at a rate i, compounded m
times in a year.

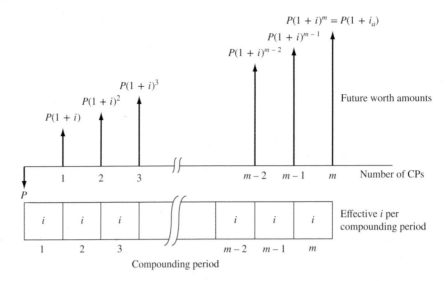

As mentioned earlier, treatment for nominal and effective interest rates parallels that of simple and compound interest. Like compound interest, an effective interest rate at any point during the year includes (compounds) the interest rate for all previous compounding periods during the year. Therefore, the derivation of an effective interest rate formula directly parallels the logic used to develop the future worth relation $F = P(1 + i)^n$. We set $P = \$1$ for simplification.

The future worth **F at the end of 1 year** is the principal P plus the interest $P(i)$ through the year. Since interest may be compounded several times during the year, use the effective annual rate symbol i_a to write the relation for F with $P = \$1$.

$$F = P + Pi_a = 1(1 + i_a)$$

Now consider Figure 4–2. The effective rate i per CP must be compounded through all m periods to obtain the total effect of compounding by the end of the year. This means that F can also be written as

$$F = 1(1 + i)^m$$

Equate the two expressions for F and solve for i_a. The **effective annual interest rate formula** for i_a is

$$i_a = (1 + i)^m - 1 \qquad \text{[4.3]}$$

Equation [4.3] calculates the effective annual interest rate i_a for any number of compounding periods per year when i is the rate for one compounding period.

If the effective annual rate i_a and compounding frequency m are known, Equation [4.3] can be solved for i to determine the *effective interest rate per compounding period.*

$$i = (1 + i_a)^{1/m} - 1 \qquad \text{[4.4]}$$

As an illustration, Table 4–2 utilizes the nominal rate of 18% per year for different compounding periods (year to week) to determine the effective annual interest rate. In each case, the effective rate i per CP is applied m times during the year. Table 4–3 summarizes the effective annual rate for frequently quoted nominal rates using Equation [4.3]. A standard of 52 weeks and 365 days per year is used throughout. The values in the continuous-compounding column are discussed in Section 4.8.

TABLE 4–2 Effective Annual Interest Rates Using Equation [4.3]

$r = 18\%$ per year, compounded CP-ly

Compounding Period, CP	Times Compounded per Year, m	Rate per Compound Period, $i\%$	Distribution of i over the Year of Compounding Periods	Effective Annual Rate, $i_a = (1 + i)^m - 1$
Year	1	18		$(1.18)^1 - 1 = 18\%$
6 months	2	9		$(1.09)^2 - 1 = 18.81\%$
Quarter	4	4.5		$(1.045)^4 - 1 = 19.252\%$
Month	12	1.5		$(1.015)^{12} - 1 = 19.562\%$
Week	52	0.34615		$(1.0034615)^{52} - 1 = 19.684\%$

TABLE 4–3   Effective Annual Interest Rates for Selected Nominal Rates

Nominal Rate, r%	Semiannually (m = 2)	Quarterly (m = 4)	Monthly (m = 12)	Weekly (m = 52)	Daily (m = 365)	Continuously (m = ∞; $e^r - 1$)
0.25	0.250	0.250	0.250	0.250	0.250	0.250
0.50	0.501	0.501	0.501	0.501	0.501	0.501
1.00	1.003	1.004	1.005	1.005	1.005	1.005
1.50	1.506	1.508	1.510	1.511	1.511	1.511
2	2.010	2.015	2.018	2.020	2.020	2.020
3	3.023	3.034	3.042	3.044	3.045	3.046
4	4.040	4.060	4.074	4.079	4.081	4.081
5	5.063	5.095	5.116	5.124	5.126	5.127
6	6.090	6.136	6.168	6.180	6.180	6.184
7	7.123	7.186	7.229	7.246	7.247	7.251
8	8.160	8.243	8.300	8.322	8.328	8.329
9	9.203	9.308	9.381	9.409	9.417	9.417
10	10.250	10.381	10.471	10.506	10.516	10.517
12	12.360	12.551	12.683	12.734	12.745	12.750
15	15.563	15.865	16.076	16.158	16.177	16.183
18	18.810	19.252	19.562	19.684	19.714	19.722
20	21.000	21.551	21.939	22.093	22.132	22.140
25	26.563	27.443	28.073	28.325	28.390	28.403
30	32.250	33.547	34.489	34.869	34.968	34.986
40	44.000	46.410	48.213	48.954	49.150	49.182
50	56.250	60.181	63.209	64.479	64.816	64.872

EXAMPLE 4.3

Janice is an engineer with Southwest Airlines. She purchased Southwest stock for $6.90 per share and sold it exactly 1 year later for $13.14 per share. She was very pleased with her investment earnings. Help Janice understand exactly what she earned in terms of (a) effective annual rate and (b) effective rate for quarterly compounding, and for monthly compounding. Neglect any commission fees for purchase and selling of stock and any quarterly dividends paid to stockholders.

Solution

(a) The effective annual rate of return i_a has a compounding period of 1 year, since the stock purchase and sales dates are exactly 1 year apart. Based on the purchase price of $6.90 per share and using the definition of interest rate in Equation [1.2],

$$i_a = \frac{\text{amount of increase per 1 year}}{\text{original price}} \times 100\% = \frac{6.24}{6.90} \times 100\% = 90.43\% \text{ per year}$$

(b) For the effective annual rates of 90.43% per year, compounded quarterly, and 90.43%, compounded monthly, apply Equation [4.4] to find corresponding effective rates on the basis of each compounding period.

Quarter: $m = 4$ times per year $i = (1.9043)^{1/4} - 1 = 1.17472 - 1 = 0.17472$

This is 17.472% per quarter, compounded quarterly.

Month: $m = 12$ times per year $i = (1.9043)^{1/12} - 1 = 1.05514 - 1 = 0.05514$

This is 5.514% per month, compounded monthly.

Comment

Note that these quarterly and monthly rates are less than the effective annual rate divided by the number of quarters or months per year. In the case of months, this would be 90.43%/12 = 7.54% per month. This computation is incorrect because it neglects the fact that compounding takes place 12 times during the year to result in the effective annual rate of 90.43%.

The spreadsheet function that displays the result of Equation [4.3], that is, the **effective annual rate** i_a, is the EFFECT function. The format is

$$= \text{EFFECT(nominal\_rate\_per\_year, compounding\_frequency)}$$

$$= \text{EFFECT}(r\%, m) \qquad\qquad\qquad\qquad\qquad\qquad \textbf{[4.5]}$$

Note that the rate entered in the EFFECT function is the **nominal annual rate $r\%$ per year**, not the effective rate $i\%$ per compounding period. The function automatically finds i for use in Equation [4.3]. As an example, assume the nominal annual rate is $r = 5.25\%$ per year, compounded quarterly, and you want to find the effective annual rate i_a. The correct input is = EFFECT(5.25%,4) to display $i_a =$ 5.354% per year. This is the spreadsheet equivalent of Equation [4.3] with $i = 5.25/4 = 1.3125\%$ per quarter with $m = 4$.

$$i_a = (1 + 0.013125)^4 - 1 = 0.05354 \quad (5.354\%)$$

The thing to remember about using the EFFECT function is that the nominal rate r entered must be expressed over the same period of time as that of the effective rate required, which is 1 year here.

The NOMINAL spreadsheet function finds the **nominal annual rate r.** The format is

$$= \text{NOMINAL(effective\_rate, compounding\_frequency\_per\_year)}$$

$$= \text{NOMINAL}(i_a\%, m) \qquad\qquad\qquad\qquad\qquad\qquad \textbf{[4.6]}$$

This function is designed to display **only** nominal **annual** rates. Accordingly, the m entered must be the number of times interest is compounded per year. For example, if the effective annual rate is 10.381% per year, compounded quarterly, and the nominal annual rate is requested, the function is = NOMINAL(10.381%,4) to display $r = 10\%$ per year, compounded quarterly. The nominal rates for shorter time periods than 1 year are determined by using Equation [4.1]. For example, the quarterly rate is $10\%/4 = 2.5\%$.

The things to remember when using the NOMINAL function are that the answer is always a nominal annual rate, the rate entered must be an effective annual rate, and the m must equal the number of times interest is compounded annually.

EXAMPLE 4.4 The Credit Card Offer Case   **PE**

In our Progressive Example, Dave is planning to accept the offer for a Chase Bank credit card that carries an APR (nominal rate) of 14.24% per year, or 1.187% per month. He will transfer a balance of $1000 and plans to pay it and the transfer fee of $30, due at the end of the first month. Let's assume that Dave makes the transfer, and only days later his employer has a 1-year assignment for him in the country of Cameroon in West Africa. Dave accepts the employment offer, and in his hurried, excited departure, he forgets to send the credit card service company a change of address. Since he is now out of mail touch, he does not pay his monthly balance due, which we calculated in Example 4.2 to be $1041.87.

(a) If this situation continues for a total of 12 months, determine the total due after 12 months and the effective annual rate of interest Dave has accumulated. Remember, the fine print on the card's interest and fee information states a penalty APR of 29.99% per year after one late payment of the minimum payment amount, plus a late payment fee of $39 per occurrence.

(b) If there were no penalty APR and no late-payment fee, what effective annual interest rate would be charged for this year? Compare this rate with the answer in part (a).

Solution

(a) Because Dave did not pay the first month's amount, the new balance of $1041.87 and all future monthly balances will accumulate interest at the higher monthly rate of

$$29.99\%/12 = 2.499\% \text{ per month}$$

Additionally, the $39 late-payment fee will be added each month, starting with the second month, and interest will be charged on these fees also each month thereafter. The first 3 months and last 2 months are detailed below. Figure 4–3 details the interest and fees for all 12 months.

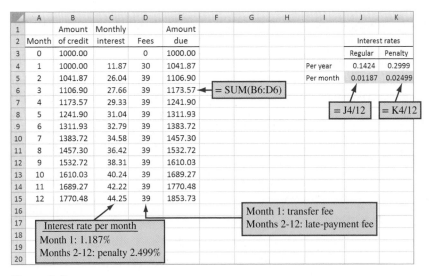

Figure 4–3
Monthly amounts due for a credit card, Example 4.4.

Month 1: $1000 + 1000(0.01187) + 30 = \$1041.87$
Month 2: $1041.87 + 1041.87(0.02499) + 39 = \$1106.91$
Month 3: $1106.91 + 1106.91(0.02499) + 39 = \$1173.57$

$$\vdots$$

Month 11: $1689.25 + 1689.25(0.02499) + 39 = \$1770.46$
Month 12: $1770.46 + 1770.46(0.02499) + 39 = \$1853.71$

The *effective monthly rate* is determined by using the F/P factor to find the i value at which \$1000 now is equivalent to \$1853.71 after 12 periods.

$$1853.71 = 1000(F/P,i,12) = 1000(1 + i)^{12}$$

$$1 + i = (1.85371)^{1/12} = 1.05278$$

$$i = 5.278\% \text{ per month}$$

Since the compounding period is 1 month, use Equation [4.3] to determine the *effective annual rate* of 85.375% per year, compounded monthly.

$$i_a = (1 + i)^m - 1 = (1.05278)^{12} - 1$$

$$= 0.85375 \quad (85.375\%)$$

(*b*) If there were no penalty fees for late payments and the nominal annual rate of 14.24% (or 1.187% per month) were applied throughout the 12 months, the effective annual rate would be 15.207% per year, compounded monthly. By Equation [4.3], with a small rounding error included,

$$i_a = (1 + i)^m - 1 = (1.01187)^{12} - 1 = 0.15207$$

First, Dave will not pay at the stated rate of 14.24%, because this is the APR (nominal rate), not the APY (effective rate) of 15.207%. Second, and much more important, is the huge difference made by (1) the increase in rate to an APR of 29.99% and (2) the monthly fees of \$39 for not making a payment. These large fees become part of the credit balance and accumulate interest at the penalty rate of 29.99% per year. The result is an effective annual rate jump from 15.207% to 85.375% per year, compounded monthly.

Comment

This is but one illustration of why the best advice to an individual or company in debt is to spend down the debt. The quoted APR by credit card, loan, and mortgage institutions can be quite deceiving; plus, the addition of penalty fees increases the effective rate very rapidly.

When Equation [4.3] is applied to find i_a, the result is usually not an integer. Therefore, the engineering economy factor cannot be obtained directly from the interest factor tables. There are alternative ways to find the factor value.

- Use the factor formula with the i_a rate substituted for i.
- Use the spreadsheet function with i_a (as discussed in Section 2.4).
- Linearly interpolate between two tabulated rates (as discussed in Section 2.4).

4.3 Effective Interest Rates for Any Time Period ● ● ●

Equation [4.3] in Section 4.2 calculates an effective interest rate per year from any effective rate over a shorter time period. We can generalize this equation to determine the **effective interest rate for any time period** (shorter or longer than 1 year).

$$\text{Effective } i \text{ per time period} = \left(1 + \frac{r}{m}\right)^m - 1 \qquad [4.7]$$

where i = effective rate for specified time period (say, semiannual)

r = nominal interest rate for same time period (semiannual)

m = number of times interest is compounded per stated time period (times per 6 months)

The term r/m is always the effective interest rate over a compounding period CP, and m is always the number of times that interest is compounded per the time period on the left of the equals sign in Equation [4.7]. Instead of i_a, this general expression uses i as the symbol for the effective interest rate, which conforms to the use of i throughout the remainder of this text. Examples 4.5 and 4.6 illustrate the use of this equation.

EXAMPLE 4.5

Tesla Motors manufactures high-performance battery electric vehicles. An engineer is on a Tesla committee to evaluate bids for new-generation coordinate-measuring machinery to be directly linked to the automated manufacturing of high-precision vehicle components. Three bids include the interest rates that vendors will charge on unpaid balances. To get a clear understanding of finance costs, Tesla management asked the engineer to determine the effective semiannual and annual interest rates for each bid. The bids are as follows:

Bid 1:   9% per year, compounded quarterly

Bid 2:   3% per quarter, compounded quarterly

Bid 3:   8.8% per year, compounded monthly

(*a*) Determine the effective rate for each bid on the basis of semiannual periods.
(*b*) What are the effective annual rates? These are to be a part of the final bid selection.
(*c*) Which bid has the lowest effective annual rate?

Solution

(*a*) Convert the nominal rates to a semiannual basis, determine m, then use Equation [4.7] to calculate the effective semiannual interest rate i. For bid 1,

$$r = 9\% \text{ per year} = 4.5\% \text{ per 6 months}$$

$$m = 2 \text{ quarters per 6 months}$$

$$\text{Effective } i\% \text{ per 6 months} = \left(1 + \frac{0.045}{2}\right)^2 - 1 = 1.0455 - 1 = 4.55\%$$

Table 4–4 (left section) summarizes the effective semiannual rates for all three bids.

TABLE 4–4	Effective Semiannual and Annual Interest Rates for Three Bid Rates, Example 4.5					
	Semiannual Rates			Annual Rates		
Bid	Nominal r per 6 Months, %	CP per 6 Months, m	Equation [4.7], Effective i, %	Nominal r per Year, %	CP per Year, m	Equation [4.7], Effective i, %
1	4.5	2	4.55	9	4	9.31
2	6.0	2	6.09	12	4	12.55
3	4.4	6	4.48	8.8	12	9.16

(*b*) For the effective annual rate, the time basis in Equation [4.7] is 1 year. For bid 1,

$$r = 9\% \text{ per year} \qquad m = 4 \text{ quarters per year}$$

$$\text{Effective } i\% \text{ per year} = \left(1 + \frac{0.09}{4}\right)^4 - 1 = 1.0931 - 1 = 9.31\%$$

The right section of Table 4–4 includes a summary of the effective annual rates.

(*c*) Bid 3 includes the lowest effective annual rate of 9.16%, which is equivalent to an effective semiannual rate of 4.48% when interest is compounded monthly.

EXAMPLE 4.6

A dot-com company plans to place money in a new venture capital fund that currently returns 18% per year, compounded daily. What effective rate is this (*a*) yearly and (*b*) semiannually?

Solution

(*a*) Use Equation [4.7], with $r = 0.18$ and $m = 365$.

$$\text{Effective } i\% \text{ per year} = \left(1 + \frac{0.18}{365}\right)^{365} - 1 = 19.716\%$$

(*b*) Here $r = 0.09$ per 6 months and $m = 182$ days.

$$\text{Effective } i\% \text{ per 6 months} = \left(1 + \frac{0.09}{182}\right)^{182} - 1 = 9.415\%$$

4.4 Equivalence Relations: Payment Period and Compounding Period ● ● ○

Now that the procedures and formulas for determining effective interest rates with consideration of the compounding period are developed, it is necessary to consider the **payment period**.

The payment period (PP) is the length of time between cash flows (inflows or outflows). It is common that the lengths of the payment period and the compounding period (CP) do not coincide. It is important to determine if PP = CP, PP > CP, or PP < CP.

If a company deposits money each month into an account that earns at the nominal rate of 8% per year, compounded semiannually, the cash flow deposits define a payment period of 1 month and the nominal interest rate defines a compounding period of 6 months. These time periods are shown in Figure 4–4. Similarly, if a person deposits a bonus check once a year into an account that compounds interest quarterly, PP = 1 year and CP = 3 months.

Figure 4–4
One-year cash flow diagram for a monthly payment period (PP) and semiannual compounding period (CP).

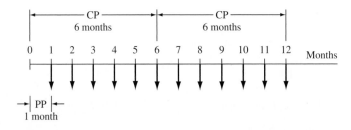

TABLE 4-5	Section References for Equivalence Calculations Based on Payment Period and Compounding Period Comparison	
Length of Time	Involves Single Amounts (*P* and *F* Only)	Involves Uniform Series or Gradient Series (*A*, *G*, or *g*)
PP = CP	Section 4.5	Section 4.6
PP > CP	Section 4.5	Section 4.6
PP < CP	Section 4.7	Section 4.7

As we learned earlier, to correctly perform equivalence calculations, an effective interest rate is needed in the factors and spreadsheet functions. Therefore, it is essential that the time periods of the interest rate and the payment period be on the same time basis. The next three sections (4.5 to 4.7) describe procedures to determine correct i and n values for engineering economy factors and spreadsheet functions. First, compare the length of PP and CP, then identify the cash flows as only single amounts (P and F) or as a series (A, G, or g). Table 4–5 provides the section reference. The section references are the same when PP = CP and PP > CP, because the procedures to determine i and n are the same, as discussed in Sections 4.5 and 4.6.

A general principle to remember throughout these equivalence computations is that when cash actually flows, it is necessary to account for the time value of money. For example, assume that cash flows occur every 6 months and that interest is compounded quarterly. After 3 months, there is no cash flow and no need to determine the effect of quarterly compounding. However, at the 6-month time point, it is necessary to consider the interest accrued during the previous two quarters.

4.5 Equivalence Relations: Single Amounts with PP ≥ CP ●●●

When only single-amount cash flows are involved, there are many correct combinations of i and n that can be used to calculate P or F, because of the following two requirements: (1) i must be expressed as an effective rate, and (2) the time period of n must be exactly the same as that of i. These two requirements are fulfilled in two equally correct ways to determine i and n for P/F and F/P factors. Method 1 is easier to apply because the interest tables in the back of the text can usually provide the factor value. Method 2 likely requires a factor formula calculation because the resulting effective interest rate is not an integer. For spreadsheets, either method is acceptable; however, method 1 is usually easier.

Method 1: Determine the effective interest rate over the **compounding period CP**, and set n equal to the number of compounding periods between P and F. The relations to calculate P and F are

$$P = F(P/F, \text{effective } i\% \text{ per CP, total number of periods } n) \qquad [4.8]$$

$$F = P(F/P, \text{effective } i\% \text{ per CP, total number of periods } n) \qquad [4.9]$$

For example, assume that the stated interest rate is a nominal 15% per year, compounded monthly. Here CP is 1 month. To find P or F over a 2-year span, calculate the effective monthly rate of $15\%/12 = 1.25\%$ and the total months of $2(12) = 24$. Then 1.25% and 24 are used in the P/F and F/P factors.

While any time period can be used to determine the effective interest rate (as shown in method 2 below), the interest rate that is associated with the CP is typically the best because it is usually a whole number. Therefore, the factor tables in the back of the text can be used.

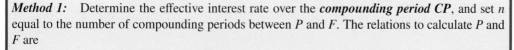

Method 2: First, determine the **effective interest rate for any** *time period t, then for the same time period,* set n equal to the total number of periods between P and F.

The P and F relations are the same as in Equations [4.8] and [4.9] with the term *effective i% per t* substituted for the interest rate. For an interest rate of 15% per year, compounded monthly, and a 2-year time span between P and F, if a time period t of 1 year is used, the effective yearly rate and the n values are

$$\text{Effective } i\% \text{ per year} = \left(1 + \frac{0.15}{12}\right)^{12} - 1 = 16.076\%$$

$$n = 2 \text{ years}$$

The P/F factor is the same by both methods: $(P/F, 1.25\%, 24) = 0.7422$ using Table 5 in the rear of the text, and $(P/F, 16.076\%, 2) = 0.7422$ using the P/F factor formula. With these calculations, you can see why method 1 is recommended when a spreadsheet is not utilized.

EXAMPLE 4.7

Over the past 10 years, Gentrack has placed varying sums of money into a special capital accumulation fund. The company sells compost produced by garbage-to-compost plants in the United States and Vietnam. Figure 4–5 is the cash flow diagram in $1000 units. Find the amount in the account now (after 10 years) at an interest rate of 12% per year, compounded semiannually.

Solution

Only P and F values are involved. Both methods are illustrated to find F in year 10.

Method 1: Use the semiannual CP to express the effective semiannual rate of 6% per 6-month period. There are $n = (2)$(number of years) semiannual periods for each cash flow. Using tabulated factor values, the future worth by Equation [4.9] is

$$F = 1000(F/P, 6\%, 20) + 3000(F/P, 6\%, 12) + 1500(F/P, 6\%, 8)$$

$$= 1000(3.2071) + 3000(2.0122) + 1500(1.5938)$$

$$= \$11,634 \quad (\$11.634 \text{ million})$$

Method 2: Express the effective annual rate, based on semiannual compounding.

$$\text{Effective } i\% \text{ per year} = \left(1 + \frac{0.12}{2}\right)^{2} - 1 = 12.36\%$$

The n value is the actual number of years. Use the factor formula $(F/P, i, n) = (1.1236)^n$ and Equation [4.9] to obtain the same answer as above.

$$F = 1000(F/P, 12.36\%, 10) + 3000(F/P, 12.36\%, 6) + 1500(F/P, 12.36\%, 4)$$

$$= 1000(3.2071) + 3000(2.0122) + 1500(1.5938)$$

$$= \$11,634 \quad (\$11.634 \text{ million})$$

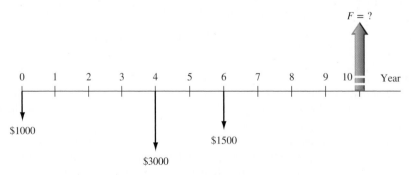

Figure 4–5
Cash flow diagram, Example 4.7.

TABLE 4–6 Examples of *n* and *i* Values Where PP = CP or PP > CP

Cash Flow Series	Interest Rate	What to Find; What Is Given	Standard Notation
$500 semiannually for 5 years	16% per year, compounded semiannually	Find *P*; given *A*	$P = 500(P/A,8\%,10)$
$75 monthly for 3 years	24% per year, compounded monthly	Find *F*; given *A*	$F = 75(F/A,2\%,36)$
$180 quarterly for 15 years	5% per quarter	Find *F*; given *A*	$F = 180(F/A,5\%,60)$
$25 per month increase for 4 years	1% per month	Find *P*; given *G*	$P = 25(P/G,1\%,48)$
$5000 per quarter for 6 years	1% per month	Find *A*; given *P*	$A = 5000(A/P,3.03\%,24)$

4.6 Equivalence Relations: Series with PP ≥ CP ●●●

When uniform or gradient series are included in the cash flow sequence, and the payment period equals or exceeds the compounding period, PP is defined by the length of time between cash flows. This also establishes the time unit of the effective interest rate. For example, if cash flows occur on a *quarterly* basis, PP is 1 *quarter* and the effective *quarterly* rate is necessary. The *n* value is the total number of *quarters*. If PP is a quarter, 5 years translates to an *n* value of 20 quarters. This is a direct application of the following general guideline:

> When cash flows involve a series (i.e., *A*, *G*, *g*) and the payment period equals or exceeds the compounding period in length:
>
> - Find the effective *i* per payment period.
> - Determine *n* as the total number of payment periods.

In performing equivalence computations for series, *only* these values of *i* and *n* can be used in interest tables, factor formulas, and spreadsheet functions. In other words, there are no other combinations that give the correct answers as there are for *single-amount cash flows*.

Table 4–6 shows the correct formulation for several cash flow series and interest rates. Note that *n* is always equal to the total number of payment periods and *i* is an effective rate expressed over the same time period as *n*.

EXAMPLE 4.8

For the past 7 years, Excelon Energy has paid $500 every 6 months for a software maintenance contract. What is the equivalent total amount after the last payment, if these funds are taken from a pool that has been returning 8% per year, compounded quarterly?

Solution

The cash flow diagram is shown in Figure 4–6. The payment period (6 months) is longer than the compounding period (quarter); that is, PP > CP. Applying the guideline, we need to determine an **effective semiannual interest rate.** Use Equation [4.7] with *r* = 4% per 6-month period and *m* = 2 quarters per semiannual period.

$$\text{Effective } i\% \text{ per 6 months} = \left(1 + \frac{0.04}{2}\right)^2 - 1 = 4.04\%$$

The effective semiannual interest rate can also be obtained from Table 4–3 by using the *r* value of 4% and *m* = 2 to get *i* = 4.04%.

The value $i = 4.04\%$ seems reasonable, since we expect the effective rate to be slightly higher than the nominal rate of 4% per 6-month period. The total number of semiannual payment periods is $n = 2(7) = 14$. The relation for F is

$$F = A(F/A,4.04\%,14)$$

$$= 500(18.3422)$$

$$= \$9171.09$$

To determine the F/A factor value 18.3422 using a spreadsheet, enter the FV function from Figure 2–9, that is, $= -FV(4.04\%,14,1)$. Alternatively, the final answer of $9171.09 can be displayed directly using the function $= -FV(4.04\%,14,500)$.

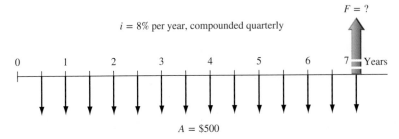

Figure 4–6
Diagram of semiannual deposits used to determine F, Example 4.8.

EXAMPLE 4.9   The Credit Card Offer Case

In our continuing credit card saga of Dave and his job transfer to Africa, let's assume he did remember that the total balance is $1030, including the $30 balance transfer fee, and he wants to set up a monthly automatic checking account transfer to pay off the entire amount in 2 years. Once he learned that the minimum payment is $25 per month, Dave decided to be sure the monthly transfer exceeds this amount to avoid any further penalty fees and the penalty APR of 29.99% per year. What amount should he ask to be transferred by the due date each month? What is the APY he will pay, if this plan is followed exactly and Chase Bank does not change the APR during the 2-year period? Also, assume he left the credit card at home and will charge no more to it.

Solution

The monthly A series is needed for a total of $n = 2(12) = 24$ payments. In this case, $PP = CP = 1$ month, and the effective monthly rate is $i = 14.24\%/12 = 1.187\%$ per month.

Solution by hand: Use a calculator or hand computation to determine the A/P factor value.

$$A = P(A/P,i,n) = 1030(A/P,1.187\%,24) = 1030(0.04813)$$

$$= \$49.57 \text{ per month for 24 months}$$

Solution by spreadsheet: Use the function $= -PMT(1.187\%,24,1)$ to determine the factor value 0.04813 to determine A for $n = 24$ payments. Alternatively use the function $= PMT (1.187\%,24,1030)$ to directly display the required monthly payment of $A = \$-49.57$.

The *effective annual interest rate or APY* is computed using Equation [4.7] with $r = 14.24\%$ per year, compounded monthly, and $m = 12$ times per year.

$$\text{Effective } i \text{ per year} = \left(1 + \frac{0.1424}{12}\right)^{12} - 1 = 1.15207 - 1$$

$$= 15.207\% \text{ per year}$$

This is the same effective annual rate i_a determined in Example 4.4b.

EXAMPLE 4.10

The Scott and White Health Plan (SWHP) has purchased a robotized prescription fulfillment system for faster and more accurate delivery to patients with stable, pill-form medication for chronic health problems, such as diabetes, thyroid, and high blood pressure. Assume this high-volume system costs $3 million to install and an estimated $200,000 per year for all materials, operating, personnel, and maintenance costs. The expected life is 10 years. An SWHP bio-medical engineer wants to estimate the total revenue requirement for each 6-month period that is necessary to recover the investment, interest, and annual costs. Find this semiannual A value both by hand and by spreadsheet, if capital funds are evaluated at 8% per year, using two different compounding periods:

Rate 1. 8% per year, compounded *semiannually*.
Rate 2. 8% per year, compounded *monthly*.

Solution

Figure 4–7 shows the cash flow diagram. Throughout the 20 semiannual periods, the annual cost occurs every other period, and the capital recovery series is sought for every 6-month period. This pattern makes the solution by hand quite involved if the P/A and A/P factors are used to find P and A values. The spreadsheet solution is recommended in cases such as this.

Solution by hand—rate 1: Steps to find the semiannual A value are as follows. First, calculate the present worth P at time 0 with $n = 10$ years using the effective annual rate.

$$\text{Effective } i \text{ per year} = (1 + 0.08/2)^2 - 1 = 8.16\%$$
$$P = 3{,}000{,}000 + 200{,}000(P/A,8.16\%,10)$$
$$= 3{,}000{,}000 + 200{,}000(6.6619)$$
$$= \$4{,}332{,}380$$

Now, find A per 6-month period.

PP = CP at 6 months; find the effective rate per semiannual period.
Effective semiannual $i = 8\%/2 = 4\%$ per 6 months, compounded semiannually.
Number of semiannual periods $n = 2(10) = 20$.

$$A = \$4{,}332{,}380(A/P,4\%,20) = \$318{,}777 \text{ per 6-months}$$

Conclusion: Revenue of $318,777 is necessary every 6 months to cover all costs and interest at 8% per year, compounded semiannually.

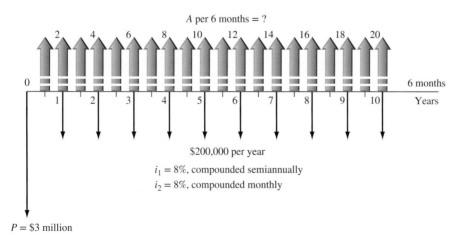

Figure 4–7
Cash flow diagram with two different compounding periods, Example 4.10.

Solution by hand—rate 2: As before, calculate P at time 0 with $n = 10$ years and an effective annual year for monthly compounding.

$$\text{Effective } i \text{ per year} = (1 + 0.08/12)^{12} - 1 = 8.30\%$$
$$P = 3{,}000{,}000 + 200{,}000(P/A,8.30\%,10)$$
$$= 3{,}000{,}000 + 200{,}000(6.6202)$$
$$= \$4{,}324{,}040$$

Now, find A per semiannual period, where

PP is 6 months, but the CP is monthly; therefore, PP > CP. To find the effective semiannual rate, the effective interest rate Equation [4.7] is applied with $r = 4\%$ and $m = 6$ months per semiannual period. The n value to determine A is 20 semiannual periods.

$$\text{Effective semiannual } i = \left(1 + \frac{0.04}{6}\right)^6 - 1 = 4.067\%$$

$$A = \$4{,}324{,}080(A/P,4.067\%,20) = \$320{,}061 \text{ per 6-months}$$

Now, \$320,061, or \$1284 more semiannually, is required to cover the more frequent compounding of the 8% per year interest. Note that all P/A and A/P factors must be calculated with factor formulas at 8.30% and 4.067%, respectively. This method is usually more calculation-intensive and error-prone than the spreadsheet solution.

Solution by spreadsheet—rates 1 and 2: Once the effective interest rates per semiannual period have been determined, a spreadsheet solution is simple. For rate 1, a single-cell PMT function, with an imbedded PV function (in italics) to determine the present worth of the \$200,000 annual series at the effective annual rate of 8.16%, displays $A = \$318{,}784$ per 6-month period. The function is:

$$= \text{PMT}(4\%,20,PV(8.16\%,10,200000) - 3000000)$$

Similarly, for rate 2, the single-cell spreadsheet function = PMT(4.067%,20,PV(8.30%,10, 200000)–3000000) displays $A = \$320{,}061$ per 6-month period. Note that the EFFECT function can be used separately or incorporated into these single-cell PMT functions to calculate the effective rates necessary to determine the correct semiannual A values.

A spreadsheet can also be used if the \$200,000 annual amounts are considered single amounts, but the spreadsheet is more involved as follows. Figure 4–8 presents a general solu-

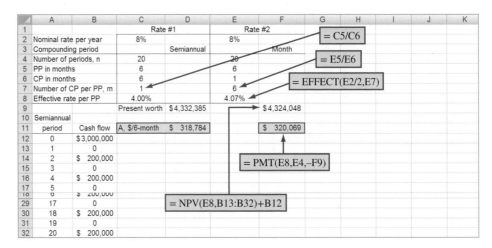

Figure 4–8
Spreadsheet solution for semiannual A series for different compounding periods, Example 4.10.

tion for the problem at both rates. (Several rows at the bottom of the spreadsheet are not printed. They continue the cash flow pattern of $200,000 every other 6 months through cell B32.) The functions in C8 and E8 are general expressions for the effective rate per PP, expressed in months. This allows some sensitivity analysis to be performed for different PP and CP values. Note the functions in C7 and E7 to determine m for the effective rate relations. This technique works well for spreadsheets once PP and CP are entered in the time unit of the CP.

Each 6-month period is included in the cash flows, including the $0 entries, so the NPV and PMT functions work correctly. The final A values in D14 ($318,784) and F14 ($320,069) are slightly different than those of the hand solution due to the number of significant digits used by spreadsheet functions.

The EFFECT function is applied in cells C8 and E8 to obtain the effective semiannual rate i. When using EFFECT for payment periods other than a year, the first entry of "nominal_rate" is the rate *per payment period*, not per year as indicated in Equation [4.5]. In the case for rate #2, the percentage is per 6-month payment period, that is, 4%, as determined in cell E8 using the relation E2/2. EFFECT correctly displays the effective semiannual rate of 4.07%.

4.7 Equivalence Relations: Single Amounts and Series with PP < CP ●●●

If a person deposits money each month into a savings account where interest is compounded quarterly, do all the monthly deposits earn interest before the next quarterly compounding time? If a person's credit card payment is due with interest on the 15th of the month, and if the full payment is made on the 1st, does the financial institution reduce the interest owed, based on early payment? The usual answers are no. However, if a monthly payment on a $10 million, quarterly compounded, bank loan were made early by a large corporation, the corporate financial officer would likely insist that the bank reduce the amount of interest due, based on early payment. These are examples of PP < CP. The timing of cash flow transactions between compounding points introduces the question of how **interperiod compounding** is handled. Fundamentally, there are two policies, one of which must be applied: interperiod cash flows earn *no interest* or they earn *compound interest*.

> For a no-interperiod-interest policy, negative cash flows (deposits or payments, depending on the perspective used for cash flows) are all regarded as *made at the end of the compounding period,* and positive cash flows (receipts or withdrawals) are all regarded as *made at the beginning.*

As an illustration, when interest is compounded quarterly, all monthly deposits are moved to the end of the quarter (no interperiod interest is earned) and all withdrawals are moved to the beginning (no interest is paid for the entire quarter). This procedure can significantly alter the distribution of cash flows before the effective quarterly rate is applied to find P, F, or A. This effectively forces the cash flows into a PP = CP situation, as discussed in Sections 4.5 and 4.6. Example 4.11 illustrates this procedure and the economic fact that, within a one-compounding-period time frame, there is no interest advantage to making payments early. Of course, noneconomic factors may be present.

EXAMPLE 4.11

Last year AllStar Venture Capital agreed to invest funds in Clean Air Now (CAN), a start-up company in Las Vegas that is an outgrowth of research conducted in mechanical engineering at the University of Nevada–Las Vegas. The product is a new filtration system used in the process of carbon capture and sequestration (CCS) for coal-fired power plants. The venture fund manager generated the cash flow diagram in Figure 4–9a in $1000 units from AllStar's perspective. Included are payments (outflows) to CAN made over the first year and receipts (inflows) from CAN to AllStar. The receipts were unexpected this first year; however, the product has great promise, and advance orders have come from eastern U.S. plants anxious to become zero-emission coal-fueled plants. The interest rate is 12% per year, compounded quarterly, and AllStar uses the no-interperiod-interest policy. How much is AllStar in the "red" at the end of the year?

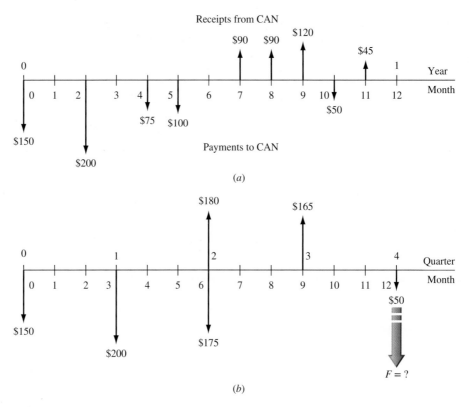

Figure 4–9
(*a*) Actual and (*b*) moved cash flows (in $1000) for quarterly compounding periods using no interperiod interest, Example 4.11.

Solution

With no interperiod interest considered, Figure 4–9*b* reflects the moved cash flows. All negative cash flows (payments to CAN) are moved to the end of the respective quarter, and all positive cash flows (receipts) are moved to the beginning of the respective quarter. Calculate the *F* value at 12%/4 = 3% per quarter.

$$F = 1000[-150(F/P,3\%,4) - 200(F/P,3\%,3) + (-175 + 180)(F/P,3\%,2)$$

$$+ 165(F/P,3\%,1) - 50]$$

$$= \$-262,111$$

AllStar has a net investment of $262,111 in CAN at the end of the year.

If PP < CP and interperiod compounding is earned, then the cash flows are not moved, and the equivalent *P*, *F*, or *A* values are determined using the effective interest rate per payment period. The engineering economy relations are determined in the same way as in the previous two sections for PP ≥ CP. The effective interest rate formula will have an *m* value less than 1 because there is only a fractional part of the CP within one PP. For example, weekly cash flows and quarterly compounding require that $m = 1/13$ of a quarter. When the nominal rate is 12% per year, compounded quarterly (the same as 3% per quarter, compounded quarterly), the effective rate per PP is

$$\text{Effective weekly } i\% = (1.03)^{1/13} - 1 = 0.228\% \text{ per week}$$

4.8 Effective Interest Rate for Continuous Compounding ● ● ●

If we allow compounding to occur more and more frequently, the compounding period becomes shorter and shorter and *m*, the number of compounding periods per payment period, increases.

Continuous compounding is present when the duration of CP, the compounding period, becomes infinitely small and m, the number of times interest is compounded per period, becomes infinite. Businesses with large numbers of cash flows each day consider the interest to be continuously compounded for all transactions.

As m approaches infinity, the effective interest rate Equation [4.7] must be written in a new form. First, recall the definition of the natural logarithm base.

$$\lim_{h \to \infty} \left(1 + \frac{1}{h}\right)^h = e = 2.71828+ \qquad\qquad [4.10]$$

The limit of Equation [4.7] as m approaches infinity is found by using $r/m = 1/h$, which makes $m = hr$.

$$\lim_{m \to \infty} i = \lim_{m \to \infty} \left(1 + \frac{r}{m}\right)^m - 1$$

$$= \lim_{h \to \infty} \left(1 + \frac{1}{h}\right)^{hr} - 1 = \lim_{h \to \infty} \left[\left(1 + \frac{1}{h}\right)^h\right]^r - 1$$

$$i = e^r - 1 \qquad\qquad [4.11]$$

Equation [4.11] is used to compute the **effective continuous interest rate,** when the time periods on i and r are the same. As an illustration, if the nominal annual $r = 15\%$ per year, the effective continuous rate per year is

$$i\% = e^{0.15} - 1 = 16.183\%$$

For convenience, Table 4–3 includes effective continuous rates for the nominal rates listed.

To find an effective or nominal interest rate for continuous compounding using the spreadsheet functions **EFFECT** or **NOMINAL,** enter a very large value for the compounding frequency m in Equation [4.5] or [4.6], respectively. A value of 10,000 or higher provides sufficient accuracy. Both functions are illustrated in Example 4.12.

EXAMPLE 4.12

(*a*) For an interest rate of 18% per year, compounded continuously, calculate the effective monthly and annual interest rates.

(*b*) An investor requires an effective return of at least 15%. What is the minimum annual nominal rate that is acceptable for continuous compounding?

Solution by Hand

(*a*) The nominal monthly rate is $r = 18\%/12 = 1.5\%$, or 0.015 per month. By Equation [4.11], the effective monthly rate is

$$i\% \text{ per month} = e^r - 1 = e^{0.015} - 1 = 1.511\%$$

Similarly, the effective annual rate using $r = 0.18$ per year is

$$i\% \text{ per year} = e^r - 1 = e^{0.18} - 1 = 19.722\%$$

(*b*) Solve Equation [4.11] for r by taking the natural logarithm.

$$e^r - 1 = 0.15$$

$$e^r = 1.15$$

$$\ln e^r = \ln 1.15$$

$$r = 0.13976$$

Therefore, a rate of 13.976% per year, compounded continuously, will generate an effective 15% per year return. The general formula to find the nominal rate, given the effective continuous rate i, is $r = \ln(1 + i)$.

Solution by Spreadsheet

(*a*) Use the EFFECT function with the nominal monthly rate $r = 1.5\%$ and annual rate $r = 18\%$ with a large m to display effective i values. The functions to enter on a spreadsheet and the responses are as follows:

Monthly:	= EFFECT(1.5%,10000)	effective $i = 1.511\%$ per month
Annual:	= EFFECT(18%,10000)	effective $i = 19.722\%$ per year

(*b*) Use the function in Equation [4.6] in the format = NOMINAL(15%,10000) to display the nominal rate of 13.976% per year, compounded continuously.

EXAMPLE 4.13

Engineers Marci and Suzanne both invest $5000 for 10 years at 10% per year. Compute the future worth for both individuals if Marci receives annual compounding and Suzanne receives continuous compounding.

Solution

Marci: For annual compounding the future worth is

$$F = P(F/P,10\%,10) = 5000(2.5937) = \$12,969$$

Suzanne: Using Equation [4.11], first find the effective i per year for use in the F/P factor.

$$\text{Effective } i\% = e^{0.10} - 1 = 10.517\%$$

$$F = P(F/P,10.517\%,10) = 5000(2.7183) = \$13,591$$

Continuous compounding causes a $622 increase in earnings. For comparison, daily compounding yields an effective rate of 10.516% ($F = \$13,590$), only slightly less than the 10.517% for continuous compounding.

For some business activities, cash flows occur throughout the day. Examples of costs are energy and water costs, inventory costs, and labor costs. A realistic model for these activities is to increase the frequency of the cash flows to become continuous. In these cases, the economic analysis can be performed for **continuous cash flow** (also called continuous funds flow) and the continuous compounding of interest as discussed above. Different expressions must be derived for the factors for these cases. In fact, the monetary differences for continuous cash flows relative to the discrete cash flow and discrete compounding assumptions are usually not large. Accordingly, most engineering economy studies do not require the analyst to utilize these mathematical forms to make a sound economic decision.

4.9 Interest Rates That Vary Over Time ●●●

Real-world interest rates for a corporation vary from year to year, depending upon the financial health of the corporation, its market sector, the national and international economies, forces of inflation, and many other elements. Loan rates may increase from one year to another. Home mortgages financed using ARM (adjustable-rate mortgage) interest is a good example. The mortgage rate is slightly adjusted annually to reflect the age of the loan, the current cost of mortgage money, etc.

When P, F, and A values are calculated using a constant or average interest rate over the life of a project, rises and falls in i are neglected. If the variation in i is large, the equivalent values will vary considerably from those calculated using the constant rate. Although an engineering economy study can accommodate varying i values mathematically, it is more involved computationally to do so.

To determine the P value for future cash flow values (F_t) at different i values (i_t) for each year t, we will assume *annual compounding*. Define

$$i_t = \text{effective annual interest rate for year } t \qquad (t = \text{years 1 to } n)$$

To determine the present worth, calculate the P of each F_t value, using the applicable i_t, and sum the results. Using standard notation and the P/F factor,

$$P = F_1(P/F,i_1,1) + F_2(P/F,i_1,1)(P/F,i_2,1) + \cdots$$
$$+ F_n(P/F,i_1,1)(P/F,i_2,1) \cdots (P/F,i_n,1) \qquad \text{[4.12]}$$

The future worth F is determined in a similar fashion. Calculate the F for each P_t value using the applicable i_t and sum the results to obtain the total F in year n.

When only single amounts are involved, that is, one P and one F in the final year n, the last term in Equation [4.12] is the expression for the present worth of the future cash flow.

$$P = F_n(P/F,i_1,1)(P/F,i_2,1) \cdots (P/F,i_n,1) \qquad \text{[4.13]}$$

If the equivalent uniform series A over all n years is needed, first find P, using either of the last two equations; then substitute the symbol A for each F_t symbol. Since the equivalent P has been determined numerically using the varying rates, this new equation will have only one unknown, namely, A. Example 4.14 illustrates this procedure.

EXAMPLE 4.14

CE, Inc. leases large earth tunneling equipment. The net profit from the equipment for each of the last 4 years has been decreasing, as shown below. Also shown are the annual rates of return on invested capital. The return has been increasing. Determine the present worth P and equivalent uniform series A of the net profit series. Take the annual variation of rates of return into account.

Year	1	2	3	4
Net Profit	$70,000	$70,000	$35,000	$25,000
Annual Rate	7%	7%	9%	10%

Solution

Figure 4–10 shows the cash flows, rates for each year, and the equivalent P and A. Equation [4.12] is used to calculate P. Since for both years 1 and 2, the net profit is $70,000 and the annual rate is 7%, the P/A factor can be used for these 2 years only.

$$P = [70(P/A,7\%,2) + 35(P/F,7\%,2)(P/F,9\%,1)$$
$$+ 25(P/F,7\%,2)(P/F,9\%,1)(P/F,10\%,1)](1000)$$
$$= [70(1.8080) + 35(0.8013) + 25(0.7284)](1000)$$
$$= \$172{,}816 \qquad \text{[4.14]}$$

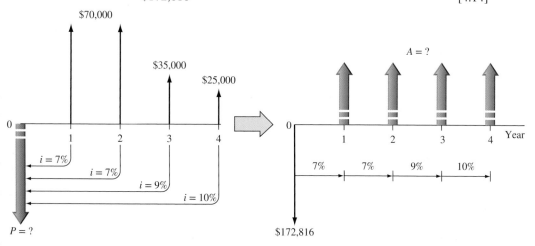

Figure 4–10
Equivalent P and A values for varying interest rates, Example 4.14.

To determine an equivalent annual series, substitute the symbol A for all net profit values on the right side of Equation [4.14], set it equal to $P = \$172,816$, and solve for A. This equation accounts for the varying i values each year. See Figure 4–10 for the cash flow diagram transformation.

$$\$172,816 = A[(1.8080) + (0.8013) + (0.7284)] = A[3.3377]$$

$$A = \$51,777 \text{ per year}$$

Comment

If the average of the four annual rates, that is, 8.25%, is used, the result is $A = \$52,467$. This is a \$690 per year overestimate of the equivalent annual net profit.

When there is a cash flow in year 0 and interest rates vary annually, this cash flow must be included to determine P. In the computation for the equivalent uniform series A over all years, including year 0, it is important to include this initial cash flow at $t = 0$. This is accomplished by inserting the factor value for $(P/F, i_0, 0)$ into the relation for A. This factor value is always 1.00. It is equally correct to find the A value using a future worth relation for F in year n. In this case, the A value is determined using the F/P factor, and the cash flow in year n is accounted for by including the factor $(F/P, i_n, 0) = 1.00$.

CHAPTER SUMMARY

Since many real-world situations involve cash flow frequencies and compounding periods other than 1 year, it is necessary to use nominal and effective interest rates. When a nominal rate r is stated, the effective interest rate i per payment period is determined by using the effective interest rate equation.

$$\text{Effective } i = \left(1 + \frac{r}{m}\right)^m - 1$$

The m is the number of compounding periods (CP) per interest period. If interest compounding becomes more and more frequent, then the length of a CP approaches zero, continuous compounding results, and the effective i is $e^r - 1$.

All engineering economy factors require the use of an effective interest rate. The i and n values placed in a factor depend upon the type of cash flow series. If only single amounts (P and F) are present, there are several ways to perform equivalence calculations using the factors. However, when series cash flows (A, G, and g) are present, only one combination of the effective rate i and number of periods n is correct for the factors. This requires that the relative lengths of PP and CP be considered as i and n are determined. The interest rate and payment periods must have the same time unit for the factors to correctly account for the time value of money.

From one year (or interest period) to the next, interest rates will vary. To accurately perform equivalence calculations for P and A when rates vary significantly, the applicable interest rate should be used, not an average or constant rate.

PROBLEMS

Nominal and Effective Rates

4.1   Identify the compounding period for the following interest statements: (*a*) 1% per week; (*b*) 2.5 % per quarter; and (*c*) 8.5% per year, compounded semi-annually.

4.2   What is the difference between APR and APY? Which one is used in interest factor formulas, tables, and spreadsheet functions?

4.3   Identify the compounding period for the following interest statements: (*a*) nominal 7% per year, compounded monthly; (*b*) effective 6.8% per year, compounded monthly; and (*c*) effective 3.4% per quarter, compounded weekly.

4.4   Determine the number of times interest is compounded in a year for the following interest statements: (*a*) 1% per quarter; (*b*) 2% per month; and (*c*) 8% per year, compounded every 2 months.

4.5 For an interest rate of 10% per year, compounded quarterly, determine the number of times interest is compounded (*a*) per quarter, (*b*) per 6 months, (*c*) per year, and (*d*) per 3 years.

4.6 For an interest rate of 1% per quarter, determine the nominal interest rate per (*a*) semiannual period, (*b*) year, and, (*c*) 2 years.

4.7 First Corp Bank advertises interest paid at 2% every 6 months. What is the APR?

4.8 For an interest rate of 9% per year, compounded every 4 months, determine the nominal interest rate per (*a*) 8 months, (*b*) 12 months, and (*c*) 2 years.

4.9 For an APR of 10% per year, if interest is compounded quarterly, determine the nominal rate per (*a*) 6 months, and (*b*) 2 years.

4.10 Identify the following interest rate statements as either nominal or effective: (*a*) 1.3% per month; (*b*) 1% per week, compounded weekly; (*c*) nominal 15% per year, compounded monthly; (*d*) effective 1.5% per month, compounded daily; and (*e*) 15% per year, compounded semiannually.

4.11 You plan to deposit $100 per week into a fund that pays interest of 6% per year, compounded quarterly. Identify the interest period, compounding period, and number of times interest is compounded per interest period.

4.12 Western Energy makes quarterly deposits into an account reserved for purchasing new equipment 2 years from now. The interest paid on the deposits is 12% per year, compounded monthly. (*a*) Identify the interest period, compounding period, and compounding frequency in the interest period. (*b*) Calculate the effective annual interest rate, that is, the APY.

4.13 Citizens Bank of Toronto advertises an APR of 12% compounded monthly for collateral loans. What is the APY? Also, write the spreadsheet function to display the APY.

4.14 A company that makes modular bevel gear drives with a tight swing ratio for optimizing fork-lift vehicles was told that the interest rate on a loan would be an effective 3.5% per quarter, compounded monthly. The owner, confused by the terminology, asked you to help. What is the (*a*) APR, and (*b*) APY? (*c*) Explain how the EFFECT and NOMINAL functions can or cannot be used to find the APR and APY.

4.15 The Premier Car Title Loan Company makes emergency loans of up to $500 for 1 month for a fee of 4% of the loan amount. If a person borrows $500, what is (*a*) the nominal interest rate per year, and (*b*) the effective rate per year?

4.16 A credit card issued by the GECU credit union has an APR of 16% and an APY of 16.64%. (*a*) What is the compounding period? (*b*) Use the EFFECT function to find the compounding period.

4.17 Eckelberger Products, Inc. makes high-speed recorders with high-speed scanning. The small company has been growing at an average rate of 75% per year for the past 4 years. The CEO asked you to convert the past growth rate into a monthly rate for its annual report. If the past growth rate was an effective rate, what was the effective growth rate per month?

Equivalence When PP ≥ CP

4.18 In order to find the future worth, F, from a present amount, P, 5 years from now at an interest rate of 12% per year, compounded quarterly, what interest rate must be used in the F/P factor, $(F/P,i\%,n)$, when n is (*a*) 5, (*b*) 10, and (*c*) 20?

4.19 Jennifer and Rex both receive a dividend from their 401(k) retirement plan every 6 months. The earning rates for this year have been 5% per year, compounded quarterly for Jennifer, and 4.85% per year, compounded monthly for Rex. Rex felt good about this because he knew the monthly compounding on his plan would make his APY higher than Jennifer's APY.
(*a*) Is Rex correct? Explain your answer.
(*b*) What is the effective rate for each plan on the basis of the payment period?

4.20 Videotech specializes in online security software development. It wants to have $65 million available in 3 years to pay stock dividends. How much money must the company set aside *now* in an account that earns interest at a rate of 12% per year, compounded quarterly? Solve using (*a*) tabulated factors, and (*b*) a single-cell spreadsheet function.

4.21 If the VF Corporation deposits $20 million of its retained earnings in an aggressive stock ownership fund for only 1 year, what will be the difference in the total amount accumulated (i.e., F) at 18% per year, compounded monthly versus 18% per year *simple* interest?

4.22 A friend tells you she has saved for 7 years and has a present sum of $10,000, which earned at the rate of 8% per year, compounded quarterly. (*a*) Determine the equivalent amount she had to start with

7 years ago. (*b*) Write the EFFECT function to find the effective rate per year.

4.23   The TerraMax truck currently being manufactured by Oshkosh Truck Co. is a driverless truck that is intended for military use. Such a truck frees personnel to more effectively perform nondriving tasks, such as reading maps, scanning for roadside explosives, or scouting for the enemy. If such trucks can result in reduced injuries to military personnel amounting to $15 million 3 years from now, determine the present worth of these benefits at an interest rate 10% per year, compounded semiannually.

4.24   Because testing of nuclear bombs was halted internationally in 1992, the Department of Energy has developed a laser system that allows engineers to simulate (in a laboratory) conditions in a thermonuclear reaction. Due to soaring cost overruns, a congressional committee undertook an investigation and discovered that the estimated development cost of the project increased at an average rate of 2% per month over a 5-year period. If the original cost was estimated to be $2.3 billion 5 years ago, what is the expected cost today?

4.25   The patriot missile, developed by Lockheed Martin, is designed to detect, identify, and shoot down aircraft and other missiles. The Patriot Advanced Capability-3 was originally promised to cost $3.9 billion, but due to extra time needed to develop computer code and scrapped tests (due to high winds) at White Sands Missile Range, the actual cost was much higher. If the total project development time was 10 years and costs increased at a rate of 0.5% per month, what was the final cost of the project?

4.26   Pollution control equipment for a pulverized coal cyclone furnace is estimated to cost $190,000 two years from now and an additional $120,000 four years from now. If Monongahela Power wants to set aside enough money now to cover these future costs, how much must be invested at an interest rate of 8% per year, compounded semiannually?

4.27   In an effort to have enough money to finance his passion for top-level off-road racing, a mechanical engineering graduate decided to sell some of his "toys." At time $t = 0$, he sold his chopper motorcycle for $18,000. Six months later, he sold his Baja pre-runner SUV for $26,000. At the end of year 1, he got $42,000 for his pro-light race truck. If he invested all of the money in a high-risk commodity hedge fund that earned 24% per year, compounded semiannually, how much did he have in the account at the end of 2 years?

4.28   Head & Shoulders shampoo insured a spokesman football player's long hair for $1 million with Lloyd's of London. The insurance payout would be triggered if he lost at least 60% of his hair during an on-field event. The insurance company placed the odds of a payout at 1% in year 5. Determine how much Head & Shoulders had to pay in a lump-sum amount for the insurance policy if Lloyd's of London wanted a rate of return of 20% per year, compounded quarterly.

4.29   Radio Frequency Identification (RFID) is a technology used by drivers with "speed passes" at toll booths and ranchers who track livestock from "farm to fork." Thrift-Mart is implementing the technology to track products within its stores. The RFID-tagged products will result in better inventory control and save the company $1.3 million per month beginning 3 months from now. How much can the company afford to spend now to implement the technology at an interest rate of 12% per year, compounded monthly, if it wants to recover its investment in 2½ years? Also, write a spreadsheet function to display the answer.

4.30   The optical products division of Panasonic is planning a $3.5 million building expansion for manufacturing its powerful Lumix DMC digital zoom camera. If the company uses an interest rate of 16% per year, compounded quarterly for all new investments, what is the uniform amount per quarter the company must make in order to recover its investment in 3 years?

4.31   Hemisphere, LLC is planning to outsource its 51-person information technology (IT) department to Dyonyx. Hemisphere's president believes this move will allow access to cutting-edge technologies and skill sets that would be cost prohibitive to obtain on its own. If it is assumed that the loaded cost of an IT employee is $100,000 per year and that Hemisphere will save 25% of this cost through outsourcing, determine the present worth of the total savings for a 5-year contract at an interest rate of 6% per year, compounded monthly.

4.32   Environmental recovery company RexChem Partners plans to finance a site reclamation project that will require a 4-year cleanup period. If the company borrows $4.1 million now, how much will the company have to get at the end of each year in order to earn 15% per year, compounded quarterly on its investment? Additionally, what is the spreadsheet function?

4.33   Northwest Iron and Steel is considering getting involved in electronic commerce. A modest e-commerce package is available for $30,000. The company wants to recover the cost in 2 years. Find the equivalent amount of new revenue that must be realized every

6 months at an interest rate of 3% per quarter using (*a*) factor formula, and (*b*) a single-cell spreadsheet function that includes an EFFECT function.

4.34 Metropolitan Water Utilities purchases surface water from Elephant Butte Irrigation District at a cost of $100,000 per month in the months of February through September. Instead of paying monthly, the utility makes a single payment of $800,000 at the end of each calendar year for the water it used. The delayed payment essentially represents a subsidy by Elephant Butte Irrigation District to the water utility. At an interest rate of 0.25% per month, what is the amount of the subsidy?

4.35 For the cash flows shown, determine the equivalent uniform annual worth in years 1 through 5 at an interest rate of 18% per year, compounded monthly.

Year	1	2	3	4	5
Cash Flow, $	0	0	350,000	350,000	350,000

4.36 Lotus Development has a software rental plan called SmartSuite that is available on the World Wide Web. A number of programs are available at $2.99 for 48 hours. If a construction company uses the service every week for an average of 48 hours, what is the future worth of the rental costs for 1 year at an interest rate of 1% per month, compounded weekly? (Assume 4 weeks per month.)

4.37 Assume you are considering the consolidation of all your electronic services with one company. By purchasing a digital phone from AT&T wireless, you can buy wireless e-mail and fax services for $6.99 per month. For $14.99 per month, you will get unlimited web access and personal organization functions. For a 2-year contract period, what is the future worth of the *difference* between the services at an interest rate of 12% per year, compounded monthly?

4.38 Thermal Systems, a company that specializes in odor control for wastewater treatment plants, made deposits of $100,000 now and $25,000 every 6 months for 2 years. Determine the future worth after 2 years of the deposits for $i = 16\%$ per year, compounded quarterly.

4.39 McMillan Company manufactures electronic flow sensors that are designed as an alternative to ball-and-tube rotometers. The company recently spent $3 million to increase the capacity of an existing production line. If the extra revenue generated by the expansion amounts to $200,000 per month, how long will it take for the company to recover its investment at an interest rate of 12% per year, compounded monthly? Solve using (*a*) tabulated factors and (*b*) a spreadsheet or calculator.

4.40 Metalfab Pump and Filter, Inc. estimates that the cost of steel bodies for pressure valves will increase by $2 every 3 months. If the cost for the first quarter is expected to be $80, what is the present worth of the costs for a 3-year time period at an interest rate of 3% per quarter?

4.41 Fieldsaver Technologies, a manufacturer of precision laboratory equipment, borrowed $2 million to renovate one of its testing labs. The loan was repaid in 2 years through quarterly payments that increased by $50,000 each time. At an interest rate of 12% per year, compounded quarterly, what was the size of the first quarterly payment?

4.42 Frontier Airlines hedged the cost of jet fuel by purchasing options that allowed the airline to buy fuel at a fixed price for 2 years. The savings in fuel costs were $140,000 in month 1, $141,400 in month 2, and amounts increasing by 1% per month through the 2-year option period. What was the present worth of the savings at an interest rate of 18% per year, compounded monthly?

4.43 Equipment maintenance costs for manufacturing explosion-proof pressure switches are projected to be $125,000 in year 1 and increase by 3% each year through year 5. What is the equivalent annual worth of the maintenance costs at an interest rate of 10% per year, compounded semiannually? Solve using (*a*) the factor formula and (*b*) a spreadsheet.

4.44 The Fairfold family decided to buy a super ski and water sports boat. They took out an $80,000, 5-year, 6% per year, compounded semiannually loan with monthly payments from First Bank and Trust (FB&T). After making only two payments, a banker friend offered to make them a better deal: a 5-year, 4.2% per year, compounded semiannually loan with no transfer fee to his bank and a complete repayment no-fee-required of the remaining principal to FB&T. The principal on the new loan will be the remaining principal from the current loan. Answer the following questions for the Fairfolds as they deliberate this new offer.
   (*a*) What is the current monthly payment on the $80,000 loan?
   (*b*) What is the current principal due on the current loan?
   (*c*) How much interest have they already paid in the first two payments?
   (*d*) What is the amount of the new monthly payment starting with month 3, if the new loan offer is accepted?

(Note: Problem 4.65 requests a spreadsheet solution for this problem, if assigned by your instructor.)

Equivalence When PP < CP

4.45 In an effort to save money for early retirement, an environmental engineering colleague plans to deposit $1200 per month, starting 1 month from now, into a fixed rate account that pays 8% per year, compounded semiannually. How much will be in the account at the end of 25 years?

4.46 Beginning 1 month from now, Thaxton Mechanical Products will set aside $1500 per month for possibly replacing its corrosion-resistant diaphragm pumps whenever it becomes necessary. The replacement isn't needed for 3 years. How much will the company have available if the hoped-for rate of return of 18% per year is achieved?

4.47 Income from recycling the paper and cardboard generated in an office building has averaged $3000 per month for the past 3 years. What is the income stream's equivalent worth now at an interest rate of 8% per year, compounded quarterly?

4.48 How much would your parents have to deposit *each month* into an account that grows at a rate of 12% per year, compounded semiannually, if they want to have $80,000 at the end of year 3 to cover part of your college expenses? Assume no interperiod compounding. Also, write the spreadsheet function to display the monthly amount.

4.49 Western Refining purchased a model MTVS peristaltic pump for injecting antiscalant at its nanofiltration water conditioning plant. The cost of the pump was $1200. If the chemical cost is $11 per day, determine the equivalent cost per month (pump plus chemicals) at an interest rate of 1% per month. Assume 30 days per month and a 4-year pump life.

4.50 An engineer deposited her annual bonus of $10,000 into an account that pays interest at 8% per year, compounded semiannually. If she withdrew $1000 in months 2, 11, and 23, what was the total value of the account at the end of 3 years? Assume no interperiod compounding. (Note: See Problem 4.65 for additional analysis of this series.)

Continuous Compounding

4.51 What effective interest rate per year, compounded continuously, is equivalent to a nominal rate of 10% per year? Solve by formula and spreadsheet function.

4.52 What effective interest rate per 6 months is equal to a nominal 2% per month, compounded continuously?

4.53 What nominal rate per quarter is equivalent to an effective rate of 12.7% per year, compounded continuously?

4.54 Corrosion problems and manufacturing defects rendered a gasoline pipeline between El Paso, TX, and Phoenix, AZ, subject to longitudinal weld seam failures. The pressure was reduced to 80% of the design value. If the reduced pressure resulted in delivery of less product valued at $100,000 per month, what will be the value of the lost revenue after a 2-year period at an interest rate of 15% per year, compounded continuously?

4.55 Because of a chronic water shortage in California, new athletic fields must use artificial turf or xeriscape landscaping. If the value of the water saved each month is $6000, how much can a private developer afford to spend now on artificial turf provided he must recover his investment in 5 years. Use an interest rate of 18% per year, compounded continuously? Write the spreadsheet function to display the answer.

4.56 A Texas chemical company had to file for bankruptcy because of a nationwide phaseout of methyl tertiary butyl ether (MTBE). The company is in the planning phase of reorganizing and expects to invest $50 million in a new ethanol production facility. Determine the revenue necessary *each month* to recover its investment in 3 years, plus interest at a rate of 2% per month, compounded continuously. Solve by (*a*) formula, and (*b*) spreadsheet function.

4.57 How long will it take for a lump-sum investment to double in value at an interest rate of 1.5% per month, compounded continuously? Solve by (*a*) formula and (*b*) spreadsheet function.

4.58 A very optimistic hedge fund investor expects his single-deposit investment to triple in value in 5 years. (*a*) What is the required effective monthly rate, compounded continuously? (*b*) What is the corresponding effective annual rate needed?

Varying Interest Rates

4.59 How much money can a production company that makes fluidized bed scrubbers spend now instead of spending $150,000 in year 5 if the interest rates are estimated to be 10% per year in years 1 to 3 and 12% per year in years 4 and 5?

4.60 Brady and Sons uses accounts receivable as collateral to borrow money for operations and payroll when revenues are low. If the company borrows $300,000 now at an interest rate of 12% per year, compounded monthly, and the rate increases to

15% per year, compounded monthly after 4 months, how much will the company owe at the end of 1 year?

4.61 Find the F, P, and A values for the negative cash flows shown in the diagram.

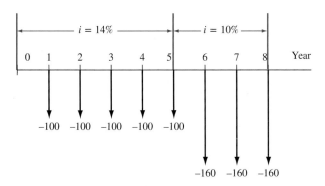

4.62 For the cash flows shown, determine the future worth in year 5.

Year	Cash Flow, $/Year	Estimated *i* Per Year
0	5000	12%
1–4	6000	12%
5	9000	20%

4.63 The Owner of Blue Bayou Café usually pays his appliance (refrigerators, dish washers, and freezers) maintenance contract by the year. If he projects the annual costs shown, find the equivalent A value for years 1 through 5.

Year	Cash Flow, $/Year	Estimated *i* Per Year
0	0	
1–3	5000	10%
4–5	7000	12%

EXERCISES FOR SPREADSHEETS

4.64 Solve Problem 4.19 by applying the EFFECT function to determine the rates.

4.65 Solve Problem 4.44 using a spreadsheet, which analyzes the loan on a boat purchased by the Fairfold family. Display the answers for parts (*a*) through (*d*) below the spreadsheet and indicate the functions that you used to obtain the answers.

4.66 An engineer deposited her annual bonus of $10,000 into an account that pays interest at 8% per year, compounded semiannually. She withdrew $1000 in months 2, 11, and 23. Now, she wants to know the total value of the account at the end of 3 years. Solve by assuming (*a*) no interperiod compounding and (*b*) that interperiod compounding is provided.

4.67 Some years ago, Penny purchased the car of her dreams for $25,000 by paying 20% down at purchase time and taking a $20,000, 5-year, 6% per year, compounded monthly loan with 60 monthly payments of $386.66 each. She is examining her loan situation and would like to have some specific information. Help her obtain the following:

(*a*) Verification of the current monthly payment amount.

(*b*) Total amount she will pay over the 5 years.

(*c*) Total interest she will pay over the 5 years and the percentage this represents of the original loan amount of $20,000.

(*d*) After she missed payment #36 at the very end of the third year, according to the loan agreement, the interest rate increased from 6% to 10% per year, compounded monthly. Based on the remaining principal immediately after the late payment, determine the new monthly payment. Verify that this increased amount is necessary to pay off the loan at the increased rate.

(*e*) Penny is now in her fourth year, has paid the increased payment for 12 payments, and wants to get rid of this loan completely. She wishes to know the remaining principal amount when payment #48 is due. There is no penalty for early repayment of principal.

ADDITIONAL PROBLEMS AND FE EXAM REVIEW QUESTIONS

4.68 An interest rate of 12% per year, compounded monthly, is nearest to:
(*a*) 12.08% per year
(*b*) 12.28% per year
(*c*) 12.48% per year
(*d*) 12.68% per year

4.69 An interest rate of 2% per month is the same as:
(*a*) 24% per year, compounded monthly
(*b*) a nominal 24% per year, compounded monthly
(*c*) an effective 24% per year, compounded monthly
(*d*) Both (*a*) and (*b*)

4.70 An interest rate of 18% per year, compounded continuously, is closest to an effective:
(a) 1.51% per quarter
(b) 4.5% per quarter
(c) 4.6% per quarter
(d) 9% per 6 months

4.71 The only time you change the original cash flow diagram in problems involving uniform series cash flows is when the:
(a) payment period is longer than the compounding period
(b) payment period is equal to the compounding period
(c) payment period is shorter than the compounding period
(d) stated interest rate is a nominal interest rate

4.72 If you make *quarterly deposits* for 3 years (beginning one quarter from now) into an account that compounds interest at 1% per month, the value of n in the F/A factor (for determining F at the end of the 3-year period) is:
(a) 3   (b) 4   (c) 12   (d) 16

4.73 Assume you make *monthly* deposits of $200 starting 1 month from now into an account that pays 6% per year, *compounded semiannually*. If you want to know how much you will have after 4 years, the value of i you should use in the F/A factor, assuming no interperiod interest, is:
(a) 0.5%   (b) 3.00%   (c) 6.0%   (d) 12.0%

4.74 An interest rate of 2% per quarter, compounded continuously, is closest to an *effective* semiannual rate of:
(a) 2.00% per semiannual period
(b) 2.02% per semiannual period
(c) 4.0% per semiannual period
(d) 4.08% per semiannual period

4.75 The present worth of a deposit of $1000 *now* and $1000 every 6 months for 10 years at an interest rate of 10% per year, compounded semiannually is represented by which of the following equations:
(a) $P = 1000(P/A,5\%,21)(F/P,5\%,1)$
(b) $P = 1000 (P/A,5\%,20)$
(c) $P = 1000 (P/A,5\%,21)$
(d) $P = 1000 + 1000(P/A,10.25\%,10)$

4.76 The cost of replacing a high-definition television production line in 6 years is estimated to be $500,000. At an interest rate of 14% per year, compounded semiannually, the uniform amount that must be deposited into a sinking fund every 6 months *beginning now* is closest to:
(a) <$21,000
(b) $21,335
(c) $24,820
(d) $27,950

4.77 You are planning to make two equal amount deposits, one now and the other 3 years from now in order to accumulate $300,000 ten years in the future. If the interest rate is 14% per year, compounded semiannually, the size of each deposit is:
(a) <$46,300   (b) $46,525
(c) $47,835   (d) >$48,200

4.78 For the cash flow diagram shown, the future worth in year 4 is closest to:
(a) <$2800   (b) $2915
(c) $3735   (d) $5219

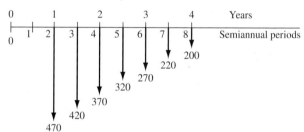

4.79 A small company plans to spend $10,000 in year 2 and $10,000 in year 5. At an interest rate of *effective* 10% per year, compounded semiannually, the equation that represents the equivalent annual worth in years 1 through 5 is:
(a) $A = 10,000(P/F10\%,2)(A/P,10\%,5) + 10,000(A/F,10\%,5)$
(b) $A = 10,000(A/P,10\%,4) + 10,000 (A/F,10\%,5)$
(c) $A = 10,000(P/F,5\%,2)(A/P,5\%,10) + 10,000 (A/F,5\%,10)$
(d) $A = [10,000(F/P,10\%,5) + 10,000] (A/F,10\%,5)$

4.80 If you deposit $P into a savings account that earns interest at a rate of $i\%$ per month for n years, the future worth in year n is represented by all of the following equations, except:
(a) $F = \$P(F/P, \text{effective } i/\text{month}, 12n)$
(b) $F = \$P(F/P, \text{effective } i/\text{quarter}, 3n)$
(c) $F = \$P(F/P, \text{effective } i/6\text{-month}, 2n)$
(d) $F = \$P(F/P, \text{effective } i/\text{year}, n)$

4.81 A midcareer engineer hopes to have $2 million available for his use when he retires 20 years from now. He plans to deposit a uniform amount semi-annually, beginning now and every 6 months thereafter through the end of year 20. If his investment account has a yearly return of 8% per year, compounded quarterly, the interest rate, i, that must be used in the A/F equation to determine the size of the uniform deposits is:

(a) 2%   (b) 8.24%   (c) 8.00%   (d) 4.04%

4.82   If you wish to accumulate $10,000 over a 5-year period by placing $200 a month, starting next

month, into a Roth IRA retirement fund that pays 6% per year, compounded quarterly with no inter-period compounding, the NPER function to determine the number of deposits is:

(a) = NPER(4.568%,−200,10000)

(b) = NPER(1.5%,200,10000)

(c) = NPER(1.5%,−600,10000)

(d) = NPER(2%,600,−10000)

CASE STUDY

IS OWNING A HOME A NET GAIN OR NET LOSS OVER TIME?

Background

The Carroltons are deliberating whether to purchase a house or continue to rent for the next 10 years. They are assured by both of their employers that no transfers to new locations will occur for at least this number of years. Plus, the high school that their children attend is very good for their college prep education, and they all like the neighborhood where they live now.

They have a total of $40,000 available now and estimate that they can afford up to $2850 per month for the total house payment.

If the Carroltons do not buy a house, they will continue to rent the house they currently occupy for $2700 per month. They will also place the $40,000 into an investment instrument that is expected to earn at the rate of 6% per year. Additionally, they will add to this investment *at the end of each year* the same amount as the monthly 15-year mortgage payments. This alternative is called the rent–don't buy plan.

Information

Two financing plans using fixed-rate mortgages are currently available. The details are as follows.

Plan	Description
A	30-year fixed rate of 5.25% per year interest; 10% down payment
B	15-year fixed rate of 5.0% per year interest; 10% down payment

Other information:

- Price of the house is $330,000.
- Taxes and insurance (T&I) are $500 per month.
- Up-front fees (origination fee, survey fee, attorney's fee, etc.) are $3000.

Any money not spent on the down payment or monthly payment will be invested and return at a rate of 6% per year (0.5% per month).

The Carroltons anticipate selling the house after 10 years and plan for a 10% increase in price, that is, $363,000 (after all selling expenses are paid).

Case Study Exercises

1. The 30-year fixed-rate mortgage (plan A) is analyzed below. No taxes are considered on proceeds from the savings or investments.

   Perform a similar analysis for the 15-year loan (plan B) and the rent–don't buy plan. The Carroltons decided to use the *largest future worth after 10 years* to select the best of the plans. Do the analysis for them and select the best plan.

Plan A analysis: 30-year fixed-rate loan

Amount of money required for closing costs:	
Down payment (10% of $330,000)	$33,000
Up-front fees (origination fee, attorney's fee, survey, filing fee, etc.)	3,000
Total	$36,000

The amount of the loan is $297,000, and equivalent monthly principal and interest (P&I) is determined at $5.25\%/12 = 0.4375\%$ per month for $30(12) = 360$ months.

$$A = 297{,}000(A/P,0.4375\%,360) = 297{,}000(0.005522)$$
$$= \$1640$$

Add the T&I of $500 for a total monthly payment of

$$\text{Payment}_A = \$2140 \text{ per month}$$

The future worth of plan A is the sum of three future worth components: remainder of the $40,000 available for the closing costs (F_{1A}); left-over money from that available for monthly payments (F_{2A}); and increase in the house value when it is sold after 10 years (F_{3A}). These are calculated here.

$$F_{1A} = (40{,}000 - 36{,}000)(F/P,0.5\%,120)$$
$$= \$7278$$

Money available each month to invest after the mortgage payment, and the future worth after 10 years, is

$$2850 - 2140 = \$710$$

$$F_{2A} = 710(F/A,0.5\%,120)$$

$$= \$116,354$$

Net money from the sale in 10 years (F_{3A}) is the difference between the net selling price ($363,000) and the remaining balance on the loan.

Loan balance

$$= 297,000(F/P,0.4375\%,120)$$
$$- 1640(F/A,0.4375\%,120)$$
$$= 297,000(1.6885) - 1640(157.3770)$$
$$= \$243,386$$

$$F_{3A} = 363,000 - 243,386 = \$119,614$$

Total future worth of plan A is

$$F_A = F_{1A} + F_{2A} + F_{3A}$$
$$= 7278 + 116,354 + 119,614$$
$$= \$243,246$$

2. Perform this analysis if all estimates remain the same, except that when the house sells 10 years after purchase, the bottom has fallen out of the housing market and the net selling price is only 70% of the purchase price, that is, $231,000.

LEARNING STAGE 2

Basic Analysis Tools

An engineering project or alternative is formulated to make or purchase a **product,** to develop a **process,** or to provide a **service** with specified results. An engineering economic analysis evaluates cash flow estimates for parameters such as initial cost, annual costs and revenues, nonrecurring costs, and possible salvage value over an estimated useful life of the product, process, or service. The chapters in this learning stage develop and demonstrate the basic tools and techniques to evaluate one or more alternatives using the factors, formulas, and spreadsheet functions learned in stage 1.

After completing these chapters, you will be able to evaluate most engineering project proposals using a well-accepted economic analysis technique, such as present worth, future worth, capitalized cost, life-cycle costing, annual worth, rate of return, or benefit/cost analysis.

The epilogue to this stage provides an approach useful in selecting the engineering economic method that will provide the best analysis for the estimates and conditions present once the mutually exclusive alternatives are defined.

Important note: If depreciation and/or after-tax analysis is to be considered along with the evaluation methods in Chapters 5 through 9, Chapter 16 and/or Chapter 17 should be covered, preferably after Chapter 6.

CHAPTER 5

Present Worth Analysis

Onoky/SuperStock

LEARNING OUTCOMES

Purpose: Utilize different present worth techniques to evaluate and select alternatives.

SECTION	TOPIC	LEARNING OUTCOME
5.1	Formulating alternatives	• Identify mutually exclusive and independent projects; define revenue and cost alternatives.
5.2	PW of equal-life alternatives	• Select the best of equal-life alternatives using present worth analysis.
5.3	PW of different-life alternatives	• Select the best of different-life alternatives using present worth analysis.
5.4	FW analysis	• Select the best alternative using future worth analysis.
5.5	CC analysis	• Select the best alternative using capitalized cost (CC) analysis.

A future amount of money converted to its equivalent value now has a present worth (PW) that is always less than that of the future cash flow because all *P/F* factors have a value less than 1.0 for any interest rate greater than zero. For this reason, present worth values are often referred to as *discounted cash flows* (DCF), and the interest rate is referred to as the *discount rate*. Besides PW, two other terms frequently used are *present value* (PV) and *net present value* (NPV). Up to this point, PW computations have been made for one project or alternative. In this chapter, techniques for comparing two or more mutually exclusive alternatives by the present worth method are treated. Two additional applications are covered here—**future worth and capitalized cost.** Capitalized costs are used for projects with very long expected lives or long planning horizons.

To understand how to organize an economic analysis, this chapter begins with a description of independent and mutually exclusive projects as well as revenue and cost alternatives.

PE

Water for Semiconductor Manufacturing Case: The worldwide contribution of semiconductor sales is about $250 billion per year, or about 10% of the world's GDP (gross domestic product). This industry produces the microchips used in many of the communication, entertainment, transportation, and computing devices we use every day. Depending upon the type and size of fabrication plant (fab), the need for ultrapure water (UPW) to manufacture these tiny integrated circuits is high, ranging from 500 to 2000 gpm (gallons per minute). Ultrapure water is obtained by special processes that commonly include reverse osmosis/deionizing resin bed technologies. Potable water obtained from purifying seawater or brackish groundwater may cost from $2 to $3 per 1000 gallons, but to obtain UPW on-site for semiconductor manufacturing may cost an additional $1 to $3 per 1000 gallons.

A fab costs upward of $2.5 billion to construct, with approximately 1% of this total, or $25 million, required to provide the UPW needed, including the necessary wastewater and recycling equipment.

A newcomer to the industry, Angular Enterprises, has estimated the cost profiles for two options to supply its anticipated fab with water. It is fortunate to have the option of desalinated seawater or purified groundwater sources in the location chosen for its new fab. The initial cost estimates for the UPW system are given below.

Source	Seawater (S)	Groundwater (G)
Equipment first cost, $M	−20	−22
AOC, $M per year	−0.5	−0.3
Salvage value, % of first cost	5	10
Cost of UPW, $ per 1000 gallons	4	5

Angular has made some initial estimates for the UPW system.

Life of UPW equipment	10 years
UPW needs	1500 gpm
Operating time	16 hours per day for 250 days per year

This case is used in the following topics (Sections) and problems of this chapter:

PW analysis of equal-life alternatives (Section 5.2)

PW analysis of different-life alternatives (Section 5.3)

Capitalized cost analysis (Section 5.5)

Problems 5.26 and 5.41.

5.1 Formulating Alternatives ●●○

The evaluation and selection of economic proposals require **cash flow estimates** over a stated period of time, mathematical techniques to calculate the **measure of worth** (review Example 1.2 for possible measures), and a **guideline for selecting the best proposal.** From all the proposals that may accomplish a stated purpose, the alternatives are formulated. This progression is detailed in Figure 5–1. Up-front, some proposals are viable from technological, economic, and/or legal perspectives; others are not viable. Once the obviously nonviable

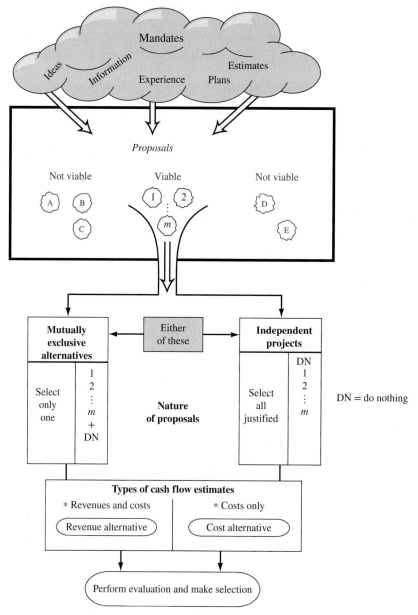

Figure 5–1
Progression from proposals to economic evaluation to selection.

ideas are eliminated, the remaining viable proposals are fleshed out to form the alternatives to be evaluated. Economic evaluation is one of the primary means used to select the best alternative(s) for implementation.

The nature of the economic proposals is always one of two types:

Mutually exclusive alternatives: Only one of the proposals can be selected. For terminology purposes, each viable proposal is called an *alternative*.

Independent projects: More than one proposal can be selected. Each viable proposal is called a *project*.

The parameters that usually must be estimated for an alternative or project were mentioned in Section 1.2 in the description of alternatives. In brief, they included first cost (commonly identified by the letter P), expected life (n), effective interest rate (i), salvage or trade-in value (SV), any anticipated major rework or upgrade costs, and annual operating costs (AOC), which may also be termed maintenance and operating (M&O) costs.

The **do-nothing (DN)** proposal is usually understood to be an option when the evaluation is performed.

> The DN alternative or project means that the **current approach is maintained;** nothing new is initiated. No new costs, revenues, or savings are generated.

Do nothing

If it is absolutely required that one or more of the defined alternatives be selected, DN is not considered. This may occur when a mandated function must be installed for safety, legal, government, or other purposes.

Mutually exclusive (ME) alternatives and independent projects are selected in completely different ways. A mutually exclusive selection takes place, for example, when an engineer must select *the* best diesel-powered engine from several available models. Only one is chosen, and the rest are rejected. If none of the alternatives are economically justified, then all can be rejected and, by default, the DN alternative is selected. For independent projects one, two, or more, in fact, all of the projects that are economically justified can be accepted, provided capital funds are available. This leads to the two following fundamentally different evaluation bases:

> Mutually exclusive alternatives **compete with one another** and are compared pairwise.
>
> Independent projects are evaluated one at a time and **compete only with the DN project.**

Any of the techniques in Chapters 5 through 9 can be used to evaluate either type of proposal—mutually exclusive or independent. When performed correctly as described in each chapter, any of the techniques will reach the same conclusion of which alternative or alternatives to select. This chapter covers the present worth method.

It is easy to develop a relationship between independent projects and mutually exclusive alternatives. Assume there are m independent projects. Zero, one, two, or more may be selected. Since each project may be in or out of the selected group of projects, there are a total of 2^m **mutually exclusive alternatives.** This number includes the DN alternative, as shown in Figure 5–1. For example, if the engineer has three diesel engine models (A, B, and C) and may select any number of them, there are $2^3 = 8$ alternatives: DN, A, B, C, AB, AC, BC, ABC. Commonly, in real-world applications, there are restrictions, such as an upper budgetary limit, that eliminate many of the 2^m alternatives. Independent project analysis without budget limits is discussed in this chapter and through Chapter 9. Chapter 12 treats independent projects with a budget limitation; this is called capital budgeting.

Finally, it is important to recognize the sense and nature of the cash flow estimates before starting the computation of a measure of worth that leads to the final selection. A positive or negative sign must be associated with every monetary estimate to identify it as an outflow (e.g., first cost or AOC), or cash inflow, such as the salvage value or a savings generated by the alternative. Cash flow estimates also determine whether the alternatives are revenue- or cost-based. All the alternatives or projects must be of the same type when the economic study is performed. Definitions for these types follow:

> **Revenue:** Each alternative generates cost (cash outflow) and revenue (cash inflow) estimates, and possibly savings, also considered cash inflows. Revenues can vary for each alternative.
>
> **Cost:** Each alternative has only cost cash flow estimates. Revenues or savings are assumed equal for all alternatives; thus, alternative selection is not dependent upon their estimation. These may also be referred to as **service alternatives.**

Revenue or cost alternative

Although the exact procedures vary slightly for revenue and cost cash flows, all techniques and guidelines covered through Chapter 9 apply to both. Differences in evaluation methodology are detailed in each chapter.

5.2 Present Worth Analysis of Equal-Life Alternatives ● ● ●

The PW comparison of alternatives with equal lives is straightforward. The present worth P is renamed PW of the alternative. The present worth method is quite popular in industry because all future costs and revenues are transformed to **equivalent monetary units NOW;** that is, all future cash flows are converted (discounted) to present amounts (e.g., dollars) at a specific rate of return, which is the MARR. This makes it very simple to determine which alternative has the best economic advantage. The required conditions and evaluation procedure are as follows:

Equal-service requirement

Project evaluation

ME alternative selection

Independent project selection

If the alternatives have the same capacities for the same time period (life), the **equal-service requirement** is met. Calculate the PW value at the stated MARR for each alternative.

For **mutually exclusive (ME)** alternatives, whether they are revenue or cost alternatives, the following guidelines are applied to justify a single project or to select one from several alternatives.

One alternative: If PW ≥ 0, the requested MARR is met or exceeded and the alternative is economically justified.

Two or more alternatives: Select the alternative with the PW that is **numerically largest,** that is, less negative or more positive. This indicates a lower PW of cost for cost alternatives or a larger PW of net cash flows for revenue alternatives.

Note that the guideline to select one alternative with the lowest cost or highest revenue uses the criterion of **numerically largest.** This is not the absolute value of the PW amount because the sign matters. The selections below correctly apply the guideline for two alternatives A and B.

PW_A	PW_B	Selected Alternative
$-2300	$-1500	B
-500	+1000	B
+2500	+2000	A
+4800	-400	A

For **independent** projects, each PW is considered separately, that is, compared with the DN project, which always has PW = 0. The selection guideline is as follows:

One or more independent projects: Select all projects with PW ≥ 0 at the MARR.

The independent projects must have positive and negative cash flows to obtain a PW value that can exceed zero; that is, they must be revenue projects.

All PW analyses require a MARR for use as the *i* value in the PW relations. The bases used to establish a realistic MARR were summarized in Chapter 1 and are discussed in detail in Chapter 10.

EXAMPLE 5.1

A university lab is a research contractor to NASA for in-space fuel cell systems that are hydrogen- and methanol-based. During lab research, three equal-service machines need to be evaluated economically. Perform the present worth analysis with the costs shown below. The MARR is 10% per year.

	Electric-Powered	Gas-Powered	Solar-Powered
First cost, $	-4500	-3500	-6000
Annual operating cost (AOC), $/year	-900	-700	-50
Salvage value S, $	200	350	100
Life, years	8	8	8

Solution

These are cost alternatives. The salvage values are considered a "negative" cost, so a + sign precedes them. (If it costs money to dispose of an asset, the estimated disposal cost has a − sign.) The PW of each machine is calculated at *i* = 10% for *n* = 8 years. Use subscripts E, G, and S.

$$PW_E = -4500 - 900(P/A,10\%,8) + 200(P/F,10\%,8) = \$-9208$$
$$PW_G = -3500 - 700(P/A,10\%,8) + 350(P/F,10\%,8) = \$-7071$$
$$PW_S = -6000 - 50(P/A,10\%,8) + 100(P/F,10\%,8) = \$-6220$$

The solar-powered machine is selected since the PW of its costs is the lowest; it has the numerically largest PW value.

EXAMPLE 5.2 Water for Semiconductor Manufacturing Case

As discussed in the introduction to this chapter, ultrapure water (UPW) is an expensive commodity for the semiconductor industry. With the options of seawater or groundwater sources, it is a good idea to determine if one system is more economical than the other. Use a MARR of 12% per year and the present worth method to select one of the systems.

Solution

An important first calculation is the cost of UPW per year. The general relation and estimated costs for the two options are as follows:

$$\textbf{UPW cost relation: } \frac{\$}{\text{year}} = \left(\frac{\text{cost in } \$}{1000 \text{ gallons}}\right)\left(\frac{\text{gallons}}{\text{minute}}\right)\left(\frac{\text{minutes}}{\text{hour}}\right)\left(\frac{\text{hours}}{\text{day}}\right)\left(\frac{\text{days}}{\text{year}}\right)$$

Seawater: $(4/1000)(1500)(60)(16)(250) = \$1.44$ M per year

Groundwater: $(5/1000)(1500)(60)(16)(250) = \$1.80$ M per year

Calculate the PW at $i = 12\%$ per year and select the option with the lower cost (larger PW value). In $1 million units:

PW relation: PW = first cost − PW of AOC − PW of UPW + PW of salvage value

$$\text{PW}_S = -20 - 0.5(P/A,12\%,10) - 1.44(P/A,12\%,10) + 0.05(20)(P/F,12\%,10)$$
$$= -20 - 0.5(5.6502) - 1.44(5.6502) + 1(0.3220)$$
$$= \$-30.64$$

$$\text{PW}_G = -22 - 0.3(P/A,12\%,10) - 1.80(P/A,12\%,10) + 0.10(22)(P/F,12\%,10)$$
$$= -22 - 0.3(5.6502) - 1.80(5.6502) + 2.2(0.3220)$$
$$= \$-33.16$$

Based on this PW analysis, the seawater option is cheaper by $2.52 M.

5.3 Present Worth Analysis of Different-Life Alternatives ●●●

When the present worth method is used to compare mutually exclusive alternatives that have different lives, the equal-service requirement must be met. The procedure of Section 5.2 is followed, with one exception:

The PW of the alternatives must be compared over the **same number of years** (or periods) and must end at the same time to satisfy the equal-service requirement.

Equal-service requirement

This is necessary, since the present worth comparison involves calculating the equivalent PW of all future cash flows for each alternative. A fair comparison requires that PW values represent cash flows associated with equal service. For cost alternatives, failure to compare equal service will always favor the shorter-lived mutually exclusive alternative, even if it is not the more economical choice, because fewer periods of costs are involved. The equal-service requirement is satisfied by using either of two approaches:

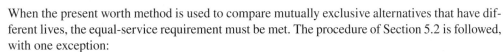

LCM: Compare the PW of alternatives over a period of time equal to the **least common multiple (LCM)** of their estimated lives.

Study period: Compare the PW of alternatives using a **specified study period of n years.** This approach does not necessarily consider the useful life of an alternative. The study period is also called the *planning horizon.*

LCM or study period

For either approach, calculate the PW at the MARR and use the same selection guideline as that for equal-life alternatives. The LCM approach makes the cash flow estimates extend to the same period, as required. For example, lives of 3 and 4 years are compared over a 12-year period.

The first cost of an alternative is reinvested at the beginning of each life cycle, and the estimated salvage value is accounted for at the end of each life cycle when calculating the PW values over the LCM period. Additionally, the LCM approach requires that some assumptions be made about subsequent life cycles.

The assumptions when using the LCM approach are that

1. The service provided will be needed over the entire LCM years or more.
2. The selected alternative can be repeated over each life cycle of the LCM in exactly the same manner.
3. Cash flow estimates are the same for each life cycle.

As will be shown in Chapter 14, the third assumption is valid only when the cash flows are expected to change by exactly the inflation (or deflation) rate that is applicable through the LCM time period. If the cash flows are expected to change by any other rate, then the PW analysis must be conducted using constant-value dollars, which considers inflation (Chapter 14).

A study period analysis is necessary if the first assumption about the length of time the alternatives are needed cannot be made. For the study period approach, a time horizon is chosen over which the economic analysis is conducted, and only those cash flows which occur during that time period are considered relevant to the analysis. All cash flows occurring beyond the study period are ignored. An estimated market value at the end of the study period must be made. The time horizon chosen might be relatively short, especially when short-term business goals are very important. The study period approach is often used in replacement analysis (Chapter 11). It is also useful when the LCM of alternatives yields an unrealistic evaluation period, for example, 5 and 9 years.

EXAMPLE 5.3

National Homebuilders, Inc. plans to purchase new cut-and-finish equipment. Two manufacturers offered the following estimates

	Vendor A	Vendor B
First cost, $	−15,000	−18,000
Annual M&O cost, $ per year	−3,500	−3,100
Salvage value, $	1,000	2,000
Life, years	6	9

(a) Determine which vendor should be selected on the basis of a PW comparison, if the MARR is 15% per year.

(b) National Homebuilders has a standard practice of evaluating all options over a 5-year period. If a study period of 5 years is used and the salvage values are not expected to change, which vendor should be selected?

Solution

(a) Since the equipment has different lives, compare them over the LCM of 18 years. For life cycles after the first, the first cost is repeated each new cycle, which is the last year of the previous cycle. These are years 6 and 12 for vendor A and year 9 for B. The cash flow diagram is shown in Figure 5–2. Calculate PW at 15% over 18 years.

$$\text{PW}_A = -15,000 - 15,000(P/F,15\%,6) + 1000(P/F,15\%,6)$$
$$-15,000(P/F,15\%,12) + 1000(P/F,15\%,12) + 1000(P/F,15\%,18)$$
$$-3,500(P/A,15\%,18)$$
$$= \$-45,036$$

$$\text{PW}_B = -18,000 - 18,000(P/F,15\%,9) + 2000(P/F,15\%,9)$$
$$+ 2000(P/F,15\%,18) - 3100(P/A,15\%,18)$$
$$= \$-41,384$$

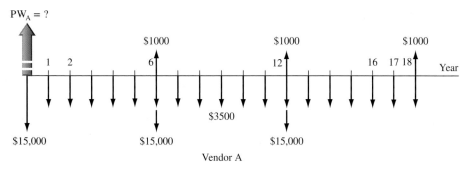

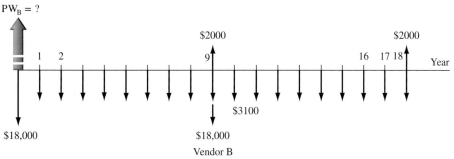

Figure 5–2
Cash flow diagram for different-life alternatives, Example 5.3a.

Vendor B is selected, since it costs less in PW terms; that is, the PW$_B$ value is numerically larger than PW$_A$.

(*b*) For a 5-year study period, no cycle repeats are necessary. The PW analysis is

$$PW_A = -15{,}000 - 3500(P/A,15\%,5) + 1000(P/F,15\%,5)$$
$$= \$-26{,}236$$

$$PW_B = -18{,}000 - 3100(P/A,15\%,5) + 2000(P/F,15\%,5)$$
$$= \$-27{,}397$$

Vendor A is now selected based on its smaller PW value. This means that the shortened study period of 5 years has caused a switch in the economic decision. In situations such as this, the standard practice of using a fixed study period should be carefully examined to ensure that the appropriate approach, that is, LCM or fixed study period, is used to satisfy the equal-service requirement.

EXAMPLE 5.4 Water for Semiconductor Manufacturing Case   **PE**

When we discussed this case in the introduction, we learned that the initial estimates of equipment life were 10 years for both options of UPW (ultrapure water)—seawater and groundwater. As you might guess, a little research indicates that seawater is more corrosive and the equipment life is shorter—5 years rather than 10. However, it is expected that, instead of complete replacement, a total refurbishment of the equipment for $10 M after 5 years will extend the life through the anticipated 10th year of service.

With all other estimates remaining the same, it is important to determine if this 50% reduction in expected usable life and the refurbishment expense may alter the decision to go with the seawater option, as determined in Example 5.2. For a complete analysis, consider both a 10-year and a 5-year option for the expected use of the equipment, regardless of the source of UPW.

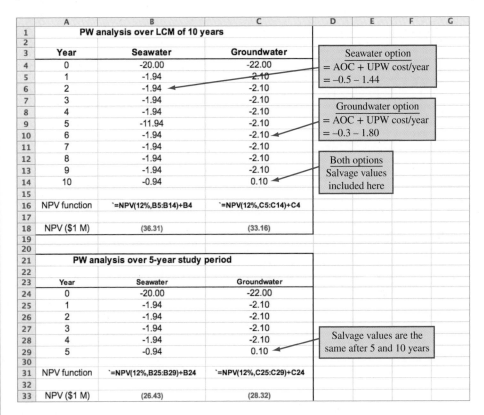

Figure 5-3
PW analyses using LCM and study period approaches for water for semiconductor manufacturing case, Example 5.4.

Solution

A spreadsheet and the NPV function are a quick and easy way to perform this dual analysis. The details are presented in Figure 5-3.

LCM of 10 years: In the top part of the spreadsheet, the LCM of 10 years is necessary to satisfy the equal-service requirement; however, the first cost in year 5 is the refurbishment cost of -10 M, not the -20 M expended in year 0. Each year's cash flow is entered in consecutive cells; the -11.94 M in year 5 accounts for the continuing AOC and annual UPW cost of -1.94 M, plus the -10 M refurbishment cost. The NPV functions shown on the spreadsheet determine the 12% per year PW values in $1 million units.

$$PW_S = \$-36.31 \qquad PW_G = \$-33.16$$

Now, the *groundwater option is cheaper*; the economic decision is reversed with this new estimate of life and year 5 refurbishment expense.

Study period of 5 years: The lower portion of Figure 5-3 details a PW analysis using the second approach to evaluating different-life alternatives, that is, a specific study period, which is 5 years in this case study. Therefore, all cash flows after 5 years are neglected.

Again the economic decision is reversed as the 12% per year PW values *favor the seawater option*.

$$PW_S = \$-26.43 \qquad PW_G = \$-28.32$$

Comments

The decision switched between the LCM and study period approaches. Both are correct answers given the decision of how the equal-service requirement is met. This analysis demonstrates how important it is to compare ME alternatives over time periods that are believable and to take the time necessary to make the most accurate cost, life, and MARR estimates when the evaluation is performed.

> *If the PW evaluation is incorrectly performed using the respective lives of the two options, the equal-service requirement is violated,* and PW values favor the shorter-lived option, that is, seawater. The PW values are
>
>   Option S: $n = 5$ years, PW_S = $\$-26.43$ M, from the bottom left calculation in Figure 5–3.
>
>   Option G: $n = 10$ years, $PW_G = \$-33.16$ M, from the top right calculation in Figure 5–3.

For **independent projects,** use of the LCM approach is unnecessary since each project is compared to the DN alternative, not to each other, and satisfying the equal-service requirement is not a problem. Simply use the MARR to determine the PW over the respective life of each project, and **select all projects with a PW ≥ 0.**

5.4 Future Worth Analysis ● ● ●

The future worth (FW) of an alternative may be determined directly from the cash flows, or by multiplying the PW value by the F/P factor, at the established MARR. The n value in the F/P factor is either the LCM value or a specified study period. Analysis of alternatives using FW values is especially applicable to large capital investment decisions when a prime goal is to maximize the *future wealth* of a corporation's stockholders.

Future worth analysis over a specified study period is often utilized if the asset (equipment, a building, etc.) might be sold or traded at some time before the expected life is reached. Suppose an entrepreneur is planning to buy a company and expects to trade it within 3 years. FW analysis is the best method to help with the decision to sell or keep it 3 years hence. Example 5.5 illustrates this use of FW analysis. Another excellent application of FW analysis is for projects that will come online at the end of a multiyear investment period, such as electric generation facilities, toll roads, airports, and the like. They are analyzed using the FW value of investment commitments made during construction.

> The selection guidelines for FW analysis are the same as for PW analysis; FW ≥ 0 means the MARR is met or exceeded. For two or more mutually exclusive alternatives, select the one with the numerically largest FW value.

ME alternative selection

EXAMPLE 5.5

A British food distribution conglomerate purchased a Canadian food store chain for £75 million 3 years ago. There was a net loss of £10 million at the end of year 1 of ownership. Net cash flow is increasing with an arithmetic gradient of £+5 million per year starting the second year, and this pattern is expected to continue for the foreseeable future. Because of the heavy debt financing used to purchase the Canadian chain, the international board of directors expects a MARR of 25% per year from any sale.

(*a*) The British conglomerate has just been offered £159.5 million by a French company wishing to get a foothold in Canada. Use FW analysis to determine if the MARR will be realized at this selling price.

(*b*) If the British conglomerate continues to own the chain, what selling price must be obtained at the end of 5 years of ownership to just make the MARR?

Solution

(*a*) Set up the FW relation in year 3 (FW_3) at $i = 25\%$ per year and an offer price of £159.5 million. Figure 5–4*a* presents the cash flow diagram in million £ units.

$$FW_3 = -75(F/P,25\%,3) - 10(F/P,25\%,2) - 5(F/P,25\%,1) + 159.5$$
$$= -168.36 + 159.5 = £-8.86 \text{ million}$$

No, the MARR of 25% will not be realized if the £159.5 million offer is accepted.

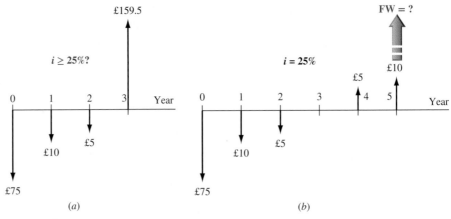

Figure 5–4
Cash flow diagrams for Example 5.5. (*a*) i = ?; (*b*) FW = ?.

(*b*) Determine the future worth 5 years from now at 25% per year. Figure 5–4*b* presents the cash flow diagram. The A/G and F/A factors are applied to the arithmetic gradient.

$$\text{FW}_5 = -75(F/P,25\%,5) - 10(F/A,25\%,5) + 5(A/G,25\%,5)(F/A,25\%,5)$$
$$= £-246.81 \text{ million}$$

The offer must be for at least £246.81 million to make the MARR. This is approximately 3.3 times the purchase price only 5 years earlier, in large part based on the required MARR of 25%.

5.5 Capitalized Cost Analysis ● ● ●

Many public sector projects such as bridges, dams, highways and toll roads, railroads, and hydro-electric and other power generation facilities have very long expected useful lives. A **perpetual or infinite life** is the effective planning horizon. Permanent endowments for charitable organizations and universities also have perpetual lives. The economic worth of these types of projects or endowments is evaluated using the present worth of the cash flows.

Capitalized Cost (CC) is the present worth of a project that has a very long life (more than, say, 35 or 40 years) or when the planning horizon is considered very long or infinite.

The formula to calculate CC is derived from the PW relation $P = A(P/A,i\%,n)$, where $n = \infty$ time periods. Take the equation for P using the P/A factor and divide the numerator and denominator by $(1 + i)^n$ to obtain

$$P = A \left[\frac{1 - \dfrac{1}{(1 + i)^n}}{i} \right]$$

As n approaches ∞, the bracketed term becomes $1/i$. We replace the symbols P and PW with CC as a reminder that this is a capitalized cost equivalence. Since the A value can also be termed AW for annual worth, the capitalized cost formula is simply

$$\text{CC} = \frac{A}{i} \quad \text{or} \quad \text{CC} = \frac{\text{AW}}{i} \tag{5.1}$$

Solving for A or AW, the amount of new money that is generated each year by a capitalization of an amount CC is

$$\text{AW} = \text{CC}(i) \tag{5.2}$$

This is the same as the calculation $A = P(i)$ for an infinite number of time periods. Equation [5.2] can be explained by considering the time value of money. If $20,000 is invested now (this is the capitalization) at 10% per year, the maximum amount of money that can be withdrawn at the end of every year for *eternity* is $2000, which is the interest accumulated each year. This leaves the original $20,000 to earn interest so that another $2000 will be accumulated the next year.

The cash flows (costs, revenues, and savings) in a capitalized cost calculation are usually of two types: *recurring,* also called periodic, and *nonrecurring.* An annual operating cost of $50,000 and a rework cost estimated at $40,000 every 12 years are examples of recurring cash flows. Examples of nonrecurring cash flows are the initial investment amount in year 0 and one-time cash flow estimates at future times, for example, $500,000 in fees 2 years hence.

The procedure to determine the CC for an infinite sequence of cash flows is as follows:

1. Draw a cash flow diagram showing all nonrecurring (one-time) cash flows and at least two cycles of all recurring (periodic) cash flows.
2. Find the PW of all nonrecurring amounts. This is their CC value.
3. Find the A value through *one life cycle* of all recurring amounts. (This is the same value in all succeeding life cycles, as explained in Chapter 6.) Add this to all other uniform amounts (A) occurring in years 1 through infinity. The result is the total equivalent uniform annual worth (AW).
4. Divide the AW obtained in step 3 by the interest rate i to obtain a CC value. This is an application of Equation [5.1].
5. Add the CC values obtained in steps 2 and 4.

Drawing the cash flow diagram (step 1) is more important in CC calculations than elsewhere because it helps separate nonrecurring and recurring amounts. In step 5, the present worths of all component cash flows have been obtained; the total capitalized cost is simply their sum.

EXAMPLE 5.6

The Haverty County Transportation Authority (HCTA) has just installed a new software to charge and track toll fees. The director wants to know the total equivalent cost of all future costs incurred to purchase the software system. If the new system will be used for the indefinite future, find the equivalent cost (*a*) now, a CC value, and (*b*) for each year hereafter, an AW value.

The system has an installed cost of $150,000 and an additional cost of $50,000 after 10 years. The annual software maintenance contract cost is $5000 for the first 4 years and $8000 thereafter. In addition, there is expected to be a recurring major upgrade cost of $15,000 every 13 years. Assume that $i = 5\%$ per year for county funds.

Solution

(*a*) The five-step procedure to find CC now is applied.
   1. Draw a cash flow diagram for two cycles (Figure 5–5).
   2. Find the PW of the nonrecurring costs of $150,000 now and $50,000 in year 10 at $i = 5\%$. Label this CC_1.

$$CC_1 = -150,000 - 50,000(P/F,5\%,10) = \$-180,695$$

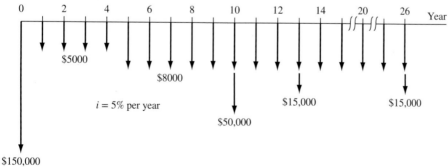

Figure 5–5
Cash flows for two cycles of recurring costs and all nonrecurring amounts, Example 5.6.

3 and 4. Convert the $15,000 recurring cost to an A value over the first cycle of 13 years, and find the capitalized cost CC_2 at 5% per year using Equation [5.1].

$$A = -15,000(A/F,5\%,13) = \$-847$$

$$CC_2 = -847/0.05 = \$-16,940$$

There are several ways to convert the annual software maintenance cost series to A and CC values. A straightforward method is to, first, consider the -5000 an A series with a capitalized cost of

$$CC_3 = -5000/0.05 = \$-100,000$$

Second, convert the step-up maintenance cost series of -3000 to a capitalized cost CC_4 in year 4, and find the PW in year 0. (Refer to Figure 5-5 for cash flow timings.)

$$CC_4 = \frac{-3000}{0.05}(P/F,5\%,4) = \$-49,362$$

5. The total capitalized cost CC_T for the Haverty County Transportation Authority is the sum of the four-component CC values.

$$CC_T = -180,695 - 16,940 - 100,000 - 49,362$$

$$= \$-346,997$$

(*b*) Equation [5.2] determines the AW value forever.

$$AW = Pi = CC_T(i) = \$346,997(0.05) = \$17,350$$

Correctly interpreted, this means Haverty County officials have committed the equivalent of $17,350 forever to operate and maintain the toll management software.

For the comparison of **two alternatives on the basis of capitalized cost,** use the procedure above to find the A value and CC_T for each alternative. Since the capitalized cost represents the total present worth of financing and maintaining a given alternative forever, the alternatives will automatically be compared for the same number of years (i.e., infinity). The alternative with the smaller capitalized cost will represent the more economical one. This evaluation is illustrated in Example 5.7 using the progressive example for this chapter.

EXAMPLE 5.7 Water for Semiconductor Manufacturing Case

Our case study has progressed (in Example 5.4) to the point that the life of the seawater option can be extended to 10 years with a major refurbishment cost after 5 years. This extension is possible only one time, after which a new life cycle would commence. In $1 million units, the estimates and PW values (from Figure 5-3) are as follows:

Seawater: $P_S = \$-20$; $AOC_S = \$-1.94$; $n_S = 10$ years; refurbishment, year 5 $= \$-10$; $S_S = 0.05(20) = \$1.00$; $PW_S = \$-36.31$

Groundwater: $P_G = \$-22$; $AOC_G = \$-2.10$; $n_G = 10$ years; $S_G = 0.10(22) = \$2.2$; $PW_G = \$-33.16$

If we assume that the UPW (ultrapure water) requirement will continue for the foreseeable future, a good number to know is the present worth of the long-term options at the selected MARR of 12% per year. What are these capitalized costs for the two options using the estimates made thus far?

Solution

Find the equivalent A value for each option over its respective life, then determine the CC value using the relation $CC = A/i$. Select the option with the lower CC. This approach satisfies the equal-service requirement because the time horizon is infinity when the CC is determined.

Seawater: $A_S = PW_S(A/P,12\%,10) = -36.31(0.17698) = \$-6.43$

$\qquad CC_S = -6.43/0.12 = \$-53.58$

Groundwater: $A_G = PW_G(A/P,12\%,10) = -33.16(0.17698) = \$-5.87$

$\qquad CC_G = -5.87/0.12 = \$-48.91$

In terms of capitalized cost, the *groundwater alternative is cheaper*.

Comment

If the seawater-life extension is not considered a viable option, the original alternative of 5 years could be used in this analysis. In this case, the equivalent A value and CC computations in \$1 million units are as follows:

$$A_{S,5\text{ years}} = -20(A/P,12\%,5) - 1.94 + 0.05(20)(A/F,12\%,5)$$
$$= \$-7.33$$
$$CC_{S,5\text{ years}} = -7.33/0.12 = \$-61.08$$

Now, the economic advantage of the groundwater option is even larger.

If a finite-life alternative (e.g., 5 years) is compared to one with an indefinite or very long life, capitalized costs can be used. To determine capitalized cost for the finite-life alternative, calculate the equivalent A value for one life cycle and divide by the interest rate (Equation [5.1]). This procedure is illustrated in Example 5.8 using a spreadsheet.

EXAMPLE 5.8

The Commissioners of several districts have mandated 100% recycling programs linked with a comprehensive waste-to-energy (WtE) program. The goal is zero waste to the landfill by 2022. Two options for the materials separation equipment are outlined below for a consortium of 32 districts within one state. The interest rate for state-mandated projects is 5% per year.

Contractor option (C): \$8 million now and \$25,000 per year will provide separation services at a maximum of 15 sites. No contract period is stated; thus the contract and services are offered for as long as needed.

Purchase option (P): Purchase equipment at each site for \$275,000 per site and expend an estimated \$12,000 in annual operating costs (AOC). Expected life of the equipment is 5 years with no salvage value.

(*a*) Perform a capitalized cost analysis for a total of 10 recycling sites.
(*b*) Determine the maximum number of sites at which the equipment can be purchased and still have a capitalized cost less than that of the contractor option.

Solution

(*a*) Figure 5–6, column B, details the solution. The contract, as proposed, has a long life. Therefore, the \$8 million is already a capitalized cost. The annual charge of $A = \$25,000$ is divided by $i = 0.05$ to determine its CC value. Summing the two values results in $CC_C = \$-8.5$ million.

For the finite, 5-year purchase alternative, column B shows the first cost (\$-275,000 per site), AOC (\$-12,000), and equivalent A value of \$-755,181, which is determined via the PMT function (cell tag). Divide A by the interest rate of 5% to determine $CC_P = \$-15.1$ million.

The *contractor option is by far more economical* for the anticipated 10 sites.

(*b*) A quick way to find the maximum number of sites for which $CC_P < CC_C$ is to use Excel's **Goal Seek** tool, introduced in Chapter 2, Example 2.10. (See Appendix A for details on how to use this tool.) The template is set up in Figure 5–6 to make the two CC values equal as the number of sites is altered (decreased). The result, shown in column C, indicates that 5.63 sites make the options economically equivalent. Since the number of sites

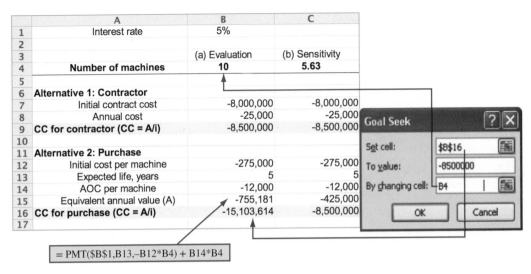

Figure 5–6
Spreadsheet solution of Example 5.8 using capitalized cost (*a*) for 10 recycling sites and (*b*) to determine the number of sites to make the alternatives economically equal.

must be an integer, *5 or fewer sites will favor purchasing the equipment* and 6 or more sites will favor contracting the separation services.

This approach to problem solution will be called *breakeven analysis* in later chapters of the text. By the way, another way to determine the number of sites is by trial and error. Enter different values in cell B4 until the CC values favor the purchase alternative.

CHAPTER SUMMARY

The present worth method of comparing alternatives involves converting all cash flows to present dollars at the MARR. The alternative with the numerically larger (or largest) PW value is selected. When the alternatives have different lives, the comparison must be made for equal-service periods. This is done by performing the comparison over either the LCM of lives or a specific study period. Both approaches compare alternatives in accordance with the equal-service requirement. When a study period is used, any remaining value in an alternative is recognized through the estimated future market value.

If the life of the alternatives is considered to be very long or infinite, capitalized cost is the comparison method. The CC value is calculated as A/i, because the P/A factor reduces to $1/i$ in the limit of $n = \infty$.

PROBLEMS

Types of Projects

5.1   When evaluating multiple alternatives or projects, against what must they be compared, if they are (*a*) independent, and (*b*) mutually exclusive?

5.2   (*a*) What is the difference between a service alternative and revenue alternative? (*b*) What is another name of a service alternative?

5.3   Define the term *capitalized cost* and give a real-world example of something that might be analyzed using a capitalized cost evaluation technique.

5.4   A rapidly growing city is dedicated to neighborhood integrity. However, increasing traffic and speed on a through street are of concern to residents. The city manager has proposed five independent traffic-calming options:

1. Stop sign at corner A
2. Stop sign at corner B
3. Low-profile speed bump at point C
4. Low-profile speed bump at point D
5. Speed dip at point E

There cannot be any of the following combinations in the final projects:

—No combination of dip and one or two bumps
—Not two bumps
—Not two stop signs

(a) If there were no restrictions on how the options can be combined, calculate how many alternatives there are to evaluate.

(b) Using the restrictions, identify the number of and list of acceptable alternatives.

Alternative Comparison—Equal Lives

5.5 When evaluating projects by the present worth method, how do you know which one(s) to select if the proposals are (a) independent, and (b) mutually exclusive?

5.6 What is meant by the term equal service alternative?

5.7 You have been asked to evaluate two alternatives, X and Y, that may increase plant capacity for manufacturing high-pressure hydraulic hoses. The parameters associated with each alternative have been estimated. Which one should be selected on the basis of a present worth comparison at an interest rate of 12% per year? Why is yours the correct choice?

Alternative	X	Y
First cost, $	−45,000	−58,000
Maintenance cost, $/year	−8,000	−4,000
Salvage value, $	2,000	12,000
Life, years	5	5

5.8 One of two methods must be used to produce expansion anchors. Method A costs $80,000 initially and will have a $15,000 salvage value after 3 years. The operating cost with this method will be $30,000 per year. Method B will have a first cost of $120,000, an operating cost of $8000 per year, and a $40,000 salvage value after its 3-year life. At an interest rate of 12% per year, which method should be used on the basis of a present worth analysis? Also, write the two spreadsheet functions to perform the PW analysis.

5.9 In order to provide drinking water as part of its 50-year plan, a west coast city is considering constructing a pipeline for importing water from a nearby community that has a plentiful supply of brackish ground water. A full-sized pipeline can be constructed at a cost of $122 million now. Alternatively, a smaller pipeline can be constructed now for $80 million and enlarged 20 years from now for another $100 million. The pumping cost will be $25,000 per year higher for the smaller pipeline during the first 20 years, but will be approximately the same thereafter. Both pipelines are expected to have the same useful life with no salvage value. (a) At an interest rate of 6% per year, which alternative is more economical? (b) Write the spreadsheet function to display the PW for the smaller pipeline alternative.

5.10 Lennon Hearth Products manufactures glass-door fireplace screens that have two types of mounting brackets for the frame. An L-shaped bracket is used for relatively small fireplace openings and a U-shaped bracket is used for all others. The company includes both types of brackets in the box with their product and the purchaser discards the one not needed. The cost of these two brackets with screws and other parts is $3.50. If the frame of the fireplace screen is redesigned, a single universal bracket can be used that will cost $1.20 to make. However, initial costs of retooling ($6000) and inventory write-downs ($8000) will be incurred immediately. If the company sells 1200 fireplace units per year, should the company keep the old brackets or go with the new ones, assuming the company uses an interest rate of 15% per year and it wants to recover its investment in 5 years. Use the present worth method for your evaluation.

5.11 Leonard, a company that manufactures explosion-proof motors, is considering two alternatives for expanding its international export capacity. Option 1 requires equipment purchases of $900,000 now and $560,000 two years from now, with annual M&O costs of $79,000 in years 1 through 10. Option 2 involves subcontracting some of the production at costs of $280,000 per year beginning now through the end of year 10. Neither option will have a significant salvage value. Use a present worth analysis to determine which option is more attractive at the company's MARR of 20% per year. (Note: Check out the spreadsheet exercises for new options that Leonard has been offered recently.)

5.12 A software package created by Navarro & Associates can be used for analyzing and designing three-sided guyed towers and three- and four-sided self-supporting towers. A single-user license will cost $4000 per year. A site license has a one-time cost of $15,000. A structural engineering consulting company is trying to decide between two alternatives: first, to buy one single-user license now and one each year for the next 4 years (which will provide 5 years of service), or second, to buy a site license now. Determine which strategy should be adopted at an interest rate of 12% per year for a 5-year planning period using present worth evaluation.

5.13   Oil from a specific type of marine microalgae can be converted into biodiesel that may serve as an alternate transportable fuel for automobiles and trucks. If lined ponds are used to grow the algae, the construction cost is $13 million and the M&O cost is estimated at $2.1 million per year. Alternatively, if long plastic tubes are used for growing the algae, the initial cost will be higher at $18 million, but less contamination will render the M&O cost lower at $0.41 million per year. At an interest rate of 10% per year and a 5-year project period, which system is better—ponds or tubes? Use a present worth analysis.

5.14   A company that makes food-friendly silicone (for use in cooking and baking pan coatings) is considering the independent projects shown, all of which can be considered to be viable for only 10 years. If the company's MARR is 15% per year, determine which should be selected on the basis of a present worth analysis. Financial values are in $1000 units.

	A	B	C	D
First cost, $	−1,200	−2,000	−5,000	−7,000
Annual net income, $/year	200	400	1100	1300
Salvage value, $	5	6	8	7

5.15   Two methods can be used to produce expansion anchors. Method A costs $80,000 initially and will have a $15,000 salvage value after 3 years. The operating cost with this method will be $30,000 in year 1, increasing by $4000 each year. Method B will have a first cost of $120,000, an operating cost of $8000 in year 1, increasing by $6500 each year, and a $40,000 salvage value after its 3-year life. At an interest rate of 12% per year, which method should be used on the basis of a present worth analysis?

5.16   Sales of bottled water in the United States totaled 34.0 gallons per person in 2014. Evian, a high-quality natural spring water, costs about 60¢ per bottle, while a local brand of purified municipal water may cost only 25¢ per bottle. On average, a local municipal water utility may provide drinkable tap water for $2.90 per 1000 gallons. If the average person drinks two bottles of water per day or uses 5 gallons per day to obtain the same amount of water from the tap, what are the present worth values per person for 1 year of drinking tap water versus bottled water using (a) Evian, and (b) local-brand bottled water?

Use an interest rate of 6% per year, compounded monthly, and 30 days per month.

Alternative Comparison—Different Lives

5.17   Dexcon Technologies, Inc. is evaluating two alternatives to produce its new plastic filament with tribological (i.e., low friction) properties for creating custom bearings for 3-D printers. The estimates associated with each alternative are shown below. Using a MARR of 20% per year, which alternative has the lower present worth?

Method	DDM	LS
First cost, $	−164,000	−370,000
M&O cost, $/year	−55,000	−21,000
Salvage value, $	0	30,000
Life, years	2	4

5.18   NASA is considering two materials for use in a space vehicle tracking station in Australia. The estimates are shown below. Which should be selected on an economic basis of present worth values at an interest rate of 10% per year?

Material	M	FF
First cost, $	−205,000	−235,000
Maintenance cost, $/year	−29,000	−27,000
Salvage value, $	2,000	20,000
Life, years	2	4

5.19   Lego Group in Bellund, Denmark manufactures Lego toy construction blocks. The company is considering two methods for producing special-purpose Lego parts. Method 1 will have an initial cost of $400,000, an annual operating cost of $140,000, and a life of 3 years. Method 2 will have an initial cost of $600,000, an operating cost of $100,000 per year, and a 6-year life. Assume 10% salvage values for both methods. Lego uses an MARR of 15% per year. (a) Which method should it select on the basis of a present worth analysis? (b) If the evaluation is *incorrectly* performed using the respective life estimates of 3 and 6 years, will Lego make a correct or incorrect economic decision? Explain your answer.

5.20   Compare two alternatives, A and B, on the basis of a present worth evaluation using $i = 10\%$ per year and a study period of 8 years.

Alternative	A	B
First cost, $	−15,000	−28,000
Annual operating cost, $/year	−6,000	−9,000
Overhaul in year 4, $	—	−2000
Salvage value, $	3,000	5,000
Life, years	4	8

5.21 An engineer is trying to decide which process to use to reduce sludge volume prior to disposal. Belt filter presses (BFP) will cost $203,000 to buy and $85,000 per year to operate. Belts will be replaced one time per year at a cost of $5500. Centrifuges (Cent) will cost $396,000 to buy and $119,000 per year to operate, but because the centrifuge will produce a thicker "cake", the sludge hauling cost to the monofill will be $37,000 per year less than for the belt presses. The useful lives are 5 and 10 years for alternatives BFP and Cent, respectively, and the salvage values are assumed to be 10% of the first cost of each process whenever they are closed down or replaced. Use PW evaluation to select the more economical process at an interest rate of 6% per year over (*a*) the LCM of lives, and (*b*) a study period of 8 years. Are the decisions the same?

5.22 The product development group of a high-tech electronics company developed five proposals for new products. The company wants to expand its product offerings, so it will undertake all projects that are economically attractive at the company's MARR of 20% per year. The cash flows (in $1000 units) associated with each project are estimated. Which projects, if any, should the company accept on the basis of a present worth analysis?

Project	A	B	C	D	E
Initial investment, $	−400	−510	−660	−820	−900
Operating cost, $/year	−100	−140	−280	−315	−450
Revenue, $/year	360	235	400	605	790
Salvage value, $	—	22	—	80	95
Life, years	3	10	5	8	4

5.23 Compare the alternatives C and D on the basis of a present worth analysis using an interest rate of 10% per year and a study period of 10 years.

Alternative	C	D
First cost, $	−40,000	−32,000
AOC, $/year	−7,000	−3,000
Annual increase in operating cost, $/year	−1,000	—
Salvage value, $	9,000	500
Life, years	10	5

5.24 You and your partner have become very interested in cross-country motorcycle racing and wish to purchase entry-level equipment. You have identified two alternative sets of equipment and gear. Package K has a first cost of $160,000, an operating cost of $7000 per quarter, and a salvage value of $40,000 after its 2-year life. Package L has a first cost of $210,000 with a lower operating cost of $5000 per quarter, and an estimated $26,000 salvage value after its 4-year life. Which package offers the lower present worth analysis at an interest rate of 8% per year, compounded quarterly?

5.25 Three different plans were presented to the GAO (General Accounting Office) by a high-tech facilities manager for operating a portable cyber-security facility.

Plan A: Renewable 1-year contracts with payments of $1 million at the beginning of each year.

Plan B: A 2-year contract that requires three payments of $600,000 each, with the first one to be made immediately and the second and third payment made at the beginning of the following two 6-month intervals; no payments required during the second year of the contract.

Plan C: A 3-year contract that entails a payment of $1.5 million now and another payment of $0.5 million 2 years from now.

Assuming the GAO can renew any of the plans under the same conditions, if it decides to do so, which plan is best on the basis of a present worth analysis at an interest rate of 6% per year, compounded semiannually? Solve using (*a*) factors and (*b*) a spreadsheet. (Hint: Construct a cash flow diagram before working this problem.)

5.26 **Water for Semiconductor Manufacturing Case** PE Throughout the present worth analyses, the decision between seawater and groundwater switched multiple times in Examples 5.2 and 5.4. A summary is given here in $1 million monetary units.

Seawater (S)			
Life *n*, years	First Cost, $	PW at 12%, $	Selected
10	−20	−30.64	**YES**
5	−20, plus −10 after 5 years	−36.31	No
5 (study period)	−20	−26.43	**YES**
Groundwater (G)			
10	−22	−33.16	No
10	−22	−33.16	**YES**
5 (study period)	−22	−28.32	No

The confusion about the recommended source for UPW has not gone unnoticed by the general manager. Yesterday, you were asked to settle the issue by determining the first cost, X_S, of the seawater option to ensure that it is the economic choice over groundwater. The study period is set by the manager as 10 years, simply because that is the time period on the lease agreement for the building where the fab will be located. Since the seawater equipment must be refurbished or replaced after 5 years, the general manager told you to assume that the equipment will be purchased anew after 5 years of use. What is the maximum first cost that Angular Enterprises should pay for the seawater option?

Future Worth Comparison

5.27   A remotely located air sampling station can be powered by solar cells or by running an electric line to the site and using conventional power. Solar cells will cost $12,600 to install and will have a useful life of 4 years with no salvage value. Annual costs for inspection, cleaning, etc. are expected to be $1400. A new power line will cost $11,000 to install, with power costs expected to be $800 per year. Since the air sampling project will end in 4 years, the salvage value of the line is considered to be zero. At an interest rate of 10% per year, which alternative should be selected on the basis of a future worth analysis?

5.28   Parker Hannifin of Cleveland, Ohio manufactures CNG fuel dispensers. It needs replacement equipment to streamline one of its production lines for a new contract, but plans to sell the equipment at or before its expected life is reached at an estimated market value for used equipment. Select between the two options using the corporate MARR of 15% per year and a future worth analysis for the expected use period. Also, write the FV spreadsheet functions that will display the correct future worth values.

Option	D	E
First cost, $	−62,000	−77,000
AOC, $ per year	−15,000	−21,000
Expected market value, $	8,000	10,000
Expected use, years	3	6

5.29   The Department of Energy is proposing new rules mandating either a 20% increase or a 35% increase in clothes washer efficiency in 3 years. The 20% increase is expected to add $100 to the current price of a washer, while the 35% increase will add $240 to the price. If the cost for energy is $80 per year with the 20% increase in efficiency and $65 per year with the 35% increase, which one of the two proposed standards is more economical on the basis of a future worth analysis at an interest rate of 10% per year? Assume a 15-year life for all washer models.

5.30   An electric switch manufacturing company is trying to decide between three different assembly methods. Method A has an estimated first cost of $40,000, an annual operating cost (AOC) of $9000, and a service life of 2 years. Method B will cost $80,000 to buy and will have an AOC of $6000 over its 4-year service life. Method C costs $130,000 initially with an AOC of $4000 over its 8-year life. Methods A and B will have no salvage value, but Method C will have equipment worth 10% of its first cost. Perform both (a) future worth, and (b) present worth analyses to select the method at $i = 10\%$ per year.

5.31   A small strip-mining coal company is trying to decide whether it should purchase or lease a new clamshell. If purchased, the "shell" will cost $150,000 and is expected to have a $65,000 salvage value after 6 years. Alternatively, the company can lease a clamshell for only $20,000 per year, but the lease payment will have to be made at the *beginning* of each year. If the clamshell is purchased, it will be leased to other strip-mining companies whenever possible, an activity that is expected to yield revenues of $12,000 per year. If the company's MARR is 15% per year, should the clamshell be purchased or leased on the basis of a future worth analysis? Assume the annual M&O cost is the same for both options.

Capitalized Cost

5.32   Determine the capitalized cost of a permanent roadside historical marker that has a first cost of $78,000 and a maintenance cost of $3500 once every 5 years. Use an interest rate of 8% per year.

5.33   The Golden Gate bridge is maintained by 38 painters and 17 ironworkers (who replace corroding steel and rivets). If the painters get an average wage of $120,000 per year (with benefits) and the ironworkers get $150,000 per year, what is the capitalized cost today of all the future wages for bridge maintenance at an interest rate of 8% per year?

5.34   What is the present worth difference between an investment of $10,000 per year for 50 years and an investment of $10,000 per year forever at an interest rate of 10% per year?

5.35   The cost of upgrading a section of Grand Loop Road in Yellowstone National Park is $1.7 million. Resurfacing and other maintenance are expected to cost $350,000 every 3 years. What is the capitalized cost of the road at an interest rate of 6% per year?

5.36 How much money was deposited 35 years ago at 10% per year interest if it is sufficient to provide a perpetual income of $10,000 per year beginning now, year 35?

5.37 Assume that 25 years ago your dad invested $200,000, plus $25,000 in years 2 through 5, and $40,000 per year from year 6 on ward. At a very good interest rate of 12% per year, determine (*a*) the CC value, and (*b*) the annual retirement amount that he can withdraw forever starting next year (year 26), if no additional investments are made.

5.38 An aggressive stockbroker claims an ability to consistently earn 12% per year on an investor's money. A client invests $10,000 now, $30,000 three years from now, and $8000 per year for 5 years starting 4 years from now. (*a*) How much money can the client withdraw every year forever, beginning 20 years from now? (*b*) What is the capitalized cost of the client's investments if the $8000 per year is expected to continue for an unspecified time into the future?

5.39 Beaver, a city in the United States, is attempting to attract a professional soccer team. Beaver is planning to build a new stadium that will cost $250 million. Annual upkeep is expected to amount to $800,000. The turf will have to be replaced every 10 years at a cost of $950,000. Painting every 5 years will cost $75,000. If the city expects to maintain the facility indefinitely, what is the estimated capitalized cost at $i = 8\%$ per year?

5.40 Compare three alternatives on the basis of their capitalized costs at $i = 10\%$ per year.

Alternative	E	F	G
First cost, $	−50,000	−300,000	−900,000
AOC, $ per year	−30,000	−10,000	−3,000
Salvage value, $	5,000	70,000	200,000
Life, years	2	4	∞

5.41 **Water for Semiconductor Manufacturing Case**—It is anticipated that the needs for UPW (ultrapure water) at the new Angular Enterprises site will continue for a long time, as much as 50 years. This is the rationale for using capitalized cost as a basis for the economic decision between desalinated seawater (S) and purified groundwater (G). These costs were determined (Example 2.7) to be $CC_S = \$-53.58$ million and $CC_G = \$-48.91$ million. Groundwater is the clear economic choice.

Yesterday, the general manager had lunch with the president of Brissa Water, who offered to supply the needed UPW at a cost of $5 million per year for the indefinite future. It would mean a dependence upon a contractor to supply the water, but the equipment, treatment, and other costly activities to obtain UPW on site would be eliminated. The manager asks you to make a recommendation about this seemingly attractive alternative under the following conditions at the same MARR of 12% per year as used for the other analyses:

(*a*) The annual cost of $5 million remains constant throughout the time it is needed.

(*b*) The annual cost starts at $5 million for the first year only, and then increases 2% per year. (This increase is above the cost of providing UPW by either of the other two methods.)

EXERCISES FOR SPREADSHEETS

5.42 Leonard Motors is trying to increase its international export business. It is considering several alternatives. Two were available earlier (per Problem 5.11), but a new one has recently been proposed by a potential internationally partnering corporation, which has funds to invest.

Option 1: Equipment costs $900,000 now and another $560,000 in 2 years
Annual M&O costs of $79,000
Life of project is 10 years
Salvage value is nil

Option 2: Subcontract production for annual payment of $280,000 for years 0 (now) through 10

Annual M&O cost is zero
Life of project is 10 years
Salvage value is nil

Option 3: Costs are $400,000 in year 1 plus an additional 5% each year through year 5
Revenues are $50,000 per year for years 6 through 10
Life of project is 10 years
Equipment salvage value is $100,000 paid in year 10 by international partner

Due to stock market pressures, Leonard plans to change its MARR from its current 20% per year to 15% per year, *compounded quarterly*. Use PW analysis to select the best of the three options.

5.43 You wish to evaluate four independent projects that all have a 10-year life at MARR = 15% per year. Preliminary estimates for first cost, annual net income, and SV have been made.

	A	B	C	D
First cost, $	−1,200	−2,000	−5,000	−7,000
Annual net income, $/year	200	400	1100	1300
Salvage value, $	5	6	8	7

(a) Accept or reject each project using a present worth analysis. On your spreadsheet, include the logical IF function to make the accept/reject decision.

(b) The preliminary estimates have changed for projects A and B as shown below. Use the same spreadsheet to reevaluate them.

	A	B
First cost, $	−1,000	−2,200
Annual net income, $/year	300	440
Salvage value, $	8	0

5.44 Use a spreadsheet and a study period of 8 years to select alternative A or B in Problem 5.20.

5.45 Charlie's Truck Repair and Service has a new contract that requires him to purchase and maintain new equipment for work on 18-wheeler trucks and heavy road equipment. Two separate vendors have made quotes and estimates for Charlie. Use the estimates and a 6% per year return requirement to find the better economic option. One problem is that Charlie does not currently know if the contract will last for 5, 8, or 10 years. He needs a recommendation for all three time periods.

Vendor	Ferguson	Halgrove
First cost P, $	−203,000	−396,000
M&O, $ per year	−90,000	−82,000
Salvage value, SV	10% of P	10% of P
Maximum life, years	5	10

5.46 Hannifin CNG Fuel Dispensers needs to purchase replacement equipment to streamline one of its production lines for a new contract, but may sell the equipment before its expected life is reached at an estimated market value for used equipment. At MARR = 15% per year, select the better option using a future worth analysis over (a) the expected usage period, and (b) the maximum life, when the salvage values are expected to be 50% of the market values for used equipment. Are the selections the same for both plans?

Option	D	E
First cost, $	−62,000	−77,000
AOC, $ per year	−15,000	−21,000
Expected market value, $	8,000	10,000
Expected use, years	3	6
Maximum life, years	4	8

5.47 Petrobras, the Brazilian energy company, has identified two alternatives to provide potable water to offshore platforms—purchase and operate the equipment, or contract long term with Manal and Associates, an international oilfield service corporation. For the estimates shown, use capitalized cost analysis at $i = 6\%$ per year to determine (a) the better economic choice for Petrobras, and (b) the maximum annual M&O cost that will cause Manal and Associates to succeed in winning the contract.

Alternative	Purchase	Contract
First cost, $	−300,000	−850,000
M&O, $ per year	−10,000, year 1 + 2% increase each year thereafter	−10,000
Salvage value, $	70,000	—
Expected use, years	8	40+

ADDITIONAL PROBLEMS AND FE EXAM REVIEW QUESTIONS

5.48 The present worth of an alternative that provides infinite service is called its:

(a) Net present value
(b) Discounted total cost
(c) Capitalized cost
(d) Perpetual annual cost

5.49 When comparing mutually exclusive alternatives that have different lives by the present worth method, it is necessary to:

(a) Always compare them over a period equal to the life of the longer-lived alternative
(b) Always compare them over a time period of equal service
(c) Always compare them over a period equal to the life of the shorter-lived alternative
(d) Find the present worth over one life cycle of each alternative

5.50   The capitalized cost of an initial investment of $200,000 and annual investments of $30,000 forever at an interest rate of 10% per year is closest to:

(a) $−230,000   (b) $−300,000
(c) $−500,000   (d) $−2,300,000

5.51   An upgraded version of a CNC machine has a first cost of $200,000, an annual operating cost of $60,000, and a salvage value of $50,000 after its 8-year life. At an interest rate of 10% per year, the capitalized cost is closest to:

(a) $−93,116   (b) $−100,060
(c) $−931,160   (d) $−1,000,600

5.52   The cost of maintaining a permanent monument is expected to be $70,000 now and $70,000 every 10 years forever. At an interest rate of 10% per year, the capitalized cost is nearest:

(a) $−11,393   (b) $−58,930
(c) $−84,360   (d) $−113,930

Problems 5.53 through 5.55 are based on the following cash flows for alternatives X and Y at an interest rate of 10% per year.

Machine	X	Y
Initial cost, $	−146,000	−220,000
AOC, $/year	−15,000	−10,000
Annual revenue, $/year	80,000	75,000
Salvage value, $	10,000	25,000
Life, years	3	6

5.53   In comparing the alternatives on a present worth basis, the PW of machine X is closest to:

(a) $23,160   (b) $40,560
(c) $58,950   (d) $72,432

5.54   In comparing the alternatives on a future worth basis, the FW of machine X is closest to:

(a) $23,160   (b) $40,560
(c) $58,950   (d) $71,860

5.55   The capitalized cost of machine Y is closest to:

(a) $17,726
(b) $86,590
(c) $177,260
(d) $207,720

5.56   An engineer analyzed four independent alternatives by the present worth method. On the basis of her results, the alternative(s) she should select are:

Alternative	A	B	C	D
Present worth, $	−5000	−2000	−3000	−1000

(a)   Only D
(b)   Can't tell from this information
(c)   All of them
(d)   None of them

Problems 5.57 through 5.61 are based on the following cash flows for alternatives A and B at an interest rate of 10% per year.

Alternative	A	B
First cost, $	−90,000	−750,000
AOC, $/year	−50,000	−10,000
Salvage value, $	8,000	2,000,000
Life, years	5	∞

5.57   You have been asked to compare the alternatives on the basis of a present worth comparison. The PW of alternative A is closest to:

(a) $−724,320   (b) $−530,520
(c) $−388,950   (d) $−72,432

5.58   The present worth of alternative B is nearest:

(a) $−85,000   (b) $−750,000
(c) $−850,000   (d) $−950,000

5.59   The capitalized cost of alternative B is nearest:

(a) $−590,000   (b) $−625,000
(c) $−734,000   (d) $−850,000

5.60   All of the following equations for calculating the capitalized cost of alternative B are correct, except:

(a) $CC_B = −750,000 − 10,000/0.10$
(b) $CC_B = −750,000 − 10,000/0.10$
        $+ 2,000,000(0.10)$
(c) $CC_B = [−750,000(0.10) − 10,000]/0.10$
(d) $CC_B = −750,000 − 10,000/0.10$
        $+ 2,000,000/(1+0.10)^∞$

5.61   The future worth of alternative B is closest to:

(a) $−85,000   (b) $−750,000
(c) $−850,000   (d)   −∞

CASE STUDY

COMPARING SOCIAL SECURITY BENEFITS

Background

When Sheryl graduated from Northeastern University in 2017 and went to work for BAE Systems, she did not pay much attention to the monthly payroll deduction for social security. It was a "necessary evil" that may be helpful in retirement years. However, this was so far in the future that she fully expected this government retirement benefit system to be broke and gone by the time she could reap any benefits from her years of contributions.

This year, Sheryl and Brad, another engineer at BAE, got married. Recently, they both received notices from the Social Security Administration of their potential retirement amounts, were they to retire and start social security benefits at preset ages. Since both of them hope to retire a few years early, they decided to pay closer attention to the predicted amount of retirement benefits and to do some analysis on the numbers.

Information

They found that their projected benefits are substantially the same, which makes sense since their salaries are very close to each other. Although the numbers were slightly different in their two mailings, the similar messages to Brad and Sheryl can be summarized as follows:

If you stop working and start receiving benefits . . .

At age 62, your payment would be about	$1400 per month
At your full retirement age (67 years), your payment would be about	$2000 per month
At age 70, your payment would be about	$2480 per month

These numbers represent a reduction of 30% for early retirement (age 62) and an increase of 24% for delayed retirement (age 70).

This couple also learned that it is possible for a spouse to take spousal benefits at the time that one of them is at full retirement age. In other words, if Sheryl starts her $2000 benefit at age 67, Brad can receive a benefit equal to 50% of hers. Then, when Brad reaches 70 years of age, he can discontinue spousal benefits and start his own. In the meantime, his benefits will have increased by 24%. Of course, this strategy could be switched with Brad taking his benefits and Sheryl receiving spousal benefits until age 70.

All these options led them to define four alternative plans.

A: Each takes early benefits at age 62 with a 30% reduction to $1400 per month.

B: Each takes full benefits at full retirement age of 67 and receives $2000 per month.

C: Each delays benefits until age 70 with a 24% increase to $2480 per month.

D: One person takes full benefits of $2000 per month at age 67, and the other person receives spousal benefits ($1000 per month at age 67) and switches to delayed benefits of $2480 at age 70.

They realize, of course, that the numbers will change over time, based on their respective salaries and number of years of contribution to the social security system by them and by their employers.

Case Study Exercises

Brad and Sheryl are the same age. Brad determined that most of their investments make an average of 6% per year. With this as the interest rate, the analysis for the four alternatives is possible. Sheryl and Brad plan to answer the following questions, but don't have time this week. Can you please help them? (Do the analysis for one person at a time, not the couple, and stop at the age of 85.)

1. How much in total (without the time value of money considered) will each plan A through D pay through age 85?

2. What is the future worth at 6% per year of each plan at age 85?

3. Plot the future worth values for all four plans on one spreadsheet graph.

4. Economically, what is the best combination of plans for Brad and Sheryl, assuming they both live to be 85 years old?

5. Develop at least one additional question that you think Sheryl and Brad may have. Answer the question.

CHAPTER 6

Annual Worth Analysis

Ingram Publishing

LEARNING OUTCOMES

Purpose: Utilize different annual worth techniques to evaluate and select alternatives.

SECTION	TOPIC	LEARNING OUTCOME
6.1	Advantages of AW	• Demonstrate that the AW value is the same for each life cycle.
6.2	CR and AW values	• Calculate and interpret the capital recovery (CR) and AW amounts.
6.3	AW analysis	• Select the best alternative using an annual worth analysis.
6.4	Perpetual life	• Evaluate alternatives with very long lives using AW analysis.
6.5	LCC analysis	• Perform a life-cycle cost (LCC) analysis using AW methods.

I n this chapter, we add to our repertoire of alternative comparison tools. In Chapter 5 we learned the present worth (PW) method and its companion, the future worth (FW) method. Here we learn the equivalent annual worth, or AW, method. AW analysis is commonly considered the more desirable of the two methods because the AW value is easy to calculate; the measure of worth—AW in monetary units per year—is understood by most individuals, and its assumptions are essentially identical to those of the PW method.

Annual worth is also known by other titles. Some are *equivalent annual worth* (EAW), *equivalent annual cost* (EAC), *annual equivalent* (AE), and *equivalent uniform annual cost* (EUAC). The alternative selected by the AW method will always be the same as that selected by the PW method, and all other alternative evaluation methods, provided they are performed correctly.

An additional application of AW analysis treated here is **life-cycle cost (LCC) analysis.** This method considers all costs of a product, process, or system from concept to phaseout.

6.1 Advantages and Uses of Annual Worth Analysis ●●◐

For many engineering economic studies, the AW method is the best to use, when compared to PW, FW, and rate of return (Chapters 7 and 8). Since the AW value is the equivalent uniform annual worth of all estimated receipts and disbursements during the life cycle of the project or alternative, AW is easy to understand by any individual acquainted with annual amounts, for example, dollars per year. The AW value, which has the same interpretation as A used thus far, is the economic equivalent of the PW and FW values at the MARR for n years. All three can be easily determined from each other by the relation

$$AW = PW(A/P,i,n) = FW(A/F,i,n) \qquad [6.1]$$

The n in the factors is the number of years for equal-service comparison. This is the LCM or the stated study period of the PW or FW analysis.

When all cash flow estimates are converted to an AW value, this value applies for every year of the life cycle and for *each additional life cycle.*

> The annual worth method offers a prime computational and interpretation advantage because the AW value needs to be calculated for **only one life cycle.** The AW value determined over one life cycle is the AW for all future life cycles. Therefore, it is **not necessary to use the LCM** of lives to satisfy the equal-service requirement.

Equal-service requirement and LCM

As with the PW method, there are three fundamental assumptions of the AW method that should be understood. When alternatives being compared have different lives, the AW method makes the assumptions that

1. The services provided are needed for at least the LCM of the lives of the alternatives.
2. The selected alternative will be repeated for succeeding life cycles in exactly the same manner as for the first life cycle.
3. All cash flows will have the same estimated values in every life cycle.

In practice, no assumption is precisely correct. If, in a particular evaluation, the first two assumptions are not reasonable, a study period must be established for the analysis. Note that for assumption 1, the length of time may be the indefinite future (forever). In the third assumption, all cash flows are expected to change exactly with the inflation (or deflation) rate. If this is not a reasonable assumption, new cash flow estimates must be made for each life cycle, and again a study period must be used. AW analysis for a stated study period is discussed in Section 6.3.

EXAMPLE 6.1

In Example 5.3, National Homebuilders, Inc. evaluated cut-and-finish equipment from vendor A (6-year life) and vendor B (9-year life). The PW analysis used the LCM of 18 years. Consider only the vendor A option now. The diagram in Figure 6–1 shows the cash flows for all three life cycles (first cost $-15,000$; annual M&O costs -3500; salvage value 1000). Demonstrate the equivalence at $i = 15\%$ of PW over three life cycles and AW over one cycle. In Example 5.3, present worth for vendor A was calculated as PW = $-45,036$.

Figure 6–1
PW and AW values for three life cycles, Example 6.1.

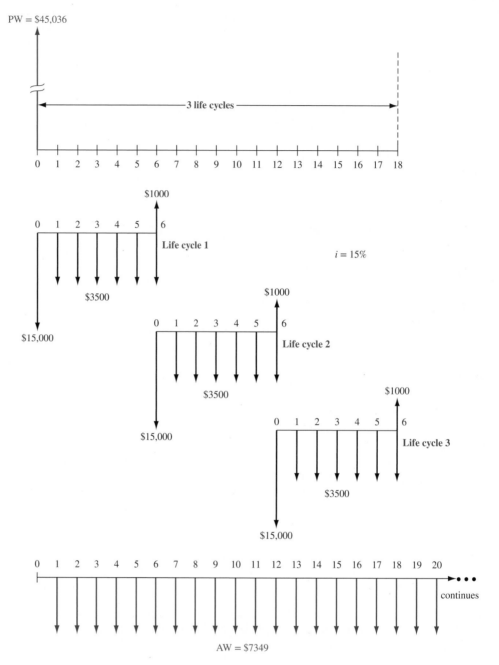

Solution

Calculate the equivalent uniform AW value for all cash flows in the first life cycle.

$$AW = -15,000(A/P,15\%,6) + 1000(A/F,15\%,6) - 3500 = \$-7349$$

When the same computation is performed on each succeeding life cycle, the AW value is $\$-7349$. Now Equation [6.1] is applied to the PW value for 18 years.

$$AW = -45,036(A/P,15\%,18) = \$-7349$$

The one-life-cycle AW value and the AW value based on 18 years are equal.

Not only is annual worth an excellent method for performing engineering economy studies, but also it is applicable in any situation where PW (and FW and benefit/cost) analysis can be utilized. The AW method is especially useful in certain types of studies: asset replacement and retention time studies to minimize overall annual costs (both covered in Chapter 11), breakeven studies and make-or-buy decisions (Chapter 13), and all studies dealing with production or manufacturing costs where a cost/unit or profit/unit measure is the focus.

If income taxes are considered, a slightly different approach to the AW method is used by some large corporations and financial institutions. It is termed *economic value added,* or EVA. This approach, covered in Chapter 17, concentrates upon the wealth-increasing potential that an alternative offers a corporation. The resulting EVA values are the equivalent of an AW analysis of after-tax cash flows.

6.2 Calculation of Capital Recovery and AW Values ●●○

As we learned in the previous chapter, an alternative should have the following cash flows estimated as accurately as possible:

Initial investment P. This is the total first cost of all assets and services required to initiate the alternative. When portions of these investments take place over several years, their PW is an equivalent initial investment. Use this amount as *P*.

Salvage value S. This is the terminal estimated value of assets at the end of their useful life. The *S* is zero if no salvage is anticipated; *S* is negative when it will cost money to dispose of the assets. For study periods shorter than the useful life, *S* is the estimated market value or trade-in value at the end of the study period.

Salvage/market value

Annual amount A. This is the equivalent annual amount (costs only for cost alternatives; costs and receipts for revenue alternatives). Often this is the annual operating cost (AOC) or M&O cost, so the estimate is already an equivalent *A* value.

The AW value for an alternative is comprised of two components: **capital recovery** for the initial investment *P* at a stated interest rate (usually the MARR) and the equivalent annual amount *A*. The symbol CR is used for the capital recovery component. In equation form,

$$AW = CR + A \qquad [6.2]$$

Both CR and *A* represent costs. The total annual amount *A* is determined from uniform recurring costs (and possibly receipts) and nonrecurring amounts. The *P/A* and *P/F* factors may be necessary to first obtain a PW amount; then the *A/P* factor converts this amount to the *A* value in Equation [6.2]. (If the alternative is a revenue project, there will be positive cash flow estimates present in the calculation of the *A* value.)

The recovery of an amount of capital *P* committed to an asset, **plus** the time value of the capital at a particular interest rate, is a **fundamental principle of economic analysis.**

Capital recovery (CR) is the equivalent annual amount that the asset, process, or system must earn (new revenue) each year to just **recover the initial investment plus a stated rate of return** over its expected life. Any expected salvage value is considered in the computation of CR.

Capital recovery

The *A/P* factor is used to convert *P* to an equivalent annual cost. If there is some anticipated positive salvage value *S* at the end of the asset's useful life, its equivalent annual value is recovered using the *A/F* factor. This action reduces the equivalent annual cost of the asset. Accordingly, CR is calculated as

$$CR = -P(A/P,i,n) + S(A/F,i,n) \qquad [6.3]$$

EXAMPLE 6.2

Lockheed Martin is increasing its booster thrust power in order to win more satellite launch contracts from European companies interested in opening up new global communications markets. A piece of earth-based tracking equipment is expected to require an investment of $13 million, with $8 million committed now and the remaining $5 million expended at the end of year 1 of the project. Annual operating costs for the system are expected to start the first year and continue at $0.9 million per year. The useful life of the tracker is 8 years with a salvage value of $0.5 million. Calculate the CR and AW values for the system, if the corporate MARR is 12% per year.

Solution

Capital recovery: Determine *P* in year 0 of the two initial investment amounts, followed by the use of Equation [6.3] to calculate the CR. In $1 million units,

$$P = 8 + 5(P/F,12\%,1) = \$12.46$$

$$\text{CR} = -12.46(A/P,12\%,8) + 0.5(A/F,12\%,8)$$
$$= -12.46(0.20130) + 0.5(0.08130)$$
$$= \$-2.47$$

The correct interpretation of this result is very important to Lockheed Martin. It means that each and every year for 8 years, the equivalent total net revenue from the tracker must be at least $2,470,000 just to recover the initial present worth investment plus the required return of 12% per year. This does not include the AOC of $0.9 million each year.

Annual worth: To determine AW, the cash flows in Figure 6–2a must be converted to an equivalent AW series over 8 years (Figure 6–2b). Since CR = $−2.47 million is an *equivalent annual cost,* as indicated by the minus sign, total AW is easily determined by adding the CR and AOC values.

$$\text{AW} = -2.47 - 0.9 = \$-3.37 \text{ million per year}$$

This is the AW for all future life cycles of 8 years, provided the costs rise at the same rate as inflation, and the same costs and services are expected to apply for each succeeding life cycle.

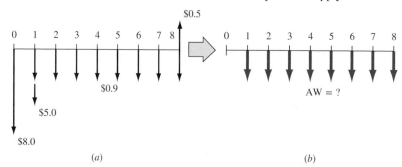

Figure 6–2
(*a*) Cash flow diagram for satellite tracker costs, and (*b*) conversion to an equivalent AW (in $1 million), Example 6.2.

There is a second, equally correct way to determine CR. Either method results in the same value. There is a relation between the A/P and A/F factors.

$$(A/F,i,n) = (A/P,i,n) - i$$

Both factors are present in the CR Equation [6.3]. Substitute for the A/F factor to obtain

$$\text{CR} = -P(A/P,i,n) + S[(A/P,i,n) - i]$$
$$= -[(P - S)(A/P,i,n) + S(i)]$$

There is a basic logic to this formula. Subtracting S from the initial investment P before applying the A/P factor recognizes that the salvage value will be recovered. This reduces CR, the annual cost of asset ownership. However, the fact that S is not recovered until year n of ownership is compensated for by charging the annual interest $S(i)$ against the CR. Although either CR relation results in the same amount, it is better to consistently use the same method. The first method, Equation [6.3], will be used in this text.

To perform an AW evaluation by spreadsheet, the PMT function can be used to determine the CR. The annual amount A is then added. The general function = PMT($i\%,n,P,F$) is rewritten using the initial investment as P and $−S$ for the salvage value. The complete format to display AW is

$$= \text{PMT}(i\%,n,P,-S) + A \qquad [6.4]$$

As an illustration, determine the AW in Example 6.2. The equivalent initial investment in year 0 is $12.46 million. The complete function for the AW amount (in $1 million units) is = PMT(12%,8,12.46,−0.5) − 0.9. The answer of $−3.37 (million) will be displayed in the spreadsheet cell.

As we learned in Section 3.1, one spreadsheet function can be embedded in another function. In the case of Example 6.2, the initial investment is distributed over a 2-year period. The one-cell PMT function, with the PV function embedded (in bold), can be written as = PMT(12%,8, **8+PV(12%,1,−5)**,−0.5) − 0.9 to display the same value AW = $−3.37.

6.3 Evaluating Alternatives by Annual Worth Analysis ●●●

The annual worth method is typically the easiest to apply of the evaluation techniques when the MARR is specified. The AW is calculated over the respective life of each alternative, and the selection guidelines are the same as those used for the PW method. For **mutually exclusive** alternatives, whether cost- or revenue-based, the guidelines are as follows:

Project evaluation

ME alternative
selection

> **One alternative:** If AW ≥ 0, the requested MARR is met or exceeded and the alternative is economically justified.
>
> **Two or more alternatives:** Select the alternative with the AW that is **numerically largest,** that is, less negative or more positive. This indicates a lower AW of cost for cost alternatives or a larger AW of net cash flows for revenue alternatives.

If any of the three assumptions in Section 6.1 is not acceptable for an alternative, a study period analysis must be used. Then the cash flow estimates over the study period are converted to AW amounts. The following two examples illustrate the AW method for one project and two alternatives.

EXAMPLE 6.3

Heavenly Pizza, which is located in Toronto, fares very well with its competition in offering fast delivery. Many students at the area universities and community colleges work part-time delivering orders made via the web. The owner, Jerry, a software engineering graduate, plans to purchase and install five portable, in-car systems to increase delivery speed and accuracy. The systems provide a link between the web order-placement software and the On-Star system for satellite-generated directions to any address in the area. The expected result is faster, friendlier service to customers, and larger income.

Each system costs $4600, has a 5-year useful life, and may be salvaged for an estimated $300. Total operating cost for all systems is $1000 for the first year, increasing by $100 per year thereafter. The MARR is 10%. Perform an annual worth evaluation for the owner that answers the following questions. Perform the solution by hand and by spreadsheet.

(a) How much new annual net income is necessary to recover only the initial investment at the MARR of 10% per year?

(b) Jerry estimates increased net income of $6000 per year for all five systems. Is this project financially viable at the MARR?

(c) Based on the answer in part (b), determine how much new net income Heavenly Pizza must have to economically justify the project. Operating costs remain as estimated.

Solution by Hand

(a) The CR amount calculated by Equation [6.3] answers the first question.

$$CR = -5[4600(A/P,10\%,5)] + 5[300(A/F,10\%,5)]$$
$$= -5[4600(0.26380)] + 5[300(0.16380)]$$
$$= \$-5822$$

The five systems must generate an equivalent annual new revenue of $5822 to recover the initial investment plus a 10% per year return.

(b) Figure 6–3 presents the cash flows over 5 years. The annual operating cost series, combined with the estimated $6000 annual income, forms an arithmetic gradient series with a base amount of $5000 and $G = \$-100$. The project is financially viable if AW ≥ 0 at $i = 10\%$ per year. Apply Equation [6.2], where A is the equivalent annual net income series.

$$AW = CR + A = -5822 + 5000 - 100(A/G,10\%,5)$$
$$= \$-1003$$

The system is not financially justified at the net income level of $6000 per year.

Figure 6–3
Cash flow diagram used to
compute AW, Example 6.3.

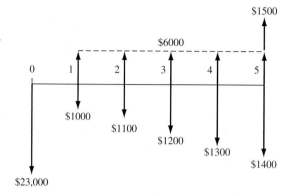

(*c*) Let the required income equal *R,* and set the AW relation equal to zero to find the minimum
income to justify the system.

$$0 = -5822 + (R - 1000) - 100(A/G,10\%,5)$$
$$R = -5822 - 1000 - 100(1.8101)$$
$$= \$7003 \text{ per year}$$

Solution by Spreadsheet

The spreadsheet in Figure 6–4 summarizes the estimates and answers the questions posed
for Heavenly Pizza with the same values that were determined in the hand solution.

Figure 6–4
Spreadsheet solution
of Example 6.3.
(*a*) Capital recovery
in cell B16, (*b*) AW in
cell E17, and (*c*) Goal
Seek template and
outcome in cell B5.

	A	B	C	D	E
1					
2	MARR	10%			
3	# of systems	5			
4	Cost/system	-4,600			
5	Income/year	$ 6,000			
6					
7		Investment	Annual	Annual	
8	Year	and salvage	costs	income	Net income
9	0	-23,000	0	0	-23,000
10	1	0	-1,000	6,000	5,000
11	2	0	-1,100	6,000	4,900
12	3	0	-1,200	6,000	4,800
13	4	0	-1,300	6,000	4,700
14	5	1,500	-1,400	6,000	6,100
15					
16	(a) CR value	-5,822			
17	(b) AW value				-1,003
18					
19		= -PMT($B$2, 5, NPV($B$2,B10:B14)+B9)			
20					
21			= -PMT($B$2, 5, NPV($B$2,E10:E14)+E9)		
22					

(*a*) and (*b*)

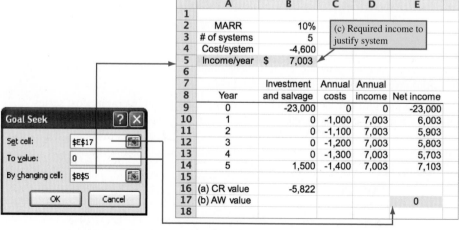

	A	B	C	D	E
1					
2	MARR	10%	(c) Required income to		
3	# of systems	5	justify system		
4	Cost/system	-4,600			
5	Income/year	$ 7,003			
6					
7		Investment	Annual	Annual	
8	Year	and salvage	costs	income	Net income
9	0	-23,000	0	0	-23,000
10	1	0	-1,000	7,003	6,003
11	2	0	-1,100	7,003	5,903
12	3	0	-1,200	7,003	5,803
13	4	0	-1,300	7,003	5,703
14	5	1,500	-1,400	7,003	7,103
15					
16	(a) CR value	-5,822			
17	(b) AW value				0
18					

Goal Seek

Set cell: $E$17
To value: 0
By changing cell: $B$5

OK   Cancel

(*c*)

Cell references are used in the spreadsheet functions to accommodate changes in estimated values.

(*a*) The capital recovery CR = $–5822 is displayed in column B using the PMT function with an embedded NPV function, as shown in the cell tag.

(*b*) The annual worth AW = $–1003 is displayed in column E using the PMT function shown. The arithmetic gradient series of costs and estimated income of $6000 in columns C and D, respectively, are added to obtain the net income necessary for the PMT function.

(*c*) The minimum required income is determined in the lower part of Figure 6–4. This is easily accomplished by setting AW = 0 (column E) in the Goal Seek tool and letting it find the income per year of $7003 to balance the AW equation.

EXAMPLE 6.4

Luby's Cafeterias is in the process of forming a separate business unit that provides meals to facilities for the elderly, such as assisted care and long-term care centers. Since the meals are prepared in one central location and distributed by trucks throughout the city, the equipment that keeps food and drink cold and hot is very important. Michele is the general manager of this unit, and she wishes to choose between two manufacturers of temperature retention units that are mobile and easy to sterilize after each use. Use the cost estimates below to select the more economic unit at a MARR of 8% per year.

	Hamilton (H)	Infinity Care (IC)
Initial cost P, $	−15,000	−20,000
Annual M&O, $/year	−6,000	−9,000
Refurbishment cost, $	0	−2,000 every 4 years
Trade-in value S, % of P	20	40
Life, years	4	12

Solution

The best evaluation technique for these different-life alternatives is the annual worth method, where AW is taken at 8% per year over the respective lives of 4 and 12 years.

$$AW_H = \text{annual equivalent of } P - \text{annual M\&O} + \text{annual equivalent of } S$$
$$= -15,000(A/P,8\%,4) - 6000 + 0.2(15,000)(A/F,8\%,4)$$
$$= -15,000(0.30192) - 6000 + 3000(0.22192)$$
$$= \$-9,863$$

$$AW_{IC} = \text{annual equivalent of } P - \text{annual M\&O} - \text{annual equivalent of refurbishment}$$
$$\qquad + \text{annual equivalent of } S$$
$$= -20,000(A/P,8\%,12) - 9000 - 2000[(P/F,8\%,4) + (P/F,8\%,8)](A/P,8\%,12)$$
$$\qquad + 0.4(20,000)(A/F,8\%,12)$$
$$= -20,000(0.13270) - 9000 - 2000[0.7350 + 0.5403](0.13270) + 8000(0.05270)$$
$$= \$-11,571$$

The Hamilton unit is considerably less costly on an annual equivalent basis.

If the projects are **independent**, the AW at the MARR is calculated. All projects with AW ≥ 0 are acceptable.

Independent project selection

6.4 AW of a Permanent Investment ● ● ●

This section discusses the annual worth equivalent of the capitalized cost introduced in Section 5.5. Evaluation of public sector projects, such as flood control dams, irrigation canals, bridges, or other large-scale projects, requires the comparison of alternatives that have such long lives that they may be considered infinite in economic analysis terms. For this type of analysis, the annual worth (and capital recovery amount) of the initial investment is the perpetual annual interest on the initial investment, that is, $A = Pi = (CC)i$. This is Equation [5.2].

Cash flows recurring at regular or irregular intervals are handled exactly as in conventional AW computations; **convert them to equivalent uniform annual amounts A for one cycle**. This automatically annualizes them for each succeeding life cycle. Add all the A values to the CR amount to find total AW, as in Equation [6.2].

EXAMPLE 6.5

The U.S. Bureau of Reclamation is considering three proposals for increasing the capacity of the main drainage canal in an agricultural region of Nebraska. Proposal A requires dredging the canal to remove sediment and weeds that have accumulated during previous years' operation. The capacity of the canal will have to be maintained in the future near its design peak flow because of increased water demand. The Bureau is planning to purchase the dredging equipment and accessories for $650,000. The equipment is expected to have a 10-year life with a $17,000 salvage value. The annual operating costs are estimated to total $50,000. To control weeds in the canal itself and along the banks, environmentally safe herbicides will be sprayed during the irrigation season. The yearly cost of the weed control program is expected to be $120,000.

Proposal B is to line the canal with concrete at an initial cost of $4 million. The lining is assumed to be permanent, but minor maintenance will be required every year at a cost of $5000. In addition, lining repairs will have to be made every 5 years at a cost of $30,000.

Proposal C is to construct a new pipeline along a different route. Estimates are an initial cost of $6 million, annual maintenance of $3000 for right-of-way, and a life of 50 years.

Compare the alternatives on the basis of AW, using an interest rate of 5% per year.

Solution

Since this is an investment for a permanent project, compute the AW for one cycle of all recurring costs. For proposals A and C, the CR values are found using Equation [6.3], with $n_A = 10$ and $n_C = 50$, respectively. For proposal B, the CR is simply $P(i)$.

Proposal A	
CR of dredging equipment:	
$-650,000(A/P,5\%,10) + 17,000(A/F,5\%,10)$	$ -82,824$
Annual cost of dredging	$-50,000$
Annual cost of weed control	$-120,000$
	$-252,824$
Proposal B	
CR of initial investment: $-4,000,000(0.05)$	$-200,000$
Annual maintenance cost	$-5,000$
Lining repair cost: $-30,000(A/F,5\%,5)$	$-5,429$
	$-210,429$
Proposal C	
CR of pipeline: $-6,000,000(A/P,5\%,50)$	$-328,680$
Annual maintenance cost	$-3,000$
	$-331,680$

Proposal B, which is a permanent solution, is selected due to its lowest AW of costs.

Comment

Note the use of the A/F factor for the lining repair cost in proposal B. The A/F factor is used instead of A/P because the lining repair cost begins in year 5, not year 0, and continues indefinitely at 5-year intervals.

If the 50-year life of proposal C is considered infinite, $CR = P(i) = \$-300,000$, instead of $\$-328,680$ for $n = 50$. This is a small economic difference. How long lives of 40 or more years are treated economically is a matter of "local" practice.

EXAMPLE 6.6

At the end of each year, all owners and employees at Bell County Utility Cooperative are given a bonus check based on the net profit of the Coop for the previous year. Bart just received his bonus in the amount of $8530. He plans to invest it in an annuity program that returns 7% per year. Bart's long-term plans are to quit the Coop job some years in the future when he is still young enough to start his own business. Part of his future living expenses will be paid from the proceeds that this year's bonus accumulates over his remaining years at the Coop.

(*a*) Use a spreadsheet to determine the amount of annual year-end withdrawal that he can anticipate (starting 1 year after he quits) that will continue forever. He is thinking of working 15 or 20 more years.

(*b*) Determine the amount Bart must accumulate after 15 and 20 years to generate $3000 per year forever.

Solution by Spreadsheet

(*a*) Figure 6–5 presents the cash flow diagram for $n = 15$ years of accumulation at 7% per year on the $8530 deposited now. The accumulated amount after $n = 15$ years is indicated as $F_{after\ 15} = ?$ and the withdrawal series starts at the end of year 16. The diagram for $n = 20$ would appear the same, except there is a 20-year accumulation period.

The spreadsheet in Figure 6–6 shows the functions and answers for $n = 15$ years in columns C and D. The FV function displays the total accumulated at 7% after 15 years as $23,535. The perpetual withdrawal is determined by viewing this accumulated amount as a P value and by applying the formula

$$A = P(i) = 23,535(0.07) = \$1647 \text{ per year}$$

The spreadsheet function = D9*B7 performs this same computation in cell reference format in column D.

Answers for $n = 20$ years are displayed in column E. At a consistent 7% per year return, Bart's perpetual income is estimated as $1647 after 15 years, or $2311 per year if he waits for 20 years.

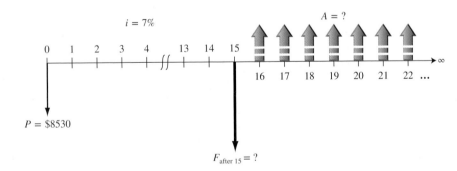

Figure 6–5
Diagram for a perpetual series starting after 15 years of accumulation, Example 6.6.

	A	B	C	D	E
1					
2					
3			Functions	Number of years, n	
4		Estimates	for n = 15 years	15	20
5					
6					
7	Interest rate, % per year	7%			
8	Amount deposited today	$8,530			
9	Amount accumulated after n years		`=-FV(B7,D4,,B8)	$ 23,535	$ 33,008
10	(a) Withdrawal forever, $ per year		`=D9*B7	$ 1,647	$ 2,311
11					
12					
13	(b) Required accumulation for $3000/year	7%	`= 3000/B13	$ 42,857	$ 42,857
14					
15	Accumulation years for $3000 per year	23.86	`= NPER(7%,,-8530,42857)		
16					

Figure 6–6
Spreadsheet solution, Example 6.6.

(*b*) To obtain a perpetual annual withdrawal of $3000, it is necessary to determine how much must be accumulated 1 year before the first withdrawal of $3000. This is an application of the relation $A = P(i)$ solved for P, or

$$P = \frac{A}{i} = \frac{3000}{0.07} = \$42,857$$

This P value is independent of how long Bart works at the Coop because he must accumulate this amount to achieve his goal. Figure 6–6, row 13, shows the function and result. Note that the number of years n does not enter into the function = 3000/B13.

Comment

The NPER function can be used to determine how many years it will take for the current amount of $8530 to accumulate the required $42,857 at 7% per year. The row 15 function indicates that Bart will have to work at the Coop for just under 24 additional years.

6.5 Life-Cycle Cost Analysis ● ● ●

The PW and AW analysis techniques discussed thus far have concentrated on estimates for first cost P, annual operating and maintenance costs (AOC or M&O), salvage value S, and predictable periodic repair and upgrade costs, plus any revenue estimates that may favor one alternative over another. There are usually a host of additional costs involved when the complete project life costs are evaluated. A life-cycle cost analysis includes these additional estimates to the extent that they can be reliably determined.

Life-cycle cost (LCC) analysis utilizes AW or PW methods to evaluate cost estimates for the entire life cycle of one or more projects. Estimates will cover the **entire life span** from the early conceptual stage, through the design and development stages, throughout the operating stage, and even the phaseout and disposal stages. Both **direct and indirect costs** are included to the extent possible, and differences in revenue and savings projections between alternatives are included.

Some typical LCC applications are life-span analysis for military and commercial aircraft, new manufacturing plants, new automobile models, new and expanded product lines, and government systems at federal and state levels. For example, the U.S. Department of Defense requires that a government contractor include an LCC budget and analysis in the originating proposal for most defense systems.

Most commonly the LCC analysis includes costs, and the AW method is used for the analysis, especially if only one alternative is evaluated. If there are expected revenue or other benefit differences between alternatives, a PW analysis is recommended. Public sector projects are usually evaluated using a benefit/cost analysis (Chapter 9), rather than LCC analysis, because estimates to the citizenry are difficult to make with much accuracy. The *direct costs* mentioned above include material, human labor, equipment, supplies, and other costs directly related to a product, process, or system. Some examples of *indirect cost components* are taxes, management, legal, warranty, quality, human resources, insurance, software, purchasing, etc. Direct and indirect costs are discussed further in Chapter 15.

Life-cycle cost analysis is most effectively applied when a substantial percentage of the life span (postpurchase) costs, relative to the initial investment, will be expended in direct and indirect operating, maintenance, and similar costs once the system is operational. For example, the evaluation of two equipment purchase alternatives with expected useful lives of 5 years and M&O costs of 5% to 10% of the initial investment does not require an LCC analysis. However, let's assume that Exxon-Mobil wants to evaluate the design, construction, operation, and support of a new type and style of tanker that can transport oil over long distances of ocean. If the initial costs are in the $100 millions with support and operating costs ranging from 25% to 35% of this amount over a 25-year life, the logic of an LCC analysis will offer a better understanding of the economic viability of the project.

To understand how an LCC analysis works, first we must understand the phases and stages of systems engineering or systems development. Many books and manuals are available on systems development and analysis. Generally, the LCC estimates may be categorized into a simplified format for the major phases of *acquisition, operation,* and *phaseout/disposal,* and their respective stages.

Acquisition phase: all activities prior to the delivery of products and services.
- Requirements definition stage—Includes determination of user/customer needs, assessing them relative to the anticipated system, and preparation of the system requirements documentation.
- Preliminary design stage—Includes feasibility study, conceptual, and early-stage plans; final go–no go decision is probably made here.
- Detailed design stage—Includes detailed plans for resources—capital, human, facilities, information systems, marketing, etc.; there is some acquisition of assets, if economically justifiable.

Operation phase: all activities are functioning, products and services are available.
- Construction and implementation stage—Includes purchases, construction, and implementation of system components; testing; preparation; etc.
- Usage stage—Uses the system to generate products and services; the largest portion of the life cycle.

Phaseout and disposal phase: covers all activities to transition to a new system; removal/recycling/disposal of old system.

EXAMPLE 6.7

In the 1860s, General Mills, Inc. and Pillsbury, Inc. both started in the flour business in the Twin Cities of Minneapolis–St. Paul, Minnesota. In the decade of 2000 to 2010, General Mills purchased Pillsbury for a combination cash and stock deal worth more than $10 billion and integrated the product lines. Food engineers, food designers, and food safety experts made many cost estimates as they determined the needs of consumers and the combined company's ability to technologically and safely produce and market new food products. At this point only cost estimates have been addressed—no revenues or profits.

Assume that the major cost estimates below have been made based on a 6-month study about two new products that could have a 10-year life span for the company. Use LCC analysis at the industry MARR of 18% to determine the size of the commitment in AW terms. (Time is indicated in product-years. Since all estimates are for costs, they are not preceded by a minus sign.)

Consumer habits study (year 0)	$0.5 million
Preliminary food product design (year 1)	0.9 million
Preliminary equipment/plant design (year 1)	0.5 million
Detail product designs and test marketing (years 1, 2)	1.5 million each year
Detail equipment/plant design (year 2)	1.0 million
Equipment acquisition (years 1 and 2)	$2.0 million each year
Current equipment upgrades (year 2)	1.75 million
New equipment purchases (years 4 and 8)	2.0 million (year 4) + 10% per purchase thereafter
Annual equipment operating cost (AOC) (years 3–10)	200,000 (year 3) + 4% per year thereafter
Marketing, year 2	$8.0 million
years 3–10	5.0 million (year 3) and −0.2 million per year thereafter
year 5 only	3.0 million extra
Human resources, 100 new employees for 2000 hours per year (years 3–10)	$20 per hour (year 3) + 5% per year
Phaseout and disposal (years 9 and 10)	$1.0 million each year

Solution

LCC analysis can get complicated rapidly due to the number of elements involved. Calculate the PW by phase and stage, add all PW values, then find the AW over 10 years. Values are in $1 million units.

Acquisition phase:
Requirements definition: consumer study

$$PW = \$0.5$$

Preliminary design: product and equipment

$$PW = 1.4(P/F,18\%,1) = \$1.187$$

Detailed design: product and test marketing, and equipment

$$PW = 1.5(P/A,18\%,2) + 1.0(P/F,18\%,2) = \$3.067$$

Operation phase:
Construction and implementation: equipment and AOC

$$PW = 2.0(P/A,18\%,2) + 1.75(P/F,18\%,2) + 2.0(P/F,18\%,4) + 2.2(P/F,18\%,8)$$

$$+ 0.2\left[\frac{1 - \left(\frac{1.04}{1.18}\right)^8}{0.14}\right](P/F,18\%,2) = \$6.512$$

Use: marketing

$$PW = 8.0(P/F,18\%,2) + [5.0(P/A,18\%,8) - 0.2(P/G,18\%,8)](P/F,18\%,2)$$
$$+ 3.0(P/F,18\%,5)$$

$$= \$20.144$$

Use: human resources: (100 employees)(2000 h/yr)($20/h) = $4.0 million in year 3

$$PW = 4.0\left[\frac{1 - \left(\frac{1.05}{1.18}\right)^8}{0.13}\right](P/F,18\%,2) = \$13.412$$

Phaseout phase:

$$PW = 1.0(P/A,18\%,2)(P/F,18\%,8) = \$0.416$$

The sum of all PW of costs is PW = $45.238 million. Finally, determine the AW over the expected 10-year life span.

$$\mathbf{AW = 45.238 \ million}(A/P,18\%,10) = \$10.066 \ \mathbf{million \ per \ year}$$

This is the LCC estimate of the equivalent annual commitment to the two proposed products.

Often the alternatives compared by LCC do not have the same level of output or amount of usage. For example, if one alternative will produce 20 million units per year and a second alternative will operate at 35 million per year, the AW values should be compared on a currency unit/unit produced basis, such as dollar/unit or euro/hour operated.

Figure 6–7 presents an overview of how costs may be distributed over an entire life cycle. For some systems, typically defense systems, operating and maintenance costs rise fast after acquisition and remain high until phaseout occurs.

The total LCC for a system is established or locked in early in the life cycle. It is not unusual to have 75% to 85% of the entire life span LCC committed during the preliminary and detail

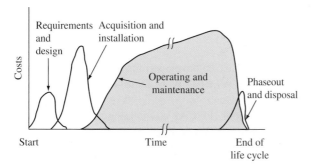

Figure 6–7

Typical distribution of life-cycle costs of the phases for one life cycle.

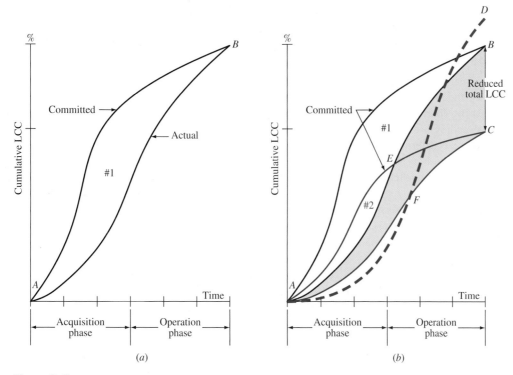

Figure 6–8

LCC envelopes for committed and actual costs: (*a*) design #1, (*b*) improved design #2.

design stages. As shown in Figure 6–8*a*, the actual or observed LCC (bottom curve *AB*) will trail the committed LCC throughout the life span (unless some major design flaw increases the total LCC of design #1 above point *B*).

The potential for significantly reducing total LCC occurs primarily during the early stages. A more effective design and more efficient equipment can reposition the envelope to design #2 in Figure 6–8*b*. Now the committed LCC curve *AEC* is below *AB* at all points, as is the actual LCC curve *AFC*. It is this lower envelope #2 we seek. The shaded area represents the reduction in actual LCC.

Even though an effective LCC envelope may be established early in the acquisition phase, it is not uncommon that unplanned cost-saving measures are introduced during the acquisition phase and early operation phase. These apparent "savings" may actually increase the total LCC, as shown by curve *AFD*. This style of ad hoc cost savings, often imposed by management early in the design stage and/or construction stage, can substantially increase costs later, especially in the after-sale portion of the use stage. For example, the use of inferior-strength concrete and steel has been the cause of structural failures many times, thus increasing the overall life-span LCC.

There have been numerous tools developed over time to help analysts in industry, business, government, and consulting better utilize the LCC approach. Some sample LCC-related tools and web-based materials are listed here.

- Life-Cycle Cost Projection (LCCP) developed for the *water industry* by the Water Environment Research Foundation (WERF). Web address is http://simple.werf.org/simple/media /LCCT/index.html.
- Life-Cycle Costing/Assessment (LCCA) methodology and several tools developed for the *building industry* as a part of the Whole Building Design Guide program of the National Institute of Building Sciences. Address: https://www.wbdg.org/resources/lcca.php.
- RealCost software developed for *pavement design* is available from the Federal Highway Administration. Address: https://www.fhwa.dot.gov/infrastructure/asstmgmt/lcca.cfm.
- There are many LCC methodologies and cost estimation tools developed for *defense-related industries.* Example models are MAAP, MOSS, COCOMO, PRICE, and many more. Information is available at the National Defense Industrial Association website www.ndia.org.

CHAPTER SUMMARY

The annual worth method of comparing alternatives is often preferred to the present worth method because the AW comparison is performed for only one life cycle. This is a distinct advantage when comparing different-life alternatives. AW for the first life cycle is the AW for the second, third, and all succeeding life cycles, under certain assumptions. When a study period is specified, the AW calculation is determined for that time period, regardless of the lives of the alternatives.

For infinite-life (perpetual) alternatives, the initial cost is annualized simply by multiplying P by i. For finite-life alternatives, the AW through one life cycle is equal to the perpetual equivalent annual worth.

Life-cycle cost analysis is appropriate for systems that have a large percentage of costs in operating and maintenance. LCC analysis helps in the analysis of all costs from design to operation to disposal phases.

PROBLEMS

Annual Worth and Capital Recovery Calculations

6.1 After you have conducted a future worth comparison of alternatives, what do you multiply the FW values by in order to obtain the AW values of the alternatives?

6.2 Assume an alternative has a 3-year life and that you calculated its AW value over its 3-year life cycle. If you were asked to provide the AW value of the alternative for a 4-year study period, is the AW value you calculated for the alternative's 3-year life cycle a valid estimate of the AW over the 4-year study period? Why or why not?

6.3 A machine used to shread cardboard boxes for composting has a first cost of $10,000, an AOC of $7000 per year, a 3-year life, and no appreciable salvage value. Assume you were told that the service provided by the machine will be needed for only 5 years, which entails a repurchase and retention for only 2 years. What salvage value, S, is required after 2 years in order to make the 2-year AW value the same as it is for its 3-year life cycle AW? Assume $i = 10\%$ per year. Solve using factors and, if assigned, using a spreadsheet and Goal Seek.

6.4 A delivery car had a first cost of $30,000, an annual operating cost of $12,000, and an estimated $4000 salvage value after its 6-year life. Due to an economic slowdown, the car will be retained for only 2 years and must be sold now as a used vehicle.
(a) At an interest rate of 10% per year, what must the market value of the 2-year-old vehicle be in order for its AW value to be the same as the AW for a full 6-year life cycle?
(b) Compare your answer in (a) with the first cost and expected salvage after 6 years. Is the required market value a reasonable one, in your opinion?

6.5 As discussed in Section 6.2, either of the following two equations can be applied to determine the amount necessary to recover invested capital and a required return:

$$CR_1 = -P(A/P,i,n) + S(A/F,i,n) \text{ or}$$
$$CR_2 = -[(P - S)(A/P,i,n) + S(i)]$$

For an alternative that has a first cost of $50,000 and a salvage value of $5,000 after its 5-year life, show that the capital recovery calculated using either of these equations is exactly the same. Use an interest rate of 10% per year.

6.6   One way to recover invested capital with interest is to collect the principal P over n years as P/n and also collect the interest on the unrecovered balance. Assume you borrowed $6000 at 10% per year interest with a repayment period of 3 years to purchase a new network controller for your company. (*a*) Use the method above to determine the amount of each of the three payments. (*b*) Determine the amount of each payment if they are all equal.

6.7   Humana Hospital Corporation installed a new MRI machine at a cost of $750,000 this year in its medical professional clinic in Cedar Park. This state-of-the-art system is expected to be used for 5 years and then sold for $125,000. Humana uses a return requirement of 25% per year for all of its medical diagnostic equipment. As a bioengineering student currently serving a coop semester on the management staff of Humana Hospital Corporation in Louisville, Kentucky, you are asked to determine the minimum revenue required each year to realize the expected recovery and return.
(*a*)   What is your answer?
(*b*)   If the AOC is expected to be $80,000 per year, what is the total revenue required to provide for recovery of capital, the 25% return, and the annual expenses?
(*c*)   Write the spreadsheet functions to display your answers.

6.8   White Oaks Properties builds strip shopping centers and small malls. The company plans to replace its refrigeration, cooking, and HVAC equipment with newer models in one entire center built 9 years ago. The original purchase price of the equipment was $638,000 nine years ago and the operating cost has averaged $240,000 per year. Determine the equivalent annual cost of the equipment if the company can now sell it for $184,000. The company's MARR is 25% per year.

6.9   In 2016, Google rented 1000 acres of a historic California airbase in the San Francisco Bay Area for $1.16 billion via a 60-year lease. Annual M&O is expected to be $6.3 million. In addition, Google

will refurbish three hangars at a cost of $2.6 million 4 years from now. Assuming the $1.16 billion represents the present worth of only the lease, what is the equivalent AW of the transaction over the 60-year period at an interest rate of 15% per year?

6.10   U.S. Steel is considering a plant expansion to produce austenitic, precipitation hardened, duplex, and martensitic stainless steel round bars that is expected to cost $13 million now and another $10 million 1 year from now. If total operating costs will be $1.2 million per year starting 1 year from now, and the estimated salvage value of the plant is virtually zero, how much must the company make annually in years 1 through 10 to recover its investment plus a return of 15% per year?

6.11   The Toro Company is expanding its El Paso, Texas plastic molding plant as it continues to transfer work from Juarez, Mexico contractors. The plant bought a $1.1 million precision injection molding machine to make plastic parts for Toro lawn mowers, trimmers, and snow blowers. The plant also spent $275,000 for three smaller plastic injection molding machines to make plastic parts for a new line of sprinkler systems. The plant expects to hire 13 people, including some engineers for the expansion. If the average loaded cost (i.e., including benefits) of each employee is $100,000 per year, determine the annual worth of the new systems and employees over a 5-year study period. Use an investment return rate of 10% per year, and assume a 25% salvage value for the new equipment. Solve using factors and a spreadsheet function.

6.12   Airodyne Wind, Inc. has wind tunnels that can operate vertically or horizontally for evaluating the effects of air flow on a component's PCB response and reliability. The company expects to build a new tunnel that will be outfitted with multiple sensor ports. For the estimates below, calculate the equivalent annual cost of the project.

First cost, $	−800,000
Replacement cost, year 2, $	−300,000
AOC, $/year	−950,000
Salvage value, $	250,000
Life, years	4
Interest rate, %	10

Alternative Comparison

6.13   The Briggs and Stratton Commercial Division designs and manufactures small engines for golf turf maintenance equipment. A robotics-based testing system with support equipment will ensure that their new signature guarantee program entitled "Always Insta-Start" does indeed work for every engine produced. (*a*) Compare the annual worth values of the two systems at MARR = 10% per year. Select the better system. (*b*) Determine the salvage value for the Push System that will make the company indifferent between the two systems.

	Pull System	Push System
Equipment first cost, $	−1,500,000	−2,250,000
AOC, $ per year	−700,000	−600,000
Salvage value, $	100,000	50,000
Estimated life, years	8	8

6.14 Polymer Molding, Inc. is considering two processes for manufacturing storm drains. Plan A involves conventional injection molding that will require making a steel mold at a cost of $2 million. The cost for inspecting, maintaining, and cleaning the molds is expected to be $60,000 per year. Since the cost of materials for this plan is expected to be the same as for the other plan, this cost is not included in the comparison. The salvage value for plan A is expected to be 10% of the first cost. Plan B involves using an innovative process known as virtual engineered composites wherein a floating mold uses an operating system that constantly adjusts the water pressure around the mold and the chemicals entering the process. The first cost to tool the floating mold is only $795,000, but because of the newness of the process, personnel and product-reject costs are expected to be higher than for a conventional process. The company expects the operating costs to be $85,000 for the first year and then decrease to $46,000 per year thereafter. There will be no salvage value with this plan. At an interest rate of 12% per year, which process should the company select on the basis of an annual worth analysis over a 3-year study period?

6.15 Two methods can be used for producing solar panels for electric power generation. Method 1 will have an initial cost of $550,000, an AOC of $160,000 per year, and $125,000 salvage value after its 3-year life. Method 2 will cost $830,000 with an AOC of $120,000, and a $240,000 salvage value after its 5-year life. Assume your boss asked you to determine which method is better, but she wants the analysis done over a 3-year planning period. You estimate the salvage value of method 2 will be 35% higher after 3 years than it is after 5 years. If the MARR is 10% per year, which method should the company select?

6.16 Your brother, a new college graduate, wants to be his own boss. He wants to open a restaurant in a small strip center or acquire and operate a food truck, an increasingly popular mode of eating, especially in larger cities. The restaurant space can be rented for $2200 per month. Modest furnishings and used equipment will have a first cost of $26,000. Income is expected to be $14,100 per month, with expenses for utilities, labor, taxes, etc. expected to average $3700 per month. Alternatively, a kitchen-ready food truck will cost $17,900 to purchase and $900 per month to operate. Income is expected to be $6200 per month. If the salvage values are assumed to be 10% of the first cost for the restaurant and 35% of the first cost of the truck after a 5-year planning period, which alternative is better on the basis of an annual worth comparison at an interest rate of 12% per year, compounded monthly? Also, write the spreadsheet functions to find the AW values.

6.17 A consulting engineering firm is considering two models of SUVs for the company principals. A GM model will have a first cost of $36,000, an operating cost of $4000, and a salvage value of $15,000 after 3 years. A Ford model will have a first cost of $32,000, an operating cost of $3100, and also have a $15,000 resale value, but after 4 years. (*a*) At an interest rate of 15% per year, which model should the consulting firm buy? Conduct an annual worth analysis. (*b*) What are the PW values for each vehicle?

6.18 An international textile company's North America Division must decide which type of fabric cutting machines it will use—straight knife or round knife. The estimates are summarized below. Compare them on the basis of annual worths values at $i = 10\%$ per year using (*a*) factors, and (*b*) single-cell spreadsheet functions.

	Round Knife	Straight Knife
First cost, $	−250,000	−170,000
AOC, $/year	−31,000	−35,000
Overhaul in year 2, $	—	−26,000
Salvage value, $	40,000	10,000
Life, years	6	4

6.19 Two types of robots (Cartesian and Articulated) with the following estimates are under consideration for a dishwasher assembly process. Using an interest rate of 10% per year, determine which one should be selected on the basis of an annual worth analysis.

Robot	Cartesian	Articulated
First cost, $	−300,000	−430,000
AOC, $/year	−60,000	−40,000
Salvage value, $	70,000	95,000
Life, years	4	6

6.20 You have two machines under consideration for an improved automated wrapping process for Snickers Fun Size candy bars as detailed below.
(a) Using an AW analysis, determine which should be selected at $i = 15\%$ per year.
(b) Assume you want machine D to be selected and are willing to extend its estimated life, if necessary. Perform this analysis to ensure D's selection using factors or a spreadsheet.

Machine	C	D
First cost, $	−40,000	−65,000
Annual cost, $/year	−10,000	−12,000
Salvage value, $	12,000	25,000
Life, years	3	6

6.21 Accurate air flow measurement requires straight, unobstructed pipe for a minimum of 10 diameters upstream and 5 diameters downstream of the measuring device. In one particular application, physical constraints compromised the pipe layout, so the engineer was considering installing the air flow probes in an elbow (Plan A), knowing that flow measurement would be less accurate but good enough for process control. This plan would be acceptable for only 2 years, after which a more accurate flow measurement system with the same costs will be available. Plan A has a first cost of $25,000 with annual M&O estimated at $4000. Plan B involves installation of a recently designed submersible air flow probe. The stainless steel probe could be installed in a drop pipe with the transmitter located in a waterproof enclosure on the handrail. The cost of this system will be $88,000, but because it is accurate, it would not have to be replaced for at least 6 years. Its maintenance cost is estimated to be $1400 per year. Neither system will have a salvage value. (a) At $i = 12\%$ per year, select A or B on the basis of an AW comparison. (b) Use future worth analysis to select between A and B.

6.22 An environmental engineer wants to evaluate three different methods for disposing of a non-hazardous chemical waste: land application, fluidized-bed incineration, and private disposal contract. Use the estimates below to help him determine which has the least cost at $i = 10\%$ per year, on the basis of an (a) annual worth, and (b) present worth evaluation.

	Land	Incineration	Contract
First cost, $	−150,000	−900,000	0
AOC, $/year	−95,000	−60,000	−140,000
Salvage value, $	25,000	300,000	0
Life, years	4	6	2

6.23 Blue Whale Moving and Storage recently purchased a warehouse building in Santiago. The manager has two good options for moving pallets of stored goods in and around the facility. Alternative 1 includes a 4000-pound capacity, electric forklift ($P = \$-30,000$; $n = 12$ years; AOC = $\$-1000$ per year; $S = \$8000$), and 500 new pallets at $10 each. The forklift operator's annual salary and indirect benefits are estimated at $32,000.

Alternative 2 involves the use of two electric pallet movers ("walkies") each with a 3000-pound capacity (for each mover, $P = \$-2000$; $n = 4$ years; AOC = $\$-150$ per year; no salvage) and 800 pallets at $10 each. The two operators' salaries and benefits will total $55,000 per year. For both options, new pallets are purchased now and every 2 years that the equipment is in use. (a) If the MARR is 8% per year, select the better alternative. (b) Rework using a spreadsheet solution.

6.24 Two mutually exclusive alternatives have the estimates shown below. Use annual worth analysis to determine which should be selected at an interest rate of 10% per year.

	Q	R
First cost, $	−42,000	−80,000
AOC, $ per year	−6,000	−7,000 in year 1, increasing by $1,000 per year thereafter
Salvage value, $	0	4,000
Life, years	2	4

6.25 A chemical engineer is considering two sizes of pipes for moving distillate from a refinery to the tank farm. A small pipeline will cost less to purchase (including valves and other appurtenances), but will have a high head loss and, therefore, a higher pumping cost. The small pipeline will cost $1.7 million installed and have an operating cost of $12,000 per month. A larger-diameter pipeline will cost $2.1 million installed, but its operating cost will be only $8000 per month. Which pipe size is more economical at an interest rate of 1% per month on the basis of an annual worth analysis? Assume the salvage value is 10% of the first cost for each pipeline at the end of the 10-year project.

Permanent Investments

6.26 Calculate the perpetual equivalent annual cost (years 1 to ∞) of $1,000,000 now and $1,000,000 three years from now at an interest rate of 10% per year.

6.27 How much must you deposit each year into your retirement account starting *now* and continuing through year 9, if you want to be able to withdraw $80,000 per year forever, beginning 30 years from now? Assume the account earns interest at (*a*) the generous rate of 10% per year, and (*b*) a more conservative rate of 4% per year. Compare the two amounts.

6.28 The State of Chiapas, Mexico, decided to fund a program for improving reading skills in elementary school students. The first cost is $300,000 now, and an update amount of $100,000 every 5 years *forever*. Determine the perpetual equivalent annual cost at an interest rate of 10% per year.

6.29 Cisco, Inc. has a proposal from the Engineering Planning Division to invest some of the Cisco retained earnings in the design, testing, and development of the next generation of smart grids useful in the Internet of Things (IoT) environment. The initial investment projection is $5,000,000 in year 0, $2,000,000 in year 10, and $100,000 in years 11 and beyond. At $i = 10$% per year, calculate the infinite-life equivalent annual cost in years 0 *through infinity* of the proposal.

6.30 Determine the difference in annual worth between an investment of $100,000 per year for 50 years and an investment of $100,000 per year forever at an interest rate of (*a*) 5% and (*b*) 10% per year. Compare the two differences.

6.31 Harmony Auto Group sells and services imported and domestic cars. The owner wants to evaluate the option of outsourcing all of its new auto warranty service work to Winslow, Inc., a private repair service that works on any make and year car. Either a 5-year contract basis or 10-year license agreement is available from Winslow. Revenue from the manufacturer will be shared with no added cost incurred by the car/warranty owner. Alternatively, Harmony can continue to do warranty work in-house for the foreseeable future. Use the estimates made by the Harmony owner to perform an annual worth evaluation at 10% per year to select the best option. Monetary values are in $ million units.

	Contract	License	In-house
First cost, $	0	−2	−20
Annual cost, $ per year	−1	−0.2	−4
Annual revenue, $ per year	2.5	1.3	8
Life, years	5	10	∞

6.32 The cost associated with maintaining rural highways follows a predictable pattern. There are usually no costs for the first 3 years, but thereafter maintenance is required for restriping, weed control, light replacement, shoulder repairs, etc. For one section of state highway S102, these costs are projected to be $6000 in year 3, $7000 in year 4,

and amounts increasing by $1000 per year through the highway's expected 30-year life.

(*a*) Assuming it is replaced with a similar roadway, determine the perpetual equivalent annual cost (years 1 to ∞) at an interest rate of 8% per year.

(*b*) Verify your factor-based answer on a spreadsheet in the most efficient way you know.

6.33 ABC Beverage, LLC purchases its 355-ml cans in large bulk from Wald-China Can Corporation. The finish on the anodized aluminum surface is produced by mechanical finishing technologies called brushing or bead blasting. Engineers at Wald are switching to more efficient, faster, and cheaper machines to supply ABC. Use the estimates and MARR = 8% per year to select between the two alternatives.

Brush alternative: $P = \$-400{,}000$; $n = 10$ years; $S = \$50{,}000$; nonlabor AOC $= \$-60{,}000$ in year 1, decreasing by $5000 annually starting in year 2

Bead blasting alternative: $P = \$-400{,}000$; n is large, assume permanent; no salvage; nonlabor AOC $= \$-70{,}000$ per year

Life Cycle Cost

6.34 An international aerospace contractor has been asked by a municipal police department to estimate and analyze the life cycle costs for a proposed drone surveillance system to monitor traffic patterns and congestion within the central thoroughfares of the city. The list of items include the following general categories: R&D costs (R&D), nonrecurring investment costs (NRI), recurring investment costs (RI), scheduled and unscheduled maintenance costs (Maint), equipment usage costs (Equip), and disposal costs (Disp). The costs (in $ million units) for the 20-year life cycle have been estimated. Calculate the annual LCC at an interest rate of 7% per year.

Year	R&D	NRI	RI	Maint	Equip	Disp
0	5.5	1.1				
1	3.5					
2	2.5					
3	0.5	5.2	1.3	0.6	1.5	
4		10.5	3.1	1.4	3.6	
5		10.5	4.2	1.6	5.3	
6–10			6.5	2.7	7.8	
11–20			2.2	3.5	8.5	
18–20						2.7

6.35 A manufacturing software engineer at a major aerospace corporation has been assigned the management responsibility of a project to design, build, test, and implement AREMSS, a new-generation automated scheduling system for routine and expedited maintenance. Reports on the disposition of each service will also be entered by field personnel, then filed and archived by the system. The initial application will be on existing Air Force in-flight refueling aircraft. The system is expected to be widely used over time for other aircraft maintenance scheduling. Once fully implemented, enhancements will have to be made, but the system is expected to serve as a worldwide scheduler for up to 15,000 separate aircraft. The engineer, who must make a presentation next week of the best estimates of costs over a 20-year life period, has decided to use the life-cycle cost approach of cost estimation. Use the following information to determine the current annual LCC at 6% per year for the AREMSS scheduling system.

| Cost Category | Cost in Year ($ millions) | | | | | | | |
	1	**2**	**3**	**4**	**5**	**6 on**	**10**	**18**
Field study	0.5							
Design of system	2.1	1.2	0.5					
Software design		0.6	0.9					
Hardware purchases			5.1					
Beta testing		0.1	0.2					
Users manual dev		0.1	0.1	0.2	0.2	0.06		
Sys implementation				1.3	0.7			
Field hardware				0.4	6.0	2.9		
Training trainers				0.3	2.5	2.5	0.7	
Software upgrades						0.6	3.0	3.7

6.36 A medium-size municipality plans to develop a software system to assist in project selection during the next 10 years. A life-cycle cost approach has been used to categorize costs into development, programming, operating, and support costs for each alternative. There are three alternatives under consideration, identified as A (tailored system), B (adapted system), and C (current system). Use an annual life-cycle cost approach to identify the best alternative at 8% per year.

Alternative	Component	Cost
A	Development	$250,000 now, $150,000 years 1–4
	Programming	$45,000 now, $35,000 years 1, 2
	Operation	$50,000 years 1–10
	Support	$30,000 years 1–5
B	Development	$10,000 now
	Programming	$45,000 year 0, $30,000 years 1–3
	Operation	$80,000 years 1–10
	Support	$40,000 years 1–10
C	Operation	$175,000 years 1–10

EXERCISES FOR SPREADSHEETS

6.37 Problem 6.9 described Google's rental agreement for an airbase in California. It included the following estimates: $P = \$1.16$ billion; M&O = $0.0063 billion per year; refurbishment cost = $0.0026 billion in year 4; $n = 60$ years; and $i = 15\%$ per year.
(a) You calculated the AW over 60 years. Now, use spreadsheet functions to display AW, PW, the capitalized cost (CC), and the difference between PW and CC.
(b) Help the CFO of Google perform a capital recovery analysis on the agreement for several retention periods, that is, n values, varying from 20 to 60 years (in 10-year increments) and infinity. Explain the meaning of the results to the CFO.

6.38 (Note: This problem requires the development of a spreadsheet and associated scatter chart to answer the economic questions.) As you came to work this morning, your department head asked you to perform an analysis of installing MAP/TOP in the manufacturing planning division of the company. (MAP/TOP is a business and manufacturing environment communications integration software system developed by Boeing Computer Services.) The entire system would be implemented and operated by one of two consultant groups, Hi Tone or Extra-S. Estimates have been developed, but the first cost (includes MAP/TOP software, hardware, implementation and training costs) that your company will agree to pay is still in question. The possible variation of the initial first cost estimate P included in the consultants' proposals is indicated below as a percentage of P.

Consultant	Hi Tone	Extra-S
First cost estimate P, $	−500,000	−750,000
Variation in P, %	From 100% to 130% of P	From 80% to 130% of P
AOC, $/year	−150,000	−120,000
Salvage value, $	125,000	240,000
Life of contract, years	3	5

(a) Develop the spreadsheet accompanied by a scatter chart that graphs the AW values versus the possible variations in first cost shown as percentage changes from 100% of P (in 10% increments). Use $i = 10\%$ per year.

(b) Select the better alternative under the following negotiated conditions:

1. Company agrees to 100% of P for Hi Tone and 110% of P for Extra-S
2. Company agrees to 110% of P for Hi Tone; Extra-S increases first cost to 130% of P
3. Company agrees to 100% of P for both consultants
4. Company agrees to 100% of P for Hi Tone; Extra-S agrees to 90% of P

6.39 Three methods to dispose of nonhazardous waste have been developed—land application, fluidized-bed incineration, and private disposal contract. Use AW analysis and an associated scatter chart of AW versus i values to select the economically best alternative for interest rates between $i = 6\%$ and $i = 24\%$ in 3% increments. (Note: This is an extension of Problem 6.22.)

	Land	Incineration	Contract
First cost, $	−150,000	−900,000	0
AOC, $/year	−95,000	−60,000	−140,000
Salvage value, $	25,000	300,000	0
Life, years	4	6	2

6.40 Cost increases imposed by design changes introduced at different times during advancing stages of the product life cycle (PLC) usually rise dramatically as the change is introduced later in the life cycle of the product. Though the numbers vary from one industry to another, one cost-increase model looks something like the following:

PLC Stage	Cost Multiplier	Example, $
Conceptual	1	100
Production planning	3	300
Prototype/testing	5	500
Manufacturing (per unit)	5–50	500–5000
Sales and after-sales	20–100	2000–10,000

Hamound Industries designed, developed, and is now manufacturing for sales an improved version of its model GR1 smoke detector system that can be used on high speed trains, such as the TGV (*Train à Grande Vitesse* in French). The original estimates, developed during the conceptual design stage, are shown below. During the testing stage, a major safety flaw was discovered and significant design changes were made, *affecting only the equipment first cost estimate*, fortunately. Now, during manufacturing, a significant problem has emerged that makes the internet-connectivity feature difficult to manufacture with the level of required reliability. This change will increase the *manufacturing cost per unit* by a factor of 5. The initial economic analysis, conducted during the conceptual stage, using the AW value at $i = 7\%$ per year, showed very promising results. However, the design changes have affected this positive analysis. Use the original estimates and the PLC cost multipliers summarized above to perform AW analyses for the four stages from conceptual through manufacturing to determine what has happened to the economic viability of the improved product.

Equipment original first cost P, $	−6,000,000
Equipment AOC, $ per year	−300,000
Manufacturing cost, $ per unit	2.75
Revenue, $ per unit	12.75
Number of units per year	500,000
Life, years	10
Salvage value, $	Constant at 10% of original P

Stockdisc

ADDITIONAL PROBLEMS AND FE EXAM REVIEW QUESTIONS

6.41 An automaton asset with a high first cost of $10 million has required capital recovery (CR) of $1,985,000 per year. The correct interpretation of this CR value is that:
  (*a*) the owner must pay an additional $1,985,000 each year to retain the asset.
  (*b*) each year of its expected life, a net revenue of $1,985,000 must be realized to recover the $10 million first cost and the required rate of return on this investment.
  (*c*) each year of its expected life, a net revenue of $1,985,000 must be realized to recover the $10 million first cost.
  (*d*) the services provided by the asset will stop if less than $1,985,000 in net revenue is reported in any year.

6.42 If you have the capitalized cost of a certain alternative that has a 5-year life, you can get its annual worth by:
  (*a*) multiplying the CC by i
  (*b*) multiplying the CC by $(A/F,i,5)$
  (*c*) multiplying the CC by $(P/A,i,5)$
  (*d*) multiplying the CC by $(A/P,i,5)$

6.43 The AW values of three cost alternatives are $-23,000 for alternative A, $-21,600 for B, and $-27,300 for C. On the basis of these results, the decision is to:
  (*a*) select alternative A
  (*b*) select alternative B
  (*c*) select alternative C
  (*d*) select the Do Nothing alternative

6.44 The initial cost of a packed-bed degassing reactor for removing trihalomethanes from potable water is $84,000. The annual operating cost for power, site maintenance, etc. is $13,000. If the salvage value of the pumps, blowers, and control systems is expected to be $9000 at the end of 10 years, the AW of the packed-bed reactor, at an interest rate of 8% per year, is closest to:
  (*a*) $-26,140
  (*b*) $-25,518
  (*c*) $-24,900
  (*d*) $-13,140

6.45 The AW amounts of three revenue alternatives are $-23,000 for alternative A, $-21,600 for B, and $-27,300 for C. On the basis of these AW values, the correct decision is to:
  (*a*) select alternative A
  (*b*) select alternative B

  (*c*) select alternative C
  (*d*) select the Do Nothing alternative

6.46 If you have the annual worth of an alternative with a 5-year life, you can calculate its perpetual annual worth by:
  (*a*) no calculation needed. The perpetual annual worth is equal to the annual worth
  (*b*) multiplying the annual worth by $(A/P,i,5)$
  (*c*) dividing the annual worth by i
  (*d*) multiplying the annual worth by i

6.47 The annual worth for years 1 through infinity of $50,000 now, $10,000 per year in years 1 through 15, and $20,000 per year in years 16 through infinity at 10% per year is closest to:
  (*a*) less than $16,900
  (*b*) $16,958
  (*c*) $17,395
  (*d*) $19,575

6.48 An alumnus of West Virginia University wishes to start an endowment that will provide scholarship money of $40,000 per year beginning in year 5 and continuing indefinitely. The donor plans to give money *now* and for each of the next 2 years. If the size of each donation is exactly the same, the amount that must be donated each year at $i = 8\%$ per year is closest to:
  (*a*) $190,820
  (*b*) $122,280
  (*c*) $127,460
  (*d*) $132,040

6.49 You will make equal deposits into your retirement account each year for 10 years starting *now* (i.e., years 0–9). If you expect to withdraw $50,000 per year forever beginning 30 years from now, and the funds will earn interest at 10% per year, the size of the 10 deposits is nearest:
  (*a*) $4239   (*b*) $4662
  (*c*) $4974   (*d*) $5471

6.50 To get the AW of a cash flow of $10,000 that occurs every 10 years forever, with the first one occurring now, it is correct to:
  (*a*) multiply the $10,000 by $(A/P,i,10)$
  (*b*) multiply the $10,000 by $(A/F,i,10)$
  (*c*) multiply the $10,000 by i
  (*d*) multiply the $10,000 by $(A/F,i,n)$ and then multiply by i

6.51 The estimates for two alternatives are to be compared on the basis of their perpetual equivalent annual worth. At an interest rate of 10% per year, the equation that represents the perpetual AW of Y1 is:

Alternative	X1	Y1
First cost, $	−50,000	−90,000
Annual cost, $/year	−10,000	−4000
Salvage value, $	13,000	15,000
Life, years	3	6

(a) $AW_{Y1} = -90,000(0.10) - 4000 + 15,000(0.10)$

(b) $AW_{Y1} = -90,000(0.10) - 4000 + 15,000(A/F,10\%,6)$

(c) $AW_{Y1} = -90,000(0.10) - 4000 - 15,000(P/F,10\%,3)(0.10) + 15,000(0.10)$

(d) $AW_{Y1} = -90,000(A/P,10\%,6) - 4000 + 15,000(A/F,10\%,6)$

Problems 6.52–6.54 are based on the following cash flows and an interest rate of 10% per year.

Alternative	X	Y
First cost, $	−200,000	−800,000
Annual cost, $/year	−60,000	−10,000
Salvage value, $	20,000	150,000
Life, years	5	∞

6.52   In comparing the alternatives by the annual worth method, the AW of X is determined by the following equation:

(a) $-200,000(0.10) - 60,000 + 20,000(0.10)$

(b) $-200,000(A/P,10\%,5) - 60,000 + 20,000(A/F,10\%,5)$

(c) $-200,000(A/P,10\%,5) - 60,000 - 20,000(A/F,10\%,5)$

(d) $-200,000(0.10) - 60,000 + 20,000(A/F,10\%,5)$

6.53   The annual worth of perpetual service for alternative X is represented by the following equation:

(a) $-200,000(0.10) - 60,000 + 20,000(0.10)$

(b) $-200,000(A/P,10\%,5) - 60,000 + 20,000(A/F,10\%,5)$

(c) $-200,000(A/P,10\%,5) - 60,000 - 20,000(A/F,10\%,5)$

(d) $-200,000(0.10) - 60,000 + 20,000(A/F,10\%,5)$

6.54   The annual worth of alternative Y is closest to:

(a) $−50,000

(b) $−76,625

(c) $−90,000

(d) $−92,000

CASE STUDY

ANNUAL WORTH ANALYSIS—THEN AND NOW

Background and Information

Harry, owner of an automobile battery distributorship in Atlanta, Georgia, performed an economic analysis 3 years ago when he decided to place surge protectors in-line for all his major pieces of testing equipment. The estimates used and the annual worth analysis at MARR = 15% are summarized below. Two different manufacturers' protectors were compared.

	PowrUp	Lloyd's
Cost and installation, $	−26,000	−36,000
Annual maintenance cost, $ per year	−800	−300
Salvage value, $	2,000	3,000
Equipment repair savings, $	25,000	35,000
Useful life, years	6	10

The spreadsheet in Figure 6–9 is the one Harry used to make the decision. Lloyd's was the clear choice due to its substantially larger AW value. The Lloyd's protectors were installed.

During a quick review this last year (year 3 of operation), it was obvious that the maintenance costs and repair savings have not followed (and will not follow) the estimates made 3 years ago. In fact, the maintenance contract cost (which includes quarterly inspection) is going from $300 to $1200 per year next year and will then increase 10% per year for the next 10 years. Also, the repair savings for the last 3 years were $35,000, $32,000, and $28,000, as best as Harry can determine. He believes savings will decrease by $2000 per year hereafter. Finally, these 3-year-old protectors are worth nothing on the market now, so the salvage in 7 years is zero, not $3000.

◇	A	B	C	D	E	F	G
1	MARR =	15%					
2							
3			**PowrUp**			**Lloyd's**	
4		Investment	Annual	Repair	Investment	Annual	Repair
5	Year	and salvage	maintenance	savings	and salvage	maintenance	savings
6	0	-26,000	0	0	-36,000	0	0
7	1	0	-800	25,000	0	-300	35,000
8	2	0	-800	25,000	0	-300	35,000
9	3	0	-800	25,000	0	-300	35,000
10	4	0	-800	25,000	0	-300	35,000
11	5	0	-800	25,000	0	-300	35,000
12	6	2,000	-800	25,000	0	-300	35,000
13	7				0	-300	35,000
14	8				0	-300	35,000
15	9				0	-300	35,000
16	10				3,000	-300	35,000
17	AW element	-6,642	-800	25,000	-7,025	-300	35,000
18	**Total AW**			$ 17,558			$ 27,675

Figure 6–9
Annual worth analysis of surge protector alternatives, case study.

Case Study Exercises

1. Plot a graph of the newly estimated maintenance costs and repair savings projections, assuming the protectors last for seven more years.

2. With these new estimates, what is the recalculated AW for the Lloyd's protectors? Use the old first cost and maintenance cost estimates for the first 3 years. If these estimates had been made 3 years ago, would Lloyd's still have been the economic choice?

3. How has the capital recovery amount changed for the Lloyd's protectors with these new estimates?

CHAPTER 7

Rate of Return Analysis: One Project

Yunus Arakon/Getty Images

LEARNING OUTCOMES

Purpose: Understand the meaning of rate of return and perform an ROR evaluation of a single project.

SECTION	TOPIC	LEARNING OUTCOME
7.1	Definition	• State and understand the meaning of rate of return.
7.2	Calculate ROR	• Use a PW or AW relation to determine the ROR of a series of cash flows.
7.3	Cautions	• State the difficulties of using the ROR method, relative to the PW and AW methods.
7.4	Multiple RORs	• Determine the maximum number of possible ROR values and their values for a specific cash flow series.
7.5	Calculate EROR	• Determine the external rate of return using the techniques of modified ROR and return on invested capital.
7.6	Bonds	• Calculate the nominal and effective rate of return for a bond investment.

T he most commonly quoted measure of economic worth for a project or alternative is its rate of return (ROR). Whether it is a corporate engineering project with cash flow estimates or an investment in a stock or bond by an individual, the ROR is a well-accepted way of determining if the project or investment is economically acceptable. Compared to the PW or AW value, the ROR is a generically different type of measure of worth, as is discussed in this chapter. Correct procedures to calculate a rate of return using a PW or AW relation are explained here, as are some cautions necessary when the ROR technique is applied to a single project's cash flows.

The ROR is known by other names such as the *internal rate of return* (IROR), which is the technically correct term, and *return on investment* (ROI). We will discuss the computation of ROI in the latter part of this chapter.

In some cases, more than one ROR value may satisfy the PW or AW equation. This chapter describes how to recognize this possibility and an approach to find the **multiple values.** Alternatively, one reliable ROR value can be obtained by using additional information established separately from the project cash flows. Two of the techniques are covered: the modified ROR technique and the ROIC (return on invested capital) technique.

Only one alternative is considered here; Chapter 8 applies these same principles to multiple alternatives. Finally, the rate of return for a bond investment is discussed.

7.1 Interpretation of a Rate of Return Value ●●○

From the perspective of someone who has borrowed money, the interest rate is applied to the *unpaid balance* so that the total loan amount and interest are paid in full exactly with the last loan payment. From the perspective of a lender of money, there is an *unrecovered balance of the principal* of the loan at each time period. The interest rate is the return on this unrecovered balance so that the total amount lent and the interest are recovered exactly with the last receipt. *Rate of return* describes both of these perspectives.

Rate of return (ROR) is the rate paid on the **unpaid balance of borrowed money,** or the rate earned on the **unrecovered balance of an investment,** so that the final payment or receipt brings the **balance to exactly zero** with interest considered.

Rate of return

The rate of return is expressed as a percent per period, for example, $i = 10\%$ per year. It is stated as a positive percentage; the fact that interest paid on a loan is actually a negative rate of return from the borrower's perspective is not considered. The numerical value of i can range from -100% to infinity, that is, $-100\% \le i < \infty$. In terms of an investment, a return of $i = -100\%$ means the entire amount is lost.

The definition above does not state that the rate of return is on the initial amount of the investment or initial principal of a loan; rather it is on the **unrecovered balance,** which **changes each time period.** Example 7.1 illustrates this difference.

EXAMPLE 7.1

To get started in a new telecommuting position with AB Hammond Engineers, Jane took out a $1000 loan at $i = 10\%$ per year for 4 years to buy home office equipment. From the lender's perspective, the investment in this young engineer is expected to produce an equivalent net cash flow (NCF) of $315.47 for each of 4 years.

$$A = \$1000(A/P,10\%,4) = \$315.47$$

This represents a 10% per year rate of return on the unrecovered balance. Compute the amount of the unrecovered investment for each of the 4 years using (*a*) the rate of return on the unrecovered balance (the correct basis) and (*b*) the return on the initial $1000 investment. (*c*) Explain why all of the initial $1000 amount is not recovered by the final payment in part (*b*). (*d*) Determine the unrecovered balance after 3 years in Table 7–1 using factors rather than detailing the amounts for each year. Also, write a single-cell spreadsheet function that will display the same result.

TABLE 7–1	Unrecovered Balances Using a Rate of Return of 10% on the Unrecovered Balance				
(1)	**(2)**	**(3) = 0.10 × (2)**	**(4)**	**(5) = (4) − (3)**	**(6) = (2) + (5)**
Year	Beginning Unrecovered Balance	Interest on Unrecovered Balance	Cash Flow	Recovered Amount	Ending Unrecovered Balance
0	—	—	$−1000.00	—	$−1000.00
1	$−1000.00	$100.00	+315.47	$215.47	−784.53
2	−784.53	78.45	+315.47	237.02	−547.51
3	−547.51	54.75	+315.47	260.72	−286.79
4	−286.79	28.68	+315.47	286.79	0
		$261.88		$1000.00	

TABLE 7–2	Unrecovered Balances Using a 10% Return on the Initial Amount				
(1)	**(2)**	**(3) = 0.10 × (2)**	**(4)**	**(5) = (4) − (3)**	**(6) = (2) + (5)**
Year	Beginning Unrecovered Balance	Interest on Initial Amount	Cash Flow	Recovered Amount	Ending Unrecovered Balance
0	—	—	$−1000.00	—	$−1000.00
1	$−1000.00	$100	+315.47	$215.47	−784.53
2	−784.53	100	+315.47	215.47	−569.06
3	−569.06	100	+315.47	215.47	−353.59
4	−353.59	100	+315.47	215.47	−138.12
		$400		$861.88	

Solution

(a) Table 7–1 shows the unrecovered balance of the principal at the end of each year in column 6 using the 10% rate on the *unrecovered balance at the beginning of the year*. After 4 years the total $1000 is recovered, and the balance in column 6 is exactly zero.

(b) Table 7–2 shows the unrecovered balance if the 10% return is always figured on the *initial $1000*. Column 6 in year 4 shows a remaining unrecovered amount of $138.12 because only $861.88 is recovered in the 4 years (column 5).

(c) As shown in column 3, a total of $400 in interest must be earned if the 10% return each year is based on the initial amount of $1000. However, only $261.88 in interest must be earned if a 10% return on the unrecovered balance is used. There is more of the annual cash flow available to reduce the remaining loan when the rate is applied to the unrecovered balance as in part (a) and Table 7–1. Figure 7–1 illustrates the correct interpretation of rate of return in Table 7–1. Each year the $315.47 receipt represents 10% interest on the unrecovered balance in column 2 plus the recovered amount in column 5.

(d) To determine the unrecovered balance after any year in the repayment schedule, simply find the future worth of the loan principal at the end of the year in question using the F/P factor and remove the equivalent future value of all payments made thus far, 3 years in this case.

For this $1000, 4-year, 10% per year loan,

$$\text{Unrecovered balance, year 3} = -1000(F/P,10\%,3) + 315.47(F/A,10\%,3)$$
$$= -1331.00 + 1044.21$$
$$= \$-286.79$$

This is the same amount shown in Table 7–1, column (6), year 3. The FV spreadsheet function will accomplish the same result. In this case, = −FV(10%,3,315.47,−1000) will display $−286.79. (The minus sign on the FV function forces the result to correspond to the signing convention of Table 7–1.)

Because rate of return is the interest rate on the unrecovered balance, the computations in *Table 7–1 for part (a) present a correct interpretation of a 10% rate of return.* Clearly, an interest

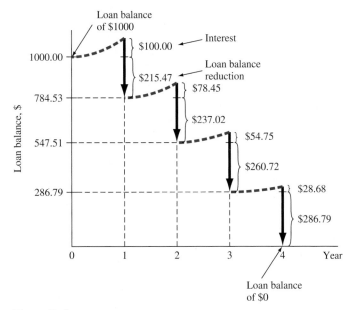

Figure 7–1
Plot of unrecovered balances and 10% per year rate of return on a $1000 amount, Table 7–1.

rate applied only to the principal represents a higher rate than is stated. In practice, a so-called add-on interest rate is frequently based on principal only, as in part (*b*). This is sometimes referred to as the *installment financing* problem.

Installment financing can be discovered in many forms in everyday finances. One popular example is a "no-interest program" offered by retail stores on the sale of major appliances, audio and video equipment, furniture, and other consumer items. Many variations are possible, but in most cases, if the purchase is not paid for in full by the time the promotion is over, usually 6 months to 1 year later, *finance charges are assessed from the original date of purchase.* Further, the program's fine print may stipulate that the purchaser use a credit card issued by the retail company, which often has a higher interest rate than that of a regular credit card, for example, 24% per year compared to 15% per year. In all these types of programs, the one common theme is more interest paid over time by the consumer. Usually, the correct definition of *i* as interest on the unpaid balance does not apply directly; *i* has often been manipulated to the financial disadvantage of the purchaser. This was demonstrated by Example 4.4 using the Credit Card Case in Chapter 4.

7.2 Rate of Return Calculation Using a PW or AW Relation ●●●

The ROR value is determined in a generically different way compared to the PW or AW value for a series of cash flows. For a moment, consider only the PW relation for a cash flow series. Using the MARR, which is established independent of any particular project's cash flows, a mathematical relation determines the PW value in actual monetary units, say, dollars or euros. For the ROR values calculated in this and later sections, **only the cash flows themselves** are used to determine an interest rate that balances the PW relation. Therefore, ROR may be considered a relative measure, while PW and AW are absolute measures. Since the resulting interest rate depends only on the cash flows themselves, the correct term is **internal rate of return (IROR);** however, the term *ROR* is used interchangeably. Another definition of rate of return is based on our previous interpretations of PW and AW.

The *rate of return* is the interest rate that makes the present worth or annual worth of a cash flow series exactly equal to 0.

Rate of return

To determine the rate of return, develop the ROR equation using either a PW or AW relation, set it equal to 0, and solve for the interest rate. Alternatively, the present worth of cash outflows (costs and disbursements) PW_O may be equated to the present worth of cash inflows (revenues and savings) PW_I. That is, solve for i using either of the relations

$$0 = PW \qquad\qquad [7.1]$$

or

$$PW_O = PW_I$$

The annual worth approach utilizes the AW values in the same fashion to solve for i.

$$0 = AW \qquad\qquad [7.2]$$

or

$$AW_O = AW_I$$

The i value that makes these equations numerically correct is called i^*. It is the root of the ROR relation. To determine if the investment project's cash flow series is viable, compare i^* with the established MARR.

The guideline is as follows:

Project evaluation

If $i^* \geq$ MARR, accept the project as economically viable.
If $i^* <$ MARR, the project is not economically viable.

The purpose of engineering economy calculations is *equivalence* in PW or AW terms for a stated $i \geq 0\%$. In ROR calculations, the objective is to *find the interest rate i^** at which the cash flows are equivalent. The calculations are the reverse of those made in previous chapters, where the interest rate was known. For example, if you deposit $1000 now and are promised payments of $500 three years from now and $1500 five years from now, the ROR relation using PW factors and Equation [7.1] is

$$1000 = 500(P/F,i^*,3) + 1500(P/F,i^*,5) \qquad\qquad [7.3]$$

The value of i^* that makes the equality correct is to be determined (see Figure 7–2). If the $1000 is moved to the right side of Equation [7.3], we have the form $0 = $ PW.

$$0 = -1000 + 500(P/F,i^*,3) + 1500(P/F,i^*,5)$$

The equation is solved for $i^* = 16.9\%$ by hand using trial and error or using a spreadsheet function. The rate of return will always be greater than zero if the total amount of cash inflow is greater than the total amount of outflow, when the time value of money is considered. Using $i^* = 16.9\%$, a graph similar to Figure 7–1 can be constructed. It will show that the unrecovered balances each year, starting with $-1000 in year 1, are exactly recovered by the $500 and $1500 receipts in years 3 and 5.

It should be evident that ROR relations are merely a rearrangement of a PW equation. That is, if the above interest rate is known to be 16.9%, and it is used to find the PW of $500 three years from now and $1500 five years from now, the PW relation is

$$PW = 500(P/F,16.9\%,3) + 1500(P/F,16.9\%,5) = \$1000$$

Figure 7–2
Cash flow for which a value of i is to be determined.

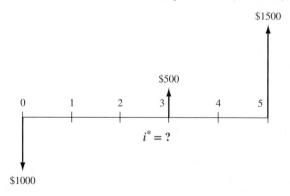

This illustrates that rate of return and present worth equations are set up in exactly the same fashion. The only differences are what is given and what is sought.

There are several ways to determine i^* once the PW relation is established: solution via trial and error by hand, using a programmable calculator, and solution by spreadsheet function. The spreadsheet is faster; the first method helps in understanding how ROR computations work. We summarize two methods here and in Example 7.2. Refer to Appendix D for the discussion about solution by calculator.

i* Using Trial and Error   The general procedure of using a PW-based equation is as follows:

1. Draw a cash flow diagram.
2. Set up the rate of return equation in the form of Equation [7.1].
3. Select values of i by trial and error until the equation is balanced.

When the trial-and-error method is applied to determine i^*, it is advantageous in step 3 to get fairly close to the correct answer on the first trial. If the cash flows are combined in such a manner that the income and disbursements can be represented by a *single factor* such as P/F or P/A, it is possible to look up the interest rate (in the tables) corresponding to the value of that factor for n years. The problem, then, is to combine the cash flows into the format of only one of the factors. This may be done through the following procedure:

1. Convert all *disbursements* into either single amounts (P or F) or uniform amounts (A) by neglecting the time value of money. For example, if it is desired to convert an A to an F value, simply multiply the A by the number of years n. The scheme selected for movement of cash flows should be the one that minimizes the error caused by neglecting the time value of money. That is, if most of the cash flow is an A and a small amount is an F, convert the F to an A rather than the other way around.
2. Convert all *receipts* to either single or uniform values.
3. Having combined the disbursements and receipts so that a P/F, P/A, or A/F format applies, use the interest tables to find the approximate interest rate at which the P/F, P/A, or A/F value is satisfied. The rate obtained is a good estimate for the first trial.

It is important to recognize that this first-trial rate is only an *estimate* of the actual rate of return, because the time value of money is neglected. The procedure is illustrated in Example 7.2.

i* by Spreadsheet   The fastest way to determine an i^* value when there is a series of equal cash flows (A series) is to apply the RATE function. This is a powerful one-cell function, where it is acceptable to have a separate P value in year 0 and a separate F value in year n. The format is

$$= \textbf{RATE}(n, A, P, F) \qquad [7.4]$$

When cash flows vary from year to year (period to period), the best way to find i^* is to enter the net cash flows into contiguous cells (including any \$0 amounts) and apply the IRR function in any cell. The format is

$$= \textbf{IRR}(\textbf{first\_cell:last\_cell,guess}) \qquad [7.5]$$

where "guess" is the i value at which the function starts searching for i^*.

The PW-based procedure for sensitivity analysis and a graphical estimation of the i^* value is as follows:

1. Draw the cash flow diagram.
2. Set up the ROR relation in the form of Equation [7.1], PW = 0.
3. Enter the cash flows onto the spreadsheet in contiguous cells.
4. Develop the IRR function to display i^*.
5. Use the NPV function to develop a PW graph (PW versus i values). This graphically shows the i^* value at which PW = 0.

EXAMPLE 7.2

Applications of green, lean manufacturing techniques coupled with value stream mapping can make large financial differences over future years while placing greater emphasis on environmental factors. Engineers with Monarch Paints have recommended to management an investment of $200,000 now in novel methods that will reduce the amount of wastewater, packaging materials, and other solid waste in their consumer paint manufacturing facility. Estimated savings are $15,000 per year for each of the next 10 years and an additional savings of $300,000 at the end of 10 years in facility and equipment upgrade costs. Determine the rate of return using hand and spreadsheet solutions.

Solution by Hand

Use the trial-and-error procedure based on a PW equation.

1. Figure 7–3 shows the cash flow diagram.
2. Use Equation [7.1] format for the ROR equation.

$$0 = -200,000 + 15,000(P/A,i^*,10) + 300,000(P/F,i^*,10) \qquad [7.6]$$

3. Use the estimation procedure to determine i for the first trial. All income will be regarded as a single F in year 10 so that the P/F factor can be used. The P/F factor is selected because most of the cash flow ($300,000) already fits this factor and errors created by neglecting the time value of the remaining money will be minimized. Only for the first estimate of i, define $P = \$200,000$, $n = 10$, and $F = 10(15,000) + 300,000 = \$450,000$. Now we can state that

$$200,000 = 450,000(P/F,i^*,10)$$
$$(P/F,i^*,10) = 0.444$$

The roughly estimated i^* is between 8% and 9%. Use 9% as the first trial because this approximate rate for the P/F factor will be lower than the true value when the time value of money is considered.

$$0 = -200,000 + 15,000(P/A,9\%,10) + 300,000(P/F,9\%,10)$$
$$0 < \$22,986$$

The result is positive, indicating that the return is more than 9%. Try $i = 11\%$.

$$0 = -200,000 + 15,000(P/A,11\%,10) + 300,000(P/F,11\%,10)$$
$$0 > \$-6002$$

Since the interest rate of 11% is too high, linearly interpolate between 9% and 11%.

$$i^* = 9.00 + \frac{22,986 - 0}{22,986 - (-6002)}(2.0)$$
$$= 9.00 + 1.58 = 10.58\%$$

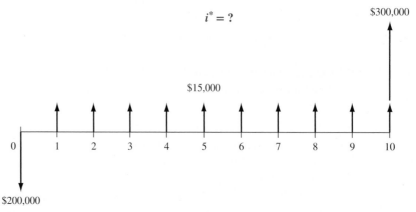

Figure 7–3
Cash flow diagram, Example 7.2.

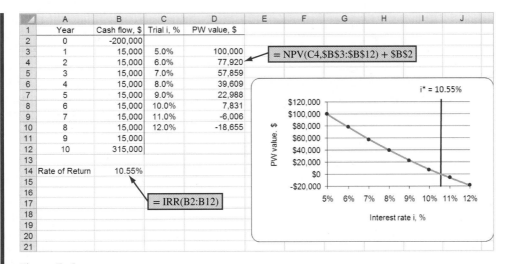

Figure 7–4
Spreadsheet to determine i^* and develop a PW graph, Example 7.2.

Solution by Spreadsheet

The fastest way to find i^* is to use the RATE function (Equation [7.4]). The entry = RATE(10,15000,−200000,300000) displays $i^* = 10.55\%$ per year. It is equally correct to use the IRR function. Figure 7–4, column B, shows the cash flows and = IRR(B2:B12) function to obtain i^*.

For a complete spreadsheet analysis, use the procedure outlined above.

1. Figure 7–3 shows cash flows.
2. Equation [7.6] is the ROR relation.
3. Figure 7–4 shows the net cash flows in column B.
4. The IRR function in cell B14 displays $i^* = 10.55\%$.
5. To graphically observe $i^* = 10.55\%$, column D displays the PW graph for different i values. The NPV function is used repeatedly to calculate PW for the scatter chart.

Just as i^* can be found using a PW equation, it may equivalently be determined using an AW relation. This method is preferred when uniform annual cash flows are involved. Solution by hand is the same as the procedure for a PW-based relation, except Equation [7.2] is used. In the case of Example 7.2, $i^* = 10.55\%$ is determined using the AW-based relation.

$$0 = -200,000(A/P,i^*,10) + 15,000 + 300,000(A/F,i^*,10)$$

The procedure for solution by spreadsheet is exactly the same as outlined above using the IRR function. Internally, IRR calculates the NPV function at different i values until NPV = 0. (There is no equivalent way to utilize the PMT function, since it requires a fixed value of i to calculate an A value.)

7.3 Special Considerations When Using the ROR Method ●●●

The rate of return method is commonly used in engineering and business settings to evaluate one project, as discussed in this chapter, and to select one alternative from two or more, as explained in the next chapter. As mentioned earlier, an ROR analysis is performed using a different basis than PW and AW analyses. The cash flows themselves determine the (internal) rate of return. As a result, there are some assumptions and special considerations with ROR analysis that must be made when calculating i^* and in interpreting its real-world meaning. A summary is provided below.

- *Multiple i* values.* Depending upon the sequence of net cash inflows and outflows, there may be more than one real-number root to the ROR equation, resulting in *more than one i* value.* This possibility is discussed in Section 7.4.
- *Reinvestment at i*.* Both the PW and AW methods assume that any net positive investment (i.e., net positive cash flows once the time value of money is considered) is reinvested at the MARR. However, the ROR method assumes reinvestment at the i^* rate. When i^* is not close to the MARR (e.g., if i^* is substantially larger than MARR), this is an unrealistic assumption. In such cases, the i^* value is not a good basis for decision making. This situation is discussed in Section 7.5.
- *Different procedure for multiple alternative evaluations.* To correctly use the ROR method to choose from two or more mutually exclusive alternatives requires an *incremental analysis* procedure that is significantly more involved than PW and AW analysis. Chapter 8 explains this procedure.

If possible, from an engineering economic study perspective, the **AW or PW method at a stated MARR should be used in lieu of the ROR method**. However, there is a strong appeal for the ROR method because rate of return values are very commonly quoted. And it is easy to compare a proposed project's return with that of in-place projects.

> When it is important to know the exact value of i^*, a good approach is to determine PW or AW at the MARR, then determine the specific i^* for the selected alternative.

As an illustration, if a project is evaluated at MARR = 15% and has PW < 0, there is no need to calculate i^*, because $i^* < 15\%$. However, if PW is positive, but close to 0, calculate the exact i^* and report it along with the conclusion that the project is financially justified.

7.4 Multiple Rate of Return Values ● ● ●

In Section 7.2, a unique rate of return i^* was determined. In the cash flow series presented thus far, the algebraic signs on the *net cash flows* changed only once, usually from minus in year 0 to plus at some time during the series. This is called a *conventional (simple or regular) cash flow series*. However, for some series the net cash flows switch between positive and negative from 1 year to another, so there is more than one sign change. Such a series is called *nonconventional (nonsimple or irregular)*. As shown in the examples of Table 7–3, each series of positive or negative signs may be one or more in length. Relatively large NCF changes in amount and sign can occur in projects that require significant spending at the end of the expected life. Nuclear plants, open-pit mines, petroleum well sites, refineries, and the like often require environmental restoration, waste disposal, and other expensive phaseout costs. The cash flow diagram will appear similar to Figure 7–5a. Plants and systems that have anticipated major refurbishment costs or upgrade investments in future years may have considerable swings in cash flow and sign changes over the years, as shown by the pattern in Figure 7–5b.

| **TABLE 7–3** | Examples of Conventional and Nonconventional Net Cash Flow for a 6-year Project |

Type of Series	0	1	2	3	4	5	6	Number of Sign Changes
	\-\-\-\-\-\-\- **Sign on Net Cash Flow by Year** \-\-\-\-\-\-\-							
Conventional	−	+	+	+	+	+	+	1
Conventional	−	−	−	+	+	+	+	1
Conventional	+	+	+	+	+	−	−	1
Nonconventional	−	+	+	+	−	−	−	2
Nonconventional	+	+	−	−	−	+	+	2
Nonconventional	−	+	−	−	+	+	+	3

When there is more than one sign change in the net cash flows, it is possible that there will be multiple i^* values in the −100% to plus infinity range. There are two tests to perform in sequence on the nonconventional series to determine if there is one unique value or possibly multiple i^* values that are real numbers.

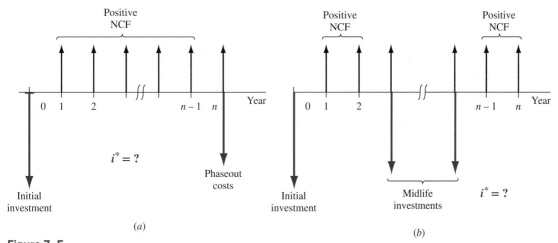

Figure 7–5
Typical cash flow diagrams for projects with (a) large restoration or remediation costs, and (b) upgrade or refurbishment costs.

> **Test 1: (Descartes') rule of signs** states that the total number of real-number roots is always less than or equal to the number of sign changes in the series.

This rule is derived from the fact that the relation set up by Equation [7.1] or [7.2] to find i^* is an nth-order polynomial. (It is possible that imaginary values or infinity may also satisfy the equation.)

> **Test 2: Cumulative cash flow sign test,** also known as *Norstrom's criterion,* states that only one sign change in a series of cumulative cash flows which *starts negatively* indicates that there is **one positive root** to the polynomial relation.

Zero values in the series are neglected when applying Norstrom's criterion. This is a more discriminating test that determines if there is one, real-number, positive i^* value. There may be negative roots that satisfy the ROR relation, but these are not useful i^* values. To perform the test, determine the series

$$S_t = \text{cumulative cash flows through period } t$$

Observe the sign of S_0 and count the sign changes in the series $S_0, S_1, \ldots, S_n$. Only if $S_0 < 0$ and signs change one time in the series is there a single, real-number, positive i^*.

With the results of these two tests, the ROR relation is solved for either the unique i^* or the multiple i^* values, using trial and error by hand, using a programmable calculator, or by spreadsheet using an IRR function that incorporates the "guess" option. Development of the PW graph is recommended, especially when using a spreadsheet. Examples 7.3 and 7.4 illustrate the tests and solution for i^*.

EXAMPLE 7.3

Sept-Îles Aluminum Company operates a bauxite mine to supply its aluminum smelter located about 2 km from the current open pit. A new branch for the pit is proposed that will supply an additional 10% of the bauxite currently available over the next 10-year period. The lease for the land will cost $400,000 immediately. The contract calls for the restoration of the land and development as part of a state park and wildlife area at the end of the 10 years. This is expected to cost $300,000. The increased production capacity is estimated to net an additional $75,000 per year for the company. Perform an ROR analysis that will provide the following information:

(a) Type of cash flow series and possible number of ROR values
(b) PW graph showing all i^* values
(c) Actual i^* values determined using the ROR relation and spreadsheet function
(d) Conclusions that can be drawn about the correct rate of return from this analysis

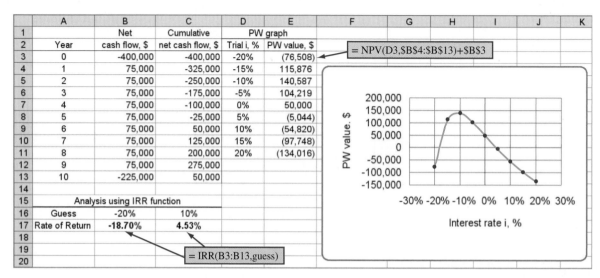

Figure 7–6
Spreadsheet determination of multiple i^* values and PW graph, Example 7.3.

Solution

(a) The net cash flows will appear like those in Figure 7–5a with an initial investment of $\$-400,000$, annual net cash flow of $\$75,000$ for years 1 through 10, and a phaseout cost of $\$-300,000$ in year 10. Figure 7–6 details the NCF series (column B) and cumulative NCF (column C) for use in the two tests for unique and multiple i^* values. The **series is nonconventional** based on the sign changes throughout the series.

Test 1: There are two sign changes in the NCF series, which indicates a **possible maximum of two roots** to the polynomial equation or i^* values for the ROR equation.

Test 2: There is one sign change in the cumulative NCF series, which indicates a unique positive root or **one positive i^* value.**

(b) Columns D and E of the spreadsheet in Figure 7–6 use i values ranging from -20% to $+20\%$ per year to plot the PW vs i curve via the NPV function. There are two times that the parabolic-shaped curve crosses the PW = 0 line; these are approximately $i_1^* = -18\%$ and $i_2^* = 5\%$.

(c) The ROR equation based on PW computations is

$$0 = -400,000 + 75,000(P/A,i^*\%,10) - 300,000(P/F,i^*\%,10) \qquad [7.7]$$

i^* values by hand   If hand solution is chosen, the same procedure used in Example 7.2 can be applied here. However, the technique to estimate the initial i value will not work as well in this case since the majority of the cash flows do not fit either the P/F or the F/P factor. In fact, using the P/F factor, the initial i value is indicated to be 1.25%. Trial-and-error solution of Equation [7.7] with various i values will approximate the correct answer of about 4.5% per year. This complies with the test results of one positive i^* value.

i^* by spreadsheet function   Use the = IRR(B3:B13,guess) function to determine the i^* value for the NCF series in column B, Figure 7–6. Entering different values in the optional "guess" field will force the function to find multiple i^* values, if they exist. As shown in row 17, two values are found.

$$i_1^* = -18.70\% \qquad i_2^* = +4.53\%$$

This result does not conflict with test results, as there is one positive value, but a negative value also balances the ROR equation.

(d) The positive $i^* = 4.53\%$ is accepted as the correct IROR for the project. The negative value is not useful in economic conclusions about the project.

EXAMPLE 7.4

The engineering design and testing group for Honda Motor Corp. does contract-based work for automobile manufacturers throughout the world. During the last 3 years, the net cash flows for contract payments have varied widely, as shown below, primarily due to a large manufacturer's inability to pay its contract fee.

Year	0	1	2	3
Net Cash Flow, $1000	+2000	−500	−8100	+6800

(a) Determine the maximum number of i^* values that may satisfy the ROR equation.
(b) Write the PW equation and approximate the i^* value(s) by plotting PW vs i.
(c) What do the i^* values mean?

Solution

(a) Table 7–4 shows the annual cash flows and cumulative cash flows. Since there are two sign changes in the cash flow sequence, the rule of signs indicates a maximum of two i^* values. The cumulative cash flow sequence starts with a positive number $S_0 = +2000$, indicating that test #2 is inconclusive. As many as two i^* values can be found.

TABLE 7–4   Cash Flow and Cumulative Cash Flow Sequences, Example 7.4

Year	Cash Flow ($1000)	Sequence Number	Cumulative Cash Flow ($1000)
0	+2000	S_0	+2000
1	−500	S_1	+1500
2	−8100	S_2	−6600
3	+6800	S_3	+200

(b) The PW relation is

$$PW = 2000 - 500(P/F,i,1) - 8100(P/F,i,2) + 6800(P/F,i,3)$$

The PW values are shown below and plotted in Figure 7–7 for several i values. The characteristic parabolic shape for a second-degree polynomial is obtained, with PW crossing the i axis at approximately $i_1^* = 8\%$ and $i_2^* = 41\%$.

$i\%$	5	10	20	30	40	50
PW ($1000)	+51.44	−39.55	−106.13	−82.01	−11.83	+81.85

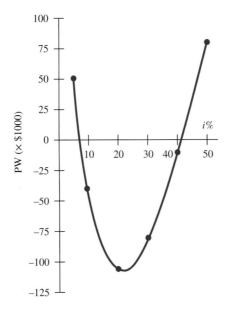

Figure 7–7
Present worth of cash flows at several interest rates, Example 7.4.

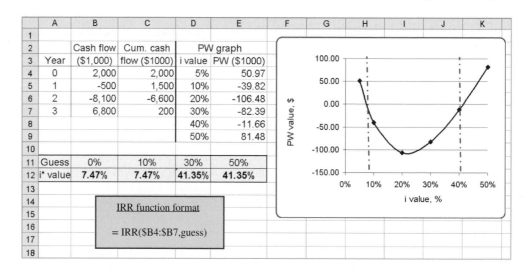

Figure 7–8
Spreadsheet solution, Example 7.4.

Figure 7–8 presents the spreadsheet PW graph with the PW curve crossing the x axis at PW = 0 two times. Also, the solution for **two positive i^* values** using the IRR function with different guess values is displayed. The values are

$$i_1^* = 7.47\% \qquad i_2^* = 41.35\%$$

(c) Since both i^* values are positive, neither can be considered the true ROR of the cash flow series. This result indicates that additional information is needed to calculate a more useful project ROR, namely, some information about the anticipated return on funds invested external to the project and the cost of capital to borrow money to continue the project. This problem is a good example of when an approach discussed in the next section should be taken. Alternatively, use of a PW, AW, or FW evaluation at a set MARR can be applied to reach an economic decision.

If the guess option is not used in the IRR function, the starting point is 10% per year. The function will find the one ROR closest to 10% that satisfies the PW relation. Entering various guess values will allow IRR to find multiple i^* values in the range of −100% to ∞, if they exist. Often, the results are unbelievable or unacceptable values that are rejected. Some helpful guidelines can be developed. Assume there are two i^* values for a particular cash flow series.

If the Results Are	What to Do
Both $i^* < 0$	Discard both values.
Both $i^* > 0$	Discard both values.
One $i^* > 0$; one $i^* < 0$	Use $i^* > 0$ as ROR.

If both i^* values are discarded, proceed to the approach discussed in the next section to determine one rate of return value for the project. However, remember the prime recommendation.

Always determine the PW or AW at the MARR first for a reliable measure of economic justification. If the PW or AW is greater than zero and the ROR is needed, then find the actual i^* of the project cash flows.

This recommendation is not to dissuade you from using the ROR method. Rather it is a recommendation that the use of the ROR technique be reserved for times when the actual i^* value is essential to the decision-making process.

7.5 Techniques to Remove Multiple Rates of Return ● ● ●

The techniques developed here are used under the following conditions:

- The PW or AW value at the MARR is determined and could be used to make the decision, but information on the ROR is deemed necessary to finalize the economic decision, and
- The two tests of cash flow sign changes (Descartes' and Norstrom's) indicate multiple roots (i^* values) are possible,

- More than one positive i^* value or all negative i^* values are obtained when the PW graph and IRR function are developed, and
- A single, reliable rate of return value is required by management or engineers to make a clear economic decision.

We will present two methods to remove multiple i^* values. The selected approach depends upon what estimates are the most reliable for the project being evaluated. An important fact to remember is the following.

> The result of follow-up analysis to obtain a single ROR value when multiple, nonuseful i^* values are present does **not determine the internal rate of return (IROR)** for nonconventional net cash flow series. The resulting rate is a function of the additional information provided to make the selected technique work, and the accuracy is further dependent upon the reliability of this information.

It is commonly accepted to refer to the resulting value as the **external rate of return (EROR)** as a reminder that it is different from the IROR obtained in all previous sections. Prior to determining the EROR, it is first and foremost important to identify the **perspective about the annual net cash flows** of a project. Take the following view: You are the project manager and the project generates cash flows each year. Some years produce positive NCF, and you want to invest the excess money at a good rate of return. We will call this the **investment rate i_i**. This can also be called the *reinvestment rate*. Other years, the NCF will be negative and you must borrow funds from some source to continue. The interest rate you pay should be as small as possible; we will call this the **borrowing rate i_b**, also referred to as the *finance rate*. Each year, you must consider the time value of money, which must utilize either the investment rate or the borrowing rate, depending upon the sign on the NCF of the preceding year. With this perspective, it is now possible to outline two approaches that rectify the multiple i^* situation. The resulting ROR value will not be the same for each method because slightly different additional information is necessary and the cash flows are treated in slightly different fashions from the time value of money viewpoint.

> **External rate of return (EROR)** is the rate of return per year for a cash flow series that considers two different situations: (1) the rate of investment for a positive net cash flow generated by the project during a year, and (2) the interest rate (independent of or external to the project, e.g., a bank loan rate) that the project must pay to continue operation, if and when the project produces a negative net cash flow, that is, a loss in a year.

External rate of return

Modified ROR (MIRR) Approach   This is the easier approach to apply, and it has a spreadsheet function that can find the single EROR value quickly. However, the investment and borrowing rates must be reliably estimated, since the results may be quite sensitive to them. The symbol i' will identify the result.

Return on Invested Capital (ROIC) Approach   Though more mathematically rigorous, this technique provides a more reliable estimate of the EROR and it requires only the investment rate i_i. The symbol i'' is used to identify the result.

At this point, it would be good to review the material in Section 7.1, including Example 7.1. Though the i' or i'' value determined here is not the ROR defined earlier in the chapter, the concepts used to make the ending cash flow balance equal to zero are used.

Before describing these two techniques to remove ambiguity about the most correct rate of return, an overview of evaluation methods for (1) different types of cash flow sequences (conventional and

Figure 7–9
Logic of ROR analysis for different types of net cash flow series and multiple rate or return values

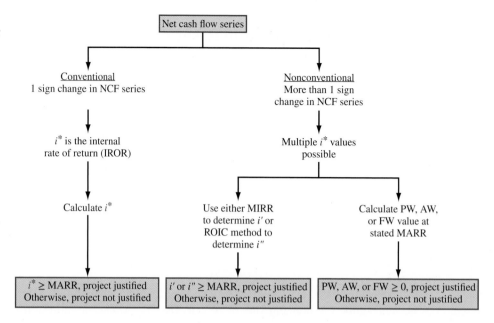

nonconventional), and (2) the number of possible i^* values will help decide upon the best evaluation approach for any situation. Figure 7–9 provides this overview. If the series is conventional (left branch), the calculation of i^* provides a correct value of the IROR, that is, the i^* is totally determined by the observed net cash flows. A comparison to the MARR cinches the decision of a justified or not justified project.

The challenge presents itself when multiple rates can solve the ROR equation mathematically (right branch). As mentioned several times, the use of PW, AW, or FW evaluation at the stated MARR on a nonconventional series is an efficient and correct way to determine economic justification. However, if an accurate and correct i^* value is needed, one of the analytical methods we are about to cover is an approach that works, provided the additional estimates that must be made are reliable.

Modified ROR Approach

The technique requires that two rates external to the project net cash flows be estimated.

- **Investment rate i_i** is the rate at which extra funds are invested in some source external to the project. This applies to all positive annual NCF. It is reasonable that the MARR is used for this rate. As noted above, this term can also be referred to as the *reinvestment rate*.
- **Borrowing rate i_b** is the rate at which funds are borrowed from an external source to provide funds to the project. This applies to all negative annual NCF. The weighted average cost of capital (WACC) can be used for this rate. This term is also called the *finance rate*.

It is possible to make all rates the same, that is, $i_i = i_b = \text{MARR} = \text{WACC}$. However, this is not a good idea as it implies that the company is willing to borrow funds and invest in projects at the same rate. This implies no profit margin over time, so the company can't survive for long using this strategy. Commonly, MARR > WACC, so usually $i_i > i_b$. (See Section 1.9 for a quick review of MARR and WACC and Chapter 10 for a more detailed discussion of WACC.)

Figure 7–10 is a reference diagram that has multiple i^* values, since the net cash flows change sign multiple times. The modified ROR method uses the following procedure to determine a single external rate of return i' and to evaluate the economic viability of the project.

1. Determine the PW value in year 0 of *all negative NCF at the borrowing rate i_b* (lightly shaded area and resulting PW_0 value in Figure 7–10).
2. Determine the FW value in year n of *all positive NCF at the investment rate i_i* (darker shaded area and resulting FW_n value in Figure 7–10).
3. Calculate the modified rate of return i' at which the *PW and FW values are equivalent* over the n years using the following relation, where i' is to be determined.

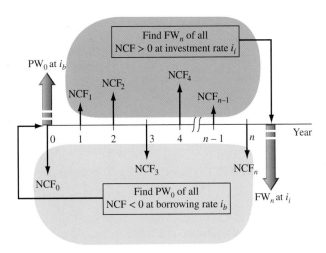

Figure 7–10
Typical cash flow diagram to determine modified rate of return i'.

$$\mathbf{FW}_n = \mathbf{PW}_0\,(F/P,i'\%,n) \qquad\qquad [7.8]$$

If using a spreadsheet rather than hand computation, the MIRR function displays i' directly with the format

$$= \mathbf{MIRR(first\_cell\!:\!last\_cell,\ i_b,\ i_i)} \qquad\qquad [7.9]$$

4. The guideline for economic decision making compares the EROR or i' to the MARR.

     If $i' >$ MARR, the project is economically justified.

     If $i' <$ MARR, the project is not economically justified.

     As in other situations, on the rare occasion that $i' =$ MARR, there is indifference to the project's economic acceptability; however, acceptance is the usual decision.

EXAMPLE 7.5

The cash flows experienced by Honda Motors in Example 7.4 are repeated below. There are two positive i^* values that satisfy the PW relation, 7.47% and 41.35% per year. Use the modified ROR method to determine the EROR value. Studies indicate that Honda has a WACC of 8.5% per year and that projects with an estimated return of less than 9% per year are routinely rejected. Due to the nature of this contract business, any excess funds generated are expected to earn at a rate of 12% per year.

Year	0	1	2	3
Net Cash Flow, $1000	+2000	−500	−8100	+6800

Solution by Spreadsheet
Using the information in the problem statement, the rate estimates are as follows:

MARR:	9% per year
Investment rate, i_i:	12% per year
Borrowing rate, i_b:	8.5% per year

The fast way to find i' is with the MIRR function. Figure 7–11 shows the result of $i' = 9.39\%$ per year. Since 9.39% > MARR, the project is economically justified.

Figure 7–11
Spreadsheet application of
MIRR function, Example 7.5.

	A	B	C	D	E	F
1	Year	NCF, $				
2	0	2000		Borrowing rate, i_b	8.5%	
3	1	-500		Investment rate, i_i	12.0%	
4	2	-8100				
5	3	6800		EROR value	**9.39%**	
6						
7				$= MIRR(B2:B5,E2,E3)$		
8						

It is vital that the interpretation be correct. The 9.39% is not the internal rate of return (IROR); it is the external ROR (EROR) based on the two external rates for investing and borrowing money.

Solution by Hand

Figure 7–10 can serve as a reference as the procedure to find i' manually is applied.

Step 1. Find PW_0 of all negative NCF at $i_b = 8.5\%$.

$$PW_0 = -500(P/F,8.5\%,1) - 8100(P/F,8.5\%,2)$$
$$= \$-7342$$

Step 2. Find FW_3 of all positive NCF at $i_i = 12\%$.

$$FW_3 = 2000(F/P,12\%,3) + 6800$$
$$= \$9610$$

Step 3. Find the rate i' at which the PW and FW are equivalent.

$$PW_0(F/P,i',3) + FW_3 = 0$$
$$-7342(1 + i')^3 + 9610 = 0$$

$$i' = \left(\frac{9610}{7342}\right)^{1/3} - 1$$

$$= 0.939 \quad (9.39\%)$$

Step 4. Since $i' > $ MARR of 9%, the project is economically justified using this EROR approach.

Comment

If either or both of the finance rate and investment rate are omitted from the MIRR function, they are assumed to be 0%. The result is the ROR as if all negative amounts were added and placed in year 0 (step 1 of the MIRR method), and all positive net cash flows were added and placed in year n (step 2). For this example, the function $= MIRR(B2:B5,,)$ results in an i' of 0.77%. If solved by hand, steps 1 through 3 appear as follows:

Step 1. $PW_0 = -500 - 8100 = \$-8600$
Step 2. $FW_3 = 2000 + 6800 = \$8800$
Step 3. $PW_0(F/P,i',3) + FW_3 = 0$
            $-8600(1 + i')^3 + 8800 = 0$
            $i' = (8800/8600)^{1/3} - 1$
            $= 0.0077 \ (0.77\%)$

This is the EROR provided it costs nothing to borrow and there is no return on any positive net cash flows that occur. As you can conclude, neither of these are good assumptions for most projects.

Return on Invested Capital Approach

The definition of ROIC should be understood before we discuss the approach.

Return on invested capital (ROIC) is a rate-of-return measure of how effectively a project utilizes the funds invested in it, that is, funds that remain internal to the project. For a corporation, ROIC is a measure of how effectively it utilizes the funds invested in its operations, including facilities, equipment, people, systems, processes, and all other assets used to conduct business.

The technique requires that the investment rate i_i be estimated for excess funds generated in any year that they are not needed by the project. The ROIC rate, which has the symbol i'', is determined using an approach called the **net-investment procedure.** It involves developing a series of future worth (F) relations moving forward 1 year at a time. In those years that the net balance of the project cash flows is positive (extra funds generated by the project), the funds are invested at the i_i rate. Usually, i_i *is set equal to the MARR.* When the net balance is negative, the ROIC rate is used, since the project keeps all of its funds internal to itself. The ROIC method uses the following procedure to determine a single external rate of return i'' and to evaluate the economic viability of the project. Remember that the perspective is that you are the project manager and when the project generates extra cash flows, they are invested external to the project at the investment rate i_i.

1. Develop a series of future worth relations by setting up the following relation for each year t ($t = 1, 2, \ldots, n$ years).

$$F_t = F_{t-1}(1 + k) + \text{NCF}_t \qquad\qquad [7.10]$$

where F_t = **future worth in year t based on previous year and time value of money**

       NCF_t = **net cash flow in year** t

$$k = \begin{cases} i_i & \text{if } F_{t-1} > 0 \quad \textbf{(extra funds available)} \\ i'' & \text{if } F_{t-1} < 0 \quad \textbf{(project uses all available funds)} \end{cases}$$

2. Set the future worth relation for the last year n equal to 0, that is, $F_n = 0$, and solve for i'' to balance the equation. The i'' value is the ROIC for the specified investment rate i_i.

   The F_t series and solution for i'' in the $F_n = 0$ relation can become involved mathematically. Fortunately, the Goal Seek spreadsheet tool can assist in the determination of i'' because there is only one unknown in the F_n relation and the target value is zero. (Both the hand and spreadsheet solutions are demonstrated in Example 7.6.)
3. The guideline for economic decision making is the same as above, namely,

   If ROIC ≥ MARR, the project is economically justified.

   If ROIC < MARR, the project is not economically justified.

It is important to remember that the **ROIC is an external rate of return dependent upon the investment rate choice.** It is not the same as the internal rate of return discussed at the beginning of this chapter, nor is it the multiple rates, nor is it the MIRR rate found by the previous method. This is a separate technique to find a single rate for the project.

EXAMPLE 7.6

Once again, we will use the cash flows experienced by Honda Motors in Example 7.4 (repeated below). Use the ROIC method to determine the EROR value. The MARR is 9% per year, and any excess funds generated by the project can earn at a rate of 12% per year.

Year	0	1	2	3
Net Cash Flow, $1000	+2000	−500	−8100	+6800

Solution by Hand
The hand solution is presented first to provide the logic of the ROIC method. Use MARR = 9% and i_i = 12% per year in the procedure to determine i'', which is the ROIC. Figure 7–12 details the cash flows and tracks the progress as each F_t is developed. Equation [7.10] is applied to develop each F_t.

Step 1. Year 0: $F_0 = \$+2000$

            Since $F_0 > 0$, externally invest in year 1 at $i_i = 12\%$.

      Year 1: $F_1 = 2000(1.12) - 500 = \$+1740$

            Since $F_1 > 0$, use $i_i = 12\%$ for year 2.

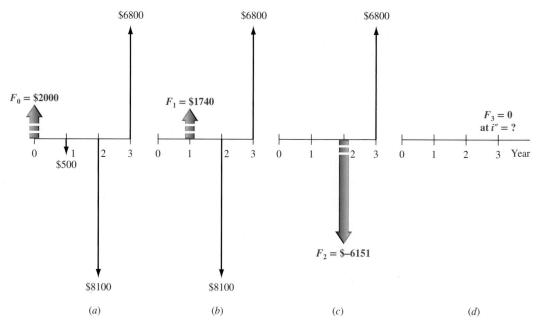

Figure 7–12
Application of ROIC method at $i_i = 12\%$ per year: (a) original cash flow; equivalent form in (b) year 1, (c) year 2, and (d) year 3.

Year 2: $F_2 = 1740(1.12) - 8100 = \$-6151$

Now $F_2 < 0$, use i'' for year 3, according to Equation [7.10].

Year 3: $F_3 = -6151(1 + i'') + 6800$

This is the last year. See Figure 7–12 for equivalent net cash flow diagrams. Go to step 2.

Step 2. Solve for $i'' = $ ROIC from $F_3 = 0$.

$$-6151(1 + i'') + 6800 = 0$$
$$i'' = 6800/6151 - 1$$
$$= 0.1055 \quad (10.55\%)$$

Step 3. Since ROIC > MARR = 9%, the project is economically justified.

Solution by Spreadsheet

Figure 7–13 provides a spreadsheet solution. The future worth values F_1 through F_3 are determined by the conditional IF statements in rows 3 through 5. The functions are shown in column D. In each year, Equation [7.10] is applied. If there are surplus funds generated by the project, $F_{t-1} > 0$ and the investment rate i_i (in cell E7) is used to find F_t. For example, because the F_1 value (in cell C3) of $1740 > 0$, the time value of money for the next year is calculated at the investment rate of 12% per year, as shown in the hand solution above for year 2.

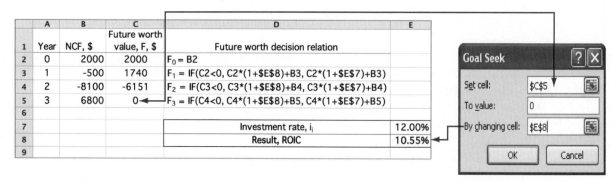

Figure 7–13
Spreadsheet application of ROIC method using Goal Seek, Example 7.6.

The Goal Seek template sets the F_3 value to zero by changing the ROIC value (cell E8). The result is $i'' = \text{ROIC} = 10.55\%$ per year. As before, since $10.55\% > 9\%$, the MARR, the project is economically justified.

Comment

Note that the rate by the ROIC method (10.55%) is different than the MIRR rate (9.39%). Also these are both different than the multiple rates determined earlier (7.47% and 41.35%). This shows how dependent the different methods are upon the additional information provided when multiple i^* rates are indicated by the two sign tests.

Now that we have learned two techniques to remove multiple i^* values, here are some connections between the multiple i^* values, the external rate estimates, and the resulting external rates (i' and i'') obtained by the two methods.

Modified ROR technique   When both the borrowing rate i_b and the investment rate i_i are exactly equal to any one of the multiple i^* values, the rate i' found by the MIRR function, or by hand solution, will equal the i^* value. That is, all four parameters have the same value.

$$\text{If any } i^* = i_b = i_i, \qquad \text{then } i' = i^*$$

ROIC technique   Similarly, if the investment rate i_i is exactly equal to any one of the multiple i^* values, the rate found by the Goal Seek tool, or by hand when the equation $F_n = 0$ is solved, will be $i'' = i^*$ value.

Finally, it is very important to remember the following fact.

None of the details of the modified ROR (MIRR) technique or the return on invested capital (ROIC) technique are necessary if the PW or AW method of project evaluation is applied at a specific MARR. When the MARR is established, this is, in effect, fixing the i^* value. Therefore, a definitive economic decision can be made directly from the PW or AW value.

7.6 Rate of Return of a Bond Investment ● ● ●

A time-tested method of raising capital funds is through the issuance of an IOU, which is financing through debt, not equity (see Chapter 1). One very common form of IOU is a bond—a long-term note issued by a corporation or a government entity (the borrower) to finance major projects. The borrower receives money now in return for a promise to pay the **face value V** of the bond on a stated maturity date. Bonds are usually issued in face value amounts of $1000, $5000, or $10,000. **Bond dividend I**, also called *bond interest*, is paid periodically between the time the money is borrowed and the time the face value is repaid. The bond dividend is paid c times per year. Expected payment periods are usually semiannually or quarterly. The amount of interest is determined using the stated dividend or interest rate, called the **bond coupon rate b.**

$$I = \frac{\textbf{(face value) (bond coupon rate)}}{\textbf{number of payment periods per year}}$$

$$I = \frac{Vb}{c} \tag{7.11}$$

There are many types or classifications of bonds. Four general classifications are summarized in Table 7–5 according to their issuing entity, fundamental characteristics, and example names or purposes. For example, *Treasury securities* are issued in different monetary amounts ($1000 and up) with varying periods of time to the maturity date (Bills up to 1 year; Notes for 2 to 10 years; and Bonds for 10 to 30 years). In the United States, Treasury securities are considered a very safe bond purchase because they are backed with the "full faith and credit of the U.S. government." The safe investment rate indicated in Figure 1–6 as the lowest level for establishing a MARR is the coupon rate on a U.S. Treasury security. Funds obtained through corporate bond issues are used for new product development, facilities upgrade, expansion into international markets, and similar business ventures.

TABLE 7–5	Classification and Characteristics of Bonds		
Classification	**Issued by**	**Characteristics**	**Examples**
Treasury securities	Federal government	Backed by faith and credit of the federal government	Bills (≤1 year) Notes (2–10 years) Bonds (10–30 years)
Municipal	Local governments	Federal tax-exempt Issued against taxes received	General obligation Revenue Zero coupon Put
Mortgage	Corporation	Backed by specified assets or mortgage Low rate/low risk on first mortgage Foreclosure, if not repaid	First mortgage Second mortgage Equipment trust
Debenture	Corporation	Not backed by collateral, but by reputation of corporation Bond rate may "float" Higher interest rates and higher risks	Convertible Subordinated Junk or high yield

EXAMPLE 7.7

General Electric just released $10 million worth of $10,000 ten-year bonds. Each bond pays dividends semiannually at a rate of 6% per year. (*a*) Determine the amount a purchaser will receive each 6 months and after 10 years. (*b*) Suppose a bond is purchased at a time when it is discounted by 2% to $9800. What are the dividend amounts and the final payment amount at the maturity date?

Solution

(*a*) Use Equation [7.11] for the dividend amount.

$$I = \frac{10,000(0.06)}{2} = \$300 \text{ per 6 months}$$

The face value of $10,000 is repaid after 10 years.

(*b*) Purchasing the bond at a discount from face value does not change the dividend or final repayment amounts. Therefore, $300 per 6 months and $10,000 after 10 years remain the amounts.

The cash flow series for a bond investment is conventional and has one unique i^*, which is best determined by solving a PW-based rate of return equation in the form of Equation [7.1], that is, $0 = \text{PW}$.

EXAMPLE 7.8

Allied Materials needs $3 million in debt capital for expanded composites manufacturing. It is offering small-denomination bonds at a discount price of $800 for a 4% $1000 bond that matures in 20 years with a dividend payable semiannually. What nominal and effective interest rates per year, compounded semiannually, will Allied Materials pay an investor?

Solution

The income that a purchaser will receive from the bond purchase is the bond dividend $I = \$20$ every 6 months plus the face value in 20 years. The PW-based equation for calculating the ROR is

$$0 = -800 + 20(P/A,i^*,40) + 1000(P/F,i^*,40)$$

Solve by the IRR function or by hand to obtain $i^* = 2.8435\%$ semiannually. The nominal interest rate per year is computed by multiplying i^* by 2.

Nominal $i = (2.8435)(2) = 5.6870\%$ per year, compounded semiannually

Using Equation [4.5], the effective annual rate is

$$i_a = (1.028435)^2 - 1 = 5.7678\%$$

EXAMPLE 7.9

Gerry is a project engineer. He took a financial risk and bought a bond from a corporation that had defaulted on its interest payments. He paid $4240 for an 8% $10,000 bond with dividends payable quarterly. The bond paid no interest for the first 3 years after Gerry bought it. If interest was paid for the next 7 years and then Gerry was able to resell the bond for $11,000, what rate of return did he make on the investment? Assume the bond is scheduled to mature 18 years after he bought it. Perform hand and spreadsheet analysis.

Solution by Hand

The bond interest received in years 4 through 10 was

$$I = \frac{(10,000)(0.08)}{4} = \$200 \text{ per quarter}$$

The effective rate of return *per quarter* can be determined by solving the PW equation developed on a per quarter basis.

$$0 = -4240 + 200(P/A, i^* \text{ per quarter}, 28)(P/F, i^* \text{ per quarter}, 12)$$
$$+ 11,000(P/F, i^* \text{ per quarter}, 40)$$

The equation is correct for $i^* = 4.1\%$ per quarter, which is a nominal 16.4% per year, compounded quarterly.

Solution by Spreadsheet

Once all the cash flows are entered into contiguous cells, the function = IRR(B2:B42) is used in Figure 7–14, row 43, to display the answer of a nominal ROR of 4.10% per quarter. (Note that many of the row entries have been hidden to conserve space.) This is the same as the nominal annual rate of

$$i^* = 4.10\%(4) = 16.4\% \text{ per year, compounded quarterly}$$

Gerry did well on his bond investment.

	A	B
1	Quarter	Cash flow, $
2	0	-4,240
3	1	0
4	2	0
8	6	0
13	11	0
14	12	0
15	13	200
21	19	200
22	20	200
29	27	200
41	39	200
42	40	11,200
43	i*/qtr	4.10%

Figure 7–14
Spreadsheet solution for a bond investment, Example 7.9.

If a bond investment is being considered and a required rate of return is stated, the same PW-based relation used to find i^* can be used to determine the maximum amount to pay for the bond now to ensure that the rate is realized. The stated rate is the MARR, and the PW computations are performed exactly as they were in Chapter 5. As an illustration, in the last example, if 12% per year, compounded quarterly, is the target MARR, the PW relation is used to find the maximum that Gerry should pay now; P is determined to be $6004. The quarterly MARR is 12%/4 = 3%.

$$0 = -P + 200(P/A, 3\%, 28)(P/F, 3\%, 12) + 11,000(P/F, 3\%, 40)$$
$$P = \$6004$$

CHAPTER SUMMARY

The rate of return of a cash flow series is determined by setting a PW-based or AW-based relation equal to zero and solving for the value of i^*. The ROR is a term used and understood by almost everybody. Most people, however, can have considerable difficulty in calculating a rate of return correctly for anything other than a conventional cash flow series. For some types of series, more than one ROR possibility exists. The maximum number of i^* values is equal to the number of changes in the sign of the NCF series (Descartes' rule of signs). Also, a single positive rate can be found if the cumulative NCF series starts negatively and has only one sign change (Norstrom's criterion).

When multiple i^* values are indicated, either of the two techniques covered in this chapter can be applied to find a single, reliable rate for the nonconventional (nonsimple or irregular) net cash flow series. (Refer to Figure 7–9 for a graphical summary.) In the case of the ROIC technique, additional information is necessary about the *investment rate* that excess project funds will realize, while the modified ROR technique requires this same information, plus the *borrowing rate* for the organization considering the project. Usually, the investment rate is set equal to the MARR, and the borrowing rate takes on the historical WACC rate. Each technique will result in slightly different rates, but they are reliable for making the economic decision, whereas the multiple rates are often not useful to decision making.

> **If an exact ROR is not necessary, it is strongly recommended that the PW or AW method at the MARR be used to decide upon economic justification.**

PROBLEMS

Understanding ROR

7.1   What is (*a*) the highest, and (*b*) lowest rate of return (in percent) possible?

7.2   A $10,000 loan amortized over 5 years at an interest rate of 10% per year requires payments of $2638 to completely remove the loan when interest is charged on the unrecovered balance of the principal. If interest is charged on the original principal instead of the unrecovered balance, what is the loan balance after 5 years, provided the same $2638 payments are made each year?

7.3   A to Z Mortgages made a home equity loan to your friend. For a 4-year loan of $10,000 at 10% per year, what annual payment must he make to pay off the entire loan in 4 years if interest is charged on (*a*) the original principal amount of $10,000, and (*b*) the unrecovered balance? (*c*) What is the difference in the annual payments between the two bases for interest? Which method requires more money to repay the loan?

7.4   Spectra Scientific of Santa Clara, CA manufactures Q-switched solid state industrial lasers for LED substrate scribing and silicon wafer dicing. The company got a $60 million loan to be repaid over a 5-year period at 8% per year interest. Determine the amount of the unrecovered balance *of the*

principal (*a*) immediately *before* the payment is made at the end of year 1, and (*b*) immediately *after* the first payment.

7.5   A small industrial contractor purchased a warehouse building for storing equipment and materials that are not immediately needed at construction job sites. The cost of the building was $100,000 and the contractor has just made an agreement with the seller to finance the purchase over a 5-year period. The agreement states that monthly payments will be made based on a 30-year repayment schedule of interest on the unrecovered balance of the principal; however, the total remaining balance of principal and interest at the end of year 5 must be paid in a lump-sum "balloon" payment. What is the size of the balloon payment, if the interest rate on the loan is 0.5% per month?

7.6   The production of polyamide from raw materials of plant origin, such as castor oil, requires 20% less fossil fuel than conventional production methods. Darvon Chemicals borrowed $6 million to implement the process. If the interest rate on the loan is 10% per year (on unrecovered balance) for 10 years, what is the amount of interest for year 2? Also, write the spreadsheet function that displays this same amount.

Determination of ROR

7.7 If a manufacturer of electronic devices invests $650,000 in equipment for making compact piezo-electric accelerometers for general purpose vibration measurement, estimate the rate of return from revenue of $225,000 per year for 10 years and $70,000 in salvage value from the used equipment sale in year 10. Solve by (a) factors, and (b) spreadsheet function.

7.8 Use factors and a spreadsheet to determine the interest rate per period from the following equation:

$$0 = -40,000 + 8000(P/A,i^*,5) + 8000(P/F,i^*,8)$$

7.9 Determine i^* per year for the cash flows shown using factors and the IRR function. (Note: See Problem 7.55 for additional spreadsheet work with this series.)

Year	0	1	2	3	4
Cash Flow, $	0	−80,000	9000	70,000	30,000

7.10 The Camino Real Landfill was required to install a plastic liner to prevent leachate from migrating into the groundwater. The fill area was 50,000 m^2 and the installed liner cost was $8 per m^2. In order to recover the investment, the owner charges to unload at the rates of $10 per pick-up, $25 per dump-truck, and $70 per compactor-truck load. The fill area is adequate for 4 years. If the annual traffic is estimated to be 2500 pick-up loads, 650 dump-truck loads, and 1200 compactor-truck loads, what rate of return will the landfill owner make on the investment?

7.11 Determine the rate of return for the following cash flow series.

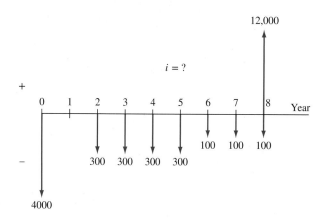

7.12 Jamison Specialties manufactures programmable incremental encoders that resist shock and vibration for use in harsh environments. Five years ago, the company invested $650,000 in an automated quality control system and recorded savings of $105,000 per year for the 5 years. The equipment now has a salvage value of $50,000.

(a) What is the rate of return per year on the investment?

(b) The owner expected to make at least 15% per year on this quality improvement investment. What annual savings were necessary to meet the 15% goal?

7.13 Over a 7-year period, the number of uninsured motorists in New Mexico dropped from 33% to 10% because an up-to-date computer database allowed the state to better police its requirements for car and truck liability insurance. Assuming the decrease occurred uniformly and compounded annually, what was the annual rate of the decrease?

7.14 What rate of return per month will an entrepreneur make over a 2½-year project period if he invested $150,000 to produce portable 12-volt air compressors? His monthly costs are $27,000 and his revenues are $33,000.

7.15 Swagelok Enterprises is a manufacturer of miniature fittings and valves. Over a 5-year period, the costs associated with one product line were as follows: first cost of $30,000 and annual costs of $18,000. Annual revenue was $27,000 and the used equipment was salvaged for $4000. What rate of return did the company make on this product?

7.16 A broadband service company borrowed $2 million for new equipment and repaid the loan in amounts of $200,000 in years 1 and 2 plus a lump-sum amount of $2.2 million at the end of year 3. What was the interest rate on the loan?

7.17 Barron Chemical uses a thermoplastic polymer to enhance the appearance of certain RV panels. The initial cost of one process was $130,000 with annual costs of $49,000 and revenues of $78,000 in year 1, increasing by $1000 per year. A salvage value of $23,000 was realized when the process was discontinued after 8 years. What rate of return did the company make on the process? Solve by trial and error and verify i^* by spreadsheet.

7.18 Steel cable barriers in highway medians are a low-cost way to improve traffic safety without overstressing department of transportation budgets. Cable barriers cost $44,000 per mile, compared with $72,000 per mile for guardrail and $419,000 per mile for concrete barriers. Furthermore, cable barriers tend to snag tractor-trailer rigs, keeping them from ricocheting back into same-direction traffic. The state of Ohio spent $4.97 million installing 113 miles of cable barriers. Answer the

following using both tabulated factors and a spreadsheet function.
(a) If the cable barriers prevent accidents totaling $1.3 million per year, what rate of return does this represent over a 10-year study period?
(b) What is the rate of return for 113 miles of guardrail if accident prevention is $1.1 million per year over a 10-year study period?

7.19 Aloma, a university graduate who started a successful business, wants to start an endowment in her name that will provide scholarships to IE students. She wants the scholarship to provide $10,000 per year and expects the first one to be awarded on the day she fulfills the endowment obligation. If Aloma plans to donate $100,000, what rate of return must the university realize in order to award the $10,000 per year scholarship forever?

7.20 A plaintiff in a successful lawsuit was awarded a judgment of $4800 per month for 5 years. The plaintiff has the need of a fairly large sum of money immediately for an investment of his own, and has offered the defendant the opportunity to pay off the award in a lump-sum amount of $110,000. If the defendant accepts the offer and pays the $110,000 now, what rate of return will the defendant have made by not paying the 60 monthly amounts?

7.21 Army Research Laboratory scientists developed a diffusion-enhanced adhesion process which is expected to significantly improve the performance of multifunction hybrid composites. NASA engineers estimate that composites made using the new process will result in savings in space exploration projects. The cash flows for one project are estimated. Determine the rate of return per year.

Year, t	Cost ($1000)	Savings ($1000)
0	−210	—
1	−150	—
2–5	—	$100 + 60(t − 2)$

7.22 An Indium-Gallium-Arsenide-Nitrogen alloy developed at Sandia National Laboratory is said to have potential uses in electricity-generating solar cells. The new material is expected to have a longer life, and it is believed to have a 40% efficiency rate, which is nearly twice that of standard silicon solar cells. The useful life of a telecommunications satellite could be extended from 10 to 15 years by using the new solar cells. What rate of return could be realized if an extra investment now of $950,000 would result in extra revenues of $450,000 in year 11, $500,000 in year 12, and amounts increasing by $50,000 per year through year 15?

Multiple ROR Values

7.23 Define the term reinvestment rate and describe how it differs for any cash flow series between (1) a PW value calculated at the MARR, and (2) the IROR value $i*$.

7.24 What cash flows are associated with Descartes' rule of signs and Norstrom's criterion?

7.25 Identify the following net cash flow series as conventional or nonconventional.

	Sign on Net Cash Flow by Year					
Year	0	1	2	3	4	5
(a)	−	−	−	−	−	+
(b)	−	−	−	+	+	−
(c)	+	+	+	+	−	−
(d)	+	−	+	−	+	−
(e)	−	−	+	+	+	+

7.26 Identify the net cash flows shown below as either conventional or nonconventional.

Year	0	1	2	3	4	5
(a)	400	300	200	100	0	−100
(b)	−50	−50	−50	40	40	40
(c)	−1000	200	300	400	300	200
(d)	−6000	−500	−500	−750	10,000	−2000
(e)	50	−10	50	50	50	50

7.27 The annual revenues (in $1000 units) associated with several large apartment complexes are $0, $350, $290, $460, $150, and $320 for years 0, 1, 2, 3, 4, and 5, respectively. Determine (1) whether each cash flow series is conventional or nonconventional, (2) the maximum number of real-number roots, and (3) if there is one, positive real-number $i*$ value. The cash flows for years 0, 1, 2, 3, 4, and 5, respectively, are:
(a) $−1500, $−90, $−40, $−85, $−60, and $−90
(b) $−1500, $−450, $−300, $−400, $−125, and $−400
(c) $1500, $−450, $−300, $−500, $−200, $−400
(d) $−1500, $−450, $−300, $−400, $−125, and $−310

7.28 According to Descartes' rule of signs, determine the possible number of $i*$ values for the net cash flow series with the following signs:
(a) −−−+++−+−
(b) −−−−−−+++++
(c) ++++−−−−−−+−+−−−
(d) −−++++−

7.29 Self-tightening wedge grips are designed for tensile testing applications up to 1200 pounds. The cash flows associated with the product are available below. Determine the following from these values: net cash flow series, cumulative cash flow series, and $i*$ displayed by the IRR spreadsheet function, if an option for "guess" is not entered.

Year	1	2	3	4
Revenue	25,000	13,000	4,000	70,000
Costs	−30,000	−7,000	−6,000	−12,000

7.30 One of your employees presented you the cash flow estimates (in $1000 units) for a new method of manufacturing box cutters for a 2-year period.
(*a*) Apply the rule of signs to determine the maximum number of possible $i*$ values at aMARR of 5% per quarter.
(*b*) Apply Norstrom's criterion to determine if there is only one positive rate of return value.
(*c*) Is it possible to determine a positive $i*$ for this net cash flow series that meets the MARR? Why or why not?

(Note: More questions about this series are included in Problem 7.57.)

Quarter	Expenses, $	Revenues, $
0	−20	0
1	−20	5
2	−10	10
3	−10	25
4	−10	26
5	−10	20
6	−15	17
7	−12	15
8	−15	2

7.31 Jenco Electric manufactures washdown adjustable speed drives in open loop, encoderless, and closed-loop servo configurations. The net cash flow associated with one phase of the production operation is shown below.
(*a*) Determine the possible rate of return values according to the rule of signs.
(*b*) Determine the changes of sign in the cumulative cash flow series. What does this mean?
(*c*) Write the PW relation to determine $i*$ values. Use the IRR function to display it.

Year	Net Cash Flow, $
0	−40,000
1	32,000
2	18,000
3	−2000
4	−1000

7.32 A cutting-edge product of Continental Fan had the following net cash flow series (in $1000 units) during its first 5-year period on the market. Find all rate of return values between 0% and 100%.

Year	Net Cash Flow, $
0	−50,000
1	+22,000
2	+38,000
3	−2000
4	−1000
5	+5000

7.33 RKI Instruments manufactures a ventilation controller designed for monitoring and controlling carbon monoxide in parking garages, boiler rooms, tunnels, etc. The net cash flow associated with one plant for the first 3 years of operation is shown. (*a*) What do the two rules about sign changes indicate concerning $i*$ values? (*b*) Find all rate of return values between 0 and 100%.

Year	Net Cash Flow, $
0	−30,000
1	20,000
2	15,000
3	−2000

7.34 A third-year college friend of yours opened Mike's Bike Repair Shop when he was a freshmen working on a sociology degree. He shared his NCF figures for the 3 years, including the $17,000 amount that it took to get started in business.
(*a*) Determine the number of possible rate of return values.
(*b*) Find all $i*$ values between −50% and 110% by plotting PW versus $i*$ for your friend. (Use 20% increments.)

Year	Net Cash Flow, $
0	−17,000
1	20,000
2	−5,000
3	8000

7.35 Arc-bot Technologies, manufacturers of six-axis, electric servo-driven robots, has experienced the cash flows shown in a shipping department. (*a*) Determine the number of possible rate of return values. (*b*) Find all $i*$ values between 0 and 100%.

Year	Expense, $	Savings, $
0	−33,000	0
1	−15,000	18,000
2	−40,000	38,000
3	−20,000	55,000
4	−13,000	12,000

7.36 Profit and loss (in $1000 units) associated with the sale of a vision-guided machine tool loading system and the resulting NCF amounts are recorded. (*a*) Use sign-change rules to determine the possible number of i^* values. (*b*) Find all i^* values between 0 and 100%. (*c*) If the required MARR for the company is 15% per year, how small can the gradient be for years 3 through 6?

Year	Cash Flow, $
0	−5,000
1	−10,100
2	4,500
3	6,500
4	8,500
5	10,500
6	12,500

Removing Multiple i^* Values

7.37 For the net cash flow series, (*a*) determine the number of possible i^* values using the two sign tests, (*b*) find the EROR using the MIRR method with an investment rate of 20% per year and a borrowing rate of 10% per year, and (*c*) use the MIRR function to find the EROR.

Year	1	2	3	4	5	6
Net Cash Flow, $	+4100	−2000	−7000	+12,000	−700	+800

7.38 For the cash flows shown, determine:
  (*a*) the number of possible i^* values
  (*b*) the i^* value displayed by the IRR function
  (*c*) the external rate of return using the MIRR method if $i_i = 18\%$ per year and $i_b = 10\%$ per year.

Year	0	1	2	3	4
Revenues, $	0	25,000	19,000	4000	18,000
Costs, $	−6000	−30,000	−7000	−6000	−12,000

7.39 For the net cash flow series shown, (*a*) apply the two rules of sign change, (*b*) find the external rate of return using the ROIC method at an investment rate of 15% per year, and (*c*) determine an i^* using the IRR function with and without the ROIC determined in part (*b*).

Year	Net Cash Flow, $
0	+48,000
1	+20,000
2	−90,000
3	+50,000
4	−10,000

7.40 A new advertising campaign by a company that manufactures products that apply biometric, surveillance, and satellite technologies resulted in the cash flows shown. Calculate unique external rate of return values using (*a*) the ROIC method with an investment rate of 30% per year, and (*b*) the MIRR approach with an investment rate of 30% and a borrowing rate of 10% per year.

Year	Cash Flow, $1000
0	2000
1	1200
2	−4000
3	−3000
4	2000

7.41 For the nonconventional net cash flow shown, determine the external rate of return using the MIRR method at a borrowing rate of 10% per year and an investment rate of (*a*) 15% per year, and (*b*) 30% per year. (Note: See Problem 7.60 for a thorough spreadsheet-based analysis of this series.)

Year	0	1	2	3
Net Cash Flow, $	−7000	+3000	+15,000	−5000

7.42 The Martian Corporation, a space vehicle development company, is starting a new division that will develop the next generation launch missile engine configuration. Use a hand application of the MIRR method to determine the EROR for the estimated net cash flows (in $1000 units) of $−50,000 in year 0, $+15,000 in years 1 through 6, and $−8000 in year 7. Assume a borrowing rate of 12% and an investment rate of 25% per year. Also, write the MIRR function to obtain i'.

7.43 A company that makes clutch disks for race cars has the annual net cash flows shown for one department.

Year	NCF, $1000
0	−65
1	30
2	84
3	−10
4	−12

  (*a*) Determine the number of positive roots to the rate of return relation.
  (*b*) Calculate the internal rate of return. Is there a negative root? How is it treated?
  (*c*) Calculate the external rate of return using the return on invested capital (ROIC) approach with an investment rate of 15% per year (as assigned by your instructor, solve by hand and/or spreadsheet).

Bonds

7.44 What is the amount and frequency of dividend payments on a bond that has a face value of

$10,000 and a coupon rate of 8% per year, payable semiannually?

7.45   If you receive a $5000 bond as a graduation present and the bond will pay you $75 interest every 3 months for 20 years, what is the bond coupon rate?

7.46   A mortgage bond issued by Automation Engineering is for sale for $8200. The bond has a face value of $10,000 with a coupon rate of 8% per year, payable annually. What rate of return will be realized if the purchaser holds the bond to maturity 5 years from now?

7.47   An engineer planning for his child's college education purchased a zero-coupon corporate bond (i.e., a bond that has no dividend payments) for $9250. The bond has a face value of $50,000 and is due in 18 years. If the bond is held to maturity, determine the i^* for the investment.

7.48   Janice V. bought a 5% $1000 twenty-year bond for $925. She received a semiannual dividend for 8 years, then sold it immediately after the 16th dividend for $800. What rate of return did she make (*a*) per semiannual period, and (*b*) per year (nominal)?

7.49   Four years ago, Valero issued $5 million worth of debenture bonds having a bond interest rate of 10% per year, payable semiannually. Market interest rates dropped and the company called the bonds (i.e., paid them off in advance) at a 10% premium on the face value. What semiannual rate of return did an investor make who purchased one $5000 bond 4 years ago and held it until it was called 4 years later?

7.50   During recessionary periods, bonds that were issued many years ago have a higher coupon rate than currently issued bonds. Therefore, they may sell at a premium, a price higher than their face value, because of currently low coupon rates. A $50,000 bond that was issued 15 years ago is for sale for $60,000. What rate of return per year will a purchaser make if the bond coupon rate is 14% per year, payable annually, and the bond is due 5 years from now? Write the ROR equation and use a single-cell spreadsheet function to display the correct answer directly.

7.51   A $10,000 mortgage bond with a bond interest rate of 8% per year, payable quarterly, was purchased for $9200. The bond was kept until it was due, a total of 7 years. What rate of return was made by the purchaser per 3 months and per year (nominal)?

7.52   A savvy investor paid $6000 for a 20-year $10,000 mortgage bond that had a bond interest rate of 8% per year, payable quarterly. Three years after he purchased the bond, market interest rates went down, so the bond increased in value. If the investor sold the bond for $11,500 three years after he bought it, what rate of return did the investor make (*a*) per quarter, and (*b*) per year (nominal)?

7.53   Ten years ago, DEWA, an electricity and water authority, issued $20 million worth of municipal bonds that carried a coupon rate of 6% per year, payable semiannually. The bonds had a maturity date of 25 years. Due to a worldwide recession, interest rates dropped significantly enough for the utility to consider paying off the bonds early at a 10% penalty to the face value. DEWA would then reissue the bonds at the same face value (i.e., $20 million) for the remaining 15 years, but at a lower coupon rate of 2% per year, payable semiannually. What would be the semiannual rate of return to DEWA, if it proceeds with this plan?

EXERCISES FOR SPREADSHEETS

7.54   Cloey has just purchased new bedroom furniture from Haverty's for a total of $10,000. She paid 20% down and, through a special promotional, she can pay off the $8000 in ten $800, no-interest payments over the next 10 months. The smaller print at the bottom of the contract states that if any of the payments is not paid in full (meaning $800) or late by one or more days, the remaining unpaid balance will revert to an interest-based loan at the APR rate of 36% per year, compounded monthly, to be paid out completely in a period of 5 months starting 1 month after any infraction of the agreement. Cloey made the first three payments on time and in the full amount; however, she could only pay $600 on time when the fourth payment was due.

(*a*)   Calculate the equal monthly payments that she must now pay, again in the full amount and on time to avoid further penalties stated in additional "fine print" of the agreement.

(*b*)   How does the total amount she paid for the furniture over the 10-month period including the APR-penalty interest compare with the original cost of $10,000? Assume Cloey makes all five interest-bearing payments in full and on time.

7.55   You need to determine i^* per year for the cash flows shown below. Since you have a new boss, you decided to perform a thorough analysis. Do the following:

(a) Use a spreadsheet to find i^*.

(b) Develop two charts on the spreadsheet: a column chart that plots cash flows versus year, and a scatter chart showing PW values versus a range of i values, one of which is i^*. (Hint: Use Figure 7–4 as a guide.)

(c) You were told when handed the estimates that the rate of return was about 25%. Determine the cash flows for years 2, 3, and 4 that are necessary to realize an i^* of 25% per year, if they remain in the same proportion to each other as initially estimated.

Year	0	1	2	3	4
Cash Flow, $	0	−80,000	9000	70,000	30,000

7.56 An Australian steel company, ASM International, claims that a savings of 40% of the cost of stainless steel threaded bar can be achieved by replacing machined threads with precision weld depositions. A U.S. manufacturer of rock bolts and grout-in-fittings plans to purchase the equipment. A mechanical engineer with the company has prepared the following estimates for additional costs and savings over the next 4 years (16 quarters). Costs decrease for some time, then increase rapidly as the equipment ages. Savings peak at $80,000 for some quarters, but then decrease as expected competition takes its toll.

The U.S. manufacturer president is not sure of the time frame to plan for the use of this new technology. The decision, in part, rests upon the rate of return achievable as the years of use progress. In the past, projects must return at least 24% per year to be retained.

(a) Determine the expected rate of return per quarter for 2 years (8 quarters) and beyond using a spreadsheet.

(b) Plot i^* versus *quarter* to graphically illustrate your results.

(c) Make a recommendation on the number of years to use the technology to economic advantage.

Quarter	Cost, $	Savings, $
0	−350,000	—
1	−50,000	10,000
2	−40,000	20,000
3	−30,000	30,000
4	−20,000	40,000
5	−10,000	50,000
6–12	0	80,000
13	−20,000	80,000
14	−40,000	40,000
15	−60,000	20,000
16	−80,000	0

7.57 Determine the following for the quarterly cash flow estimates. (a) i^* value or values; (b) if an MARR of 5% per quarter is achievable; and (c) the minimum revenue in quarter 8 that will generate an i^* that meets the MARR.

Quarter	Expenses, $	Revenues, $
0	−20	0
1	−20	5
2	−10	10
3	−10	25
4	−10	26
5	−10	20
6	−15	17
7	−12	15
8	−15	2

7.58 Charles Enterprises got into the drone manufacturing business at the inception of the technology. Net cash flows (in $ million units) for years 0 through 5 had its ups and downs as shown below. However, a few years ago, Charles started selling 3-D printable, disposable drones that can be printed, assembled, and launched with different types of sensors in remote locations—ships, isolated land areas, etc. The net cash flows for the last 5 years have improved considerably. With the assistance of the "guess" option in the IRR function, perform a thorough analysis of the NCF series using the following questions.

(a) Find all i^* values between −100% and +100% for the NCF series for years 0 through 5.

(b) Find all i^* values for the years 0 through 10.

(c) Did the i^* analysis reflect the two sign change rule predictions with this range of i^* for the 5-year period? For the 10-year period?

(d) Plot i^* versus PW for years 0–5 and years 0–10 using two scatter charts.

Year	NCF, $	Year	NCF, $
0	−40	6	−15
1	32	7	0
2	18	8	5
3	10	9	10
4	−10	10	15
5	−8		

7.59 Ten years ago, JD and his colleagues resigned from a major aerospace corporation, after many years of salaried employment, to form JRG Solar, Inc. with the intent to make a significant impact on the international renewable energy market based on years of design and test work in neighborhood garages of the three partners. Additionally, they hoped to make at least 10% per year on their own and other investors' financial commitment to JRG Solar. They were successful in obtaining $12 million in capital funding to start the business. JD, who has always been the one most interested in the economic return side of the business, recently looked at the 10-year NCF series of JRG Solar (shown below in $1000 units). When he did a quick analysis on a spreadsheet, he saw that the IRR over the 10 years was 11.26% per year. He was very pleased with this return; however, he noticed that the NCF and cumulative cash flow (CCF) series both indicated multiple i^* values.

During this 10-year start-up phase of JRG Solar, the average investment rate for positive cash flows has averaged 4% per year. Since the investment rate is quite low for corporate retained earnings, and because the two rules-of-sign-changes indicate multiple roots to the ROR equation, JD wants to understand the results of an ROR analysis beyond that of the simple ROR result ($i^* = 11.26\%$) that he obtained using the IRR function on the NCF series. Perform an ROIC analysis for JD to determine if JRG Solar's owners are realizing the MARR of 10% per year that they anticipated.

Year	0	1	2	3	4	5	6	7	8	9	10
NCF, $1000	−12,000	4000	−3000	−7000	15,000	1000	4000	−2000	−5000	8000	10,000
CCF, $1000	−12,000	−8000	−11,000	−18,000	−3000	−2000	2000	0	−5000	3000	13,000

7.60 (Note to instructor: This is a comprehensive evaluation using most of the ROR analysis techniques in the chapter. Except for (f), parts of the problem may be assigned independently of others.) For the nonconventional net cash flow series experienced over the first 3 years of operation by Viking, Inc., an Internet-based sports boat and ski equipment sales company, perform a thorough ROR analysis for the owners, Julie Merkel and Carl Upton, to include the following:
   (a) Application of sign tests to determine the number and nature of the roots to the ROR equation.
   (b) All real-number i^* values between −100% and +100% using the IRR function.
   (c) A plot of PW versus i values indicating the i^* values found in part (b).
   (d) The EROR values using the MIRR method at an investment rate of 10% and various

borrowing rates ranging from 4% to 14%, in 2% increments (Viking does not know currently what it will cost to borrow additional funds, if needed).
   (e) The EROR value using the ROIC method at the same 10% per year investment rate.
   (f) Before you started your analysis, the owners told you they expected to realize at least a 25% per year return. With this MARR and your results, develop a short written summary for Julie and Carl's review and understanding of the different ROR values and interpretations, that is, for all of the i^*, i', and i'' values. Tell them if they are meeting their MARR.

Year	0	1	2	3
Net Cash Flow, $1000	−7000	+3000	+15,000	−5000

ADDITIONAL PROBLEMS AND FE EXAM REVIEW QUESTIONS

7.61 According to Descartes' rule of signs, the possible number of rate of return values for the net cash flow series ++++−−−−−−+−+−−−++ is
   (a) 2   (b) 4   (c) 6   (d) 8

7.62 According to Norstrom's criterion, there is only one positive rate of return value in a cash flow series when:
   (a) The cumulative cash flow starts out positive and changes sign only once
   (b) The cumulative cash flow starts out negative and changes sign only once
   (c) The cumulative cash flow total is greater than zero
   (d) The cumulative cash flow total is less than zero

7.63 When positive net cash flows are generated before the end of a project, and when these cash flows are

reinvested at an interest rate that is greater than the internal rate of return,
   (a) The resulting rate of return is equal to the internal rate of return.
   (b) The resulting rate of return is less than the internal rate of return.
   (c) The resulting rate of return is equal to the reinvestment rate of return.
   (d) The resulting rate of return is greater than the internal rate of return.

7.64 As part of your inheritance, you received a bond that will pay interest of $700 every 6 months for 15 years. If the coupon rate is 7% per year, the face value of the bond is
   (a) $10,000   (b) $20,000
   (c) $30,000   (d) $40,000

7.65 A municipal bond with a face value of $10,000 was issued today with an interest rate of 6% per year payable semiannually. The bond matures 20 years from now. If an investor paid $9,000 for the bond and holds it to maturity, all of the following equations will yield the correct semiannual rate of return except:

(a) $0 = -9000 + 300(P/A,i^*,40) + 10,000(P/F,i^*,40)$

(b) $0 = -9000(F/P,i^*,40) + 300(F/A,i^*,40) + 10,000$

(c) $0 = -9000(A/P,i^*,40) + 300 + 10,000(A/F,i^*,40)$

(d) $0 = -9000 + 600(P/A,i^*,40) + 10,000(P/F,i^*,40)$

7.66 A $20,000 collateral bond has a coupon rate of 7% per year payable quarterly. The bond matures 30 years from now. At a market interest rate of 7% per year, compounded semiannually, the amount and frequency of the bond interest payments is:

(a) $1400 per year
(b) $1400 per quarter
(c) $350 per year
(d) $350 per quarter

7.67 A $10,000 mortgage bond that is due in 20 years pays interest of $250 every 6 months. The bond coupon rate is:

(a) 2.5% per year, payable quarterly
(b) 5.0% per year, payable quarterly
(c) 5% per year, payable semiannually
(d) 10% per year, payable quarterly

7.68 For the net cash flow and cumulative cash flows shown, the value of x is nearest:

(a) $-8,000   (b) $-16,000
(c) $16,000   (d) $41,000

Year	1	2	3	4	5
Net Cash Flow, $	+13,000	−29,000	−25,000	50,000	x
CCF, $	+13,000	−16,000	−41,000	+9000	+1000

7.69 Basset, a furniture manufacturing company, borrowed $1 million and repaid the loan through monthly payments of $20,000 for 2 years plus a single lump-sum payment of $1 million at the end of 2 years. The interest rate on the loan was closest to:

(a) 0.5% per month
(b) 2% per month
(c) 2% per year
(d) 8% per year

7.70 An investment of $60,000 ten years ago resulted in uniform income of $10,000 per year for the 10-year period. The rate of return on the investment was closest to:

(a) 10.6% per year
(b) 14.2% per year
(c) 16.4% per year
(d) 18.6% per year

7.71 A bulk materials hauler purchased a used dump truck for $50,000 two years ago. The operating costs have been $5000 per month and revenues have averaged $7500 per month. The truck was just sold for $11,000. The rate of return is closest to:

(a) 2.6% per month
(b) 2.6% per year
(c) 3.6% per month
(d) 15.6% per year

7.72 Five years ago, an alumnus of a university donated $50,000 to establish a permanent endowment for scholarships. The first scholarships were awarded 1 year after the contribution. If the amount awarded each year, that is, the interest on the endowment, is $4500, the rate of return earned on the fund is closest to:

(a) 7.5% per year
(b) 8.5% per year
(c) 9% per year
(d) 10% per year

7.73 When applying the MIRR approach to determine the external rate of return of a project, which of the following statements is true?

(a) The borrowing rate, i_b, is usually equal to the MARR.
(b) The borrowing rate, i_b, is usually greater than the investment rate, i_i.
(c) The borrowing rate, i_b, is usually less than the investment rate, i_i.
(d) The borrowing rate, i_b, and investment rate, i_i, are usually equal.

7.74 For the nonconventional net cash flow series shown, the external rate of return per year using the MIRR method, with an investment rate of 20% per year and a borrowing rate of 8% per year, is closest to:

(a) 10.8%   (b) 12.0%
(c) 14.8%   (d) 16.7%

Year	0	1	2	3	4
NCF, $	−40,000	+13,000	−29,000	+25,000	+50,000

CASE STUDY

DEVELOPING AND SELLING AN INNOVATIVE IDEA

Background

Three engineers who worked for Mitchell Engineering, a company specializing in public housing development, went to lunch together several times a week. Over time they decided to work on solar energy production ideas. After a lot of weekend time over several years, they had designed and developed a prototype of a low-cost, scalable solar energy plant for use in multifamily dwellings on the low end and medium-sized manufacturing facilities on the upper end. For residential applications, the collector could be mounted along side a TV dish and be programmed to track the sun. The generator and additional equipment are installed in a closet-sized area in an apartment or on a floor for multiple-apartment supply. The system serves as a supplement to the electricity provided by the local power company. After some 6 months of testing, it was agreed that the system was ready to market and reliably state that an electricity bill in high-rises could be reduced by approximately 40% per month. This was great news for low-income dwellers on government subsidy that are required to pay their own utility bills.

Information

With a hefty bank loan and $200,000 of their own capital, they were able to install demonstration sites in three cities in the sunbelt. Net cash flow after all expenses, loan repayment, and taxes for the first 4 years was acceptable; $55,000 at the end of the first year, increasing by 5% each year thereafter. A business acquaintance introduced them to a potential buyer of the patent rights and current subscriber base with an estimated $500,000 net cashout after only these 4 years of ownership. However, after serious discussion replaced the initial excitement of the sales offer, the trio decided to not sell at this time. They wanted to stay in the business for a while longer to develop some enhancement ideas and to see how much revenue may increase over the next few years.

During the next year, the fifth year of the partnership, the engineer who had received the patents upon which the collector and generator designs were based became very displeased with the partnering arrangements and left the trio to go into partnership with an international firm in the energy business. With new research and development funds and the patent rights, a competing design was soon on the market and took much of the business away from the original two developers. Net cash flow dropped to $40,000 in year 5 and continued to decrease by $5000 per year. Another offer to sell in year 8 was presented, but it was only for $100,000 net cash. This was considered too much of a loss, so the two owners did not accept. Instead, they decided to put $200,000 more of their own savings into the company to develop additional applications in the housing market.

It is now 12 years since the system was publicly launched. With increased advertising and development, net cash flow has been positive the last 4 years, starting at $5000 in year 9 and increasing by $5000 each year until now.

Case Study Exercises

It is now 12 years after the products were developed, and the engineers invested most of their savings in an innovative idea. However, the question of "When do we sell?" is always present in these situations. To help with the analysis, determine the following:

1. The rate of return at the end of year 4 for two situations: (*a*) the business is sold for the net cash amount of $500,000 and (*b*) no sale.
2. The rate of return at the end of year 8 for two situations: (*a*) the business is sold for the net cash amount of $100,000 and (*b*) no sale.
3. The rate of return now at the end of year 12.
4. Consider the cash flow series over the 12 years. Is there any indication that multiple rates of return may be present? If so, use the spreadsheet already developed to search for ROR values in the range ±100% other than the one determined in exercise 3 above.
5. Assume you are an investor with a large amount of ready cash, looking for an innovative solar energy product. What amount would you be willing to offer for the business at this point (end of year 12) if you require a 12% per year return on all your investments and, if purchased, you plan to own the business for 12 additional years? To help make the decision, assume the current NCF series continues increasing at $5000 per year for the years you would own it. **Explain your logic for offering this amount.**

CHAPTER 8

Rate of Return Analysis: Multiple Alternatives

PhotoLink/Getty Images

LEARNING OUTCOMES

Purpose: Select the best alternative on the basis of incremental rate of return analysis.

SECTION	TOPIC	LEARNING OUTCOME
8.1	Incremental analysis	• State why the ROR method of comparing alternatives requires an incremental cash flow analysis.
8.2	Incremental cash flows	• Calculate the incremental cash flow series for two alternatives.
8.3	Meaning of Δi^*	• Interpret the meaning of the incremental ROR (Δi^*) determined from the incremental cash flow series.
8.4	Δi^* from PW relation	• Based on a PW relation, select the better of two alternatives using incremental ROR analysis or a breakeven ROR value.
8.5	Δi^* from AW relation	• Select the better of two alternatives using incremental ROR analysis based on an AW relation.
8.6	More than two alternatives	• Select the best from several alternatives using incremental ROR analysis.
8.7	All-in-one spreadsheet	• Use a single spreadsheet to perform PW, AW, ROR, and incremental ROR analyses for mutually exclusive and independent alternatives.

This chapter presents the methods by which two or more alternatives can be evaluated using a rate of return (ROR) comparison based on the methods of the previous chapter. The ROR evaluation, correctly performed, will result in the same selection as the PW and AW analyses, but the computational procedure is considerably different for ROR evaluations. The ROR analysis evaluates the increments between two alternatives in pairwise comparisons. As the cash flow series becomes more complex, spreadsheet functions help speed computations.

8.1 Why Incremental Analysis Is Necessary ●●●

When two or more mutually exclusive alternatives are evaluated, engineering economy can identify the one alternative that is the best economically. As we have learned, the PW and AW techniques can be used to do so, and are the recommended methods. Now the procedure using ROR to identify the best is presented.

Let's assume that a company uses a MARR of 16% per year, that the company has $90,000 available for investment, and that two alternatives (A and B) are being evaluated. Alternative A requires an investment of $50,000 and has an internal rate of return i_A^* of 35% per year. Alternative B requires $85,000 and has an i_B^* of 29% per year. Intuitively we may conclude that the better alternative is the one that has the larger return, A in this case. However, this is not necessarily so. While A has the higher projected return, its initial investment ($50,000) is much less than the total money available ($90,000). What happens to the investment capital that is left over? It is generally assumed that excess funds will be invested at the company's MARR, as we learned in previous chapters. Using this assumption, it is possible to determine the consequences of the two alternative investments. If alternative A is selected, $50,000 will return 35% per year. The $40,000 left over will be invested at the MARR of 16% per year. The rate of return on the total capital available, then, will be the weighted average. Thus, if alternative A is selected,

$$\text{Overall ROR}_A = \frac{50,000(0.35) + 40,000(0.16)}{90,000} = 26.6\%$$

If alternative B is selected, $85,000 will be invested at 29% per year, and the remaining $5000 will earn 16% per year. Now the weighted average is

$$\text{Overall ROR}_B = \frac{85,000(0.29) + 5000(0.16)}{90,000} = 28.3\%$$

These calculations show that even though the i^* for alternative A is higher, alternative B presents the better overall ROR for the $90,000. If either a PW or AW comparison is conducted using the MARR of 16% per year as i, alternative B will be chosen.

This simple example illustrates a major fact about the rate of return method for ranking and comparing alternatives:

> Under some circumstances, project ROR values do not provide the same ranking of alternatives as do PW and AW analyses. This situation does not occur if we conduct an **incremental ROR analysis** (discussed below).

When **independent projects** are evaluated, no incremental analysis is necessary between projects. Each project is evaluated separately from others, and more than one can be selected. Therefore, the only comparison is with the do-nothing alternative for each project. The project ROR can be used to accept or reject each one.

Independent project selection

8.2 Calculation of Incremental Cash Flows for ROR Analysis ●●●

To conduct an incremental ROR analysis, it is necessary to calculate the **incremental cash flow** series over the lives of the alternatives. Based upon the equivalence relations (PW and AW), ROR evaluation makes the equal-service assumption.

TABLE 8–1	Format for Incremental Cash Flow Tabulation		
	Cash Flow		Incremental Cash Flow (3) = (2) − (1)
Year	Alternative A (1)	Alternative B (2)	
0			
1			
.			
.			
.			

Equal-service requirement

> The incremental ROR method requires that the equal-service requirement be met. Therefore, the **LCM (least common multiple) of lives for each pairwise comparison** must be used. All the assumptions of equal service present for PW analysis are necessary for the incremental ROR method.

A format for hand or spreadsheet solutions is helpful (Table 8–1). Equal-life alternatives have n years of incremental cash flows, while unequal-life alternatives require the LCM of lives for analysis. At the end of each life cycle, the salvage value and initial investment for the next cycle must be included for the LCM case.

When a **study period** is established, only this number of years is used for the evaluation. All incremental cash flows outside the period are neglected. As we learned earlier, using a study period, especially one shorter than the life of either alternative, can change the economic decision from that rendered when the full lives are considered.

Only for the purpose of simplification, use the convention that between two alternatives, the one with the *larger initial investment* will be regarded as *alternative B*. Then, for each year in Table 8–1,

$$\text{Incremental cash flow} = \text{cash flow}_B - \text{cash flow}_A \qquad [8.1]$$

The initial investment and annual cash flows for each alternative (excluding the salvage value) are one of the types identified in Chapter 5:

Revenue alternative, where there are both negative and positive cash flows

Cost alternative, where all cash flow estimates are negative

Revenue or cost alternative

In either case, Equation [8.1] is used to determine the incremental cash flow series with the sign of each cash flow carefully determined.

EXAMPLE 8.1

Liberty Pharmacy must improve its point-of-sale (POS) speed for customers to remain competitive with large chains such as Walgreen's and CVS. Liberty owners plan to replace all scanner and cash register equipment in their six locations throughout Austin, Texas. With the decreasing costs and improved technology, two options are available. New systems will cost a total of $21,000; however, slightly used equipment is available for $15,000. Because the new systems are more sophisticated, their annual operating cost is expected to be $7000 per year, while the used systems are expected to require $8200 per year. The expected life is 10 years for both systems with a 5% salvage value. Tabulate the incremental cash flow series.

Solution

Incremental cash flow is tabulated in Table 8–2. The subtraction performed is (new – used) since the new system has a larger initial cost. The salvage values in year 10 are separated from ordinary cash flow for clarity. When disbursements are the same for a number of consecutive years, *for hand solution only*, it saves time to make a single cash flow listing, as is done for years 1 to 10. However, remember that several years were combined when performing the analysis. This approach *cannot be used for spreadsheets,* when the IRR or NPV function is used, as each year's cash flow must be entered separately, even if it is zero.

TABLE 8–2	Cash Flow Tabulation for Example 8.1		
	Cash Flow		**Incremental Cash Flow**
Year	**Used Press**	**New Press**	**(New – Used)**
0	$−15,000	$−21,000	$−6,000
1–10	−8,200	−7,000	+1,200
10	+750	+1,050	+300

EXAMPLE 8.2

A sole-source vendor can supply a new industrial park with large transformers suitable for underground utilities and vault-type installation. Type A has an initial cost of $70,000 and a life of 8 years. Type B has an initial cost of $95,000 and a life expectancy of 12 years. The annual operating cost for type A is expected to be $9000, while the AOC for type B is expected to be $7000. If the salvage values are $5000 and $10,000 for type A and type B, respectively, tabulate the incremental cash flow using their LCM for hand and spreadsheet solutions.

Solution by Hand

The LCM of 8 and 12 is 24 years. In the incremental cash flow tabulation for 24 years (Table 8–3), note that the reinvestment and salvage values are shown in years 8 and 16 for type A and in year 12 for type B.

Solution by Spreadsheet

Figure 8–1 shows the incremental cash flows for the LCM of 24 years. As in the hand tabulation, reinvestment is made in the last year of each intermediate life cycle. The incremental values in column D are the result of subtractions of column B from C.

   Note that the final row includes a summation check. The total incremental cash flow should agree in both the column D total and the subtraction C29 − B29. Also note that the incremental values change signs three times, indicating the possibility of multiple i^* values, per Descartes' rule of signs. This possible dilemma is discussed later in the chapter.

TABLE 8–3	Incremental Cash Flow Tabulation, Example 8.2		
	Cash Flow		**Incremental Cash Flow**
Year	**Type A**	**Type B**	**(B − A)**
0	$ −70,000	$ −95,000	$−25,000
1–7	−9,000	−7,000	+2,000
8	$\begin{cases}-70{,}000\\-9{,}000\\+5{,}000\end{cases}$	−7,000	+67,000
9–11	−9,000	−7,000	+2,000
12	−9,000	$\begin{cases}-95{,}000\\-7{,}000\\+10{,}000\end{cases}$	−83,000
13–15	−9,000	−7,000	+2,000
16	$\begin{cases}-70{,}000\\-9{,}000\\+5{,}000\end{cases}$	−7,000	+67,000
17–23	−9,000	−7,000	+2,000
24	$\begin{cases}-9{,}000\\+5{,}000\end{cases}$	$\begin{cases}-7{,}000\\+10{,}000\end{cases}$	+7,000
	$−411,000	$−338,000	$+73,000

Figure 8–1
Spreadsheet computation of incremental cash flows for unequal-life alternatives, Example 8.2.

	A	B	C	D	E	F	G
1				Incremental			
2		Net cash flows		cash flow			
3	Year	Type A	Type B	(B - A)			
4	0	-70,000	-95,000	-25,000			
5	1	-9,000	-7,000	2,000			
6	2	-9,000	-7,000	2,000			
7	3	-9,000	-7,000	2,000			
8	4	-9,000	-7,000	2,000			
9	5	-9,000	-7,000	2,000			
10	6	-9,000	-7,000	2,000			
11	7	-9,000	-7,000	2,000			
12	8	-74,000	-7,000	67,000			
13	9	-9,000	-7,000	2,000			
14	10	-9,000	-7,000	2,000			
15	11	-9,000	-7,000	2,000			
16	12	-9,000	-92,000	-83,000			
17	13	-9,000	-7,000	2,000			
18	14	-9,000	-7,000	2,000			
19	15	-9,000	-7,000	2,000			
20	16	-74,000	-7,000	67,000			
21	17	-9,000	-7,000	2,000			
22	18	-9,000	-7,000	2,000			
23	19	-9,000	-7,000	2,000			
24	20	-9,000	-7,000	2,000			
25	21	-9,000	-7,000	2,000			
26	22	-9,000	-7,000	2,000			
27	23	-9,000	-7,000	2,000			
28	24	-4,000	3,000	7,000			
29	Totals	-411,000	-338,000	73,000			
30							
31							

Starting new life cycle for A
= initial cost + AOC + salvage
= − 70,000 − 9,000 + 5,000

Check on summations
Incremental column should equal difference of columns

8.3 Interpretation of Rate of Return on the Extra Investment ● ● ●

The incremental cash flows in year 0 of Tables 8–2 and 8–3 reflect the *extra investment* or *cost* required if the alternative with the larger first cost is selected. This is important in an incremental ROR analysis in order to determine the ROR earned on the extra funds expended for the larger-investment alternative. If the incremental cash flows of the larger investment don't justify it, we must select the cheaper one. If there is no cheaper, justified alternative, the DN alternative is the best economic choice. In Example 8.1, the new systems require an extra investment of $6000 (Table 8–2). If the new systems are purchased, there will be a "savings" of $1200 per year for 10 years, plus an extra $300 in year 10. The decision to buy the used or new systems can be made on the basis of the profitability of investing the extra $6000 in the new systems. If the equivalent worth of the savings is greater than the equivalent worth of the extra investment at the MARR, the extra investment should be made (i.e., the larger first-cost proposal should be accepted). On the other hand, if the extra investment is not justified by the savings, select the lower-investment proposal.

It is important to recognize that the rationale for making the selection decision is the same as if only *one alternative* were under consideration, that alternative being the one represented by the incremental cash flow series. When viewed in this manner, it is obvious that unless this investment yields a rate of return equal to or greater than the MARR, the extra investment should not be made. As further clarification of this extra investment rationale, consider the following: The rate of return attainable through the incremental cash flow is an alternative to investing at the MARR. Section 8.1 states that any excess funds not invested in the alternative are assumed to be invested at the MARR. The conclusion is clear:

> If the rate of return available through the incremental cash flow equals or exceeds the MARR, the alternative associated with the extra investment should be selected.

ME alternative selection

Not only must the return on the extra investment meet or exceed the MARR, the return on the investment that is common to both alternatives must meet or exceed the MARR. Accordingly, prior to performing an incremental ROR analysis, it is advisable to determine the internal rate of return i^* for each alternative. This can be done only for *revenue alternatives* because cost alternatives have only cost (negative) cash flows and no i^* can be determined. The guideline is as follows:

> **For multiple revenue alternatives,** calculate the internal rate of return i^* for each alternative, and eliminate all alternatives that have an $i^* <$ MARR. Compare the remaining alternatives incrementally.

As an illustration, if the MARR = 15% and two alternatives have i^* values of 12% and 21%, the 12% alternative can be eliminated from further consideration. With only two alternatives, it is obvious that the second one is selected. If both alternatives have $i^* <$ MARR, no alternative is justified and the do-nothing alternative is the best economically. When three or more alternatives are evaluated, it is usually worthwhile, but not required, to calculate i^* for each alternative for preliminary screening. Alternatives that cannot meet the MARR may be eliminated from further evaluation using this option. This option is especially useful when performing the analysis by spreadsheet. The IRR function applied to each alternative's cash flow series can quickly indicate unacceptable alternatives, as demonstrated in Section 8.6.

Let's consider the ROR on the incremental cash flow series between two alternatives B and A. The incremental ROR value, identified as i^*_{B-A}, and discussed further below, has a predictable relation to individual alternative ROR values in weighted average calculations like those presented in Section 8.1. In brief, if alternative B has a larger initial investment than A, and if $\text{ROR}_B < \text{ROR}_A$, then $i^*_{B-A} < \text{ROR}_B$. Furthermore, if $\text{ROR}_B > \text{ROR}_A$, then $i^*_{B-A} > \text{ROR}_B$.

When **independent projects** are evaluated, there is no comparison on the extra investment. The ROR value is used to accept all projects with $i^* \geq$ MARR, assuming there is no budget limitation. For example, assume MARR = 10%, and three independent projects are available with ROR values of

Independent project selection

$$i^*_A = 12\% \qquad i^*_B = 9\% \qquad i^*_C = 23\%$$

Projects A and C are selected, but B is not because $i^*_B <$ MARR.

8.4 Rate of Return Evaluation Using PW:
Incremental and Breakeven (Two Alternatives) ●●●

In this section, we discuss the primary approach to making mutually exclusive alternative selections by the incremental ROR method. A PW-based relation is developed for the incremental cash flows and set equal to zero. Use hand solution or spreadsheet functions to find Δi^*_{B-A}, the internal ROR for the series. Placing Δ (delta) before i^*_{B-A} distinguishes it from the overall ROR values i^*_A and i^*_B. (Δi^* may replace Δi^*_{B-A} when only two alternatives are present.)

Since incremental ROR requires equal-service comparison, the LCM of lives must be used in the PW formulation. Because of the reinvestment requirement for PW analysis for different-life assets, the incremental cash flow series may contain several sign changes, indicating multiple Δi^* values. Though incorrect, this indication is usually neglected in actual practice. The correct approach is to follow one of the techniques of Section 7.5. This means that the single external ROR ($\Delta i'$ or $\Delta i''$) for the incremental cash flow series is determined and the alternative selection is based on this result.

These three elements—**incremental cash flow series, LCM, and multiple roots**—are the primary reasons that the ROR method is often applied incorrectly in engineering economy analyses of multiple alternatives. As stated earlier, it is always possible, and generally advisable, to use a PW or AW analysis *at an established MARR* in lieu of the ROR method when multiple rates are indicated.

The complete procedure for hand or spreadsheet solution for an incremental ROR analysis for two *cost* alternatives is as follows:

1. Order the alternatives by initial investment or cost, starting with the smaller one, called A. The one with the larger initial investment is in the column labeled B in Table 8–1.
2. Develop the cash flow and incremental cash flow series using the LCM of years, assuming reinvestment in alternatives.
3. Draw an incremental cash flow diagram, if needed.
4. Count the number of sign changes in the incremental cash flow series to determine if multiple rates of return may be present. If necessary, use Norstrom's criterion to determine if a single positive root exists.
5. Set up the PW = 0 equation and determine Δi^*_{B-A}.
6. Select the economically better alternative as follows:

If Δi^*_{B-A} < MARR, select alternative A.

If $\Delta i^*_{B-A} \geq$ MARR, the extra investment is justified; select alternative B.

If Δi^* is exactly equal to or very near the MARR, noneconomic considerations help in the selection of the "better" alternative.
    For revenue alternatives in step 2, first calculate the i^* for each alternative to ensure that it is economically acceptable relative to the MARR.

ME alternative
selection

In step 5, if trial and error by hand are used, time may be saved if the Δi^*_{B-A} value is bracketed, rather than approximated by a point value using linear interpolation, provided that a single ROR value is not needed. For example, if the MARR is 15% per year and you have established that Δi^*_{B-A} is in the 15% to 20% range, an exact value is not necessary to accept B since you already know that $\Delta i^*_{B-A} \geq$ MARR.

The IRR function on a spreadsheet will normally determine one Δi^* value. Multiple guess values can be input to find multiple roots in the range −100% to ∞ for a nonconventional series, as illustrated in Example 7.4. If this is not the case, to be correct, the indication of multiple roots in step 4 requires that one of the techniques of Section 7.5 be applied to find an EROR.

EXAMPLE 8.3

As Ford Motor Company of Europe retools an old assembly plant in the United Kingdom to produce a fuel-efficient economy model automobile, Ford and its suppliers are seeking additional sources for light, long-life transmissions. Automatic transmission component manufacturers use highly finished dies for precision forming of internal gears and other moving parts. Two international vendors make the required dies. Use the per unit estimates below and a MARR of 12% per year to select the more economical vendor bid. Show both hand and spreadsheet solutions.

	A	B
Initial cost, $	−8,000	−13,000
Annual costs, $ per year	−3,500	−1,600
Salvage value, $	0	2,000
Life, years	10	5

Solution by Hand

These are cost alternatives, since all cash flows are costs. Use the procedure described above to determine Δi^*_{B-A}.

1. Alternatives A and B are correctly ordered with the higher first-cost alternative in column 2 of Table 8–4.
2. The cash flows for the LCM of 10 years are tabulated.

			Incremental
TABLE 8–4	Incremental Cash Flow Tabulation, Example 8.3		
Year	Cash Flow A (1)	Cash Flow B (2)	Cash Flow (3) = (2) − (1)
0	$ −8,000	$−13,000	$ −5,000
1–5	−3,500	−1,600	+1,900
5	—	{ +2,000 / −13,000	−11,000
6–10	−3,500	−1,600	+1,900
10	—	+2,000	+2,000
	$−43,000	$−38,000	$ +5,000

3. The incremental cash flow diagram is shown in Figure 8–2.
4. There are three sign changes in the incremental cash flow series, indicating as many as three roots. There are also three sign changes in the cumulative incremental series, which starts negatively at $S_0 = $−5000$ and continues to $S_{10} = $+5000$, indicating that more than one positive root may exist.
5. The rate of return equation based on the PW of incremental cash flows is

$$0 = -5000 + 1900(P/A,\Delta i^*,10) - 11,000(P/F,\Delta i^*,5) + 2000(P/F,\Delta i^*,10) \qquad [8.2]$$

In order to resolve any multiple-root problem, we can assume that the investment rate i_i in the ROIC technique will equal the Δi^* found by trial and error. Solution of Equation [8.2] for the first root discovered results in Δi^* between 12% and 15%. By interpolation $\Delta i^* = 12.65\%$.
6. Since the rate of return of 12.65% on the extra investment is greater than the 12% MARR, the higher-cost vendor B is selected.

Comment

In step 4, the presence of up to three i^* values is indicated. The preceding analysis finds one of the roots at 12.65%. When we state that the incremental ROR is 12.65%, we assume that any positive net cash flows are reinvested at 12.65%. If this is not a reasonable assumption, the ROIC or modified ROR technique (Section 7.5) must be applied to find a different single $\Delta i'$ or $\Delta i''$ to compare with MARR = 12%.

The other two roots are very large positive and negative numbers, as the IRR function reveals. So they are not useful to the analysis.

Solution by Spreadsheet

Steps 1 through 4 are the same as above.

5. Figure 8–3 includes the same incremental net cash flows from Table 8–4 calculated in column D. Cell D15 displays the Δi^* value of 12.65% using the IRR function.
6. Since the rate of return on the extra investment is greater than the 12% MARR, the higher-cost vendor B is selected.

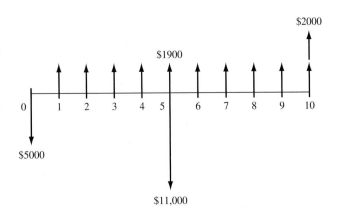

Figure 8–2
Diagram of incremental cash flows, Example 8.3.

	A	B	C	D	E
1					
2		Vendor A	Vendor B	Incremental	
3	Year	cash flow, $	cash flow, $	cash flow, $	
4	0	-8,000	-13,000	-5,000	
5	1	-3,500	-1,600	1,900	
6	2	-3,500	-1,600	1,900	
7	3	-3,500	-1,600	1,900	
8	4	-3,500	-1,600	1,900	
9	5	-3,500	-12,600	-9,100	
10	6	-3,500	-1,600	1,900	
11	7	-3,500	-1,600	1,900	
12	8	-3,500	-1,600	1,900	
13	9	-3,500	-1,600	1,900	
14	10	-3,500	400	3,900	
15	Incremental i*			**12.65%**	
16					
17	Check: PW @ 12%			$137.67	
18					
19			= NPV(12%,D5:D14) + D4		
20					

Figure 8–3
Spreadsheet solution using LCM of lives and IRR function, Example 8.3.

Comment

Once the spreadsheet is set up, there are a wide variety of analyses that can be performed. For example, row 17 uses the NPV function to verify that the present worth is positive at MARR=12%. Charts such as PW versus Δi and PW versus i help graphically interpret the situation.

The rate of return determined for the incremental cash flow series or the actual cash flows can be interpreted as a **breakeven rate of return** value.

Breakeven ROR

> The **breakeven rate of return** is the incremental $i*$ value, $\Delta i*$, at which the PW (or AW) value of the *incremental cash flows* is exactly zero. This is the same as stating that the breakeven ROR is the i value, $i*$, at which the PW (or AW) values of two alternatives' actual cash flows are exactly equal to each other.

If the incremental cash flow ROR ($\Delta i*$) is greater than the MARR, the larger-investment alternative is selected. For example, if the PW versus Δi graph for the incremental cash flows in Table 8–4 (and spreadsheet Figure 8–3) is plotted for various interest rates, the graph shown in Figure 8–4 is obtained. It shows the $\Delta i*$ breakeven at 12.65%. The conclusions are that

- For MARR < 12.65%, the extra investment for B is justified.
- For MARR > 12.65%, the opposite is true—the extra investment in B should not be made, and vendor A is selected.
- If MARR is exactly 12.65%, the alternatives are equally attractive.

Figure 8–5, which is a breakeven graph of PW versus i for the cash flows (not incremental) of each alternative in Example 8.3, provides the same results. Since all net cash flows are negative (cost alternatives), the PW values are negative. Now, the same conclusions are reached using the following logic:

- If MARR < 12.65%, select B since its PW of cost cash flows is smaller (numerically larger).
- If MARR > 12.65%, select A since its PW of costs is smaller.
- If MARR is exactly 12.65%, either alternative is equally attractive.

Example 8.4 illustrates incremental ROR evaluation and breakeven rate of return graphs for revenue alternatives. More of breakeven analysis is covered in Chapter 13.

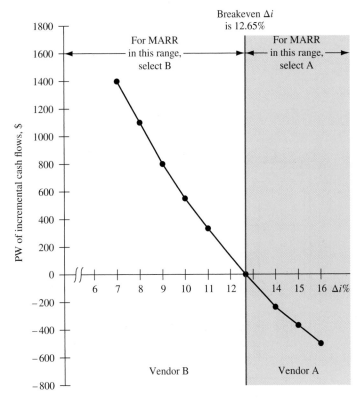

Figure 8–4
Plot of present worth of incremental cash flows for Example 8.3 at various Δi values.

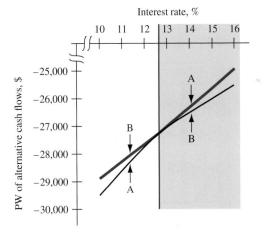

Figure 8–5
Breakeven graph of Example 8.3 cash flows (not incremental).

EXAMPLE 8.4

New filtration systems for commercial airliners are available that use an electric field to remove up to 99.9% of infectious diseases and pollutants from aircraft air. This is vitally important, as many of the flu germs, viruses, and other contagious diseases are transmitted through the systems that recirculate aircraft air many times per hour. Investments in the new filtration equipment can cost from $100,000 to $150,000 per aircraft, but savings in fuel, customer complaints, legal actions, etc., can also be sizable. Use the estimates below (in $100 units) from two suppliers provided to an international carrier to do the following, using a spreadsheet and an MARR of 15% per year.

- Plot two graphs: PW versus i values for both alternatives' cash flows and PW versus Δi values for incremental cash flows.
- Estimate the breakeven ROR values from both graphs, and use this estimate to select one alternative.

	Air Cleanser (Filter 1)	Purely Heaven (Filter 2)
Initial cost per aircraft, $	−1000	−1500
Estimated savings, $ per year	375	700 in year 1, decreasing by 100 per year thereafter
Estimated life, years	5	5

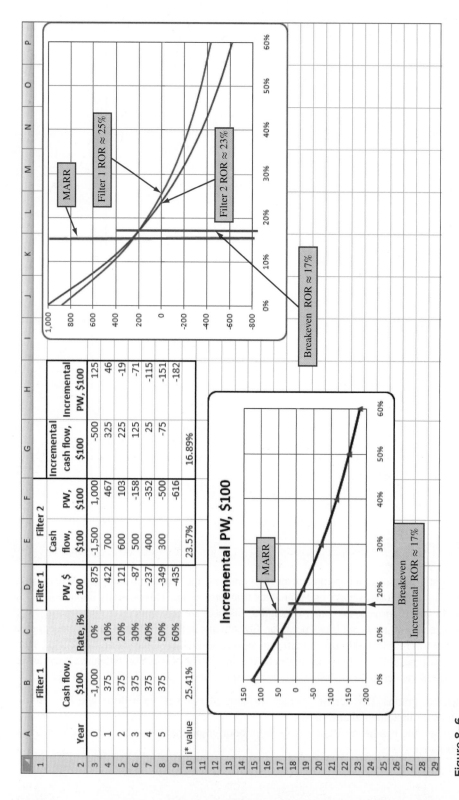

Figure 8–6
PW versus *i* graph and PW versus incremental *i* graph, Example 8.4.

Solution by Spreadsheet

Refer to Figure 8–6 as the solution is explained. For information, row 10 shows the ROR values calculated using the IRR function for filter 1 and filter 2 cash flows and the incremental cash flow series (filter 2 − filter 1).

The cash flow sign tests for each filter indicate no multiple rates. The incremental cash flow sign test does not indicate the presence of a unique positive root; however, the second rate is an extremely large and useless value. The PW values of filter 1 and filter 2 cash flows are plotted on the right side for i values ranging from 0% to 60%. Since the PW curves cross each other at approximately $i^* = 17\%$, the rate does exceed the MARR of 15%. The higher-cost filter 2 (Purely Heaven) is selected.

The PW curve for the incremental cash flows series (column G) is plotted at the bottom left of Figure 8–6. As expected, the curve crosses the PW = 0 line at approximately 17%, indicating the same economic conclusion of filter 2.

Figure 8–6 provides an excellent opportunity to see why the ROR method can result in selecting the wrong alternative when only i^* values are used to select between two alternatives. This is sometimes called the **ranking inconsistency problem** of the ROR method. *The inconsistency occurs when the MARR is set less than the breakeven rate between two revenue alternatives.* Since the MARR is established based on conditions of the economy and market, MARR is established external to any particular alternative evaluation. In Figure 8–6, the incremental breakeven rate is 16.89%, and the MARR is 15%. The MARR is lower than breakeven; therefore, the *incremental ROR analysis* results in correctly selecting filter 2. But if only the i^* values were used, filter 1 would be wrongly chosen because its i^* exceeds that of filter 2 (25.41% > 23.57%). This error occurs because the rate of return method assumes reinvestment at the alternative's ROR value, while PW and AW analyses use the MARR as the reinvestment rate. The conclusion is simple:

If the ROR method is used to evaluate two or more alternatives, use the **incremental cash flows and Δi^*** to make the decision between alternatives.

8.5 Rate of Return Evaluation Using AW ●●○

Comparing alternatives by the ROR method (correctly performed) always leads to the same selection as PW and AW analyses, whether the ROR is determined using a PW-based or an AW-based relation. However, for the AW-based technique, there are two equivalent ways to perform the evaluation: (1) using the *incremental cash flows* over the LCM of alternative lives, just as for the PW-based relation (Section 8.4), or (2) finding the AW for each alternative's *actual cash flows* and setting the difference of the two equal to zero to find the Δi^* value. There is no difference between the two approaches if the alternative lives are equal. Both methods are summarized here.

Since the ROR method requires comparison for equal service, the **incremental cash flows must be evaluated over the LCM of lives.** There may be no real computational advantage to using AW, as was found in Chapter 6. The same six-step procedure of the previous section (for PW-based calculation) is used, except in step 5, the AW-based relation is developed.

Equal-service requirement

The second AW-based method mentioned above takes advantage of the AW technique's assumption that the equivalent AW value is the same for each year of the first and all succeeding life cycles. Whether the lives are equal or unequal, set up the *AW relation for the cash flows of each alternative,* form the relation below, and solve for i^*.

$$0 = AW_B - AW_A \qquad\qquad [8.3]$$

For both methods, all equivalent values are on an AW basis, so the i^* that results from Equation [8.3] is the same as the Δi^* found using the first approach. Example 8.5 illustrates ROR analysis using AW-based relations for unequal lives.

EXAMPLE 8.5

Compare the alternatives of vendors A and B for Ford in Example 8.3, using an AW-based incremental ROR method and the same MARR of 12% per year.

Solution

For reference, the PW-based ROR relation, Equation [8.2], for the incremental cash flow in Example 8.3 shows that vendor B should be selected with $\Delta i^* = 12.65\%$.

For the AW relation, there are two equivalent solution approaches. Write an AW-based relation on the *incremental* cash flow series over the *LCM of 10 years*, or write Equation [8.3] for the *two actual* cash flow series over *one life cycle* of each alternative.

For the incremental method, the AW equation is easily developed from Table 8–4, column (3).

$$0 = -5000(A/P,\Delta i^*,10) - 11{,}000(P/F,\Delta i^*,5)(A/P,\Delta i^*,10) + 2000(A/F,\Delta i^*,10) + 1900$$

It is easy to enter the incremental cash flows onto a spreadsheet, as in Figure 8–3, column D, and use the = IRR(D4:D14) function to display $\Delta i^* = 12.65\%$.

For the second method, the ROR is found using the actual cash flows and the respective lives of 10 years for A and 5 years for B.

$$AW_A = -8000(A/P,i,10) - 3500$$
$$AW_B = -13{,}000(A/P,i,5) + 2000(A/F,i,5) - 1600$$

Now develop $0 = AW_B - AW_A$.

$$0 = -13{,}000(A/P,i^*,5) + 2000(A/F,i^*,5) + 8000(A/P,i^*,10) + 1900$$

Solution again yields $i^* = 12.65\%$.

Comment

It is very important to remember that when an incremental ROR analysis using an AW-based equation is made on the *incremental cash flows,* the LCM must be used.

8.6 Incremental ROR Analysis of Multiple (More than Two) Alternatives ●●●

This section treats selection from multiple alternatives that are mutually exclusive, using the incremental ROR method. Acceptance of one alternative automatically precludes acceptance of any others. The analysis is based upon PW (or AW) relations for incremental cash flows between two alternatives at a time.

When the incremental ROR method is applied, the entire investment must return at least the MARR. When the i^* values on several alternatives exceed the MARR, incremental ROR evaluation is required. (For revenue alternatives, if not even one $i^* \geq$ MARR, the do-nothing alternative is selected.) For all alternatives (revenue or cost), the incremental investment must be separately justified. If the return on the extra investment equals or exceeds the MARR, then the extra investment should be made in order to maximize the total return on the money available, as discussed in Section 8.1.

For ROR analysis of multiple, mutually exclusive alternatives, the following criteria are used.

ME alternative
selection

Select the one alternative
That requires the **largest investment,** and
Indicates that the **extra investment** over another acceptable alternative **is justified.**

An important rule to apply when evaluating multiple alternatives by the incremental ROR method is that *an alternative should never be compared with one for which the incremental investment is not justified.*

The incremental ROR evaluation procedure for multiple, equal-life alternatives is summarized below. Step 2 applies only to revenue alternatives because the first alternative is compared to DN only when revenue cash flows are estimated. The terms **defender** and **challenger** are dynamic in that they refer, respectively, to the alternative that is currently selected (the defender) and the one that is challenging it for acceptance based on Δi^*. In every pairwise evaluation, there is one of each. The steps for solution by hand or by spreadsheet are as follows:

1. Order the alternatives from **smallest to largest initial investment.** Record the annual cash flow estimates for each equal-life alternative.
2. *Revenue alternatives only:* Calculate i^* for the first alternative. In effect, this makes DN the defender and the first alternative the challenger. If $i^* <$ MARR, eliminate the alternative and go to the next one. Repeat this until $i^* \geq$ MARR for the first time, and define that alternative as the defender. The next alternative is now the challenger. Go to step 3. (*Note:* This is where solution by spreadsheet can be a quick assist. Calculate the i^* for all alternatives first using the IRR function, and select as the defender the first one for which $i^* \geq$ MARR. Label it the defender and go to step 3.)
3. Determine the incremental cash flow between the challenger and defender, using the relation

   Incremental cash flow = challenger cash flow − defender cash flow

   Set up the ROR relation.
4. Calculate Δi^* for the incremental cash flow series using a PW- or AW-based equation. (PW is most commonly used.)
5. If $\Delta i^* \geq$ MARR, the challenger becomes the defender and the previous defender is eliminated. Conversely, if $\Delta i^* <$ MARR, the challenger is removed, and the defender remains against the next challenger.
6. Repeat steps 3 to 5 until only one alternative remains. It is the selected one.

Note that only two alternatives are compared at any one time. It is vital that the correct alternatives be compared, or the wrong alternative may be selected.

EXAMPLE 8.6

Caterpillar Corporation wants to build a spare parts storage facility in the Phoenix, Arizona, vicinity. A plant engineer has identified four different location options. The initial cost of earthwork and prefab building and the annual net cash flow estimates are detailed in Table 8–5. The annual net cash flow series vary due to differences in maintenance, labor costs, transportation charges, etc. If the MARR is 10%, use incremental ROR analysis to select the one economically best location.

TABLE 8–5	Estimates for Four Alternative Building Locations, Example 8.6			
	A	**B**	**C**	**D**
Initial cost, $	−200,000	−275,000	−190,000	−350,000
Annual cash flow, $ per year	+22,000	+35,000	+19,500	+42,000
Life, years	30	30	30	30

Solution
All sites have a 30-year life, and they are revenue alternatives. The procedure outlined above is applied.

1. The alternatives are ordered by increasing initial cost in Table 8–6.
2. Compare location C with the do-nothing alternative. The ROR relation includes only the P/A factor.

$$0 = -190{,}000 + 19{,}500(P/A,i^*,30)$$

Table 8–6, column 1, presents the calculated $(P/A, \Delta i^*, 30)$ factor value of 9.7436 and $\Delta i_c^* = 9.63\%$. Since $9.63\% < 10\%$, location C is eliminated. Now the comparison is A to DN, and column 2 shows that $\Delta i_A^* = 10.49\%$. This eliminates the do-nothing alternative; the defender is now A and the challenger is B.

TABLE 8–6	Computation of Incremental Rate of Return for Four Alternatives, Example 8.6			
	C **(1)**	**A** **(2)**	**B** **(3)**	**D** **(4)**
Initial cost, $	−190,000	−200,000	−275,000	−350,000
Cash flow, $ per year	+19,500	+22,000	+35,000	+42,000
Alternatives compared	C to DN	A to DN	B to A	D to B
Incremental cost, $	−190,000	−200,000	−75,000	−75,000
Incremental cash flow, $	+19,500	+22,000	+13,000	+7,000
Calculated $(P/A,\Delta i^*,30)$	9.7436	9.0909	5.7692	10.7143
$\Delta i^*,\%$	9.63	10.49	17.28	8.55
Increment justified?	No	Yes	Yes	No
Alternative selected	DN	A	B	B

3. The incremental cash flow series, column 3, and Δi^* for *B-to-A comparison* are determined from

$$0 = -275,000 - (-200,000) + (35,000 - 22,000)(P/A,\Delta i^*,30)$$
$$= -75,000 + 13,000(P/A,\Delta i^*,30)$$

4. From the interest tables, look up the P/A factor at the MARR, which is $(P/A,10\%,30) =$ 9.4269. Now, any P/A value greater than 9.4269 indicates that the Δi^* will be less than 10% and is unacceptable. The P/A factor is 5.7692, so B is acceptable. For reference purposes, $\Delta i^* = 17.28\%$.
5. Alternative B is justified incrementally (new defender), thereby eliminating A.
6. Comparison D-to-B (steps 3 and 4) results in the PW relation $0 = -75,000 +$ $7000(P/A,\Delta i^*,30)$ and a P/A value of 10.7143 ($\Delta i^* = 8.55\%$). Location D is eliminated, and **only alternative B remains; it is selected.**

Comment

An alternative must *always* be incrementally compared with an acceptable alternative, and the do-nothing alternative can end up being the only acceptable one. Since C was not justified in this example, location A was not compared with C. Thus, *if* the B-to-A comparison had not indicated that B was incrementally justified, then the D-to-A comparison would be correct instead of D-to-B.

To demonstrate how important it is to apply the ROR method correctly, consider the following. If the i^* of each alternative is computed initially, the results by ordered alternatives are

Location	C	A	B	D
i^*, %	9.63	10.49	12.35	11.56

Now apply *only* the first criterion stated earlier; that is, make the largest investment that has a MARR of 10% or more. Location D is selected. But, as shown above, this is the wrong selection, because the extra investment of $75,000 over location B will not earn the MARR. In fact, it will earn only 8.55%. This is another example of the *ranking inconsistency problem of the ROR method* mentioned in Section 8.4.

For cost alternatives, the incremental cash flow is the difference between costs for two alternatives. There is no do-nothing alternative and no step 2 in the solution procedure. Therefore, the lowest-investment alternative is the initial defender against the next-lowest investment (challenger). This procedure is illustrated in Example 8.7 using a spreadsheet solution.

EXAMPLE 8.7

Globally, automobile manufacturers are working to reduce recalls caused by safety problems due to design and manufacturing flaws. Efforts to gain the 5-Star Safety Rating, which evaluates crashworthiness and rollover safety of a vehicle for a consumer, require that the public reputation of each model improves each year. One major automaker is in the process of developing an 8-year campaign to achieve 5-Star Safety Rating on all of its sedans and SUVs. One important aspect of the program is to perform current recalls reliably, fast, efficiently, and without repeats. The strategy is to locate in each major city, one dealer that will meet strict standards of performance and quality. Four dealers in Los Angeles, California are in the final competition for the selection. The average cost estimates to perform a successful recall repair are shown in Table 8–7 for each dealer. The close-out value is the expected bonus that may be paid to dealers that perform in a superior manner during the 8-year campaign. Use MARR = 13% per year and a spreadsheet-based ROR analysis to select the economically best dealer.

TABLE 8–7	Costs for Four Alternative Dealers, Example 8.7			
Dealer	**1**	**2**	**3**	**4**
First cost, $	−5,000	−6,500	−10,000	−15,000
Annual average cost per repair, $	−3,500	−3,200	−3,000	−1,400
Close-out value, $	+500	+900	+700	+1,000
Life, years	8	8	8	8

Solution by Spreadsheet

Follow the procedure for incremental ROR analysis. The spreadsheet in Figure 8–7 contains the complete solution.

1. The alternatives are already ordered by increasing first costs.
2. These are cost alternatives, so there is no comparison to DN, since i^* values cannot be calculated.
3. Dealer 2 is the first challenger to dealer 1; the incremental cash flows for the 2 versus 1 comparison are in column D.
4. The 2 versus 1 comparison results in $\Delta i^* = 14.57\%$ by applying the IRR function.
5. This return exceeds MARR = 13%; dealer 2 is the new defender (cell D17).

The comparison continues for 3 versus 2 in column E, where the return is negative at $\Delta i^* = -18.77\%$; dealer 2 is retained as the defender. Finally the 4 versus 2 comparison has an incremental ROR of 13.60%, which is larger than MARR = 13%. The conclusion is to select dealer 4 because the extra investment is justified.

	A	B	C	D	E	F
1		Year	Dealer 1	Dealer 2	Dealer 3	Dealer 4
2	Initial investment, $		-5,000	-6,500	-10,000	-15,000
3	Average cost per repair, $/year		-3,500	-3,200	-3,000	-1,400
4	Close-out value, $		500	900	700	1,000
5	Incremental comparison			2 vs. 1	3 vs. 2	4 vs. 2
6	Incremental investment, $	0		-1,500	-3,500	-8,500
7	Incremental cash flow, $ per year	1		300	200	1,800
8		2		300	200	1,800
9		3		300	200	1,800
10		4		300	200	1,800
11		5		300	200	1,800
12		6		300	200	1,800
13		7		300	200	1,800
14		8		700	0	1,900
15	Incremental i* (Δi*)			14.57%	-18.77%	13.60%
16	Increment justified?			Yes	No	Yes
17	Alternative selected			2	2	4

$= IRR(D6:D14)$

Figure 8–7
Spreadsheet solution to select from multiple cost alternatives, Example 8.7.

Comment

As mentioned earlier, it is not possible to generate a PW versus i graph for each cost alternative because all cash flows are negative. However, it is possible to generate PW versus Δi graphs for the incremental series in the same fashion as we have demonstrated previously. The curves will cross the PW = 0 line at the Δi^* values determined by the IRR functions.

Selection from multiple, mutually exclusive alternatives with **unequal lives** using Δi^* values requires that the incremental cash flows be evaluated over the LCM of the two alternatives being compared. This is another application of the principle of equal-service comparison. The spreadsheet application in the next section illustrates the computations.

It is always possible to rely on a PW or AW analysis of the incremental cash flows at the MARR to make the selection. In other words, don't find Δi^* for each pairwise comparison; find PW or AW at the MARR instead. However, it is still necessary to make the comparison over the LCM number of years for an incremental analysis to be performed correctly.

8.7 All-in-One Spreadsheet Analysis ● ● ●

For professors and students who like to pack a spreadsheet, Example 8.8 combines many of the economic analysis techniques we have learned so far—(internal) ROR analysis, incremental ROR analysis, PW analysis, and AW analysis. Now that the IRR, NPV, and PV functions are mastered, it is possible to perform a wide variety of evaluations for multiple alternatives on a single spreadsheet. No cell tags are provided in this example. A nonconventional cash flow series for which multiple ROR values may be found, and selection from both mutually exclusive alternatives and independent projects are included in this example.

EXAMPLE 8.8

Constant improvements for in-flight texting and Internet connections provided at airline passenger seats are an expected service by many customers. Cathay Pacific Airlines knows it will have to replace 15,000 to 24,000 units in the next few years on its Boeing 777, 787, and its Airbus A300 and A380 aircraft. Four optional data handling features that build upon one another are available from the manufacturer, but at an added cost per unit. Besides costing more, the higher-end options (e.g., satellite-based plug-in video service) are estimated to have longer lives before the next replacement is forced by new, advanced features expected by flyers. All four options are expected to boost annual revenues by varying amounts. Figure 8–8 spreadsheet rows 2 through 6 include all the estimates for the four options.

(a) Using MARR = 15%, perform ROR, PW, and AW evaluations to select the one level of options that is the most promising economically.

(b) If more than one level of options can be selected, consider the four that are described as independent projects. If no budget limitations are considered at this time, which options are acceptable if the MARR is increased to 20% when more than one option may be implemented?

Solution by Spreadsheet

(a) The spreadsheet (Figure 8–8) is divided into six sections:

Section 1 (rows 1, 2): MARR value and the alternative names (A through D) are in increasing order of initial cost.

Section 2 (rows 3 to 6): Per-unit net cash flow estimates for each alternative. These are revenue alternatives with unequal lives.

Section 3 (rows 7 to 20): Actual and incremental cash flows are displayed here.

Section 4 (rows 21, 22): Because these are all revenue alternatives, i^* values are determined by the IRR function. If an alternative passes the MARR test ($i^* > 15\%$), it is

	A	B	C	D	E	F	G	H
1	MARR =	15%						
2	**Alternative**		**A**	**B**	**C**		**D**	
3	Initial cost, $		-6,000	-7,000	-9,000		-17,000	
4	Annual cash flow, $ per year		2,000	3,000	3,000		3,500	
5	Salvage value, $		0	200	300		1,000	
6	Life, years	Year	3	4	6		12	
7	**Incr. ROR comparison**		**Actual CF**	**Actual CF**	**Actual CF**	**C to B**	**Actual CF**	**D to C**
8	Incremental investment, $	0	-6,000	-7,000	-9,000	-2,000	-17,000	-8,000
9	Incremental cash flow	1	2,000	3,000	3,000	0	3,500	500
10	over the LCM, $ per year	2	2,000	3,000	3,000	0	3,500	500
11		3	2,000	3,000	3,000	0	3,500	500
12		4		3,200	3,000	6,800	3,500	500
13		5			3,000	0	3,500	500
14		6			3,300	-8,700	3,500	9,200
15		7				0	3,500	500
16		8				6,800	3,500	500
17		9				0	3,500	500
18		10				0	3,500	500
19		11				0	3,500	500
20		12				100	4,500	1,200
21	Overall i*		0.00%	26.32%	24.68%		17.87%	
22	Retain or eliminate?		Eliminate	Retain	Retain		Retain	
23	Incremental i* (Δi^*)					19.42%		11.23%
24	Increment justified?					Yes		No
25	**Alternative selected**					C		C
26	AW at MARR		-628	588	656		398	
27	PW at MARR		-3,403	3,188	3,557		2,159	
28	Alternative selected		No	No	Yes		No	

Figure 8–8
Spreadsheet analysis using ROR, PW, and AW methods for unequal-life, revenue alternatives, Example 8.8.

retained and a column is added to the right of its actual cash flows so the incremental cash flows can be determined. Columns F and H were inserted to make space for the incremental evaluations. Alternative A does not pass the i^* test.

Section 5 (rows 23 to 25): The IRR functions display the Δi^* values in columns F and H. Comparison of C to B takes place over the LCM of 12 years. Since $\Delta i^*_{C-B} = 19.42\% > 15\%$, eliminate B; alternative C is the new defender and D is the next challenger. The final comparison of D to C over 12 years results in $\Delta i^*_{D-C} = 11.23\% < 15\%$, so D is eliminated. Alternative C is the chosen one.

Section 6 (rows 26 to 28): These include the AW and PW analyses. The AW value over the life of each alternative is calculated using the PMT function at the MARR with an embedded NPV function. Also, the PW value is determined from the AW value for 12 years using the PV function. For both measures, alternative C has the numerically largest value, as expected.

Conclusion: All methods result in the same, correct choice of alternative C.

(b) Since each option is independent of the others, and there is no budget limitation at this time, each i^* value in row 21 of Figure 8–8 is compared to MARR = 20%. This is a comparison of each option with the do-nothing alternative. Of the four, options B and C have $i^* > 20\%$. They are acceptable; the other two are not.

Comment

In part (a), we should have applied the two multiple-root sign tests to the incremental cash flow series for the C-to-B comparison. The series itself has three sign changes, and the cumulative cash flow series starts negatively and also has three sign changes. Therefore, up to three real-number roots may exist. The IRR function is applied in cell F23 to obtain $\Delta i^*_{C-B} = 19.42\%$ without using a supplemental (Section 7.5) procedure. This means that the investment assumption of 19.42% for positive cash flows is a reasonable one. If the MARR = 15%, or some other earning rate were more appropriate, the ROIC procedure could be applied to determine a single rate, which would be different from 19.42%. Depending upon the investment rate chosen, alternative C may or may not be incrementally justified against B. Here, the assumption is made that the Δi^* value is reasonable, so C is justified.

CHAPTER SUMMARY

Just as present worth and annual worth methods find the best alternative from among several, incremental rate of return calculations can be used for the same purpose. In using the ROR technique, it is necessary to consider the **incremental cash flows** when selecting between mutually exclusive alternatives. The incremental investment evaluation is conducted between only two alternatives at a time, beginning with the lowest initial investment alternative. Once an alternative has been eliminated, it is not considered further.

Rate of return values have a natural appeal to management, but the ROR analysis is often more difficult to set up and complete than the PW or AW analysis using an established MARR. Care must be taken to perform an ROR analysis correctly on the incremental cash flows; otherwise it may give incorrect results.

If there is no budget limitation when independent projects are evaluated, the ROR value of each project is compared to the MARR. Any number, or none, of the projects can be accepted.

PROBLEMS

Understanding Incremental ROR Values

8.1 Assume alternatives A and B are being evaluated by the rate of return method against a MARR of 15% per year. Alternative B requires a higher initial investment than A and the i^* values are $i_A^* = 20\%$ and $i_B^* = 16\%$ per year. Under what circumstance is alternative B the preferred choice?

8.2 Schneeberger, Inc. is considering two alternatives to increase the acceleration of its linear motor actuators. The initial investment required in alternative X is $200,000 and in Y is $150,000. The MARR = 20% per year; a total of $200,000 is available for investment; and the rates of return are $i_X^* = 22\%$ and $i_Y^* = 25\%$ per year. (*a*) Will the rate of return on the increment of investment between alternatives X and Y be larger or smaller than i_X^*? larger or smaller than i_Y^*? (*b*) What is the expected i_{X-Y}^*?

8.3 If the sum of the incremental cash flows is negative, what is known about the rate of return on the incremental investment?

8.4 Incremental cash flow is calculated as (cash flow$_B$ − cash flow$_A$), where B represents the alternative with the larger initial investment. If the two cash flows were switched wherein B represents the one with the *smaller* initial investment, which alternative should be selected if the incremental rate of return is 20% and the MARR is 15%? Explain.

8.5 A food processing company is considering two types of moisture analyzers. Only one can be selected. The company expects an infrared model to yield a rate of return of 27% per year. A more expensive microwave model will yield a rate of return of 19% per year. If the company's MARR is 19% per year, can you determine which model should be

purchased solely on the basis of the overall rate of return information provided? Why or why not?

8.6 If alternative A has $i_A^* = 10\%$ and alternative B has $i_B^* = 18\%$ per year, what is known about the rate of return *on the increment* between A and B if the investment required in B is (*a*) larger than that required for A, and (*b*) smaller than that required for A?

8.7 Why is an incremental analysis necessary when conducting a rate of return analysis for cost alternatives?

8.8 What is the overall rate of return on a $100,000 investment that returns 20% on the first $30,000 and 14% on the remaining $70,000?

8.9 A small construction company has $100,000 set aside in a capital improvement fund to purchase new equipment. If $30,000 is invested at 30%, $20,000 at 25%, and the remaining $50,000 at 20% per year, what is the overall rate of return on the entire $100,000?

8.10 Assume you have a total of $200,000 to invest in two corporate stocks identified as Z1 and Z2. The overall rate of return you require on the $200,000 is 26% per year.
   (*a*) If $40,000 is invested in Z2 with an estimated i_{Z2}^* of 14% per year, what value must i_{Z1}^* exceed to realize at least 26% per year?
   (*b*) If the best return expected from the Z1 stock is 27%, determine the threshold level of investment in Z2 to maintain an overall ROR of 26% per year. Solve by hand or using Goal Seek, as instructed.

8.11 Amigo Mobility, which manufactures battery-powered mobility scooters, has $700,000 to invest.

The company is considering three different battery projects that will yield the following rates of return:

Deep cycle = 28%
Wet/flooded = 42%
Lithium ion = 19%

The initial investment required for each project is $200,000, $100,000, and $400,000, respectively. If Amigo's MARR is 15% per year and it invests in all three projects, what rate of return will the company make?

8.12 A total of $50,000 was allocated to a project to detect and reduce insider theft in the ZipCar auto parts store. Two alternatives identified as Y and Z are under consideration. The overall ROR on the $50,000 is expected to be 40%, with the rate of return on the $20,000 increment between Y and Z at 15% per year. If Z is the higher first-cost alternative, (a) what is the size of the investment required in Y, and (b) what is the rate of return on Y?

8.13 For each of the following scenarios, state whether an incremental ROR analysis is required to select an alternative and state why or why not. Assume that alternative Y requires a larger initial investment than alternative X and that the MARR is 20% per year.
   (a) X has a rate of return of 22% per year, and Y has a rate of return of 20% per year.
   (b) X has a rate of return of 19% per year, and Y has a rate of return of 21% per year.
   (c) X has a rate of return of 16% per year, and Y has a rate of return of 19% per year.
   (d) X has a rate of return of 25% per year, and Y has a rate of return of 23%.
   (e) X has a rate of return of 20%, and Y has a rate of return of 22% per year.

8.14 For the cash flows shown, determine the incremental cash flow between machines B and A (a) in year 0, (b) in year 3, and (c) in year 6.

Machine	A	B
First cost, $	−15,000	−25,000
AOC, $ per year	−1,600	−400
Salvage value, $	3,000	6,000
Life, years	3	6

8.15 For the two alternatives, demonstrate that the sum of the incremental cash flow series (Z − X) over the LCM is equal to the difference in the sums of the individual cash flow series for X and Z.

System	X	Z
First cost, $	−40,000	−95,000
AOC, $ per year	−12,000	−5,000
Salvage value, $	6,000	14,000
Life, years	3	6

8.16 Certain parts for NASA's reusable space exploration vehicle can either be anodized (A) or powder coated (P). Some of the costs for each process are shown in the table below. The incremental cash flow future worth equation associated with (P−A) is:

$$0 = -53{,}000(F/P,i^*_{P-A},5) + 21{,}000(F/A,i^*_{P-A},5) + 8000$$

Determine the following: (a) first cost for P; (b) M&O for A; and (c) resale value of P.

Process	Anodize, A	Powder Coated, P
First cost, $	−30,000	?
M&O, $/year	?	−11,000
Resale value, $	4000	?
Life, years	5	5

8.17 Lesco Chemical is considering two processes for making a cationic polymer. Process A has a first cost of $100,000 and an AOC of $60,000 per year. Process B's first cost is $165,000. If both process will be adequate for 4 years and the rate of return on the increment between the alternatives is 25%, what is the amount of the AOC for process B?

Incremental ROR Comparison (Two Alternatives)

8.18 The PW-based relation for the incremental cash flow series to find Δi^* between the lower first-cost alternative X and alternative Y has been developed.

$$0 = -40{,}000 + 9000(P/A,\Delta i^*,10) - 2000(P/F,\Delta i^*,10)$$

Determine the highest MARR value for which Y is preferred over X. Write the single-cell spreadsheet function that displays Δi^*.

8.19 A consulting engineering firm's CFO wants to purchase either Ford Explorers or Toyota 4Runners for company principals. The two models under consideration cost $30,900 for the Ford and $36,400 for the Toyota. When considering life-cycle costs, the AOC of the Explorer is expected to be $600 per year more than that of the 4Runner. The trade-in values after 3 years are estimated to be 50% of the first cost for the Explorer and 60% for the 4Runner. (a) What is the incremental ROR between the two vehicles? (b) Provided the firm's MARR is 18% per year, which vehicle should it buy?

8.20 EP Electric has identified two new methods to treat its cooling water. Alternative I (for inflow) would treat the raw water with a conventional reverse osmosis system so that the cycles of concentration could be increased from 5 to 20. This will result in water cost savings of $360,000 per year and chemical cost savings of $56,000 per year. The initial cost of the equipment will be $2.3 million with an operating cost of $125,000 per year.

Alternative B (for blowdown) will treat the cooling tower blowdown water using a high-pressure seawater reverse osmosis system to recover most of the water that is sent to an evaporation pond. This option will result in water savings of $270,000 per year. The cost of the system will be $1.2 million with an operating cost of $105,000 per year. Assuming one of the two methods must be installed, determine which is preferred on the basis of the incremental ROR value using MARR of 5% per year, which is a typically low return expected of government projects. Use a 10-year study period with no salvage value for either system. (Note: See Problem 8.41 for more on this situation.)

8.21 Polytec Chemical, Inc. must decide between two additives to improve the dry-weather stability of its low-cost acrylic paint. Additive A will have an equipment and installation cost of $125,000 and an annual cost of $55,000. Additive B will have an installation cost of $175,000 and an annual cost of $35,000. If the company uses a 5-year recovery period for paint products and a MARR of 20% per year, which process is favored on the basis of an incremental rate of return analysis? Also, write the function to display Δi^*.

8.22 A company that manufactures amplified pressure transducers wishes to decide between the machines shown—variable speed (VS) and dual speed (DS). Compare them on the basis of rate of return and determine which should be selected if the MARR = 15% per year.

	VS	DS
First cost, $	−250,000	−225,000
AOC, $ per year	−231,000	−235,000
Overhaul in year 3, $	—	−26,000
Overhaul in year 4, $	−39,000	—
Salvage value, $	50,000	10,000
Life, years	6	6

8.23 A process control manager is considering two robots to improve materials handling capacity in the production of rigid shaft couplings that mate dissimilar drive components. Robot X has a first cost of $84,000, an annual M&O cost of $31,000, a $40,000 salvage value, and will improve revenues by $96,000 per year. Robot Y has a first cost of $146,000, an annual M&O cost of $28,000, a $47,000 salvage value, and will increase revenues by $119,000 per year. The company's MARR is 15% per year and it uses a 3-year study period for economic evaluations. Which one should the manager select (a) on the basis of

ROR values, and (b) on the basis of the incremental ROR value? (c) Which is the correct selection basis? Perform the analysis by hand or spreadsheet, as instructed.

8.24 The manager of a canned-food processing plant has two labeling machine options. On the basis of a rate of return analysis with a MARR of 20% per year, determine (a) which model is economically better, and (b) if the selection changes, provided both options have a 4-year life and all other estimates remain the same.

Model	105	200
First cost, $	−15,000	−25,000
AOC, $ per year	−1,600	−400
Salvage value, $	3,000	4,000
Life, years	2	4

8.25 A solid-waste recycling plant is considering two types of storage bins using an MARR of 10% per year. (a) Use ROR evaluation to determine which should be selected. (b) Confirm the selection using the regular AW method at MARR = 10% per year.

Storage Bin	P	Q
First cost, $	−18,000	−35,000
AOC, $ per year	−4000	−3600
Salvage value, $	1000	2700
Life, years	3	6

8.26 Konica Minolta plans to sell a copier that prints documents on both sides simultaneously, cutting in half the time it takes to complete big commercial jobs. The faster copier is expected to increase profit by $2,500,000 per year, regardless of which of the following rollers the company uses in its copiers. The estimated costs associated with chemically treated vinyl rollers and fiber-impregnated rubber rollers are shown below. Determine which of the roller types should be selected on the basis of an ROR analysis assuming the company's MARR is 25% per year.

Roller Type	Treated	Impregnated
First cost, $1000	−5000	−6500
Annual cost, $1000/year	−1000	−650
Salvage value, $1000	100	200
Life, years	5	5

8.27 Alternative R has a first cost of $100,000, annual M&O costs of $50,000, and a $20,000 salvage value after 5 years. Alternative S has a first cost of $175,000 and a $40,000 salvage value after 5 years, but its annual M&O costs are not known. Determine the M&O costs for alternative S that would yield a required incremental rate of return of 20%. Solve (*a*) by hand, and (*b*) using the Goal Seek tool or RATE function on a spreadsheet.

8.28 Two roadway designs are under consideration for access to a permanent suspension bridge. Design 1A will cost $3 million to build and $100,000 per year to maintain. Design 1B will cost $3.5 million to build and $40,000 per year to maintain. Both designs are assumed to be permanent. Use an AW-based rate of return equation to determine (*a*) the breakeven ROR, and (*b*) which design is preferred at a MARR of 10% per year.

8.29 One of Suzanne's project engineers e-mailed her the screen-shot (below) of his spreadsheet analysis of the annual costs and savings for two equivalent systems considered for installation as a major upgrade of the fire and safety systems that reduce the possibility of human injury on the company's prime assembly line. He included the following statement in the e-mail: "Since the company's MARR is 20% per year, the extra investment in FirstSafe (FS) is not justified with a breakeven incremental ROR of 16.86%. However, with an exception to reduce the required return to 16%, or slightly above this, for this case only, FirstSafe is economically acceptable over FireWall (FW). Besides, I like the features on FirstSafe much better than those of the other, less expensive FW system. Can I go ahead and order the FirstSafe system?"

From all that you have learned in these last two chapters, would you recommend that Suzanne accept this analysis and respond "Yes" to the request to purchase FirstSafe? If not satisfied, what additional information would you suggest that Suzanne request of her engineer?

The rate of return on each increment of investment was less than the MARR. Which alternative should be selected?

8.31 A metal plating company is considering four different methods for recovering by-product heavy metals from a manufacturing site's liquid waste. The investment costs and annual net incomes associated with each method have been estimated. All methods have an 8-year life; the MARR is 11% per year; and an AW-based ROR analysis is required. (*a*) If the methods are independent (because they can be implemented at different plants), which ones are acceptable? (*b*) If the methods are mutually exclusive, determine which one should be selected.

Method	First Cost, $	Salvage Value, $	Annual Net Income, $/Year
A	−30,000	+1,000	+4,000
B	−36,000	+2,000	+5,000
C	−41,000	+500	+8,000
D	−53,000	−2000	+10,500

8.32 Old Southwest Canning Co. has determined that any one of four machines can be used in its chili-canning operation. The cost of the machines are estimated below, and all machines have a 5-year life. If the minimum attractive rate of return is 25% per year, determine which machine should be selected on the basis of a rate of return analysis.

Machine	First Cost, $	AOC, $
1	−28,000	−20,000
2	−51,000	−12,000
3	−32,000	−19,000
4	−33,000	−18,000

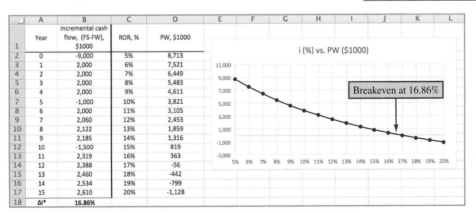

Multiple Alternative (More than Two) Evaluation

8.30 Four mutually exclusive service alternatives are under consideration for automating a manufacturing operation. The alternatives were ranked in order of increasing initial investment and then compared by incremental rate of return analysis.

8.33 A recent graduate who wants to start an excavation/earth-moving business is trying to determine which size of used dump truck to buy. He knows that as the bed size increases, the net income increases, but he is uncertain whether the incremental expenditure required for the larger trucks is justified. The cash flows associated with each size truck are estimated below. The contractor has established a MARR of 18% per year, and all trucks are expected to have a remaining economic life of 5 years. (*a*) Determine which size truck he should purchase. (*b*) If two trucks are to be purchased, what should be the size of the second truck? (Note: Problem 8.44 requests a spreadsheet solution of these alternatives.)

Truck Bed Size, m³	Initial Investment, $	M&O, $/Year	Salvage Value, $	Annual Revenue, $/Year
8	−30,000	−14,000	+2,000	+26,500
10	−34,000	−15,500	+2,500	+30,000
15	−38,000	−18,000	+3,000	+33,500
20	−48,000	−21,000	+3,500	+40,500
25	−57,000	−26,000	+4,600	+49,000

8.34 For the four revenue alternatives below, use the ROR method results to determine:
(*a*) Which one(s) to select, if MARR = 17% per year and the proposals are independent.
(*b*) Which one to select, if MARR = 14.5% per year and the alternatives are mutually exclusive.
(*c*) Which one to select, if MARR = 10% per year and the alternatives are mutually exclusive.

			Δ*i**% When Compared with Alternative		
Alternative	Initial Investment, $	Overall ROR, *i**%	A	B	C
A	−60,000	11.7	–	–	–
B	−90,000	22.2	43.3	–	–
C	−140,000	17.9	22.5	10.0	–
D	−190,000	15.8	17.8	10.0	10.0

8.35 You are considering five projects, all of which can be considered to last indefinitely. If the company's MARR is 15% per year, determine which should be selected if they are (*a*) independent projects, and (*b*) mutually exclusive alternatives.

Alternative	First Cost, $	Net Annual Income, $	ROR, %
A	−20,000	+3,000	15.0
B	−10,000	+2,000	20.0
C	−15,000	+2,800	18.7
D	−70,000	+10,000	14.3
E	−50,000	+6,000	12.0

8.36 A small manufacturing company could expand its operation by adding new products. Any or all of the products shown below can be added. If the company uses a MARR of 15% per year and a

5-year project period, which products, if any, should the company introduce?

Product	1	2	3	4
Initial cost, $	−340,000	−500,000	−570,000	−620,000
Annual cost, $/year	−70,000	−64,000	−48,000	−40,000
Annual revenue, $/year	180,000	190,000	220,000	205,000

8.37 The four alternatives described below are being evaluated by the rate of return method.
(*a*) If the proposals are independent, which should be selected at a MARR of 16% per year?
(*b*) If the proposals are mutually exclusive, which one should be selected at a MARR of 9% per year?
(*c*) If the proposals are mutually exclusive, which one should be selected when the MARR is 12% per year?

			Δ*i**% , When Compared with Alternative		
Alternative	Initial Investment, $	*i**%	A	B	C
A	−40,000	29	–	–	–
B	−75,000	15	1	–	–
C	−100,000	16	7	20	–
D	−200,000	14	10	13	12

8.38 Four different machines were under consideration for materials flow improvement on a drug bottling line. An engineer performed the economic analysis to select the best machine, but some of his calculations were removed from the report by a disgruntled employee. All machines are assumed to have a 10-year life.
(*a*) Fill in the missing values in the comparison table.
(*b*) Select the best machine at MARR = 18% per year, provided one must be chosen.

Machine	1	2	3	4
Initial cost, $	?	−60,000	−72,000	−98,000
Annual cost, $ per year	−70,000	−64,000	−61,000	−58,000
Annual savings, $ per year	+80,000	+80,000	+80,000	+82,000
Overall ROR, *i**, %	18.6%	?	23.1%	20.8%
Machines compared		2 vs. 1	3 vs. 2	4 vs. 3
Incremental investment, $		−16,000	?	−26,000
Incremental cash flow, $ per year		+6,000	+3,000	?
ROR on increment, Δ*i**, %		35.7%	?	?

8.39 A rate of return analysis was initiated for the infinite-life alternatives shown below.
  (*a*) Fill in the 10 blanks in the incremental rate of return (Δi^*) columns.
  (*b*) How much revenue is associated with each alternative?
  (*c*) Which alternative should be selected if they are mutually exclusive and MARR is 16% per year?
  (*d*) Which alternative should be selected if they are mutually exclusive and MARR is 11% per year?
  (*e*) Select the two best alternatives at MARR = 19% per year.

| | | | $\Delta i^*\%$ on Incremental Cash Flows when Compared with Alternative | | | |
Alternative	Initial Investment, $	i^*, %	E	F	G	H
E	−20,000	20	–		–	–
F	−30,000	35		–	–	–
G	−50,000	25			–	11.7
H	−80,000	20			11.7	–

EXERCISES FOR SPREADSHEETS

8.40 Kleen Corp., a privately owned and operated single-stream recycling facility, has annual contracts with several cities in the Tri-County Metropolitan Area. Kleen Corp. wants to add a new set of sensors to its existing machinery that will separate plastics and metals from paper and glass materials earlier in the separation process. Two versions of the sensor equipment are available from the Green Corporation. Model 400 has a first cost of $700,000, while Model 1000 costs $1 million. Both have an expected 10% salvage value after their respective useful lives of 6 and 3 years.

  Assume you work for Kleen Corp. as a project engineer. You have made first-cut estimates of the annual savings (with no annual increases for efficiency) and expenses (AOC with no annual decreases or increases) for both models.
  (*a*) Perform an ROR analysis using MARR = 5% per year to recommend one of the two models to your president. You know that your president likes dollar-per-year figures, and will want to see a plot of the annual worth amounts for different rate of return values when you visit with her.
  (*b*) Before you finalize your recommendation, ensure that there is no ranking inconsistency present with these two alternatives. If there is not, no problem. If there is, be prepared to provide the logic of your recommendation.

Model	Year	1	2	3	4	5	6
400	Savings, $1000/year	180	180	180	180	180	180
	Expenses, $1000/year	−40	−40	−40	−40	−40	−40
1000	Savings, $1000/year	410	410	410			
	Expenses, $1000/year	−60	−60	−60			

8.41 In Problem 8.20, two methods of cooling water treatment are analyzed. These are revenue alternatives which generate annual savings against their costs. (*a*) Perform the complete procedure for incremental ROR analysis for revenue alternatives to determine which method to select at MARR = 5% per year. (*b*) If a MARR of 8% per year were required, what recommendation would you make?

8.42 Ryan has received cost and salvage value estimates for two competing fire sprinkler systems to be installed in his office building. System A has a first cost of $100,000, annual M&O costs of $10,000, and a $20,000 salvage value after 5 years. System B has a first cost of $175,000, M&O costs of $8,000, and a $40,000 salvage value after 10 years. (*a*) Plot the breakeven ROR point between the two systems using PW values for two situations: incremental cash flows and alternative cash flows. (Hint: Use Figure 8–6 as a model.) (*b*) State which system is economically preferred if the MARR is larger than this value.

8.43 A mechanical engineer at Anode Metals is considering five equivalent projects, some of which have different life expectations. Salvage value is nil for all alternatives. Assuming that the company's MARR is 13% per year, determine which should be selected (*a*) if they are independent, and (*b*) if they are mutually exclusive. (*c*) Explain why your selection in part (*b*) is correct.

	First Cost, $	Net Annual Income, $/Year	Life, Years
A	−20,000	+5,500	4
B	−10,000	+2,000	6
C	−15,000	3,800	6
D	−60,000	+11,000	12
E	−80,000	+9,000	12

8.44 Perform the required analysis of truck-bed sizes in Problem 8.33 using a spreadsheet.

ADDITIONAL PROBLEMS AND FE EXAM REVIEW QUESTIONS

8.45 Alternative A has a rate of return of 14% and alternative B has a rate of return of 17%. If the investment required in B is larger than that required for A, the rate of return on the increment of investment between A and B is:
  (a) Larger than 14%
  (b) Larger than 17%
  (c) Between 14% and 17%
  (d) Smaller than 14%

8.46 The rate of return for alternative X is 18% per year and for alternative Y is 17%, with Y requiring a larger initial investment. If a company has a minimum attractive rate of return of 16%:
  (a) The company should select alternative X
  (b) The company should select alternative Y
  (c) The company should conduct an incremental analysis between X and Y in order to select the better alternative
  (d) The company should select the do-nothing alternative

8.47 A company that manufactures high-strength epoxys is considering investing $100,000 in two new adhesives identified as X and Z. The investment in X is $20,000 and is expected to yield a rate of return of 40% per year. Your supervisor asked you to determine what rate of return would be required on the remaining $80,000 in order for the total return to be at least 25%. You responded that the return would have to be at least:
  (a) 10.4%
  (b) 16.8%
  (c) 21.3%
  (d) 24.1%

8.48 When conducting an ROR analysis of mutually exclusive cost alternatives:
  (a) All of the projects must be compared against the do-nothing alternative
  (b) More than one project may be selected
  (c) An incremental investment analysis is necessary to identify the best one
  (d) The project with the highest incremental ROR should be selected

8.49 A chemical engineer working for a large chemical products company was asked to make a recommendation about which of three mutually exclusive *revenue* alternatives should be selected for improving the marketability of personal care products used for conditioning hair, cleansing skin, removing wrinkles, etc. The alternatives (X, Y, and Z) were ranked in order of increasing initial investment and then compared by incremental rate of return analysis. The rate of return on each increment of investment was less than the company's MARR of 17% per year. The alternative to select is:
  (a) DN
  (b) Alternative X
  (c) Alternative Y
  (d) Alternative Z

8.50 When comparing two mutually exclusive alternatives by the ROR method, if the rate of return on the alternative with the higher first cost is less than that of the lower first-cost alternative:
  (a) The rate of return on the increment between the two is greater than the rate of return for the lower first-cost alternative
  (b) The rate of return on the increment is less than the rate of return for the lower first-cost alternative
  (c) The higher first-cost alternative may be the better of the two alternatives
  (d) The lower first-cost alternative should be selected

8.51 The incremental cash flow between two alternatives is shown below. The equation that can be used to correctly solve for the incremental rate of return is:
  (a) $0 = -20{,}000 + 3000(A/P,\Delta i^*,10) + 400(P/F,\Delta i^*,10)$
  (b) $0 = -20{,}000 + 3000(A/P,\Delta i^*,10) + 400(A/F,\Delta i^*,10)$
  (c) $0 = -20{,}000(A/P,\Delta i^*,10) + 3000 + 400(P/F,\Delta i^*,10)$
  (d) $0 = -20{,}000(F/P,\Delta i^*,10) + 3000(F/A,\Delta i^*,10) + 400$

Year	Incremental Cash Flow, $
0	−20,000
1–10	+3,000
10	+400

8.52 Standby power for pumps at water distribution booster stations can be provided by either gasoline- or diesel-powered engines. The estimates for the gasoline engines are as follows:

	Gasoline
First cost, $	−150,000
M&O, $ per year	−41,000
Salvage value, $	23,000
Life, years	15

If the incremental PW-based ROR equation associated with (Diesel – Gasoline) is $0 = -40,000 + 11,000(P/A,\Delta i^*,15) + 16,000(P/F,\Delta i^*,15)$, the first cost of the diesel engine is closest to:
- (a) –190,000
- (b) –110,000
- (c) –60,000
- (d) –40,000

Questions 8.53 through 8.55 are based on the following five alternatives that are evaluated by the rate of return method.

Alternative	Initial Investment, $	Alternative s^*, %	Incremental ROR, %, When Compared with Alternative				
			A	B	C	D	E
A	–25,000	9.6	–	27.3	19.4	35.3	25.0
B	–35,000	15.1		–	0	38.5	24.4
C	–40,000	13.4			–	46.5	27.3
D	–60,000	25.4				–	26.8
E	–75,000	20.2					–

8.53 If the alternatives are independent and the MARR is 15% per year, the one(s) to select is (are):
- (a) Only D
- (b) Only D and E
- (c) Only B, D, and E
- (d) Only E

8.54 If the alternatives are mutually exclusive and the MARR is 15% per year, the alternative to select is:
- (a) Either B, D, or E
- (b) Only B
- (c) Only D
- (d) Only E

8.55 If the alternatives are mutually exclusive and the MARR is 25% per year, the alternative to select is:
- (a) B
- (b) D
- (c) E
- (d) None of them

8.56 Jewel-Osco evaluated three different pay-by-touch systems that identify a customer by a finger scan and automatically deduct the amount of the bill directly from their checking account. The revenue alternatives were ranked according to increasing initial investment and identified as alternatives A, B, and C.

Comparison	i^*, %	Δi^*, %
A to DN	23.4	
B to DN	8.1	
C to DN	16.6	
B vs. A		–5.1
C vs. A		12.0
C vs. B		83.9

Based on the alternative ROR values (i^*), the incremental ROR values (Δi^*), and the company's MARR of 16% per year, the alternative to select is:
- (a) A
- (b) B
- (c) C
- (d) DN

8.57 Five mutually exclusive cost alternatives that have *infinite lives* are under consideration for decreasing the fruit-bruising rates of a thin skin-fruit grading and packing operation (peaches, pears, apricots, etc.). The initial costs and cash flows of each alternative are available. If the MARR is 15% per year, the one alternative to select is:

Alternative	A	B	C	D	E
Initial cost, $	–11,000	–12,000	–9,000	–14,000	–15,000
Cash flow, $ per year	–1000	–900	–1400	–700	–300

- (a) A
- (b) B
- (c) D
- (d) E

CASE STUDY

PEFORMING ROR ANALYSIS FOR 3D PRINTER AND IIoT TECHNOLOGY

Background

Software used in 3D printer systems manufactured by Spectrum LASER Corp. is currently able to automatically develop hollows, shells, and "tree supports" to save development time. Network printing for use on multiple machines and at different locations globally is available. New software, called

JIT II, is being beta tested that will allow IIoT-type (Industrial Internet of Things) connections between machines at distant locations. When onboard monitoring sensors detect that a part is about to fail, or it is time to replace one of several high-profile parts, the JIT II software will automatically queue the part's manufacturing code onto a 3D printer that is detected to be locally idle or ready. The company believes it can design and produce the JIT II software, the 3D printers, and the accompanying computer equipment necessary to place this technology in a wide variety of harsh environments—chemically toxic to humans, in flight, underground in mines, on the ocean floor, in war-torn areas, etc.

Information

For this case, we will analyze the available computers that can provide the server function necessary to make this interlinking available and successful. The first cost and other parameter estimates, including expected contribution to annual net cash flow, are summarized below.

	Server 1	Server 2
First cost, $	100,000	200,000
Net cash flow, $/year	35,000	50,000 year 1, plus 5000 per year for years 2, 3, and 4 (gradient)
		70,000 maximum for years 5 on, even if the server is replaced
Life, years	3 or 4	5 or 8

The life estimates were developed by two different individuals: a design engineer and a manufacturing manager. They have asked that, at this stage of the project, all analyses be performed using both life estimates for each system.

Case Study Exercises

Use spreadsheet analysis to determine the following:

1. If the MARR = 12%, which server should be selected? Use the PW or AW method to make the selection.
2. Use incremental ROR analysis to decide between the servers at MARR = 12%.
3. Use any method of economic analysis to display on the spreadsheet the value of the incremental ROR between server 2 with a life estimate of 5 years and a life estimate of 8 years.

CASE STUDY

HOW A NEW ENGINEERING GRADUATE CAN HELP HIS FATHER[1]

Background

"I don't know whether to sell it, expand it, lease it, or what. But I don't think we can keep doing the same thing for many more years. What I really want to do is to keep it for 5 more years, then sell it for a bundle," Elmer Kettler said to his wife, Janise, their son, John Kettler, and new daughter-in-law, Suzanne Gestory, as they were gathered around the dinner table. Elmer was sharing thoughts on Gulf Coast Wholesale Auto Parts, a company he has owned and operated for 25 years on the southern outskirts of Houston, Texas. The business has excellent contracts for parts supply with several national retailers operating in the area—NAPA, AutoZone, O'Reilly, and Advance. Additionally, Gulf Coast operates a rebuild shop serving these same retailers for major automobile components, such as carburetors, transmissions, and air conditioning compressors.

At his home after dinner, John decided to help his father with an important and difficult decision: What to do with his business? John graduated just last year with an engineering degree from a major state university in Texas, where he completed a course in engineering economy. Part of his job at Energcon Industries is to perform basic rate of return and present worth analyses on energy management proposals.

Information

Over the next few weeks, John outlined five options, including his dad's favorite of selling in 5 years. John summarized all the estimates over a 10-year horizon. The options and estimates were given to Elmer, and he agreed with them.

Option 1: *Remove rebuild.* Stop operating the rebuild shop and concentrate on selling wholesale parts. The removal of the rebuild operations and the switch to an "all-parts house" are expected to cost $750,000 in the first year. Overall revenues will drop to $1 million the first year with an expected 4% increase per year thereafter. Expenses are projected at $0.8 million the first year, increasing 6% per year thereafter.

Option 2: *Contract rebuild operations.* To get the rebuild shop ready for an operations contractor to take over will cost $400,000 immediately. If

[1]Based upon a study by Alan C. Stewart, Consultant, Communications and High Tech Solutions Engineering, Accenture LLP.

expenses stay the same for 5 years, they will average $1.4 million per year, but they can be expected to rise to $2 million per year in year 6 and thereafter. Elmer thinks revenues under a contract arrangement can be $1.4 million the first year and can rise 5% per year for the duration of a 10-year contract.

Option 3: *Maintain status quo and sell out after 5 years* (Elmer's personal favorite). There is no cost now, but the current trend of negative net profit will probably continue. Projections are $1.25 million per year for expenses and $1.15 million per year in revenue. Elmer had an appraisal last year, and the report indicated Gulf Coast Wholesale Auto Parts is worth a net $2 million. Elmer's wish is to sell out completely after 5 more years at this price, and to make a deal that the new owner pay $500,000 per year at the end of year 5 (sale time) and the same amount for the next 3 years.

Option 4: *Trade-out.* Elmer has a close friend in the antique auto parts business who is making a "killing," so he says, with e-commerce. Although the possibility is risky, it is enticing to Elmer to consider a whole new line of parts, but still in the basic business that he already understands. The trade-out would cost an estimated $1 million for Elmer immediately. The 10-year horizon of annual expenses and revenues is considerably higher than for his current business. Expenses are estimated at $3 million per year and revenues at $3.5 million each year.

Option 5: *Lease arrangement.* Gulf Coast could be leased to some turnkey company with Elmer remaining the owner and bearing part of the expenses for building, delivery trucks, insurance, etc. The first-cut estimates for this option are $1.5 million to get the business ready now, with annual expenses at $500,000 per year and revenues at $1 million per year for a 10-year contract.

Case Study Exercises

Help John with the analysis by doing the following:

1. Develop the actual cash flow series and incremental cash flow series (in $1000 units) for all five options in preparation for an incremental ROR analysis.
2. Discuss the possibility of multiple rate of return values for all the actual and incremental cash flow series. Find any multiple rates in the range of 0% to 100%.
3. If John's father insists that he make 25% per year or more on the selected option over the next 10 years, what should he do? Use all the methods of economic analysis you have learned so far (PW, AW, ROR) so John's father can understand the recommendation in one way or another.
4. Prepare plots of the PW versus i for each of the five options. Estimate the breakeven rate of return between options.
5. What is the minimum amount that must be received in each of years 5 through 8 for option 3 (the one Elmer wants) to be best economically? Given this amount, what does the sale price have to be, assuming the same payment arrangement as presented above?

David Sucsy/Getty Images

CHAPTER 9

Benefit/Cost Analysis and Public Sector Economics

LEARNING OUTCOMES

Purpose: Understand public sector projects and select the best alternative on the basis of incremental benefit/cost analysis.

SECTION	TOPIC	LEARNING OUTCOME
9.1	Public sector	• Explain some of the fundamental differences between private and public sector projects.
9.2	B/C for single project	• Calculate the benefit/cost ratio and use it to evaluate a single project.
9.3	Incremental B/C	• Select the better of two alternatives using the incremental B/C ratio method.
9.4	More than two alternatives	• Based on the incremental B/C ratios, select the best of multiple alternatives.
9.5	Service projects and CEA	• Explain service sector projects and use cost-effectiveness analysis (CEA) to evaluate projects.
9.6	Ethical considerations	• Explain the major aspects of public project activities, and describe how ethical compromise may enter public sector project analysis.

T he evaluation methods of previous chapters are usually applied to alternatives in the private sector, that is, for-profit and not-for-profit corporations and businesses. This chapter introduces **public sector and service sector alternatives** and their economic consideration. In the case of public projects, the owners and users (beneficiaries) are the citizens and residents of a government unit—city, county, state, province, or nation. Government units provide the mechanisms to raise capital and operating funds. Public-private partnerships have become increasingly common, especially for large infrastructure projects such as major highways, power generation plants, water resource developments, health care facilities, and the like.

The benefit/cost (B/C) ratio introduces objectivity into the economic analysis of public sector evaluation, thus reducing the effects of politics and special interests. The different formats of B/C analysis, and associated disbenefits of an alternative, are discussed here. The B/C analysis can use equivalency computations based on PW, AW, or FW values. Performed correctly, the benefit/cost method will always select the same alternative as PW, AW, FW, and ROR analyses.

This chapter also introduces **service sector projects** and discusses how their economic evaluation is different from that for other projects. Finally, there is a discussion on **professional ethics** and ethical dilemmas in the public sector.

PE

Water Treatment Facility #3 Case: Allen Water Utilities has planned for the last 25 years to construct a new drinking water treatment facility that will supply the rapidly growing north and northwest areas of the city. An expectation of over 100,000 new residents in the next several years and 500,000 by 2040 prompted development and construction of the facility in the 2014–17 time frame. The supply is from a large surface lake currently used to provide water to all of Allen and the surrounding communities. The project is termed WTF3, and its initial capital investment is $540 million for the treatment plant and two large steel-pipe transmission mains (84- and 48-inch) that will be installed via tunneling approximately 100 to 125 feet (approximately 30 to 38 meters) under suburban areas of the city to reach current reservoirs.

Tunneling was selected after geotechnical borings indicated that open trenching was not supportable by the soil and based upon a large public outcry against trenching in the living areas along the selected transmission routes. Besides the treatment plant construction on the 95-acre site, there must be at least three large vertical shafts (25 to 50 feet in diameter) bored along each transmission main to gain underground access for equipment and debris removal during the tunneling operations.

The stated criteria used to make decisions for WTF3 and the transmission mains were economics, environment, community impact, and constructability.

There are major long-term benefits for the new facility. These are some mentioned by city engineers:

- It will meet projected water needs of the city for the next 50 years.

- The new treatment plant is at a higher elevation than the current two plants, allowing gravity flow to replenish reservoirs, thereby using little or no electric pumping.

- There will be an increase in the diversity and reliability of supply as other plants age.

- It will provide a water quality that is more consistent due to the location of the raw water intakes.

- The facility uses water supplies already purchased; therefore, there is no need to negotiate additional allowances.

The disbenefits are mostly short-term during the construction of WTF3 and transmission mains. Some of these are mentioned by citizen groups and one retired city engineer:

- There will be disruption of habitat for some endangered species of birds, lizards, and trees not found in any other parts of the country.

- Large amounts of dust and smoke will enter the atmosphere in a residential area during the 3½ years of construction, tunneling, and transmission main completion.

- Noise pollution and traffic congestion will result during an estimated 26,000 truck trips to remove debris from the plant site and tunnel shafts, in addition to the problems from regular construction traffic.

- Natural landscape in plant and tunnel shaft sites will be destroyed.

- Safety will be compromised for children in a school where large trucks will pass about every 5 minutes for approximately 12 hours per day, 6 days per week for 2½ years.

- There may be delays in fire and ambulance services in emergencies, since many neighborhood streets are country-road width and offer only single ingress/egress streets for neighborhoods along the indicated routes.

- The need for the facility has not been proved, as the water will be sold to developers outside the city limits, not provided to residences within Allen.

- Newly generated revenues will be used to pay off the capital funding bonds approved for the plant's construction.

Last year, the city engineers did a benefit/cost analysis for this massive public sector project; none of the results were publicized. Public and elected official intervention has now caused some of the conclusions using the criteria mentioned above to be questioned by the general manager of Allen Water Utilities.

This case is used in the following topics of this chapter:

Public sector projects (Section 9.1)

Incremental B/C analysis, two alternatives (Section 9.3)

Incremental B/C analysis, more than two alternatives (Section 9.4)

9.1 Public Sector Projects ● ● ●

Virtually all the examples and problems of previous chapters have involved the private sector, where products, systems, and services are developed and offered by corporations and businesses for use by individual customers and clients, the government, or other companies. (Notable exceptions are the long-life alternatives discussed in Chapters 5 (PW) and 6 (AW) where capitalized cost analysis was applied.) Now we will explore projects that concentrate on government units and the citizens they serve. These are called public sector projects.

A **public sector project** is a product, service, or system used, financed, and owned by the citizens of any government level. The primary purpose is to **provide service to the citizenry for the public good at no profit.** Areas such as public health, criminal justice, safety, transportation, welfare, and utilities are publically owned and require economic evaluation.

Upon reflection, it is surprising how much of what we use on a daily or as-needed basis is publicly owned and financed to serve us—the citizenry. These are some public sector examples:

Hospitals and clinics	Economic development projects
Parks and recreation	Convention centers
Utilities: water, electricity, gas, sewer, sanitation	Sports arenas
Schools: primary, secondary, community colleges, universities	Transportation: highways, bridges, waterways
Police and fire protection	Public housing
Courts and prisons	Emergency relief
Food stamp and rent relief programs	Codes and standards
Job training	

There are significant differences in the characteristics of private and public sector alternatives. They are summarized here.

Characteristic	Public Sector	Private Sector
Size of investment	Large	Some large; more medium to small

Often alternatives developed to serve public needs require large initial investments, possibly distributed over several years. Modern highways, public transportation systems, universities, airports, and flood control systems are examples.

Characteristic	Public Sector	Private Sector
Life estimates	Longer (30–50+years)	Shorter (2–25 years)

The long lives of public projects often prompt the use of the capitalized cost method, where infinity is used for n and annual costs are calculated as $A = P(i)$. As n gets larger, especially over 30 years, the differences in calculated A values become small. For example, at $i = 7\%$, there will be a very small difference in 30 and 50 years, because $(A/P,7\%,30) = 0.08059$ and $(A/P,7\%,50) = 0.07246$.

Characteristic	Public Sector	Private Sector
Annual cash flow estimates	No profit; costs, benefits, and disbenefits are estimated	Revenues contribute to profit; costs are estimated

Public sector projects (also called publicly owned) do not have profits; they do have costs that are paid by the appropriate government unit; and they benefit the citizenry. Public sector projects often have undesirable consequences, as interpreted by some sectors of the public, such as self-formed citizen groups and organized, funded special-interest groups. It is these consequences that can cause public controversy about the projects. The economic analysis should consider these consequences in monetary terms to the degree estimable. (Often in private sector analysis, undesirable consequences are not considered, or they may be directly addressed as costs.) To perform a benefit/cost economic analysis of public alternatives, the costs (initial and annual), the benefits, and the disbenefits, if considered, must be estimated as accurately as possible in *monetary units*.

Costs—estimated expenditures *to the government entity* for construction, operation, and maintenance of the project, less any expected salvage value.

Benefits—advantages to be experienced *by the owners, the public.*

Disbenefits—expected undesirable or negative consequences *to the owners* if the alternative is implemented. Disbenefits may be indirect economic disadvantages of the alternative.

It is difficult to estimate and agree upon the economic impact of benefits and disbenefits for a public sector alternative. For example, assume a short bypass around a congested area in town is recommended. How much will it benefit a driver in *dollars per driving minute* to be able to bypass five traffic lights while averaging 35 miles per hour, as compared to currently driving through the lights averaging 20 miles per hour and stopping at an average of two lights for an average of 45 seconds each? The bases and standards for benefits estimation are always difficult to establish and verify. Relative to revenue cash flow estimates in the private sector, benefit estimates are much harder to make, and vary more widely around uncertain averages. (The inability to make economic estimates for benefits *may be overcome* by using the evaluation technique discussed in Section 9.5.) And the disbenefits that accrue from an alternative are even harder to estimate. In fact, the disbenefit itself may not be known at the time the evaluation is performed.

Characteristic	Public Sector	Private Sector
Funding	Taxes, fees, bonds, private funds	Stocks, bonds, loans, individual owners

The capital used to finance public sector projects is commonly acquired from taxes, bonds, and fees. Taxes are collected from those who are the owners—the citizens (e.g., federal gasoline taxes for highways are paid by all gasoline users, and health care costs are covered by insurance premiums). This is also the case for fees, such as toll road fees for drivers. Bonds are often issued: U.S. Treasury bonds, municipal bond issues, and special-purpose bonds, such as utility district bonds. Private lenders can provide up-front financing. Also, private donors may provide funding for museums, memorials, parks, and garden areas through gifts.

Characteristic	Public Sector	Private Sector
Interest rate	Lower	Higher, based on cost of capital

Because many of the financing methods for public sector projects are classified as *low-interest,* the interest rate is virtually always lower than for private sector alternatives. Government agencies are exempt from taxes levied by higher-level units. For example, municipal projects do not have to pay state taxes. (Private corporations and individual citizens do pay taxes.) Many loans are very low-interest, and grants with no repayment requirement from federal programs may share project costs. This results in interest rates in the 3% to 7% range. As a matter of standardization, directives to use a specific rate are beneficial because different government agencies are able to obtain varying types of funding at different rates. This can result in projects of the same type being rejected in one state or city but accepted in another. Standardized rates tend to increase the consistency of economic decisions and to reduce gamesmanship.

The determination of the interest rate for public sector evaluation is as important as the determination of the MARR for a private sector analysis. The public sector interest rate is identified as i; however, it is referred to by other names to distinguish it from the private sector rate. The most common terms are *discount rate* and *social discount rate.*

Characteristic	Public Sector	Private Sector
Alternative selection criteria	Multiple criteria	Primarily based on rate of return

Multiple categories of users, economic as well as noneconomic interests, and special-interest political and citizen groups make the selection of one alternative over another much more difficult in public sector economics. Seldom is it possible to select an alternative on the sole basis of a criterion such as PW or ROR. It is important to describe and itemize the criteria and selection method prior to the analysis. This helps determine the perspective or viewpoint when the evaluation is performed. Viewpoint is discussed below. (Multiple criteria analysis, including noneconomic attributes, is discussed in Chapter 10, Section 10.6 and may be covered now or at a later point.)

Characteristic	Public Sector	Private Sector
Environment of the evaluation	Politically inclined	Primarily economic

There are often public meetings and debates associated with public sector projects to accommodate the various interests of citizens (owners). Elected officials commonly assist with the selection, especially when pressure is brought to bear by voters, developers, environmentalists, and others. The selection process is not as "clean" as in private sector evaluation.

The **viewpoint of a public sector analysis** must be determined before cost, benefit, and disbenefit estimates are made and before the evaluation is formulated and performed. There are several viewpoints for any situation, and the different perspectives may alter how a cash flow estimate is classified.

Some example perspectives commonly taken are the citizen; the tax base; number of students in the school district; creation and retention of jobs; economic development potential; a particular industry interest (agriculture, banking, electronics manufacturing); even the reelection of a public officeholder (often termed *pork projects*) can be the viewpoint of the analysis. In general, the viewpoint of the analysis should be as broadly defined as those who will bear the costs of the project and reap its benefits. Once established, the viewpoint assists in categorizing cost, benefit, and disbenefit estimates of each alternative. This is illustrated in Example 9.1.

EXAMPLE 9.1 Water Treatment Facility #3 Case   **PE**

The situation with the location and construction of the new WTF3 and associated transmission mains described in the chapter's introduction has reached a serious level because of recent questions posed by some city council members and self-formed citizen groups. Before going public to the city council with the analysis performed last year, the director of Allen Water Utilities has asked an engineering management consultant to review it and determine if it was an acceptable analysis and correct economic decision, then and now. The lead consultant, Joel Whiterson, took engineering economy as a part of his B.S. education and has previously worked on economic studies in the government sector, but never as the lead person.

Within the first hour of checking background notes, Joel found several initial estimates (shown below) from last year for expected consequences if WTF3 were built. He realized that no viewpoint of the study was defined, and, in fact, the estimates were never classified as costs, benefits, or disbenefits. He did determine that disbenefits were considered at some point in the analysis, though the estimates for them are very sketchy.

Joel defined two viewpoints: a *citizen* of Allen and the Allen Water Utilities *budget*. He wants to identify each of the estimates as a cost, benefit, or disbenefit from each viewpoint. Please help with this classification.

Economic Dimension	Monetary Estimate
1. Cost of water: 10% annual increase to Allen households	Average of $29.7 million (years 1–5, steady thereafter)
2. Bonds: Annual debt service at 3% per year on $540 million	$16.2 million (years 1–19); $516.2 million (year 20)
3. Use of land: Payment to Parks and Recreation for shaft sites and construction areas	$300,000 (years 1–4)
4. Property values: Loss in value, sales price, and property taxes	$4 million (years 1–5)
5. Water sales: Increases in sales to surrounding communities	$5 million (year 4) plus 5% per year (years 5–20)
6. M&O: Annual maintenance and operations costs	$300,000 plus 4% per year increase (years 1–20)
7. Peak load purchases: Savings in purchases of treated water from secondary sources	$500,000 (years 5–20)

Solution

The perspective of each viewpoint is identified and estimates are classified. (How this classification is done will vary depending upon who does the analysis. This solution offers only one logical answer.)

Viewpoint 1: Citizen of the city of Allen. Goal: Maximize the quality of life and wellness of citizens with family and neighborhood as prime concerns.

> Costs: 1, 2, 4, 6   Benefits: 5, 7   Disbenefits: 3

Viewpoint 2: Allen Water Utilities budget. Goal: Ensure the budget is balanced and of sufficient size to fund rapidly growing city service demands.

> Costs: 2, 3, 6   Benefits: 1, 5, 7   Disbenefits: 4

Citizens view costs in a different light than a city budget employee does. For example, the loss of property values (item 4) is considered a real cost to a citizen, but is an unfortunate disbenefit from the city budget perspective. Similarly, the Allen Water Utilities budget interprets estimate

3 (payment for use of land to Parks and Recreation) as a real cost; but a citizen might interpret this as merely a movement of funds between two municipal budgets—therefore, it is a disbenefit, not a real cost.

Comment

The inclusion of disbenefits can easily change the economic decision. However, agreement on the disbenefits and their monetary estimates is difficult (to impossible) to develop, often resulting in the exclusion of any disbenefits from the economic analysis. Unfortunately, this usually transfers the consideration of disbenefits to the noneconomic (i.e., political) realm of public project decision making.

Most of the large public sector projects are developed through public-private partnerships (PPPs). A partnership is advantageous in part because of the greater efficiency of the private sector and in part because of the sizable cost to design, construct, and operate such projects. Full funding by the government unit may not be possible using traditional means—fees, taxes, and bonds. Some examples of the projects are as follows:

Project	Some Purposes of the Project
Mass transportation	Reduce transit time; reduce congestion; improve environment; decrease road accidents
Bridges and tunnels	Speed traffic flows; reduce congestion; improve safety
Ports and harbors	Increase cargo capacity; support industrial development; increase tourism
Airports	Increase capacity; improve passenger safety; support development
Water resources	Desalination and brackish-water purification for drinking water; meet irrigation and industrial needs; improve wastewater treatment

In these joint ventures, the public sector (government) is responsible for the funding and service to the citizenry, and the private sector partner (corporation) is responsible for varying aspects of the projects as detailed below. The government unit cannot make a profit, but the corporation(s) involved can realize a reasonable profit; in fact, the profit margin is usually written into the contract that governs the design, construction, and operation of the project.

Traditional methods of contracting were *fixed-price* (historically called lump-sum) and *cost reimbursable* (also called cost-plus). In these formats, a government unit took responsibility for funding and possibly some of the design elements, and later all operation activities, while the contractor did not share in the risks involved—liability, natural disasters, funding shortfalls, etc. More recently, the PPP has become the arrangement of choice for most large public projects. Commonly these contracts are called **design-build (DB),** under which contractors take on more and more of the functions from design to operation. Details about different types of DB contracts may be found on the website of The National Council for Public-Private Partnerships (www.ncppp.org). The most reliance is placed upon a contractor or contractors with a DBOMF contract, as described below.

The **Design-Build-Finance-Operate-Maintain (DBFOM) contract** is considered a turnkey approach to a project. It requires the contractor(s) to perform all the DBFOM activities with collaboration and approval of the owner (the government unit). The activity of **financing** is the *management of cash flow to support project implementation* by a contracting firm. Although a contractor may assist in some instances, the **funding** (obtaining the capital funds) remains the government's responsibility through bonding, commercial loans, taxation, grants, and gifts.

When the financing activity is not managed by a contractor, the contract is a DBOM; it is also common to develop a design-build contract. In virtually all cases, some forms of design-build arrangements for public projects are made because they offer several advantages to the government and citizens served:

- Cost and time savings in the design, build, and operate phases
- Earlier and more reliable (less variable) cost estimates

- Reduced administrative responsibilities for the owner
- Better efficiency of resource allocation by private enterprise
- Environmental, liability, and safety issues addressed by the private sector, where there usually is greater expertise

Many of the projects in international settings and in developing countries utilize the public-private partnership. There are, of course, disadvantages to this arrangement. One risk is that the amount of funding committed to the project may not cover the actual build cost because it is considerably higher than estimated. Another risk is that a reasonable profit may not be realized by the private corporation due to low usage of the facility during the operate phase. To prevent such problems, the original contract may provide for special subsidies and loans guaranteed by the government unit. The subsidy may cover costs plus (contractually agreed-to) profit if usage is lower than a specified level. The level used may be the breakeven point with the agreed-to profit margin considered.

9.2 Benefit/Cost Analysis of a Single Project ●●●

The benefit/cost ratio is relied upon as a fundamental analysis method for public sector projects. The B/C analysis was developed to introduce greater objectivity into public sector economics, and as one response to the U.S. Congress approving the Flood Control Act of 1936. There are several variations of the B/C ratio; however, the fundamental approach is the same. All cost and benefit estimates must be converted to a common equivalent monetary unit (PW, AW, or FW) at the discount rate (interest rate). The B/C ratio is then calculated using one of these relations:

$$B/C = \frac{\text{PW of benefits}}{\text{PW of costs}} = \frac{\text{AW of benefits}}{\text{AW of costs}} = \frac{\text{FW of benefits}}{\text{FW of costs}} \qquad [9.1]$$

Present worth and annual worth equivalencies are preferred to future worth values. The sign convention for B/C analysis is positive signs; **costs are preceded by a + sign.** Salvage values and additional revenues to the government, when they are estimated, are subtracted from costs in the denominator. Disbenefits are considered in different ways depending upon the model used. Most commonly, **disbenefits are subtracted from benefits** and placed in the numerator. The different formats are discussed below.

The decision guideline is simple:

If B/C ≥ 1.0, accept the project as economically justified for the estimates and discount rate applied.

If B/C < 1.0, the project is not economically acceptable.

Project evaluation

If the B/C value is exactly or very near 1.0, noneconomic factors will help make the decision.

The *conventional B/C ratio,* probably the most widely used, is calculated as follows:

$$B/C = \frac{\text{benefits} - \text{disbenefits}}{\text{costs}} = \frac{B - D}{C} \qquad [9.2]$$

In Equation [9.2], disbenefits are subtracted from benefits, not added to costs. The B/C value could change considerably if disbenefits are regarded as costs. For example, if the numbers 10, 8, and 5 are used to represent the PW of benefits, disbenefits, and costs, respectively, the correct procedure results in B/C = (10 − 8)/5 = 0.40. The incorrect placement of disbenefits in the denominator results in B/C = 10/(8 + 5) = 0.77, which is approximately twice the correct B/C value of 0.40. Clearly, then, the method by which disbenefits are handled affects the magnitude of the B/C ratio. However, regardless of whether disbenefits are (correctly) subtracted from the numerator or (incorrectly) added to costs in the denominator, a B/C ratio of less than 1.0 by the first method will always yield a B/C ratio less than 1.0 by the second method, and vice versa.

The *modified B/C ratio* includes all the estimates associated with the project, once operational. Annual operating costs (AOC) and maintenance and operation (M&O) costs are placed in the numerator and treated in a manner similar to disbenefits. The denominator includes only the initial investment. Once all amounts are expressed in PW, AW, or FW terms, the modified B/C ratio is calculated as

$$\text{Modified B/C} = \frac{\text{benefits} - \text{disbenefits} - \text{M\&O costs}}{\text{initial investment}} \quad [9.3]$$

Salvage value is usually included in the denominator as a negative cost. The modified B/C ratio will obviously yield a different value than the conventional B/C method. However, as with disbenefits, *the modified procedure can change the magnitude of the ratio but not the decision to accept or reject the project.* The decision guideline for the modified B/C ratio is the same as that for the conventional B/C ratio.

The ***benefit and cost difference*** measure of worth, which does not involve a ratio, is based on the difference between the PW, AW, or FW of benefits and costs, that is, $B - C$. If $(B - C) \geq 0$, the project is acceptable. This method has the advantage of eliminating the discrepancies noted above when disbenefits are regarded as costs because B represents *net benefits.* Thus, for the numbers 10, 8, and 5, the same result is obtained regardless of how disbenefits are treated.

Subtracting disbenefits from benefits: $B - C = (10 - 8) - 5 = -3$
Adding disbenefits to costs: $B - C = 10 - (8 + 5) = -3$

Before calculating the B/C ratio by any formula, check whether the alternative with the larger AW or PW of costs also yields a larger AW or PW of benefits. It is possible for one alternative with larger costs to generate lower benefits than other alternatives, thus making it unnecessary to further consider the larger-cost alternative.

By the very nature of benefits and especially disbenefits, monetary estimates are difficult to make and will vary over a wide range. The extensive use of **sensitivity analysis** on the more questionable parameters helps determine how sensitive the economic decision is to estimate variation. This approach assists in determining the **economic and public acceptance risk** associated with a defined project. Also, the use of sensitivity analysis can alleviate some of the public's concerns commonly expressed that people (managers, engineers, consultants, contractors, and elected officials) designing (and promoting) the public project are narrowly receptive to different approaches to serving the public's interest.

EXAMPLE 9.2

In the past, the Afram Foundation has awarded many grants to improve the living and medical conditions of people in war-torn and poverty-stricken countries throughout the world. In a proposal for the foundation's board of directors to construct a new hospital and medical clinic complex in a deprived central African country, the project manager has developed some estimates. These are developed, so she states, in a manner that does not have a major negative effect on prime agricultural land or living areas for citizens.

Award amount:	$20 million (end of) first year, decreasing by $5 million per year for 3 additional years; local government will fund during the first year only
Annual costs:	$2 million per year for 10 years, as proposed
Benefits:	Reduction of $8 million per year in health-related expenses for citizens
Disbenefits:	$0.1 to $0.6 million per year for removal of arable land and commercial districts

Use the conventional and modified B/C methods to determine if this grant proposal is economically justified over a 10-year study period. The foundation's discount rate is 6% per year.

Solution

Initially, determine the AW for each parameter over 10 years. In $1 million units,

Award:	$20 - 5(A/G,6\%,4) = \$12.864$ per year
Annual costs:	$2 per year
Benefits:	$8 per year
Disbenefits:	Use $0.6 for the first analysis

The *conventional B/C analysis* applies Equation [9.2].

$$B/C = \frac{8.0 - 0.6}{12.864 + 2.0} = 0.50$$

The *modified B/C analysis* uses Equation [9.3].

$$\text{Modified } B/C = \frac{8.0 - 0.6 - 2.0}{12.864} = 0.42$$

The proposal is not justified economically since both measures are less than 1.0. If the low disbenefits estimate of \$0.1 million per year is used, the measures increase slightly, but not enough to justify the proposal.

It is possible to develop a direct formula connection between the B/C of a public sector and B/C of a private sector project that is a **revenue alternative;** that is, both revenues and costs are estimated. Further, we can identify a direct correspondence between the modified B/C relation in Equation [9.3] and the PW method we have used repeatedly. (The following development also applies to AW or FW values.) Let's concentrate on the net cash flow (NCF) of the project for year 1 through its expected life. For the private sector, the PW for project cash flows is

$$\text{PW of NCF} = \text{PW of revenues} - \text{PW of costs}$$

Since private sector revenues are approximately the same as public sector benefits minus disbenefits $(B - D)$, the modified B/C relation in Equation [9.3] may be written as

$$\text{Modified } B/C = \frac{\text{PW of } (B - D) - \text{PW of } C}{\text{PW of initial investment}}$$

This relation can be slightly rewritten to form the **profitability index (PI)**, which is useful in evaluating revenue projects in the public or private sector.

$$PI = \frac{\text{PW of NCF series}}{\text{PW of initial investment}} \qquad [9.4]$$

Note that the denominator includes only first cost (initial investment) items, while the numerator has only cash flows that result from the project for years 1 through its life. The PI measure of worth provides a sense of getting the most for the investment dollar (euro, yen, etc.). That is, the result is in PW units per PW of money invested at the beginning. This is a "bang for the buck" measure. When used solely for a private sector project, the disbenefits are usually omitted, whereas they should be estimated and included in the modified B/C version of this measure for a public project.

The evaluation guideline for a single project using the PI is the same as for the conventional B/C or modified B/C.

If PI $\geq$ 1.0, the project is economically acceptable at the discount rate.
If PI $<$ 1.0, the project is not economically acceptable at the discount rate.

Project evaluation

Remember, the computations for PI and modified B/C are essentially the same, except the PI is usually applied without disbenefits estimated. The PI has another name: the **present worth index (PWI)**. It is often used to rank and assist in the selection of independent projects when the capital budget is limited. This application is discussed in Chapter 12, Section 12.5.

EXAMPLE 9.3

The Georgia Transportation Directorate is considering a public-private partnership with Young Construction as the prime contractor using a DBFOM contract for a new 22.51-mile toll road on the outskirts of Atlanta's suburban area. The design includes three 4-mile-long commercial/retail corridors on both sides of the toll road. Highway construction is expected to require 5 years at an average cost of \$3.91 million per mile. The discount rate is 4% per year, and the study period is 30 years. Evaluate the economics of the proposal using (*a*) the modified B/C analysis from the State of Georgia perspective, and (*b*) the profitability index from the Young corporate viewpoint in which disbenefits are not included.

Initial investment: $88 million distributed over 5 years; $4 million now and in year 5; and $20 million in each of years 1 through 4.

Annual M&O cost: $1 million per year, plus an additional $3 million each fifth year, including year 30.

Annual revenue/benefits: Include tolls and retail/commercial growth; start at $2 million in year 1, increasing by a constant $0.5 million annually through year 10, and then increasing by a constant $1 million per year through year 20 and remaining constant thereafter.

Estimable disbenefits: Include loss of business income, taxes, and property value in surrounding areas; start at $10 million in year 1, decrease by $0.5 million per year through year 21, and remain at zero thereafter.

Solution

The PW values in year 0 for all estimates must be developed initially usually by hand, calculator, or spreadsheet computations. If the 30 years of estimates are entered into a spreadsheet and NPV functions at 4% are applied, the results in $1 million units are obtained. All values are positive because of the sign convention for B/C and PI measures.

$$\text{PW of investment} = \$71.89 \qquad \text{PW of benefits} = \$167.41$$
$$\text{PW of costs} = \$26.87 \qquad \text{PW of disbenefits} = \$80.12$$

(*a*) From the public project perspective, the State will apply Equation [9.3].

$$\text{Modified B/C} = \frac{167.41 - 80.12 - 26.87}{71.89} = 0.84$$

The toll road proposal is *not economically acceptable*, since $B/C < 1.0$.

(*b*) From the private corporation viewpoint, Young Construction will apply Equation [9.4].

$$\text{PI} = \frac{167.41 - 26.87}{71.89} = 1.95$$

The proposal is clearly justified without the disbenefits, since PI > 1.0. The private project perspective predicts that every investment dollar will return an equivalent of $1.95 over 30 years at a 4% per year discount rate.

Comment

The obvious question that arises concerns the correct measure to use. When PI is used in the private project setting, there is no problem, since disbenefits are virtually never considered in the economic analysis. The public project setting will commonly use some form of the B/C ratio with disbenefits considered. When a public-private partnership is initiated, there should be some agreement beforehand that establishes the economic measure acceptable for analysis and decision making throughout the project. Then the numerical dilemma presented above should not occur.

ME alternative
selection

9.3 Incremental B/C Analysis (Two Alternatives) ● ● ●

The technique to compare two mutually exclusive alternatives using benefit/cost analysis is virtually the same as that for incremental ROR in Chapter 8. The incremental (conventional) B/C ratio, which is identified as $\Delta B/C$, is determined using PW, AW, or FW calculations. The higher-cost alternative is justified if $\Delta B/C$ is equal to or larger than 1.0. The selection rule is as follows:

> If $\Delta B/C \geq 1.0$, choose the higher-cost alternative, because its extra cost is economically justified.
>
> If $\Delta B/C < 1.0$, choose the lower-cost alternative.

To perform a correct incremental B/C analysis, it is required that each alternative be compared only with another alternative for which the incremental cost is already justified. This same rule was used for incremental ROR analysis.

There are two dimensions of an incremental B/C analysis that differ from the incremental ROR method in Chapter 8. We already know the first, all costs have a positive sign in the B/C ratio. The second, and significantly more important, concerns the ordering of alternatives prior to incremental analysis.

Alternatives are **ordered by increasing equivalent total costs,** that is, PW or AW of all cost estimates that will be utilized in the denominator of the B/C ratio. When not done correctly, the incremental B/C analysis may reject a justified higher-cost alternative.

If two alternatives, A and B, have equal initial investments and lives, but B has a larger equivalent annual cost, then B must be incrementally justified against A. (This is illustrated in Example 9.4 below.) If this convention is not correctly followed, it is possible to get a negative cost value in the denominator, which can incorrectly make $B/C < 1$ and reject a higher-cost alternative that is actually justified.

As with the incremental ROR method, $\Delta B/C$ has a predictable relation to individual alternative B/C values in weighted average calculations. In brief, if alternative B has a larger total cost than A, and if $B/C_B < B/C_A$, then $\Delta B/C < B/C_B$. Furthermore, if $B/C_B > B/C_A$, then $\Delta B/C > B/C_B$.

Follow these steps to correctly perform a conventional B/C ratio analysis of two alternatives. Equivalent values can be expressed in PW, AW, or FW terms.

1. Determine the equivalent total costs for both alternatives.
2. Order the alternatives by equivalent total cost: first smaller, then larger. Calculate the incremental cost (ΔC) for the larger-cost alternative. This is the denominator in $\Delta B/C$.
3. Calculate the equivalent total benefits and any disbenefits estimated for both alternatives. Calculate the incremental benefits (ΔB) for the larger-cost alternative. This is $\Delta(B - D)$ if disbenefits are considered.
4. Calculate the $\Delta B/C$ ratio using Equation [9.2], $(B - D)/C$.
5. Use the selection guideline to select the higher-cost alternative if $\Delta B/C \geq 1.0$.

When the B/C ratio is determined for the lower-cost alternative, it is a comparison with the do-nothing (DN) alternative. If $B/C < 1.0$, then DN should be selected and compared to the second alternative. If neither alternative has an acceptable B/C value and one of the alternatives does not have to be selected, the DN alternative must be selected. In public sector analysis, the DN alternative is usually the current condition.

EXAMPLE 9.4

The city of Garden Ridge, Florida, has received designs for a new patient room wing to the municipal hospital from two architectural consultants. One of the two designs must be accepted in order to announce it for construction bids. The costs and benefits are the same in most categories, but the city financial manager decided that the estimates below should be considered to determine which design to recommend at the city council meeting next week and to present to the citizenry in preparation for an upcoming bond referendum next month.

	Design A	Design B
Construction cost, $	10,000,000	15,000,000
Building maintenance cost, $/year	35,000	55,000
Patient usage copay, $/year	450,000	200,000

The patient usage copay is an estimate of the amount paid by patients over the insurance coverage generally allowed for a hospital room. The discount rate is 5%, and the life of the building is estimated at 30 years.

(*a*) Use incremental B/C analysis to select design A or B.

(*b*) Once the two designs were publicized, the privately owned hospital in the directly adjacent city of Forest Glen lodged a complaint that design A will reduce its own municipal hospital's income by an estimated $500,000 per year because some of the day-surgery features of design A duplicate its services. Subsequently, the Garden Ridge merchants' association argued that design B could reduce its annual revenue by an estimated $400,000 because it will eliminate an entire parking lot used by their patrons for short-term parking. The city financial manager stated that these concerns would be entered into the evaluation as disbenefits of the respective designs. Redo the B/C analysis to determine if the economic decision is still the same as when disbenefits were not considered.

Solution

(*a*) Since most of the cash flows are already annualized, the incremental B/C ratio will use AW values. No disbenefit estimates are considered. Follow the steps of the procedure above:

1. The AW of costs is the sum of construction and maintenance costs.

$$AW_A = 10,000,000(A/P,5\%,30) + 35,000 = \$685,500$$
$$AW_B = 15,000,000(A/P,5\%,30) + 55,000 = \$1,030,750$$

2. Design B has the larger AW of costs, so it is the alternative to be incrementally justified. The incremental cost is

$$\Delta C = AW_B - AW_A = \$345,250 \text{ per year}$$

3. The AW of benefits is derived from the patient usage copays, since these are consequences to the public. The benefits for the ΔB/C analysis are not the estimates themselves, but the *difference* if design B is selected. The lower usage copay is a positive benefit for design B.

$$\Delta B = copay_A - copay_B = \$450,000 - \$200,000 = \$250,000 \text{ per year}$$

4. The incremental B/C ratio is calculated by Equation [9.2].

$$\Delta B/C = \frac{\$250,000}{\$345,250} = 0.72$$

5. The B/C ratio is less than 1.0, indicating that the extra costs associated with design B are not justified. Therefore, design A is selected for the construction bid.

(*b*) The revenue loss estimates are considered disbenefits. Since the disbenefits of design B are $100,000 less than those of A, this positive difference is added to the $250,000 benefits of B to give it a total benefit of $350,000. Now

$$\Delta B/C = \frac{\$350,000}{\$345,250} = 1.01$$

Design B is slightly favored. In this case, the inclusion of disbenefits has reversed the previous economic decision. This has probably made the situation more difficult politically. New disbenefits will surely be claimed in the near future by other special-interest groups.

Like other methods, incremental B/C analysis requires **equal-service comparison** of alternatives. Usually, the expected useful life of a public project is long (25 or 30 or more years), so alternatives generally have equal lives. However, when alternatives do have unequal lives, the use of PW or AW to determine the equivalent costs and benefits requires that the LCM of lives be used to calculate $\Delta B/C$. As with ROR analysis of two alternatives, this is an excellent opportunity to use the AW equivalency of estimated (not incremental) costs and benefits, if the implied assumption that the project could be repeated is reasonable. Therefore, use AW-based analysis of actual costs and benefits for B/C ratios when different-life alternatives are compared.

EXAMPLE 9.5   Water Treatment Facility #3 Case

As our case unfolds, the consultant, Joel Whiterson, has pieced together some of the B/C analysis estimates for the 84-inch Jolleyville transmission main study completed last year. The two options for constructing this main were open trench (OT) for the entire 6.8-mile distance or a combination of trenching and bore tunneling (TT) for a shorter route of 6.3 miles. One of the two options had to be selected to transport approximately 300 million gallons per day (gpd) of treated water from the new WTF3 to an existing aboveground reservoir.

The general manager of Allen Water Utilities has stated publicly several times that the trench-tunnel combination option was selected over the open-trench alternative based on analysis of both quantitative and nonquantitative data. He stated the equivalent annual costs in an internal e-mail some months ago, based on the expected construction periods of 24 and 36 months, respectively, as equivalent to

$$AW_{OT} = \$1.20 \text{ million per year}$$

$$AW_{TT} = \$2.37 \text{ million per year}$$

This analysis indicated that the open-trench option was economically better, at that time. The planning horizon for the transmission mains is 50 years; this is a reasonable study period, Joel concluded. Use the estimates below that Joel has unearthed to perform a correct incremental B/C analysis and comment on the results. The interest (discount) rate is 3% per year, compounded annually, and 1 mile is 5280 feet.

	Open Trench (OT)	Trench-Tunnel (TT)
Distance, miles	6.8	6.3
First cost, $ per foot	700	Trench for 2.0 miles: 700
		Tunnel for 4.3 miles: 2100
Time to complete, months	24	36
Construction support costs, $ per month	250,000	175,000
Ancillary expenses, $ per month:		
Environmental	150,000	20,000
Safety	140,000	60,000
Community interface	20,000	5,000

Solution

One of the alternatives must be selected, and the construction lives are unequal. Since it is not reasonable to assume that this construction project will be repeated many cycles in the future, it is incorrect to conduct an AW analysis over the respective completion periods of 24 and 36 months, or the LCM of these time periods. However, the study period of 50 years is a reasonable evaluation time frame, since the mains are considered permanent installations. We can assume that the construction first costs are a present worth value in year 0, but the equivalent PW and 50-year AW of other monthly costs must be determined.

$$PW_{OT} = \text{PW of construction} + \text{PW of construction support costs}$$

$$= 700(6.8)(5280) + 250,000(12)(P/A,3\%,2)$$

$$= \$30,873,300$$

$$AW_{OT} = 30,873,300(A/P,3\%,50)$$

$$= \$1.20 \text{ million per year}$$

$$PW_{TT} = [700(2.0) + 2100(4.3)](5280) + 175,000(12)(P/A,3\%,3)$$

$$= \$61,010,460$$

$$AW_{TT} = 61,010,460(A/P,3\%,50)$$

$$= \$2.37 \text{ million per year}$$

The *trench-tunnel (TT) alternative has a larger equivalent cost*; it must be justified against the OT alternative. The incremental cost is

$$\Delta C = AW_{TT} - AW_{OT} = 2.37 - 1.20 = \$1.17 \text{ million per year}$$

The difference between ancillary expenses defines the incremental benefit for TT.

$$PW_{OT\text{-anc}} = 310,000(12)(P/A,3\%,2)$$

$$= \$7,118,220$$

$$AW_{OT\text{-anc}} = 7,118,220(A/P,3\%,50)$$

$$= \$276,685 \text{ per year}$$

$$PW_{TT\text{-}anc} = 85{,}000(12)(P/A,3\%,3)$$
$$= \$2{,}885{,}172$$

$$AW_{TT\text{-}anc} = 2{,}885{,}172(A/P,3\%,50)$$
$$= \$112{,}147 \text{ per year}$$

$$\Delta B = AW_{OT\text{-}anc} - AW_{TT\text{-}anc} = 276{,}685 - 112{,}147 = \$164{,}538 \text{ per year} \quad (\$0.16 \text{ million})$$

Calculate the incremental B/C ratio.

$$\Delta B/C = 0.16/1.17 = 0.14$$

Since $\Delta B/C \ll 1.0$, the trench-tunnel option is not economically justified. Joel can now conclude that the general manager's earlier comment that the TT option was selected based on quantitative *and* nonquantitative data must have had heavy dependence on nonquantitative information not yet discovered.

9.4 Incremental B/C Analysis of Multiple (More than Two) Alternatives ● ● ●

The procedure to select one from three or more mutually exclusive alternatives using incremental B/C analysis is essentially the same as that of Section 9.3. The procedure also parallels that for incremental ROR analysis in Section 8.6. The selection guideline is as follows:

> Choose the largest-cost alternative that is justified with an incremental $B/C \geq 1.0$ when this selected alternative has been compared with another justified alternative.

There are two types of benefit estimates—estimation of **direct benefits,** and **implied benefits based on usage cost** estimates. The previous two examples (9.4 and 9.5) are good illustrations of the second type of implied benefit estimation. *When direct benefits are estimated,* the B/C ratio for each alternative may be calculated first as an initial screening mechanism to eliminate unacceptable alternatives. At least one alternative must have $B/C \geq 1.0$ to perform the incremental B/C analysis. If all alternatives are unacceptable, the DN alternative is the choice. (This is the same approach as that of step 2 for "revenue alternatives only" in the ROR procedure of Section 8.6. However, the term *revenue alternative* is not applicable to public sector projects.)

As in the previous section when comparing two alternatives, selection from multiple alternatives by incremental B/C ratio utilizes equivalent total costs to initially order alternatives from smallest to largest. Pairwise comparison is then undertaken. Also, remember that all costs are considered positive in B/C calculations. The terms *defender* and *challenger alternative* are used in this procedure, as in an ROR-based analysis. The procedure for incremental B/C analysis of multiple, mutually exclusive alternatives is as follows:

1. Determine the equivalent total cost for all alternatives. Use AW, PW, or FW equivalencies.
2. Order the alternatives by equivalent total cost, smallest first.
3. Determine the equivalent total benefits (and any disbenefits estimated) for each alternative.
4. *Direct benefits estimation only:* Calculate the B/C for the first ordered alternative. If $B/C < 1.0$, eliminate it. By comparing each alternative to DN, we eliminate all that have $B/C < 1.0$. The lowest-cost alternative with $B/C \geq 1.0$ becomes the defender and the next higher-cost alternative is the challenger in the next step. (For analysis by spreadsheet, determine the B/C for all alternatives initially and retain only acceptable ones.)
5. Calculate incremental costs (ΔC) and benefits (ΔB) using the relations

$$\Delta C = \text{challenger cost} - \text{defender cost}$$

$$\Delta B = \text{challenger benefits} - \text{defender benefits}$$

If relative **usage costs** are estimated for each alternative, rather than direct benefits, ΔB may be found using the relation

$$\Delta B = \text{defender usage costs} - \text{challenger usage costs} \qquad [9.5]$$

6. Calculate the ΔB/C for the first challenger compared to the defender.

$$\Delta B/C = \Delta B/\Delta C \qquad\qquad [9.6]$$

If $\Delta B/C \geq 1.0$ in Equation [9.6], the challenger becomes the defender and the previous defender is eliminated. Conversely, if $\Delta B/C < 1.0$, remove the challenger and the defender remains against the next challenger.

7. Repeat steps 5 and 6 until only one alternative remains. It is the selected one.

In all the steps above, incremental disbenefits may be considered by replacing ΔB with $\Delta(B - D)$.

EXAMPLE 9.6

Schlitterbahn Waterparks of Texas, a very popular water and entertainment park headquartered in New Braunfels, has been asked by four different cities outside of Texas to consider building a park in their area. All the offers include some version of the following incentives:

- Immediate cash incentive (year 0)
- A 10% of first-year incentive as a direct property tax reduction for 8 years
- Sales tax rebate sharing plan for 8 years
- Reduced entrance (usage) fees for area residents for 8 years

Table 9–1 (top section) summarizes the estimates for each proposal, including the present worth of the initial construction cost and anticipated annual revenue. The annual M&O costs are expected to be the same for all locations. Use incremental B/C analysis at 7% per year and an 8-year study period to advise the board of directors if they should consider any of the offers to be economically attractive.

Solution

The viewpoint is that of Schlitterbahn, and the benefits are direct estimates. Develop the AW equivalents over 8 years, and use the procedure detailed above. The results are presented in Table 9–1.

1. AW of total costs and an example for city 1 are determined in $1 million units.

$$\text{AW of costs} = \text{first cost}(A/P,7\%,8) + \text{entrance fee reduction to residents}$$
$$= 38.5(0.16747) + 0.5$$
$$= \$6.948 \quad (\$6,948,000 \text{ per year})$$

2. The four alternatives are correctly ordered by increasing equivalent total cost in Table 9–1.

TABLE 9–1	Incremental B/C Analysis of Water Park Proposals, Example 9.6			
	City 1	City 2	City 3	City 4
First cost, $ million	38.5	40.1	45.9	60.3
Entrance fee costs, $/year	500,000	450,000	425,000	250,000
Annual revenue, $ million/year	7.0	6.2	10.0	10.4
Initial cash incentive, $	250,000	350,000	500,000	800,000
Property tax reduction, $/year	25,000	35,000	50,000	80,000
Sales tax sharing, $/year	310,000	320,000	320,000	340,000
AW of total costs, $ million/year	6.948	7.166	8.112	10.348
AW of total benefits, $ million/year	7.377	6.614	10.454	10.954
Overall B/C	1.06	0.92	1.29	1.06
Alternatives compared	1 vs. DN	B/C < 1.0	3 vs. 1	4 vs. 3
Incremental costs ΔC, $/year	6.948		1.164	2.236
Incremental benefits ΔB, $/year	7.377		3.077	0.50
$\Delta B/C$	1.06		2.64	0.22
Increment justified?	Yes	Eliminated	Yes	No
City selected	1		3	3

3. AW of total benefits and an example for city 1 are also determined in $1 million units.

$$AW \text{ of benefits} = \text{revenue} + \text{initial incentive}(A/P,7\%,8)$$
$$+ \text{property tax reduction} + \text{sales tax sharing}$$
$$= 7.0 + 0.25(0.16747) + 0.025 + 0.31$$
$$= \$7.377 \quad (\$7{,}377{,}000 \text{ per year})$$

4. Since benefits are directly estimated (and no disbenefits are included), determine the overall B/C for each alternative using Equation [9.1]. In the case of city 1,

$$B/C_1 = 7.377/6.948 = 1.06$$

City 2 is eliminated with $B/C_2 = 0.92$; the rest are initially acceptable.

5. The ΔC and ΔB values are the actual estimates for the 1 vs. DN comparison.
6. The overall B/C is the same as $\Delta B/C = 1.06$, using Equation [9.6]. City 1 is economically justified and becomes the defender.
7. Repeat steps 5 and 6. Since city 2 is eliminated, the 3 vs. 1 comparison results in

$$\Delta C = 8.112 - 6.948 = 1.164$$
$$\Delta B = 10.454 - 7.377 = 3.077$$
$$\Delta B/C = 3.077/1.164 = 2.64$$

City 3 is well justified and becomes the defender against city 4. From Table 9–1, $\Delta B/C = 0.22$ for the 4 vs. 3 comparison. City 4 is clearly eliminated and **city 3** is the one to recommend to the board. Note that the DN alternative could have been selected had no proposal met the B/C or $\Delta B/C$ requirements.

Independent project selection

> When two or more ***independent projects*** are evaluated using B/C analysis and there is no budget limitation, no incremental comparison is necessary. The only comparison is between each project separately with the do-nothing alternative. The project B/C values are calculated, and those with $B/C \geq 1.0$ are accepted.

This is the same procedure as that used to select from independent projects using the ROR method (Chapter 8). When a budget limitation is imposed, the capital budgeting procedure discussed in Chapter 12 must be applied.

When the lives of mutually exclusive alternatives are so long that they can be considered infinite, the capitalized cost is used to calculate the equivalent PW or AW values for costs and benefits. As discussed in Section 5.5, Capitalized Cost Analysis, the relation $A = P(i)$ is used to determine the equivalent AW values in the incremental B/C analysis. Example 9.7 illustrates this using the progressive example and a spreadsheet.

EXAMPLE 9.7 Water Treatment Facility #3 Case

Land for Water Treatment Facility #3 was initially purchased in the year 2010 for $19.3 million; however, when it was publicized, influential people around Allen spoke strongly against the location. We will call this location 1. Some of the plant design had already been completed when the general manager announced that this site was not the best choice anyway, and that it would be sold and a different, better site (location 2) would be purchased for $28.5 million. This was well over the budget amount of $22.0 million previously set for land acquisition. As it turns out, there was a third site (location 3) available for $35.0 million that was never seriously considered.

In his review and after much resistance from Allen Water Utilities staff, the consultant, Joel, received a copy of the estimated costs and benefits for the three plant location options. The revenues, savings, and sale of bulk water rights to other communities are estimated as increments from a base amount for all three locations. Using the assumption of a very long life for the WTF3 facility and the established discount rate of 3% per year, determine what Joel discovered when he did the B/C analysis. Was the general manager correct in concluding that location 2 was the best, all said and done?

	Location 1	Location 2	Location 3
Land cost, $ million	19.3	28.5	35.0
Facility first cost, $ million	460.0	446.0	446.0
Benefits, $ million per year:			
Pumping cost savings	5	3	0
Sales to area communities	12	10	8
Added revenue from Allen	6	6	6
Total benefits, $ million per year	23	19	14

Solution

A spreadsheet can be very useful when performing an incremental B/C analysis of three or more alternatives. Figure 9–1a presents the analysis with the preliminary input of AW values for costs using the relation $A = P(i)$ and annual benefits. Figure 9–1b details all the functions used in the analysis. Logical IF statements indicate alternative elimination and selection decisions. In $1 million units,

$$\text{Location 1:} \quad \text{AW of costs} = (\text{land cost} + \text{facility first cost})i$$
$$= (19.3 + 460.0)(0.03)$$
$$= \$14.379 \text{ per year}$$
$$\text{AW of benefits} = \$23$$

$$\text{Location 2:} \quad \text{AW of costs} = \$14.235 \quad \text{AW of benefits} = \$19$$

$$\text{Location 3:} \quad \text{AW of costs} = \$14.430 \quad \text{AW of benefits} = \$14$$

Though the AW of cost values are close to one another, the increasing order is locations 2, 1, and 3 to determine $\Delta B/C$ values. The benefits are direct estimates; therefore, the overall B/C ratios indicate that location 3 (row 5; $B/C_3 = 0.97$) is not economically justified at the outset. It is eliminated, and one of the remaining locations must be selected. Location 2 is justified against the DN alternative ($B/C_2 = 1.33$); the only remaining comparison is 1-to-2 as detailed in column C of Figure 9–1. **Location 1** is a clear winner with $\Delta B/C = 27.78$.

In conclusion, Joel has learned that location 1 is indeed the best and that, from the economic perspective, the general manager was incorrect in stating that location 2 was better. However, given the original evaluation criteria listed in the introduction—*economics, environment, community impact, and constructability*—location 2 is likely a good compromise selection.

	A	B	C	D
1				
2	Order of analysis	Location 2	Location 1	Location 3
3	AW of cost, $M/year	14.235	14.379	14.430
4	Annual benefits, $M/year	19.0	23.0	14.0
5	Overall B/C	1.33	1.60	0.97
6	Acceptable	Yes	Yes	No
7				
8	Comparison		1-to-2	Eliminated
9	ΔC $/year		0.144	
10	ΔB, $/year		4.0	
11	ΔB/C		27.78	
12	Increment justified?		Yes	
13	Selection		Location 1	

(a)

	A	B	C	D
1				
2	Order of analysis	Location 2	Location 1	Location 3
3	AW of cost, $M/year	14.235	14.379	14.43
4	Annual benefits, $M/year	19	23	14
5	Overall B/C	= B4/B3	= C4/C3	= D4/D3
6	Acceptable	= IF(B5<1,"No","Yes")	= IF(C5<1,"No","Yes")	= IF(D5<1,"No","Yes")
7				
8	Comparison		1-to-2	Eliminated
9	ΔC $/year		= C3-B3	
10	ΔB, $/year		= C4-B4	
11	ΔB/C		= C10/C9	
12	Increment justified?		= IF(C11<1,"No","Yes")	
13	Selection		= IF(C12="Yes",C2,"No")	

(b)

Figure 9–1
Incremental B/C analysis for WTF3 case: (a) numerical results and (b) functions developed for the analysis.

Comment

This is an actual situation with changed names and values. Location 1 was initially purchased and planned for WTF3. However, the presence of political, community, and environmental stress factors changed the decision to location 2, when all was said and done.

9.5 Service Sector Projects and Cost-Effectiveness Analysis ● ● ●

Much of the GDP of the United States and some countries in Europe and Asia is generated by what has become known as the service sector of the economy. A large percentage of service sector projects are generated by and dependent upon the private sector (corporations, businesses, and other for-profit institutions). However, many projects in the public sector are also service sector projects.

> A **service sector project** is a process or system that provides services to individuals, businesses, or government units. The economic value is developed primarily by the **intangibles** of the process or system, **not the physical entities** (buildings, machines, and equipment). Manufacturing and construction activities are commonly not considered a service sector project, though they may support the theme of the service provided.

Service projects have a tremendous range of variety and purpose; to name a few: health care systems, health and life insurance, airline reservation systems, credit card services, police and court systems, security programs, safety training programs, and all types of consulting projects. The intangible and intellectual work done by engineers and other professionals is often a part of a service sector project.

The economic evaluation of a service project is difficult to a great degree because the cost and benefit estimates are not accurate and often not within an acceptable degree of error. In other words, undue risk may be introduced into the decision because of poor monetary estimates. For example, consider the decision to place red-light cameras at stop lights to ticket drivers who run the red light. This is a public and a service project, but its (*economic*) *benefits are quite difficult to estimate*. Depending upon the viewpoint, benefits could be in terms of accidents averted, deaths prevented, police personnel released from patrolling the intersection, or, from a more mercenary viewpoint, amount of fines collected. In all but the last case, benefits in monetary terms will be poor estimates. These are examples where B/C analysis does not work well and a different form of analysis is needed.

In service and public sector projects, as expected, it is the benefits that are the more difficult to estimate. An evaluation method that combines monetary cost estimates with nonmonetary benefit estimates is **cost-effectiveness analysis (CEA)**. The CEA approach utilizes a **cost-effectiveness measure** or the **cost-effectiveness ratio (CER)** as a basis of ranking projects and selecting the best of independent projects or mutually exclusive alternatives. The CER ratio is defined as

$$\text{CER} = \frac{\textbf{equivalent total costs}}{\textbf{total of effectiveness measure}} = \frac{C}{E} \qquad [9.7]$$

In the red-light camera example, the effectiveness measure (the benefit) may be one of the samples mentioned earlier, accidents averted or deaths prevented. Different from the B/C ratio of costs to benefits, CER places the PW or AW of total costs in the numerator and the effectiveness measure in the denominator. (The reciprocal of Equation [9.7] can also be used as the measure of worth, but we will use CER as defined above.) With costs in the numerator, smaller ratio values are more desirable for the same value of the denominator, since smaller ratio values indicate a lower cost for the same level of effectiveness.

Like ROR and B/C analysis, cost-effectiveness analysis requires the ordering (ranking) of alternatives prior to selection and the use of incremental analysis for mutually exclusive alternative selection. Cost-effectiveness analysis utilizes a different ranking criterion than ROR or BC analysis. The *ordering criteria* are as follows:

> **Independent projects:** Initially rank projects by **CER value.**
>
> **Mutually exclusive alternatives:** Initially rank alternatives by **effectiveness measure,** then perform an incremental CER analysis.

Return again to the public/service project of red-light cameras. If the CER is defined as "*cost per total accidents averted*" and the projects are independent, increasing CER value is the ranking basis. If the projects are mutually exclusive, "*total accidents averted*" is the correct ranking basis and an incremental analysis is necessary.

There are significantly different analysis procedures for independent and mutually exclusive proposals. To select some from several (independent) projects, a budget limit, termed *b*, is inherently necessary once ordering is complete. However, for selecting one from several (mutually exclusive) alternatives, a pairwise incremental analysis is necessary and selection is made on the basis of ΔC/E ratios. The procedures and examples follow.

> For **independent projects,** the procedure is as follows:
>
> 1. Determine the equivalent total cost *C* and effectiveness measure *E*, and calculate the CER measure for each project.
> 2. Order projects from the *smallest to the largest CER value.*
> 3. Determine cumulative cost for each project and compare with the budget limit *b*.
> 4. The selection criterion is to fund all projects such that *b* is not exceeded.

Independent project
selection

EXAMPLE 9.8

Recent research indicates that corporations throughout the world need employees who demonstrate creativity and innovation for new processes and products. One measure of these talents is the number of patents approved each year through the R&D efforts of a company. Rollings Foundation for Innovative Thinking has allocated $1 million in grant funds to award to corporations that enroll their top R&D personnel in a 1- to 2-month professional training program in their home state that has a historically proven track record over the last 5 years in helping individuals earn patents.

Table 9–2 summarizes data for six corporations that submitted proposals. Columns 2 and 3 give the proposed number of attendees and cost per person, respectively, and column 4 provides the historical track record of program graduates in patents per year. Use cost-effectiveness analysis to select the corporations and programs to fund.

Solution

We assume that across all programs and all patent awards there is equal quality. Use the procedure for independent projects and *b* = $1 million to select from the proposals.

TABLE 9–2	Data for Programs to Increase Patents Used for CEA		
Program (1)	Total Personnel (2)	Cost/Person, $ (3)	5-Year History, Patents/Graduate/Year (4)
1	50	5000	0.5
2	35	4500	3.1
3	57	8000	1.9
4	24	2500	2.1
5	12	5500	2.9
6	87	3800	0.6

1. Using Equation [9.7], the effectiveness measure E is patents per year, and the CER is

$$\text{CER} = \frac{\text{program cost per person}}{\text{patents per graduate}} = \frac{C}{E}$$

The program cost C is a PW value, and the E values are obtained from the proposals.
2. The CER values are shown in Table 9–3 in increasing order, column 5.
3. Cost per course, column 6, and cumulative costs, column 7, are determined.
4. Programs 4, 2, 5, 3, and 6 (68 of the 87 people) are selected to not exceed $1 million.

TABLE 9–3	Programs Ordered by CER Value, Example 9.8					
Program (1)	Total Personnel (2)	Cost/Person C, $ (3)	Patents per Year E (4)	CER, $ per Patent (5) = (3)/(4)	Program Cost, $ (6) = (2)(3)	Cumulative Cost, $ (7) = Σ(6)
4	24	2,500	2.1	1,190	60,000	60,000
2	35	4,500	3.1	1,452	157,500	217,500
5	12	5,500	2.9	1,897	66,000	283,500
3	57	8,000	1.9	4,211	456,000	739,500
6	87	3,800	0.6	6,333	330,600	1,070,100
1	50	5,000	0.5	10,000	250,000	1,320,100

Comment

This is the first time that a budget limit has been imposed for the selection among independent projects. This is often referred to as *capital budgeting*, which is discussed further in Chapter 12.

For **mutually exclusive** alternatives and no budget limit, the alternative with the highest effectiveness measure E is selected without further analysis. Otherwise, an incremental CER analysis is necessary and the budget limit is applied to the selected alternative(s). The analysis is based on the incremental ratio $\Delta C/E$, and the procedure is similar to that we have applied for incremental ROR and B/C, except now the concept of *dominance* is utilized.

ME alternative
selection

> **Dominance** occurs when the incremental analysis indicates that the challenger alternative offers an improved incremental CER measure compared to the defender's CER, that is,
>
> $$(\Delta C/E)_{\text{challenger}} < (C/E)_{\text{defender}}$$
>
> Otherwise, no dominance is present, and both alternatives remain in the analysis.

For **mutually exclusive alternatives,** the selection procedure is as follows:

1. Order the alternatives from *smallest to largest effectiveness measure E*. Record the cost for each alternative.
2. Calculate the CER measure for the first alternative. This, in effect, makes DN the defender and the first alternative the challenger. This CER is a baseline for the next incremental comparison, and the first alternative becomes the new defender.
3. Calculate incremental costs (ΔC) and effectiveness (ΔE) and the incremental measure $\Delta C/E$ for the new challenger using the relation

$$\Delta C/E = \frac{\text{cost of challenger} - \text{cost of defender}}{\text{effectiveness of challenger} - \text{effectiveness of defender}} = \frac{\Delta C}{\Delta E}$$

4. If $\Delta C/E < C/E_{\text{defender}}$, the challenger dominates the defender and it becomes the new defender; the previous defender is eliminated. Otherwise, no dominance is present and both alternatives are retained for the next incremental evaluation.
5. *Dominance present:* Repeat steps 3 and 4 to compare the next ordered alternative (challenger) and new defender. Determine if dominance is present.

Dominance not present: The current challenger becomes the new defender, and the next alternative is the new challenger. Repeat steps 3 and 4 to compare the new challenger and new defender. Determine if dominance is present.

6. Continue steps 3 through 5 until only one alternative or only nondominated alternatives remain.
7. Apply the budget limit (or other criteria) to determine which of the remaining alternative(s) can be funded.

EXAMPLE 9.9

One of the corporations not selected for funding in Example 9.8 decided to fund its 50 R&D personnel to attend one of the innovation and creativity programs at its own expense. One criterion is that the program must have a historical average for a graduate of at least 2.0 patents per year. Use the data in Table 9–2 to select the best program.

TABLE 9–4	Mutually Exclusive Alternatives Evaluated by Cost-Effectiveness Analysis, Example 9.9				
Program (1)	Total Personnel (2)	Cost/Person C, $ (3)	Patents per Year E (4)	CER, $ per Patent (5) = (3)/(4)	Program Cost, $ (6) = (2)(3)
4	50	2,500	2.1	1190	125,000
5	50	5,500	2.9	1897	275,000
2	50	4,500	3.1	1452	225,000

Solution

From Table 9–2, three programs—2, 4, and 5—have a historical record of at least two patents per graduate per year. Since only one program will be selected, these are now *mutually exclusive alternatives*. Use the procedure to perform the incremental analysis.

1. The alternatives are ranked by increasing patents per year in Table 9–4, column 4.
2. The CER measure for program 4 is compared to the DN alternative.

$$C/E_4 = \frac{\text{program cost per person}}{\text{patents per graduate}} = \frac{2500}{2.1} = 1190$$

3. Program 5 is now the challenger.

$$\text{5-to-4 comparison: } \Delta C/E = \frac{\Delta C}{\Delta E} = \frac{5500 - 2500}{2.9 - 2.1} = 3750$$

4. In comparison to $C/E_4 = 1190$, it costs $3750 per additional patent if program 5 is chosen over 4. Program 5 is more expensive for more patents; however, clear dominance is not present; both programs are retained for further evaluation.
5. *Dominance not present:* Program 5 becomes the new defender, and program 2 is the new challenger. Perform the 2-to-5 comparison.

$$\text{2-to-5 comparison: } \Delta C/E = \frac{\Delta C}{\Delta E} = \frac{4500 - 5500}{3.1 - 2.9} = -5000$$

Compared to $C/E_5 = 1897$, this increment is much cheaper—more patents for less money per person. Dominance is present; eliminate program 5 and compare 2 to 4.

6. Repeat steps 3 through 5 and compare $\Delta C/E$ to $C/E_4 = 1190$.

$$\text{2-to-4 comparison: } \Delta C/E = \frac{\Delta C}{\Delta E} = \frac{4500 - 2500}{3.1 - 2.1} = 2000$$

This does not represent dominance of program 2 over 4. The conclusion is that both programs are eligible for funding, that is, CEA in this case does not indicate only one program. This occurs when there is not lower cost and higher effectiveness of one alternative over another; that is, one alternative does not dominate all the others.

7. Now the budget and other considerations (probably noneconomic) are brought to bear to make the final decision. The fact that program 4 costs $125,000—significantly less than program 2 at $225,000—will likely enter into the decision.

Cost-effectiveness analysis is a form of multiattribute decision making in which economic and noneconomic dimensions are integrated to evaluate alternatives from several perspectives by different decision makers. See Chapter 10 for further discussion of alternative analysis using multiple attributes.

9.6 Ethical Considerations in the Public Sector ● ● ●

Usual expectations by citizens of their elected officials—locally, nationally, and internationally—are that they make decisions for the good of the public, ensuring safety and minimizing risk and cost to the public. Above these is the long-standing expectation that public servants have **integrity.**

Similarly, the expectations of engineers employed by government departments, and those serving as consultants to government agencies, are held to high standards. **Impartiality,** consideration of a **wide range of circumstances,** and the use of **realistic assumptions** are but three of the foundation elements upon which engineers should base their recommendations to decision makers. This implies that engineers in public service avoid

- Self-serving, often greedy individuals and clients with goals of excessive profits and future contract awards
- Using a politically favorable perspective that compromises the results of a study
- Narrowly defined assumptions that serve special interest groups and subcommunities potentially affected by the findings

Many people are disappointed and discouraged with government when elected officials and public employees (engineers and others, alike) do not have real commitment to integrity and unbiasedness in their work.

Engineers are routinely involved in two of the major aspects of public sector activities:

Public policy making—the **development of strategy** for public service, behavior, fairness, and justice. This may involve literature study, background discovery, data collection, opinion giving, and hypothesis testing. An example is *transportation management.* Engineers make virtually all the recommendations based on data and long-standing decision algorithms for policy items such as capacity of roads, expansion of highways, planning and zoning rules, traffic signal usage, speed limit corridors, and many related topics in transportation policy. Public officials use these findings to establish public transportation policy.

Public planning—the **development of projects** that implement strategy and affect people, the environment, and financial resources in a variety of ways. Consider *traffic control,* where the use and placement of traffic control signs, signals, speed limits, parking restrictions, etc. are detailed based upon established policy and current data. (In effect, this is systems engineering, that is, an application of the life-cycle phases and stages explained in Section 6.5.)

Whether in the arena of policy making or public planning, engineers can find ethical compromise a possibility when working with the public sector. A few circumstances are summarized here.

- *Use of technology*   Many public projects involve the use of new technology. The public risks and safety factors are not always known for these new advances. It is common and expected that engineers make every attempt to apply the latest technology while ensuring that the public is not exposed to undue risk.
- *Scope of study*   A client may pressure the engineers to limit the range of options, the assumption base, or the breadth of alternative solutions. These restrictions may be based on financial reasons, politically charged topics, client-favored options, or a wide variety of other reasons. To remain impartial, it is the responsibility of the engineer to submit a fully unbiased analysis, report, and recommendation, even though it may jeopardize future contract possibilities, promote public disfavor, or generate other negative consequences.
- *Negative community impact*   It is inevitable that public projects will adversely affect some groups of people, or the environment, or businesses. The intentional silencing of these projected effects is often the cause of strong public outcries against what may be a project that is in the best interest of the community at large. Engineers who find (stumble onto) such negative impacts may be pressured by clients, managers, or public figures to overlook them, though the Code of Ethics for Engineers dictates a full and fair analysis and report. For example,

TABLE 9–5	Some Ethical Considerations When Performing B/C and CEA Analysis	
What the Study Includes	**Ethical Dimension**	**Example**
Audience for study	Is it ethical to select a specific group of people affected by the project and neglect possible effects on other groups?	Construct children's health care clinics for city dwellers, but neglect rural families with poor transportation means.
Impact time of decision	Is it ethical to decide now for future generations who may be adversely and economically affected by the current project decision?	Accomplish financial bailouts of corporations when future generations' taxes will be significantly higher to recover the costs, plus interest and inflation effects.
Greater good for community as a whole	Vulnerable minority groups, especially economically deprived ones, may be disproportionally affected. Is this ethical if the impact is predictable?	Allow a chemical plant that is vital to the community's employment to pollute a waterway when a minority group is known to eat fish from the water that is predictably contaminated.
Reliance on economic measures only	Is it acceptable to reduce all costs and benefits to monetary estimates for a decision, then subjectively impute nonquantified factors in the final decision?	Softening of building codes can improve the financial outlook for home builders; however, increased risks of fire loss, storm and water damage to structures, and reduced future resale values are considered only in passing as a new subdivision is approved by the planning and zoning committee.
Scope of disbenefits estimated and evaluated	Is it ethical to disregard any disbenefits in the B/C study or use indirect effectiveness measures in a CEA study based on the difficulty to estimate some of them?	Noise and air pollution caused by a planned open-pit quarry will have a negative effect on area ranchers, residents, wildlife, and plant life; but the effectiveness measure considers only suburban residents due to estimation difficulty of effects on other constituencies.

a planned rerouting of a city street may effectively cut off a section of international citizens' businesses, thus resulting in a clearly predictable economic downturn. Considering this outcome in the recommendation to the transportation department should be a goal of the analyzing engineers, yet pressure to bias the results may be quite high.

The results of a B/C or CEA analysis are routinely depended upon by public officials and government staff members to assist in making public planning decisions. As discussed earlier, estimations for benefits, disbenefits, effectiveness measures, and costs can be difficult and inaccurate, but these analysis tools are often the best available to structure a study. Some examples of ethically oriented challenges that may be confronted during B/C and CEA analyses are summarized in Table 9–5.

CHAPTER SUMMARY

The benefit/cost method is used primarily to evaluate alternatives in the public sector. When one is comparing mutually exclusive alternatives, the incremental B/C ratio must be greater than or equal to 1.0 for the incremental equivalent total cost to be economically justified. The PW, AW, or FW of the initial costs and estimated benefits can be used to perform an incremental B/C analysis. For independent projects, no incremental B/C analysis is necessary. All projects with $B/C \geq 1.0$ are selected provided there is no budget limitation. It is usually quite *difficult to make accurate estimates of benefits* for public sector projects. The characteristics of public sector projects are substantially different from those of the private sector: initial costs are larger; expected life is longer; additional sources of capital funds include taxation, user fees, and government grants; and interest (discount) rates are lower.

Service projects develop economic value largely based on the intangibles of the services provided to users, not the physical items associated with the process or system. Evaluation by B/C analysis can be difficult with no good way to make monetary estimates of benefits. Cost-effectiveness analysis (CEA) combines cost estimates and a nonmonetary effectiveness measure (the benefit) to evaluate independent or mutually exclusive projects using procedures that are similar to incremental ROR and B/C analysis. The concept of dominance is incorporated into the procedure for comparing mutually exclusive alternatives.

As a complement to the discussion on professional ethics in Chapter 1, some potential ethical challenges in the public sector for engineers, elected officials, and government consultants are discussed here. Examples are included.

PROBLEMS

Understanding B/C Concepts

9.1   What is the primary purpose of a public sector project?

9.2   In conducting a B/C analysis, (a) why is it usually necessary to take a specific viewpoint in categorizing cost, benefit, disbenefit estimates; and (b) what are two specific viewpoints that you can identify if the situation is a financial transaction between you and another person? Between your company and an international customer?

9.3   Identify the following projects as primarily public or private sector.
(a)   Bridge across Ohio River
(b)   Coal mine expansion
(c)   Baja 1000 race team
(d)   Consulting engineering firm
(e)   New county courthouse building
(f)   Flood control project
(g)   Endangered species designation
(h)   Freeway lighting (lumen increase)
(i)   Antarctic cruise for you and your spouse
(j)   Crop dusting airplane purchase

9.4   Identify the following *funding sources* as primarily public or private.
(a)   Municipal bonds
(b)   Retained earnings
(c)   Sales taxes
(d)   Automobile license fees
(e)   Bank loans
(f)   Savings accounts
(g)   An engineer's IRA (Individual Retirement Account)
(h)   State fishing license revenues
(i)   Entrance fees to Tokyo Disneyland
(j)   State park entrance fees

9.5   Explain how the viewpoint established before a public sector analysis is started can turn an estimate from being categorized as a disbenefit to a cost, or vice versa.

9.6   Identify the viewpoint (e.g., budget, government unit, citizen, business owner), and categorize the following cash flows as a benefit, disbenefit, or cost:
(a) $600,000 annual income to area businesses from tourism created by new freshwater reservoir/recreation area
(b) $450,000 per year for repainting of bridge across the Mississippi River
(c) $800,000 per year maintenance by container-ship port authority
(d)   Loss of $1.6 million in salaries for border residents because of strict enforcement of immigration laws
(e)   Reduction of $600,000 per year in car repairs because of improved roadways
(f)   Expenditure of $350,000 for guardrail replacement on freeway
(g) $1.8 million loss of revenue by farmers because of highway right-of-way purchases

9.7   Identify the following as primarily private or public sector characteristics:
(a)   Large investment
(b)   No profits
(c)   Funding from fees
(d)   MARR-based selection criteria
(e)   Low interest rate
(f)   Short project life estimate
(g)   Disbenefits

9.8   List two advantages of a public-private partnership (PPP).

Project B/C Value

9.9   Where is the salvage value placed in a conventional B/C ratio? Why? How is it handled mathematically?

9.10 The estimated annual cash flows for a proposed municipal government project are costs of $750,000 per year, benefits of $900,000 per year, and disbenefits of $225,000 per year. Calculate the conventional B/C ratio at an interest rate of 6% per year and determine if it is economically justified.

9.11 Officials from the City of Galveston and State of Texas gathered to celebrate the start of a beach restoration project that involves dumping sand and adding antierosion structures. The first cost of the project is $30 million with annual maintenance estimated at $340,000. If the restored/expanded beaches attract visitors who will spend $6.2 million per year, what is the conventional B/C ratio at the social discount rate of 8% per year? Assume the State wants to recover the investment in 20 years.

9.12 The National Environmental Protection Agency has established that 2.5% of the median household income is a reasonable amount to pay for safe drinking water. For a median household income of $45,000 per year, what would the health benefits have to be (in dollars per household per year) for the B/C ratio to be equal to 1.5?

9.13 As part of the rehabilitation of the downtown area of a southern U.S. city, the Parks and Recreation Department expects to develop the space below several overpasses into basketball, handball, miniature golf, and tennis courts. The estimates are: initial cost of $190,000, life of 20 years, and annual M&O costs of $21,000. The department expects 20,000 people per year to use the facilities an average of 2 hours each. The value of the recreation has been conservatively set at $1.00 per hour. At a discount rate of 6% per year, what is the B/C ratio for the project?

9.14 The B/C ratio for a flood control project along the Swanee River was calculated to be 1.3. If the benefits were $500,000 per year and the maintenance costs were $200,000 per year, determine the initial cost of the project at an interest rate of 7% per year and a 50-year life.

9.15 The State Legislative Budget Board approved adding 4000 surveillance cameras along the 800-mile Texas-Mexico border from El Paso to Brownsville at a cost of $300 per camera. In addition to the cost of the cameras, manpower and other resources are expected to cost $3.2 million per year. The benefits associated with interception of people and drugs are estimated to be $5.1 million per year. What is the conventional B/C ratio at an interest rate of 6% per year over a 10-year project period?

9.16 The State Highway Department is considering a bypass loop that is expected to save motorists $820,000 per year in gasoline and other automobile-related expenses. However, local businesses will experience revenue losses estimated to be $135,000 each year. The cost of the loop will be $9,000,000. (*a*) Calculate the conventional B/C ratio using an interest rate of 6% per year and a 20-year project period. (*b*) Calculate the conventional B/C ratio without considering the disbenefits. Is the project economically justified with and without considering the revenue losses? (*c*) Develop the single-cell spreadsheet functions that will answer the two questions above.

9.17 A rural, agriculture-based city that has 17,000 households is required to install treatment systems for the removal of arsenic and other harmful chemicals from its drinking water. The annual cost is projected to be $150 per household per year. Assume that one life will be saved every 3 years as a result of the removal systems. (*a*) What is the B/C ratio, if a human life is valued at $4.8 million? Use an interest rate of 8% per year and assume the life is saved at the end of each 3-year period. (*b*) What justifies the project?

9.18 An Army Corps of Engineers project for improving navigation on the Ohio River will have an initial cost of $6,500,000 and annual maintenance of $130,000. Benefits for barges and paddle wheel touring boats are estimated at $820,000 per year. The project is assumed to be permanent and the discount rate is 8% per year. Determine if the Corps should proceed with the project.

9.19 The following estimates (in $1000 units) have been developed for a security system upgrade at Chicago's O'Hare Airport. (*a*) Calculate the conventional B/C ratio at a discount rate of 10% per year. Is the project justified? (*b*) Determine the minimum first cost that is possible to render the project just economically *unjustified*.

Item	Cash Flow
First cost, $	13,000
AW of benefits, $ per year	3,800
FW of disbenefits, year 20, $	6,750
M&O costs, $ per year	400
Life, years	20

9.20 A consultant, after 3 months of work, reported that the modified B/C ratio for a city-owned hospital heliport project is 1.7. If the initial cost is $1 million and the annual benefits are $150,000, what is the amount of the annual M&O costs used in the calculation? The report stated that a discount rate of 6% per year and an estimated life of 30 years were used.

9.21 The B/C ratio for a mosquito control program proposed by the Harris County Department of Health is reported to be 2.1. The person who prepared the report stated that the annual health benefits were estimated to be $400,000, and that disbenefits of $25,000 per year were used in the calculation. He also stated that the costs for chemicals, machinery, maintenance, and labor were estimated at $150,000 per year, but he forgot to list the cost for initiating the program (trucks, pumps, tanks, etc.). If the initial cost was amortized over a 10-year period at 8% per year, what is the estimated initial cost?

9.22　Calculate the B/C ratio for the following cash flow estimates at a discount rate of 10% per year. Is the project justified?

Item	Estimate
PW of benefits, $	3,800,000
AW of disbenefits, $/year	45,000
First cost, $	1,200,000
M&O costs, $/year	300,000
Life, years	20

9.23　From the following data for a PPP project, calculate the (a) conventional, and (b) modified benefit/cost ratios using an interest rate of 6% per year and an infinite project period.

	To the People		To the Government
Benefits:	$100,000 per year beginning now	Costs:	$1.8 million now and $200,000 every 3 years
Disbenefits:	$60,000 per year	Savings:	$90,000 per year

9.24　With an installed capacity in excess of 11,000 MW, the Belo Monte dam complex on the Xingu River in Brazil will be the world's third largest hydroelectric dam. The project is expected to cost $16 billion and will begin producing electricity upon completion by the end of 2019 after the 5-year construction period. Even though the dam will provide clean energy for millions of people, environmentalists are sharply opposed. They say it will devastate wildlife and the livelihoods of 40,000 people who live in the area to be flooded.

　　Assume that the funding will occur evenly over the 5-year construction period (i.e., $3.2 billion per year), and that the disbenefits will be $66,000 for each displaced person and $1 billion for wildlife destruction. Further, assume the disbenefits will occur evenly through the 5-year construction period and the benefits will begin in 2020 and continue indefinitely. At a discount rate of 8% per year, use the B/C ratio to determine the minimum annual benefits required to economically justify the project.

9.25　In United States airspace, the average number of commercial aircraft in the sky on an average morning is 4000. There are another 16,000 planes on the ground. Aerospace company Rockwell-Collins developed what it calls a "digital parachute," a panic-button technology that will land any plane in a pinch at the closest airport, no matter what the weather or geography, and without the help of a pilot. The technology could be applied where a pilot is no longer capable of flying the plane, or is panicked and confused about what to do in an emergency. Assume the cost of retrofitting 20,000 commercial airplanes is $55,000 each and the plane stays in service for 15 years. The technology could save an estimated average of 40 lives per year; the value of a human life is estimated at $4,000,000. Use an interest rate of 8% per year to determine the B/C ratio.

9.26　When red light cameras are installed at high-risk intersections, rear-end collisions go up, but all other types of accidents go down, including those involving pedestrians. Analysis of traffic accidents in a northwestern city revealed that the total number of collisions at a group of selected photo-enhanced intersections decreased from 33 per month to 18. At the same time, the number of traffic tickets issued for red light violations averaged 1100 per month, at a cost to violators of $85 per citation. The cost for the contractor to install the basic camera system at the selected intersections was $750,000. For analysis purposes, use an interest rate of 0.5% per month and a 3-year study period. The cost of a collision is estimated at $41,000 and traffic ticket costs are considered as disbenefits.
(a)　Calculate the B/C ratio for the camera system from the perspective of the city.
(b)　The contractor has proposed to operate and maintain the system, plus provide collected violation data to the police department each 24-hour period for a fee of 50% of the estimated cost of a collision. Calculate the profitability index (PI) from the perspective of the contractor.

9.27　Dickinson, a large oil and gas drilling and operating corporation, has invested over the past 6 years in the installation and operation of a FOUNDATION Fieldbus H1 (FF H1) system developed by Pepperl+Fuchs of Germany. A project engineer has collected information on annual net cash flow increases (ΔNCF) generated by the FF H1 system and the annual investments made by Dickinson in the system. At an interest rate of 10% per year, determine the PI of this endeavor. Has it proven to be economically worthwhile? (Note: Problem 9.54 explores the fieldbus implementation further using a spreadsheet.)

Year	0	1	2	3	4	5	6
ΔNCF, $10,000 per year	0	5	7	9	11	13	20
Investment, $10,000	15	8	10	0	0	5	10

Two-Alternative Comparison

9.28 Two alternatives, identified as X and Y, are evaluated using the B/C method. Alternative Y has a higher total cost than X. If the B/C ratios are 1.2 and 1.0 for alternatives X and Y, respectively, which alternative should be selected? Why?

9.29 One of two alternatives will be selected to reduce flood damage in a rural community in central Arizona. The estimates associated with each alternative are available. Use B/C analysis at a discount rate of 8% per year over a 20-year study period to determine which alternative should be selected. For analysis purposes only, assume the flood damage would be prevented in years 3, 9, and 18 of the study period.

	Retention Pond	Channel
Initial cost, $	880,000	2,900,000
Annual maintenance, $/year	92,000	30,000
Reduced flood damage, $	200,000	600,000

9.30 The city engineer and economic development directors are evaluating two sites for construction of a multipurpose sports arena. The sites are downtown (DT) and southwest (SW) of the metropolitan area. The city already owns enough land at the DT site; however, the land for a parking garage will cost $1 million, and construction costs will be $10 million for the parking garage, infrastructure relocation, and drainage. The SW site is 30 km from downtown, but the land will be donated by a developer who knows that an arena at this site will dramatically increase the value of the remainder of his land holdings.

Because of its centralized location, there will be greater attendance at most of the events held at the DT site. This will result in more revenue to vendors and local merchants in the amount of $700,000 per year. Additionally, the average attendee will not have to travel as far, resulting in annual benefits of $400,000 per year. All other costs and revenues are expected to be the same at either site. If the city uses a discount rate of 6% per year, and will construct at one site or the other, which site should be selected? Use a 30-year study period.

9.31 Two relatively inexpensive alternatives are available for reducing potential earthquake damage at a top secret government research site. The cash flow estimates for each alternative are given below. At an interest rate of 8% per year, use the B/C ratio method to determine which one should be selected.

Use a 20-year study period, and assume the damage costs would occur in the middle of the study period, that is, in year 10.

	Alternative 1	Alternative 2
Initial cost, $	600,000	1,100,000
Annual maintenance, $/year	50,000	70,000
Potential damage costs, $	950,000	250,000

9.32 A state agency is considering two mutually exclusive alternatives for upgrading the skills of its technical staff. Alternative 1 involves purchasing software that will reduce the time required to collect background information on each client. The total cost for the purchase, installation, and training associated with the new software is $840,900. The present worth of the benefits from increased efficiency is expected to be $1,020,000.

Alternative 2 involves developing multimedia training to improve the performance of the staff technicians. The total cost to develop, install, and train the technicians will be $1,780,000. The present worth of the benefits from increased performance due to training is expected to be $1,850,000. Use a B/C analysis to determine which alternative, if any, the agency should undertake.

9.33 There are two potential locations to construct an urgent care walk-in clinic to serve rural residents. Use B/C analysis to determine which location, if any, is better at an interest rate of 8% per year. (Note: See Problem 9.55 for more on this analysis.)

Location	1	2
Initial cost, $	1,200,000	2,000,000
Annual M&O cost, $/year	80,000	75,000
Annual benefits, $/year	520,000	580,000
Annual disbenefits, $/year	90,000	140,000
Site suitability, years	10	20

9.34 Locations under consideration for a border patrol station have their costs estimated by the federal government. Use the B/C ratio method at an interest rate of 6% per year to determine which location to select, if any.

Location	North, N	South, S
Initial cost, $	1.1×10^6	2.9×10^6
Annual cost, $ per year	480,000	390,000
Disbenefits, $ per year	70,000	40,000
Life, years	∞	∞

9.35 Solar and conventional power alternatives are available to provide energy for monitoring equipment at a remote irrigation canal site. The estimates associated with each alternative have been developed. Use the B/C method to select one at an interest rate of 7% per year over a 5-year study period. One of the alternatives must be selected.

Alternative	Conventional	Solar
Initial cost, $	200,000	1,300,000
Annual power cost, $/year	80,000	9,000
Salvage value, $	—	150,000

9.36 A public utility in a medium-size city is considering two cash rebate programs to achieve water conservation. Program 1, which is expected to cost an average of $60 per household, provides a rebate of 75% of the purchase and installation costs of an ultra-low-flush toilet ($100 max). This program is projected to achieve a 5% reduction in overall household water use over a 5-year evaluation period. This will benefit the citizenry to the extent of $1.25 per household per month. Program 2 involves turf replacement with xeric (low water need) landscaping. The program is expected to cost $500 per household, but it will result in reduced water cost estimated at $8 per household per month (on the average). Use the B/C method at a discount rate of 0.5% per month to determine which program, if any, the utility should undertake. The programs may be (a) mutually exclusive, or (b) independent.

9.37 A public-private partnership has been formed between a city, county, and construction/management company to attract a professional athletic team to the area. Assume you are the engineer with the company and are assisting with the benefit/cost analysis. The primary options are to construct a domed arena (DA) or a conventional stadium (CS), one of which will definitely be built. The DA option will cost $200 million to construct, have a useful life of 50 years, and require M&O costs of $360,000 the first year, increasing by $10,000 per year thereafter. In 25 years, a remodeling expenditure of $4,800,000 is predicted.

The CS option will cost only $50 million to construct, have a useful life of 50 years, and require M&O costs of $175,000 the first year, increasing by $8000 per year. Periodic costs for repainting and resurfacing for the stadium are estimated at $100,000 every 10 years, except in year 50.

Revenue from DA is expected to be greater than that from CS by $10,900,000 the first year, with amounts increasing by $200,000 per year through year 15. Thereafter, the extra revenue from the dome is expected to remain constant at $13.7 million per year. Assuming that both structures will have a salvage value of $5 million, use an interest rate of 8% per year and a B/C analysis to determine which structure should be built. Solve by hand or spreadsheet, as instructed.

Multiple (More than Two) Alternatives

9.38 Three alternatives identified as X, Y, and Z were evaluated by the B/C method. The analyst calculated project B/C values of 0.92, 1.34, and 1.29. The alternatives are listed in the order of increasing equivalent total costs. The analyst is not sure whether an incremental analysis is needed.
    (a) What do you think? If no incremental analysis is needed, why not; if so, which alternatives must be compared incrementally?
    (b) For what type of projects is incremental analysis never necessary? If X, Y, and Z are all this type of project, which alternatives should be selected based on the calculated B/C values?

9.39 The Department of Defense is considering three sites in the National Wildlife Preserve for extraction of rare metals. The cash flows associated with each site are summarized. The extraction period is limited to 5 years and the interest rate is 10% per year. Use the B/C method to determine which site, if any, is acceptable. The monetary unit is $ million. (Note: Problem 9.56 includes further analysis for this situation.)

Site	A	B	C
Initial cost, $	50	90	200
Annual cost, $/year	3	4	6
Annual benefits, $/year	20	29	61
Annual disbenefits, $/year	0.5	2	2.1

9.40 Four mutually exclusive, direct benefit alternatives are compared using the B/C method. Which alternative, if any, should be selected?

Alternative	Initial Investment, $ Million	B/C Ratio	ΔB/C Ratio When Compared with Alternative			
			J	K	L	M
J	20	1.10	—			
K	25	0.96	0.40	—		
L	33	1.22	1.42	2.14	—	
M	45	0.89	0.72	0.80	0.08	—

9.41 The city of Valley View, California, is considering various proposals regarding the disposal of used tires. All proposals involve shredding, but the benefits differ in each plan. An incremental B/C analysis was initiated, but the engineer conducting the study left recently. Using a 20-year study period and an interest rate of 8% per year, (a) fill in the blanks in the incremental B/C columns of the table. (b) Which alternative should be selected?

Alternative	PW of Cost, $ Million	B/C Ratio	ΔB/C Ratio When Compared with Alternative			
			P	Q	R	S
P	10	1.1	—	2.83		
Q	40	2.4	2.83	—		
R	50	1.4			—	
S	80	1.8				—

9.42 From the information shown for possible location of campgrounds and lodging at a national park, determine which project, if any, should be selected from the six mutually exclusive projects. Selected incremental B/C ratios are included. If the proper comparisons have not been made, state which one(s) are needed.

	Location Identifier					
	G	H	I	J	K	L
Total cost, $	20,000	45,000	50,000	35,000	85,000	70,000
B/C ratio	1.15	0.89	1.10	1.11	0.94	1.06

Comparison	ΔB/C
G vs. H	0.68
G vs. I	0.73
H vs. J	0.10
I vs. J	1.07
J vs. G	1.07
H vs. K	1.00
H vs. L	1.36
J vs. K	0.82
J vs. L	1.00
K vs. L	0.40
G vs. L	1.02

9.43 You have been requested to compare four mutually exclusive alternatives using the B/C method and the results shown. Which alternative, if any, do you recommend?

Alternative	Equivalent Total Cost, $ Million	Direct B/C Ratio	ΔB/C Ratio When Compared with Alternative			
			X	Y	Z	Q
X	20	0.75	—			
Y	30	1.07	1.70	—		
Z	50	1.20	1.50	1.40	—	
Q	90	1.11	1.21	1.13	1.00	—

9.44 In order to safeguard the public health, environment, public beaches, water quality, and economy of south San Diego County, California, and Tijuana, Mexico, federal agencies in the United States and Mexico developed four alternatives for treating wastewater prior to discharge into the ocean. The project will minimize untreated wastewater flows that have caused chronic and substantial pollution in the Tijuana River Valley, the Tijuana River National Estuarine Research Reserve, coastal areas used for agriculture and public recreation, and areas designated as critical habitat for federal- and state-listed endangered species. For the costs and benefits estimated, which alternative should be selected on the basis of a B/C analysis at 6% per year and a 40-year project period? All monetary amounts are in $ million units.

	Pond System	Expand Plant	Advanced Primary	Partial Secondary
Capital cost, $	58	76	2	48
M&O cost, $/year	5.5	5.3	2.1	4.4
Benefits, $/year	11.1	12.0	2.7	8.3

9.45 Several alternatives are under consideration to enhance security at a county jail. Since the alternatives serve different areas of the facility, all that are economically attractive will be implemented. Determine which one(s) should be selected, based on a B/C analysis using an interest rate of 7% per year and a 10-year study period.

	Extra Cameras (EC)	New Sensors (NS)	Steel Tubing (ST)	Access Control (AC)
First cost, $	38,000	87,000	99,000	61,000
M&O, $/year	49,000	64,000	42,000	38,000
Benefits, $/year	110,000	160,000	74,000	52,000
Disbenefits, $/year	26,000	21,000	32,000	14,000

Cost-Effectiveness Analysis

9.46 Various techniques have been proposed to curb cross-border drug smuggling into a country. The costs of implementing each strategy along a particularly rugged section of the border are indicated below. The table also includes a score that is compiled based on deterrence, interdiction, and apprehension, with a higher score indicating better performance. For a budget of $60 million, determine which techniques should be employed on the basis of a cost-effectiveness analysis.

Activity	Cost, $ Millions	Score
Tethered aerostats	3.8	8
Boots-on-the-ground	31.4	52
Fence	18.7	12
Motion sensors	9.8	7
Seismic sensors	8.3	5
Drones	12.1	26

9.47 Productivity improvement is a primary goal for the new owners of a plastic pipe extrusion plant.

The CEO decided to undertake, on a trial basis, a series of actions directed toward improving employee morale as a first step toward productivity-increasing programs. Six separate strategies were implemented, such as increased employee autonomy, flexible work schedules, improved training, company picnics, and better work environment. Periodically, the company surveyed the employees to measure the change in morale. The measure of effectiveness is the difference between the number of employees who rate their job satisfaction as "very high" and those who rate it "very low." The per-employee cost of each strategy (identified as A through F) and the resultant measurement score are shown below for the six programs.

The CEO has set aside $60 per employee to spend on the permanent implementation of as many of the strategies as are justified from both the effectiveness and economical viewpoints. Determine which strategies are the best to implement. Use hand or spreadsheet solution, as instructed.

Strategy	Cost/Employee, $	Measurement Score
A	5.20	50
B	23.40	182
C	3.75	40
D	10.80	75
E	8.65	53
F	15.10	96

9.48 There are a number of techniques to help people stop smoking, but their cost and effectiveness vary widely. One accepted measure of effectiveness of a program is "percentage of enrollees quitting." The table in this problem shows several techniques touted as effective stop-smoking methods, some historical data on the approximate cost of each program per person, and the percentage of people smoke-free 3 months after the program ended.

The Cancer Society provides annual cost-offset funding to cancer patients so more people can afford these programs. A large clinic in St. Louis has the capacity to treat each year the number of people shown. If the clinic plans to submit a proposal to the Cancer Society to treat a specified number of people annually, estimate the amount of money the clinic should ask for in its proposal to do the following:

(a) Conduct programs at the capacity level for the technique with the lowest cost-effectiveness ratio.

(b) Offer programs using as many techniques as possible to treat up to 1100 people per year using the most cost-effective techniques.

Technique	Cost, $/Enrollee	% Quitting	Treatment Capacity per Year
Acupuncture	700	9	250
Subliminal message	150	1	500
Aversion therapy	1700	10	200
Out-patient clinic	2500	39	400
In-patient clinic	2800	41	550
Nicotine replacement therapy (NRT)	1300	20	100

9.49 The annual cost and an effectiveness measure of items salvaged per year for four mutually exclusive, service sector alternatives have been collected. Calculate (a) the cost effectiveness ratio for each alternative, and (b) use the CER to identify the best alternative.

Alternative	Cost C, $/Year	Salvaged Items/Year, E
W	355	20
X	208	17
Y	660	41
Z	102	7

9.50 An engineering student has only 45 minutes before the final exam in her Engineering Economy class. She needs help in understanding cost-effectiveness analysis because she knows from the instructor's review session that a CEA problem will be on the exam. There is time for using only one method of assistance before the exam; she must select well. In a rapid process of estimation, she determines how many minutes it would take for each method of assistance and how many points it might gain for her on the final. The method and estimates follow. Where should she seek help to be most effective?

Assistance from	Minutes for Assistance	Points Gained
Teaching assistant (TA)	15	20
Slides on web	20	10
Friend in class	10	5
Instructor	20	15

Public Sector Ethics

9.51 In most developed countries, laws that prohibit texting-while-driving (TWD) have been enacted. Hambara and Associates, a systems engineering consulting company, has won the contract to assist the Callaghan Police Department in implementing projects that will have a zero-tolerance policy for TWD, with the expressed goal to not increase the average number of violation tickets issued per month. This is a highly publicized goal because the Police Chief is keenly aware of the public's negativity toward police officers as they issue tickets for any violation, especially for TWD. Assume that you work for Hambara and have been identified as the lead project engineer on the Callaghan contract. Immediately, you recognize this as a public planning effort, since Hambara will be responsible for developing and implementing projects. You will be meeting the other four members of the Hambara team and two assigned liaison officers from the Callaghan Police Department tomorrow morning. They expect you to present a general overview of your approach to the project.

Because of the sensitivity of the public, the Chief's announced goal of no increase in ticketing, and your own concerns about possible ethical dimensions within the police department and your own Hambara team members, you need some structure to understand any ethical considerations that may arise in the future. Use the format of Table 9–5, especially the dimensions suggested in the leftmost column, to identify *for yourself* the ethical considerations of this project. This material will likely not be shared directly at the initial meeting of the group.

9.52 During the design and development stages of a remote meter reading system for residential electricity use, the two engineers working on the project for the City of Forest Ridge noted something different than they expected. The first, an electrical/software engineer, noted that their city liaison staff member provided all the information on the software options, but only one option, the one from Lorier Software, was ever discussed and detailed. The second designer, an industrial systems engineer, further noted that all the hardware specifications provided to them by this same liaison all came from the same distributor, namely, Delsey Enterprises. Coincidently, at a weekend family picnic for city employees, to which the engineers had been invited, they met a couple named Don Delsey and Susan Lorier. Upon review, they learned that the man Don is the son-in-law of the city liaison and Susan is his step-daughter. Based on these observations, before they proceed further with development of the system, what should the two engineers do, if anything, about their suspicions that the city liaison person is trying to bias the design to favor the use of his relative's software and hardware businesses?

9.53 Since transportation via automobile was introduced, drivers throughout the country of Yalturia in Eastern Europe have driven on the left side of the road. Recently, the Yalturian National Congress passed a law that within 3 years, a right-hand driving convention will be adopted and implemented throughout the country. This is a major policy change for the country and will require significant public planning and project development to implement successfully and safely. Assume you are the lead engineering consultant from Halcrow Engineers, responsible for developing and describing many of these major projects. Identify six of the projects you deem necessary. State a name and provide a one or two sentence description of each project.

EXERCISES FOR SPREADSHEETS

9.54 For the FOUNDATION Fieldbus H1 installation analyzed in Problem 9.27, do the following: (Data is repeated below.)
 (a) Find the PI for the first 6 years of operation.
 (b) Find the required ΔNCF for next year if Dickinson plans to invest an additional $150,000 in year 7 and wants to realize an increase in profitability such that PI = 1.20.

Year	0	1	2	3	4	5	6
ΔNCF, $10,000 per year	0	5	7	9	11	13	20
Investment, $10,000	15	8	10	0	0	5	10

9.55 There are two potential locations to construct an urgent care walk-in clinic to serve rural residents. The DN alternative is an option since the clinic is not a requirement; it would be a convenience for future patients. When incremental B/C analysis was applied at a discount rate of 8% per year, using the estimates summarized below, the results were:

Rank by increasing total cost: DN, 1, 2
Location 1 vs. DN: $B/C = 1.66$ Conclusion: eliminate DN
Location 2 vs.1: $\Delta B/C = 0.50$ Conclusion: select location 1

Location	1	2
Initial cost, $	1,200,000	2,000,000
Annual M&O cost, $/year	80,000	75,000
Annual benefits, $/year	520,000	580,000
Annual disbenefits, $/year	90,000	140,000
Site suitability, years	10	20

Dr. Thompson, the designated lead physician of the clinic, wanted location 2 to be selected. He has now challenged the estimates, especially the initial cost and disbenefits estimated for location 2. Besides, he calculated the B/C ratio for location 2 itself and obtained a value of 1.58, which he contends is very close to B/C = 1.66 reported for location 1, thus making the selection of location 1 a marginal decision. Answer the following questions for the group of people who must decide if and where to construct the clinic:

(a) Are the B/C ratios for the two locations correct as reported?
(b) Can we select location 2 and still be economically justified?
(c) How much less does location 2 have to cost initially to make it the better location?
(d) If the initial cost of location 2 can't be changed, is it possible to increase the benefits to select location 2?
(e) Is the result the same if the disbenefits are completely neglected?

9.56 Use the estimates from Problem 9.39 (repeated below) to answer the questions asked concerning the extraction of rare metals in a National Wildlife Preserve. Utilize the B/C method, a 5-year study period, and a discount rate of 10% per year. The monetary unit is $ million.

Site	A	B	C
Initial cost, $	50	90	200
Annual cost, $/year	3	4	6
Annual benefits, $/year	20	29	61
Annual disbenefits, $/year	0.5	2	2.1

(a) Which site is the best economically?
(b) What is the minimum benefits estimate necessary to make site B the best of the three options? Assume that $\Delta B/C$ must be at least 1.01 and that all other estimates remain as quoted.
(c) For the original estimates, if site A is eliminated for environmental reasons, is site B or C acceptable?

9.57 Consider the four independent projects outlined below. At 5% per year and a 15-year study period, (a) determine which projects are acceptable, and (b) the value of the annual M&O cost to make any nonacceptable project acceptable with a B/C ratio of at least 1.0. (Note: Utilize a logical IF function somewhere in your analysis.)

Project	M	N	O	P
First cost, $	38,000	87,000	99,000	61,000
M&O, $/year	49,000	64,000	42,000	38,000
Benefits, $/year	110,000	160,000	74,000	52,000
Disbenefits, $/year	26,000	21,000	32,000	14,000

ADDITIONAL PROBLEMS AND FE EXAM REVIEW QUESTIONS

9.58 When comparing different-life alternatives by the B/C method, the alternatives should be compared over:
(a) The life of the shorter-lived alternative
(b) The life of the longer-lived alternative
(c) The LCM of their lives
(d) An infinite time period

9.59 When a B/C analysis is conducted, the benefits and costs:
(a) Must be expressed in terms of their present worth values
(b) Must be expressed in terms of their annual worth values
(c) Must be expressed in terms of their future worth values
(d) Can be expressed in terms of PW, AW, or FW

9.60 In a conventional B/C ratio:
(a) Disbenefits and M&O costs are subtracted from benefits
(b) Disbenefits are subtracted from benefits, and M&O costs are subtracted from costs

(c) Disbenefits are subtracted from benefits, and M&O costs are added to costs
(d) Disbenefits are added to costs, and M&O costs are subtracted from benefits

9.61 All of the following are primarily associated with public sector projects, except:
(a) Low interest rates
(b) Stocks
(c) Benefits
(d) Infinite life

9.62 All of the following cash flows can be identified as benefits, except:
(a) Longer tire life because of smooth pavement
(b) $200,000 annual income to local businesses because of tourism created by water reservoir
(c) Expenditure of $20 million for construction of a highway
(d) Fewer highway accidents because of improved lighting

9.63 An alternative has the following cash flows: benefits = $50,000 per year; disbenefits = $27,000 per year; costs = $25,000 per year. The B/C ratio is closest to:

(a) 0.92   (b) 0.96
(c) 1.04   (d) 2.00

9.64 In evaluating three mutually exclusive alternatives by the B/C method, the alternatives were ranked A, B, and C, respectively, in terms of increasing cost, and the following results were obtained for overall B/C ratios: 1.1, 0.9, and 1.3. On the basis of these results, it is correct to:

(a) Select only alternative A
(b) Select only alternative C
(c) Compare A and C incrementally
(d) Select alternatives A and C

9.65 All of the following are examples of unethical behavior, except:

(a) Deception
(b) Withholding information
(c) Lying
(d) Offering engineering services at a lower cost than a competitor

9.66 If two mutually exclusive alternatives have B/C ratios of 1.4 and 1.5 for the lower and higher cost alternatives, respectively:

(a) The B/C ratio on the increment between them is equal to 1.5
(b) The B/C ratio on the increment between them is between 1.4 and 1.5
(c) The B/C ratio on the increment between them is less than 1.5
(d) The higher cost alternative is definitely better economically

9.67 From the PW, AW, and FW values shown, the conventional B/C ratio is closest to:

(a) 1.27   (b) 1.33
(c) 1.54   (d) 2.76

	PW, $	AW, $/Year	FW, $
First cost	100,000	16,275	259,370
M&O cost	68,798	11,197	178,441
Benefits	245,784	40,000	637,496
Disbenefits	30,723	5,000	79,687

9.68 If two mutually exclusive alternatives have B/C ratios of 1.5 and 1.4 for the lower first-cost and higher first-cost alternatives, respectively:

(a) The B/C ratio on the increment between them is greater than 1.4
(b) The B/C ratio on the increment between them is between 1.4 and 1.5
(c) The B/C ratio on the increment between them is less than 1.4
(d) The lower-cost alternative is the better one

9.69 All of the following cash flows should be classified as disbenefits, except:

(a) Cost of fish from hatchery to stock lake at state park
(b) Decrease in property values due to closure of a government research lab
(c) School overcrowding because of military base expansion
(d) Revenue loss to local motels because of shortened national park season

9.70 If benefits are $10,000 per year forever starting in year 1, and costs are $50,000 at time zero and $50,000 at the end of year 2, the B/C ratio at $i = 10\%$ per year is closest to:

(a) 0.93   (b) 1.10
(c) 1.24   (d) 1.73

9.71 The first cost of grading and spreading gravel on a short rural road is expected to be $700,000. The road will have to be maintained at a cost of $25,000 per year. Even though the new road is not very smooth, it allows access to an area that previously could only be reached with off-road vehicles. This improved accessibility has increased the property values along the road from $400,000 to $600,000. The conventional B/C ratio at an interest rate of 6% per year for a 10-year study period is closest to:

(a) 1.3   (b) 1.7
(c) 2.1   (d) 2.8

9.72 A permanent flood-control dam is expected to have an initial cost of $2.8 million and an annual M&O cost of $20,000. In addition, minor reconstruction will be required every 5 years at a cost of $200,000. As a result of the dam, flood damage will be reduced by an average of $310,000 per year. Using an interest rate of 6% per year, the conventional B/C ratio will be closest to:

(a) 0.46   (b) 1.02
(c) 1.40   (d) 1.96

9.73 For the two *independent* projects shown, determine which, if any, should be funded at $i = 10\%$ per year using the B/C ratio method:

Alternative	X	Y
Annualized first cost, $/year	60,000	90,000
Annual M&O cost, $/year	45,000	35,000
Annual benefits, $/year	110,000	150,000
Annual disbenefits, $/year	20,000	45,000
Life, years	∞	∞

(a) Fund neither   (b) Fund X
(c) Fund Y   (d) Fund both X and Y

9.74 The estimated first cost of a permanent national monument is $2 million with annual benefits and disbenefits estimated at $360,000 and $42,000, respectively. The B/C ratio at 6% per year is closest to:
  (a) 0.16   (b) 0.88
  (c) 1.73   (d) 2.65

9.75 Cost-effectiveness analysis (CEA) differs from benefit/cost analysis (B/C) in that:
  (a) CEA cannot handle multiple alternatives
  (b) CEA compares alternatives on the basis of a specific outcome rather than solely on monetary units
  (c) CEA cannot handle independent alternatives
  (d) CEA is more time consuming and resource intensive

9.76 The statements contained in a code of ethics are variously known as all of the following, except:
  (a) Canons   (b) Norms
  (c) Standards   (d) Laws

9.77 Of the following words, the one *not* related to ethics is:
  (a) Virtuous   (b) Honest
  (c) Lucrative   (d) Proper

CASE STUDY

HIGHWAY LIGHTING OPTIONS TO REDUCE TRAFFIC ACCIDENTS

Background

This case study compares benefit/cost analysis and cost-effectiveness analysis on the same information about highway lighting and its role in accident reduction.

Poor highway lighting may be one reason that proportionately more traffic accidents occur at night. Traffic accidents are categorized into six types by severity and value. For example, an accident with a *fatality* is valued at approximately $4 million, while an accident in which there is *property damage* (to the car and contents) is valued at $6000. One method by which the impact of lighting is measured compares day and night accident rates for lighted and unlighted highway sections with similar characteristics. Observed reductions in accidents seemingly caused by too low lighting can be translated into either monetary estimates of the benefits B of lighting or used as the effectiveness measure E of lighting.

Information

Freeway accident data were collected in a 5-year study. The property damage category is commonly the largest based on the accident rate. The number of accidents recorded on a section of highway is presented here.

| Accident Type | Number of Accidents Recorded[1] | | | |
| | Unlighted | | Lighted | |
	Day	Night	Day	Night
Property damage	379	199	2069	839

[1]Portion of data reported in Michael Griffin, "Comparison of the Safety of Lighting on Urban Freeways," *Public Roads,* vol. 58, pp. 8–15, 1994.

The ratios of night-to-day accidents involving property damage for the unlighted and lighted freeway sections are $199/379 = 0.525$ and $839/2069 = 0.406$, respectively. These results indicate that the lighting was beneficial. To quantify the benefit, the accident rate ratio from the unlighted section will be applied to the lighted section. This will yield the number of accidents that were prevented. Thus, there would have been $(2069)(0.525) = 1086$ accidents instead of 839 if there had not been lights on the freeway. This is a difference of 247 accidents. At a cost of $6000 per accident, this results in a net annual benefit of

$$B = (247)(\$6000) = \$1,482,000$$

For an effectiveness measure of number of accidents prevented, this results in $E = 247$.

To determine the cost of the lighting, it will be assumed that the light poles are center poles 67 meters apart with two bulbs each. The bulb size is 400 watts, and the installation cost is $3500 per pole. Since these data were collected over 87.8 kilometers of lighted freeway, the installed cost of the lighting is (with number of poles rounded off):

$$\text{Installation cost} = \$3500 \left(\frac{87.8}{0.067} \right)$$
$$= 3500(1310)$$
$$= \$4,585,000$$

There are a total of $87.8/0.067 = 1310$ poles, and electricity costs $0.10 per kWh. Therefore,

Annual power cost

$$= 1310 \text{ poles}(2 \text{ bulbs/pole})(0.4 \text{ kilowatt/bulb})$$
$$\times (12 \text{ hours/day})(365 \text{ days/year})$$
$$\times (\$0.10/\text{kilowatt-hour})$$
$$= \$459,024 \text{ per year}$$

The data were collected over a 5-year period. Therefore, the annualized cost C at $i = 6\%$ per year is

$$\text{Total annual cost} = \$4,585,000(A/P,6\%,5)$$
$$+ 459,024$$
$$= \$1,547,503$$

If a benefit/cost analysis is the basis for a decision on additional lighting, the B/C ratio is

$$B/C = \frac{1,482,000}{1,547,503} = 0.96$$

Since $B/C < 1.0$, the lighting is not justified. Consideration of other categories of accidents is necessary to obtain a better basis for decisions. If a cost-effectiveness analysis (CEA) is applied, due to a judgment that the monetary estimates for lighting's benefit is not accurate, the C/E ratio is

$$C/E = \frac{1,547,503}{247} = 6265$$

This can serve as a base ratio for comparison when an incremental CEA is performed for additional accident reduction proposals.

These preliminary B/C and C/E analyses prompted the development of four lighting options:

W)   Implement the plan as detailed above; light poles every 67 meters at a cost of $3500 per pole.

X)   Install poles at twice the distance apart (134 meters). This is estimated to cause the accident prevention benefit to decrease by 40%.

Y)   Install cheaper poles and surrounding safety guards, plus slightly lowered lumen bulbs (350 watts) at a cost of $2500 per pole; place the poles 67 meters apart. This is estimated to reduce the benefit by 25%.

Z)   Install cheaper equipment for $2500 per pole with 350-watt lightbulbs and place them 134 meters apart. This plan is estimated to reduce the accident prevention measure by 50% from 247 to 124.

Case Study Exercises

Determine if a definitive decision on lighting can be determined by doing the following:

1. Use a *benefit/cost analysis* to compare the four alternatives to determine if any are economically justified.

2. Use a cost-effectiveness analysis to compare the four alternatives.

From an understanding viewpoint, consider the following:

3. How many property-damage accidents could be prevented on the unlighted portion if it were lighted?

4. What would the lighted, night-to-day accident ratio have to be to make alternative Z economically justified by the B/C ratio?

5. Discuss the analysis approaches of B/C and C/E. Does one seem more appropriate in this type of situation than the other? Why? Can you think of other bases that might be better for decisions for public projects such as this one?

Selecting the Basic Analysis Tool

In the previous five chapters, several equivalent evaluation techniques have been discussed. Any method—PW, AW, FW, ROR, or B/C—can be used to select one alternative from two or more and obtain the same, correct answer. Only one method is needed to perform the engineering economy analysis because any method, correctly performed, will select the same alternative. Yet different information about an alternative is available with each different method. The selection of a method and its correct application can be confusing.

Table LS2–1 gives a recommended evaluation method for different situations, if it is not specified by the instructor in a course or by corporate practice in professional work. The primary criteria for selecting a method are speed and ease of performing the analysis. Interpretation of the entries in each column follows.

Evaluation period: Most private sector alternatives (revenue and cost) are compared over their equal or unequal estimated lives, or over a specific period of time. Public sector projects are commonly evaluated using the B/C ratio and usually have long lives that may be considered infinite for economic computation purposes.

Type of alternatives: Private sector alternatives have cash flow estimates that are revenue-based (includes income and cost estimates) or cost-based (cost estimates only). For cost alternatives, the revenue cash flow series is assumed to be equal for all alternatives. For public sector projects, the difference between costs and timing is used to select one alternative over another. *Service sector projects* for which benefits are estimated using a nonmonetary effectiveness measure are usually evaluated with a method such as cost-effectiveness analysis. This applies to all evaluation periods.

Recommended method: Whether an analysis is performed by hand, calculator, or spreadsheet, the method(s) recommended in Table LS2–1 will correctly select one alternative from two or more as

TABLE LS2–1	Recommended Method to Compare Mutually Exclusive Alternatives, Provided the Method Is Not Preselected		
Evaluation Period	**Type of Alternatives**	**Recommended Method**	**Series to Evaluate**
Equal lives of alternatives	Revenue or cost	AW or PW	Cash flows
	Public sector	B/C, based on AW or PW	Incremental cash flows
Unequal lives of alternatives	Revenue or cost	AW	Cash flows
	Public sector	B/C, based on AW	Incremental cash flows
Study period	Revenue or cost	AW or PW	Updated cash flows
	Public sector	B/C, based on AW or PW	Updated incremental cash flows
Long to infinite	Revenue or cost	AW or PW	Cash flows
	Public sector	B/C, based on AW	Incremental cash flows

rapidly as possible. Any other method can be applied subsequently to obtain additional information and, if needed, verification of the selection. For example, if lives are unequal and the rate of return is needed, it is best to first apply the AW method at the MARR and then determine the selected alternative's $i*$ using the same AW relation with i as the unknown.

Series to evaluate: The estimated cash flow series for one alternative and the incremental series between two alternatives are the only two options for present worth or annual worth evaluation. For spreadsheet analyses, this means that the NPV or PV function (for present worth) or the PMT function (for annual worth) is applied. The word *updated* is added as a reminder that a study period analysis requires that cash flow estimates (especially salvage/market values) be reexamined and updated before the analysis is performed.

Once the evaluation method is selected, a specific procedure must be followed. These procedures were the primary topics of the last five chapters. Table LS2–2 summarizes the important elements of the procedure for each method—PW, AW, ROR, and B/C. FW is included as an extension of PW. The meaning of the entries in Table LS2–2 follows.

TABLE LS2–2	Characteristics of an Economic Analysis of Mutually Exclusive Alternatives Once the Evaluation Method Is Determined					
Evaluation Method	**Equivalence Relation**	**Lives of Alternatives**	**Time Period for Analysis**	**Series to Evaluate**	**Rate of Return; Interest Rate**	**Decision Guideline: Select[1]**
Present worth	PW	Equal	Lives	Cash flows	MARR	Numerically largest PW
	PW	Unequal	LCM	Cash flows	MARR	Numerically largest PW
	PW	Study period	Study period	Updated cash flows	MARR	Numerically largest PW
	CC	Long to infinite	Infinity	Cash flows	MARR	Numerically largest CC
Future worth	FW	Same as present worth for equal lives, unequal lives, and study period				Numerically largest FW
Annual worth	AW	Equal or unequal	Lives	Cash flows	MARR	Numerically largest AW
	AW	Study period	Study period	Updated cash flows	MARR	Numerically largest AW
	AW	Long to infinite	Infinity	Cash flows	MARR	Numerically largest AW
Rate of return	PW or AW	Equal	Lives	Incremental cash flows	Find $\Delta i*$	Last $\Delta i* \geq$ MARR
	PW or AW	Unequal	LCM of pair	Incremental cash flows	Find $\Delta i*$	Last $\Delta i* \geq$ MARR
	AW	Unequal	Lives	Cash flows	Find $\Delta i*$	Last $\Delta i* \geq$ MARR
	PW or AW	Study period	Study period	Updated incremental cash flows	Find $\Delta i*$	Last $\Delta i* \geq$ MARR
Benefit/cost	PW	Equal or unequal	LCM of pairs	Incremental cash flows	Discount rate	Last $\Delta B/C \geq$ 1.0
	AW	Equal or unequal	Lives	Incremental cash flows	Discount rate	Last $\Delta B/C \geq$ 1.0
	AW or PW	Long to infinite	Infinity	Incremental cash flows	Discount rate	Last $\Delta B/C \geq$ 1.0

[1]Lowest equivalent cost or largest equivalent income.

Equivalence relation The basic equation written to perform any analysis is either a PW or an AW relation. The capitalized cost (CC) relation is a PW relation for infinite life, and the FW relation is likely determined from the PW equivalent value. Additionally, as we learned in Chapter 6, AW is simply PW times the A/P factor over the LCM or study period.

Lives of alternatives and time period for analysis The length of time for an evaluation (the n value) will always be one of the following: equal lives of the alternatives, LCM of unequal lives, specified study period, or infinity because the lives are very long.

- PW analysis always requires the LCM of compared alternatives.
- Incremental ROR and B/C methods require the LCM of the two alternatives being compared.
- The AW method allows analysis over the respective alternative lives.
- CC analysis has an infinite time line and uses the relation $P = A/i$.

The one exception is for the incremental ROR method for unequal-life alternatives using an AW relation for *incremental cash flows*. The LCM of the two alternatives compared must be used. This is equivalent to using an AW relation for the *actual cash flows* over the respective lives. Both approaches find the incremental rate of return Δi^*.

Series to evaluate Either the estimated cash flow series or the incremental series is used to determine the PW value, the AW value, the i^* value, or the B/C ratio.

Rate of return (interest rate) The MARR value must be stated to complete the PW, FW, or AW method. This is also correct for the discount rate for public sector alternatives analyzed by the B/C ratio. The ROR method requires that the incremental rate be found in order to select one alternative. It is here that the dilemma of multiple rates appears, if the sign tests indicate that a unique, real number root does not necessarily exist for a nonconventional series.

Decision guideline The selection of one alternative is accomplished using the general guideline in the rightmost column. Always select the alternative with the **numerically largest PW, FW, or AW value.** This is correct for both revenue and cost alternatives. The incremental cash flow methods—ROR and B/C—require that the largest initial cost and incrementally justified alternative be selected, provided it is justified against an alternative that is itself justified. This means that the Δi^* exceeds MARR, or the ΔB/C exceeds 1.0.

EXAMPLE LS2–1

Read through the problem statement of the following examples, neglecting the evaluation method used in the example. Determine which evaluation method is probably the fastest and easiest to apply. Is this the method used in the example? (*a*) 8.6, (*b*) 6.5, (*c*) 5.8, (*d*) 5.4.

Solution
Referring to the contents of Table LS2–1, the following methods should be applied first:

(*a*) Example 8.6 involves four revenue alternatives with equal lives. Use the AW or PW value at the MARR of 10%. The incremental ROR method was applied in the example.

(*b*) Example 6.5 requires selection between three public sector alternatives with unequal lives, one of which is 50 years and another is infinite. The B/C ratio of AW values is the best choice. This is how the problem was solved.

(*c*) Since Example 5.8 involves two cost alternatives with one having a long life, either AW or PW can be used. Since one life is long, capitalized cost, based on $P = A/i$, is best in this case. This is the method applied in the example.

(*d*) Example 5.4 is in the series of progressive examples. It involves 5-year and 10-year cost alternatives. The AW method is the best to apply in this case. The PW method for the LCM of 10 years and a study period of 5 years were both presented in the example.

LEARNING STAGE 3

Making Better Decisions

M ost of the evaluations in the real world involve more than a simple economic selection of new assets or projects. The chapters in this stage introduce information-gathering and techniques that make decisions better. For example, **noneconomic parameters** can be introduced into the project analysis study through multiple attribute evaluation, and the **appropriate MARR** for a corporation or type of alternative can tailor and improve the economic decision.

The future is certainly not exact. However, techniques such as **replacement/retention** studies, **breakeven analysis,** and **payback analysis** help make informed decisions about future uses of existing assets and systems.

After completing these chapters, you will be able to go beyond the basic alternative analysis tools of the previous chapters. The techniques covered in this learning stage take into consideration the moving targets of change over time.

Important note: If asset depreciation and taxes are to be considered by an *after-tax analysis,* Chapters 16 and 17 should be covered before or in conjunction with these chapters.

© Floresco Productions/age fotostock

CHAPTER 10

Project Financing and Noneconomic Attributes

LEARNING OUTCOMES

Purpose: Explain debt and equity financing, select the appropriate MARR, and consider multiple attributes when comparing alternatives.

SECTION	TOPIC	LEARNING OUTCOME
10.1	COC and MARR	• Explain the relation between cost of capital and the MARR; explain why MARR values vary.
10.2	D-E mix and WACC	• Understand debt-to-equity mix and calculate the weighted average cost of capital.
10.3	Cost of debt capital	• Estimate the cost of debt capital, considering tax advantages.
10.4	Cost of equity capital	• Estimate the cost of equity capital and describe its relation to MARR and WACC.
10.5	High D-E mixes	• Demonstrate the connection between high D-E mixes and financial risk for a corporation or an individual.
10.6	Multiple attributes	• Develop weights for multiple attributes used in alternative evaluation and selection.
10.7	Additive weights	• Apply the weighted attribute method to select an alternative when economic and noneconomic attributes are considered.

his chapter discusses the different ways to finance a project through debt and equity sources and explains how the MARR is established. The descriptions here complement the introductory material of Chapter 1 on the same topics. Some of the parameters specified earlier are unspecified here, and in future chapters. As a result, some of the textbook aspects apparent in previous chapters are removed, thus coming closer to treating the more complex, real-world situations in which professional practice and decision making occur.

Until now, only one dimension—the economic one—has been the basis for judging the economic viability of one project, or the selection basis from two or more alternatives. In this chapter, guidelines and techniques explain the determination and use of multiple (noneconomic) attributes helpful in selecting between alternatives.

10.1 MARR Relation to the Cost of Capital ●●●

The MARR value used in alternative evaluation is one of the most important parameters of a study. In Chapter 1, the MARR was described relative to the weighted costs of debt and equity capital. This and the next four sections explain how to establish a MARR under varying conditions.

To form the basis for a realistic MARR, the types and cost of each source of project financing should be understood and estimated. There is a strong connection between the costs of debt and equity capital and the MARR used to evaluate one or more alternatives, whether they are mutually exclusive or independent. There are several terms and relationships important to the understanding of project financing and the MARR that is specified to evaluate projects using PW, AW, FW, or B/C methods. (Reference to Section 1.9 will complement the following material.)

The **cost of capital** is the weighted average interest rate paid based on the proportion of investment capital from *debt* and *equity sources*.

The **MARR** is then set relative to the cost of capital. The MARR can be set for one project, a series of projects, a division of a corporation, or the entire company. MARR values *change over time* due to changing circumstances.

When no specific MARR is established, the estimated net cash flows and available capital establish an inherent MARR. This rate is determined by finding the ROR (i^*) value of the project cash flows. This rate is utilized as the **opportunity cost**, which, in terms of an interest rate, is the ROR of the first project not funded due to the lack of capital funds.

Cost of capital

MARR

Opportunity cost

Before we discuss cost of capital, let's be sure we understand the two primary sources of capital.

Debt capital represents borrowing from outside the company, with the principal repaid at a stated interest rate following a specified time schedule. Debt financing includes borrowing via *bonds, loans,* and *mortgages*. The lender does not share in the profits made using the debt funds, but there is **risk** in that the borrower could default on part of or all the borrowed funds. The amount of outstanding debt financing is indicated in the liabilities section of the corporate balance sheet. (See Appendix B.)

Equity capital is corporate money comprised of the *funds of owners* and *retained earnings*. Owners' funds are further classified as money obtained from the sale of stocks or owners' capital for a private (non-stock-issuing) company. Retained earnings are funds previously retained in the corporation for capital investment. The amount of equity is indicated in the net worth section of the corporate balance sheet.

To illustrate the relation between cost of capital and MARR, assume a municipal power utility plans to install a state-of-the-art Internet of Things-based (IoT-based) vibration control system on gas-powered turbines. The IoT system will be completely financed by a $2.5 million bond issue (100% debt financing). Further, assume the dividend rate on the bonds is 8%. Therefore, the cost of debt capital is 8% as shown in Figure 10–1. This 8% is the minimum for MARR. Management may increase this MARR in increments that reflect its desire for added return and its perception of risk. For example, management may add an amount for all capital commitments in this area. Suppose this amount is 2%. This increases the expected return to 10% (Figure 10–1). Also, if the risk associated with the investment is considered substantial enough to warrant an additional 1% return requirement, the final MARR is 11%.

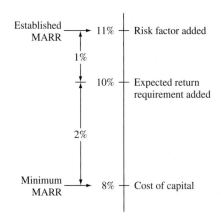

Figure 10–1
A fundamental relation between cost of capital
and MARR used in practice.

The recommended approach does not follow the logic presented above. Rather, the cost of capital (8% here) should be the established MARR. Then the i^* value is determined from the estimated net cash flows. Using this approach, suppose the control system is estimated to return 11%. Now, additional return requirements and risk factors are considered to determine if 3% above the MARR of 8% is sufficient to justify the capital investment. After these considerations, if the project is not funded, the effective MARR is now 11%. This is the opportunity cost discussed previously—the unfunded project i^* has established the effective MARR for the IoT system at 11%, not 8% per year.

Clearly, the setting of the MARR for an economy study is not an exact process. The debt and equity capital mix changes over time and between projects. Also, the MARR is not a fixed value established corporatewide. It is altered for different opportunities and types of projects. For example, a corporation may use a MARR of 10% for evaluating the purchase of assets (equipment, cars) and a MARR of 20% for expansion investments, such as acquiring smaller companies.

The effective MARR varies from one project to another and through time because of factors such as the following:

Project risk. Where there is greater risk (perceived or actual) associated with proposed projects, the tendency is to set a higher MARR. This is encouraged by the higher cost of debt capital for projects considered risky. This usually means that there is some concern that the project will not realize its projected revenue requirements.

Investment opportunity. If management is determined to expand in a certain area, the MARR may be lowered to encourage investment with the hope of recovering lost revenue in other areas. This common reaction to investment opportunity can create havoc when the guidelines for setting a MARR are too strictly applied. Flexibility becomes very important.

Government intervention. Depending upon the state of the economy, international relations, and a host of other factors, the federal government (and possibly lower levels) can dictate the forces and direction of the free market. This may occur through price limits, subsidies, import tariffs, and limitation on availability. Both short-term and long-term government interventions are commonly present in different areas of the economy. Examples are steel imports, foreign capital investment, car imports, and agricultural product exports. During the time that such government actions are in force, there is a strong impact to increase or decrease taxes, prices, etc., thus tending to move the MARR up or down.

Tax structure. If corporate taxes are rising (due to increased profits, capital gains, local taxes, etc.), pressure to increase the MARR is present. Use of after-tax analysis may assist in eliminating this reason for a fluctuating MARR, since accompanying business expenses will tend to decrease taxes and after-tax costs.

Limited capital. As debt and equity capital become limited, the MARR is increased. If the demand for limited capital exceeds supply, the MARR may tend to be set even higher. The opportunity cost has a large role in determining the MARR actually used.

Market rates at other corporations.   If the MARR increases at other corporations, especially competitors, a company may alter its MARR upward in response. These variations are often based on changes in interest rates for loans, which directly impact the cost of capital.

If the details of after-tax analysis are not of interest, but the effects of income taxes are important, the MARR may be increased by incorporating an effective tax rate using the formula

$$\text{Before-tax MARR} = \frac{\text{after-tax MARR}}{1 - \text{tax rate}} \qquad [10.1]$$

The total or effective tax rate, including federal, state, and local taxes, for most corporations is in the range of 30% to 50%. If an after-tax rate of return of 10% is required and the effective tax rate is 35%, the MARR for the before-tax economic analysis is $10\%/(1 - 0.35) = 15.4\%$.

EXAMPLE 10.1

Twin brother and sister, Carl and Christy, graduated several years ago from college. Carl, an architect, has worked in home design with Bulte Homes since graduation. Christy, a civil engineer, works with Butler Industries in structural components and analysis. They both reside in Richmond, Virginia. They have started a creative e-commerce network through which Virginia-based builders can buy their "spec home" plans and construction materials much more cheaply. Carl and Christy want to expand into a regional e-business corporation. They have gone to the Bank of America (BA) in Richmond for a business development loan. Identify some factors that might cause the loan rate to vary when BA provides the quote. Also, indicate any impact on the established MARR when Carl and Christy make economic decisions for their business.

Solution

In all cases, the direction of the loan rate and the MARR will be the same. Using the six factors mentioned above, some loan rate considerations are as follows:

Project risk: The loan rate may increase if there has been a noticeable downturn in housing starts, thus reducing the need for the e-commerce connection.

Investment opportunity: The rate could increase if other companies offering similar services have already applied for a loan at other BA branches regionally or nationwide.

Government intervention: The loan rate may decrease if the federal government has recently offered Federal Reserve loan money at low rates to banks. The intervention may be designed to boost the housing economic sector in an effort to offset a significant slowdown in new home construction.

Taxes: If the state recently removed house construction materials from the list of items subject to sales tax, the rate might be lowered slightly.

Capital limitation: Assume the computer equipment and software rights held by Carl and Christy were bought with their own funds and there are no outstanding loans. If additional equity capital is not available for this expansion, the rate for the loan (debt capital) should be lowered.

Market loan rates: The local BA branch probably receives its development loan money from a large national pool. If market loan rates to this BA branch have increased, the rate for this loan will likely increase because money is becoming "tighter."

10.2 Debt-Equity Mix and Weighted Average Cost of Capital ●●●

The **debt-to-equity (D-E) mix** identifies the percentages of debt and equity financing for a corporation. A company with a 40–60 D-E mix has 40% of its capital originating from debt capital sources (bonds, loans, and mortgages) and 60% derived from equity sources (stocks and retained

Figure 10–2
General shape of different
cost of capital curves.

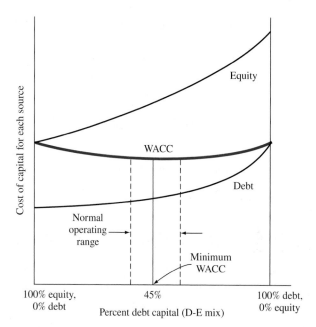

earnings). Most projects are funded with a combination of debt and equity capital made available specifically for the project or taken from a corporate *pool of capital*. The **weighted average cost of capital (WACC)** of the pool is estimated by the relative fractions from debt and equity sources. If known exactly, these fractions are used to estimate WACC; otherwise the historical fractions for each source are used in the relation

$$\text{WACC} = \text{(equity fraction)(cost of equity capital)}$$
$$+ \text{(debt fraction)(cost of debt capital)} \qquad [10.2]$$

The two *cost* terms are expressed as percentage interest rates.

Since virtually all corporations have a mixture of capital sources, the WACC is a value between the debt and equity costs of capital. If the fraction of each type of equity financing—common stock, preferred stock, and retained earnings—is known, Equation [10.3] results. The fraction of debt financing—bonds and loans—can also be separated in the WACC calculation.

$$\text{WACC} = \text{(common stock fraction)(cost of common stock capital)}$$
$$+ \text{(preferred stock fraction)(cost of preferred stock capital)}$$
$$+ \text{(retained earnings fraction)(cost of retained earnings capital)}$$
$$+ \text{(debt fraction)(cost of debt capital)} \qquad [10.3]$$

Figure 10–2 indicates the usual shape of cost of capital curves. If 100% of the capital is derived from equity or 100% is from debt sources, the WACC equals the cost of capital of that source of funds. There is virtually always a mixture of capital sources involved for any capitalization program. As an illustration only, Figure 10–2 indicates a minimum WACC at about 45% debt capital. Most firms operate over a range of D-E mixes. For example, a range of 30% to 50% debt financing for some companies may be very acceptable to lenders, with no increases in risk or MARR. However, another company may be considered "risky" with only 20% debt capital. It takes knowledge about management ability, current projects, and the economic health of the specific industry to determine a reasonable operating range of the D-E mix for a particular company.

EXAMPLE 10.2

Historically, Hong Kong has imported over 95% of its fresh vegetables each day. In an effort to develop sustainable and renewable vegetable sources, a new commercial vertical crop technology is being installed through a public-private partnership with Valcent Products.[1] For

[1]*Source:* "Valcent Announces Agreement to Supply Verticrop™ Vertical Farming Technology to Hong Kong's VF Innovations Ltd.," www.verticrop.com.

illustration purposes, assume that the present worth of the total system cost is $20 million with financing sources and costs as follows.

Commercial loan for debt financing	$10 million at 6.8% per year
Retained earnings from partnering corporations	$4 million at 5.2% per year
Sale of stock (common and preferred)	$6 million at 5.9% per year

There are three existing international vertical farming projects with capitalization and WACC values as follows:

Project 1:	$5 million with $WACC_1 = 7.9\%$
Project 2:	$30 million with $WACC_2 = 10.2\%$
Project 3:	$7 million with $WACC_3 = 4.8\%$

Compare the WACC for the Hong Kong (HK) project with the WACC of the existing projects.

Solution

To apply Equation [10.3] to this new project, the fraction of equity (stock and retained earnings) and debt financing is needed. These are 0.3 for stock ($6 out of $20 million), 0.2 for retained earnings, and 0.5 for debt ($10 out of $20 million).

$$WACC_{HK} = 0.3(5.9\%) + 0.2(5.2\%) + 0.5(6.8\%) = 6.210\%$$

To correctly weight the other three project WACCs by size, determine the fraction in each one of the $42 million in total capital: project 1 has $5 million/$42 million = 0.119; project 2 has 0.714; project 3 has 0.167. The WACC weighted by project size is $WACC_W$.

$$WACC_W = 0.119(7.9\%) + 0.714(10.2\%) + 0.167(4.8\%) = 9.025\%$$

The Hong Kong project has a considerably lower cost of capital than the weighted average of other projects, considering all sources of funding.

The WACC value can be computed using before-tax or after-tax values for cost of capital. The after-tax method is the correct one since debt financing has a distinct tax advantage, as discussed in Section 10.3 below. Approximations of after-tax or before-tax cost of capital are made using the effective tax rate T_e in the relation

> **After-tax cost of debt capital = (before-tax cost)$(1 - T_e)$**   **[10.4]**

The effective tax rate is a combination of federal, state, and local tax rates. They are reduced to a single number T_e to simplify computations. Equation [10.4] may be used to approximate the cost of debt capital separately or inserted into Equation [10.2] for an after-tax WACC rate. Chapter 17 treats taxes and after-tax economic analysis in detail.

10.3 Determination of the Cost of Debt Capital ●●●

Debt financing includes borrowing, primarily via bonds and loans. (We learned about bonds in Section 7.6.) In most industrialized countries, bond dividends and loan interest payments are tax-deductible as a corporate expense. This reduces the income base upon which taxes are calculated, with the end result of less taxes paid. The cost of debt capital is, therefore, reduced because there is an annual *tax savings* of the expenses times the effective tax rate T_e. This tax savings is subtracted from corporate expenses in order to calculate the net cash flow (NCF). In formula form,

> **Tax savings = (expenses) (effective tax rate) = expenses (T_e)**   **[10.5]**
> **Net cash flow = expenses − tax savings = expenses $(1 - T_e)$**   **[10.6]**

To find the resulting cost of debt capital, develop a PW- or AW-based relation of the NCF series with i^* as the unknown. Find i^* by trial and error, by calculator, or by the RATE or IRR function on a spreadsheet. This is the cost of debt capital used in the WACC computation, Equation [10.2].

EXAMPLE 10.3

Boeing Aerospace will generate $5 million in debt capital by issuing five thousand $1000, 8% per year, 10-year bonds. If the effective tax rate of the company is 30% and the bonds are discounted 2%, compute the cost of debt capital (a) before taxes, and (b) after taxes from the company perspective. Obtain the answers by hand and spreadsheet. (c) Approximate the after-tax cost of debt capital and compare it with the actual cost.

Solution by Hand

(a) The annual bond dividend is $1000(0.08) = $80, and the 2% discounted sales price is $980 now. Using the company perspective, find the i^* in the PW relation

$$0 = 980 - 80(P/A, i^*,10) - 1000(P/F, i^*,10)$$
$$i^* = 8.3\%$$

The before-tax cost of debt capital is $i^* = 8.3\%$, which is slightly higher than the 8% bond interest rate because of the 2% sales discount.

(b) With the allowance to reduce taxes by deducting the bond dividend, Equation [10.5] shows a tax savings of $80(0.3) = $24 per year. The bond dividend amount for the PW relation is now $80 − 24 = $56. Solving for i^* after taxes reduces the cost of debt capital to 5.87%.

(c) By Equation [10.4], the approximation of the cost of debt capital is 8.3%(1 − 0.3) = 5.81%. This approximation is a slight underestimate compared to the actual cost of 5.87% from part (b).

Solution by Spreadsheet

Figure 10–3 is a spreadsheet image for both before-tax (column B) and after-tax (column C) analysis using the IRR function. The after-tax net cash flow is calculated using Equation [10.6] with $T_e = 0.3$.

	A	B	C	D	E	F	G
1		Before-tax	After-tax				
2	Year	cash flow	cash flow				
3	0	980	980				
4	1	-80	-56				
5	2	-80	-56	Bond dividend before taxes			
6	3	-80	-56	= −1000*0.08			
7	4	-80	-56				
8	5	-80	-56				
9	6	-80	-56				
10	7	-80	-56	Bond dividend after taxes			
11	8	-80	-56	= (−1000*0.08)*(1 − 0.3)			
12	9	-80	-56				
13	10	-1,080	-1,056				
14	**Cost of debt capital**	8.30%	5.87%	= IRR(C3:C13)			
15							
16							
17							

Figure 10–3
Use of IRR function to determine cost of debt capital before taxes and after taxes, Example 10.3.

EXAMPLE 10.4

LST Trading Company will purchase a $20,000 ten-year-life asset. Company managers have decided to put $10,000 down now from retained earnings and borrow $10,000 at an interest rate of 6%. The simplified loan repayment plan is $600 in interest each year, with the entire $10,000 principal paid in year 10. (a) What is the after-tax cost of debt capital if the effective tax rate is 42%? (b) How are the interest rate and cost of debt capital used to calculate WACC?

Solution

(a) The after-tax net cash flow for interest on the $10,000 loan is an annual amount of $600(1 - 0.42) = \$348$ by Equation [10.6]. The loan repayment is $10,000 in year 10. PW is used to estimate a cost of debt capital of 3.48%.

$$0 = 10,000 - 348(P/A, i^*, 10) - 10,000(P/F, i^*, 10)$$

(b) The 6% annual interest on the $10,000 loan is not the WACC because 6% is paid only on the borrowed funds. Nor is 3.48% the WACC, since it is only the cost of debt capital. The cost of the $10,000 equity capital is needed to determine the WACC.

10.4 Determination of the Cost of Equity Capital and the MARR ●●●

Equity capital is usually obtained from the following sources:

Sale of preferred stock

Sale of common stock

Use of retained earnings

Use of owner's private capital

The cost of each type of financing is estimated separately and entered into the WACC computation. A summary of one commonly accepted way to estimate each source's cost of capital is presented here. One additional method for estimating the cost of equity capital via common stock is presented. *There are no tax savings for equity capital because dividends paid to stockholders and owners are not tax-deductible.*

Issuance of **preferred stock** carries with it a commitment to pay a stated dividend annually. The cost of capital is the stated dividend percentage, for example, 10%, or the dividend amount divided by the price of the stock. Preferred stock may be sold at a discount to speed the sale, in which case the actual proceeds from the stock should be used as the denominator. For example, if a 10% dividend preferred stock with a value of $200 is sold at a 5% discount for $190 per share, there is a cost of equity capital of $(\$20/\$190) \times 100\% = 10.53\%$.

Estimating the cost of equity capital for **common stock** is more involved. The dividends paid are not a true indication of what the stock issue will actually cost in the future. Usually a valuation of the common stock is used to estimate the cost. If R_e is the cost of equity capital, that is, common stock (in decimal form),

$$R_e = \frac{\text{first-year dividend}}{\text{price of common stock}} + \text{expected dividend growth rate}$$

$$= \frac{DV_1}{P} + g \qquad \qquad [10.7]$$

The growth rate g is an estimate of the annual increase in returns that the shareholders receive. Stated another way, it is the compound growth rate on dividends that the company believes is required to attract stockholders. For example, assume a U.S.-based corporation plans to raise capital through its international subsidiary for a new plant in South America by selling $2,500,000 worth of common stock valued at $20 each. If a 5% or $1 dividend is planned for the first year and an appreciation of 2% per year is anticipated for future dividends, the cost of capital for this common stock issue from Equation [10.7] is 7%.

$$R_e = \frac{1}{20} + 0.02 = 0.07$$

The **retained earnings** and **owner's funds** cost of equity capital are usually set equal to the common stock cost, since it is the shareholders and owners who will realize any returns from projects in which these funds are invested.

Figure 10–4
Expected return on common stock issue using CAPM.

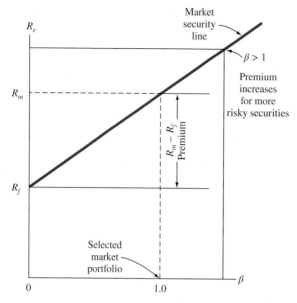

Once the cost of capital for all planned equity sources is estimated, the WACC is calculated using Equation [10.3].

A second method used to estimate the **cost of common stock capital** is the **capital asset pricing model (CAPM)**. Because of the fluctuations in stock prices and the higher return demanded by some corporations' stocks compared to others, this valuation technique is commonly applied. The cost of equity capital from common stock R_e, using CAPM, is

$$R_e = \text{risk-free return} + \text{premium above risk-free return}$$
$$= R_f + \beta(R_m - R_f) \hspace{3cm} [10.8]$$

where β = volatility of a company's stock relative to other stocks in the market ($\beta = 1.0$ is the norm)

R_m = return on stocks in a defined market portfolio measured by a prescribed index

R_f = risk-free interest rate, usually the rate for government bonds

The term $(R_m - R_f)$ is the premium paid above the "safe investment" or risk-free rate. The risk-free rate, which is commonly in the range of 2.5% to 3.5% per year, is determined by bonds such as the U.S. Treasury long-term bond coupon rate. The coefficient β (beta) indicates how the stock is expected to vary compared to a selected portfolio of stocks in the same general market area, often the Standard and Poor's 500 stock index. If $\beta < 1.0$, the stock is less volatile, so the resulting premium can be smaller; when $\beta > 1.0$, larger price movements are expected, so the premium is increased.

Security is a word that identifies a stock, bond, or any other financial instrument used to develop capital. To better understand how CAPM works, consider Figure 10–4. This is a plot of a market security line, which is a linear fit by regression analysis to indicate the expected return for different β values. When $\beta = 0$, the risk-free return R_f is acceptable (no premium). As β increases, the premium return requirement grows. Beta values are published periodically for most stock-issuing corporations. Once complete, this estimated cost of common stock equity capital can be included in the WACC computation.

EXAMPLE 10.5

The lead software engineer at SafeSoft, a food industry service corporation, has convinced the president to develop new software technology for the meat and food safety industry. It is envisioned that processes for prepared meats can be completed more safely and faster using this automated control software. A common stock issue is a possibility to raise capital if the cost of equity capital is below 9%. SafeSoft, which has a historical beta value of 1.09, uses CAPM to determine the premium of its stock compared to other software corporations. The security market line indicates that a 5% premium above the risk-free rate is desirable. If U.S. Treasury bills are paying 2%, estimate the cost of common stock capital.

Solution

The premium of 5% represents the term $R_m - R_f$ in Equation [10.8].

$$R_e = 2.0 + 1.09(5.0) = 7.45\%$$

Since this cost is lower than 9%, SafeSoft should issue common stock to finance this new venture.

In theory, a correctly performed engineering economy study uses a MARR equal to the cost of the capital committed to the specific alternatives in the study. Of course, such detail is not known. For a combination of debt and equity capital, the calculated WACC sets the minimum for the MARR. The most rational approach is to set MARR between the cost of equity capital and the corporation's WACC. The risks associated with an alternative should be treated separately from the MARR determination, as stated earlier. This supports the guideline that the MARR should not be arbitrarily increased to account for the various types of risk associated with the cash flow estimates. The MARR is often set above the WACC because, in practice, management insists on accommodating risk by an increase in the MARR for a project with perceived extra risk.

EXAMPLE 10.6

The Engineering Products Division of 4M Corporation has two mutually exclusive alternatives A and B with ROR values of $i_A^* = 9.2\%$ and $i_B^* = 5.9\%$. The financing scenario is yet unsettled, but it will be one of the following: plan 1—use all equity funds, which are currently earning 8% for the corporation; plan 2—use funds from the corporate capital pool which is 25% debt capital costing 14.5% and the remainder from the same equity funds mentioned above. The cost of debt capital is currently high because the company has narrowly missed its forecasted revenue target for the last 2 years, and banks have increased the borrowing rate for 4M. Make the economic decision on alternative A versus B under each financing scenario. The MARR is set equal to the calculated WACC.

Solution

The capital is available for one of the two mutually exclusive alternatives. For plan 1, 100% equity, the financing is specifically known, so the cost of equity capital is the MARR, that is, 8%. Only alternative A is acceptable; alternative B is not since the estimated return of 5.9% does not exceed this MARR.

Under financing plan 2, with a D-E mix of 25–75,

$$WACC = 0.25(14.5) + 0.75(8.0) = 9.625\%$$

Now, neither alternative is acceptable since both ROR values are less than MARR = WACC = 9.625%. The selected alternative should be to do nothing. If one alternative absolutely must be selected, noneconomic attributes must be considered.

10.5 Effect of Debt-Equity Mix on Investment Risk ●●○

The D-E mix was introduced in Section 10.2. As the proportion of debt capital increases, the calculated cost of capital decreases due to the tax advantages of debt capital.

The leverage offered by larger debt capital percentages increases the riskiness of projects undertaken by the company. When large debts are already present, additional financing using debt (or equity) sources gets more difficult to justify, and the corporation can be placed in a situation where it owns a smaller and smaller portion of itself. This is sometimes referred to as a **highly leveraged corporation**.

Inability to obtain operating and investment capital means increased difficulty for the company and its projects. Thus, a reasonable balance between debt and equity financing is important for the financial health of a corporation. Example 10.7 illustrates the disadvantages of unbalanced D-E mixes.

EXAMPLE 10.7

Three auto parts manufacturing companies have the following debt and equity capital amounts and D-E mixes. Assume all equity capital is in the form of common stock.

Company	Amount of Capital		D-E Mix (%–%)
	Debt ($ in Millions)	Equity ($ in Millions)	
A	10	40	20–80
B	20	20	50–50
C	35	15	70–30

Assume the annual revenue is $15 million for each one and that after interest on debt is considered, the net incomes are $14.4, $12.0, and $10.0 million, respectively. Compute the return on equity for each company, and comment on the return relative to their D-E mixes.

Solution

Divide the net income by the stock (equity) amount to compute the common stock return. In million dollars,

$$\text{Return}_A = \frac{14.4}{40} = 0.36 \quad (36\%)$$

$$\text{Return}_B = \frac{12.0}{20} = 0.60 \quad (60\%)$$

$$\text{Return}_C = \frac{10.0}{15} = 0.67 \quad (67\%)$$

As expected, the return is larger for higher-leveraged companies. For C, where only 30% of the company is in the hands of the ownership, the return is excellent. However, the risk associated with this firm is high compared to A, where the D-E mix has only 20% debt.

The use of *large percentages of debt financing greatly increases the risk* taken by lenders and stock owners. Long-term confidence in the corporation diminishes, no matter how large the short-term return on stock.

The leverage of large D-E mixes does increase the return on *equity capital,* as shown in previous examples; but it can also work against the owners and investors. A decrease in asset value will more negatively affect a highly debt-leveraged company compared to one with small leveraging. Example 10.8 illustrates this fact.

EXAMPLE 10.8

During the last several years, the U.S. airline industry has had financial problems, in part due to high fuel costs, fewer customers, security problems, government regulations, aging aircraft, and union dissatisfaction. As a consequence, the D-E mixes of the large companies have become larger on the debt side than is historically acceptable. Meanwhile, the D-E mixes of low-cost airlines have suffered, but not to the same extent. In an effort to reduce costs, assume that three airlines joined forces to cooperate on a range of services (baggage handling, onboard food preparation, ticket services, and software development) by forming a new company called FullServe, Inc. This required $5 billion ($5 B) up-front funding from each airline.

Table 10–1 summarizes the D-E mixes and the total equity capitalization for the three airlines *after* its share of $5 B was removed from available equity funds. The percentage of the $5 B obtained as debt capital was the same proportion as the debt in the company's D-E mix.

TABLE 10–1	Debt and Equity Statistics, Example 10.8		
Airline Company	**Corporate D-E Mix, %**	**Amount Borrowed, $ B**	**Equity Capital Available, $ B**
National	30–70	1.50	5.0
Transglobal	65–35	3.25	3.7
Continental	91–9	4.55	6.7

For example, National had 30% of its capitalization in debt capital; therefore, 30% of $5 B was borrowed, and 70% was provided from National's equity fund.

Unfortunately, after a short time, it was clear that the three-way collaborative effort was a complete failure due to an inability to productively cooperate, and FullServe was dissolved and its assets were distributed or sold for a total of $3.0 billion, only 20% of its original value. A total of $1.0 billion in equity capital was returned to each airline. The commercial banks that provided the original loans then required that the airlines each pay back the entire borrowed amount now, since FullServe was dissolved and no profit from the venture could be realized. Assuming the loan and equity amounts are the same as shown in Table 10–1, determine the resulting equity capital situation for each airline after it pays off the loan from its own equity funds. Also, describe one impact on each company as a result of this failure.

Solution

Determine the level of post-FullServe equity capital using the following relation, in $ billions.

$$\text{Equity capital} = \text{pre-FullServe level} + \text{returned capital} - \text{loan repayment}$$

National: Equity capital = 5.0 + 1.0 − 1.50 = $4.50
Transglobal: Equity capital = 3.7 + 1.0 − 3.25 = $1.45
Continental: Equity capital = 6.7 + 1.0 − 4.55 = $3.15

Comparing the equity capital levels (Table 10–1) with the levels above indicates that the FullServe effort *reduced equity amounts* by 10% for National, 60% for Transglobal, and 53% for Continental. The debt capital to fund the failed FullServe effort has affected National airlines the least, in large part due to its low D-E mix of 30%–70%. However, Transglobal and Continental are in much worse shape financially, and they now must maintain business with a significantly lower ownership level and a reduced ability to obtain future capital—debt or equity.

The same principles discussed above for corporations are equally applicable to individuals as they manage their debt. The person who is highly leveraged has large debts in terms of credit card balances, personal loans, car loans, and a house mortgage. As an example, assume two successful engineers each have an annual take-home amount of $60,000 after all income tax, social security, and insurance premiums are deducted from their annual salaries. Further, assume that the cost of the debt (money borrowed via credit cards and loans) averages 15% per year and that the total debt is being repaid in equal amounts over 20 years with required interest paid each year. If Sherry has a total debt of $25,000 and Linda owes $100,000, the remaining amount of the annual take-home pay is calculated below. As you see, Sherry has 91.7% of her base salary available for the year, while Linda has a much smaller 66.7% available.

	Total Debt, $ (1)	Interest on Debt at 15%, $/Year (2) = (1) × 0.15	Repayment of Debt, $/Year (3) = (1)/20	Total Paid, $/Year (4) = (2) + (3)	Remaining from Salary	
					Amount, $ (5) = 60,000 − (4)	Percent, % (6) = (5)/60,000
Sherry	25,000	3,750	1,250	5,000	55,000	91.7
Linda	100,000	15,000	5,000	20,000	40,000	66.7

10.6 Multiple Attribute Analysis: An Introduction ● ● ●

In Chapter 1, the fundamentals of engineering economy were explored. The decision-making process explained in that chapter (Figure 1–1) included the seven steps listed on the right side of Figure 10–5. Step 4 is to identify the one or multiple attributes (criteria) upon which the selection will be based. In all prior evaluations, only one attribute—the economic one—has been identified and used to select the best alternative. The criterion has been maximization of PW, AW, FW, ROR, B/C ratio, or the CER value. As we all know, virtually all evaluations and decisions in industry, business, engineering—and in our own personal lives—are based on multiple factors (attributes), many are not purely economic in nature. These factors, labeled as noneconomic in step 5 of Figure 1–1, tend to be intangible and often difficult, if not impossible, to quantify with economic scales. Nonetheless, among the many attributes that can be identified, there are key ones that must be considered in earnest before the alternative selection process is complete. This section and the next describe some of the techniques that accommodate multiple attributes in an engineering study.

Multiple attributes enter into the decision-making process in many studies. Public and service sector projects are excellent examples of multiple-attribute environments. For example, the proposal to construct a dam to form a lake or to widen the catch basin of a river usually has several purposes, such as flood control, hydroelectric power generation, drinking water, industrial use, downstream irrigation, commercial development, recreation, nature conservation, and possibly other less obvious purposes. High levels of complexity are introduced into the selection process by the multiple attributes thought to be important in selecting an alternative for the dam's location, design, environmental impact, etc.

The left side of Figure 10–5 expands steps 4 and 5 to consider multiple attributes. The discussion below concentrates on the expanded step 4 and the next section focuses on the evaluation measure and alternative selection of step 5.

4-1 Attribute Identification   Attributes to be considered in the evaluation methodology can be identified and defined by several methods, some much better than others depending upon the nature of the study. To seek input from individuals other than the analyst is important; it helps focus the study on key attributes. The following list is a sample of ways in which key attributes are identified.

- Comparison with similar studies that include multiple attributes
- Input from experts with relevant past experience
- Surveys of stakeholders (customers, employees, managers)
- Small group discussions using approaches such as focus groups, brainstorming, or nominal group technique
- Delphi method, a progressive procedure used to develop reasoned consensus

Figure 10–5
Expansion of the decision-making process to include multiple attributes.

Consider multiple attributes

4-1. Identify the attributes for decision making.
4-2. Determine the relative importance (weights) of attributes.
4-3. For each alternative, determine each attribute's value rating.

5. Evaluate each alternative using a multiple-attribute technique. Use sensitivity analysis for key attributes.

Emphasis on one attribute

1. Understand the problem; define the objective.
2. Collect relevant information; define alternatives.
3. Make estimates.

4. Identify the selection criteria (one or more attributes).

5. Evaluate each alternative; use sensitivity and risk analysis.

6. Select the best alternative.
7. Implement the solution and monitor results.

As an illustration, assume that Singapore Airlines has decided to purchase five Boeing 787-10 Dreamliners for overseas flights, primarily between the North American west coast and Asian cities, principally Singapore, Hong Kong, and Tokyo. There are approximately 8000 options for each plane that must be decided upon by engineering, purchasing, maintenance, and marketing personnel before the order to Boeing is placed. Options range *in scope* from the material and color of the plane's interior to the type of latching devices used on the engine cowlings, and *in function* from maximum engine thrust to pilot instrument design. An economic study based on the equivalent AW of the estimated passenger income per trip has determined that 150 of these options are clearly advantageous. But other noneconomic attributes are to be considered before some of the more expensive options are specified. A Delphi study was performed using input from 25 individuals. Concurrently, options from another airline's recent order were shared with Singapore personnel. From these two studies, it was determined that there are four strategic, economic, and noneconomic attributes for options selection. They are

- *Repair time:* mean time to repair or replace (MTTR) if the option is a flight-critical component.
- *Safety:* mean time to failure (MTTF) of flight-critical components.
- *Economic:* estimated extra revenue for the option. (Basically, this is the attribute evaluated by the economic study already performed.)
- *Crewmember needs:* some measure of the necessity and benefits of the option judged by crewmembers—pilots and attendants.

Of course, there are many other attributes that can be, and are, used. However, the point is that the economic study directly addresses only one of the key attributes vital to alternative decision making.

A vitally important attribute routinely identified is **risk**.

> Risk is the **variation** in a parameter (i.e., an attribute, variable, or factor) from an expected, desired, or predicted value that may be detrimental to the intended outcome(s) of the product, process, or system. Risk represents deviation from certainty. Risk is present when there are two or more observable values of a parameter. Determination of the **chance** that each value may occur is called risk estimation.

Risk

Actually, risk is not a stand-alone attribute, because it is a part of every attribute in one form or another. Considerations of variation, probabilistic estimates, etc. in the decision-making process are treated in Chapters 18 and 19. Formalized sensitivity analysis, expected values, simulation, and decision trees are some of the techniques useful in handling risk.

4-2 Importance (Weights) for the Attributes   Determination of the **extent of importance** for each attribute i ($i = 1, 2, \ldots, m$ attributes) results in a weight W_i that is incorporated into the evaluation. The weight, a number between 0 and 1, is based upon the experienced opinion of one individual or a group of persons familiar with the attributes, and possibly the alternatives. If a group is utilized to determine the weights, there must be consensus among the members for each weight. Otherwise, some averaging technique must be applied to arrive at a single W_i for each attribute.

Table 10–2 is a tabular layout of attributes and alternatives used to perform a multiple attribute evaluation. Weights W_i for each attribute are entered on the left side. The remainder of the table is discussed as we proceed through steps 4 and 5 of the expanded decision-making process.

Attribute weights are usually normalized such that their sum over all the alternatives is 1.0. This normalizing implies that each attribute's importance score is divided by the sum S over all attributes. Expressed in formula form, these two properties of weights for attribute i ($i = 1, 2, \ldots, m$) are

$$\text{Normalized weights: } \sum_{i=1}^{m} W_i = 1.0 \qquad [10.9]$$

$$\text{Weight calculation: } W_i = \frac{\text{importance score}_i}{\sum_{i=1}^{m} \text{importance score}_i} = \frac{\text{importance score}_i}{S} \qquad [10.10]$$

TABLE 10–2	Tabular Layout of Attributes and Alternatives Used for Multiple Attribute Evaluation					
		Alternatives				
Attributes	**Weights**	**1**	**2**	**3**	**...**	**n**
1	W_1					
2	W_2					
3	W_3		Value ratings V_{ij}			
$\vdots$	$\vdots$					
m	W_m					

Of the many procedures developed to assign weights to an attribute, an analyst is likely to rely upon one that is relatively simple, such as equal weighting, rank order, or weighted rank order. Pairwise comparison is another technique. Each is briefly presented below.

Equal Weighting   All attributes are considered to be of approximately the **same importance**, or there is no rationale to distinguish the more important from the less important attribute. This is the default approach. Each weight in Table 10–2 will be $1/m$, according to Equation [10.10]. Alternatively, the normalizing can be omitted, in which case each weight is 1 and their sum is m. Now, the final evaluation measure for an alternative is the sum over all attributes.

Rank Order   The m attributes are ordered (ranked) by increasing importance with a score of 1 assigned to the least important and m assigned to the most important. By Equation [10.10], the weights follow the pattern $1/S, 2/S, \ldots, m/S$. With this method, the difference in weights between attributes of **increasing importance is constant**.

Weighted Rank Order   The m attributes are again placed in the order of increasing importance; however, **differentiation between attributes** is possible. The most important attribute is assigned a score, usually 100, and all other attributes are scored relative to it between 100 and 0. Now, define the score for each attribute as s_i, and Equation [10.10] takes the form

$$W_i = \frac{s_i}{\sum\limits_{i=1}^{m} s_i} \qquad [10.11]$$

This is a very practical method to determine weights because one or more attributes can be heavily weighted if they are significantly more important than the remaining ones, and Equation [10.11] automatically normalizes the weights. For example, suppose the four key attributes in the previous aircraft purchase example are ordered: safety, repair time, crewmember needs, and economic. If repair time is only one-half as important as safety, and the last two attributes are each one-half as important as repair time, the scores and weights are as follows:

Attribute	Score	Weights
Safety	100	100/200 = 0.50
Repair time	50	50/200 = 0.25
Crewmember needs	25	25/200 = 0.125
Economic	25	25/200 = 0.125
Sum of scores and weights	200	1.000

Pairwise Comparison   Each attribute is compared to each other attribute in a pairwise fashion using a rating scale that indicates the importance of one attribute over the other. Assume the three

TABLE 10–3	Pairwise Comparison of Three Attributes to Determine Weights		
Attribute i	1 = Cost	2 = Constructability	3 = Environment
Cost	—	0	1
Constructability	2	—	1
Environment	1	1	—
Sum of scores	3	1	2
Weight W_i	0.500	0.167	0.333

attributes upon which a public works project decision is based are *cost, constructability,* and *environmental impact.* Define the importance comparison scale as follows:

   0   if attribute is *less important* than one compared to

   1   if attribute is *equally important* as one compared to

   2   if attribute is *more important* than one compared to

Set up a table listing attributes across the top and down the side, and perform the pairwise comparison for each column attribute with each row attribute. Table 10–3 presents a comparison with importance scores included. The arrow to the right of the table indicates the direction of comparison, that is, column with row attribute. For example, cost is judged more important than constructability, thus a score of 2. The complement score of 0 is placed in the reverse comparison of constructability with cost. The weights are determined by normalizing the scores using Equation [10.11], where the sum for each column is s_i. For the first attribute, cost $i = 1$.

$$s_1 = 3$$
$$\sum s_i = 3 + 1 + 2 = 6$$
$$\text{Cost weight } W_1 = 3/6 = 0.500$$

Similarly, the other weights are $W_2 = 1/6 = 0.167$ and $W_3 = 2/6 = 0.333$.

    There are other attribute weighting techniques, especially for group processes, such as utility functions, and the Dunn-Rankin procedure. These become increasingly sophisticated, but they are able to provide an advantage that these simple methods do not afford the analyst: *consistency of ranks and scores* between attributes and between individuals. If this consistency is important in that several decision makers with diverse opinions about attribute importance are involved in a study, a more sophisticated technique may be warranted. There is substantial literature on this topic.

4-3 Value Rating of Each Alternative by Attribute   This is the final step prior to calculating the evaluation measure. Each alternative j is awarded a value rating V_{ij} for each attribute i. These are the entries within the cells in Table 10–2. The ratings are appraisals by decision makers of how well an alternative will perform as each attribute is considered.

    The scale for the value rating can vary depending upon those doing the rating. A scale of 0 to 100 can be used for attribute importance scoring. However, the most popular is a scale of 4 or 5 gradations about the perceived ability of an alternative to accomplish the intent of the attribute. This is called a *Likert scale,* which can have descriptions for the gradations (e.g., very poor, poor, good, very good), or numbers assigned between 0 and 10, or −1 to +1, or −2 to +2. The last two scales can give a negative impact to the evaluation measure for poor alternatives. An example Likert scale of 0 to 10 is as follows:

If You Value the Alternative as	Give It a Rating between the Numbers
Very poor	0–2
Poor	3–5
Good	6–8
Very good	9–10

It is preferable to have a Likert scale with four choices (an even number) so that the central tendency of "fair" is not overrated.

TABLE 10–4	Completed Layout for Four Attributes and Three Alternatives for Multiple Attribute Evaluation			
			Alternatives	
Attributes	**Weights**	**1**	**2**	**3**
Safety	0.50	6	4	8
Repair	0.25	9	3	1
Crew needs	0.125	5	6	6
Economic	0.125	5	9	7

Returning to the earlier illustration of airplane purchases by Singapore Airlines, we can now include value ratings. The cells are filled with ratings awarded by a decision maker. Table 10–4 includes example ratings V_{ij} and the weights W_i determined above. Initially, there will be one such table for each decision maker. Prior to calculating a final evaluation measure, the ratings can be combined in some fashion; or a different measure can be calculated using each decision maker's ratings. Determination of an evaluation measure is discussed below.

10.7 Evaluation Measure for Multiple Attributes ● ● ●

The evaluation measure that accommodates multiple attributes and multiple alternatives (step 5 of Figure 10–5, left side) can be one that attempts to retain all of the ratings, values, and complexity of assessments made by multiple decision makers, or it can reduce these inputs to a single-dimension measure. This section introduces a single-dimension measure that is widely applied.

A **single-dimension measure** effectively combines the different aspects addressed by the attribute importance weights W_i and the alternative value ratings V_{ij}. The resulting evaluation measure is a formula that calculates an aggregated measure for use in selecting from two or more alternatives. The approach applied in this process is called the *rank-and-rate method*.

This reduction process removes much of the complexity of trying to balance the different attributes; however, it also eliminates much of the robust information captured by the process of ranking attributes for their importance and rating each alternative's performance against each attribute.

There are additive, multiplicative, and exponential measures, but by far the most commonly applied is the additive model. The most used additive model is the **weighted attribute method**, also called the **additive weight technique**. The evaluation measure, symbolized by R_j for each alternative j, is defined as

$$R_j = \text{sum of (weight} \times \text{value rating)}$$
$$= \sum_{i=1}^{m} W_i \times V_{ij} \qquad\qquad [10.12]$$

The W_i numbers are the attribute importance weights, and V_{ij} is the value rating by attribute i for each alternative j. If the attributes are of equal weight (also called *unweighted*), all $W_i = 1/m$, as determined by Equation [10.10]. This means that W_i can be moved outside of the summation in the formula for R_j. (If an equal weight of $W_i = 1.0$ is used for all attributes, in lieu of $1/m$, then the R_j value is simply the sum of all ratings for the alternative.)

The selection guideline is as follows:

Choose the alternative with the **largest R_j value.** This measure assumes that as an attribute's importance increases, its W_i value increases. Similarly for increasing ratings V_{ij} for improved performance of an alternative against an attribute.

ME alternative selection

Sensitivity analysis for any score, weight, or value rating is used to determine sensitivity of the decision to it. The case study in Chapter 18 includes an example of multiple attribute sensitivity analysis.

EXAMPLE 10.9

Tenneco requested bids to replace equipment on one of its offshore platforms. The spreadsheet in Figure 10–6, left two columns, presents the attributes and normalized weights W_i published for use in selecting one of the vendors presenting proposals. Four acceptable proposals were received. The next four columns (C through F) include value ratings between 0 and 100 developed by a group of Tenneco decision makers when the details of each proposal were evaluated against each attribute. For example, proposal 2 received a perfect score of 100 on delivery date, but life cycle costs were considered too high (rating of 20) and the purchase price was considered relatively high (rating of 55). Use these weights and ratings to determine which proposal to pursue first.

© Digital Vision/PunchStock

Solution

Assume an additive weighting model is appropriate and apply the weighted attribute method. Equation [10.12] determines the R_j measure for the four alternatives. As an illustration, for proposal 3,

$$R_3 = 0.30(95) + 0.15(60) + 0.20(70) + 0.15(85) + 0.20(80)$$
$$= 28.5 + 9.0 + 14.0 + 12.8 + 16.0$$
$$= 80.3$$

The four totals in Figure 10–6 (columns G through J, row 8) indicate that proposal 4 is the overall best choice.

Comment

Any economic measure can be incorporated into a multiple attribute evaluation using this method. All measures of worth—PW, AW, ROR, B/C, and C/E—can be included; however, their impact on the final selection will vary relative to the importance placed on the noneconomic attributes.

	A	B	C	D	E	F	G	H	I	J
1			Value rating V_{ij} (0 to 100 basis)				Evaluation measure R_i			
2	Attribute, i	Normalized weight, W_i	Proposal 1	Proposal 2	Proposal 3	Proposal 4	Proposal 1	Proposal 2	Proposal 3	Proposal 4
3	Construction quality	0.30	95	60	95	95	28.5	18.0	28.5	28.5
4	Delivery date	0.15	60	100	60	60	9.0	15.0	9.0	9.0
5	Equipment reliability	0.20	100	80	70	100	20.0	16.0	14.0	20.0
6	Purchase price	0.15	20	55	85	65	3.0	8.3	12.8	9.8
7	Life cycle costs	0.20	100	20	80	100	20.0	4.0	16.0	20.0
8	Total	1.00					80.5	61.3	80.3	87.3

Figure 10–6

Attributes, weights, ratings, and evaluation measure for offshore platform proposals, Example 10.9.

CHAPTER SUMMARY

The interest rate at which the MARR is established depends principally upon the cost of capital and the mix between debt and equity financing. The MARR is strongly influenced by the *weighted average* cost of capital (WACC). Risk, profit, and other factors can be considered after the AW, PW, or ROR analysis is completed and prior to final alternative selection. A high debt-to-equity mix can significantly increase the riskiness of a project and make further debt financing difficult to acquire for the corporation.

If multiple attributes, which include more than the economic dimension of a study, are to be considered in making the alternative decision, first the attributes must be identified and their relative importance assessed. Then each alternative can be value-rated for each attribute. The evaluation measure is determined using a model such as the weighted attribute method, where the measure is calculated by Equation [10.12]. The largest value indicates the best alternative.

PROBLEMS

Working with MARR

10.1 Identify the two primary sources of capital and state what is meant by each.

10.2 State how the opportunity cost sets the MARR when, because of limited capital, only one alternative can be selected from two or more.

10.3 For each of the following factors, state if it will raise or lower the MARR:
(a) Higher risk
(b) Company wants to expand into a competitor's area
(c) Higher corporate taxes
(d) Limited availability of capital
(e) Increased market interest rates
(f) Government imposition of price controls

10.4 After 15 years of employment in the airline industry, John started his own consulting company to use physical and computer simulation in the analysis of commercial airport accidents on runways. He estimates his average cost of new capital at 8% per year for physical simulation projects, that is, where he physically reconstructs the accident using scale versions of planes, buildings, vehicles, etc. He has established 12% per year as his MARR. What *net* rate of return on capital investments for physical simulation does he expect?

10.5 Lodi Enterprises uses an after-tax MARR of 15% per year. If the company's effective tax rate (federal, state, and local taxes) is 38%, determine the company's before-tax MARR.

10.6 The owner of a small pipeline construction company wants to determine how much he should bid in his attempt to win his first "big" contract. He estimates that his cost to complete the project will be $7.2 million in PW equivalency. He wants to bid an amount that will generate an after-tax rate of return of 15% per year; however, he doesn't know how much to bid on a before-tax basis. He told you that his effective state tax rate is 12% and his effective federal tax rate is 22% per year.
(a) The equation for determining the overall effective tax rate is:

state rate + (1 − state rate)(federal rate)

Determine his before-tax MARR in order to realize an after-tax MARR of 15% per year.
(b) How much should he bid?

10.7 State whether each of the following involves debt financing or equity financing:
(a) A bond issue for $3,500,000 by a city-owned utility

(b) An initial public offering (IPO) of $35,000,000 in common stock for a dot-com company
(c) $31,000 taken from your retirement account to pay cash for a new car
(d) A homeowner's equity loan for $40,000

10.8 Five independent projects were ranked in decreasing order by two measures—rate of return (ROR) and present worth (PW)—to determine which should be funded with the total initial investment not to exceed $30 million. (a) Use the results below to determine the opportunity cost in ROR terms for each measure. (b) If a MARR of 15% per year is a firm requirement, how does the opportunity cost help in selecting projects to fund?

Measures for Each Project				Ranking by ROR		Ranking by PW	
Project	ROR, %	PW at 15%, $1,000	Initial Investment, $1,000	Project	Cumulative Investment, $1,000	Project	Cumulative Investment, $1,000
A	44.5	7,138	8,000	A	8,000	A	8,000
B	12.8	−1,162	15,000	E	13,000	C	16,000
C	20.4	1,051	8,000	C	21,000	E	21,000
D	9.6	−863	8,000	B	36,000	D	29,000
E	26.0	936	5,000	D	44,000	B	44,000

10.9 Tom, the owner of Burger Palace, determined that his weighted average cost of capital is 8%. He expects a return of 4% per year on all of his investments. A proposal presented by the owner of the Dairy Choice next door seems quite risky to Tom, but it is an intriguing partnership opportunity. Tom has determined that the proposal's "risk factor" will require an additional 3% per year return for him to accept it.
(a) Use the recommended approach to determine the MARR that Tom should use and explain how the 3% risk factor is compensated for in this MARR.
(b) Determine the effective MARR for his business if Tom turns down the proposal.

D-E Mix and WACC

10.10 A new cross-country, trans-mountain water pipeline needs to be built at an estimated first cost of $200,000,000. The consortium of cooperating companies has not fully decided the financial arrangements of this adventurous project. The WACC for similar projects has averaged 10% per year. (a) Two financing options have been identified. The first requires an investment of 60% equity funds at 12% and a loan for the balance at an interest rate of 9% per year. The second option requires only 20% equity funds and the balance obtained by a massive international loan estimated to carry an interest rate of 12.5% per year, which is, in part, based on the geographic location of the pipeline. Which financing plan will result in the smaller average cost of capital? (*b*) If the consortium CFOs have decided that the WACC must not exceed the 5-year historical average of 10% per year, what is the maximum acceptable loan interest rate for each financing option?

10.11 Nucor Corp. manufactures generator coolers for nuclear and gas turbine power plants. The company completed a plant expansion through financing that had a debt/equity mix of 40%–60%. If $15 million came from mortgages and bond sales, what was the total amount of the financing?

10.12 Nano-Technologies bought out RT-Micro using financing as follows: $16 million from mortgages, $4 million from retained earnings, $12 million from cash on hand, and $30 million from bonds. Determine the debt-to-equity mix.

10.13 Tiffany Baking Co. wants to arrange for $37.5 million in capital for manufacturing a new baked potato chip product line. The current financing plan is 60% equity and 40% debt capital. Calculate the expected WACC for the following financing scenario:

Equity capital: 60%, or $22.5 million, via common stock sales for 40% of this amount that will pay dividends at a rate of 5% per year, and the remaining 60% from retained earnings, which currently earn 9% per year.

Debt capital: 40%, or $15 million, obtained through two sources: bank loans for $10 million borrowed at 8% per year, and the remainder in convertible bonds at a coupon rate estimated to be 10% per year.

10.14 Seven different financing plans with their D-E mixes and costs of debt and equity capital for a new innovations project are summarized below. Use the data to determine what mix of debt and equity capital will result in the lowest WACC. (Note: Problem 10.51 explores these financing plans more deeply using a spreadsheet.)

	Debt Capital		Equity Capital	
Plan	Percentage	Rate, %	Percentage	Rate, %
1	100	14.5	–	–
2	70	13.0	30	7.8
3	65	12.0	35	7.8
4	50	11.5	50	7.9
5	35	9.9	65	9.8
6	20	12.4	80	12.5
7	–	–	100	12.5

10.15 A public corporation in which you own common stock reported a WACC of 10.7% for the year in its annual report to stockholders. The common stock that you own has averaged a total return of 6% per year over the last 3 years. The annual report also mentions that projects within the corporation are 80% funded by its own capital. Estimate the company's cost of debt capital. Does this seem like a reasonable rate for borrowed funds?

10.16 Alpha Engineering invested $30 million using a D-E mix of 65%–35% for the development, marketing, and delivery of a web-based training program for project management. Determine the return on the company's equity if the net income from the sale of the program for the first year was $4 million from total revenue of $6 million.

10.17 Business and engineering seniors are comparing methods of financing their college education during their senior year. The business student has $30,000 in student loans that comes due at graduation. Interest is an effective 4% per year. The engineering senior owes $50,000: 50% from his parents with no interest due, and 50% from a credit union loan. This latter amount is also due at graduation with an effective rate of 7% per year.
(*a*) What is the D-E mix for each student?
(*b*) If their grandparents pay the loans in full at graduation, what are the amounts on the checks they write for each graduate?
(*c*) When grandparents pay the full amount at graduation, what percent of the principal does the interest represent?

10.18 Two public corporations, First Engineering and Midwest Development, each show capitalization of $175 million in their annual reports. The balance sheet for First indicates total debt of $87 million, and that of Midwest indicates net worth of $62 million. Determine the D-E mix for each company.

10.19 Determine the WACC for Delta Corporation, which manufactures miniature triaxial accelerometers for space-restricted applications. The financing profile, with interest rates, is as follows: $3 million in stock sales at 8% per year, $4 million in bonds at 9%, and $6 million in retained earnings at 11% per year.

10.20 To understand the advantage of debt capital from a tax perspective in the United States, determine the before-tax and approximated after-tax weighted average costs of capital if a project is funded 40%–60% with debt capital borrowed at 9% per year. A recent study indicates that corporate equity funds earn 12% per year and that the effective tax rate is 35% for the year.

10.21 BASF will invest $14 million this year to upgrade its ethylene glycol processes. This chemical is used to produce polyester resins to manufacture products varying from construction materials to aircraft, and from luggage to home appliances. Equity capital costs 14.5% per year and will supply 65% of the capital funds. Debt capital costs 10%

per year before taxes. The effective tax rate for BASF is 36%.

(a) Determine the amount of annual revenue after taxes that is consumed in covering the interest on the project's initial cost.

(b) If the corporation does not want to use 65% of its own funds, the financing plan may include 75% debt capital. Determine the amount of annual revenue needed to cover the interest with this plan, and explain the effect it may have on the corporation's ability to borrow in the future.

Cost of Debt Capital

10.22 In order to finance a new project, a company borrowed $4,000,000 at 8% per year with the stipulation that the company would repay the loan plus all interest at the end of one year. Assume the company's effective tax rate is 39%. What was the company's cost of debt capital (a) before taxes, and (b) after taxes? (c) Compare the calculated after-tax cost with the approximated cost using Equation [10.4].

10.23 The cash flow plan associated with a debt financing transaction allowed a company to receive $2,800,000 now in lieu of future interest payments of $196,000 per year for 10 years plus a lump sum of $2,800,000 in year 10. If the company's effective tax rate is 33%, determine its cost of debt capital (a) before taxes, and (b) after taxes.

10.24 Engineers at a semiconductor company developed an improved front-end-of-line (FEOL) formulation process that requires an investment of $6 million. The company plans to issue $6 million worth of 10-year bonds that will pay interest of 6% per year, payable annually. If the company's effective tax rate is 40%, what is the after-tax cost (i.e., interest rate) of the debt financing?

10.25 A company that makes several different types of skateboards, Jennings Outdoors, incurred interest expenses of $1,200,000 per year from various types of debt financing. The company received $19,000,000 in year 0 through the sale of discounted bonds with a face value of $20,000,000. The company repaid the principal of the loans in year 15 in a lump sum payment of $20,000,000. If the company's effective tax rate is 29%, what was Jennings' cost of debt capital (a) before taxes, and (b) after taxes? (c) Write a single-cell RATE function to display the rate for each debt capital cost requested.

10.26 Tri-States Gas Producers expects to borrow $800,000 for field engineering improvements. Two methods of debt financing are possible—borrow it all from a bank or issue debenture bonds. The company will pay an effective 8% per year to the bank for 8 years. The principal on the loan will be reduced

uniformly over the 8 years, with the remainder of each annual payment going toward interest. The bond issue will be for 800 ten-year bonds of $1000 each that require a 6% per year dividend payment.

(a) Which method of financing is cheaper after an effective tax rate of 40% is considered?

(b) Which is the cheaper method using a before-tax analysis? Is it the same as the after-tax choice?

10.27 An international pharmaceutical company is initiating a new project that requires $2.5 million in debt capital. The current plan is to sell 20-year bonds that pay 4.2% per year, payable quarterly, at a 3% discount on the face value. The company has an effective tax rate of 35% per year. Determine (a) the total face value of the bonds required to obtain $2.5 million, and (b) the effective annual after-tax cost of debt capital using two methods—factors and spreadsheet functions.

Cost of Equity Capital

10.28 Which form of financing has the lower after-tax cost, debt or equity? Why?

10.29 Harris International currently pays a dividend of $3.24 per share on its preferred stock that sells for $54 per share. In order to raise capital to purchase a smaller competitor, the company plans to issue 2.7 million shares of preferred stock at a 10% discount to its current price. Determine (a) the amount of funding that Harris will realize through the stock offering, and (b) the cost of equity financing.

10.30 BBK Industries plans to sell 2 million shares of its common stock for $80 per share, with an annual dividend of $1.90 per share. Determine the lower cost of equity capital under the following conditions: (a) The company expects a dividend growth rate of 3% per year, and (b) a 5% discount is offered to attract stock purchases and a much lower dividend growth rate of 1% per year is anticipated.

10.31 What dividend growth rate would be required to produce a cost of equity capital of 8% when the common stock price is $140 per share and the dividend is $4.76 per share?

10.32 When the risk-free return is 3.5%, what is the cost of equity capital for a company whose stock has a historical beta factor of 0.92 and the security market indicates that the premium above the risk-free rate is 5%?

10.33 H2W Technologies is considering raising capital to expand its offerings of 2-phase and 4-phase linear stepper motors. The beta value for its stock is high at 1.41. Use the capital asset pricing model and a 3.8% premium above the risk-free return to determine the cost of equity capital. The risk-free return is 3.2% per year.

10.34 Common stock issued by Meggitt Sensing Systems paid stockholders an initial dividend of $0.93 per share on an average price of $18.80 last year. The company expects to grow the dividend rate at a maximum of 1.5% per year. The stock volatility is 1.19, and other stocks in the same industry are paying an average of 4.95% per year dividend. U.S. Treasury bills are returning 2.0%. Determine Meggitt's cost of equity capital last year using (*a*) the dividend method, and (*b*) the CAPM. (*c*) To what amount could the initial year dividend have decreased before the CAPM estimate would have exceeded the dividend method estimate?

10.35 Last year a Japanese engineering materials corporation, Yamachi Inc., purchased U.S. Treasury bonds that returned an average of 4% per year. Now, Euro bonds are being purchased with a realized average return of 3.9% per year. The volatility factor of Yamachi stock last year was 1.10; but, it has increased this year to 1.18. Other publicly traded stocks in this same business arena are paying an average dividend of 5.1% per year. Determine the cost of equity capital for each year and explain why the increase or decrease seems to have occurred.

10.36 The engineering manager at FXO Plastics wants to complete an alternative evaluation study. She asked the finance manager for the corporate MARR. The finance manager gave her some data on the project and stated that all projects must clear their average (pooled) cost by at least 4%. Use the data to determine the minimum before-tax MARR.

Source of Funds	Amount, $	Average Cost, %
Retained earnings	4,000,000	7.4
Stock sales	6,000,000	4.8
Long-term loans	5,000,000	9.8

Different D-E Mixes

10.37 Why is it financially unhealthy for an individual to maintain a large percentage of debt financing over a long period of time, that is, to be highly leveraged?

10.38 Mosaic Software has an opportunity to invest $10,000,000 in a new engineering remote-control system for offshore drilling platforms in partnership with two other companies. Financing for Mosaic will be split between common stock sales ($5,000,000) and a loan with an 8% per year interest rate. Mosaic's share of the annual net cash flow is estimated to be $1,115,000 for each of the next 6 years. Mosaic is about to initiate CAPM as its common stock evaluation model. Recent analysis shows that it has a volatility rating of 1.05 and is paying a premium of 5% common stock dividend. Risk-free government bond investments are currently paying 4% per year. Is the venture financially attractive if the MARR equals (*a*) the cost of equity capital, and (*b*) the WACC? (Note: Refer to problem 10.52 for further analysis of this investment opportunity.)

10.39 Halifax Technologies primarily relies on 100% equity financing to fund projects. A good opportunity is available that will require $250,000 in capital. The Halifax owner can supply the money from personal investments that currently earn an average of 8.5% per year. The annual net cash flow from the project is estimated at $30,000 for the next 15 years. Alternatively, 60% of the required amount can be borrowed for 15 years at 9% per year. Using a before-tax analysis and setting the MARR equal to the WACC, determine which plan, if either, is better.

10.40 Mrs. McKay's Nutrition Products has different methods by which a $600,000 project can be funded using debt and equity capital. A net cash flow of $90,000 per year is estimated for 7 years.

Type of Financing	Financing Plan, %			Cost per Year, %
	1	2	3	
Debt	20	50	60	10.0
Equity	80	50	40	7.5

Determine the rate of return for each plan, and identify the ones that are economically acceptable if (*a*) MARR equals the cost of equity capital, (*b*) MARR equals the WACC, and (*c*) MARR is halfway between the cost of equity capital and the WACC. (*d*) Do the decisions for the three financing plans support the fact that a highly leveraged project is more likely to be acceptable in that the rate of return on equity capital is higher? Explain the basis of your answer.

10.41 A new annular die process is to be installed for extruding pipes, tubes, and tubular films. The phase I installed price for the dies and machinery is $2,000,000. The manufacturer has not decided how to finance the system. The WACC over the last 5 years has averaged 9.5% per year.
(*a*) Two financing alternatives have been defined. The first requires an investment of 40% equity funds at 9% and a loan for the balance at an interest rate of 10% per year. The second alternative requires only 25% equity funds and the balance borrowed at 10.5% per year. Which approach will result in the smaller average cost of capital?
(*b*) Yesterday, the corporate finance committee decided that the WACC for all new projects must not exceed the 5-year historical average of 9.5% per year. With this restriction, what is the maximum loan interest rate that can be incurred for each of the financing alternatives?

10.42 Deavyanne Johnston, the engineering manager at TZO Chemicals, is conducting an evaluation of alternatives based on ROR. She was given the following data and told that due to the unusually large number of investment opportunities the company now has, all future projects must have a ROR that is at least 12.5% above the company's weighted average cost of capital on an after-tax basis. If the company's effective tax rate is 32%, what is the *after-tax* MARR she should use in her evaluation?

Funds Source	Amount, $	Cost, %
Retained earnings	4 million	7.4
Stock sales	6 million	4.8
Long-term loans	5 million	10.4
Budgeted funds for project	15 million	

10.43 Two friends each invested $20,000 of their own (equity) funds. Stan, being more conservative, purchased utility and manufacturing corporation stocks. Theresa, being a risk taker, leveraged the $20,000 and purchased a $100,000 condo for rental property. Considering no taxes, dividends, or revenues, analyze these two purchases by doing the following for 1 year after the funds were invested.

  (a) Determine the year-end values of their equity funds if there was a 10% *increase* in the value of the stocks and the condo.

  (b) Determine the year-end values of their equity funds if there was a 10% *decrease* in the value of the stocks and the condo.

  (c) Use your results to explain why leverage can be financially risky.

Multiple, Noneconomic Attributes

10.44 Three alternatives are being evaluated based on six different attributes, all of which are considered of equal importance. (*a*) Determine the weight to assign to each attribute. (*b*) Write the equation used to calculate the weighted attribute measure R_j for each alternative.

10.45 A company executive assigned importance values between 0 and 100 to five attributes included in an alternative evaluation process. Determine the weight of each attribute using the importance scores.

Attribute, i	Importance Score, s_i
1. Safety	40
2. Cost	60
3. Production rate	70
4. Environmental	30
5. Maintainability	50

10.46 Jill rank-ordered 10 attributes in increasing importance and identified them as A, B, ..., J, with a value of 1 assigned to A, 2 to B, etc. (*a*) What is the sum of the scores? (*b*) What is the weight for attribute D? (*c*) Answer the two questions above if Jill decides that attribute D has the same importance as J, but all other scores stay the same.

10.47 A committee of four people submitted the following statements about the attributes to be used in a weighted attribute evaluation. Use the statements to determine the normalized weights of each attribute if scores are assigned between 0 and 10, with 10 indicating the most important factor.

Attribute	Statement
1. Flexibility	The most important factor
2. Safety	50% as important as uptime
3. Uptime	One-half as important as flexibility
4. Speed	As important as uptime
5. Rate of return	Twice as important as safety

10.48 Different types and capacities of crawler hoes are being considered for use in a significant excavation project to bury fiber-optic cable in Argentina. Several supervisors who have experience with similar projects have identified key attributes and their view of relative importance. Determine the weighted rank order (0 to 10 scale) and the normalized weights.

Attribute	Comment
1. Truck vs. hoe height	90% as important as trenching speed
2. Type of topsoil	Only 10% of most important attribute
3. Type of subsoil	30% as important as trenching speed
4. Hoe cycle time	Twice as important as type of subsoil
5. Hoe trenching speed	Most important attribute
6. Cable-laying speed	80% as important as hoe cycle time

10.49 John, who works for Dumas Jewelers, has decided to use the weighted attribute method to compare three systems of cutting diamonds for setting into rings, earrings, necklaces, and bracelets. Once the final inspector and cutting manager scored each of three attributes in terms of importance to them, John placed an evaluation from 0 to 100 on each system for the three attributes. John's *ratings* for each system follow:

Attribute	System 1	System 2	System 3
1. ROR > MARR	50	70	100
2. Throughput rate	100	60	30
3. Accuracy	100	40	50

(*a*) Use the *weights* below to evaluate the systems. (*b*) Are the results the same for both individuals' weights? Why?

Importance Score	Inspector	Manager
1. ROR > MARR	20	100
2. Throughput rate	80	80
3. Accuracy	100	20

10.50 An airport Baggage Handing Department has evaluated two proposals for baggage delivery conveyor systems. A present worth analysis at $i = 15\%$ per year of estimated revenues and costs resulted in $PW_A = \$460,000$ and $PW_B = \$395,000$. In addition to this economic measure, three more attributes were independently assigned a relative importance score from 0 to 100 by the department manager and a senior team supervisor.

Attribute	Importance Scores	
	Manager	Supervisor
1. Economics	80	25
2. Durability	35	80
3. Safety	30	100
4. Maintainability	20	90

Separately, you have used the four attributes to value rate the two proposals on a scale of 0 to 1.0 as shown in the following table. (The economics attribute was rated using the PW values.)

Attribute	Value Rating	
	Proposal A	Proposal B
1. Economics	1.00	0.90
2. Durability	0.35	0.50
3. Safety	1.00	0.20
4. Maintainability	0.25	1.00

Select the better proposal using each of the following methods:
(a) Weighted evaluation of the department manager
(b) Weighted evaluation of the team supervisor
(c) Present worth

EXERCISES FOR SPREADSHEETS

10.51 Financing plans for a project are summarized below for Encore Productions.
  (a) Plot the WACC and indicate the D-E mix with the lowest WACC.
  (b) Yesterday, the president of Angkor Bank, the bank that usually makes loans to Encore, informed the CFO that interest rates on all loans will increase by 1% per year immediately due to economic instability. Encore's CFO does not want the WACC to exceed 9.9% for this project. Update the rates for debt capital, determine the D-E mix with the lowest WACC, and determine for this specific D-E mix the percentage debt capital that must not be exceeded to ensure that the WACC is at 9.9%. Determine this maximum limit on debt capital using Goal Seek.

Plan	Debt Capital		Equity Capital	
	Percentage	Rate, %	Percentage	Rate, %
1	100	14.5	–	–
2	70	13.0	30	7.8
3	65	12.0	35	7.8
4	50	11.5	50	7.9
5	35	9.9	65	9.8
6	20	12.4	80	12.5
7	–	–	100	12.5

10.52 In Problem 10.38, Mosaic Software could invest $10,000,000 over a 6-year period with a net cash flow estimate of $1,115,000 per year. The equity portion of the investment will cost 9.25% per year; however, the debt portion can vary from 20% to 80% of the total amount, and the required loan rate may change with increasing amounts of debt financing. After an important meeting with the loan officers of the two prime lending banks, the CFO

of Mosaic formulated scenarios Bank 1 and Bank 2 for funding the project with different D-E mixes. Develop the WACC curves using a spreadsheet for each scenario and determine if the project is economically justified based on equity financing provided the MARR is set equal to the WACC (a) for the D-E mix of 50%–50%, and (b) for each funding scenario.

Project	Cost of Debt Capital, %	
D-E mix	Bank 1	Bank 2
20–80	8%	7.5 %
30–70	8%	7.5%
40–60	8%	8.0%
50–50	8%	8.0%
60–40	8%	8.5%
70–30	8%	9.0%
80–20	8%	9.5%

10.53 TrvlSafe, a manufacturer of air-freightable pet crates with imbedded chips to monitor the health of the pet, has identified two projects that have relatively high risk; however, they are expected to move the company into new revenue markets. Utilize a spreadsheet to determine:
  (a) Which of the two projects, if either, are acceptable when the MARR is equal to the after-tax WACC.
  (b) If the same projects are acceptable, provided the risk factors are significant and warrant an additional return of 2% per year above the established MARR.

Project	Initial Investment, $	After-tax Cash Flow, $/Year	Life, Years
Wildlife (W)	–250,000	48,000	10
Reptiles (R)	–125,000	30,000	5

Financing will be developed using a D-E mix of 60%–40% with equity funds costing 7.5% per year. Debt financing will be developed from $10,000, 5% per year, quarterly dividend, 10-year bonds. The effective tax rate is 30% per year.

10.54 To work this problem via spreadsheet, please refer to the data in Problem 10.50.
   (a)   Determine which proposal, A or B, to select using the weighted attribute method and the importance scores for the manager and the supervisor.
   (b)   You have given proposal B a very low rating, V_{ij}, for safety, only 20%. Determine the minimum value rating that you would have to assign to proposal B for the safety attribute to select B with an R_B that is at least 10% more than that determined previously for proposal A.

ADDITIONAL PROBLEMS AND FE EXAM REVIEW QUESTIONS

10.55 All of the following are examples of debt capital, except:
   (a)   Mortgage on equipment
   (b)   Long-term bonds
   (c)   Short-term loan from a bank
   (d)   Preferred stock

10.56 For a 60–40 D-E mix of investment capital, the maximum cost for debt capital that would yield a WACC of 10% when the cost of equity capital is 4% is closest to:
   (a)   8%
   (b)   12%
   (c)   14%
   (d)   16%

10.57 If a company finances an expansion in its production facilities by issuing $6 million in preferred stock, using $3.5 million in retained earnings, and obtaining $15 million via a secured loan, it will have a D-E mix closest to:
   (a)   60–40
   (b)   50–50
   (c)   40–60
   (d)   30–70

10.58 All of the following are factors that affect the effective MARR of a project, except:
   (a)   Project risk
   (b)   Product selling price
   (c)   Availability of capital
   (d)   Attractiveness of other investment opportunities

10.59 Gonzales, Inc. financed a new product as follows: $5 million in stock sales at 13.7% per year, $2 million in retained earnings at 8.9% per year, and $3 million through convertible bonds at 7.8% per year. The company's WACC is closest to:
   (a)   9% per year
   (b)   10% per year
   (c)   11% per year
   (d)   12% per year

10.60 If the after-tax rate of return for a cash flow series is 11.2% and the corporate effective tax rate is 39%, the approximated before-tax rate of return is closest to:
   (a)   6.8%
   (b)   5.4%
   (c)   18.4%
   (d)   28.7%

10.61 Medzyme Pharmaceuticals has maintained a 50–50 D-E mix for capital investments. Equity capital has cost 11%; however, debt capital that has historically cost 9% is now 20% higher than that. If Medzyme does not want to exceed its historical weighted average cost of capital (WACC), and it is forced to go to a D-E mix of 75–25, the maximum acceptable cost of equity capital is closest to:
   (a)   7.6%
   (b)   9.2%
   (c)   9.8%
   (d)   10.9%

10.62 The importance values (0 to 100) for five attributes are shown below. The weight to assign to attribute 1 is:
   (a)   0.16   (b)   0.20
   (c)   0.22   (d)   0.55

Attribute	Importance Score
1	55
2	45
3	85
4	30
5	60

10.63 All of the following are acceptable attribute identification approaches, except:
   (a)   Employing small group discussions
   (b)   Using the same attributes that competing entities use
   (c)   Getting input from experts with relevant experience
   (d)   Surveying the stakeholders

10.64 Three attributes are first cost, safety, and environmental concerns. Assigned importance scores are 100, 75, and 50, respectively. The weight for environmental concerns is closest to:
(*a*) 0.44   (*b*) 0.33
(*c*) 0.22   (*d*) 0.11

10.65 Alternative locations for an advanced wastewater recycling plant are being evaluated using four attributes identified as attributes 1, 2, 3, and 4 with weights of 0.4, 0.3, 0.2, and 0.1, respectively. If the value rating scale is from 1 to 10, and the ratings are 3, 7, 2, and 10 for attributes 1, 2, 3, and 4, respectively, the weighted attribute measure R is closest to:
(*a*) 3.7   (*b*) 3.9
(*c*) 4.0   (*d*) 4.7

10.66 Ten noneconomic attributes are identified as *A*, *B*, *C*, ..., *J*. If they are rank-ordered in terms of *decreasing* importance with a value of 10 assigned to *A*, 9 to *B*, etc., the weighting of attribute *B* is closest to:
(*a*) 0.24   (*b*) 0.17
(*c*) 0.08   (*d*) 0.04

CASE STUDY

EXPANDING A BUSINESS—DEBT VS. EQUITY FINANCING?

Background

Hormel Foods is in the process of initiating a new meal service called Sheila's In-home Meals. Whatsapp, the cross-platform instant messaging service for smartphones, will be used to communicate with a person in the family to select a meal design from several offered each day. The entire evening meal, including dishes, drinks, etc., will be delivered and placed on the table for the family and/or guests to enjoy at an agreed-upon time. Later, the truck returns to remove any leftovers and the utensils. To provide this service, Hormel is about to purchase 100 vans with custom interiors for a total of $1.5 million. Each van is expected to be used for 10 years and have a $2000 salvage value.

A feasibility study completed last year indicated that this expanded business could realize an estimated annual net cash flow of $300,000 before taxes. After-tax considerations would have to take into account an effective tax rate of 35%.

An engineer with Hormel Farms Distribution Division has worked with the corporate finance office to determine how to best develop the $1.5 million capital needed for the purchase of vans. There are two viable financing plans.

Information

Plan A is debt financing for 50% of the capital ($750,000) with the 8% per year compound interest loan repaid over 10 years with uniform year-end payments. (A simplifying assumption that $75,000 of the principal is repaid with each annual payment can be made.)

Plan B is 100% equity capital raised from the sale of $15 per share common stock. The financial manager informed the engineer that stock is paying $0.50 per share in dividends and that this dividend rate has been increasing at an average of 5%

each year. This dividend pattern is expected to continue, based on the current financial environment.

Case Study Exercises

1. What values of MARR should the engineer use to determine the better financing plan?
2. The engineer must make a recommendation on the financing plan by the end of the day. He does not know how to consider all the tax angles for the debt financing in plan A. However, he does have a handbook that gives these relations for equity and debt capital about taxes and cash flows:

   Equity capital: no income tax advantages

   After-tax net cash flow
       = (before-tax net cash flow)(1 − tax rate)

   Debt capital: income tax advantage comes from interest paid on loans

   After-tax net cash flow = before-tax net cash flow
       − loan principal
       − loan interest − taxes

   Taxes = (taxable income)(tax rate)
   Taxable income = net cash flow
       − loan interest

   He decides to forget any other tax consequences and use this information to prepare a recommendation. Is A or B the better plan?
3. The division manager would like to know how much the WACC varies for different D-E mixes, especially about 15% to 20% on either side of the 50% debt financing option in plan A. Plot the WACC curve and compare its shape with that of Figure 10–2.

Royalty-Free/CORBIS

CHAPTER 11

Replacement and Retention Decisions

LEARNING OUTCOMES

Purpose: Perform a replacement/retention study between an in-place asset, process, or system and one that could replace it.

SECTION	TOPIC	LEARNING OUTCOME
11.1	Replacement study basics	• Explain the fundamental approach and terminology of replacement analysis.
11.2	Economic service life	• Determine the ESL that minimizes the total AW for estimated costs and salvage value.
11.3	Replacement analysis	• Perform a replacement/retention study between a defender and the best challenger.
11.4	Additional considerations	• Understand the approach to special situations in a replacement study.
11.5	Study period analysis	• Perform a replacement/retention study over a specified number of years.
11.6	Replacement value	• Calculate the minimum trade-in (breakeven) value required to make the challenger economically attractive.

O ne of the most common and important issues in industrial practice is that of replacement or retention of an asset, process, or system that is currently installed. This differs from previous situations where all the alternatives were new. The fundamental question answered by a replacement study (also called a replacement/retention study) about a currently installed system is, *Should it be replaced now or later*? When an asset is currently in use and its function is needed in the future, it will be replaced at some time. In reality, a replacement study answers the question of **when, not if,** to replace.

A replacement study is usually designed to first make the economic decision to retain or replace *now*. If the decision is to replace, the study is complete. If the decision is to retain, the cost estimates and decision can be revisited periodically to ensure that the decision to retain is still economically correct. This chapter explains how to perform the initial-year and follow-on year replacement studies.

A replacement study is an application of the AW method of comparing unequal-life alternatives, first introduced in Chapter 6. In a replacement study with no specified study period, the AW values are determined by a technique called the **economic service life (ESL)** analysis. If a study period is specified, the replacement study procedure is slightly different.

If asset depreciation and taxes are to be considered in an *after-tax replacement analysis,* Chapters 16 and 17 should be covered before or in conjunction with this chapter. After-tax replacement analysis is included in Chapter 17.

PE

Keep or Replace the Kiln Case: B&T Enterprises manufactures and sells high-melting-temperature ceramics and high-performance metals to other corporations. The products are sold to a wide range of industries from the nuclear and solar power industry to sports equipment manufacturers of specialty golf and tennis gear, where kiln temperatures up to approximately 1700°C are needed. For years, B&T has owned and been very satisfied with Harper International pusher-plate tunnel kilns. Two are in use currently at plant locations on each coast of the country; one kiln is 10 years old, and the second was purchased only 2 years ago and serves, primarily, the ceramics industry needs on the west coast. This newer kiln can reach temperatures of 2800°C.

During the last two or three quarterly maintenance visits, the Harper team leader and the head of B&T quality have discussed the ceramic and metal industry needs for higher temperatures. In some cases the temperatures are as high as 3000°C for emerging nitride, boride, and carbide transition metals that form very high-melting-temperature oxides. These may find use in hypersonic vehicles, engines, plasma arc electrodes, cutting tools, and high-temperature shielding.

A looming question on the mind of the senior management and financial officers of B&T revolves around the need to seriously consider a new graphite hearth kiln, which can meet higher temperature and other needs of the current and projected customer base. This unit will have lower operating costs and significantly greater furnace efficiency in heat time, transit, and other crucial parameters. Since virtually all of this business is on the west coast, the graphite hearth kiln would replace the newer of the two kilns currently in use.

For identification, let

PT identify the currently installed pusher-plate tunnel kiln (defender)

GH identify the proposed new graphite hearth kiln (challenger)

Relevant estimates follow in $ millions for monetary units.

	PT	GH
First cost, $ M	$25; 2 years ago	$38; with no trade-in
AOC, $ M per year	year 1: $5.2; year 2: $6.4	starts at $3.4, increases 10%/year
Life, years	6 (remaining)	12 (estimated)
Heating element, $ M	—	$2.0 every 6 years

This case is used in the following topics of this chapter:

Economic service life (Section 11.2)

Replacement study (Section 11.3)

Replacement study with study period specified (Section 11.5)

Replacement value (Section 11.6)

Problems 11.18 and 11.38

11.1 Basics of a Replacement Study ● ● ●

The need for a replacement study can develop from several sources:

Reduced performance. Because of physical deterioration, the ability to perform at an expected level of *reliability* (being available and performing correctly when needed) or *productivity* (performing at a given level of quality and quantity) is not present. This usually results in increased costs of operation, higher scrap and rework costs, lost sales, reduced quality, diminished safety, and larger maintenance expenses.

Altered requirements. New requirements of accuracy, speed, or other specifications cannot be met by the existing equipment or system. Often the choice is between complete replacement or enhancement through retrofitting or augmentation.

Obsolescence. International competition and rapidly changing technology make currently used systems and assets perform acceptably but less productively than equipment coming available. The ever-decreasing development cycle time to bring new products to market is often the reason for *premature replacement* studies, that is, studies performed before the estimated useful or economic life is reached.

Replacement studies use some terminology that is closely related to terms in previous chapters.

Defender and **challenger** are the names for two mutually exclusive alternatives. The defender is the currently installed asset, and the challenger is the potential replacement. A replacement study compares these two alternatives. The challenger is the "best" challenger because it has been selected as the best one to possibly replace the defender. (This is the same terminology used earlier for incremental ROR and B/C analysis, but both alternatives were new.)

Salvage/market value

Market value is the current value of the installed asset if it were sold or traded on the open market. Also called *trade-in value*, this estimate is obtained from professional appraisers, resellers, or liquidators familiar with the industry. As in previous chapters, **salvage value** is the estimated value at the end of the expected life. In replacement analysis, the salvage value at the end of one year is used as the market value at the beginning of the next year.

AW values are used as the primary economic measure of comparison between the defender and challenger. The term *equivalent uniform annual cost* (*EUAC*) may be used in lieu of AW because often only costs are included in the evaluation; revenues generated by the defender or challenger are assumed to be equal. (Since EUAC calculations are exactly the same as for AW, we use the term AW.) Therefore, all values will be negative when only costs are involved. Salvage or market value is an exception; it is a cash inflow and carries a plus sign.

Economic service life

Economic service life (ESL) for an alternative is the *number of years at which the lowest AW of cost occurs*. The equivalency calculations to determine ESL establish the life n for the best challenger and the lowest cost life for the defender in a replacement study. The next section explains how to find the ESL.

Defender first cost is the initial investment amount P used for the defender. The *current market value* (*MV*) is the correct estimate to use for P for the defender in a replacement study. The estimated salvage value at the end of one year becomes the market value at the beginning of the next year, provided the estimates remain correct as the years pass. It is incorrect to use the following as MV for the defender first cost: trade-in value that *does not represent a fair market value*, or the depreciated book value taken from accounting records. If the defender must be upgraded or augmented to make it equivalent to the challenger (in speed, capacity, etc.), this cost is added to the MV to obtain the estimated defender first cost.

Challenger first cost is the amount of capital that must be recovered when replacing a defender with a challenger. This amount is almost always equal to P, the first cost of the challenger.

If an unrealistically high trade-in value is offered for the defender compared to its fair market value, the *net* cash flow required for the challenger is reduced, and this fact should be considered in the analysis. The correct amount to recover and use in the economic analysis for the challenger is its first cost minus the difference between the trade-in value (TIV) and market value (MV) of the defender. In equation form, this is $P - (TIV - MV)$. This amount represents the actual cost to the company because it includes both the opportunity cost (i.e., market value of the defender) and the out-of-pocket cost (i.e., first cost − trade-in) to acquire the challenger. Of course, when the trade-in and market values are the same, the challenger P value is used in all computations.

As an illustration, assume an installed piece of equipment has a MV of \$50,000; however, a TIV of \$75,000 is offered provided a newer model (the challenger) is purchased for \$300,000. The amount to recover is \$275,000, if the challenger is acquired, based on the difference of \$25,000 between the TIV and MV estimates.

The challenger first cost is the estimated initial investment necessary to acquire and install it. Sometimes, an analyst or manager will attempt to *increase* this first cost by an amount equal to the *unrecovered capital* remaining in the defender, as shown on the accounting records for the asset. This incorrect treatment of capital recovery is observed most often when the defender is working well and in the early stages of its life, but technological obsolescence, or some other reason, has forced consideration of a replacement. This leads us to identify two additional characteristics of replacement analysis, in fact, of any economic analysis: *sunk costs* and *nonowner's viewpoint*.

An amount of money that has been expended in the past and cannot be recovered now or in the future is a **sunk cost.** The replacement alternative for an asset, system, or process should not be burdened with this cost in any direct fashion; sunk costs should be handled in a realistic way using tax laws and write-off allowances.

Sunk cost

A sunk cost is the difference between an asset's book value (determined using accounting procedures discussed in chapter 16) and its current market value. The amount of the sunk cost should never be added to the challenger's first cost because it will make the *challenger appear to be more costly than it actually is*. For example, assume an asset costing \$100,000 two years ago has a depreciated value of \$80,000 on the corporate books. It must be replaced prematurely due to rapidly advancing technology. If the replacement alternative (challenger) has a first cost of \$150,000, the \$80,000 from the current asset is a sunk cost were the challenger purchased. For the purposes of an economic analysis, it is *incorrect to increase* the challenger's first cost to \$230,000 or any number between this and \$150,000.

The second characteristic is the perspective taken when conducting a replacement study. You, the analyst, are a consultant from outside the company.

The **nonowner's viewpoint**, also called the *outsider's viewpoint* or *consultant's viewpoint,* provides the greatest objectivity in a replacement study. This viewpoint performs the analysis without bias; it means the analyst owns neither the defender nor the challenger. Additionally, it assumes the services provided by the defender can be purchased now by making an "initial investment" equal to the market value of the defender.

Nonowner's viewpoint

Besides being unbiased, this perspective is correct because the defender's market value is a forgone opportunity of cash inflow were the replacement not selected, and the defender chosen.

As mentioned in the introduction, a replacement study is an application of the annual worth method. As such, the fundamental assumptions for a replacement study parallel those of an AW analysis. If the *planning horizon is unlimited*, that is, a study period is not specified, the assumptions are as follows:

1. The services provided are needed for the indefinite future.
2. The challenger is the best challenger available now and in the future to replace the defender. When this challenger replaces the defender (now or later), it can be repeated for succeeding life cycles.
3. Cost estimates for every life cycle of the defender and challenger will be the same as in their first cycle.

As expected, none of these assumptions is precisely correct. We discussed this previously for the AW method (and the PW method). When the intent of one or more of the assumptions becomes unacceptable, the estimates for the alternatives must be updated and a new replacement study conducted. The replacement procedure discussed in Section 11.3 explains how to do this. When the planning horizon is limited to a specified study period, the assumptions above do not hold. The procedure of Section 11.5 discusses replacement analysis over a fixed study period.

EXAMPLE 11.1

Only 2 years ago, Techtron purchased for $275,000 a fully loaded SCADA (supervisory control and data acquisition) system including hardware and software for a processing plant operating on the Houston ship channel. When it was purchased, a life of 5 years and salvage of 20% of first cost were estimated. Actual M&O costs have been $25,000 per year, and the book value is $187,000. There has been a series of insidious malware infections targeting Techtron's command and control software. Additionally, some next-generation hardware marketed recently could greatly reduce Techtron's competitiveness in several of its product lines. Given these factors, the system is likely worth nothing if kept in use for the final 3 years of its anticipated useful life.

Model K2-A1, a new replacement turnkey system, can be purchased for $300,000 net cash, that is, $400,000 first cost and a $100,000 trade-in for the current system. A 5-year life, salvage value of 15% of stated first cost or $60,000, and an M&O cost of $15,000 per year are good estimates for the new system. The current system was appraised this morning, and a market value of $100,000 was confirmed for today; however, with the current virus discovery, the appraiser anticipates that the market value will fall rapidly to the $80,000 range once the virus problem and new model are publicized.

Using the above values as the best possible today, state the correct defender and challenger estimates for P, M&O, S, and n in a replacement study to be performed *TODAY*.

Solution

Defender: Use the current market value of $100,000 as the first cost for the defender. All others—original cost of $275,000, book value of $187,000, and trade-in value of $100,000—are irrelevant to a replacement study conducted today. The estimates are as follows:

First cost	$P = \$-100,000$
M&O cost	$A = \$-25,000$ per year
Expected life	$n = 3$ years
Salvage value	$S = 0$

Challenger: The $400,000 stated first cost is the correct one to use for P, because the trade-in and market values are equal.

First cost	$P = \$-400,000$
M&O cost	$A = \$-15,000$ per year
Expected life	$n = 5$ years
Salvage value	$S = \$60,000$

Comment

If the replacement study is conducted next week when estimates will have changed, the defender's first cost will be $80,000, the new market value according to the appraiser. The challenger's first cost will be $380,000, that is, $P - (\text{TIV} - \text{MV}) = 400,000 - (100,000 - 80,000)$.

11.2 Economic Service Life ●●●

Until now the estimated life n of an alternative or asset has been stated. In reality, the best life estimate to use in the economic analysis is not known initially. When a replacement study or an analysis between new alternatives is performed, the best value for n should be determined using current cost estimates. The best life estimate is called the *economic service life*.

Economic service life

> The **economic service life (ESL)** is the number of years n at which the equivalent uniform annual worth (AW) of costs is the minimum, considering the most current cost estimates over all possible years that the asset may provide a needed service.

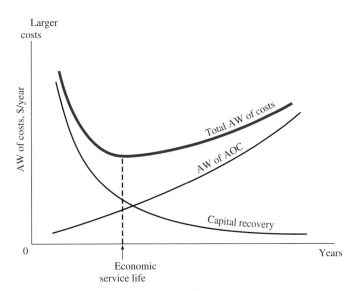

The ESL is also referred to as the *economic life* or *minimum cost life*. Once determined, the ESL should be the estimated life for the asset used in an engineering economy study, if only economics are considered. When n years have passed, the ESL indicates that the asset should be replaced to minimize overall costs. To perform a replacement study correctly, it is important that the ESL of the challenger and the ESL of the defender be determined, since their n values are usually not preestablished.

The ESL is determined by calculating the total AW of costs if the asset is in service 1 year, 2 years, 3 years, and so on, up to the last year the asset is considered useful. Total AW of costs is the sum of capital recovery (CR), which is the AW of the initial investment and any salvage value, and the AW of the estimated annual operating cost (AOC), that is,

> **Total AW = capital recovery − AW of annual operating costs**
> **= CR − AW of AOC**   [11.1]

The ESL is the n value for the smallest total AW of costs. (Remember: These AW values are *cost* estimates, so the AW values are negative numbers. Therefore, $-200 is a lower cost than $-500.) Figure 11–1 shows the characteristic shape of a total AW of cost curve. The CR component of total AW decreases, while the AOC component increases, thus forming the concave shape. The two AW components are calculated as follows.

Decreasing cost of capital recovery. The capital recovery is the AW of investment; it decreases with each year of ownership. Capital recovery is calculated by Equation [6.3], which is repeated here. The salvage value S, which usually decreases with time, is the estimated market value (MV) in that year.

$$\text{Capital recovery} = -P(A/P,i,n) + S(A/F,i,n) \qquad [11.2]$$

Capital recovery

Increasing cost of AW of AOC. Since the AOC (or M&O cost) estimates usually increase over the years, the AW of AOC increases. To calculate the AW of the AOC series for 1, 2, 3, . . . years, determine the present worth of each AOC value with the P/F factor, then redistribute this PW value using the A/P factor.

The complete equation for total AW of costs over k years ($k = 1, 2, 3, \dots$) is

> $$\text{Total AW}_k = -P(A/P,i,k) + S_k(A/F,i,k) - \left[\sum_{j=1}^{j=k} \text{AOC}_j(P/F,i,j)\right](A/P,i,k) \quad [11.3]$$

where P = initial investment or current market value
         S_k = salvage value or market value after k years
     AOC_j = annual operating cost for year j ($j = 1$ to k)

The current MV is used for P when the asset is the defender, and the estimated future MV values are substituted for the S values in years 1, 2, 3, ... Plotting the AW_k series as in Figure 11–1 clearly indicates where the ESL is located and the trend of the AW_k curve on each side of the ESL. When several total AW values are approximately equal, the curve will be flat over several periods. This indicates that the ESL is relatively insensitive to costs.

To determine ESL by spreadsheet, the PMT function (with embedded NPV functions as needed) is used repeatedly for each year to calculate capital recovery and the AW of AOC. Their sum is the total AW for k years of ownership. The PMT function formats for the capital recovery and AOC components for each year k ($k = 1, 2, 3, \ldots$) are as follows:

> **Capital recovery for the challenger:** PMT($i\%$,years,P,−MV_in_year_k)
> **Capital recovery for the defender:** PMT($i\%$,years,current_MV,−MV_in_year_k)   [11.4]
> **AW of AOC:** −PMT($i\%$,years,NPV($i\%$,year_1_AOC:year_k_AOC)+0)

When the spreadsheet is developed, it is recommended that the PMT functions in year 1 be developed using cell-reference format; then drag down the function through each column. A final column summing the two PMT results displays total AW. Augmenting the table with a scatter or line chart graphically displays the cost curves in the general form of Figure 11–1, and the ESL is easily identified. Example 11.2 illustrates ESL determination by hand and by spreadsheet.

EXAMPLE 11.2

A 3-year-old heavy-duty transport vehicle is being considered for early replacement. Its current market value is $20,000. Estimated future market values and annual operating costs for the next 5 years are given in Table 11–1, columns 2 and 3. What is the economic service life of this defender if the interest rate is 10% per year? Solve by hand and by spreadsheet.

Solution by Hand

Equation [11.3] is used to calculate total AW_k for $k = 1, 2, \ldots, 5$. Table 11–1, column 4, shows the capital recovery for the $20,000 current market value ($j = 0$) plus 10% return. Column 5 gives the equivalent AW of AOC for k years. As an illustration, the computation of total AW for $k = 3$ from Equation [11.3] is

$$
\begin{aligned}
\text{Total } AW_3 &= -P(A/P,i,3) + MV_3(A/F,i,3) - [\text{PW of } AOC_1, AOC_2, \text{ and } AOC_3](A/P,i,3) \\
&= -20{,}000(A/P,10\%,3) + 6000(A/F,10\%,3) - [5000(P/F,10\%,1) \\
&\quad + 6500(P/F,10\%,2) + 8000(P/F,10\%,3)](A/P,10\%,3) \\
&= -6230 - 6405 = \$-12{,}635
\end{aligned}
$$

A similar computation is performed for each year 1 through 5. The lowest equivalent cost (numerically largest AW value) occurs at $k = 3$. Therefore, the defender ESL is $n = 3$ years, and the AW value is $\$-12{,}635$. In the replacement study, this AW will be compared with the best challenger AW determined by a similar ESL analysis.

TABLE 11–1   Computation of Economic Service Life

Year j (1)	MV_j, $ (2)	AOC_j, $ (3)	Capital Recovery, $ (4)	AW of AOC, $ (5)	Total AW_k, $ (6) = (4) + (5)
1	10,000	−5,000	−12,000	−5,000	−17,000
2	8,000	−6,500	−7,714	−5,714	−13,428
3	6,000	−8,000	−6,230	−6,405	−12,635
4	2,000	−9,500	−5,878	−7,072	−12,950
5	0	−12,500	−5,276	−7,961	−13,237

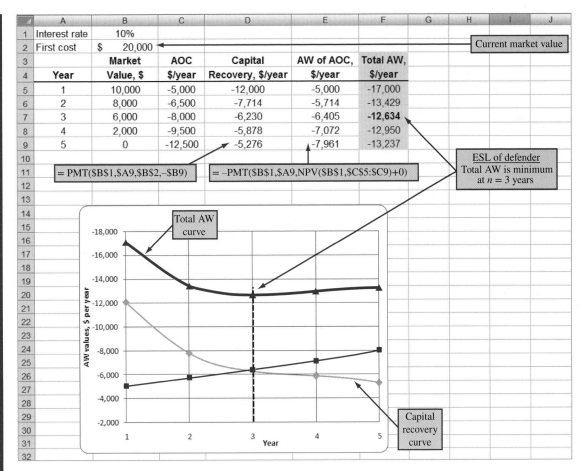

Figure 11–2
Determination of ESL and plot of curves, Example 11.2.

Solution by Spreadsheet

See Figure 11–2 for the spreadsheet image and graph that indicates the ESL is $n = 3$ years and AW = $\$-12{,}634$. (This format is a template for any ESL analysis; simply change the estimates and add rows for more years.) Contents of columns D and E are described below. The PMT functions apply the formats as described in Equation [11.4]. Cell tags show detailed cell-reference format for year 5. The $ symbols are included for absolute cell referencing, needed when the entry is dragged down the column.

Column D: Capital recovery is the AW of the $20,000 investment in year 0 for each year 1 through 5 with the estimated MV in that year. For example, in actual numbers, the cell-reference PMT function in year 5 shown on the spreadsheet reads = PMT (10%,5,20000, −0), resulting in $−5276. This series is plotted in Figure 11–2.

Column E: The NPV function embedded in the PMT function obtains the present worth in year 0 of all AOC estimates through year k. Then PMT calculates the AW of AOC over k years. For example, in year 5, the PMT in numbers is = −PMT(10%,5,NPV (10%,C5:C9)+0). The 0 is the AOC in year 0; it is optional. The graph plots the AW of AOC curve, which constantly increases in cost because the AOC estimates increase each year.

Comment

The *capital recovery curve* in Figure 11–2 (middle curve) is not the expected shape (see year 4) because the estimated market value changes each year. If the same MV were estimated for each year, the curve would appear like Figure 11–1.

It is reasonable to ask about the difference between the ESL analysis above and the AW analyses performed in previous chapters. Previously we had a *specific life estimated to be n years* with associated other estimates: first cost in year 0, possibly a salvage value in year n, and an AOC that remained constant or varied each year. For all previous analyses, the calculation of AW using these estimates determined the AW over n years. This is the economic service life when n is fixed. Also, in all previous cases, there were no year-by-year market value estimates. Therefore, we can conclude the following:

> When the **expected life n is known and specified** for the challenger or defender, no ESL computations are necessary. Determine the AW over n years, using the first cost or current market value, estimated salvage value after n years, and AOC estimates. This AW value is the correct one to use in the replacement study.

However, when n is not fixed, the following is useful. First the market/salvage series is needed. It is not difficult to estimate this series for a new or current asset. For example, an asset with a first cost of P can lose market value of, say, 20% per year, so the market value series for years 0, 1, 2, . . . is P, $0.8P$, $0.64P$, . . . , respectively. If it is reasonable to predict the MV series on a year-by-year basis, it can be combined with the AOC estimates to produce what is called the *marginal costs* for the asset.

> **Marginal costs (MC)** are year-by-year estimates of the costs to own and operate an asset for that year. Three components are added to determine the marginal cost:
> - Cost of ownership (loss in market value is the best estimate of this cost)
> - Forgone interest on the market value at the beginning of the year
> - AOC for each year

Once the marginal costs are estimated for each year, their equivalent AW value is calculated. *The sum of the AW values of the first two of these components is the capital recovery amount.* Now, it should be clear that the total AW of all three marginal cost components over k years is the same value as the total annual worth for k years calculated in Equation [11.3]. That is, the following relation is correct.

$$\text{AW of marginal costs} = \text{total AW of costs} \qquad [11.5]$$

Therefore, there is no need to perform a separate, detailed marginal cost analysis when yearly market values are estimated. The ESL analysis presented in Example 11.2 is sufficient in that it results in the same numerical values. This is demonstrated in Example 11.3 using the progressive example.

EXAMPLE 11.3 Keep or Replace the Kiln Case

In our progressive example, B&T Enterprises is considering the replacement of a 2-year-old kiln with a new one to meet emerging market needs. When the current tunnel kiln was purchased 2 years ago for $25 million, an ESL study indicated that the minimum cost life was between 3 and 5 years of the expected 8-year life. The analysis was not very conclusive because the total AW cost curve was flat for most years between 2 and 6, indicating insensitivity of the ESL to changing costs. Now, the same type of question arises for the proposed graphite hearth model that costs $38 million new: What are the ESL and the estimated total AW of costs? The Manager of Critical Equipment at B&T estimates that the market value after only 1 year will drop to $25 million and then retain 75% of the previous year's value over the 12-year expected life. Use this market value series and $i = 15\%$ per year to illustrate that an ESL analysis and marginal cost analysis result in exactly the same total AW of cost series.

Solution

Figure 11–3 is a spreadsheet screen shot of the two analyses in $ million units. The market value series is detailed in column B starting at $25 (million) and decreasing by 25% per year. A brief description of each analysis follows.

	A	B	C	D	E	F	G
1	Interest rate	15%	**ESL analysis**		First cost, $	38.00	
2		**Market**	**AOC**	**Capital**	**AW of AOC,**	**Total AW,**	
3	**Year**	**Value, $**	**$/year**	**Recovery, $/year**	**$/year**	**$/year**	
4	1	25.00	-3.40	-18.70	-3.40	-22.10	
5	2	18.75	-3.74	-14.65	-3.56	-18.21	
6	3	14.06	-4.11	-12.59	-3.72	-16.31	
7	4	10.55	-4.53	-11.20	-3.88	-15.08	
8	5	7.91	-4.98	-10.16	-4.04	-14.21	Two AW series are identical
9	6	5.93	-7.48	-9.36	-4.43	-13.80	
10	7	4.45	-6.02	-8.73	-4.58	-13.31	
11	8	3.34	-6.63	-8.23	-4.73	-12.95	
12	9	2.50	-7.29	-7.81	-4.88	-12.69	
13	10	1.88	-8.02	-7.48	-5.03	-12.51	
14	11	1.41	-8.82	-7.20	-5.19	-12.39	
15	12	1.06	-9.70	-6.97	-5.35	-12.32	
16	Formulas, year 12	= B14*0.75	= C14*1.1	= PMT($B$1, $A15,$F$1,-$B15)	= -PMT($B$1,$A15, NPV($B$1,$C$4:$C15)+0)	= E15+D15	
17							
18	Interest rate	15%	**Marginal cost analysis**		Current MV, $	38.00	
19		**Market**	**Loss in**	**Lost Interest**		**Marginal Cost**	**AW of Marginal**
20	**Year**	**Value, $**	**Market Value, $**	**on MV for Year, $**	**AOC, $**	**for the Year, $**	**Cost, $/year**
21	1	25.00	-13.00	-5.70	-3.40	-22.10	-22.10
22	2	18.75	-6.25	-3.75	-3.74	-13.74	-18.21
23	3	14.06	-4.69	-2.81	-4.11	-11.61	-16.31
24	4	10.55	-3.52	-2.11	-4.53	-10.15	-15.08
25	5	7.91	-2.64	-1.58	-4.98	-9.20	-14.21
26	6	5.93	-1.98	-1.19	-7.48	-10.64	-13.80
27	7	4.45	-1.48	-0.89	-6.02	-8.40	-13.31
28	8	3.34	-1.11	-0.67	-6.63	-8.41	-12.95
29	9	2.50	-0.83	-0.50	-7.29	-8.62	-12.69
30	10	1.88	-0.63	-0.38	-8.02	-9.02	-12.51
31	11	1.41	-0.47	-0.28	-8.82	-9.57	-12.39
32	12	1.06	-0.35	-0.21	-9.70	-10.26	-12.32
33	Formulas, year 12	= B31*0.75	= B32-B31	= -$B$1*$B31	= E31*1.1	= $C32+$D32+$E32	= -PMT($B$1,$A32, NPV($B$1,$F$21:$F$32)+0)
34							

Figure 11–3

Comparison of annual worth series resulting from ESL analysis and marginal cost analysis, Example 11.3.

ESL analysis: Equation [11.4] is applied repeatedly for $k = 1, 2, \ldots, 12$ years (columns C, D, and E) in the top of Figure 11–3. Row 16 details the spreadsheet functions for year 12. The result in column F is the total AW series that is of interest now.

Marginal cost (MC): The functions in the bottom of Figure 11–3 (columns C, D, and E) develop the three components added to obtain the MC series. Row 33 details the functions for year 12. The resulting AW of marginal costs (column G) is the series to compare with the corresponding ESL series above (column F).

The two AW series are identical, thus demonstrating that Equation [11.5] is correct. Therefore, either an ESL or a marginal cost analysis will provide the same information for a replacement study. In this case, the results show that the new kiln will have a minimum AW of costs of $–12.32 million at its full 12-year life.

We can draw two important conclusions about the n and AW values to be used in a replacement study. These conclusions are based on the extent to which detailed annual estimates are made for the market value.

1. **Year-by-year market value estimates are made.** Use them to perform an ESL analysis, and determine the n value with the lowest total AW of costs. These are the best n and AW values for the replacement study.

2. **Yearly market value estimates are not available.** The only estimates available are the current market value (salvage value) of the defender and its salvage value in year n. Use them to calculate the AW over n years. These are the n and AW values to use; however, they may not be the "best" values in that they may not represent the best equivalent total AW of cost when compared to the results of a full ESL analysis.

Upon completion of the ESL analysis (item 1 above), the replacement study procedure in Section 11.3 is applied using the values

Challenger alternative (C): AW_C for n_C years

Defender alternative (D): AW_D for n_D years

11.3 Performing a Replacement Study ● ● ●

Replacement studies are performed in one of two ways: without a study period specified or with one defined. Figure 11–4 gives an overview of the approach taken for each situation. The procedure discussed in this section applies when no study period (planning horizon) is specified. If a specific number of years is identified for the replacement study, for example, over the next 5 years, with no continuation considered after this time period in the economic analysis, the procedure in Section 11.5 is applied.

A replacement study determines when a challenger replaces the in-place defender. The complete study is finished if the challenger (C) is selected to replace the defender (D) now. However, if the defender is retained now, the study may extend over a number of years equal to the life of the defender n_D, after which a challenger replaces the defender. Use the annual worth and life values for C and D determined in the ESL analysis in the following procedure. Assume the services provided by the defender could be obtained at the AW_D amount.

The replacement study procedure is:

New replacement study:
1. On the basis of the better AW_C or AW_D value, select the challenger C or defender D. When the challenger is selected, replace the defender now, and expect to keep the challenger for n_C years. This replacement study is complete. If the defender is selected, plan to retain it for up to n_D more years. (This is the leftmost branch of Figure 11–4.) Next year, perform the following steps.

One-year-later analysis:
2. Determine if all estimates are still current for both alternatives, especially first cost, market value, and AOC. If any are not current, proceed to step 3. If this is year n_D and the challenger's estimates are still acceptable, replace the defender. If this is not year n_D, retain the defender for another year and repeat this same step. This step may be repeated several times.
3. Whenever the estimates have changed, update them and determine new AW_C and AW_D values. Initiate a new replacement study (step 1).

Figure 11–4
Overview of replacement study approaches.

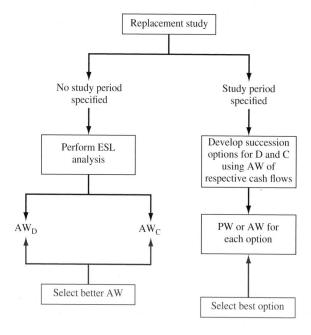

If the defender is selected initially (step 1), estimates may need updating after 1 year of retention (step 2). Possibly there is a new best challenger to compare with D. Either significant changes in defender estimates or availability of a new challenger indicates that a new replacement study is to be performed. In actuality, a replacement study can be performed each year or more frequently to determine the advisability of replacing or retaining any defender, provided a competitive challenger is available.

Example 11.4 illustrates the application of ESL analysis for a challenger and defender, followed by the use of the replacement study procedure. The planning horizon is unspecified in this example.

EXAMPLE 11.4

Two years ago, Toshiba Electronics made a $15 million investment in new assembly line machinery. It purchased approximately 200 units at $70,000 each and placed them in plants in 10 different countries. The equipment sorts, tests, and performs insertion-order kitting of components in preparation for the assembly of special-purpose circuit boards. This year, new international industry standards will require a $16,000 retrofit on each unit, in addition to the expected operating cost. Due to the new standards, coupled with rapidly changing technology, a new system is challenging the retention of these 2-year-old machines. Since the chief engineer at Toshiba USA realizes that the economics must be considered, he has asked that a replacement study be performed this year and each year in the future, if need be. The i is 10% and the estimates are below.

Challenger: First cost: $50,000
                 Future market values: decreasing by 20% per year
                 Estimated retention period: no more than 10 years
                 AOC estimates: $5000 in year 1 with increases of $2000 per year
                     thereafter

Defender: Current international market value: $15,000
                 Future market values: decreasing by 20% per year
                 Estimated retention period: no more than 3 more years
                 AOC estimates: $4000 next year, increasing by $4000 per year
                     thereafter, plus the $16,000 retrofit next year

(a) Determine the AW values and economic service lives necessary to perform the replacement study.
(b) Perform the replacement study now.
(c) After 1 year, it is time to perform the follow-up analysis. The challenger is making large inroads to the market for electronic components assembly equipment, especially with the new international standards features built in. The expected market value for the defender is still $12,000 this year, but it is expected to drop to virtually nothing in the future—$2000 next year on the worldwide market and zero after that. Also, this prematurely outdated equipment is more costly to keep serviced, so the estimated AOC next year has been increased from $8000 to $12,000 and to $16,000 two years out. Perform the follow-up replacement study analysis.

Solution

(a) The results of the ESL analysis, shown in Figure 11–5, include all the MV and AOC estimates in columns B and C. For the challenger, note that $P = \$50,000$ is also the MV in year 0. The total AW of costs is for each year, should the challenger be placed into service for that number of years. As an example, the year $k = 4$ amount of $\$-19,123$ is determined using Equation [11.3], where the A/G factor accommodates the arithmetic gradient series in the AOC.

$$\text{Total AW}_4 = -50,000(A/P,10\%,4) + 20,480(A/F,10\%,4)$$
$$- [5000 + 2000(A/G,10\%,4)]$$
$$= \$-19,123$$

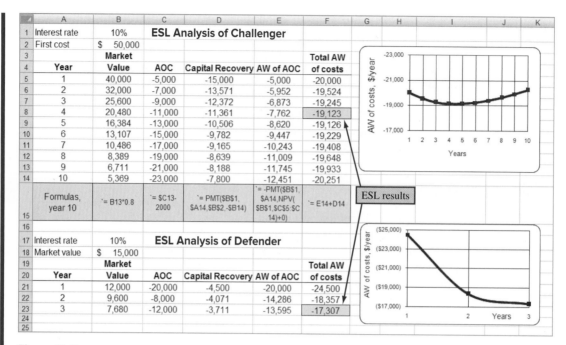

Figure 11–5
ESL analysis of challenger and defender, Example 11.4.

For spreadsheet-based ESL analysis, this same result is achieved in cell F8 using Equation [11.4]. The functions are

$$\text{Total AW}_4 = \text{PMT}(10\%,4,50000,-20480) - \text{PMT}(10\%,4,\text{NPV}(10\%,\text{C5:C8})+0)$$
$$= -11{,}361 - 7{,}762$$
$$= \$ -19{,}123$$

The defender costs are analyzed in the same way up to the maximum retention period of 3 years.

The lowest AW cost (numerically largest) values for the replacement study are as follows:

Challenger: $AW_C = \$-19{,}123$ for $n_C = 4$ years
Defender: $AW_D = \$-17{,}307$ for $n_D = 3$ years

The challenger total AW of cost curve (Figure 11–5) is classically shaped and relatively flat between years 3 and 6; there is virtually no difference in the total AW for years 4 and 5. For the defender, note that the estimated AOC values change substantially over 3 years, and they do not constantly increase or decrease.

(*b*) To perform the replacement study now, apply only the first step of the procedure. Select the defender because it has the better AW of costs ($-17,307), and expect to retain it for 3 more years. Prepare to perform the one-year-later analysis 1 year from now.

(*c*) One year later, the situation has changed significantly for the equipment Toshiba retained last year. Apply the steps for the one-year-later analysis:

  2. After 1 year of defender retention, the challenger estimates are still reasonable, but the defender market value and AOC estimates are substantially different. Go to step 3 to perform a new ESL analysis for the defender.

  3. The defender estimates in Figure 11–5 are updated below for the ESL analysis. New AW values are calculated using Equation [11.3]. There is now a maximum of 2 more years of retention, 1 year less than the 3 years determined last year.

Year k	Market Value, $	AOC, $	Total AW If Retained k More Years, $
0	12,000	—	—
1	2,000	−12,000	−23,200
2	0	−16,000	−20,819

The AW and n values for the new replacement study are as follows:

Challenger: unchanged at $AW_C = \$-19{,}123$ for $n_C = 4$ years
Defender: new $AW_D = \$-20{,}819$ for $n_D = 2$ more years

Now select the challenger based on its favorable AW value. Therefore, replace the defender now, not 2 years from now. Expect to keep the challenger for 4 years, or until a better challenger appears on the scene.

EXAMPLE 11.5   Keep or Replace the Kiln Case

We continue with the progressive example of possibly replacing a kiln at B&T Enterprises. A marketing study revealed that the improving business activity on the west coast implies that the revenue profile between the installed kiln (PT) and the proposed new one (GH) would be the same, with the new kiln possibly bringing in new revenue within the next couple of years. The president of B&T decided it was time to do a replacement study. Assume you are the lead engineer and that you previously completed the ESL analysis on the challenger (Example 11.3). It indicates that for the GH system the ESL is its expected useful life.

Challenger: ESL $n_{GH} = 12$ years with total equivalent annual cost $AW_{GH} = \$-12.32$ million

The president asked you to complete the replacement study, stipulating that, due to the rapidly rising annual operating costs (AOC), the defender would be retained a maximum of 6 years. You are expected to make the necessary estimates for the defender (PT) and perform the study at a 15% per year return.

Solution

After some data collection, you have good evidence that the market value for the PT system will stay high, but that the increasing AOC is expected to continue rising about $1.2 million per year. The best estimates for the next 6 years in $ million units are these:

Year	1	2	3	4	5	6
Market value, $ M	22.0	22.0	22.0	20.0	18.0	18.0
AOC, $ M per year	−5.2	−6.4	−7.6	−8.8	−10.0	−11.2

You developed a spreadsheet and performed the analysis in Figure 11–6. As an illustration, total AW computation for 3 years of retention, in $ million units, is

$$
\begin{aligned}
\text{Total AW}_3 &= -22.0(A/P,15\%,3) + 22.0(A/F,15\%,3) - [5.2(P/F,15\%,1) + 6.4(P/F,15\%,2) \\
&\quad + 7.6(P/F,15\%,3)](A/P,15\%,3) \\
&= -9.63 + 6.34 - [14.36](0.43798) \\
&= \$-9.59 \text{ per year}
\end{aligned}
$$

Though the system could be retained up to 6 years, the ESL is much shorter at 1 year.

Defender: ESL $n_{PT} = 1$ year with total equivalent annual cost $AW_{PT} = \$-8.50$ million

	A	B	C	D	E	F
1			**ESL Analysis of PT**			
2	Interest rate	15%			Market value	$ 22.00
3						
4		Market				Total AW
5	Year	Value	AOC	Capital Recovery	AW of AOC	of costs
6	1	22.00	-5.20	-3.30	-5.20	-8.50
7	2	22.00	-6.40	-3.30	-5.76	-9.06
8	3	22.00	-7.60	-3.30	-6.29	-9.59
9	4	20.00	-8.80	-3.70	-6.79	-10.49
10	5	18.00	-10.00	-3.89	-7.27	-11.16
11	6	18.00	-11.20	-3.76	-7.72	-11.47
12	Formula, year 3	22.00	-7.60	`= PMT($B$2, $A8,$F$2,-$B8)`	`= -PMT($B$2,$A8, NPV($B$2,$C$6:$C8)+0)`	`= D8+E8`
13						

Figure 11–6
ESL analysis of defender kiln PT for progressive example, Example 11.5.

To make the replacement/retention decision, apply step 1 of the procedure. Since $AW_{PT} =$ $\$-8.50$ million per year is considerably less than $AW = \$-12.32$ million, you should recommend keeping the current kiln only 1 more year and doing another study during the year to determine if the current estimates are still reliable.

Comment

An observation of the trends of the two final AW series in this problem is important. A comparison of Figure 11–3 (top), column F, and Figure 11–6, column F, shows us that the largest total AW of the current system ($\$-11.47$ M for 6 years) is still below the smallest total AW of the proposed system ($\$-12.32$ M for 12 years). This indicates that the graphite hearth system (the challenger) will not be chosen on an economic basis, if the decision to consider the installed kiln as a defender in the future were made. It would take some significant estimate changes to justify the challenger.

11.4 Additional Considerations in a Replacement Study ● ● ●

There are several additional aspects of a replacement study that may be introduced. Three of these are identified and discussed in turn.

- Future-year replacement decisions at the time of the initial replacement study
- Opportunity cost versus cash flow approaches to alternative comparison
- Anticipation of improved future challengers

In most cases when management initiates a replacement study, the question is best framed as "Replace now, 1 year from now, 2 years from now, etc.?" The procedure above does answer this question, provided the estimates for C and D do not change as each year passes. In other words, *at the time it is performed, step 1 of the procedure does answer the replacement question for multiple years.* It is only when estimates change over time that the decision to retain the defender may be prematurely reversed (prior to n_D years) in favor of the then-best challenger.

The first costs (P values) for the challenger and defender have been correctly taken as the initial investment for the challenger C and current market value for the defender D. This is called the **opportunity cost approach** because it recognizes that a cash inflow of funds equal to the market value is forgone if the defender is selected. This approach, also called the conventional approach, is correct for every replacement study. A second approach, called the **cash flow approach,** recognizes that when C is selected, the market value cash inflow for the defender is received and, in effect, immediately reduces the capital needed to invest in the challenger. *Use of the cash flow approach is strongly discouraged* for at least two reasons: possible violation of the equal-service requirement and incorrect capital recovery value for C. As we are aware, all economic evaluations must compare alternatives with equal service. Therefore, the cash flow approach can work only when challenger and defender lives are exactly equal. This is commonly not the case; in fact, the ESL analysis and the replacement study procedure are designed to compare two mutually exclusive, *unequal-life* alternatives via the annual worth method. If this equal-service comparison reason is not enough to avoid the cash flow approach, consider what happens to the challenger's capital recovery amount when its first cost is decreased by the market value of the defender. The capital recovery (CR) terms in Equation [11.3] will decrease, resulting in a falsely low value of CR for the challenger, were it selected. From the vantage point of the economic study itself, the decision for C or D will not change; but when C is selected and implemented, this CR value is not reliable. The conclusion is simple:

Use the initial investment of C and the market value of D as the first costs in the ESL analysis and in the replacement study.

A basic premise of a replacement study is that some challenger will replace the defender at a future time, provided the service continues to be needed and a worthy challenger is available. The expectation of ever-improving challengers can offer strong encouragement to retain the defender until some situational elements—technology, costs, market fluctuations, contract negotiations, etc.— stabilize. This was the case in the previous two examples. A large expenditure on equipment when the standards changed soon after purchase forced an early replacement consideration and a large loss of invested capital. The replacement study is no substitute for forecasting challenger availability. *It is important to understand trends, new advances, and competitive pressures that can complement the economic outcome of a good replacement study.* It is often better to compare a challenger with an augmented defender in the replacement study. Adding needed features to a currently installed defender may prolong its useful life and productivity until challenger choices are more appealing.

It is possible that a significant tax impact may occur when a defender is traded early in its expected life. If taxes should be considered, proceed now, or after the next section, to Chapter 17 and the after-tax replacement analysis in Section 17.7.

11.5 Replacement Study over a Specified Study Period ● ● ●

When the time period for the replacement study is limited to a specified study period or planning horizon, for example, 6 years, the ESL analysis is not performed.

> The AW values for the challenger and for the remaining life of the defender are not based on the economic service life; the AW is calculated over the study period only. What happens to the alternatives after the study period is not considered in the replacement analysis.

This means that the defender or challenger is not needed beyond the study period. In fact, a study period of fixed duration does not comply with the three assumptions stated in Section 11.1—service needed for indefinite future, best challenger available now, and estimates will be identical for future life cycles.

When performing a replacement study over a fixed study period, it is crucial that the estimates used to determine the AW values be accurate and used in the study. This is especially important for the defender. Failure to do the following violates the requirement of equal-service comparison.

> When the defender's remaining life is **shorter than the study period,** the cost of providing the defender's services from the end of its expected remaining life to the end of the study period must be estimated as accurately as possible and included in the replacement study.

Study period

The right branch of Figure 11–4 presents an overview of the replacement study procedure for a stated study period.

1. *Succession options and AW values.* Develop all the viable ways to use the defender and challenger during the study period. There may be only one option or many options; the longer the study period, the more complex this analysis becomes. The AW values for the challenger and defender cash flows are used to build the equivalent cash flow values for each option.

2. *Selection of the best option.* The PW or AW for each option is calculated over the study period. Select the option with the lowest cost, or highest income if revenues are estimated. (As before, the best option will have the numerically largest PW or AW value.)

The following examples use this procedure and illustrate the importance of making cost estimates for the defender alternative when its remaining life is less than the study period.

EXAMPLE 11.6

Claudia works with Lockheed-Martin (LMCO) in the aircraft maintenance division. She is preparing for what she and her boss, the division chief, hope to be a new 10-year defense contract with the Air Force on long-range cargo aircraft. A key piece of equipment for maintenance operations is an avionics circuit diagnostics system. The current system was purchased 7 years ago on an earlier contract. It has no capital recovery costs remaining, and the following are reliable estimates: current market value = $70,000, remaining life of 3 more years, no salvage value, and AOC = $30,000 per year. The only options for this system are to replace it now or retain it for the full 3 additional years.

Claudia has found that there is only one good challenger system. Its cost estimates are: first cost = $750,000, life = 10 years, *S* = 0, and AOC = $50,000 per year.

Realizing the importance of accurate defender alternative cost estimates, Claudia asked the division chief what system would be a logical follow-on to the current one 3 years hence, if LMCO wins the contract. The chief predicted LMCO would purchase the very system she had identified as the challenger because it is the best on the market. The company would keep it for the entire 10 additional years for use on an extension of this contract or some other application that could recover the remaining 3 years of invested capital. Claudia interpreted the response to mean that the last 3 years would also be capital recovery years, but on some project other than this one. Claudia's estimate of the first cost of this same system 3 years from now is $900,000. Additionally, the $50,000 per year AOC is the best estimate at this time.

The division chief mentioned any study had to be conducted using the interest rate of 10%, as mandated by the Office of Management and Budget (OMB). Perform a replacement study for the fixed contract period of 10 years.

Solution

The study period is fixed at 10 years, so the intent of the replacement study assumptions is not present. This means the defender follow-on estimates are very important to the analysis. Further, *any analyses to determine the ESL values are unnecessary* since alternative lives are already set and no projected annual market values are available. The first step of the replacement study procedure is to define the options. Since the defender will be replaced now or in 3 years, there are only two options:

1. Challenger for all 10 years.
2. Defender for 3 years, followed by the challenger for 7 years.

Cash flows are diagrammed in Figure 11–7. For option 1, the challenger is used for all 10 years. Equation [11.3] is applied to calculate AW using the following estimates:

Challenger: $P = \$-750,000$ $AOC = \$-50,000$
                          $n = 10$ years $S = 0$

$$AW_C = -750,000(A/P,10\%,10) - 50,000 = \$-172,063$$

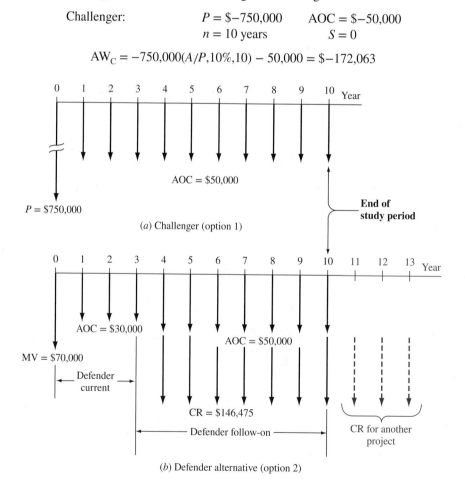

(a) Challenger (option 1)

(b) Defender alternative (option 2)

Figure 11–7
Cash flow diagrams for a 10-year study period replacement study, Example 11.6.

The second option has more complex cost estimates. The AW for the in-place system is calculated over the first 3 years. Added to this is the capital recovery for the defender follow-on for the next 7 years. *However in this case, the CR amount is determined over its full 10-year life.* (It is not unusual for the recovery of invested capital to be moved between projects, especially for contract work.) Refer to the AW components as AW_{DC} (subscript DC for defender current) and AW_{DF} (subscript DF for defender follow-on). The final cash flows are shown in Figure 11–7b.

Defender current: market value = \$−70,000   AOC = \$−30,000
$n = 3$ years $S = 0$

$$AW_{DC} = [−70,000 − 30,000(P/A,10\%,3)](A/P,10\%,10) = \$−23,534$$

Defender follow-on: $P = \$−900,000$, $n = 10$ years for capital recovery calculation only, AOC = \$−50,000 for years 4 through 10, $S = 0$.

The CR and AW for all 10 years are

$$CR_{DF} = −900,000(A/P,10\%,10) = \$−146,475 \qquad [11.6]$$
$$AW_{DF} = (−146,475 − 50,000)(F/A,10\%,7)(A/F,10\%,10) = \$−116,966$$

Total AW_D for the defender is the sum of the two annual worth values above. This is the AW for option 2.

$$AW_D = AW_{DC} + AW_{DF} = −23,534 − 116,966 = \$−140,500$$

Option 2 has a lower cost (\$−140,500 vs. \$−172,063). Retain the defender now and expect to purchase the follow-on system 3 years hence.

Comment

The capital recovery cost for the defender follow-on will be borne by some yet-to-be-identified project for years 11 through 13. If this assumption were not made, its capital recovery cost would be calculated over 7 years, not 10, in Equation [11.6], increasing CR to \$−184,869. This raises the annual worth to $AW_D = \$−163,357$. The defender alternative (option 2) is still selected.

EXAMPLE 11.7

Three years ago, Chicago's O'Hare Airport purchased a new fire truck. Because of flight increases, new fire-fighting capacity is needed once again. An additional truck of the same capacity can be purchased now, or a double-capacity truck can replace the current fire truck. Estimates are presented below. Compare the options at 12% per year using (*a*) a 12-year study period and (*b*) a 9-year study period.

	Presently Owned	New Purchase	Double Capacity
First cost P, \$	−151,000 (3 years ago)	−175,000	−190,000
AOC, \$	−1,500	−1,500	−2,500
Market value, \$	70,000	—	—
Salvage value, \$	10% of P	12% of P	10% of P
Life, years	12	12	12

Solution

Identify option 1 as retention of the presently owned truck and augmentation with a new same-capacity vehicle. Define option 2 as replacement with the double-capacity truck.

	Option 1		Option 2
	Presently Owned	Augmentation	Double Capacity
P, \$	−70,000	−175,000	−190,000
AOC, \$	−1,500	−1,500	−2,500
S, \$	15,100	21,000	19,000
n, years	9	12	12

(a) For a full-life 12-year study period of option 1,

$$AW_1 = \text{(AW of presently owned)} + \text{(AW of augmentation)}$$
$$= [-70{,}000(A/P,12\%,9) + 15{,}100(A/F,12\%,9) - 1500]$$
$$\quad + [-175{,}000(A/P,12\%,12) + 21{,}000(A/F,12\%,12) - 1500]$$
$$= -13{,}616 - 28{,}882$$
$$= \$-42{,}498$$

This computation assumes the equivalent services provided by the current fire truck can be purchased at $\$-13{,}616$ per year for years 10 through 12.

$$AW_2 = -190{,}000(A/P,12\%,12) + 19{,}000(A/F,12\%,12) - 2500$$
$$= \$-32{,}386$$

Replace now with the double-capacity truck (option 2) at an advantage of $\$10{,}112$ per year.

(b) The analysis for an abbreviated 9-year study period is identical, except that $n = 9$ in each factor; that is, 3 fewer years are allowed for the augmentation and double-capacity trucks to recover the capital investment plus a 12% per year return. The salvage values remain the same since they are quoted as a percentage of P for all years.

$$AW_1 = \$-46{,}539 \qquad AW_2 = \$-36{,}873$$

Option 2 is again selected.

The previous two examples indicate an important consideration for setting the length of the study period for a replacement analysis. It involves the capital recovery amount for the challenger, when the strict definition of a study period is applied.

Study period

Capital recovery

> When a study period shorter than the life of the challenger is defined, the challenger's capital recovery amount **increases** in order to recover the initial investment plus a return in this **shortened time period**. Highly abbreviated study periods tend to disadvantage the challenger because no consideration of time beyond the end of the study period is made in calculating the challenger's capital recovery amount.

If there are several options for the number of years that the defender may be retained before replacement with the challenger, the first step of the replacement study—succession options and AW values—must include all the viable options. For example, if the study period is 5 years and the defender will remain in service 1 year, or 2 years, or 3 years, cost estimates must be made to determine AW values for each defender retention period. In this case, there are four options; call them W, X, Y, and Z.

Option	Defender Retained, Years	Challenger Serves, Years
W	3	2
X	2	3
Y	1	4
Z	0	5

The respective AW values for defender retention and challenger use define the cash flows for each option. Example 11.8 illustrates the procedure using the progressive example.

EXAMPLE 11.8   Keep or Replace the Kiln Case

We have progressed to the point that the replacement study between the defender PT and challenger GH was completed (Example 11.5). The defender was the clear choice with a much smaller AW value ($\$-8.50$ M) than that of the challenger ($\$-12.32$ M). Now the management of B&T is in a dilemma. They know the current tunnel kiln is much cheaper than the new graphite hearth, but the prospect of future new business should not be dismissed.

TABLE 11–2	Replacement Study Options and Total AW Values, Example 11.8			
	Defender PT		**Challenger GH**	
Option	**Years Retained**	**AW, $ M/Year**	**Years Retained**	**AW, $ M/Year**
A	1	−8.50	5	−14.21
B	2	−9.06	4	−15.08
C	3	−9.59	3	−16.31
D	4	−10.49	2	−18.21
E	5	−11.16	1	−22.10
F	6	−11.47	0	—

The president asked, "Is it possible to determine when it is economically the cheapest to purchase the new kiln, provided the current one is kept at least 1 year, but no more than 6 years, its remaining expected life?" The chief financial officer answered, yes, of course. (*a*) Determine the answer for the president. (*b*) Discuss the next step in the analysis based on the conclusion reached here.

Solution

(*a*) Actually, this is a quite easy question to answer, because all the information has been determined previously. We know the MARR is 15% per year, the study period has been established at 6 years, and the defender PT will stay in place between 1 and 6 years. Therefore, the challenger GH will be considered for 0 to 5 years of service. The total AW values were determined for the defender in Example 11.5 (Figure 11–6) and for the challenger in Example 11.3 (Figure 11–3). They are repeated in Table 11–2 for convenience. Use the procedure for a replacement study with a fixed study period.

Step 1: Succession options and AW values. There are six options in this case; the defender is retained from 1 to 6 years while the challenger is installed from 0 to 5 years. We will label them A through F. Figure 11–8 presents the options and the AW series for each option from Table 11–2. No consideration of the fact that the challenger has an expected life of 12 years is made since the study period is fixed at 6 years.

Step 2: Selection of the best option. The PW value for each option is determined over the 6-year study period in column J of Figure 11–8. The conclusion is clearly to **keep the defender** in place for 6 more years.

(*b*) Every replacement analysis has indicated that the defender should be retained for the near future. If the analysis is to be carried further, the possibility of increased revenue based on services of the challenger's high-temperature and operating efficiency should be considered next. In the introductory material, new business opportunities were mentioned. A revenue increase for the challenger will reduce its AW of costs and possibly make it more economically viable.

	A	B	C	D	E	F	G	H	I	J
1		Time in Service, Years								Option
2		Defender	Challenger	AW Cash Flows for Each Option, $M per year						PW at 15%,
3	Option	PT	GH	1	2	3	4	5	6	$M
4	A	1	5	-8.50	-14.21	-14.21	-14.21	-14.21	-14.21	-48.81
5	B	2	4	-9.06	-9.06	-15.08	-15.08	-15.08	-15.08	-47.28
6	C	3	3	-9.59	-9.59	-9.59	-16.31	-16.31	-16.31	-46.38
7	D	4	2	-10.49	-10.49	-10.49	-10.49	-18.21	-18.21	-46.87
8	E	5	1	-11.16	-11.16	-11.16	-11.16	-11.16	-22.10	-46.96
9	F	6	0	-11.47	-11.47	-11.47	-11.47	-11.47	-11.47	**-43.41**
10										
11					Conclusion: Keep defender all 6 years					
12										
13										

Figure 11–8
PW values for 6-year study period replacement analysis, Example 11.8.

11.6 Replacement Value ● ● ●

Often it is helpful to know the minimum market value of the defender necessary to make the challenger economically attractive. If a realizable market value or trade-in of at least this amount can be obtained, from an economic perspective the challenger should be selected immediately. This is a *breakeven value* between AW_C and AW_D; it is referred to as the **replacement value (RV)**. Set up the relation $AW_C = AW_D$ with the market value for the defender identified as RV, which is the unknown. The AW_C is known, so RV can be determined. The selection guideline is as follows:

> If the actual market trade-in **exceeds** the breakeven **replacement value**, the challenger is the better alternative and should replace the defender now.

Determination of the RV for a defender is an excellent opportunity to utilize the Goal Seek tool. The target cell is the current market value, and the AW_D value is forced to equal the AW_C amount. Example 11.9 discusses a replacement value of the progressive example.

EXAMPLE 11.9   Keep or Replace the Kiln Case

As one final consideration of the challenger kiln, you decide to determine what the trade-in amount would have to be so that the challenger is the economic choice next year. This is based on the ESL analysis that concluded the following (Examples 11.3 and 11.5):

> Defender: ESL $n_{PT} = 1$ year with $AW_{PT} = \$-8.50$ million
> Challenger: ESL $n_{GH} = 12$ years with $AW_{GH} = \$-12.32$ million

The original defender price was $25 million, and a current market value of $22 million was estimated earlier (Figure 11–6). Since the installed kiln is known for retention of its market value (MV), you are hopeful the difference between RV and estimated MV may not be so significant. What will you discover RV to be? The MARR is 15% per year.

Solution

Set the AW relation for the defender for the ESL time of 1 year equal to $AW_{GH} = \$-12.32$ and solve for RV. The estimates for AOC and MV next year are in Figure 11–6; they are, in $ million,

$$\text{Year 1:}\quad AOC = \$-5.20 \qquad MV = \$22.0$$
$$-12.32 = -RV(A/P,15\%,1) + 22.00(A/F,15\%,1) - 5.20$$
$$1.15RV = 12.32 + 22.00 - 5.20$$
$$RV = \$25.32$$

Though the RV is larger than the defender's estimated MV of $22 million, some flexibility in the trade-in offer or the challenger's first cost may cause the challenger to be economically justifiable.

Comment

To find RV using a spreadsheet, return to Figure 11–6. In the Goal Seek template, the "set" cell is the AW for 1 year (currently $-8.50), and the required value is $-12.32, the AW_{GH} for its ESL of 12 years. The "changing" cell is the current market value (cell F2), currently $22.00. When "OK" is touched, $25.32 is displayed as the breakeven market value. This is the RV.

CHAPTER SUMMARY

It is important in a replacement study to compare the best challenger with the defender. *Best (economic) challenger is described as the one with the lowest annual worth (AW) of costs for the time period under consideration in the analysis.* If the expected remaining life of the defender and the estimated life of the challenger are specified, the AW values over these years are determined and the replacement study proceeds. However, if reasonable estimates of the expected market value (MV) and AOC for each year of ownership can be made, these year-by-year (marginal) costs help determine the best challenger.

The economic service life (ESL) analysis is designed to determine the best challenger's years of service and the resulting lowest total AW of costs. The resulting n_C and AW_C values are used in the replacement study. The same analysis can be performed for the ESL of the defender.

Replacement studies in which no study period (planning horizon) is specified utilize the annual worth method of comparing two unequal-life alternatives. The better AW value determines how long the defender is retained before replacement.

When a study period is specified for the replacement study, it is vital that the market value and cost estimates for the defender be as accurate as possible. When the defender's remaining life is shorter than the study period, it is critical that the cost for continuing service be estimated carefully. All the viable options for using the defender and challenger are enumerated, and their AW equivalent cash flows are determined. For each option, the PW or AW value is used to select the best option. This option determines how long the defender is retained before replacement.

PROBLEMS

Foundations of Replacement

11.1 Briefly explain what is meant by the defender/challenger concept.

11.2 List three reasons why a replacement study might be needed.

11.3 In a replacement analysis, what number should be used as the first cost for the currently owned asset? How is this value best obtained?

11.4 You, an engineer, and an attorney friend, Rob, started a small business 3 years ago to do energy audits for small businesses. A piece of equipment that costs $25,000 then has become prematurely obsolete and needs to be replaced with a solid state version that has a purchase price of $20,000 and the current equipment has a nil salvage value. Your company accountant shows the book value of the equipment to be $10,000. (*a*) If you buy the solid state equipment, how should the difference between the cost of the new equipment and the value of the old equipment be considered? (*b*) Your partner thinks of this difference as an added cost to the new equipment, effectively making its purchase price $30,000. Is she correct?

11.5 In conducting a replacement study of assets with different lives, can the annual worth values over the asset's own life cycle be used in the comparison, if the study period is (*a*) unlimited, (*b*) limited, and the study period *is not* an even multiple of asset lives, and (*c*) limited wherein the period *is* a multiple of asset lives? Explain your answers for each part.

11.6 A machine purchased 1 year ago for $85,000 costs more to operate than anticipated. When purchased, the machine was expected to be used for 10 years with annual maintenance costs of $22,000 and a $10,000 salvage value. However, last year, it cost the company $35,000 to maintain it, and these costs are expected to escalate to $36,500 this year and increase by $1500 each year thereafter. The market value is now estimated to be $85,000 − $10,000$k$, where k is the number of years since the machine was purchased. It is now estimated that this machine will be useful for a maximum of 5 more years. A replacement study is to be performed now. Determine the values of P, n, AOC, and S of this defender.

11.7 A mechanical engineer who designs and sells equipment that automates manual labor processes is offering a machine/robot combination that will significantly reduce labor costs associated with manufacturing garage-door opener transmitters. The equipment has a first cost of $170,000, an estimated annual operating cost of $54,000, a maximum useful life of 5 years, and a $20,000 salvage value anytime it is replaced. The existing equipment was purchased 12 years ago for $65,000 and has an annual operating cost of $78,000. At most, the currently owned equipment can be used 2 more years, at which time it will be auctioned off for an expected amount of $6000, less 33% paid to the company handling the auction. The same scenario will occur if the currently owned equipment is replaced now. Determine the defender and challenger estimates of P, n, S, and AOC in conducting a replacement analysis today at an interest rate of 20% per year.

11.8 A machine tool purchased 2 years ago for $40,000 has a market value best described by the relation $40,000 − 3000k$, where k is the number of years from time of purchase. Experience with this type of asset has shown that its annual operating cost is described by the relation $30,000 + 1000k$. The asset's salvage value was originally estimated to be $10,000 after a predicted 10-year useful life. Determine the current estimates for P, S, and AOC for a replacement study, assuming it will be kept only 1 more year, which will be the third year of ownership.

11.9 Equipment that was purchased by Newport Corporation for making pneumatic vibration isolators cost $90,000 two years ago. It has a market value that can be described by the relation $90,000 - 8000k$, where k is the years from time of purchase. The operating cost for the first 5 years is $65,000 per year, after which it increases by $6300 per year. The asset's salvage value was originally estimated to be $7000 after a predicted 10-year useful life. Determine the values of P, S, and AOC if (a) a replacement study is done now and it is assumed that the equipment will be kept a maximum of only 1 more year, and (b) a replacement study is done 1 year from now and it is assumed that the equipment will be kept a maximum of only 1 more year after that.

Economic Service Life

11.10 The AW values for retaining a presently owned machine for additional years are shown in the table. Note that the values represent the AW amount for *each of the n years* that the asset is kept, that is, if it is kept 5 more years, the annual worth is $95,000 for each of the 5 years. Assume that future costs remain as estimated for the replacement study and that used machines like the one presently owned will always be available.

(a) What is the ESL and associated AW of the defender at a MARR of 12% per year?

(b) A challenger with an ESL of 7 years and an $AW_C = \$ -89,500$ per year has been identified. Which AW will be less for the respective ESL periods?

Retention Period, Years	AW Value, $ per Year
1	-92,000
2	-88,000
3	-85,000
4	-89,000
5	-95,000

11.11 From the data shown, determine the economic service lives of the defender and challenger. (Note: Values in the table are AW values, not individual year-end values.)

Years Retained	AW of Defender, $	AW of Challenger, $
1	-145,000	-136,000
2	-96,429	-126,000
3	-63,317	-92,000
4	-39,321	-53,000
5	-49,570	-38,000

11.12 From the data shown, determine the ESL of the asset. (Note: Values in the table are AW values, not individual year-end values.)

Years Retained	AW of First Cost, $	AW of Operating Cost, $	AW of Salvage Value, $
1	-165,000	-36,000	99,000
2	-86,429	-36,000	38,095
3	-60,317	-42,000	18,127
4	-47,321	-43,000	6,464
5	-39,570	-48,000	3,276

11.13 When determining the economic service life of a new piece of equipment, an engineer made the calculations below; however, the annual worth of the salvage value for 2 years of retention was omitted. Complete the analysis by determining the following to make the ESL equal 2 years: (a) minimum AW of the salvage value; (b) estimated salvage value that would produce the AW calculated in part (a) at $i = 10\%$ per year. (Note: Values in the table are AW values, not individual year-end values.)

Years Retained	AW of First Cost, $	AW of Operating Cost, $	AW of Salvage Value, $
1	-88,000	-45,000	50,000
2	-46,095	-46,000	?
3	-32,169	-51,000	6,042
4	-25,238	-59,000	3,232
5	-21,104	-70,000	1,638

11.14 A piece of onboard equipment has a first cost of $600,000, an annual cost of $92,000, and a salvage value that decreases to zero by $150,000 each year of the equipment's maximum useful life of 5 years. Assume the company's MARR is 10% per year.

(a) Determine the ESL by hand.

(b) Use a spreadsheet with a graph indicating the capital recovery, AOC, and total AW per year to determine the ESL.

11.15 An injection molding system has a first cost of $180,000, and an annual operating cost of $84,000 in years 1 and 2, increasing by $5000 per year thereafter. The salvage value of the system is 25% of the first cost regardless of when the system is retired within its maximum useful life of 5 years. Using a MARR of 15% per year, determine the ESL and the respective AW value of the system (a) by hand solution, and (b) via spreadsheet that graphs the total AW and its components.

11.16 A large, standby electricity generator in a hospital operating room has a first cost of $70,000 and may be used for a maximum of 6 years. Its salvage value, which decreases by 15% per year, is described by the equation $S = 70,000(1 - 0.15)^n$, where n is the number of years after purchase. The operating cost of the generator will be constant at $75,000 per year. At an interest rate of 12% per year, what is the economic service life and associated AW value?

11.17 A piece of equipment has a first cost of $150,000, a maximum useful life of 7 years, and a market (salvage) value described by the relation $S = 120,000 - 20,000k$, where k is the number of years since it was purchased. The salvage value cannot go below zero. The AOC series is estimated using $AOC = 60,000 + 10,000k$. The interest rate is 15% per year. Determine the economic service life (a) by hand solution, using regular AW computations, and (b) by spreadsheet, using annual marginal cost estimates.

11.18 **Keep or Replace the Kiln Case**   **PE**

In Example 11.3, the market value (salvage value) series of the proposed $38 million replacement kiln (GH) dropped to $25 million in only 1 year and then retained 75% of the previous year's market value through the remainder of its 12-year expected life. Based on the experience with the current kiln, and the higher temperature capability of the replacement, the new kiln is actually expected to retain only 50% of the previous year's value starting in year 5. Additionally, the heating element replacement in year 6 will probably cost $4 million, not $2 million. And finally, the maintenance costs will be considerably higher as the kiln ages. Starting in year 5, the AOC is expected to increase by 25% per year, not 10% as predicted earlier. The Manager of Critical Equipment is now very concerned that the ESL will be significantly decreased from the 12 years calculated earlier (Example 11.3).
(a)   Determine the new ESL and associated AW value.
(b)   In percentage changes, estimate how much these new cost estimates may affect the minimum-cost life and AW of cost estimate.

Replacement Study

11.19 What is the fundamental difference between the AW analysis that you learned in Chapter 6 and the AW-based replacement analysis that you are learning in this chapter?

11.20 If a replacement study is performed and the defender is selected for retention for n_D years, explain what should be done 1 year later if a new challenger is identified.

11.21 What is meant by the opportunity-cost approach for a replacement study?

11.22 State what is meant by the cash flow approach and list two of its limitations in a replacement study.

11.23 An engineer with Calahan Technologies calculated the AW of cost values shown for a presently owned machine, using estimates she obtained from the vendor and company records.

Retention Period, Years	AW of Costs, $ per Year
1	-92,000
2	-81,000
3	-87,000
4	-89,000
5	-95,000

A challenger has an economic service life of 7 years with an AW of $ -86,000 per year. Assume that all future costs remain as estimated and the challenger's technology will definitely replace that of the defender within 5 years. (a) When should the company purchase the challenger? (b) If the AW for the challenger is $ -81,000 per year, when should the challenger be purchased?

11.24 Three years ago, Witt Gas Controls purchased equipment for $80,000 that was expected to have a useful life of 5 years with a $9000 salvage value. Increased demand necessitated an upgrade costing $30,000 one year ago. Technology changes now require that the equipment be upgraded again for another $25,000 so that it can be used for 3 more years. Its annual operating cost will be $48,000 and it will have a $19,000 salvage value after 3 years. Alternatively, it can be replaced with new equipment that will cost $68,000 with operating costs of $35,000 per year and a salvage value of $21,000 after 3 years. If replaced now, the existing equipment will be sold for $12,000.
(a)   Calculate the annual worth of the defender at an interest rate of 10% per year.
(b)   Determine the AW of the challenger over 3 years and select it or the defender.

11.25 A presently owned machine can last 3 more years, if properly maintained at a cost of $15,000 per year. Its AOC is $31,000 per year. After 3 years, it can be sold for an estimated $9000. A replacement costs $80,000 with a $10,000 salvage value after 3 years and an operating cost of $19,000 per year. Different vendors have offered $10,000 and $20,000, respectively, for the current system as trade-in for the replacement machine. At $i = 12$% per year, perform a replacement study and determine whether the defender should be retained or replaced.

11.26 State-of-the-art digital imaging equipment purchased 2 years ago for $50,000 had an expected useful life of 5 years and a $5000 salvage value. After its installation the performance was poor, and it was upgraded for $20,000 one year ago. Increased demand now requires another upgrade for an additional $22,000 so that it can be used for 3 more years. Its new annual operating cost will be $27,000 with a $12,000 salvage after the 3 years. Alternatively, it can be replaced with new equipment costing $65,000, an estimated AOC of $14,000, and an expected salvage of $23,000 after 3 years. If replaced now, the existing equipment can be traded for only $7000. Use a MARR of 10% per year. (*a*) Determine whether the company should retain or replace the defender now. (*b*) Based on the poor experience with the current equipment, assume the person doing this analysis decides the challenger may be kept for only 2 years, not 3, with the same AOC and salvage estimates for the 2 years. What is the decision?

11.27 The plant manager has asked you to do a cost analysis to determine when currently owned equipment should be replaced. The manager stated that under no circumstances will the existing equipment be retained longer than two more years and that once it is replaced, a contractor will provide the same service from then on at a cost of $97,000 per year. The salvage value of the currently owned equipment is estimated to be $37,000 now, $30,000 in 1 year, and $19,000 two years from now. The operating cost is expected to be $85,000 per year. Using an interest rate of 10% per year, determine when the defending equipment should be retired.

11.28 A small company that manufactures vibration isolation platforms is trying to decide whether it should replace the current assembly system (D), which is rather labor intensive, now or 1 year from now with a system that is more automated (C). Some components of the current system can be sold immediately for $9000, but they will be worthless hereafter. The operating cost of the existing system is $192,000 per year. System C will cost $320,000 with a $50,000 salvage value after 4 years. Its operating cost will be $68,000 per year. If you are told to do a replacement analysis using an interest rate of 10% per year, which system do you recommend?

System	D	C
Market value, $	9,000	320,000
AOC, $ per year	−192,000	−68,000
Salvage value, $	0	50,000
Life, years	1	4

11.29 A textile processing company is evaluating whether it should retain the current bleaching process that uses chlorine dioxide or replace it with a proprietary oxypure process. The relevant information for each process is shown. Use an interest rate of 15% per year to perform the replacement

study. Additionally, write the PMT functions that display the information necessary to make the decision.

Process	Current	Oxypure
Original cost 6 years ago, $	450,000	—
Investment cost now, $	—	600,000
Current market value, $	25,000	—
Annual operating cost, $/year	190,000	70,000
Remaining life, years	3	10
Salvage value, $	0	50,000

11.30 A presently owned machine has the projected market value and M&O costs shown below. An outside vendor of services has offered to provide the service of the existing machine at a fixed price per year. If the presently owned machine is replaced now, the cost of the fixed-price contract will be $33,000 per year. If the presently owned machine is replaced next year or any time after that, the contract price will be $35,000 per year. Determine if and when the defender should be replaced with the outside vendor using an interest rate of 10% per year. Assume used equipment similar to the defender will always be available, but that the current equipment will not be retained more than three additional years.

Retention Year	Market Value, $	M&O Cost, $ per Year
0	32,000	—
1	25,000	24,000
2	14,000	25,000
3	10,000	26,000
4	8,000	—

11.31 Randall-Rico Consultants, 5 years ago, purchased for $45,000 a microwave signal graphical plotter for corrosion detection in concrete structures. It is expected to have the market values and annual operating costs shown for its remaining useful life of up to 3 years. It could be traded now at an appraised market value of $8000.

Year	Market Value at End of Year, $	AOC, $ per Year
1	6000	−50,000
2	4000	−53,000
3	1000	−60,000

A replacement plotter with new Internet-based, digital technology costs $125,000, has an estimated salvage value of 8% of the purchase price after its 5-year life, and an AOC of 20% of its purchase price. At an interest rate of 15% per year, determine how many more years Randall-Rico should retain the present plotter. (Problem 11.45 continues the analysis using a spreadsheet.)

Replacement Study over a Specified Study Period

11.32 The estimated future market values and M&O costs for an in-place backup generator at MediCare Hospital and a possible replacement are shown below. Mrs. Jamison, Hospital Director, told you that she is interested only in what happens over the next 3 years. If the current generator is to be replaced, it must be done now or kept in-place for the 3 years. Using an interest rate of 10% per year and a 3-year study period, determine whether or not the replacement is economically advantageous now.

	In-Place		Replacement	
Year	Market Value, $	M&O Cost, $	Market Value, $	M&O Cost, $
0	40,000		80,000	
1	32,000	−55,000	65,000	−37,000
2	23,000	−55,000	39,000	−37,000
3	11,000	−55,000	20,000	−37,000
4			19,000	−38,000
5			11,000	−39,000

11.33 An electric company must decide between two options for managing the blowdown water from its cooling tower. Option 1 is to continue the lease on 50 acres of land for another 5-year period and dispose of the water by spray irrigation. The landowner will move the pipe around as necessary and maintain the spray nozzles and valves. The previous lease cost $125,000 per year with payments due midway through each year. Now the landowner will require *beginning of year payments* of $180,000 each year.

Option 2, which releases the 50 acre tract of land, involves purchasing a treatment system that will allow the recycling of most of the blowdown water. This system will have an initial cost of $1,600,000 and an AOC of $58,000 per year. However, the company will save $220,000 per year because it will not have to purchase as much make-up water as with option 1. At the end of 5 years, the company will be able to sell the equipment back to the local equipment supplier for 30% of the first cost.

If the electric company uses a MARR of 15% per year, should it continue to lease (defender) or purchase the treatment system (challenger)?

11.34 Searching for ways to cut costs and increase profit, one of the industrial engineers at Home Comfort Furniture Manufacturers, Inc. determined that the equivalent annual worth of an existing machine over its remaining useful life of 2 years is $70,000 per year. The IE also determined that used machines like the one currently in use

are not available any longer, but the defender can be replaced with a challenger that is more advanced. It will have an AW of $80,000 if it is kept for 2 years or less, $75,000 if it is kept between 3 and 4 years, and $65,000 if it is kept for 5 to 10 years. If the company uses a specified 3-year planning horizon and an interest rate of 15% per year, determine (a) when the company should replace the machine, and (b) the AW for the next 3 years.

11.35 In the process of performing a replacement study, an engineer at a fiber optics manufacturing company has two options to reduce costs on a production line. The currently owned Robot X can be sold now for $82,000. If kept, it will have an annual M&O cost of $30,000, and salvage values of $50,000, $42,000, and $35,000 after 1, 2, and 3 years, respectively. A challenger, Robot Y, will have a first cost of $97,000, an annual M&O cost of $27,000, and salvage values of $66,000, $56,000, and $42,000 after 1, 2, and 3 years, respectively. Which robot should be selected if a 2-year study period is used at an interest rate of 12% per year? Solve by two ways: by hand and by spreadsheet.

11.36 A machine purchased 3 years ago for $140,000 is now too slow to satisfy the demand of customers. It can be upgraded now for $79,000 or sold to a smaller company internationally for $40,000. The upgraded machine will have an annual operating cost of $85,000 per year and a $30,000 salvage value in 3 years. If upgraded, the presently owned machine will be retained for only 3 more years, then replaced with a machine to be used in the manufacture of several other product lines. The replacement machine, which will serve the company now and for a maximum of 8 years, costs $220,000. Its salvage value will be $50,000 for years 1 through 5; $20,000 after 6 years; and $10,000 thereafter. It will have an estimated operating cost of $45,000 per year.

Your boss asks you to perform an economic analysis at 15% per year using a specified 3-year planning horizon (a) to determine if the current machine should be replaced now or 3 years from now, and (b) once decided, to determine the equivalent AW for the next 3 years. (c) Write the spreadsheet functions that determine the AW values.

(d) After you return to your office, your boss texts the following question and wants an answer immediately. What do you tell her and what is the spreadsheet function used to determine the answer? "Hi, what is the equivalent annual worth of the replacement machine if we keep it for its maximum expected life and how much does this reduce the capital recovery each year?"

11.37 In order to replace an old process for producing a polymer that reduces friction loss in engines, two current-technology machines have been identified. Process K will have a first cost of $160,000, an operating cost of $7000 per month, and a salvage value of $50,000 after 1 year and $40,000 after its maximum 2-year life.

Process L will have a first cost of $210,000, an operating cost of $5000 per month, and salvage values of $100,000 after 1 year, $70,000 after 2 years, $45,000 after 3 years, and $26,000 after its maximum 4-year life. You have been asked to determine which process is better using a study period of (a) 1 year, (b) 2 years, and (c) 3 years. The company's MARR is 12% per year compounded monthly. (Note: Problem 11.46 continues this analysis via spreadsheet with the added option of upgrading the in-place process.)

11.38 **Keep or Replace the Kiln Case**

In Example 11.8, the in-place kiln and replacement kiln (GH) were evaluated using a fixed study period of 6 years. This is a significantly shortened period compared to the expected 12-year life of the challenger. Use the best estimates available throughout this case to determine the impact on the capital recovery amount for the replacement kiln (GH) of shortening the evaluation time from 12 to 6 years.

Replacement Value

11.39 Huntington Medical Center purchased a used low-field MRI scanner 2 years ago for $445,000. Its operating cost is $272,000 per year and it can be sold for $150,000 anytime in the next 3 years. The Center's director is considering replacing the presently owned MRI scanner with a state-of-the-art 3 Tesla machine that will cost $2.2 million. The operating cost of the new machine will be $340,000 per year, but it will generate *extra revenue* that is expected to amount to $595,000 per year. The new unit can probably be sold for $800,000 three years from now. You have been asked to determine how much the presently owned scanner would have to be worth on the open market for the AW values of the two machines to be the same over a 3-year planning period. The Center's MARR is 20% per year.

11.40 A company that makes micro motion compact coriolis meters purchased a new packaging system for $600,000. The estimated salvage value was $28,000 after 10 years. Currently the expected remaining life is 7 years with an AOC of $27,000 per year and an estimated salvage value of $40,000. The company is considering early replacement of the system with one that costs $370,000, has a 12-year economic service life, a $22,000 salvage value, and an estimated AOC of $50,000 per year. The MARR for the corporation is 12% per year. (a) Determine the minimum trade-in value necessary now to make the replacement economically advantageous. (b) Write the spreadsheet functions necessary to determine RV using Goal Seek.

11.41 With the estimates shown below, Sarah needs to determine the trade-in (replacement) value of machine X that will render its AW equal to that of machine Y at an interest rate of 8% per year. Determine the RV (a) by hand solution, and (b) using a spreadsheet.

	Machine X	Machine Y
Market value, $	?	80,000
Annual cost, $ per year	−60,000	−40,000 year 1, increasing by 2000 per year thereafter
Salvage value, $	15,000	20,000
Life, years	3	5

11.42 Machine A was purchased 5 years ago for $90,000. Its operating cost is higher than expected, so it will be used for only 4 more years. Its operating cost this year will be $40,000, increasing by $2000 per year through the end of its useful life. The challenger, machine B, will cost $150,000 with a $50,000 salvage value after its 10-year ESL. Its operating cost is expected to be $10,000 for year 1, increasing by $500 per year thereafter. What is the market value for machine A that would make the two machines equally attractive at an interest rate of 12% per year. Solve by hand and spreadsheet. (Hint: Be sure you check the RV value carefully.)

11.43 Because it fumes at room temperatures, hydrochloric acid creates a very corrosive work environment. A machine, working in that environment, is deteriorating fast and can be used for only 1 more year, at which time it will be scrapped with no salvage value. It was purchased 3 years ago for $88,000; its operating cost for the next year is expected to be $53,000. A more corrosion-resistant challenger will cost $226,000 with an operating cost of $48,000 per year. It is expected to have a $60,000 salvage value after its 10-year ESL. At an interest rate of 15% per year, what minimum replacement value would render the challenger attractive?

EXERCISES FOR SPREADSHEETS

11.44 Halcrow, Inc. expects to replace a downtime tracking system currently installed on CNC machines. The challenger system has a first cost of $70,000, an estimated AOC of $20,000 the first year increasing by 20% per year thereafter, a maximum useful life of 10 years, and a $10,000 market value after 1 year deceasing by 10% per year thereafter. (*a*) At an interest rate of 8% per year, determine the ESL and corresponding AW value for the challenger. (*b*) What happens to the ESL and the shape of the total AW curve if the company is able to reduce the annual AOC percentage increase to 10%?

11.45 First, develop the spreadsheet to confirm the replacement study decision conducted for Randall-Rico Consultants in Problem 11.31. Then answer the questions below posed by Mr. Randall in a meeting when he was informed of the results of the replacement study.
- (*a*) I want to buy the new graphical plotter. What is the required purchase price of the new plotter to make me indifferent between buying it and keeping the old one?
- (*b*) How much of a trade-in do I have to get for the old plotter to make me indifferent between retaining it (with the current ESL) and buying the new one? Is this the economically correct decision?

11.46 An existing process involving polymers that reduce friction loss in engines is described in Problem 11.37. Assume it is possible to do an augmentation and upgrade to the existing process machinery that will provide a 2-year extension to its useful life. Call this Process M. The upgrade will cost $50,000 now, have an increased operating cost of $9,000 per month, and no salvage value after the 2-year extension. Using the same criteria mentioned in Problem 11.37(*b*), that is, effective $i = 1\%$ per month and a 2-year study period, is it economically advantageous to upgrade (option M) or to purchase the equipment for Process K or L?

ADDITIONAL PROBLEMS AND FE EXAM REVIEW QUESTIONS

11.47 In a replacement study, the correct value to use when determining the purchase price of the challenger is:
- (*a*) Its first cost when it was purchased
- (*b*) Its first cost minus the trade-in value of the defender
- (*c*) Its first cost plus the trade-in value of the defender
- (*d*) The book value of the defender

11.48 A replacement analysis is most objectively conducted from the viewpoint of:
- (*a*) An outsider
- (*b*) A consultant
- (*c*) A nonowner
- (*d*) Any of the above

11.49 The economic service life of an asset is:
- (*a*) The length of time required to recover the first cost of the asset
- (*b*) The time when the operating cost is at a minimum
- (*c*) The time when the salvage value goes below 25% of the first cost
- (*d*) The time when the AW of the asset is at a minimum

11.50 In a one-year-later replacement analysis, if all estimates are still current and the year is n_D, the action that should be taken is:
- (*a*) Keep the defender 1 more year
- (*b*) Replace the defender with the challenger
- (*c*) Look for a new challenger and calculate its AW
- (*d*) Keep the defender until its market value is equal to the estimated salvage value of the challenger

11.51 A milling machine with enhanced CNC controls that allow for high-speed machining of free-form parts was purchased 2 years ago for $195,000. The company wants to purchase a recently available faster model with control up to 8 axes for $240,000. The presently owned machine can be sold today for $105,000. Its operating cost over the past 2 years has been $30,000 per year. The value that should be used as *P* for the presently owned machine is:
- (*a*) $240,000
- (*b*) $195,000
- (*c*) $105,000
- (*d*) $30,000

11.52 The equivalent annual worth of an existing machine at American Semiconductor is estimated to be $ −70,000 per year over its remaining useful life of 3 years. It can be replaced now or later with a machine that will have an AW of $ −90,000 per year if it is kept for 2 years or less, $75,000 if it is kept between 3 and 5 years, and $ −65,000 if it is kept for 6 to 8 years. The company wants an analysis of what it should do for a 3-year study period at

an interest rate of 15% per year. You should recommend that the existing machine be replaced:
(a)   Now
(b)   One year from now
(c)   Two years from now
(d)   Do not replace it

11.53 The annual worth values for a defender, which can be replaced with a similar used asset, and a challenger are estimated below. The economic service life of the challenger is
(a)   2 years   (b)   3 years
(c)   4 years   (d)   5 years

	AW Value, $ per Year	
Number of Years Retained	Defender	Challenger
1	−14,000	−21,000
2	−13,700	−18,000
3	−16,900	−13,100
4	−17,000	−15,600
5	−18,000	−17,500

11.54 From the data shown below, the economic service life of the asset is:
(a)   2 years   (b)   3 years
(c)   4 years   (d)   5 years

Years Retained	AW of First Cost, $	AW of Operating Cost, $	AW of Salvage Value, $
1	−165,000	−36,000	99,000
2	−86,429	−36,000	39,095
3	−60,317	−42,000	18,127
4	−47,321	−43,000	5,464
5	−39,570	−48,000	3,276

11.55 Assume the MARR is 10% per year for this analysis. A presently owned machine that was purchased 8 years ago for $450,000 is under consideration for replacement. It has an annual operating cost of $120,000 per year and a salvage value of $40,000 whenever it is replaced. The challenger has a first cost of $670,000, an expected annual operating cost of $94,000, and a salvage value of $60,000 after its 10-year economic life. The breakeven market value of the presently owned machine required to make the AW values of the two machines the same, if the presently owned machine is kept for 5 more years and then replaced with the challenger that has the same AW, is closest to:
(a) $196,340   (b) $255,390
(c) $325,360   (d) $394,770

11.56 A presently owned machine has the projected market value and M&O costs shown below. At an interest rate of 10% per year, the AW for keeping the machine 2 more years is closest to:

(a) $29,650   (b) $32,840
(c) $35,770   (d) $39,720

Year	Market Value, $	M&O Cost, $
0	32,000	—
1	25,000	24,000
2	14,000	24,000
3	10,000	26,000
4	8,000	29,000

11.57 The cost characteristics of a CO testing machine that was purchased 5 years ago for $100,000 are shown below. The equation to determine the AW of retaining the tester one more year and then replacing it is:
(a) $AW = -15,000(A/P,i,1) - 54,000 + 10,000(A/F,i,1)$
(b) $AW = -100,000(A/P,i,6) - 54,000 + 10,000(A/F,i,6)$
(c) $AW = -60,000(A/P,i,1) - 42,000 + 40,000(A/F,i,1)$
(d) $AW = -100,000(A/P,i,5) - 42,000 + 15,000(A/F,i,5)$

Machine Age, Years from Purchase	M&O Costs, $ per Year	Salvage Value at End of Year, $
1	42,000	60,000
2	47,000	40,000
3	49,000	31,000
4	50,000	24,000
5	52,000	15,000
6	54,000	10,000
7	63,000	10,000
8	67,000	10,000

11.58 The annual worth values for a defender, which can be replaced with a similar used asset, and a challenger are estimated. The defender should be replaced:
(a)   Now   (b)   1 year from now
(c)   2 years from now   (d)   3 years from now

	AW Value, $ per Year	
Number of Years Retained	Defender	Challenger
1	−14,000	−21,000
2	−13,700	−18,000
3	−16,900	−13,800
4	−17,000	−15,600
5	−18,000	−17,500

CASE STUDY

A PUMPER SYSTEM WITH AN ESL PROBLEM

Background

New pumper system equipment is under consideration by a Gulf Coast chemical processing plant. One crucial pump moves highly corrosive liquids from specially lined tanks on intercoastal barges into storage and preliminary refining facilities dockside. Because of the variable quality of the raw chemical and the high pressures imposed on the pump chassis and impellers, a close log is maintained on the number of hours per year that the pump operates. Safety records and pump component deterioration are considered critical control points for this system. As currently planned, rebuild and M&O cost estimates are increased accordingly when cumulative operating time reaches the 6000-hour mark.

Information

You are the safety engineer at the plant. Estimates made for this pump are as follows:

First cost: $800,000

Rebuild cost: $150,000 whenever 6000 cumulative hours are logged. Each rework will cost 20% more than the previous one. A maximum of three rebuilds are allowed.

M&O costs: $25,000 for each year 1 through 4 $40,000 per year starting the year after the first rebuild, plus 15% per year thereafter

MARR: 10% per year

Based on previous logbook data, the current estimates for number of operating hours per year are as follows:

Year	Hours per Year
1	500
2	1500
3 on	2000

Case Study Questions

1. Determine the economic service life of the pump. How does the ESL compare with the maximum allowed rebuilds?

2. The *plant superintendent* told you, the safety engineer, that only one rebuild should be planned for, because these types of pumps usually have their minimum-cost life before the second rebuild. Determine a market value for this pump that will force the ESL to be 6 years. Comment on the practicality of ESL = 6 years, given the MV calculated.

3. In a separate conversation, the *line manager* told you to not plan for a rebuild after 6000 hours because the pump will be replaced after a total of 10,000 hours of operation. The line manager wants to know what the base AOC in year 1 can be to make the ESL 6 years. He also told you to assume now that the 15% growth rate applies from year 1 forward. How does this base AOC value compare with the rebuild cost after 6000 hours?

4. What do you think of these suggestions from the plant superintendent and the line manager?

CHAPTER 12

Independent Projects with Budget Limitation

Ariel Skelley/Blend Images/Getty Images

LEARNING OUTCOMES

Purpose: Select independent projects for funding when there is a limitation on the amount of capital available for investment.

SECTION	TOPIC	LEARNING OUTCOME
12.1	Capital rationing	• Explain how a capital budgeting problem is approached.
12.2	Equal-life projects	• Use PW-based capital budgeting to select from several equal-life independent projects.
12.3	Unequal-life projects	• Use PW-based capital budgeting to select from several unequal-life independent projects.
12.4	Linear programming	• Set up the linear programming model and use the Solver spreadsheet tool to select projects.
12.5	Ranking options	• Use the internal rate of return (IROR) and profitability index (PI) to rank and select from independent projects.

I n most of the previous economic comparisons, the alternatives have been mutually exclusive; only one could be selected. If the projects are not mutually exclusive, they are categorized as independent of one another, as discussed at the beginning of Chapter 5. Now we learn techniques to select from several independent projects. It is possible to select any number of projects from none (do nothing) to all viable projects.

There is virtually always some upper limit on the amount of capital available for investment in new projects. This limit is considered as each independent project is economically evaluated. The techniques applied are called *capital budgeting methods,* also referred to as capital rationing. They determine the economically best rationing of initial investment capital among independent projects based upon different measures, such as PW, ROR, and the profitability index. These three are discussed here.

12.1 An Overview of Capital Rationing among Projects ● ● ●

Investment capital is a scarce resource for all corporations; thus there is virtually always a limited amount to be distributed among competing investment opportunities. When a corporation has several options for placing investment capital, a "reject or accept" decision must be made for each project. Effectively, each option is independent of other options, so the evaluation is performed on a project-by-project basis. Selection of one project does not impact the selection decision for any other project. This is the fundamental difference between mutually exclusive alternatives and independent projects.

The term **project** is used to identify each independent option. We use the term **bundle** to identify a collection of independent projects. The term *mutually exclusive alternative* continues to identify a project when only one may be selected from several.

There are two exceptions to purely independent projects: A *contingent project* is one that has a condition placed upon its acceptance or rejection. Two examples of contingent projects A and B are as follows: A cannot be accepted unless B is accepted; and A can be accepted in lieu of B, but both are not needed. A *dependent project* is one that must be accepted or rejected based on the decision about another project(s). For example, B must be accepted if both A and C are accepted. In practice, these complicating conditions can be bypassed by forming packages of related projects that are economically evaluated themselves as independent projects along with the remaining, unconditioned projects.

A **capital budgeting study** has several fundamental characteristics:

- Several independent projects are identified, and net cash flow estimates are available.
- Each project is either selected entirely or not selected; that is, partial investment in a project is not possible.
- A stated budgetary constraint restricts the total amount available for investment. Budget constraints may be present for the first year only or for several years. This investment limit is identified by the symbol b.
- The objective is to maximize the return on the investments using a measure of worth, such as the PW value.

Independent project selection

Limited budget

By nature, independent projects are usually quite different from one another. For example, in the public sector, a city government may develop several projects to choose from: drainage, city park, metro rail, and an upgraded public bus system. In the private sector, sample projects may be a new warehousing facility, expanded product base, improved quality program, an upgraded IT system, automation, and acquisition of another firm. The size of the study can range from only four or five projects to a complex study involving 50 to 100 projects. The typical capital budgeting problem is illustrated in Figure 12–1. For each independent project, there is an initial investment, project life, and estimated net cash flows that can include a salvage value.

Present worth analysis using the capital budgeting process is the recommended method to select projects. The general selection guideline is as follows:

Accept projects with the **best PW values** determined at the MARR over the project life, provided the capital investment limit is not exceeded.

Figure 12–1
Basic characteristics of a
capital budgeting study.

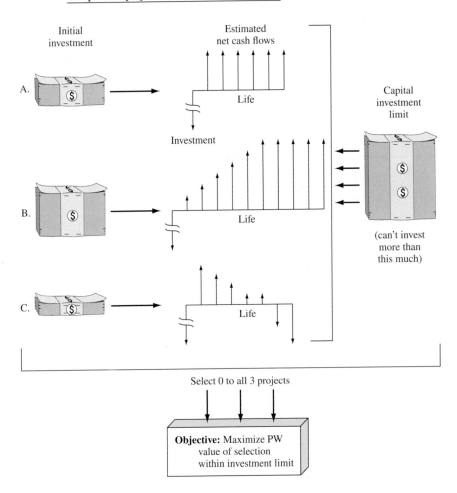

Independent projects and net cash flow estimates

This guideline is not different from that used for selection in previous chapters for independent projects. As before, each project is compared with the do-nothing project; that is, incremental analysis between projects is not necessary. The primary difference now is that the amount of money available to invest is limited, thus the title *capital budgeting* or *rationing*. A specific solution procedure that incorporates this constraint is needed.

Previously, PW analysis had the requirement of equal service between alternatives. This assumption is not necessary for capital rationing because there is no life cycle of a project beyond its estimated life. Rather, the selection guideline has the following implied assumption.

Equal-service requirement

> When the present worth at the MARR over the respective project life is used to select projects, the reinvestment assumption is that all positive net cash flows are reinvested at the MARR from the time they are realized until the **end of the longest-lived project**.

This fundamental assumption is demonstrated to be correct at the end of Section 12.3, which treats PW-based capital rationing for unequal-life projects.

Another dilemma of capital rationing among independent projects concerns the flexibility of the capital investment limit *b*. The limit may marginally disallow an acceptable project that is next in line for acceptance. For example, assume project A has a positive PW value at the MARR. If A will cause the capital limit of $5,000,000 to be exceeded by only $1000, should A be included in the PW analysis? Commonly, a capital investment limit is somewhat flexible, so project A would usually be included. Since the examples included here are for learning and illustration purposes, we will not exceed a stated investment limit.

As we learned earlier (Sections 1.9 and 10.1), the rate of return on the first unfunded project is an **opportunity cost**. The lack of capital to fund the next project defines the ROR level that is forgone. The opportunity cost will vary with each set of independent projects evaluated, but over time it provides information to fine-tune the MARR and other measures used by the company in future evaluations.

Opportunity cost

It is common to rank independent projects and select them based on measures other than PW at the MARR. Two are the *internal rate of return (IROR)*, discussed in Chapter 7, and the *profitability index (PI)*, also called the *present worth index (PWI)*, introduced in Chapter 9. Neither of these measures guarantees an optimal PW-based selection. The capital budgeting process, covered in the next three sections, does find the optimal solution for PW values. We recommend use of this PW-based technique; however, it should be recognized that both PI and IROR usually provide excellent, near-optimal selections, and they both work very well when the number of projects is large. Application of these two measures is presented in Section 12.5.

The PW-based selection method can be correctly applied (for equal or unequal life projects) using a hand solution approach or a spreadsheet solution using a linear programming approach. Both of these techniques are discussed in the next three sections.

12.2 Capital Rationing Using PW Analysis of Equal-Life Projects ● ● ●

To select from projects that have the same expected life while investing no more than the limit b, first formulate all **mutually exclusive bundles**—one project at a time, two at a time, etc. Each feasible bundle must have a total investment that does not exceed b. One of these bundles is the do-nothing (DN) project. The total number of bundles for m projects is 2^m. The number increases rapidly with m. For $m = 4$, there are $2^4 = 16$ bundles, and for $m = 16$, $2^{16} = 65,536$ bundles. Then the PW of each bundle is determined at the MARR. The bundle with the largest PW value is selected.

To illustrate the development of mutually exclusive bundles, consider these four projects with equal lives.

Project	Initial Investment, $
A	−10,000
B	−5,000
C	−8,000
D	−15,000

If the investment limit is $b = \$25,000$, of the 16 bundles, there are 12 feasible ones to evaluate. The bundles ABD, ACD, BCD, and ABCD have investment totals that exceed $25,000. The viable bundles are shown below.

Projects	Total Initial Investment, $	Projects	Total Initial Investment, $
A	−10,000	AD	−25,000
B	−5,000	BC	−13,000
C	−8,000	BD	−20,000
D	−15,000	CD	−23,000
AB	−15,000	ABC	−23,000
AC	−18,000	Do nothing	0

The procedure to conduct a capital budgeting study using PW analysis is as follows:

1. Develop all mutually exclusive bundles with a total initial investment that does not exceed the capital limit b.
2. Sum the net cash flows NCF_{jt} for all projects in each bundle j ($j = 1, 2, \ldots, 2^m$) and each year t ($t = 1, 2, \ldots, n_j$). Refer to the initial investment of bundle j at time $t = 0$ as NCF_{j0}.
3. Compute the present worth value PW_j for each bundle at the MARR.

> $\text{PW}_j = \text{PW of bundle net cash flows} - \text{initial investment}$
>
> $$= \sum_{t=1}^{t=n_j} \text{NCF}_{jt}(P/F,i,t) - \text{NCF}_{j0} \qquad [12.1]$$

4. Select the bundle with the (numerically) largest PW_j value.

Selecting the maximum PW_j means that this bundle has a PW value larger than any other bundle. Any bundle with $\text{PW}_j < 0$ is discarded because it does not produce a return of at least the MARR.

EXAMPLE 12.1

The projects review committee of Microsoft has $20 million to allocate next year to new software product development. Any or all of five projects in Table 12–1 may be accepted. All amounts are in $1000 units. Each project has an expected life of 9 years. Select the project(s) if a 15% return is expected.

TABLE 12–1   Five Equal-Life Independent Projects ($1000 Units)

Project	Initial Investment, $	Annual Net Cash Flow, $	Project Life, Years
A	−10,000	2870	9
B	−15,000	2930	9
C	−8,000	2680	9
D	−6,000	2540	9
E	−21,000	9500	9

Solution

Use the procedure above with $b = \$20{,}000$ to select one bundle that maximizes present worth. Remember the units are in $1000.

1. There are $2^5 = 32$ possible bundles. The eight bundles that require no more than $20,000 in initial investments are described in columns 2 and 3 of Table 12–2. The $21,000 investment for E eliminates it from all bundles.

TABLE 12–2   Summary of Present Worth Analysis of Equal-Life Independent Projects ($1000 Units)

Bundle j (1)	Projects Included (2)	Initial Investment NCF_{j0}, $ (3)	Annual Net Cash Flow NCF_j, $ (4)	Present Worth PW_j, $ (5)
1	A	−10,000	2,870	+3,694
2	B	−15,000	2,930	−1,019
3	C	−8,000	2,680	+4,788
4	D	−6,000	2,540	+6,120
5	AC	−18,000	5,550	+8,482
6	AD	−16,000	5,410	+9,814
7	CD	−14,000	5,220	+10,908
8	Do nothing	0	0	0

2. The bundle net cash flows, column 4, are the sum of individual project net cash flows.
3. Use Equation [12.1] to compute the present worth for each bundle. Since the annual NCF and life estimates are the same for a bundle, PW_j reduces to

$$PW_j = NCF_j(P/A,15\%,9) - NCF_{j0}$$

4. Column 5 of Table 12–2 summarizes the PW_j values at $i = 15\%$. Bundle 2 does not return 15% since $PW_2 < 0$. The largest is $PW_7 = \$10{,}908$; therefore, invest $14 million in C and D. This leaves $6 million uncommitted.

This analysis assumes that the $6 million not used in this initial investment will return the MARR by placing it in some other, unspecified investment opportunity.

12.3 Capital Rationing Using PW Analysis of Unequal-Life Projects ● ● ●

Usually independent projects do not have the same expected life. The PW method for solution of the capital budgeting problem assumes that each project will last for the period of the longest-lived project n_L. Additionally, reinvestment of any positive net cash flows is assumed to be at the MARR from the time they are realized until the end of the longest-lived project, that is, from year n_j through year n_L. Therefore, use of the **LCM of lives is not necessary**, and it is correct to use Equation [12.1] to select bundles of unequal-life projects by PW analysis using the procedure of the previous section.

EXAMPLE 12.2

For MARR = 15% per year and b = \$20,000, select from the following independent projects. Solve by hand and by spreadsheet.

Project	Initial Investment, \$	Annual Net Cash Flow, \$	Project Life, Years
A	−8,000	3870	6
B	−15,000	2930	9
C	−8,000	2680	5
D	−8,000	2540	4

Solution by Hand

The unequal-life values make the net cash flows vary over a bundle's life, but the selection procedure is the same as above. Of the $2^4 = 16$ bundles, 8 are economically feasible. Their PW values by Equation [12.1] are summarized in Table 12–3. As an illustration, for bundle 7:

$$PW_7 = -16,000 + 5220(P/A,15\%,4) + 2680(P/F,15\%,5) = \$235$$

Select bundle 5 (projects A and C) for a \$16,000 investment.

TABLE 12–3	Present Worth Analysis for Unequal-Life Independent Projects, Example 12.2					

Bundle J (1)	Project (2)	Initial Investment, NCF$_{j0}$, \$ (3)	Net Cash Flows		Present Worth PW$_j$, \$ (6)
			Year t (4)	NCF$_{jt}$, \$ (5)	
1	A	−8,000	1–6	3,870	+6,646
2	B	−15,000	1–9	2,930	−1,019
3	C	−8,000	1–5	2,680	+984
4	D	−8,000	1–4	2,540	−748
5	AC	−16,000	1–5	6,550	+7,630
			6	3,870	
6	AD	−16,000	1–4	6,410	+5,898
			5–6	3,870	
7	CD	−16,000	1–4	5,220	+235
			5	2,680	
8	Do nothing	0		0	0

Solution by Spreadsheet

Figure 12–2 presents a spreadsheet with the same information as in Table 12–3. It is necessary to initially develop the mutually exclusive bundles manually and total net cash flows each year using each project's NCF. The NPV function is used to determine PW for each bundle j over its respective life. Bundle 5 (projects A and C) has the largest PW value (row 16).

	A	B	C	D	E	F	G	H	I
1	MARR =	15%						= D7+E7	
2									
3	Bundle	1	2	3	4	5	6	7	8
4	Projects	A	B	C	D	AC	AD	CD	Do nothing
5	Year				Net cash flows, NCF, $				
6	0	-8,000	-15,000	-8,000	-8,000	-16,000	-16,000	-16,000	0
7	1	3,870	2,930	2,680	2,540	6,550	6,410	5,220	0
8	2	3,870	2,930	2,680	2,540	6,550	6,410	5,220	0
9	3	3,870	2,930	2,680	2,540	6,550	6,410	5,220	0
10	4	3,870	2,930	2,680	2,540	6,550	6,410	5,220	0
11	5	3,870	2,930	2,680		6,550	3,870	2,680	0
12	6	3,870	2,930			3,870	3,870		0
13	7		2,930						0
14	8		2,930						0
15	9		2,930						0
16	PW @ 15%	6,646	-1,019	984	-748	7,630	5,898	235	0
17									
18					= NPV($B$1,F7:F15) + F6				
19									

Figure 12–2
Computation of PW values for independent project selection, Example 12.2.

The rest of this section will help you understand why solution of the capital budgeting problem by PW evaluation using Equation [12.1] is correct. The following logic verifies the assumption of reinvestment at the MARR for all net positive cash flows when project lives are unequal. Refer to Figure 12–3, which uses the general layout of a two-project bundle. Assume each project has the same net cash flow each year. The P/A factor is used for PW computation. Define n_L as the life of the longer-lived project. At the end of the shorter-lived project, the bundle has a total future worth of $NCF_j(F/A,MARR,n_j)$ as determined for each project. Now, assume reinvestment at the MARR from year n_{j+1} through year n_L (a total of $n_L - n_j$ years). The assumption of the return at the MARR is important; this PW approach does not necessarily select the correct projects if the return is not at the MARR. The results are the two future worth arrows in year n_L in Figure 12–3. Finally,

Figure 12–3
Representative cash flows used to compute PW for a bundle of two independent unequal-life projects by Equation [12.1].

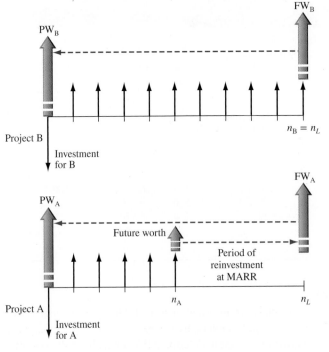

Bundle PW = PW_A + PW_B

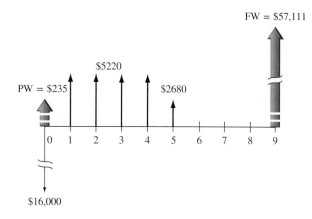

Figure 12–4
Initial investment and cash flows for bundle 7, projects C and D, Example 17.2.

compute the bundle PW value in the initial year. This is the bundle $PW = PW_A + PW_B$. In general form, the bundle j present worth is

$$PW_j = NCF_j(F/A,MARR,n_j)(F/P,MARR,n_L-n_j)(P/F,MARR,n_L) \qquad [12.2]$$

Substitute the symbol i for the MARR, and use the factor formulas to simplify.

$$PW_j = NCF_j \frac{(1+i)^{n_j}-1}{i}(1+i)^{n_L-n_j}\frac{1}{(1+i)^{n_L}}$$

$$= NCF_j\left[\frac{(1+i)^{n_j}-1}{i(1+i)^{n_j}}\right] \qquad [12.3]$$

$$= NCF_j(P/A,i,n_j)$$

Since the bracketed expression in Equation [12.3] is the $(P/A,i,n_j)$ factor, computation of PW_j for n_j years assumes reinvestment at the MARR of all positive net cash flows until the longest-lived project is completed in year n_L.

To demonstrate numerically, consider bundle $j = 7$ in Example 12.2. The evaluation is in Table 12–3, and the net cash flow is pictured in Figure 12–4. Calculate the future worth in year 9, which is the life of the longest-lived project (B).

$$FW = 5220(F/A,15\%,4)(F/P,15\%,5) + 2680(F/P,15\%,4) = \$57,111$$

The present worth at the initial investment time is

$$PW = -16,000 + 57,111(P/F,15\%,9) = \$235$$

The PW value is the same as PW_7 in Table 12–3 and Figure 12–2. This demonstrates the reinvestment assumption for positive net cash flows. If this assumption is not realistic, the PW analysis must be conducted using the *LCM of all project lives.*

12.4 Capital Budgeting Problem Formulation
Using Linear Programming ● ● ●

The procedure discussed above requires the development of mutually exclusive bundles one project at a time, two projects at a time, etc., until all 2^m bundles are developed and each one is compared with the capital limit b. As the number of independent projects increases, this process becomes prohibitively cumbersome and unworkable. Fortunately, the capital budgeting problem can be stated in the form of a linear programming model. The problem is formulated using the integer linear programming (ILP) model, which means simply that all relations are linear and that the variable x can take on only integer values. In this case, the variables can only take on the values 0 or 1, which makes it a special case called the 0-or-1 ILP model. The formulation in words follows.

Maximize: Sum of PW of net cash flows of independent projects.

Constraints:
- Capital investment constraint is that the sum of initial investments must not exceed a specified limit.
- Each project is completely selected or not selected.

For the math formulation, define b as the capital investment limit, and let x_k ($k = 1$ to m projects) be the variables to be determined. If $x_k = 1$, project k is completely selected; if $x_k = 0$, project k is not selected. Note that the subscript k *represents each independent project*, not a mutually exclusive bundle.

If the sum of PW of the net cash flows is Z, the math programming formulation is as follows:

$$\text{Maximize:} \quad \sum_{k=1}^{k=m} PW_k x_k = Z$$

$$\text{Constraints:} \quad \sum_{k=1}^{k=m} NCF_{k0} x_k \le b \qquad \text{[12.4]}$$

$$x_k = 0 \text{ or } 1 \qquad \text{for } k = 1, 2, \ldots, m$$

The PW_k of each project is calculated using Equation [12.1] at MARR $= i$.

$$PW_k = \text{PW of project net cash flows for } n_k \text{ years}$$

$$= \sum_{t=1}^{t=n_k} NCF_{kt}(P/F,i,t) - NCF_{k0} \qquad \text{[12.5]}$$

Computer solution is accomplished by a linear programming software package which treats the ILP model. Also, Excel and its optimizing tool **Solver** can be used to develop the formulation and select the projects. The Solver tool is similar to Goal Seek with significantly more capabilities. For example, Solver allows the target cell to be maximized, minimized, or set to a specific value. This means that the function Z in Equation [12.4] can be maximized. Also, multiple changing cells can be identified, so the 0 or 1 value of the unknowns can be determined. Additionally, with the added capability to include constraints, the investment limit b and 0-or-1 requirement on the unknowns in Equation [12.4] can be accommodated. Solver is explained in detail in Appendix A, and Example 12.3 illustrates its use.

EXAMPLE 12.3

Review Example 12.2. (*a*) Formulate the capital budgeting problem using the math programming model presented in Equation [12.4]. (*b*) Select the projects using Solver.

Solution

(*a*) Define the subscript $k = 1$ through 4 for the four projects, which are relabeled as 1, 2, 3, and 4. The capital investment limit is $b = \$20,000$ in Equation [12.4].

$$\text{Maximize:} \quad \sum_{k=1}^{k=4} PW_k x_k = Z$$

$$\text{Constraints:} \quad \sum_{k=1}^{k=4} NCF_{k0} x_k \le 20,000$$

$$x_k = 0 \text{ or } 1 \qquad \text{for } k = 1, 2, 3, 4$$

Now, substitute the PW_k and NCF_{k0} values from Table 12–3 into the model. Plus signs are used for all values in the budget constraint. We have the complete 0-or-1 ILP formulation.

Maximize: $\qquad 6646x_1 - 1019x_2 + 984x_3 - 748x_4 = Z$

Constraints: $\qquad 8000x_1 + 15,000x_2 + 8000x_3 + 8000x_4 \le 20,000$

$x_1, x_2, x_3,$ and $x_4 = 0$ or 1

The maximum PW is \$7630, and the solution from Example 12.2 is written as

$$x_1 = 1 \qquad x_2 = 0 \qquad x_3 = 1 \qquad x_4 = 0$$

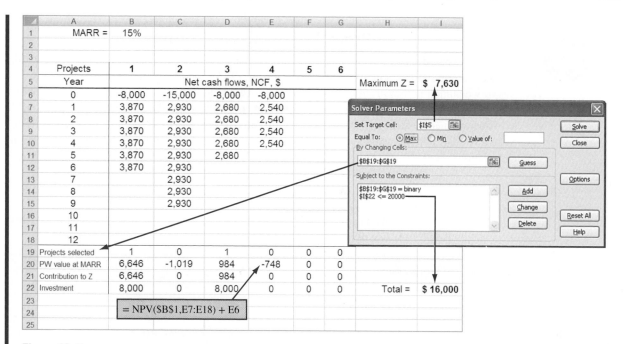

The spreadsheet shows:

	A	B	C	D	E	F	G	H	I
1	MARR =	15%							
2									
3									
4	Projects	1	2	3	4	5	6		
5	Year		Net cash flows, NCF, $					Maximum Z =	$ 7,630
6	0	-8,000	-15,000	-8,000	-8,000				
7	1	3,870	2,930	2,680	2,540				
8	2	3,870	2,930	2,680	2,540				
9	3	3,870	2,930	2,680	2,540				
10	4	3,870	2,930	2,680	2,540				
11	5	3,870	2,930	2,680					
12	6	3,870	2,930						
13	7		2,930						
14	8		2,930						
15	9		2,930						
16	10								
17	11								
18	12								
19	Projects selected	1	0	1	0	0	0		
20	PW value at MARR	6,646	-1,019	984	-748	0	0		
21	Contribution to Z	6,646	0	984	0	0	0		
22	Investment	8,000	0	8,000	0	0	0	Total =	$ 16,000

$$= NPV(\$B\$1,E7:E18) + E6$$

Solver Parameters dialog:
- Set Target Cell: $I$5
- Equal To: ⦿ Max ○ Min ○ Value of:
- By Changing Cells: $B$19:$G$19
- Subject to the Constraints:
  - $B$19:$G$19 = binary
  - $I$22 <= 20000
- Buttons: Solve, Close, Guess, Options, Add, Change, Delete, Reset All, Help

Figure 12–5

Spreadsheet and Solver template configured to solve a capital budgeting problem, Example 12.3.

(*b*) Figure 12–5 presents a spreadsheet template developed to select from six or fewer independent projects with 12 years or less of net cash flow estimates per project. The spreadsheet template can be expanded in either direction if needed. Figure 12–5 (inset) shows the Solver parameters set to solve this example for four projects and an investment limit of $20,000. The descriptions below and the cell tag identify the contents of the rows and cells in Figure 12–5, and their linkage to Solver parameters.

> Rows 4 and 5: Projects are identified by numbers to distinguish them from spreadsheet column letters. Cell I5 is the expression for Z, the sum of the PW values for the projects. This is the target cell for Solver to maximize.
>
> Rows 6 to 18: These are initial investments and net cash flow estimates for each project. Zero values that occur after the life of a project need not be entered; however, any $0 estimates that occur during a project's life must be entered.
>
> Row 19: The entry in each cell is 1 for a selected project and 0 if not selected. These are the changing cells for Solver. Since each entry must be 0 or 1, a binary constraint is placed on all row 19 cells in Solver, as shown in Figure 12–5. When a problem is to be solved, it is best to initialize the spreadsheet with 0s for all projects. Solver will find the solution to maximize Z.
>
> Row 20: The NPV function is used to find the PW for each net cash flow series. The NPV functions are developed for any project with a life up to 12 years at the MARR entered in cell B1.
>
> Row 21: When a project is selected, the contribution to the Z function is shown by multiplying rows 19 and 20.
>
> Row 22: This row shows the initial investment for the selected projects. Cell I22 is the total investment. This cell has the budget limitation placed on it by the constraint in Solver. In this example, the constraint is I22 < = $20,000.

To solve the example, set all values in row 19 to 0, set up the Solver parameters as described above, and click on Solve. (Since this is a linear model, the Solver options choice "Assume Linear Model" may be checked, if desired.) If needed, further directions on saving the solution, making changes, etc., are available in Appendix A, Section A.5, and on the Excel help function.

For this problem, the selection is projects 1 and 3 with Z = $7630, the same as determined previously, and $16,000 of the $20,000 limit is invested.

12.5 Additional Project Ranking Measures ● ● ●

The PW-based method of solving a capital budgeting problem covered in Sections 12.2 to 12.4 provides an optimal solution that maximizes the PW of the competing projects. However, it is very common in industrial, professional, and government settings to learn that the rate of return is the basis for ranking projects. The **internal rate of return (IROR),** as we learned in Section 7.2, is determined by setting a PW or AW relation equal to zero and solving for i^*—the IROR. Using a PW basis and the estimated net cash flow (NCF) series for each project j, solve for i^* in the relation

$$0 = \sum_{t=1}^{t=n_j} NCF_{jt}(P/F,i^*,t) - NCF_{j0} \qquad [12.6]$$

This is the same as Equation [12.1] set equal to zero with $i = i^*$ as the unknown. The spreadsheet function RATE or IRR will provide the same answer. The selection guideline is as follows:

Independent project selection

> Once the project ranking by IROR is complete, select all projects in order without exceeding the investment limit b.
>
> If there is no budget limit, select all projects that have IROR ≥ MARR.

The ordering of projects using *IROR ranking may differ from the PW-based ranking* we used in previous sections. This can occur because IROR ranking maximizes the overall rate of return, not necessarily the PW value. The use of IROR ranking is illustrated in the next example.

Another common ranking method is the **profitability index (PI)** that we learned in Section 9.2. This is a "bang for the buck" measure that provides a sense of getting the most from the initial investment over the life of the project. (Refer to Section 9.2 for more details.) When utilized for project ranking, it is often called the **present worth index (PWI);** however, we will use the PI term for consistency, and because there are other ways to mathematically define measures also referred to as a PW index. The PI measure is defined as

$$\frac{\text{PW of net cash flows}}{\text{PW of initial investment}} = \frac{\sum\limits_{t=1}^{t=n_j} NCF_{jt}(P/F,i,t)}{|NCF_{j0}|} \qquad [12.7]$$

Note that the denominator includes only the initial investment, and it is its *absolute value* that is used. The numerator has only cash flows that result from the project for years 1 through its life n_j. Salvage value, if there is one estimated, is incorporated into the numerator. Similar to the previous case, the selection guideline is as follows:

Independent project selection

> Once the project ranking by PI is complete, select all projects in order without exceeding the investment limit b.
>
> If there is no budget limit, select all projects that have PI ≥ 1.0.

Depending upon the project NCF estimates, the *PI ranking can differ from the IROR ranking*. Example 12.4 compares results using the different ranking methods—IROR, PI, and PW values. None of these results are incorrect; they simply maximize different measures, as you will see. The use of IROR, PI, or other measures is common when there are a large number of projects because the PW basis (when solved by hand, Solver, or ILP software) becomes increasingly cumbersome as the number of possible mutually exclusive bundles grows using the formula 2^m. Additionally, greater complexity is introduced when dependent and contingent projects are involved.

EXAMPLE 12.4

Georgia works as a financial analyst in the Management Science Group of General Electronics. She has been asked to recommend which of the five projects detailed in Table 12–4 should be funded if the MARR is 15% per year and the investment budget limit for next year is a firm

TABLE 12–4	IROR, PI, and PW Values for Five Projects, Example 12.4				
Projects	**1**	**2**	**3**	**4**	**5**
Investment, NCF_0, $1000	−8,000	−15,000	−8,000	−8,000	−5,000
NCF, $1000 per year	4,000	2,900	2,700	2,500	2,600
Life n_j, years	6	9	5	4	3
IROR, %	44.5	12.8	20.4	9.6	26.0
PI	1.89	0.92	1.13	0.89	1.19
PW at 15%, $1000	7,138	−1,162	1,051	−863	936

$18 million. She has confirmed the computations and is ready to do the ranking and make the selection. Help her by doing the following:

(a) Use the IROR measure to rank and select projects.
(b) Use the PI measure to rank and select projects.
(c) Use the PW measure to rank and select projects.
(d) Compare the selected projects by the three methods and determine which one will maximize the overall ROR value of the $18 million budget.

Solution

Refer to Table 12–5 for the ranking, cumulative investment for each project, and selection based on the ranking and budget limit $b = 18 million.

(a) Ranking by overall IROR values indicates that projects 1 and 5 should be selected with $13 million of the $18 million budget expended. The remaining $5 million is assumed to be invested at the MARR of 15% per year.
(b) As an example, the PI for project 1 is calculated using Equation [12.7].

$$PI_1 = \frac{4000(P/A,15\%,6)}{|-8000|}$$
$$= 15,138/8,000$$
$$= 1.89$$

Ranking and projects selected are the same as in the IROR-based analysis. Again, the remaining $5 million is assumed to generate a return of MARR = 15% per year.

(c) Ranking by PW value results in a different selection from that of IROR and PI rankings. Projects 1 and 3, rather than 1 and 5, are selected for a total PW = $8.189 million and an investment of $16 million. The remaining $2 million is assumed to earn 15% per year.
(d) IROR and PI rankings result in projects 1 and 5 being selected. PW ranking results in the selection of projects 1 and 3. With MARR = 15%, the $18 million will earn at the following ROR values. In $1000 units,

Projects 1 and 5	NCF, year 0:	$−13,000
	NCF, years 1–3:	$6,600
	NCF, years 4–6:	$4,000

$$0 = -13,000 + 6600(P/A,i,3) + 4000(P/A,i,3)(P/F,i,3)$$

TABLE 12–5	Ranking of Projects by Different Measures, Example 12.4 (monetary units in $1000)							
Ranking by IROR			**Ranking by PI**			**Ranking by PW**		
IROR, % (1)	Project (2)	Cumulative Investment, $ (3)	PI (4)	Project (5)	Cumulative Investment, $ (6)	PW, $ (7)	Project (8)	Cumulative Investment, $ (9)
44.5	1	8,000	1.89	1	8,000	7,138	1	8,000
26.0	5	13,000	1.19	5	13,000	1,051	3	16,000
20.4	3	21,000	1.13	3	21,000	936	5	21,000
12.8	2		0.92	2		−863	4	
9.6	4		0.89	4		−1,162	2	

By IRR function, the rate of return is 39.1%. The overall return on the entire budget is

$$ROR = [39.1(13{,}000) + 15.0(5000)]/18{,}000$$
$$= 32.4\%$$

Projects 1 and 3   NCF, year 0: $\$-16{,}000$
                                NCF, years 1–5: $\$6{,}700$
                                  NCF, year 6: $\$4{,}000$

$$0 = -16{,}000 + 6700(P/A,i,5) + 4000(P/F,i,6)$$

By IRR function, the rate of return is 33.5%. The overall return on the entire budget is

$$ROR = [33.5(16{,}000) + 15.0(2000)]/18{,}000$$
$$= 31.4\%$$

In conclusion, the IROR and PI selections maximize the overall rate of return at 32.4%. The PW selection maximizes the PW value at $8.189 (i.e., 7.138 + 1.051) million, as determined from Table 12–5, column 7.

CHAPTER SUMMARY

Investment capital is always a scarce resource, so it must be rationed among competing projects using economic and noneconomic criteria. Capital budgeting involves proposed projects, each with an initial investment and net cash flows estimated over the life of the project. The fundamental capital budgeting problem has specific characteristics (Figure 12–1).

- Selection is made from among independent projects.
- Each project must be accepted or rejected as a whole.
- Maximizing the present worth of the net cash flows is the objective.
- The total initial investment is limited to a specified maximum.

The present worth method is used for evaluation. To start the procedure, formulate all mutually exclusive bundles that do not exceed the investment limit, including the do-nothing bundle. There are a maximum of 2^m bundles for m projects. Calculate the PW at MARR for each bundle, and select the bundle with the largest PW value. Reinvestment of net positive cash flows at the MARR is assumed for all projects with lives shorter than that of the longest-lived project.

The capital budgeting problem may be formulated as a linear programming problem to select projects directly in order to maximize the total PW. The Solver tool solves this problem by spreadsheet.

Projects can be ranked and selected on bases other than PW values. Two measures are the internal rate of return (IROR) and the profitability index (PI), also called the PW index. The ordering of projects may differ between the various ranking bases since different measures are optimized. When there are a large number of projects, the IROR basis is commonly applied in industry and business.

PROBLEMS

(Note: Exercises for Spreadsheets *are integrated with the problems in this and most future chapters.)*

Capital Rationing and Independent Projects

12.1   List at least three characteristics of a capital budgeting study.

12.2   State the difference between a *contingent project* and a *dependent project*.

12.3   How many mutually exclusive bundles can be formed from the following number of different independent projects: (*a*) 5; (*b*) 8; (*c*) 13? (*d*) By what amount is each total reduced if the DN project is not an option?

12.4   Develop all of the mutually exclusive bundles from independent projects identified as A, B, and C.

12.5 Six projects have been identified for possible implementation by a company that makes dry ice blasters, machines that propel tiny dry ice pellets at supersonic speeds so they flash freeze and then lift grime, paint, rust, mold, asphalt, and other contaminants off of in-place machines and a wide range of surfaces. The present worth of each project has been determined. Identify all acceptable mutually exclusive bundles if the budget limitation is $31,000.

Project	Project PW at 15%, $
L	29,000
M	11,000
N	41,000
O	35,000
P	6,000
Q	2,000

12.6 Develop all (a) acceptable, and (b) nonacceptable mutually exclusive bundles for four independent projects if the investment limit is $350 and the following project selection restriction applies: Project 2 can be selected only if project 3 is selected.

Project	Initial Investment, $
1	250
2	150
3	75
4	235

12.7 Five independent projects (1, 2, 3, 4, and 5) will be evaluated for investment by General Electric, Energy Division. Develop all of the acceptable and nonacceptable mutually exclusive bundles based on the following two selection restrictions developed by the Division Manager:
1. Project 2 can be selected only if projects 4 and 5 are selected.
2. Projects 1 and 5 should not both be selected because they are essentially duplicates.

Selecting Independent Projects Using PW Analysis

12.8 An engineer at Suncore Micro, LLC calculated the present worth of mutually exclusive bundles, each comprised of one or more independent projects. (a) Select the acceptable bundle if the capital investment limit is $50,000 and the MARR is 15% per year. (b) What happens to any leftover capital if not all of the $50,000 is committed to the selected bundle?

Project Bundle	Initial Investment, $	PW, $
1	−27,000	2,400
2	−33,000	9,200
3	−44,000	7,300
4	−51,000	11,400
5	−66,000	10,800

12.9 The CFO for Woodsome Appliance Company Plant #A14 in Mexico City has five independent projects she can fund this year to improve surface durability on stainless steel products. The project investments and 18%-per-year PW values are as shown. What projects should be accepted if the investment limit is (a) no limit, and (b) $55,000?

Project	Initial Investment, $	PW at 18% per Year, $
1	−15,000	−400
2	−25,000	8500
3	−20,000	500
4	−40,000	−5600
5	−52,000	9800

12.10 The capital fund for new investment at Systems Corporation is limited to $75,000 for next year. You have been asked to recommend one or more of three projects as economically acceptable for investment at the corporate MARR of 15% per year. Perform the analysis in two ways: (a) by hand, and (b) using NPV functions and a spreadsheet.

Project	Initial Investment, $	Annual NCF, $/Year	Life, Years	Salvage Value, $
A	−25,000	6,000	4	4,000
B	−30,000	9,000	4	−1,000
C	−50,000	15,000	4	20,000

12.11 Abel and Family Perfumes wants to add one or more of four new products to its current line of colognes. Historically, Abel has used a 5-year project recovery period and a MARR of 20% per year. (a) Determine which of the four options the company should undertake on the basis of a present worth analysis, provided the total amount of investment capital available is $800,000. Use a hand solution, unless assigned otherwise. (b) What products are selected if the investment limit is increased to $900,000? Use a spreadsheet, unless assigned otherwise. (All cash flows are in $1000 units.)

Product Line	R1	S2	T3	U4
Investment, $	−200	−400	−500	−700
M&O cost, $/year	−50	−200	−300	−400
Revenue, $/year	150	450	520	770

12.12 Determine which of the following independent projects should be selected for investment if a maximum of $240,000 is available and the MARR is 10% per year. Use the PW method to evaluate mutually exclusive bundles to perform your analysis.

Project	Investment, $	NCF, $/Year	Life, Years
A	−100,000	50,000	8
B	−125,000	24,000	8
C	−120,000	75,000	8
E	−220,000	39,000	8
F	−200,000	82,000	8

12.13 Feng Seawater Desalination Systems has established a capital investment limit of $800,000 for next year for projects that target improved recovery of highly brackish groundwater. All projects have a 4-year life and the MARR is 10% per year. (*a*) Select any or all of the projects. (*b*) Project Z is a favorite of Mr. Feng. Determine the minimum NCF necessary to make Z the economically best project. (*c*) Use a spreadsheet to answer the two parts above.

Project	Initial Investment, $	Net Cash Flow, $/Year	Salvage Value, $
X	−250,000	50,000	45,000
Y	−300,000	90,000	−10,000
Z	−550,000	150,000	100,000

12.14 State two assumptions when doing capital rationing using a PW analysis for unequal-life projects.

12.15 Dwayne has four independent vendor proposals to contract the nationwide oil recycling services for the Ford Corporation manufacturing plants. All combinations are acceptable, except that vendors B and C cannot both be chosen. Revenue sharing of recycled oil sales with Ford is a part of the requirement. Develop all possible mutually exclusive bundles under the additional following restrictions and select the best projects. The corporate MARR is 10% per year. Write both the hand solution PW relations and spreadsheet functions necessary to perform the analysis.
(*a*) A maximum of $4 million can be invested.
(*b*) A larger budget of $5.5 million is allowed, but no more than two vendors can be selected.
(*c*) There is no limit on spending.

Vendor	Initial Investment, $	Life, Years	Annual Net Revenue, $/Year
A	−1.5 million	8	360,000
B	−3.0 million	10	600,000
C	−1.8 million	5	620,000
D	−2.0 million	5	520,000

12.16 The chief engineer at Clean Water Engineering has established a capital investment limit of $710,000 for next year for projects related to concentrate management. (*a*) Select any or all of the following independent projects, using a MARR of 10% per year. (*b*) Use a spreadsheet to determine the minimum annual NCF necessary to select any of the other viable bundles besides the one selected in part (*a*).

Project	Initial Investment, $	Annual NCF, $/Year	Life, Years	Salvage Value, $
A	−140,000	50,000	4	45,000
B	−300,000	90,000	4	−10,000
C	−590,000	150,000	6	100,000

12.17 The PW of five independent projects have been calculated at an MARR of 12% per year. Select the best combination at a capital investment limit of (*a*) $25,000, (*b*) $49,000, and (*c*) unlimited.

Project	Initial Investment, $	Life, Years	PW, $
S	−15,000	6	8,540
A	−26,000	8	12,100
M	−10,000	6	3,000
E	−25,000	4	10
H	−40,000	12	15,350

12.18 The independent project estimates below have been developed by the engineering and finance managers at Golphanen Enterprises. The corporate MARR is 8% per year, and the capital investment limit set by the CFO is $4 million. As a new employee in the Engineering Department, you have been asked to recommend the economically best projects. Use (*a*) hand solution, and (*b*) spreadsheet solution to determine your recommendation.

Project	Project Cost, $ Millions	Life, Years	NCF, $ per Year
1	−1.5	8	360,000
2	−3.5	10	600,000
3	−1.8	5	520,000
4	−2.0	4	820,000

12.19 Use *hand solution* and the PW method to evaluate four independent projects. Select as many as three of the four projects. The MARR is 12% per year, and up to $15,000 in capital investment funds are available.

Project	1	2	3	4
Investment, $	−5000	−7000	−9000	−10,000
Life, years	5	5	3	4

	Project			
	1	**2**	**3**	**4**
Year	**NCF Estimates, $ per Year**			
1	1000	500	5000	0
2	1700	500	5000	0
3	2400	500	2000	0
4	3000	500		17,000
5	3800	10,500		

12.20 Using a *spreadsheet*, (*a*) work Problem 12.19 as stated, and (*b*) expand the bundles to allow up to three projects and no more than $25,000 invested.

12.21 A capital-budgeting problem is defined as follows: three projects are to be evaluated at a MARR of 12.5% per year. No more than $3.0 million can be invested.
  (*a*) Use a spreadsheet to select from the independent projects.
  (*b*) After some study, it was determined that the life of project 3 can be increased from 5 to 8 years for the same $1 million investment; however, the estimated annual NCF would remain steady from year 5 on. Use Goal Seek to determine the NCF in year 1 and beyond for project 3 (as a bundle of one project) to have the same PW as the best bundle in part (*a*). All other estimates remain the same. With these new NCF and life estimates, what are the best projects for investment?

	Investment,	Life,	Estimated NCF, $ per Year	
Project	**$ Millions**	**Years**	**Year 1**	**Gradient after Year 1**
1	−0.9	6	250,000	−5000
2	−2.1	10	385,000	+5000
3	−1.0	5	200,000	+25%

Linear Programming and Capital Budgeting

12.22 Formulate the linear programming model, develop a spreadsheet, and solve the capital rationing problem in Example 12.1 with an investment limit of (*a*) $20 million, as presented, (*b*) $13 million, and (*c*) $30 million.

12.23 Use linear programming, a spreadsheet, and the Solver tool to select from the independent, unequal-life projects in Problem 12.18.

12.24 Solve the capital budgeting problem in Problem 12.21(*a*) using the linear programming model and a spreadsheet.

Problems 12.25 and 12.26 utilize the following estimates.

For some years, the City of Armstrong Power Authority has used 4-20 mA technology to measure parameters such as pressure, temperature, flow rates, and speed of various machinery and systems installed in its generation facilities. It now plans to use the same technology to measure vibration on all of its rotating machinery with the goal of evaluating continuously the health of the equipment, to include friction buildup, breakdown prediction, and preventive maintenance needs. Depending upon the vendors chosen, the useful lives and anticipated annual net savings will vary, as reflected in the estimates shown.

Vendor	**1**	**2**	**3**	**4**
Investment, $1000	−500	−700	−900	−1000
Life, years	5	5	3	4

	Vendor			
	1	**2**	**3**	**4**
Year	**Savings Estimates, $1000 per Year**			
1	100	50	500	0
2	170	50	500	0
3	240	50	200	0
4	300	50		1800
5	380	1000		

12.25 Develop the linear programming model and use a spreadsheet to select the vendor(s) that offer the best opportunity economically. The city's MARR is 5% per year and the investment limit is $1,600,000.

12.26 Use repeated spreadsheet solutions of the capital budgeting problem with investment limits ranging from $0.5 million to $2.5 million (in $0.25 million increments, i.e., $250 increments since all estimated values are in $1000 units) to determine the best selections. Once done, develop a graph that plots the investment limit versus the value of Z. Use MARR = 5% per year.

Other Ranking Measures

12.27 Rincon, LLC is considering a project that will require an initial investment of $750,000 with estimated net income of $135,000 per year for 10 years. (*a*) Determine the IROR, PI, and PW values at MARR = 12% per year. (*b*) For which of these measures is the project economically justified? (*c*) Reflect on the answers above and the breakeven i^*. Is there any MARR value that will cause any of the three measures to result in different conclusions about the economic viability of the project? Explain your answer.

12.28 The Board of Directors at Senico Systems is considering investment in five independent projects, all of which can be considered to last indefinitely. The MARR is 12% per year.

(a) Determine which projects should be selected on the basis of IROR if the investment limitation is $60,000.

(b) Determine the overall rate of return if the funds not invested in a project are assumed to earn a rate of return equal to the MARR and the investment limitation is $60,000.

Project	Investment, $	Income, $/Year	IROR, % per Year
A	−30,000	7,000	23.3
B	−10,000	1,900	19.0
C	−15,000	2,600	17.3
D	−55,000	9,000	16.4
E	−5,000	6,000	12.0

12.29 Determine the IROR and profitability index at 12% per year for an industrial smart-grid system that has a first cost of $400,000, an AOC of $75,000 per year, estimated annual savings of $192,000, and a salvage value of 20% of its first cost after a 5-year useful life.

12.30 There are five independent projects that must be evaluated on the basis of the profitability index. (a) Given an investment limit of $35 million and a MARR of 10% per year, select from the projects. (b) Write the spreadsheet function for a project, specifically for project 5, which will correctly display the PI value used to rank the project.

Project	First Cost, $1000	NCF, $1000/Year	Life, Years
1	−4,000	900	7
2	−7,000	1,900	10
3	−17,000	2,900	15
4	−15,000	3,600	10
5	−30,000	5,000	8

12.31 Selex Aerospace is considering five independent projects to improve net revenue as estimated below. (All estimates have been divided by $1000.) The company's MARR is 15% per year. Use a *hand-based solution* and a *spreadsheet-based analysis* to determine the following:

(a) The projects that should be undertaken on the basis of IROR if the investment limit is $120,000.

(b) The overall rate of return if the funds not invested in a project earns a rate of return equal to the MARR.

Project	Initial Cost, $	Net Revenue Increase, $/Year	Life, Years
X	−30,000	9,000	10
Y	−15,000	4,900	10
Z	−45,000	11,100	10
A	−70,000	19,000	10
B	−40,000	10,000	10

12.32 An estimated 6 billion gallons of clean drinking water disappear each day across the United States due to aging, leaky pipes, and water mains before it gets to the consumer or industrial user. The American Society of Civil Engineers (ASCE) has teamed with municipalities, counties, and several excavation companies to develop robots that can travel through mains, detect leaks, and repair many of them immediately. Four proposals have been received for funding. There is a $100 million ($100 M) limit on capital funding and the MARR is established at 12% per year.

(a) Use the IROR to rank and determine which of the four independent projects should be funded. *(Solve by hand or spreadsheet as instructed.)*

(b) Determine the rate of return for the projects selected in part (a).

(c) Determine the overall rate of return for the projects selected in part (a), assuming that excess funds are invested at the MARR. Is the overall return economically acceptable?

Project	First Cost, $M	Estimated Annual Savings, $M per Year	Project Life, Years
W	−12	5.0	3
X	−25	7.3	4
Y	−45	12.1	6
Z	−60	9.0	8

12.33 If the capital budget limit is $120,000, the MARR is 10% per year, and all projects have a 10-year life, rank and select from the independent projects using the (a) PI measure, (b) IROR measure, and (c) PW at the MARR. (d) Are different projects selected using the three methods?

Project	First Cost, $	Net Income, $ per Year	IROR, %	PW at 10%, $
A	−18,000	4,000	18.0	6,578
B	−15,000	2,800	13.3	2,205
C	−35,000	12,600	34.1	42,422
D	−60,000	13,000	17.3	19,879
E	−50,000	8,000	9.6	−843

12.34 The six independent projects shown below are presented to you as an engineer with Peyton Packing. There are capital budget constraints because the company always has more projects to engage in than it has capital to fund them. Therefore, Peyton uses a relatively high MARR of 25% per year. Since all projects are considered long-term ventures, the company uses an infinite period for their life. Determine which projects to fund and the total investment for a limited capital budget of (*a*) $700,000, and (*b*) $600,000. The selection should be based on three measures: (1) IROR, (2) PI, and (3) PW.

Project	First Cost, $	Estimated Annual Income, $ per Year
F	−200,000	54,000
G	−120,000	21,000
H	−250,000	115,000
I	−370,000	205,000
J	−50,000	26,000
K	−9000	2,100

ADDITIONAL PROBLEMS AND FE EXAM REVIEW QUESTIONS

12.35 A dependent project is one whose acceptance or rejection is based on:
(*a*) Its rate of return with respect to the MARR
(*b*) The amount of capital available for investment
(*c*) A decision about another project
(*d*) The amount receivable in salvage value of the existing asset

12.36 In capital budgeting, when the present worth over the respective life of each project is used to select independent projects, all positive net cash flows are assumed to be:
(*a*) Reinvested at the MARR through the end of the shortest-lived project
(*b*) Reinvested at the internal rate of return of the project through the end of the longest-lived project
(*c*) Reinvested at the MARR through the least common multiple of years of all of the projects
(*d*) Reinvested at the MARR through the end of the longest-lived project

12.37 The profitability index at 10% per year for a project that has an initial investment of $400,000, an AOC of $80,000 per year, revenue of $170,000 per year, and a salvage value of $60,000 after its 5-year life is closest to:
(*a*) 0.95 (*b*) 1.06 (*c*) 1.38 (*d*) 0.86

12.38 Listed below are bundles, each comprised of three independent proposals for which the PW has been estimated. For a capital investment limit of $45,000 and a MARR of 9% per year, the bundle selected is:
(*a*) 1 (*b*) 2 (*c*) 3 (*d*) 4

Bundle	Investment, for Bundle, $	PW at 9%, $
1	−18,000	−1,400
2	−26,000	8,500
3	−34,000	7,100
4	−41,000	10,500

12.39 From the five independent projects shown below, the one(s) to select with a capital limitation of $35,000 is/are:
(*a*) P (*b*) PT (*c*) T (*d*) Q

Project	Initial Investment $,	Life, Years	PW at 12% per Year, $
P	−12,000	5	8,000
Q	−25,000	9	12,600
R	−17,000	4	3,500
S	−35,000	6	2,000
T	−20,000	10	15,800

12.40 When there are a total of four projects involved in a capital budgeting study, the maximum number of bundles that can be formulated is:
(*a*) 6   (*b*) 8
(*c*) 15   (*d*) 16

12.41 Determine the number of acceptable mutually exclusive bundles for the four independent projects described below if the investment limit is $400,000 and the following project selection restriction applies: Project 1 can be selected only if projects 3 and 4 are selected.
(*a*) <7 (*b*) 7 (*c*) 10 (*d*) 12

Project	Initial Investment, $
1	−100,000
2	−150,000
3	−75,000
4	−235,000

12.42 All of the following are correct when a capital budgeting problem is solved using the 0-1 integer linear programming model, except:
(*a*) Partial investment in a project is acceptable.
(*b*) The objective is to maximize the present worth of the investments.

(c) Budget constraints may be present for the first year only or for several years.

(d) Contingent and dependent project restrictions may be considered.

12.43 All of the following are true when formulating mutually exclusive bundles of independent projects, except:

(a) One of the bundles is the do-nothing project.

(b) A bundle may consist of only one project.

(c) The capital limit may be exceeded as long as it is exceeded by less than 3%.

(d) A bundle may include contingent and dependent projects.

12.44 The independent projects shown below are under consideration for possible implementation by Renishaw, Inc. If the company's MARR is 14% per year and it uses the IROR method of capital budgeting, the projects it should select under a budget limitation of $105,000 are:

(a) A, B, and C   (b) A, B, and D
(c) B, C, and D   (d) A, C, and D

Project	First Cost, $	Annual Income, $/Year	Rate of Return, %
A	−20,000	4,000	20.0
B	−10,000	1,900	19.0
C	−15,000	2,600	17.3
D	−70,000	10,000	14.3
E	−50,000	6,000	12.0

CHAPTER 13

Breakeven and Payback Analysis

Randy Allbritton/Getty Images

LEARNING OUTCOMES

Purpose: Determine the breakeven for one or two alternatives and calculate the payback period with and without a return required.

SECTION	TOPIC	LEARNING OUTCOME
13.1	Breakeven point	• Determine the breakeven point for one parameter.
13.2	Two alternatives	• Calculate the breakeven point of a parameter and use it to select between two alternatives.
13.3	Payback period	• Determine the payback period of a project at $i = 0\%$ and $i > 0\%$. Illustrate the cautions when using payback analysis.
13.4	Spreadsheet	• Answer breakeven and payback questions that are best resolved by spreadsheet and Goal Seek tools.

B reakeven analysis is performed to determine the value of a parameter of a project or alternative that makes two elements equal, for example, the sales volume that will equate revenues and costs. A breakeven study is performed for two alternatives to determine when either alternative is equally acceptable. Breakeven analysis is commonly applied in **make-or-buy decisions** when a decision is needed about the source for manufactured components, services, etc.

Payback analysis determines the required minimum life of an asset, process, or system to recover the initial investment. There are two types of payback: return ($i > 0\%$) and no return ($i = 0\%$). Payback analysis should not be considered the final decision maker; it is used as a **screening tool** or to provide **supplemental information** for a PW, AW, or other analysis. These aspects are discussed in depth in this chapter.

Breakeven and payback studies use estimates that are considered to be certain; that is, if the estimated values are expected to vary enough to possibly change the outcome, another study is necessary using different estimates. If the parameter of interest is allowed to vary, the approaches of sensitivity analysis (Chapter 18) should be used. Additionally, if probability and risk assessment are considered, the tools of simulation (Chapter 19) can be used to supplement the static nature of a breakeven or payback study.

13.1 Breakeven Analysis for a Single Project ●●●

When the estimate of one of the engineering economy parameters—$P, F, A, i,$ or n—is not known or the estimate is considered to be inaccurate, a breakeven quantity can be determined by setting an equivalence relation for PW or AW equal to zero. This form of breakeven analysis has been used many times so far. For example, we have solved for the rate of return i^*, found the replacement value for a defender, and determined the $P, F, A,$ or salvage value S at which a series of cash flow estimates return a specific MARR. Methods used to determine the quantity include

Direct solution by hand if only one factor is present (say, P/A) or only single amounts are estimated (e.g., P and F)

Trial and error by hand or calculator when multiple factors are present

Spreadsheet when cash flow and other estimates are entered into cells and used in resident functions (PV, FV, RATE, IRR, NPV, PMT, and NPER) or tools (Goal Seek and Solver).

We now concentrate on the determination of the **breakeven quantity Q_{BE} for one parameter or decision variable**. For example, the variable may be a design element to minimize cost or the production level needed to realize revenues that exceed costs by a stated percent.

> Breakeven analysis finds the value of a parameter that **makes two elements equal**. The breakeven point Q_{BE} is determined from mathematical relations, for example, product revenue and costs or materials supply and demand or other parameters that involve the parameter Q. Breakeven analysis is fundamental to evaluations such as make-buy decisions.

Breakeven

The unit of the parameter Q may vary widely: units per year, cost per kilogram, hours per month, percentage of full plant capacity, etc.

Figure 13–1a presents different shapes of a revenue relation identified as R. A linear revenue relation is commonly assumed, but a nonlinear relation is often more realistic. It can model an increasing per unit revenue with larger volumes (curve 1 in Figure 13–1a) or a decreasing per unit revenue that usually prevails at higher quantities (curve 2).

Costs, which may be linear or nonlinear, usually include two components—fixed and variable—as indicated in Figure 13–1b.

Fixed costs (FC). These include costs such as buildings, insurance, fixed overhead, some minimum level of labor, equipment capital recovery, and information systems.

Variable costs (VC). These include costs such as direct labor, materials, indirect costs, contractors, marketing, advertisement, and warranty.

The **fixed-cost** component is essentially constant for all values of the variable, so it does not vary for a large range of operating parameters, such as production level or workforce size. Even if no units are produced, fixed costs are incurred at some threshold level. Of course, this situation cannot last long before the plant management must adjust to reduce fixed costs. Fixed costs are

Figure 13–1
Linear and nonlinear
revenue and cost relations.

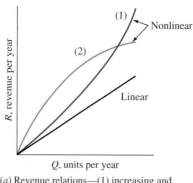

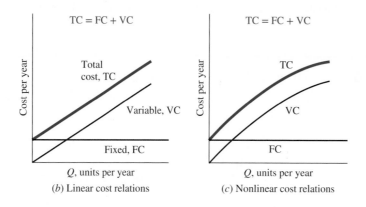

(a) Revenue relations—(1) increasing and
(2) decreasing revenue per unit

(b) Linear cost relations

(c) Nonlinear cost relations

reduced through the upgrading of equipment using new technology, advanced information systems, workforce efficiencies, less costly fringe benefit packages for employees, subcontracting specific functions, and so on.

Variable costs change with production level, workforce size, and other parameters. It is usually possible to decrease variable costs through better product design, manufacturing efficiency, improved quality and safety, and higher sales volume.

When FC and VC are added, they form the **total cost relation TC.** Figure 13–1*b* illustrates the TC relation for linear fixed and variable costs. Figure 13–1*c* shows a general TC curve for a nonlinear VC in which unit variable costs decrease as the quantity level rises.

At a specific but unknown value *Q* of the decision variable, the revenue *R* and total cost TC relations will intersect to identify the breakeven point Q_{BE} (Figure 13–2). As a general guideline, for linear *R* and VC relations, the greater the quantity, the larger the profit.

A simple relation for the breakeven point may be derived when revenue and total cost are *linear functions of quantity Q* by setting the relations for *R* and TC equal to each other, indicating a profit of zero.

$$R = TC$$
$$rQ = FC + vQ$$

where   *r* = revenue per unit
         v = variable cost per unit

Solve for the breakeven quantity $Q = Q_{BE}$ for linear *R* and TC functions.

$$Q_{BE} = \frac{FC}{r - v} \qquad [13.1]$$

The profit at any quantity level *Q* is

$$\text{Profit} = \text{revenue} - \text{total cost}$$
$$= R - (FC + VC)$$
$$= rQ - FC - vQ$$
$$= (r - v)Q - FC \qquad [13.2]$$

If $Q > Q_{BE}$, there is a profit; if $Q < Q_{BE}$, there is a loss.

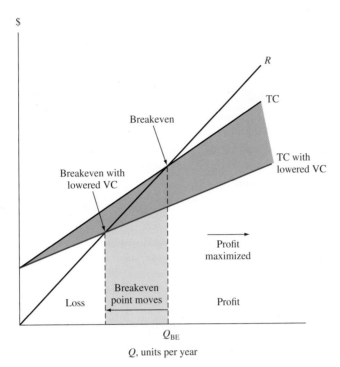

Figure 13–2
Effect on the breakeven point when the variable cost per unit is reduced.

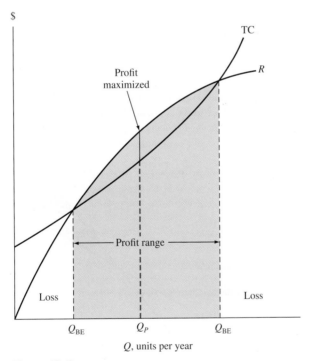

Figure 13–3
Breakeven points and maximum profit point for a nonlinear analysis.

The breakeven graph is an important management tool because it is easy to understand and may be used in decision making in a variety of ways. For example, if the variable cost per unit is reduced, then the TC line has a smaller slope (Figure 13–2) and the breakeven point will decrease. This is an advantage because the smaller the value of Q_{BE}, the greater the profit for a given amount of revenue. A similar analysis is possible for fixed VC and increased levels of production, as shown in the next example.

If nonlinear R or TC models are used, there may be more than one breakeven point. Figure 13–3 presents this situation for two breakeven points. The **maximum profit** occurs at Q_P between the two breakeven points where the distance between the R and TC relations is greatest.

Of course, no static R and TC relations—linear or nonlinear—are able to estimate exactly the revenue and cost amounts over an extended period of time. But the breakeven point is an excellent target for planning purposes.

EXAMPLE 13.1

Indira Industries is a major producer of diverter dampers used in the gas turbine power industry to divert gas exhausts from the turbine to a side stack, thus reducing the noise to acceptable levels for human environments. Normal production level is 60 diverter systems per month, but due to significantly improved economic conditions in Asia, production is at 72 per month. The following information is available.

Fixed costs	FC = $2.4 million per month
Variable cost per unit	v = $35,000
Revenue per unit	r = $75,000

(a) How does the increased production level of 72 units per month compare with the current breakeven point?
(b) What is the current profit level per month for the facility?
(c) What is the difference between the revenue and variable cost per damper that is necessary to break even at a significantly reduced monthly production level of 45 units, if fixed costs remain constant?

Solution

(a) Use Equation [13.1] to determine the breakeven number of units. All dollar amounts are in $1000 units.

$$Q_{BE} = \frac{FC}{r - v}$$

$$= \frac{2400}{75 - 35} = 60 \text{ units per month}$$

Figure 13–4 is a plot of R and TC lines. The breakeven value is 60 damper units. The increased production level of 72 units is above the breakeven value.

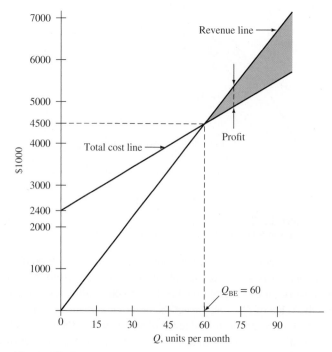

Figure 13–4
Breakeven graph, Example 13.1.

(*b*) To estimate profit (in $1000 units) at $Q = 72$ units per month, use Equation [13.2].

$$\text{Profit} = (r - v)Q - \text{FC} \qquad [13.3]$$
$$= (75 - 35)72 - 2400$$
$$= \$480$$

There is a profit of $480,000 per month currently.

(*c*) To determine the required difference $r - v$, use Equation [13.3] with profit = 0, $Q = 45$, and FC = $2.4 million. In $1000 units,

$$0 = (r - v)(45) - 2400$$
$$r - v = \frac{2400}{45} = \$53.33 \text{ per unit}$$

The spread between r and v must be $53,330. If v stays at $35,000, the revenue per damper must increase from $75,000 to $88,330 (i.e., 35,000 + 53,330) just to break even at a production level of $Q = 45$ per month.

In some circumstances, breakeven analysis performed on a per unit basis is more meaningful. The value of Q_{BE} is still calculated using Equation [13.1]; however, the relations for R and TC are divided by Q. In the case of TC, the expression for cost per unit, also termed *average cost per unit* C_u, is:

$$C_u = \frac{\text{TC}}{Q} = \frac{\text{FC} + vQ}{Q} = \frac{\text{FC}}{Q} + v \qquad [13.4]$$

At the breakeven quantity $Q = Q_{BE}$, the revenue per unit is exactly equal to the cost per unit. If graphed, the FC per unit term in Equation [13.4] takes on the shape of a hyperbola.

The breakeven point for a project with one unknown variable can always be determined by equating revenue and total cost. This is the same as setting profit equal to zero in Equation [13.2]. It may be necessary to perform some dimensional analysis initially to obtain the correct revenue and total cost relations in order to use the same dimension for both relations, for example, $ per unit, miles per month, or units per year.

13.2 Breakeven Analysis Between Two Alternatives ● ● ●

Now we consider breakeven analysis between two mutually exclusive alternatives.

> Breakeven analysis determines the value of a common variable or parameter between two alternatives. Equating the two PW or AW relations determines the breakeven point. Selection of the alternative is different depending upon two facts: **slope of the variable cost curve** and the **parameter value relative to the breakeven point.**

Breakeven

The parameter can be the interest rate i, first cost P, annual operating cost (AOC), or any parameter. We have already performed breakeven analysis between alternatives on several parameters. For example, the incremental ROR value (Δi^*) is the breakeven rate between alternatives. If the MARR is lower than Δi^*, the extra investment of the larger investment alternative is justified. In Section 11.6, the replacement value (RV) of a defender was determined. If the market value is larger than RV, the decision should favor the challenger.

Often breakeven analysis involves revenue or cost variables common to both alternatives, such as price per unit, operating cost, cost of materials, or labor cost. Figure 13–5 illustrates the breakeven concept for two alternatives with linear cost relations. The fixed cost of alternative 2 is greater than that of alternative 1. However, alternative 2 has a smaller variable cost, as indicated by its lower slope. The intersection of the total cost lines locates the breakeven point, and the variable cost establishes the slope. Thus, if the number of units of the common variable is greater than the breakeven amount, alternative 2 is selected, since the total cost will be lower. Conversely, an anticipated level of operation below the breakeven point favors alternative 1.

Figure 13–5
Breakeven between two
alternatives with linear
cost relations.

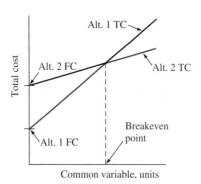

Instead of plotting the total costs of each alternative and estimating the breakeven point graphically, it may be easier to calculate the breakeven point numerically using engineering economy expressions for the PW or AW at the MARR. The AW is preferred when the variable units are expressed on a yearly basis, and AW calculations are simpler for alternatives with unequal lives. The following steps determine the breakeven point of the common variable and the slope of a linear total cost relation:

1. Define the common variable and its dimensional units.
2. Develop the PW or AW relation for each alternative as a function of the common variable.
3. Equate the two relations and solve for the breakeven value of the variable.

Selection between alternatives is based on this guideline:

> If the anticipated level of the common variable is **below** the breakeven value, select the alternative with the higher variable cost (larger slope).
>
> If the level is **above** the breakeven point, select the alternative with the lower variable cost. (Refer to Figure 13–5.)

EXAMPLE 13.2

Devon Products produces a superior quality, high-gloss, nonskid surface concrete stone primarily used as flooring in kitchens and baths. The equipment necessary to complete the nonskid surface operations can be a fully automated or semiautomated machine. The fully automated machine has an initial cost of $23,000, an estimated salvage value of $4000, and a predicted life of 10 years. One person will operate the machine at a total cost of $40 per hour. The expected output is 8 tons per hour. Annual maintenance and operating cost is expected to be $3500.

The semiautomatic machine has a first cost of $8000, no expected salvage value, a 5-year life, and an output of 10 tons per hour; however, an operator with additional skills is required at the rate of $60 per hour. The machine will have an annual maintenance and operation cost of $1500. All projects are expected to generate a return of 10% per year. How many tons of finished stone per year must be produced to justify the higher purchase cost of the fully automatic machine?

Solution

Use the steps above to calculate the breakeven point between the two alternatives.

1. Let x represent the number of tons per year.
2. For the fully automatic machine, the annual variable cost is

$$\text{Annual VC} = \frac{\$40}{\text{hour}} \frac{1 \text{ hour}}{8 \text{ tons}} \frac{x \text{ tons}}{\text{year}}$$
$$= 5x$$

The VC is developed in dollars per year. The AW expression is

$$\text{AW}_{\text{fully}} = -23{,}000(A/P,10\%,10) + 4000(A/F,10\%,10) - 3500 - 5x$$
$$= \$-6992 - 5x$$

Similarly, the annual variable cost and AW for the semiautomatic machine are

$$\text{Annual VC} = \frac{\$60}{\text{hour}} \frac{1 \text{ hour}}{10 \text{ tons}} \frac{x \text{ tons}}{\text{year}}$$

$$= 6x$$

$$\text{AW}_{\text{semi}} = -8000(A/P,10\%,5) - 1500 - 6x$$

$$= \$-3610 - 6x$$

3. Equate the two cost relations and solve for x.

$$\text{AW}_{\text{fully}} = \text{AW}_{\text{semi}}$$

$$-6992 - 5x = -3610 - 6x$$

$$x = 3382 \text{ tons per year}$$

If the output is expected to exceed 3382 tons per year, purchase the fully automatic machine, since its VC slope of 5 is smaller than the semiautomatic VC slope of 6.

The breakeven analysis approach is commonly used for **make-or-buy decisions.** This means the company contracts to *buy* the product or service from the *outside,* or *makes* it *within* the company. The alternative to buy usually has no fixed cost and a larger variable cost than the option to make. Where the two cost relations cross is the make-buy decision quantity. Amounts above this indicate that the item should be made, not purchased outside.

EXAMPLE 13.3

Guardian is a national manufacturing company of home health care appliances. It is faced with a make-or-buy decision. A newly engineered lift can be installed in a car trunk to raise and lower a wheelchair. The steel arm of the lift can be purchased internationally for $3.50 per unit or made in-house. If manufactured on site, two machines will be required. Machine A is estimated to cost $18,000, have a life of 6 years, and have a $2000 salvage value; machine B will cost $12,000, have a life of 4 years, and have a $−500 salvage value (carry-away cost). Machine A will require an overhaul after 3 years costing $3000. The annual operating cost for machine A is expected to be $6000 per year and for machine B is $5000 per year. A total of four operators will be required for the two machines at a rate of $12.50 per hour per operator. In a normal 8-hour period, the operators and two machines can produce parts sufficient to manufacture 1000 units. Use a MARR of 15% per year to determine the following:

(*a*) Number of units to manufacture each year to justify the in-house (make) option.
(*b*) The maximum capital expense justifiable to purchase machine A, assuming all other estimates for machines A and B are as stated. The company expects to produce 10,000 units per year.

Solution

(*a*) Use steps 1 to 3 stated previously to determine the breakeven point.

1. Define x as the number of lifts produced per year.
2. There are variable costs for the operators and fixed costs for the two machines for the make option.

$$\text{Annual VC} = (\text{cost per unit})(\text{units per year})$$

$$= \frac{4 \text{ operators}}{1000 \text{ units}} \frac{\$12.50}{\text{hour}} (8 \text{ hours})x$$

$$= 0.4x$$

The annual fixed costs for machines A and B are the AW amounts.

$$\text{AW}_A = -18{,}000(A/P,15\%,6) + 2000(A/F,15\%,6)$$
$$-6000 - 3000(P/F,15\%,3)(A/P,15\%,6)$$

$$\text{AW}_B = -12{,}000(A/P,15\%,4) - 500(A/F,15\%,4) - 5000$$

Total cost is the sum of AW_A, AW_B, and VC.

3. Equating the annual costs of the buy option ($3.50x$) and the make option yields

$$-3.50x = AW_A + AW_B - VC$$
$$= -18{,}000(A/P,15\%,6) + 2000(A/F,15\%,6) - 6000$$
$$-3000(P/F,15\%,3)(A/P,15\%,6) - 12{,}000(A/P,15\%,4)$$
$$-500(A/F,15\%,4) - 5000 - 0.4x \qquad\qquad [13.5]$$
$$-3.10x = -20{,}352$$
$$x = 6565 \text{ units per year}$$

A minimum of 6565 lifts must be produced each year to justify the make option, which has the lower variable cost of $0.4x$.

(*b*) Substitute 10,000 for x and P_A for the to-be-determined first cost of machine A (currently $18,000) in Equation [13.5]. Solution yields $P_A = \$58{,}295$. This is approximately three times the estimated first cost of $18,000, because the production of 10,000 per year is considerably larger than the breakeven amount of 6565.

Even though the preceding examples treat only two alternatives, the same type of analysis can be performed for three or more alternatives. To do so, compare the alternatives in pairs to find their respective breakeven points. The results are the ranges through which each alternative is more economical. For example, in Figure 13–6, if the output is less than 40 units per hour, alternative 1 should be selected. Between 40 and 60, alternative 2 is more economical; and above 60, alternative 3 is favored.

If the variable cost relations are nonlinear, analysis is more complicated. If the costs increase or decrease uniformly, mathematical expressions that allow direct determination of the breakeven point can be developed.

13.3 Payback Analysis ●●●

Payback analysis is another use of the present worth technique. It is used to determine the amount of time, usually expressed in years, required to recover the first cost of an asset or project. Payback is allied with breakeven analysis; this is illustrated later in the section. The **payback period,** also called *payback* or *payout period,* has the following definition and types.

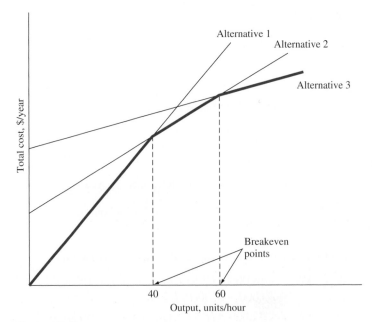

Figure 13–6
Breakeven points for three alternatives.

The **payback period** n_p is an estimated time for the revenues, savings, and any other monetary benefits to completely recover the **initial investment plus a stated rate of return** i.

There are two types of payback analysis as determined by the required return.

No return; $i = 0\%$: Also called *simple payback*, this is the recovery of only the initial investment with no interest.

Discounted payback; $i > 0\%$: The time value of money is considered in that some return, for example, 10% per year, must be realized in addition to recovering the initial investment.

Payback period

An example application of payback may be a corporate senior manager who insists that every proposal return the initial cost and some stated return within 3 years. Using payback as an initial screening tool, no proposal with $n_p > 3$ years can become a viable alternative. The payback period should be determined using a required $i > 0\%$. Unfortunately in practice, no-return payback is used too often to make economic decisions. After the formulas are presented, a couple of cautions about payback usage are provided.

The terminology is P for the initial investment in the asset, project, contract, etc., and NCF for the estimated annual net cash flow. Using Equation [1.5], annual NCF is

$$\text{NCF} = \text{cash inflows} - \text{cash outflows}$$

To calculate the payback period for $i = 0\%$ or $i > 0\%$, determine the pattern of the NCF series. Note that n_p is usually not an integer. The equations that determine a payback period differ for no-return and discounted analyses. For $t = 1, 2, \ldots , n_p$,

No return, $i = 0\%$; NCF$_t$ **varies annually:** $0 = -P + \sum_{t=1}^{t=n_p} \text{NCF}_t$ **[13.6]**

No return, $i = 0\%$; **annual uniform NCF:** $n_p = \dfrac{P}{\text{NCF}}$ **[13.7]**

Discounted, $i > 0\%$; NCF$_t$ **varies annually:** $0 = -P + \sum_{t=1}^{t=n_p} \text{NCF}_t(P/F,i,t)$ **[13.8]**

Discounted, $i > 0\%$; **annual uniform NCF:** $0 = -P + \text{NCF}(P/A,i,n_p)$ **[13.9]**

After n_p years, the cash flows will recover the investment in year 0 plus the required return of $i\%$. If the alternative is used more than n_p years, with the same or similar cash flows, a larger return results. If the estimated life is less than n_p years, there is not enough time to recover the investment and $i\%$ return. It is important to understand that payback analysis *neglects all cash flows after the payback period of n_p years.* Consequently, it is preferable to use payback as an **initial screening method** or **supplemental tool** rather than as the primary means to select an alternative. The reasons for this caution are that

- No-return payback **neglects the time value of money,** since no return on an investment is required.
- Either type of payback **disregards all cash flows occurring after the payback period.** These cash flows may increase the return on the initial investment.

Payback analysis utilizes a significantly different approach to alternative evaluation than the primary methods of PW, AW, ROR, and B/C. It is possible for payback analysis to select a different alternative than these techniques. However, the information obtained from discounted payback analysis performed at an appropriate $i > 0\%$ can be very useful in that a sense of the **risk** involved in undertaking an alternative is provided. For example, if a company plans to utilize a machine for only 3 years and payback is 6 years, indication is that the equipment should not be obtained. Even here, the 6-year payback is considered supplemental information and does not replace a complete economic analysis.

EXAMPLE 13.4

The board of directors of Halliburton International has just approved an $18 million worldwide engineering construction design contract. The services are expected to generate new annual net cash flows of $3 million. The contract has a potentially lucrative repayment clause to Halliburton of $3 million at any time that the contract is canceled by either party during the 10 years of the contract period. (*a*) If $i = 15\%$, compute the payback period. (*b*) Determine the no-return payback period and compare it with the answer for $i = 15\%$. This is an initial check to determine if the board made a good economic decision. Show both hand and spreadsheet solutions.

Solution by Hand

(*a*) The net cash flow each year is $3 million. The single $3 million payment (call it CV for cancellation value) could be received at any time within the 10-year contract period. Equation [13.9] is altered to include CV.

$$0 = -P + NCF(P/A,i,n) + CV(P/F,i,n)$$

In $1,000,000 units,

$$0 = -18 + 3(P/A,15\%,n) + 3(P/F,15\%,n)$$

The 15% payback period is $n_p = 15.3$ years, found by trial and error. During the period of 10 years, the contract will not deliver the required return.

(*b*) If Halliburton requires absolutely no return on its $18 million investment, Equation [13.6] results in $n_p = 5$ years, as follows (in $ million).

$$0 = -18 + 5(3) + 3$$

There is a very significant difference in n_p for 15% and 0%. At 15% this contract would have to be in force for 15.3 years, while the no-return payback period requires only 5 years. A longer time is always required for $i > 0\%$ for the obvious reason that the time value of money is considered.

Solution by Spreadsheet

Enter the function = NPER(15%,3,−18,3) to display 15.3 years. Change the rate from 15% to 0% to display the no-return payback period of 5 years.

If two or more alternatives are evaluated using payback periods to indicate that one may be better than the other(s), the second shortcoming of payback analysis (neglect of cash flows after n_p) may lead to an economically incorrect decision. When cash flows that occur after n_p are neglected, it is possible to favor short-lived assets even when longer-lived assets produce a higher return. In these cases, PW (or AW) analysis should always be the primary selection method. Comparison of short- and long-lived assets in Example 13.5 illustrates this situation.

EXAMPLE 13.5

Two equivalent pieces of quality inspection equipment are being considered for purchase by Square D Electric. Machine 2 is expected to be versatile and technologically advanced enough to provide net income longer than machine 1.

	Machine 1	**Machine 2**
First cost, $	12,000	8,000
Annual NCF, $	3,000	1,000 (years 1–5), 3,000 (years 6–14)
Maximum life, years	7	14

The quality manager used a return of 15% per year and software that incorporates Equations [13.8] and [13.9] to recommend machine 1 because it has a shorter payback period of 6.57 years at $i = 15\%$. The computations are summarized here.

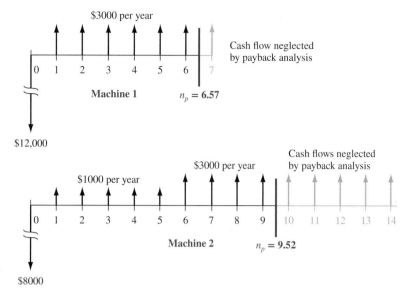

Figure 13–7
Illustration of payback periods and neglected net cash flows, Example 13.5.

Machine 1: $n_p = 6.57$ years, which is less than the 7-year life.

Equation used: $\qquad\qquad 0 = -12,000 + 3000(P/A,15\%,n_p)$

Machine 2: $n_p = 9.52$ years, which is less than the 14-year life.

Equation used: $\qquad 0 = -8000 + 1000(P/A,15\%,5)$
$$+ 3000(P/A,15\%,np-5)(P/F,15\%,5)$$

Recommendation: Select machine 1.

Now, use a 15% PW analysis to compare the machines and comment on any difference in the recommendation.

Solution

For each machine, consider the net cash flows for all years during the estimated (maximum) life. Compare them over the LCM of 14 years.

$$PW_1 = -12,000 - 12,000(P/F,15\%,7) + 3000(P/A,15\%,14) = \$663$$
$$PW_2 = -8000 + 1000(P/A,15\%,5) + 3000(P/A,15\%,9)(P/F,15\%,5)$$
$$= \$2470$$

Machine 2 is selected since its PW value is numerically larger than that of machine 1 at 15%. This result is the opposite of the payback period decision. The PW analysis accounts for the increased cash flows for machine 2 in the later years. As illustrated in Figure 13–7 (for one life cycle for each machine), payback analysis neglects all cash flow amounts that may occur after the payback time has been reached.

Comment

This is a good example of why payback analysis is best used for initial screening and supplemental risk assessment. Often a shorter-lived alternative evaluated by payback analysis may appear to be more attractive, when the longer-lived alternative has cash flows later in its life that make it more economically attractive.

As mentioned in the introduction to this section, breakeven and payback analyses are allied. They can be used in conjunction to determine the payback period when a desired level of breakeven is specified. The reverse is also possible; when a desired payback period is established, the breakeven value with or without a return requirement can be determined. By working together in this fashion, better economic decisions can be made. Example 13.6 illustrates the second of the situations mentioned above.

EXAMPLE 13.6

The president of a local company expects a product to have a profitable life of between 1 and 5 years. Help her determine the breakeven number of units that must be sold annually (without any return) to realize payback for each of the time periods 1 year, 2 years, and so on up to 5 years. The cost and revenue estimates are as follows:

Fixed costs: Initial investment of $80,000 with $1000 annual operating cost.

Variable cost: $8 per unit.

Revenue: Twice the variable cost for the first 5 years and 50% of the variable cost thereafter.

Solution by Hand

Define X_{BE} as the breakeven quantity and n_p as the payback period. Since values of X_{BE} are sought for $n_p = 1, 2, 3, 4, 5$, solve for breakeven by substituting each payback period. First develop the FC, r, and v terms.

Fixed cost, FC $\qquad \dfrac{80,000}{n_p} + 1000$

Revenue per unit, r $\qquad$ $16 $\qquad$ (years 1 through 5 only)

Variable cost per unit, v $\qquad$ $8

The breakeven relation from Equation [13.2] is

$$X_{BE} = \frac{80,000/n_p + 1000}{8} \qquad\qquad [13.10]$$

Insert n_p values and solve for X_{BE}, the breakeven value.

n_p, payback years	1	2	3	4	5
X_{BE}, units per year	10,125	5125	3458	2625	2125

Solution by Spreadsheet

Figure 13–8 presents a spreadsheet solution for the breakeven values. Equation [13.10] is encoded to display the answers in column C. The breakeven values are the same as those above. For example, sell 5125 units per year to pay back in 2 years. The breakeven curve rapidly flattens out as shown in the accompanying chart.

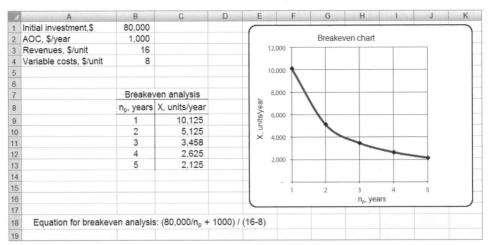

Figure 13–8
Breakeven number of units for different payback periods, Example 13.6.

13.4 More Breakeven and Payback
Analysis on Spreadsheets ● ● ●

The Goal Seek tool that we have used previously is an excellent tool to perform breakeven and payback analysis. Examples 13.7 and 13.8 demonstrate the use of Goal Seek for both types of problems.

EXAMPLE 13.7

The Naruse brake-accelerator pedal (www.narusepedal.com) is designed to minimize the chances that a driver will accidently step on the accelerator pedal of the car when the brake pedal is the intended target. The design is based on the fact that a person naturally steps downward on his or her foot when surprised, shocked, or struck with a medical emergency. In this pedal design, downward motion of the foot will always engage the brake, never the accelerator. Assume that for the manufacture of pedal components, two equally qualified machines have been identified and estimates made.

	Machine 1	Machine 2
First cost, $	−80,000	−110,000
Net cash flow, $ per year	25,000	22,000
Salvage value, $	2,000	3,000
Life, years	4	6

Using an AW analysis at MARR = 10%, the spreadsheet screen shot in Figure 13–9 indicates that machine 1 is the economic choice with a positive AW value of $193. However, the automated controls, safety features, and ergonomic design of machine 2 make it a better choice for the plant in the opinion of the project engineer. Use breakeven analysis to find the threshold values for each of several parameters that will make machine 2 equally qualified economically. The parameters to concentrate on are (*a*) first cost, (*b*) net cash flow, and (*c*) life of machine 2, if all other estimates remain the same.

	A	B	C	D
1	MARR =	10%		
2		Net cash flows, $/year		
3	Year	Machine 1	Machine 2	
4	0	-80,000	-110,000	
5	1	25,000	22,000	
6	2	25,000	22,000	
7	3	25,000	22,000	
8	4	27,000	22,000	
9	5		22,000	
10	6		25,000	
11	AW @ MARR	193	-2,868	
12				
13	= -PMT($B$1,4,NPV($B1,B5:B8)+B4)			
14				

Figure 13–9
AW values for two machines, Example 13.7.

Solution

Figure 13–10 shows the spreadsheet and Goal Seek templates that determine breakeven values for first cost and NCF.

(*a*) Figure 13–10*a*: By forcing the AW for machine 2 to equal $193, Goal Seek finds a breakeven of $96,669. If the first cost can be negotiated down to this cost from $110,000, machine 2 will be economically equivalent to machine 1.

(*b*) Figure 13–10*b*: (Remember to reset the first cost to $−110,000 on the spreadsheet.) By setting all NCFs equal to the value in year 1 (using the function = $C$5), Goal Seek determines a breakeven of $25,061 per year. Therefore, if the NCF estimate can realistically be increased from $22,000 to $25,061, again machine 2 will be economically equivalent.

(*c*) Finding an extended life estimate for machine 2 is a payback question; Goal Seek is not needed. The easiest approach is to use the NPER function to find the payback period. Entering = NPER(10%,22,000−110000,3000) displays $n_p = 7.13$ years. Therefore, extending the estimated life from 6 to between 7 and 8 years and retaining the salvage value of $3000 will select machine 2 over 1.

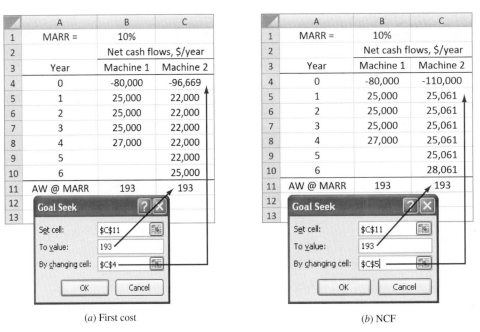

(a) First cost   *(b)* NCF

Figure 13–10
Breakeven values for *(a)* first cost and *(b)* annual net cash flow using Goal Seek, Example 13.7.

EXAMPLE 13.8

Chris and her father just purchased a small office building for $160,000 that is in need of a lot of repairs, but is located in a prime commercial area of the city. The estimated costs each year for repairs, insurance, etc. are $18,000 the first year, increasing by $1000 per year thereafter. At an expected 8% per year return, use spreadsheet analysis to determine the payback period if the building is *(a)* kept for 2 years and sold for $290,000 sometime beyond year 2, or *(b)* kept for 3 years and sold for $370,000 sometime beyond 3 years.

Solution

Figure 13–11 shows the annual costs (column B) and the sales prices if the building is kept 2 or 3 years (columns C and E, respectively). The NPV function is applied (columns D and F) to determine when the PW changes sign from plus to minus. These results bracket the payback period for each retention period and sales price. When PW > 0, the 8% return is exceeded.

(a) The 8% return payback period is between 3 and 4 years (column D). If the building is sold after exactly 3 years for $290,000, the payback period was not exceeded; but after 4 years it is exceeded.

(b) At a sales price of $370,000, the 8% return payback period is between 5 and 6 years (column F). If the building is sold after 4 or 5 years, the payback is not exceeded; however, a sale after 6 years is beyond the 8%-return payback period.

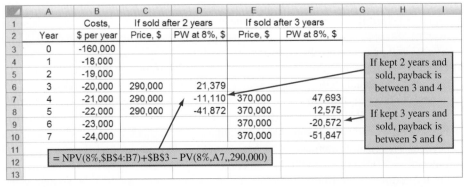

Figure 13–11
Payback period analysis, Example 13.8.

CHAPTER SUMMARY

The breakeven point for a variable for one project is expressed in terms such as units per year or hours per month. At the breakeven amount Q_{BE}, there is indifference to accept or reject the project. Use the following decision guideline:

Single Project   (Refer to Figure 13–2.)

Estimated quantity is *larger* than Q_{BE} → accept project

Estimated quantity is *smaller* than Q_{BE} → reject project

For two or more alternatives, determine the breakeven value of the common variable. Use the following guideline to select an alternative:

Two Alternatives   (Refer to Figure 13–5.)

Estimated level is *below* breakeven → select alternative with higher
variable cost (larger slope)

Estimated level is *above* breakeven → select alternative with lower
variable cost (smaller slope)

Payback analysis estimates the amount of time, for example, number of years, necessary to recover the initial investment plus a stated rate of return. This is a supplemental analysis technique used primarily for initial screening prior to a full evaluation by PW or some other method. The technique has some drawbacks, especially for no-return payback analysis, where $i = 0\%$ is the stated return.

PROBLEMS

Breakeven Analysis for a Project

13.1   The fixed cost at Harley Motors is $5 million annually. The main product has revenue of $89 per unit and $45 variable cost. Estimate the following: (*a*) Breakeven quantity per year; and (*b*) annual profit if 100,000 units are sold, and if 200,000 units are sold.

13.2   A manufacturing process at Simplicity XP has a fixed cost of $40,000 per month. A total of 100 units can be produced in 1 day at a cost of $3000 for materials and labor for the day. If the company's MARR is 12% per year, compounded monthly, how many units must be sold each month at $50 per unit for the company to just break even?

13.3   A professional photographer who specializes in wedding-related activities paid $48,000 for equipment that will have a $2000 salvage value after 5 years. He estimates that his costs associated with each event amount to $65 per day. If he charges $300 per day for his services, how many days per year must he be employed in order to break even at an interest rate of 8% per year?

13.4   An independent over-the-road (OTR) truck driver-owner paid $68,000 for a used tractor-trailer. The

salvage value of the rig after 5 more years of use is expected to be $36,000. The operating cost is $0.50 per mile and the base mileage rate (i.e., revenue) is $0.61 per mile.

(*a*)   How many miles per year must the owner drive just to break even at an interest rate of 10% per year?

(*b*)   If the owner drives 600 miles per day, how many days per year will be required for break even?

13.5   Handheld fiber optic meters with white light polarization interferometry are useful for measuring temperature, pressure, and strain in electrically noisy environments. The fixed costs associated with manufacturing are $800,000 per year. If variable costs are $290 per unit and the company sells 4000 units per year, at what selling price per unit will the company break even?

13.6   The costs and revenue projections for a new product are estimated. What is the estimated profit at a production rate of 20% above breakeven?

Fixed cost = $500,000 per year

Production cost per unit = $200

Revenue per unit = $250

13.7 Freeport McMoRan engineers estimated that the capital investment necessary for recovering valuable metals (nickel, silver, gold, etc.) from a copper refinery's wastewater would be $150 million. The equipment is expected to have a useful life of 10 years with no salvage value. The amount of metal currently discharged in the wastewater is 12,500 pounds per year. The recovered metals are expected to have a selling price of $250 per pound. The efficiency relation of the recovery operation is represented by $X^{0.5}$, where X represents the efficiency in percentage. What value of X is necessary for the company to break even? Assume $i = 10\%$ per year.

13.8 A company that manufactures automatic blowdown control valves (for applications where boilers are operated unsupervised for 24 to 36 hours) has fixed cost of $160,000 per year and variable cost of $400 per valve. If the company expects to sell 12,000 valves per year, determine the selling price in order for the company to (*a*) break even, and (*b*) make a profit of $400,000 per year.

13.9 If the cost of gasoline is $2.50 per gallon, how far out of your way (one way) could you afford to drive (in miles) to a service station that has gasoline at $2.25 per gallon if your vehicle gets 30 miles per gallon and you will fill up with 20 gallons of fuel? Consider only the cost of gasoline in vehicle driving cost. (Hint: Since you are going out of your way, you have to drive back the same distance.)

13.10 The landfill for Wellsburg has an area of 30 acres available for receiving waste from the city of 40,000 people who generate 25,000 tons of municipal solid waste (MSW) per year. If the fixed cost of the landfill is $300,000 per year and the operating cost is $12 per ton, how much must the landfill charge per ton of MSW to break even?

13.11 A financial services consulting company bought an office building for $900,000. The company has 10 professional staff members. Monthly expenses for salaries, utilities, grounds maintenance, etc. are $1.1 million. The average billing rate per professional is $90 per hour. Use an interest rate of 1% per month and assume the building will have a resale value of $1.5 million after 10 years. (*a*) How many hours per month must be billed in order to make a profit of $15,000 per month? (*b*) How many hours per professional per month must be billed? (*c*) There are 260 eight-hour workdays per year. Of the total work hours available per month, what percentage does the hours per professional in part (*b*) represent?

13.12 You work for Bellevue Window Products. While performing an analysis for a new window product, you found a report from last year that provided the following information regarding the manufacture of a similar product: annual production rate = 40,000 units; selling price = $70 per unit; fixed production cost = $240,000 per year; variable production cost = $1,700,000 per year; variable selling expenses = $96,000 per year. As a first-cut, you decide to use this information to estimate (*a*) the breakeven production rate per year, (*b*) the company's profit last year, and (*c*) the annual production rate that would generate a profit of $1,000,000 per year. What are your estimates?

13.13 An automobile company is investigating the advisability of converting a plant that manufactures economy cars into one that will make retro-sports cars. The initial cost for equipment conversion will be $250 million with a 20% salvage value anytime within a 5-year period. The cost of producing a car will be $25,000, but it is expected to have a selling price of $43,000 (to dealers). The production capacity for the first year will be 4000 units. At an interest rate of 12% per year, by what uniform amount will production have to increase each year in order for the company to recover its investment in 3 years? Assume the production cost and selling price will remain constant for the 3-year period. Solve (*a*) by hand, and (*b*) using a spreadsheet.

Breakeven Analysis Between Alternatives

13.14 A rotational molding operation has fixed costs of $10,000 per year and variable costs of $50 per unit. If the process is automated via conveyor, its fixed cost will be $22,800 per year but its variable cost will be only $10 per unit. Determine the number of units each year necessary for the two operations to break even.

13.15 A two-lane road surface can be finished with concrete or asphalt. Concrete will cost $2.3 million per mile (excluding right-of-way, environmental mitigation, or soil and site conditions) and will last for 20 years. If signing, mowing, and winter maintenance are not included, the basic maintenance costs for concrete and asphalt roadways average $486 and $774 per mile per year, respectively. The interest rate is 8% per year. (*a*) What is the maximum amount that can be spent on asphalt if it lasts 10 years? (*b*) *Use a spreadsheet* to graph the breakeven curve for the cost of asphalt, if different quality levels of asphalt can be used with expected lives of 5, 8, 10, 12, and 15 years.

13.16 Pure gasoline has an energy density of 115,600 BTU per gallon while ethanol has an energy density of 75,670 BTU per gallon. If gasoline costs $3.50 per gallon, (*a*) what would the cost of pure ethanol have to be in order for the energy costs of the two fuels to break even? (*b*) What would the price of E85 (85% ethanol, 15% gasoline) have to be for its energy cost to be the same as that of pure gasoline?

13.17 A construction company can purchase a used backhoe for $90,000 and spend $450 per day in operating costs. The equipment will have a 5-year life with no salvage value. Alternatively, the company can lease the equipment for $800 per day. How many days per year must the company use the equipment in order to justify its purchase at an interest rate of 8% per year.

13.18 Process A has a fixed cost of $160,000 per year and a variable cost of $50 per unit. For Process B, 10 units can be produced in 1 day at a cost of $200. If the company's MARR is 10% per year, what will the annual fixed cost have to be for Process B in order for the two alternatives to have the same annual total cost at a production rate of 1000 units per year?

13.19 There are a number of reasons for not paving a rural road, including low traffic volume, property owners who do not want it paved, speed control, and political issues. On the other hand, paved roads have lower road-user operating costs and are less costly to maintain. In Washington County, a report prepared by the county engineer showed that the construction cost for a stabilized gravel road that will last 3 years is $1,025,000, with a maintenance cost of $355,000 per year. A bituminous road will cost $3,525,000, but it will last for 10 years. Use $i = 8\%$ per year. (*a*) At what maintenance cost per year (for the bituminous road) will the equivalent annual costs of the two roads be the same? (*b*) If the road is not paved and the annual maintenance cost of the stabilized gravel surface is actually 30% per year higher than estimated, how many years must this surface last for the two to break even, assuming the paved road maintenance would be the breakeven value determined above? Answer these questions using manual and *spreadsheet solution*, as your instructor requests.

13.20 A land development company is considering the purchase of earth-moving equipment. This equipment will have an estimated first cost of $190,000, a salvage value of $70,000, a life of 10 years, a maintenance cost of $40,000 per year, and an operating cost of $260 per day. Alternatively, the company can rent the necessary equipment for $1100 per day and hire a driver at $180 per day. (*a*) If the company's MARR is 10% per year, how many days per year must the company need the equipment in order to justify its purchase? (*b*) When approached to rent for the breakeven number of days, the equipment owner indicated that the minimum rental is for 100 days per year; however, he might consider a lower daily rental cost. What is the daily rental cost to justify renting over purchasing? If the equipment was purchased, assume it would be used for the breakeven number of days. *Use the Goal Seek or Solver tool* to determine the required rental cost per day.

13.21 Microsurfacing is part of a pavement restoration and maintenance program that seals the surface of a street that has minor cracking to prevent water from penetrating into the base material. The annual cost of the equipment (truck, tank, valves, etc.) is $109,000 per year and the material cost is $2.75 per square yard. Alternatively, regular street resurfacing requires equipment that has a first cost of $225,000 with a 15-year life and no salvage value. The variable cost for regular resurfacing is $13 per square yard. At an interest rate of 8% per year, how many square yards per year must be resurfaced for the two methods to break even?

13.22 Two membrane systems are under consideration for treating cooling tower blowdown to reduce its volume. A low-pressure seawater reverse osmosis (SWRO) system will operate at 500 psi and produce 720,000 gallons of permeate per day. It will have a fixed cost of $465 per day and an operating cost of $485 per day. A higher-pressure SWRO system operating at 800 psi will produce 950,000 gallons per day at a cost of $1280 per day. The fixed cost of the high-pressure SWRO system will be only $328 per day because fewer membranes will be required. How many gallons of blowdown water will require treatment each day for the two systems to break even?

13.23 Three methods can be used for producing heat sensors for high-temperature furnaces. The interest rate for the economic evaluation is 10% per year.

Method A: Fixed cost of $140,000 per year
Production cost of $62 per part

Method B: Fixed cost of $210,000 per year
           Production cost of $28 per part

Method C: Equipment first cost of $500,000
           Life of 5 years
           Salvage value of 25% of first cost
           Production cost of $53 per part

(a) Determine the breakeven annual production rate between the two lowest-cost methods.

(b) Develop the *spreadsheet graph* of the annual cost for all three methods and estimate the breakeven production rate(s) from the graph.

(c) Use your *spreadsheet and graph* to determine the required annual fixed cost of method A that forces a breakeven between methods A and C at the rate of 2000 parts per year.

13.24 An online commercial directory service must decide between composing the ads for its clients in-house or paying a production company to compose them. To develop the ads in-house, the company will have to purchase computers, printers, and a database management system at an estimated cost of $42,000. This equipment will have a useful life of 3 years, after which it will be sold for $2000. The employee who creates the ads will be paid $55,000 per year. In addition, each ad will have an average cost of $10. Alternatively, the company can outsource development at a flat fee of $21 per ad. At an interest rate of 10% per year, how many ads must the company sell each year for the options to just break even?

13.25 James, the plant manager at Global Foundries (GF), received estimates from two contractors to improve traffic flow and repave the parking areas at his production facility. Proposal A includes new curbs, grading, and paving at an initial cost of $250,000. The expected life of the parking lot surface is 4 years with annual costs for maintenance and repainting of strips of $3000. According to proposal B, the pavement has a higher quality and an expected life of 12 years. The annual maintenance cost will be negligible for the parking area, but the markings will have to be repainted every 2 years at a cost of $5000. The MARR is 12% per year. (a) How much can GF afford to spend on proposal B for the two proposals to break even? (b) Proposal B first cost came in at $700,000. Now, what is the breakeven initial cost for proposal A, if all other estimates are correct?

13.26 Alfred Home Construction is considering the purchase of five dumpsters and the transport truck to store and transfer construction debris from building sites. The entire rig is estimated to have an initial cost of $125,000, a life of 8 years, a $5000 salvage value, an operating cost of $40 per day, and an annual maintenance cost of $2000. Alternatively, Alfred can obtain the same services from the city as needed at each construction site for an initial delivery cost of $125 per dumpster per site and a daily charge of $20 per day per dumpster. An estimated 45 construction sites will need debris storage throughout the average year. If the minimum attractive rate of return is 12% per year, how many days per year must the equipment be required to justify its purchase?

13.27 Process X is estimated to have a fixed cost of $40,000 per year and a variable cost of $60 per unit in year 1, decreasing by $5 per unit per year thereafter. Process Y will have a fixed cost of $70,000 per year and a variable cost of $10 per unit in year 1, increasing by $1 per unit per year thereafter. At an interest rate of 12% per year, how many units must be produced *in year 3* for the two processes to break even?

13.28 The Ecology Group wishes to purchase a piece of equipment for recycling of various metals. Machine 1 costs $123,000, has a life of 10 years, an annual cost of $5000, and requires one operator at a cost of $24 per hour. It can process 10 tons per hour. Machine 2 costs $70,000, has a life of 6 years, an annual cost of $2500, and requires two operators at a cost of $24 per hour each to process 6 tons per hour. Assume $i = 7\%$ per year and 2080 hours per work year. Determine the annual breakeven tonnage of scrap metal at $i = 7\%$ per year and select the better machine for a processing level of 1500 tons per year.

13.29 Claris Water Company makes and sells filters for water drinking fountains for the public. The filter sells for $50. Recently a make/buy analysis was done based on the need for new manufacturing equipment. The equipment first cost of $200,000 and $25,000 annual operation cost comprise the fixed cost, while Claris's variable cost is $20 per filter. The equipment has a 5-year life, no salvage value, and the MARR is 6% per year. The decision to make the filter was based on the breakeven point and the historical sales level of 5000 filters per year.

(a) Determine the breakeven point.

(b) An engineer at Claris learned that an outsourcing firm offered to make the filters for $30 each, but this offer was rejected by the president as entirely too expensive. Perform the breakeven analysis of the two options and determine if the "make" decision was correct.

(c) Develop and use the profit relations for both options to verify the preceding answers.

(d) Use a *spreadsheet* to verify the answers to parts (b) and (c) above by plotting the profit lines.

Payback Analysis

13.30 State why payback analysis should be used only as a supplemental analysis tool when an economic study is performed.

13.31 Darnell Enterprises constructed an addition to its building at a cost of $70,000. Extra annual expenses are expected to be $1850, but extra income will be $14,000 per year. How long will it take for the company to recover its investment at an interest rate of 10% per year? Also, write the spreadsheet function that determines n_p.

13.32 (a) How long would it take to recover an investment of $245,000 in enhanced CNC controls that include axis control to eight axes on the milling model, if the associated income is $92,000 per year, expense is $38,000 per year, and the salvage value is estimated to be 15% of the first cost? Use a MARR of 15% per year. (b) Also, write the spreadsheet function that determines n_p.

13.33 Accusoft Systems is offering small business owners a software package that keeps track of many accounting functions from bank transactions to sales invoices. The site license will cost $39,000 to install and will require a fee of $6000 every 3 months. If your company can save $13,500 every quarter and have the security of managing its books in-house, how long will it take for you to recover the investment at an interest rate of 10% per quarter?

13.34 (a) What is the approximate number of years you would have to sell a mobile phone app to break even if income is estimated to be $50,000 per year, expense is $15,000 per year, your initial investment is $280,000, and your MARR is 10% per year? (b) Find the exact payback period using a spreadsheet function.

13.35 The price of a car you want is $42,000 today. Its price is expected to increase by $1000 each year. You now have $25,000 in an investment account, which is earning 10% per year. How many years will it be before you have enough to buy the car without borrowing any money? Solve by (a) trial and error, and (b) by spreadsheet.

13.36 A process for producing the mosquito repellant Deet has an initial investment of $200,000 with annual costs of $50,000. Income is expected to be $90,000 per year. (a) What is the payback period at $i = 0\%$ per year? A $i = 12\%$ per year? (Note: Round your answers to the nearest integer.) (b) What is the annual breakeven production quantity for both payback periods (determined above) if net profit, that is, income minus cost, is $10 per gallon?

13.37 Two machines can be used to produce a part from titanium. The costs and other cash flows associated with each alternative are estimated. The salvage values are constant regardless of when the machines are replaced. Determine which alternative(s) should be selected for further analysis if alternatives must have a payback of 5 years or less. Perform the analysis with (a) $i = 0\%$, and (b) $i = 10\%$ per year.

Machine	Semiautomatic	Automatic
First cost, $	−40,000	−90,000
Net annual income, $ per year	10,000	15,000
Maximum life, years	10	10
Salvage value, $	0	0

13.38 A manufacturer of diaphragm seals has identified the cash flows shown for manufacturing and sales functions. Determine the no-return payback period.

First cost of equipment, $	−130,000
Annual expenses, $ per year	−45,000
Annual revenue, $ per year	75,000

13.39 In desalting groundwater that contains a significant amount of sulfates, the concentrate that is generated during the desalting process can sometimes be tough on equipment, so the equipment's useful life is uncertain. For a treatment train that has an initial cost of $90,000 and an estimated operating cost (OC) between $15,000 and $20,000 per month, use a *spreadsheet* to determine how many months the equipment must last to recover the investment at $i = 0.5\%$ per month at each OC value? Assume the income from the sale of calcium sulfate is $22,000 per month. (Note: Solve using increments of $1000 OC per month.)

13.40 A multinational engineering consulting firm that wants to provide resort accommodations to special clients is considering the purchase of a three-bedroom lodge in upper Montana that will cost $250,000. The property in that area is rapidly appreciating in value because people anxious to get away from urban developments are bidding up the prices. If the company spends an average of $500 per month for utilities and the investment increases at a rate of 0.75% per month, how long would it be before the company could sell the property for $100,000 more than it has invested in it?

13.41 A window frame manufacturer is searching for ways to improve revenue from its triple-insulated sliding windows, sold primarily in the far northern areas of the United States. Alternative A is an increase in TV and radio marketing. A total of $300,000 spent now is expected to increase revenue by $60,000 per year. Alternative B requires the same investment for enhancements to the in-plant manufacturing process that will improve the temperature retention properties of the seals around each glass pane. New revenues start slowly for this alternative at an estimated $10,000 the first year, with growth of $15,000 per year as the improved product gains reputation among builders. The MARR is 8% per year and the maximum projection period is 10 years for either alternative. Use (*a*) payback analysis, and (*b*) present worth analysis (for 10 years) to select the more economical alternative. State the reason(s) for any difference in the alternative chosen between the two analyses. (*c*) Find the payback and present worth values using a *spreadsheet*.

EXERCISES FOR SPREADSHEETS

13.42 Benjamin used regression analysis to fit quadratic relations to monthly revenue, R, and total cost, TC, data with the following results, where Q is quantity.

$$R = -0.008Q^2 + 32Q$$
$$TC = 0.005Q^2 + 2.2Q + 10$$

(*a*) Plot R and TC. Estimate the quantities Q_{BE} and Qp, where the maximum profit should occur. Estimate the amount of profit at this quantity.

(*b*) The profit relation $P = R - TC$ and calculus can be used to determine the quantity Q_p at which the maximum profit will occur and the amount of this profit. The equations are:

$$\text{Profit} = aQ^2 + bQ + c$$
$$Q_p = -b/2a$$
$$\text{Maximum profit} = -b^2/4a + c$$

Use these relations to confirm your graphical estimate of Q_P. (Your instructor may ask you to derive the relations above.)

13.43 The National Potato Cooperative purchased a deskinning machine last year for $150,000. Revenue for the first year was $50,000. Over the total estimated life of 8 years, use a spreadsheet to estimate the remaining equivalent annual revenues (years 2 through 8) to ensure breakeven by recovering the investment and a return of 10% per year. Costs are expected to be constant at $42,000 per year and a salvage value of $10,000 is anticipated.

Problems 13.44 and 13.45 are based upon the following information.

Wilson Partners manufactures thermocouples for remote temperature monitoring of electronics applications. The current system has a fixed cost of $400,000 per year and a variable cost of $10 per unit. Wilson sells the units for $14 each. A newly proposed system will add on-board features that allow the revenue to increase to $16 per unit, but the fixed cost will now be $600,000 per year. The variable cost of the new system will be based on a $48 per hour rate with 0.2 hour dedicated to produce each unit.

13.44 Determine the annual breakeven quantity for the (*a*) current system, and (*b*) proposed system.

13.45 Plot the two profit relations and estimate graphically the breakeven quantity between the current and proposed systems. Comment on your estimate.

Problems 13.46 through 13.49 are based on the following information.

Mid-Valley Industrial Extension Service, a state-sponsored agency, provides water quality sampling services to all business and industrial firms in a 10-county region. Last month, the service purchased all necessary lab equipment for full in-house testing and analysis. Now, an outsourcing company has offered to take over this function on a per-sample basis. Data and quotes for the two options have been collected. The MARR for government projects is 5% per year and a study period of 8 years is chosen.

In-house: Equipment and supplies initially cost $125,000 for a life of 8 years, an AOC of $15,000, and annual salaries of $175,000. Sample costs average $25. There is no significant salvage value for the equipment and supplies currently owned.

Outsourced: Cost averages $100 per sample for the first 5 years, increasing to $125 per sample for years 6 through 8.

13.46 Determine the breakeven number of tests between the two options.

13.47 Use a spreadsheet to graph the AW curves for both options for test loads between 0 and 5000 per year in increments of 1000 tests. What is the estimated breakeven quantity?

13.48 The service director has asked the outsource company to reduce the per sample costs by 25% across the board over the 8 years. Will this increase or decrease the breakeven point? (Hint: Look carefully at your graph from the previous problem before answering.) Determine the breakeven point *by hand and using your spreadsheet*.

13.49 Assume the Extension Service can reduce its annual salaries from $175,000 to $100,000 per year and the per sample cost from $25 to $20. Outsourced samples will cost $100 per sample. What will this do to the breakeven point? (Hint: Again, look carefully at your graph from the previous problems before answering.) Use your *spreadsheet and hand solution* to estimate the new annual breakeven test quantity.

ADDITIONAL PROBLEMS AND FE EXAM REVIEW QUESTIONS

13.50 In linear breakeven analysis, if process A has a variable cost of $45 per unit and process B has a variable cost of $31 per unit, which alternative would be preferred if the breakeven point is 7400 units and production is expected to be 6200 units?
   (*a*) Process A
   (*b*) Process B
   (*c*) Process B if its fixed cost is lower than the fixed cost of Process A
   (*d*) Cannot tell; need more information

13.51 An automobile leasing company has a contract with a new car dealer to do major repairs for $720 per car. The leasing company estimates that for $400,000, it could buy equipment to service their own cars at a cost of $300 per car. If the equipment will have a salvage value of 10% of its first cost after 15 years, the minimum number of cars that must require major servicing each year to justify the equipment at a MARR of 10% per year is closest to:
   (*a*) 88   (*b*) 122   (*c*) 128   (*d*) 143

13.52 For the following two AW relations, the breakeven point Q_{BE} in miles per year is closest to:

$$AW_1 = -23,000(A/P,10\%,10) + 4000(A/F,10\%,10) - 5000 - 4X$$
$$AW_2 = -8,000(A/P,10\%,4) - 2000 - 6X$$

   (*a*) 1984   (*b*) 1224   (*c*) 1090   (*d*) 655

13.53 A milling process is accomplished using either method X or Y. Method X has fixed costs of $10,000 per year with a variable cost of $50 per unit. Method Y is an automated process with a fixed cost of $5,000 per year and a variable cost of only $30 per unit. The number of units that must be produced each year in order for method Y to be favored is closest to:
   (*a*) Y will be favored for any level of production
   (*b*) 125
   (*c*) 375
   (*d*) X will be favored for any level of production

13.54 You saw that there was a gasoline price war going on in a small town 27 miles from where you live (54 miles round trip), so you are thinking about driving there to fill the tank in your pick-up truck, which gets 18 miles per gallon. You think it will take 22 gallons to fill your tank. If the cost of gasoline at your usual station is $3.60 per gallon, the required price at the out-of-town station to just breakeven is closest to:
   (*a*) $3.63
   (*b*) $2.82
   (*c*) $2.95
   (*d*) $3.11

13.55 Revcon Products has two subcontractor bids to automate a composite winding process. Process A will have fixed costs of $42,000 per year and will require two workers at $80 per day each. Together, these workers can generate 100 units. Process B will have fixed costs of $56,000 per year, but with this process, three workers will generate 200 units of product. If *x* is the number of units, the variable cost (VC) per year for B is best represented as:
   (*a*) VC per Year = [2(80)/100]*x*
   (*b*) VC per Year = [3(80)/200]*x*
   (*c*) VC per Year = [3(80)/200]*x* + 56,000
   (*d*) VC per Year = [2(80)/100]*x* + 42,000

13.56 The pieces of equipment required to place 160 cubic yards of concrete in 1 day by experienced workmen are two gasoline engine vibrators and one concrete pump. If the vibrators cost $76 per day and the concrete pump costs $580 per day, the variable cost of the equipment per cubic yard of concrete is closest to:
   (*a*) $0.95   (*b*) $3.63   (*c*) $4.10   (*d*) $4.60

13.57 A jalapeno canning company is faced with a make/buy decision. Cardboard shipping cartons can be purchased for $0.60 each or made in-house. If manufactured, two machines will be required. Machine X will cost $20,000 and have a life of 6 years with a $2000 salvage value. Machine Y will cost $11,000 and have a life of 4 years with no salvage value. The annual maintenance cost for machines X and Y are $6000 and $5000 per year, respectively. A total of four operators will be required for the two machines at a rate of $22.50 per hour per person. In a normal

8-hour day, the four operators and two machines can produce 1000 cartons. The variable cost per carton associated with the in-house option is closest to:

(a) $0.0625
(b) $0.10
(c) $0.72
(d) $0.81

13.58 A wireless monitored process has fixed cost of $40,000 per year and variable cost of $30 per unit. An IoT-based process using "Internet of Things" technology has fixed costs of $80,000 per year. If the two processes break even at a production level of 4000 units per year, the variable cost of the IoT-based process is closest to:

(a) $10 per unit
(b) $20 per unit
(c) $30 per unit
(d) $40 per unit

13.59 A mixing process for laboratory-grade sodium phosphate has an estimated first cost of $320,000 with annual costs of $40,000. Income is expected to be $98,000 per year. At a MARR of 20% per year, the payback period is closest to:

(a) 3 years
(b) 15 years
(c) 19 years
(d) It will never pay off

13.60 Ma Bryan sells homemade preserves. The profit relation for the following estimates at a quantity that is 10% above breakeven is closest to:

Fixed cost = $500,000 per year
Cost per 100 units = $200
Revenue per 100 units = $250

(a) Profit = 200(11,000) − 250(11,000) − 500,000
(b) Profit = 250(11,000) − 500,000 − 200(11,000)
(c) Profit = 250(11,000) − 200(11,000) + 500,000
(d) Profit = 250(10,000) − 200(10,000) − 500,000

13.61 The cost of equipment to apply epoxy paint to a warehouse floor is $1400. Materials and supplies (masking tape, surface repair material, solvents, cleanup supplies, etc.) will cost $2.03 per ft^2. Alternatively, a contractor can be hired annually to do the job for $3.25 per ft^2. If the cost of the equipment will be recovered at $i = 10\%$ per year over a 3-year period with no salvage value, the number of square feet that must be treated each year to justify the equipment purchase is closest to:

(a) 460 ft^2
(b) 953 ft^2
(c) 1148 ft^2
(d) 2457 ft^2

13.62 Two methods to control newly discovered poisonous weeds in bar-ditches on the sides of county roads in New Farmendale are under consideration. Method A involves use of a 20-year life lining at an initial cost of $14,000 and an annual maintenance cost of $3 per kilometer (km). Method B involves spraying a chemical that costs $40 per liter. One liter will treat 8 km, but the treatment must be applied four times per year. In determining the number of km per year that would result in breakeven, the variable cost for method B is closest to:

(a) $5 per km
(b) $15 per km
(c) $20 per km
(d) $40 per km

CASE STUDY

WATER TREATMENT PLANT PROCESS COSTS

Background

Aeration and sludge recirculation have been practiced for many years at municipal and industrial water treatment plants. Aeration is used primarily for the physical removal of gases or volatile compounds, while sludge recirculation can be beneficial for turbidity removal and hardness reduction.

When the advantages of aeration and sludge recirculation in water treatment were first recognized, energy costs were so low that such considerations were seldom of concern in treatment plant design and operation. With the huge increases in electricity cost that have occurred in some localities, however, it became necessary to review the cost-effectiveness of all water treatment processes that consume

significant amounts of energy. This study was conducted at a municipal water treatment plant for evaluating the cost-effectiveness of the pre-aeration and sludge recirculation practices.

Information

This study was conducted at a 106 m^3/min water treatment plant where, under normal operating circumstances, sludge from the secondary clarifiers is returned to the aerator and subsequently removed in the primary clarifiers. Figure 13–12 is a schematic of the process.

To evaluate the effect of sludge recirculation, the sludge pump was turned off, but aeration was continued. Next, the

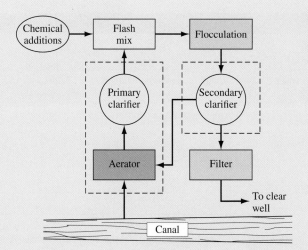

Figure 13–12
Schematic of water treatment plant.

sludge pump was turned back on, and aeration was discontinued. Finally, both processes were discontinued. Results obtained during the test periods were averaged and compared to the values obtained when both processes were operational.

The results obtained from the four operating modes showed that the hardness decreased by 4.7% when both processes were in operation (i.e., sludge recirculation and aeration). When only sludge was recirculated, the reduction was 3.8%. There was no reduction due to aeration only, or when there was neither aeration nor recirculation. For turbidity, the reduction was 28% when both recirculation and aeration were used. The reduction was 18% when *neither* aeration nor recirculation was used. The reduction was also 18% when aeration alone was used, which means that aeration alone was of no benefit for turbidity reduction. With sludge recirculation alone, the turbidity reduction was only 6%, meaning that sludge recirculation alone actually resulted in an *increase* in turbidity—the difference between 18% and 6%.

Since aeration and sludge recirculation did cause readily identifiable effects on treated water quality (some good and others bad), the cost-effectiveness of each process for turbidity and hardness reduction was investigated. The calculations are based on the following data:

Aerator motor = 40 hp

Aerator motor efficiency = 90%

Sludge recirculation motor = 5 hp

Recirculation pump efficiency = 90%

Electricity cost = 9 ¢/kWh (previous analysis)

Lime cost = 7.9 ¢/kg

Lime required = 0.62 mg/L per mg/L hardness

Coagulant cost = 16.5 ¢/kg

Days/month = 30.5

As a first step, the costs associated with aeration and sludge recirculation were calculated. In each case, costs are independent of flow rate.

Aeration cost:

$$40 \text{ hp} \times 0.75 \text{ kW/hp} \times 0.09 \text{ \$/kWh} \times 24 \text{ h/day}$$
$$\div\, 0.90 = \$72 \text{ per day or } \$2196 \text{ per month}$$

Sludge recirculation cost:

$$5 \text{ hp} \times 0.75 \text{ kW/hp} \times 0.09 \text{ \$/kWh} \times 24 \text{ h/day}$$
$$\div\, 0.90 = \$9 \text{ per day or } \$275 \text{ per month}$$

The estimates appear in columns 1 and 2 of the cost summary in Table 13–1.

Costs associated with turbidity and hardness removal are a function of the chemical dosage required and the water flow rate. The calculations below are based on a design flow of 53 m^3/min.

As stated earlier, there was less turbidity reduction through the primary clarifier without aeration than there was with it (28% vs. 6%). The extra turbidity reaching the flocculators could require further additions of the coagulating chemical. If it is assumed that, as a worst case, these chemical additions would be proportional to the extra turbidity, then 22% more coagulant would be required. Since the average dosage before discontinuation of aeration was 10 mg/L, the *incremental chemical cost* incurred because of the increased turbidity in the clarifier effluent would be

$$(10 \times 0.22) \text{ mg/L} \times 10^{-6} \text{ kg/mg} \times 53 \text{ m}^3/\text{min}$$
$$\times\, 1000 \text{ L/m}^3 \times 0.165 \text{ \$/kg} \times 60 \text{ min/h}$$
$$\times\, 24 \text{ h/day} = \$27.70/\text{day or } \$845/\text{month}$$

Similar calculations for the other operating conditions (i.e., aeration only, and neither aeration nor sludge recirculation) reveal that the additional cost for turbidity removal would be $469 per month in each case, as shown in column 5 of Table 13–1.

Changes in hardness affect chemical costs by virtue of the direct effect on the amount of lime required for water softening. With aeration and sludge recirculation, the average hardness reduction was 12.1 mg/L (i.e., 258 mg/L × 4.7%). However, with sludge recirculation only, the reduction

TABLE 13–1	Cost Summary in Dollars per Month							
Alt. I.D.	Alternative Description	Savings from Discontinuation of		Total Savings (3) = (1) + (2)	Extra Cost for Removal of		Total Extra Cost (6) = (4) + (5)	Net Savings (7) = (3) − (6)
		Aeration (1)	Recirculation (2)		Hardness (4)	Turbidity (5)		
1	Sludge recirculation and aeration	Normal operating condition						
2	Aeration only	—	275	275	1380	469	1849	−1574
3	Sludge recirculation only	2196	—	2196	262	845	1107	+1089
4	Neither aeration nor sludge recirculation	2196	275	2471	1380	469	1849	+622

was 9.8 mg/L, resulting in a difference of 2.3 mg/L attributed to aeration. The *extra cost of lime* incurred because of the discontinuation of aeration, therefore, was

$$2.3 \text{ mg/L} \times 0.62 \text{ mg/L lime} \times 10^{-6} \text{ kg/mg}$$
$$\times 53 \text{m}^3/\text{min} \times 1000 \text{ L/m}^3 \times 0.079 \text{ \$/kg}$$
$$\times 60 \text{ min/h} \times 24 \text{ h/day} = \$8.60/\text{day or}$$
$$\$262/\text{month}$$

When sludge recirculation was discontinued, there was no hardness reduction through the clarifier, so that the extra lime cost would be $1380 per month.

The total savings and total costs associated with changes in plant operating conditions are tabulated in columns 3 and 6 of Table 13–1, respectively, with the net savings shown in column 7. Obviously, the optimum condition is represented by "sludge recirculation only." This condition would result in a net savings of $1089 per month, compared to a net savings of $622 per month when both processes are discontinued and a net *cost* of $1574 per month for aeration only. Since the calculations made here represent worst-case conditions, the actual savings that resulted from modifying the plant operating procedures were greater than those indicated.

In summary, the commonly applied water treatment practices of sludge recirculation and aeration can significantly affect the removal of some compounds in the primary clarifier. However, increasing energy and chemical costs warrant

continued investigations on a case-by-case basis of the cost-effectiveness of such practices.

Case Study Exercises

1. What will be the monthly savings in electricity from discontinuation of aeration if the cost of electricity is now 12 ¢/kWh?
2. Does a decrease in the efficiency of the aerator motor make the selected alternative of sludge recirculation only more attractive, less attractive, or the same as before?
3. If the cost of lime were to increase by 50%, would the cost difference between the best alternative and second-best alternative increase, decrease, or remain the same?
4. If the efficiency of the sludge recirculation pump were reduced from 90% to 70%, would the net savings difference between alternatives 3 and 4 increase, decrease, or stay the same?
5. If hardness removal were to be discontinued at the treatment plant, which alternative would be the most cost-effective?
6. If the cost of electricity decreased to 8 ¢/kWh, which alternative would be the most cost-effective?
7. At what electricity cost would the following alternatives just break even? (*a*) Alternatives 1 and 2, (*b*) alternatives 1 and 3, (*c*) alternatives 1 and 4.

LEARNING STAGE 4

Rounding Out the Study

This stage includes topics to enhance your ability to perform a thorough engineering economic study of one project or several alternatives. The effects of **inflation, depreciation, income taxes** in all types of studies, and **indirect costs** are incorporated into the methods of previous chapters. Techniques of **cost estimation** to better predict cash flows are treated in order to base alternative selection on more accurate estimates. The last two chapters include additional material on the use of engineering economics in decision making. An expanded version of **sensitivity analysis** is developed to examine parameters that vary over a predictable range of values. The use of **decision trees** and an introduction to **real options** are included. Finally, the elements of **risk** and **probability** are explicitly considered using expected values, probabilistic analysis, and spreadsheet-based Monte Carlo simulation.

Several of these topics can be covered earlier in the text, depending on the objectives of the course. Use the chart in the Preface to determine appropriate points at which to introduce the material in Learning Stage 4.

CHAPTER 14

Effects of Inflation

Photodisc/Getty Images

LEARNING OUTCOMES

Purpose: Consider the effects of inflation when performing an engineering economy evaluation.

SECTION	TOPIC	LEARNING OUTCOME
14.1	Inflationary impact	• Demonstrate the difference that inflation makes on money now and money in the future; also, explain deflation.
14.2	PW with inflation	• Calculate the PW of cash flows with an adjustment made for inflation.
14.3	FW with inflation	• Determine the real interest rate and calculate the inflation-adjusted FW with different interpretations of future worth values.
14.4	CR with inflation	• Calculate capital recovery of an investment using the AW value with inflation considered.

his chapter concentrates upon understanding and calculating the effects of inflation in time value of money computations. Inflation is a reality that we deal with nearly everyday in our professional and personal lives.

The annual inflation rate is closely watched and historically analyzed by government units, businesses, and industrial corporations. An engineering economy study can have different outcomes in an environment in which inflation is a serious concern compared to one in which it is of minor consideration. In the first 15+ years of the 21st century, inflation has not been a major concern in the United States or most industrialized nations; however, it has caused serious economic problems in many less developed countries. The rate of inflation is sensitive to real, as well as perceived, factors of the economy. Factors such as the cost of energy, interest rates, availability and cost of skilled people, scarcity of materials, political stability, and other, less tangible factors have short-term and long-term impacts on the inflation rate. In some industries, it is vital that the effects of inflation be integrated into an economic analysis. The basic techniques to do so are covered here.

14.1 Understanding the Impact of Inflation ●●○

Inflation

We are all very well aware that $20 now does not purchase the same amount as $20 did in 2015 and purchases significantly less than in 2010 or 2000. Why? Primarily this is due to inflation and its impact on the purchasing power of money.

> **Inflation** is an increase in the amount of money necessary to obtain the **same amount** of goods or services before the inflated price was present.
>
> **Purchasing power,** or **buying power,** measures the value of a currency in terms of the quantity and quality of goods or services that one unit of money will purchase. Inflation decreases the purchasing ability of money in that **less** goods or services **can be purchased** for the same one unit of money.

Inflation occurs because the value of the currency has changed—it has gone down in value. The value of money has decreased, and as a result, it takes more money for the same amount of goods or services. This is a **sign of inflation.** To make comparisons between monetary amounts that occur in different time periods, the different-valued money first must be converted to constant-value money in order to represent the same **purchasing power over time.** This is especially important when future sums of money are considered, as is the case with all alternative evaluations.

Money in one period of time t_1 can be brought to the same value as money in another period of time t_2 by using the equation

$$\text{Amount in period } t_1 = \frac{\text{amount in period } t_2}{1 + \text{inflation rate between } t_1 \text{ and } t_2} \qquad [14.1]$$

Using dollars as the currency, dollars in period t_1 are called **constant-value dollars** or **today's dollars.** Dollars in period t_2 are called **future dollars** or **then-current dollars** and have inflation taken into account. If f represents the inflation rate per period (year) and n is the number of time periods (years) between t_1 and t_2, Equation [14.1] is

$$\text{Constant-value dollars} = \frac{\text{future dollars}}{(1+f)^n} \qquad [14.2]$$

$$\text{Future dollars} = \text{constant-value dollars}(1+f)^n \qquad [14.3]$$

We can express future dollars in terms of constant-value (CV) dollars, and vice versa, by applying the last two equations. This is how the Consumer Price Index (CPI) and cost estimation indices (of Chapter 15) are determined. As an illustration, use the price of a cheese pizza.

$$\$8.99 \qquad \text{March 2017}$$

If inflation on food prices averaged 5% during the last year, in *CV 2016 dollars,* this cost is last year's equivalent of

$$\$8.99/1.05 = \$8.56 \qquad \text{March 2016}$$

A predicted price next year, according to Equation [14.3], is

$$\$8.99(1.05) = \$9.44 \qquad \text{March 2018}$$

The price of $9.44 in 2018 buys exactly the same cheese pizza as $8.56 did in 2016. If inflation averages 5% per year over the next 9 years, Equation [14.3] is used to predict a price in 2025 based on 2016.

$$\$8.56(1.05)^9 = \$13.28 \qquad \text{March 2025}$$

This is a 55% increase over the 2016 price at 5% inflation for prepared food prices, which is generally not considered excessive. In some areas of the world, hyperinflation may average 25% to 50% per year. At 25% annual inflation, in such a stressed economy, the cheese pizza in 5 years rises from the dollar equivalent of $8.99 to $27.44–just think of the huge increase in a single, everyday food item. This is why countries experiencing hyperinflation must devalue the currency by factors of 100 and 1000 when unacceptable inflation rates persist.

Placed into an industrial or business context, at a reasonably low inflation rate averaging 4% per year, equipment or services with a first cost of $209,000 will increase by 48% to $309,000 over a 10-year span. This is before any consideration of the rate of return requirement is placed upon the equipment's revenue-generating ability. **Make no mistake: Inflation is a formidable force in our economy.**

There are three different rates that are important to understanding inflation: the real interest rate (i), the market interest rate (i_f), and the inflation rate (f). Only the first two are interest rates.

Real or inflation-free interest rate i. This is the rate at which interest is earned when the effects of changes in the value of currency (inflation) have been removed. Thus, the real interest rate presents an actual gain in purchasing power. (The equation used to calculate i, with the influence of inflation removed, is derived later in Section 14.3.) The real rate of return that generally applies for individuals is approximately 2.5% to 3.5% per year. This is the risk-free or "safe investment" rate discussed in Section 10.4. The required real rate for corporations (and many individuals) is set above this rate when a MARR is established without adjustment for inflation.

Inflation-adjusted or market interest rate i_f. As its name implies, this is the interest rate that has been adjusted to take inflation into account. This is the interest rate we hear every day. It is a combination of the real interest rate i and the inflation rate f, and, therefore, it changes as the inflation rate changes. It is also known as the *inflated interest rate.* A company's MARR adjusted for inflation is referred to as the inflation-adjusted or market MARR. The determination of this value is discussed in Section 14.3.

Inflation rate f. As described above, this is a measure of the rate of change in the value of the currency.

Deflation is the opposite of inflation in that when deflation is present, the purchasing power of the monetary unit is greater in the future than at present. That is, it will take fewer dollars in the future to buy the same amount of goods or services as it does today. Inflation occurs much more commonly than deflation, especially at the national economy level. In deflationary economic conditions, the market interest rate is always less than the real interest rate.

Temporary price deflation may occur in specific sectors of the economy due to the introduction of improved products, cheaper technology, or imported materials or products that force current prices down. In normal situations, prices equalize at a competitive level after a short time. However, deflation over a short time in a specific sector of an economy can be orchestrated through **dumping.** An example of dumping may be the importation of materials, such as steel, cement, or cars, into one country from international competitors at very low prices compared to current market prices in the targeted country. The prices will go down for the consumer, thus forcing domestic manufacturers to reduce their prices in order to compete for business. If domestic manufacturers are not in good financial condition, they may fail, and the imported items replace the domestic supply. Prices may then return to normal levels and, in fact, become inflated over time, if competition has been significantly reduced.

On the surface, having a moderate rate of deflation sounds good when inflation has been present in the economy over long periods. However, if deflation occurs at a more general level, say nationally, it is likely to be accompanied by the lack of money for new capital. Another result is that individuals and families have less money to spend due to fewer jobs, less credit, and fewer loans available; an *overall "tighter" money* situation prevails. As money gets tighter, less is available to be committed to industrial growth and capital investment. In the extreme case, this can evolve over time into a deflationary spiral that disrupts the entire economy. This has happened on occasion, notably in the United States during the Great Depression of the 1930s.

Engineering economy computations that consider deflation use the same relations as those for inflation. For basic equivalence between constant-value dollars and future dollars, Equations [14.2] and [14.3] are used, except the deflation rate is a $-f$ value. For example, if deflation is estimated to be 2% per year, an asset that costs $10,000 today would have a first cost 5 years from now determined by Equation [14.3].

$$10,000(1 - f)^n = 10,000(0.98)^5 = 10,000(0.9039) = \$9039$$

14.2 Present Worth Calculations Adjusted for Inflation ● ● ●

When the amounts of currency in different time periods are to be expressed in *constant-value (CV) amounts,* the equivalent present and future amounts must be determined using the *real interest rate i.* The calculations involved in this procedure are illustrated in Table 14–1, where the inflation rate is 4% per year. Column 2 shows the inflation-driven increase for each of the next 4 years for an item that has a cost of $5000 today. Column 3 shows the cost in future dollars, and column 4 verifies the cost in CV dollars via Equation [14.2]. When the future dollars of column 3 are converted to CV dollars (column 4), the cost is always $5000, the same as the cost at the start. This is predictably true when the costs are increasing by an amount *exactly equal* to the inflation rate. The actual cost (in inflated dollars) of the item 4 years from now will be $5849, but in CV dollars the cost in 4 years will still amount to $5000. We have added column 5, which shows the present worth of amounts of $5000 in each of the four future years at a real interest rate of $i = 10\%$ per year.

A couple of conclusions that can be drawn are: At $f = 4\%$, $5000 today inflates to $5849 in 4 years. And $5000 four years in the future has a PW of only $3415 now in constant-value dollars at a real interest rate of 10% per year.

Figure 14–1 graphs the differences over a 4-year period of the constant-value amount of $5000, the future-dollar costs at 4% inflation, and the present worth at 10% real interest with inflation considered. The effect of compounded inflation and interest rates can be large, as you can see by the shaded area.

An alternative, less complicated method of accounting for inflation in a present worth analysis involves adjusting the interest formulas themselves to account for inflation. Consider the P/F formula, where i is the real interest rate.

$$P = F \frac{1}{(1 + i)^n}$$

TABLE 14–1	Inflation Calculations Using Constant-Value Dollars ($f = 4\%$, $i = 10\%$)			
Year n (1)	Cost Increase due to 4% Inflation, $ (2)	Cost in Future Dollars, $ (3)	Future Cost in Constant-Value Dollars, $ (4) = (3)/1.04^n	Present Worth at Real $i = 10\%$, $ (5) = (4)(P/F,10\%,n)
0		5000	5000	5000
1	5000(0.04) = 200	5200	$5200/(1.04)^1 = 5000$	4545
2	5200(0.04) = 208	5408	$5408/(1.04)^2 = 5000$	4132
3	5408(0.04) = 216	5624	$5624/(1.04)^3 = 5000$	3757
4	5624(0.04) = 225	5849	$5849/(1.04)^4 = 5000$	3415

Figure 14–1
Comparison of constant-value dollars, future dollars, and their present worth values.

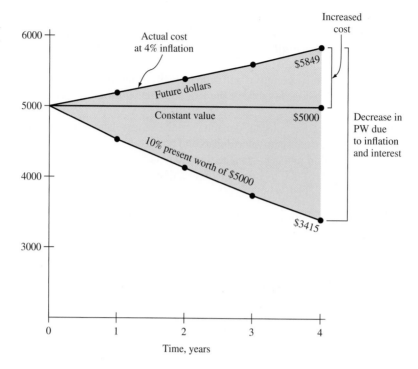

The F, which is a future-dollar amount with inflation built in, can be converted to CV dollars by using Equation [14.2].

$$P = \frac{F}{(1+f)^n} \frac{1}{(1+i)^n}$$

$$= F \frac{1}{(1+i+f+if)^n} \qquad [14.4]$$

If the term $i + f + if$ is defined as i_f, the equation becomes

$$P = F \frac{1}{(1+i_f)^n} = F(P/F, i_f, n) \qquad [14.5]$$

As described earlier, i_f is the **inflation-adjusted or market interest rate** and is defined as

$$i_f = i + f + if \qquad [14.6]$$

where i = real interest rate
        f = inflation rate

For a real interest rate of 10% per year and an inflation rate of 4% per year, Equation [14.6] yields a market interest rate of 14.4%.

$$i_f = 0.10 + 0.04 + 0.10(0.04) = 0.144$$

Table 14–2 illustrates the use of $i_f = 14.4\%$ in PW calculations for $5000 now, which inflates to $5849 in future dollars 4 years hence. As shown in column 4, the present worth for each year is the same as column 5 of Table 14–1.

The present worth of any series of cash flows—uniform, arithmetic gradient, or geometric gradient—can be found similarly. That is, either i or i_f is introduced into the P/A, P/G, or P_g factors, depending upon whether the cash flow is expressed in constant-value (today's) dollars or future dollars, respectively.

If a cash flow series is expressed in today's (constant-value) dollars, then its PW is the discounted value using the real interest rate i.

If the cash flow is expressed in future dollars, the PW value is obtained using i_f.

It is always acceptable to first convert all future dollars to constant-value dollars using Equation [14.2] and then find the PW at the real interest rate i.

TABLE 14–2	Present Worth Calculation Using an Inflated Interest Rate		
Year *n* (1)	Cost in Future Dollars, $ (2)	(*P/F*,14.4%,*n*) (3)	PW, $ (4) = (2)(3)
0	5000	1	5000
1	5200	0.8741	4545
2	5408	0.7641	4132
3	5624	0.6679	3757
4	5849	0.5838	3415

EXAMPLE 14.1

Bicycle tire sales are a highly competitive business, especially over the Internet, plus retailers must contend with competition from companies that practice international dumping. Two Internet retailers, Sprang and Biker-U-R, collaborated in performing a web-based study of a popular tire size's retail price over time. The study indicated that prices set earlier at $16 per tire were reduced to an average, non-sale price of $12 per tire. Assume that when the price was set at $16, it was expected to increase to $19 over the next 5 years, not to decrease. However, now the longer-term price is estimated to be $10 per tire, further reducing revenue prospects. Use this information to perform the following analysis:

(*a*) Determine the annual rate of inflation over 5 years to increase the price from $16 to $19.

(*b*) Using the same annual rate determined above as the rate at which the price continues to *decline* from the new $12 price, calculate the expected price in 5 years. Compare this result with $10 per tire predicted to be the longer-term price.

(*c*) Provided the two retailers were somehow able to recover the same market share as they had previously, and the same inflation rate was applied to the reduced $12 per tire price, determine the price 5 years in the future and compare it with the price of $16 per tire.

(*d*) Determine the market interest rate that must be used in economic equivalence computations if inflation is considered and an 8% per year *real* return is expected by the Internet retailers.

Solution

The first three parts involve inflation only—no return on investments.

(*a*) Solve Equation [14.2] for the annual inflation rate *f* with known constant-value and future amounts.

$$16 = 19(P/F,f,5) = \frac{19}{(1+f)^5}$$

$$1 + f = (1.1875)^{0.2}$$

$$f = 0.035 \quad (3.5\% \text{ per year})$$

(*b*) If the price deflation rate is 3.5% per year, find the *F* value in 5 years with *P* = $12.

$$F = P(F/P,-3.5\%,5) = 12(1 - 0.035)^5$$
$$= 12(0.8368)$$
$$= \$10.04$$

The price will fall to exactly $10 per tire after 5 years, as predicted.

(*c*) Five years in the future, at 3.5% per year inflation, the price will be

$$F = P(F/P,3.5\%,5) = 12(1.035)^5$$
$$= 12(1.1877)$$
$$= \$14.25$$

After 5 years of recovery at the same level as historically experienced, the price will still be considerably lower than it was at the pre-dumping point ($14.25 versus $16).

(*d*) With inflation at 3.5% per year and a real return of 8% per year, Equation [14.6] results in a market rate of 11.78% per year.

$$i_f = 0.08 + 0.035 + (0.08)(0.035)$$
$$= 0.1178 \quad (11.78\% \text{ per year})$$

EXAMPLE 14.2

A 15-year $50,000 bond that has a dividend rate of 10% per year, payable semiannually, is currently for sale. If the expected rate of return of the purchaser is 8% per year, compounded semiannually, and if the inflation rate is expected to be 2.5% each 6-month period, what is the bond worth now (*a*) without an adjustment for inflation, and (*b*) when inflation is considered? Show both hand and spreadsheet solutions.

Solution by Hand

(*a*) *Without inflation adjustment*: The semiannual dividend is $I = [(50,000)(0.10)]/2 = \$2500$. At a nominal 4% per 6 months for 30 periods,

$$PW = 2500(P/A,4\%,30) + 50,000(P/F,4\%,30) = \$58,645$$

(*b*) *With inflation*: Use the inflated rate i_f.

$$i_f = 0.04 + 0.025 + (0.04)(0.025) = 0.066 \text{ per semiannual period}$$
$$PW = 2500(P/A,6.6\%,30) + 50,000(P/F,6.6\%,30)$$
$$= 2500(12.9244) + 50,000(0.1470)$$
$$= \$39,660$$

Solution by Spreadsheet

Both (*a*) and (*b*) require simple, single-cell functions on a spreadsheet (Figure 14–2). Without an inflation adjustment, the PV function is developed at the nominal 4% rate for 30 periods; with inflation considered the rate is $i_f = 6.6\%$, as determined above.

Comment

The $18,985 difference in PW values illustrates the tremendous negative impact made by only 2.5% inflation each 6 months (5.06% per year). Purchasing the $50,000 bond means receiving $75,000 in dividends over 15 years and the $50,000 principal in year 15. Yet, this is worth only $39,660 in constant-value dollars.

	A	B	C
1		Function	PW value, $
2	(a) PW without inflation	`= PV(4%,30,-2500,-50,000)`	58,646
3			
4	(b) PW with inflation adjustment	`= PV(6.6%,30,-2500,-50,000)`	39,660
5			
6			

Figure 14–2
PW computation of a bond purchase (*a*) without and (*b*) with an inflation adjustment, Example 14.2.

EXAMPLE 14.3

A self-employed chemical engineer is on contract with Dow Chemical, currently working in a relatively high-inflation country in Central America. She wishes to calculate a project's PW with estimated costs of $35,000 now and $7000 per year for 5 years beginning 1 year from now with increases of 12% per year thereafter for the next 8 years. Use a real interest rate of 15% per year to make the calculations (a) without an adjustment for inflation, and (b) considering inflation at a rate of 11% per year.

Solution

(a) Figure 14–3 presents the cash flows. The PW without an adjustment for inflation is found using $i = 15\%$ and $g = 12\%$ in Equations [2.34] and [2.35] for the geometric series.

$$PW = -35,000 - 7000(P/A,15\%,4)$$

$$- \left\{ \frac{7000\left[1 - \left(\frac{1.12}{1.15}\right)^9\right]}{0.15 - 0.12} \right\}(P/F,15\%,4)$$

$$= -35,000 - 19,985 - 28,247$$

$$= \$-83,232$$

In the P/A factor, $n = 4$ because the $7000 cost in year 5 is the A_1 term in Equation [2.34].

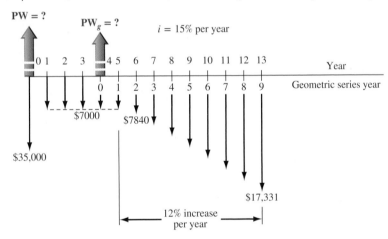

Figure 14–3
Cash flow diagram, Example 14.3.

(b) To adjust for inflation, calculate the inflated interest rate by Equation [14.6] and use it to calculate PW.

$$i_f = 0.15 + 0.11 + (0.15)(0.11) = 0.2765$$

$$PW = -35,000 - 7000(P/A,27.65\%,4)$$

$$- \left\{ \frac{7000\left[1 - \left(\frac{1.12}{1.2765}\right)^9\right]}{0.2765 - 0.12} \right\}(P/F,27.65\%,4)$$

$$= -35,000 - 7000(2.2545) - 30,945(0.3766)$$

$$= \$-62,436$$

Comment

This result demonstrates that in a high-inflation economy, when negotiating the amount of the payments to repay a loan, it is economically advantageous for the borrower to use future (inflated) dollars whenever possible to make the payments. The present value of future inflated dollars is significantly less when the inflation adjustment is included. And the higher the inflation rate, the larger the discounting because the P/F and P/A factors decrease in size.

Examples 14.2 and 14.3 above add credence to the "buy now, pay later" philosophy. However, at some point, the debt-ridden company or individual will have to pay off the debts and the accrued interest with the inflated dollars. If cash is not readily available at that time, the debts cannot be repaid. This can happen, for example, when a company unsuccessfully launches a new product, when there is a serious downturn in the economy, or when an individual loses a salary. In the longer term, this buy now, pay later approach must be tempered with sound financial practices now, and in the future.

14.3 Future Worth Calculations Adjusted for Inflation ● ● ●

In future worth calculations, a future amount F can have any one of several interpretations:

Case 1. The number of *future (then-current) dollars required* at time n to maintain the *same purchasing power* as today; that is, inflation is considered; interest is not.

Case 2. The amount of money required at time n to *maintain purchasing power and earn a stated real interest rate.*

Case 3. The *purchasing power* of the actual amount accumulated at time n, but stated in today's (CV) amount.

Depending upon which interpretation is intended, the F value is calculated differently, as described below. Each case is illustrated.

Case 1: Future Amount Required, No Interest   This first case recognizes that prices increase when inflation is present. Simply put, future dollars are worth less, so more are needed. No interest rate is considered in this case—only inflation. This is the situation if someone asks, "How much will a car cost in 5 years if its current cost is \$20,000 and its price will increase by the inflation rate of 6% per year?" (The answer is \$26,765.) No interest rate—only inflation—is involved. To find the future cost, substitute f for the interest rate in the F/P factor or use the FV spreadsheet function with f as the rate, that is, $= \text{FV}(f\%,A,P)$.

$$F = P(1 + f)^n = P(F/P,f,n) \tag{14.7}$$

For example, consider \$1000 now. At an inflation rate of 4% per year, the future amount required 7 years from now will be

$$F = 1000(F/P,4\%,7) = \$1316$$

Case 2: Inflation and Real Interest   This is the case applied when a market MARR is established. Maintaining purchasing power and earning interest must account for both increasing prices (case 1) and the time value of money. If the growth of capital is to keep up, funds must grow at a rate equal to or above the real interest rate i plus the inflation rate f. For example, to make a *real rate of return of 5.77%* when the inflation rate is 4%, i_f is the market (inflation-adjusted) rate that must be used. For the same \$1000 amount considered above,

$$i_f = 0.0577 + 0.04 + 0.0577(0.04) = 0.10$$
$$F = 1000(F/P,10\%,7) = \$1948$$

This calculation shows that \$1948 seven years in the future will be equivalent to \$1000 now with a real return of $i = 5.77\%$ per year and inflation of $f = 4\%$ per year.

Case 3: Constant-Value Dollars with Purchasing Power   The purchasing power of future dollars is determined by first using the market rate i_f to calculate F and then deflating the future amount through division by $(1 + f)^n$.

$$F = \frac{P(1 + i_f)^n}{(1 + f)^n} = \frac{P\,(F/P, i_f, n)}{(1 + f)^n} \tag{14.8}$$

This relation, in effect, recognizes the fact that inflated prices mean $1 in the future purchases less than $1 now. The percentage loss in purchasing power is a measure of how much less. As an illustration, consider the same $1000 now, and a 10% per year *market rate,* which includes an inflation rate of 4% per year. In 7 years, the purchasing power has risen, but only to $1481.

$$F = \frac{1000\ (F/P,\ 10\%,\ 7)}{(1.04)^7} = \frac{\$1948}{1.3159} = \$1481$$

This is $467 (or 24%) less than the $1948 actually accumulated at 10% (case 2). Therefore, we conclude that 4% inflation over 7 years reduces the purchasing power of money by 24%.

Also for case 3, the future amount of money accumulated with today's purchasing power could equivalently be determined by calculating the real interest rate and using it in the F/P factor to compensate for the decreased purchasing power. This **real interest rate** is the i in Equation [14.6].

$$i_f = i + f + if$$
$$= i(1 + f) + f$$

$$i = \frac{i_f - f}{1 + f} \qquad\qquad [14.9]$$

The real interest rate i represents the rate at which today's dollars expand with their *same purchasing power* into equivalent future dollars. An inflation rate larger than the market interest rate leads to a negative real interest rate. The use of this interest rate is appropriate for calculating the future worth of an investment (such as a savings account or money market fund) when the effect of inflation must be removed. For the example of $1000 in today's dollars from Equation [14.9]

$$i = \frac{0.10 - 0.04}{1 + 0.04} = 0.0577 \quad (5.77\%)$$

$$F = 1000(F/P, 5.77\%, 7) = \$1481$$

The market interest rate of 10% per year has been reduced to a real rate that is less than 6% per year because of the erosive effects of 4% per year inflation.

Table 14–3 summarizes which rate is used in the equivalence formulas for the different interpretations of F. The calculations in this section verify the following statements:

- The amount of $1000 now at a market rate of 10% per year will accumulate to $1948 in 7 years.
- The $1948 will have the purchasing power of $1481 of today's dollars if $f = 4\%$ per year.
- An item with a cost of $1000 now will cost $1316 in 7 years at an inflation rate of 4% per year.
- It will take $1948 of future dollars to be equivalent to $1000 now at a real interest rate of 5.77% plus inflation considered at 4% per year.

TABLE 14–3	Calculation Methods for Various Future Worth Interpretations	
Future Worth Desired	**Method of Calculation**	**Example for $P = \$1000, n = 7$, $i_f = 10\%, f = 4\%$**
Case 1: Dollars required for same purchasing power	Use f in place of i in equivalence formulas	$F = 1000(F/P, 4\%, 7)$
Case 2: Future dollars to maintain purchasing power **and** to earn a return	Calculate i_f and use in equivalence formulas	$F = 1000(F/P, 10\%, 7)$
Case 3: Purchasing power of accumulated dollars in terms of constant-value dollars	Use market rate i_f in equivalence and divide by $(1 + f)^n$ or	$F = \dfrac{1000\ (F/P, 10\%, 7)}{(1.04)^7}$ or
	Use real i	$F = 1000(F/P, 5.77\%, 7)$

Most corporations evaluate alternatives at a MARR large enough to cover inflation plus some return greater than their cost of capital, and significantly higher than the safe investment return of approximately 2.5% to 3.5% mentioned earlier. Therefore, for case 2 (inflation plus real interest), the resulting MARR will normally be higher than the market rate i_f. Define the symbol $MARR_f$ as the inflation-adjusted or market MARR, which is calculated in a fashion similar to i_f.

$$MARR_f = i + f + i(f) \qquad [14.10]$$

The real rate of return i used here is the required rate for the corporation relative to its cost of capital. Now the future worth F, or FW, is calculated as

$$F = P(1 + MARR_f)^n = P(F/P,MARR_f,n)$$

For example, if a company has a WACC (weighted average cost of capital) of 10% per year and requires that a project return 3% per year above its WACC, the real return is $i = 13\%$. The inflation-adjusted MARR is calculated by including the inflation rate of, say, 4% per year. Then the project PW, AW, or FW will be determined at the rate obtained from Equation [14.10].

$$MARR_f = 0.13 + 0.04 + 0.13(0.04) = 0.1752 \qquad (17.52\%)$$

EXAMPLE 14.4

Abbott Mining Systems wants to determine whether it should upgrade a piece of equipment used in deep mining operations in one of its international operations now or later. If the company selects plan A, the upgrade will be purchased now for $200,000. However, if the company selects plan I, the purchase will be deferred for 3 years when the cost is expected to rise to $300,000. Abbott is ambitious; it expects a real MARR of 12% per year. The inflation rate in the country has averaged 3% per year. From only an economic perspective, determine whether the company should purchase now or later (*a*) when inflation is not considered, and (*b*) when inflation is considered.

Solution

(*a*) *Inflation not considered*: The real rate, or MARR, is $i = 12\%$ per year. The cost of plan I is $300,000 three years hence. Calculate the FW value for plan A 3 years from now and select the lower cost.

$$FW_A = -200,000(F/P,12\%,3) = \$-280,986$$
$$FW_I = \$-300,000$$

Select plan A (purchase now).

(*b*) *Inflation considered*: This is case 2; the real rate (12%), and inflation of 3% must be accounted for. First, compute the inflation-adjusted MARR by Equation [14.10].

$$MARR_f = 0.12 + 0.03 + 0.12(0.03) = 0.1536$$

Use $MARR_f$ to compute the FW value for plan A in future dollars.

$$FW_A = -200,000(F/P,15.36\%,3) = \$-307,040$$
$$FW_I = \$-300,000$$

Purchase later (plan I) is now selected because it requires fewer equivalent future dollars. The inflation rate of 3% per year has raised the equivalent future worth of costs by 9.3% from $280,986 to $307,040. This is the same as an increase of 3% per year, compounded over 3 years, or $(1.03)^3 - 1 = 9.3\%$.

   Most countries have inflation rates in the range of 2% to 8% per year, but **hyperinflation** is a problem in countries where political instability, turmoil and insurrection, overspending by the government, weak international trade balances, etc., are present. Hyperinflation rates may be very high—10% to 20% *per month*. In these cases, the government may take drastic actions: re-define the currency in terms of the currency of another country, control banks and corporations, and control the flow of capital into and out of the country in order to decrease inflation.

   In a hyperinflated environment, people usually spend all their money immediately since the cost of goods and services will be much higher the next month, week, or day. To appreciate the disastrous effect of hyperinflation on a company's ability to keep up, we can rework Example 14.4*b* using an inflation rate of 10% per month, that is, a nominal 120% per year (not considering the compounding effect of inflation). The FW_A amount skyrockets and plan I is a clear choice. Of course, in such an environment, the $300,000 purchase price for plan I 3 years hence would obviously not be assured, so the entire economic analysis is unreliable. Good economic decisions in a hyperinflated economy are very difficult to make using traditional engineering economy methods, since the estimated future values are totally unreliable and the future availability of capital is uncertain.

14.4 Capital Recovery Calculations Adjusted for Inflation ●●●

It is particularly important in capital recovery (CR) calculations used for AW analysis to include inflation because current capital dollars must be recovered with future inflated dollars. Since future dollars have less buying power than today's dollars, it is obvious that more dollars will be required to recover the present investment. This suggests the use of the inflated interest rate in the A/P formula. For example, if **$1000 is invested today** at a real interest rate of 10% per year when the inflation rate is 8% per year, the equivalent amount that must be recovered each year for 5 years in future dollars is

$$A = 1000(A/P,18.8\%,5) = \$325.59$$

On the other hand, the decreased value of dollars through time means that investors can spend fewer present (higher-value) dollars to accumulate a specified amount of future (inflated) dollars. This suggests the use of a higher interest rate, that is, the i_f rate, to produce a lower A value in the A/F formula. The annual equivalent (with adjustment for inflation) of $F =$ **$1000 five years from now** in future dollars is

$$A = 1000(A/F,18.8\%,5) = \$137.59$$

   For comparison, the equivalent annual amount to accumulate $F = \$1000$ at a real $i = 10\%$ (without adjustment for inflation) is $1000(A/F,10\%,5) = \$163.80$. When F is a future known cost, uniformly distributed payments should be spread over as long a time period as possible so that the leveraging effect of inflation will reduce the effective annual payment ($137.59 versus $163.80 here).

EXAMPLE 14.5

What annual deposit is required for 5 years to accumulate an amount of money with the same purchasing power as $680.58 today, if the market interest rate is 10% per year and inflation is 8% per year?

Solution

First, find the actual number of inflated dollars required 5 years in the future that is equivalent to $680.58 today. This is case 1; Equation [14.7] applies.

$$F = (\text{present purchasing power})(1 + f)^5 = 680.58(1.08)^5 = \$1000$$

The actual amount of the annual deposit is calculated using the market interest rate of 10%. This is case 2 where A is calculated for a given F.

$$A = 1000(A/F,10\%,5) = \$163.80$$

Comment

The real interest rate is $i = 1.85\%$ as determined using Equation [14.9]. To put these calculations into perspective, if the inflation rate is zero when the real interest rate is 1.85%, the future amount of money with the same purchasing power as \$680.58 today is obviously \$680.58. Then the annual amount required to accumulate this future amount in 5 years is $A = 680.58(A/F,1.85\%,5) = \$131.17$. This is \$32.63 lower than the \$163.80 calculated above for $f = 8\%$. This difference is due to the fact that during inflationary periods, dollars deposited now have more purchasing power than the dollars returned at the end of the period. To make up the purchasing power difference, more higher-value dollars are required. That is, to maintain equivalent purchasing power at $f = 8\%$ per year, an extra \$32.63 per year is required.

The logic discussed here explains why, in times of increasing inflation, lenders of money (credit card companies, mortgage companies, and banks) tend to further increase their market interest rates. People tend to pay off less of their incurred debt at each payment because they use any excess money to purchase additional items before the price is further inflated. Also, the lending institutions must have more money in the future to cover the expected higher costs of lending money. All this is due to the spiraling effect of increasing inflation. Breaking this cycle is difficult to do at the individual level and much more difficult to alter at a national level.

CHAPTER SUMMARY

Inflation, treated computationally as an interest rate, makes the cost of the same product or service increase over time due to the decreased value of money. There are several ways to consider inflation in engineering economy computations in terms of today's (constant-value) dollars and in terms of future dollars. Some important relations are the following:

Inflated interest rate: $i_f = i + f + if$

Real interest rate: $i = (i_f - f)/(1 + f)$

PW of a future amount with inflation considered: $P = F(P/F, i_f, n)$

Future worth in constant-value dollars of a present amount with the same purchasing power:
$F = P(F/P, i, n)$

Future amount to cover a current amount with inflation only: $F = P(F/P, f, n)$

Future amount to cover a current amount with inflation and interest: $F = P(F/P, i_f, n)$

Annual equivalent of a future amount: $A = F(A/F, i_f, n)$

Annual equivalent of a present amount in future dollars: $A = P(A/P, i_f, n)$

Hyperinflation implies very high f values. Available funds are expended immediately because costs increase so rapidly that larger cash inflows cannot offset the fact that the currency is losing value. This can, and usually does, cause a national financial disaster when it continues over extended periods of time.

PROBLEMS

Adjusting for Inflation

14.1   How do you convert a unit of inflated currency (e.g., a dollar) into a constant-value currency statement?

14.2   If a 3D printer increased in cost at exactly the inflation rate, what was the inflation rate if the printer costs exactly twice as much now as it did 10 years ago?

14.3   In an inflationary period, what is the difference between (a) inflated dollars and "then-current" future dollars, and (b) "then-current" future dollars and constant-value future dollars?

14.4   How many future dollars 10 years from now will have the same buying power as \$10,000 today?

The market interest rate is 12% per year and the inflation rate is 7% per year.

14.5 Determine today's purchasing power of $1,000,000 thirty years in the future, if $i_f = 15\%$ per year and $f = 5\%$ per year.

14.6 In an effort to reduce pipe breakage, water hammer, and product agitation, a French chemical company plans to install several chemically resistant pulsation dampeners. The cost of the dampeners today is €106,000, but the chemical company has to wait until a permit is approved for its bidirectional port-to-plant product pipeline. The permit approval process will take at least 2 years because of the time required for preparation of an environmental impact statement. Because of intense foreign competition, the manufacturer plans to increase the price only by the inflation rate of 3% each year. Determine the cost of the dampeners in 2 years in terms of (*a*) then-current euros, and (*b*) constant-value euros.

14.7 The lease cost for a specialized highway design software package is estimated to be $13,000 for each of years 1, 2, and 3 (future dollars). (*a*) Calculate the CV amount today (year 0) of each future cost estimate at the inflation rate of 6% per year. (*b*) Develop a spreadsheet and graph for inflation rates of 3%, 6%, and 8% per year that show the CV today.

14.8 For many years, college cost (including tuition, fees, room, and board) increases have been higher than the inflation rate, averaging 5% to 8% per year. According to the College Board's *Trends in College Pricing*, the average total costs in 2015 dollars were $19,548 for students attending in-state 4-year public colleges and universities and $43,921 for students at 4-year private colleges and universities. Assume an additional $4000 per year for textbooks, supplies, transportation, and other expenses. Using a 7% per year inflation rate, (*a*) how much can a sophomore high-school student expect to spend on in-state tuition, fees, room, and board for the freshman year (3 years from now) at a 4-year public university, and (*b*) what is the estimated total cost for the second year at the university, if textbooks, supplies, etc. also increase at 7% per year?

14.9 The federal-level Pell Grant program provides financial aid to needy college students. The grant increases annually to account for inflation. The maximum total Pell Grant award increased from $5550 to $5,730 from 2013 to 2015. (*a*) What was the average inflation rate per year in this prior 2-year period? (*b*) Beginning in 2018, grants will no longer increase with inflation. What will be the maximum award starting in 2018 and thereafter, assuming the inflation rate in 2016 and 2017 is 2.5% per year?

14.10 According to NACE's April 2014 Salary Survey, engineering majors held 8 of the top 10 spots on the list of top-paid majors for the Class of 2014 bachelor's degree graduates. Graduates in petroleum engineering were the highest paid at $93,500, earning $28,000 more than the next highest paid at $67,300. Civil engineering graduates were the tenth highest paid at $62,100. Assuming the starting salary of petroleum engineering graduates stays the same for the next 10 years, what annual increase in pay would be required for civil engineering graduates to receive the same starting salary as that of petroleum engineering graduates?

14.11 A UK-based life insurance company will pay a cash-value sum of £500,000 when the insured reaches the age of 65. The insured will be 65 years old 27 years from today. (*a*) Determine the cash value of the £500,000 in CV purchasing power, assuming inflation remains constant at 3% per year. (*b*) Write the spreadsheet single-cell function that displays the answer. (*c*) Convert your answer to U.S. dollars and euros using current exchange rates.

14.12 An engineer who is now 65 years old began planning for retirement 40 years ago. At that time, he thought that if he had $1 million when he retired, he would have more than enough money to live his remaining life in luxury. Assume the inflation rate over the 40-year time period averaged a constant 4% per year. (*a*) What is the CV purchasing power of his $1 million at age 65? (Hint: Use the day he started 40 years ago as the base year.) (*b*) How many future dollars should he have accumulated over the 40 years to have a CV purchasing power equal to $1 million at his current age of 65?

14.13 Jack has tracked the annual inflation rate (shown below) for a top-rated sportsman-level fishing outfit that he has always hoped to own. It cost $1000 ten years ago, but he could not afford it over the intervening years. (*a*) Estimate the cost now, which is the end of year 10. (*b*) Is the cost estimate the same using an average inflation rate of 5% per year over the 10-year period? Why or why not?

Year	Inflation Rate per Year
1, 3, 5, 7, 9	10%
2, 4, 6, 8, 10	0%

14.14 The Bureau of Labor Statistics has a website (www.bls.gov) that contains a Consumer Price Index inflation calculator that uses the average CPI to adjust the purchasing power of money over different periods of time. The CPI index value has been calculated every year since 1913. The calculator indicated that $1 million in 1913 would have the same purchasing power as $23,930,909 in 2016. What was the average inflation rate over this 103-year time period?

14.15 The Bank of England has an inflation calculator website that uses the UK price index to show how the cost of goods and services has changed since 1750. The website reveals that something that cost £10,000 in 1920 cost £7984 in 1940. What was the average annual inflation rate over that 20-year time period in the early-to-mid 20th century?

14.16 Emissions of heat-trapping carbon dioxide (CO_2) reached an all-time high of 36.9 gigatons in 2014, a 2.5% increase over 2013. The International Energy Agency said this further reduces the chance that the world could avoid a dangerous rise in global average temperature by 2020. (*a*) If the discharge of CO_2 continues at the same 2.5% rate per year for the next 6 years, how many gigatons will be released in 2020, and (*b*) what will be the total percentage increase between 2014 and 2020?

14.17 In 2015, a retired person in the United States who received maximum social security benefits got $2699 per month, which represented a 1.7% cost of living adjustment (COLA) over 2014. (*a*) Determine the monthly benefit in 2018 if the COLA adjustments are 1.5%, 2.1%, and 2.7%, respectively, for the next 3 years. (*b*) Assuming the average annual inflation is 2.0% per year, determine if the monthly benefits in 2018 have more or less purchasing power in current-value terms assuming the 3 years of COLA adjustments above.

14.18 During periods of hyperinflation, prices increase rapidly over short periods of time. One of the worst cases of hyperinflation in history occurred in Zimbabwe in 2008 where prices doubled *every day*. If that rate continued for just 1 week, how much would a French baguette cost after 7 days if the cost at the beginning of the week was $1.25?

Present Worth Calculations with Inflation

14.19 For years, Jake has practiced the "buy now, pay later" philosophy of money management. For example, he purchased a car 3 years ago with a 6-year (72 month) loan at 7% per year interest. He has refinanced the loan each year at a higher interest rate; 7.5% last year and 8.5% this year. With a serious recession anticipated in his line of work, he was told yesterday by his boss that his salary would

be cut by 25% for the next 2 to 4 years. When Jake tried to refinance his car loan yet once again, the bank loan officer said that due to his multiple loan applications, his credit rating had been lowered significantly and that his current loan must be paid off in the next 6 months to recover his credit rating in the future. Provide Jake with some examples of what he could do to get his finances and credit rating in better order.

14.20 α-β, Inc., a high-tech company in San Diego, whose stock trades on the NYSE exchange, uses a MARR of 25% per year. If the chief financial officer (CFO) said the company expects to make a *real rate of return* of 20% per year on its investments over the next 3-year period, what is the company expecting the annual inflation rate to be over that time period?

14.21 Cellgene Biometrics, a small biotech company, uses a MARR of 40% per year when evaluating new investments. If the inflation rate is 9% per year, what will the real rate of return be, provided it realizes its optimistic MARR objective?

14.22 The new CEO of a high-tech incubator company wants to entice venture capitalists by promising a growth rate of 40% per year for at least 3 years. Therefore, the company's MARR was set at 40%. If this ROR was actually realized, but the CEO did not account for the observed 8% per year inflation rate, what was the real growth rate?

14.23 Calculate the inflation-adjusted interest rate when the annualized inflation rate is 7% per year and the real interest rate is 4% per year.

14.24 Because retail supermarket corporations operate on a low ROR, it is common to use a market MARR of about 3% per year. What real return rate is implied from a market interest rate of 3% per year when the annual inflation rate is 4% per year? Explain your answer.

14.25 Find the present worth of earthmoving equipment that has a first cost today of $150,000, an annual operating cost of $60,000, and a salvage value of 20% of the first cost after 5 years; these estimates being in future dollars. Assume the interest rate is 10% per year and that inflation has averaged 7% per year. Solve *by hand* and *spreadsheet* with inflation (*a*) not accounted for, and (*b*) accounted for.

14.26 Some of the following future cash flows have been expressed in then-current (future) dollars and others in CV dollars. Use an interest rate of 10% per year and an inflation rate of 6% per year. (*a*) Find the present worth. (*b*) Use a spreadsheet to find the

PW value using one NPV function with all cash flows expressed as future dollars.

Year	Cash Flow, $	Expressed as
0	16,000	CV
3	40,000	Then-current
4	12,000	Then-current
7	26,000	CV

14.27 The company you work for is considering a new product line projected to have the net cash flows (NCF) shown. The values are in future dollars, which have been inflated by 5% per year. Because the plant manager is unsure how the present worth is calculated, he asked you to do it in two ways over the 4-year planning horizon: (1) using the company's *market* rate of 20% per year, and (2) converting all of the estimates into CV dollars and using the company's *real* rate. You said both ways will provide the same answer, but he asked you to show him the calculations. What is the present worth by method (1) and method (2)? Solve by *hand* and *spreadsheet*.

Year	0	1	2	3	4
NCF, $1000	−10,000	2000	5000	5000	5000

14.28 A regional infrastructure building and maintenance contractor must decide to buy a new compact horizontal directional drilling (HDD) machine now, or wait and buy it 2 years from now when a large pipeline contract will require the new equipment. The HDD machine will include an innovative pipeloader design and a maneuverable undercarriage system. The cost of the system is $68,000 if purchased now or an estimated $81,000 if purchased 2 years from now. At $i = 10\%$ per year and $f = 5\%$ per year, determine if the contractor should buy now or later (*a*) without any adjustment for inflation, and (*b*) with inflation considered.

14.29 Joan, the project manager, asks you to evaluate alternatives A and B on the basis of their PW values using a real interest rate of 10% per year and an inflation rate of 3% per year (*a*) without any adjustment for inflation, and (*b*) with inflation considered. Also, write the spreadsheet functions that will display the correct PW values. (*c*) Joan clearly wants alternative A to be selected. If inflation is steady at 3% per year, what real return i would machine A have to generate each year to make the choice between A and B indifferent? What is the required return with inflation considered?

Machine	A	B
First cost, $	−31,000	−48,000
AOC, $ per year	−28,000	−19,000
Salvage, $	5,000	7,000
Life, years	5	5

14.30 Compare the alternatives below on the basis of their capitalized costs with adjustments made for inflation. Use $i = 12\%$ per year and $f = 3\%$ per year.

Alternative	X	Y
First cost, $	−18,500,000	−9,000,000
AOC, $ per year	−25,000	−10,000
Salvage value, $	105,000	82,000
Life, years	∞	10

14.31 A grandfather is planning to leave his only granddaughter well off when she reaches the age of 25. He plans to deposit a lump sum now, which is her second birthday, such that she will have enough money to live comfortably without working. He wants her to have an amount that would have the same purchasing power as $2 million today. If he can invest the money now and earn an average market interest rate of 8% per year while the inflation rate averages 4% per year, what amount must he deposit?

14.32 A doctor is on contract to a medium-sized oil company to provide medical services at remotely located, widely separated refineries. The doctor is considering the purchase of a private plane to reduce the total travel time between refineries. The doctor can buy a used Learjet 31A now for $1.1 million or wait for a new very light jet (VLJ) that will be available 3 years from now. The cost of the VLJ will be $2.1 million, payable when the plane is delivered in 3 years. The doctor has asked you, his friend, to determine the present worth of the VLJ so that he can decide to buy the used Learjet now or wait for the VLJ. The MARR is 15% per year and the inflation rate is projected to be 3% per year. (*a*) What is the present worth of the VLJ with inflation considered? (*b*) Which plane should he buy?

14.33 A salesman from Industrial Water Services (IWS), who is trying to get his foot in the door of Westco Refining, offered electro-dialysis equipment for $2.5 million. This is $800,000 more than the price offered by a competing saleswoman from AG Enterprises. However, IWS said Westco won't have to pay for the equipment until the 2-year warranty runs out. IWS will also offer an *extended* 2-year warranty for $100,000, payable 2 years from now. If Westco does want the extended warranty, determine which offer is better using Westco's real return requirement of 15% per year and an assumed inflation rate of 3.5% per year.

14.34 As an innovative way to pay for various software packages that your company sells, a high-tech service company has offered to pay your company in any one of three ways: (1) pay $450,000 now;

(2) pay $1.1 million 5 years from now; or (3) pay $200,000 now and $400,000 two years from now. You want to earn a real return of 10% per year and the inflation rate is 6% per year in the specialized software market. (*a*) Use PW analysis to determine which offer you should accept with inflation considered. (*b*) Determine the market return rates at which all three methods have the same PW value.

Future Worth and Other Calculations with Inflation

14.35 The company you work for signed a contract with a security firm to control access to company grounds and corporate offices. The contract amount was for $140,000 for the first year, renewable each year for up to a total of 5 years at the same cost plus a percentage increase equal to the inflation rate for the preceding year. Your boss asked you to calculate the expected cost in the final year of the contract (year 5), assuming the inflation rate is 3% for the next 3 years and 5% for the last one.

14.36 A plant manager is not sure whether he will get the approval to buy new equipment for automating an engine assembly line now or at some future time within the next 3 years. In order to have the money whenever he is given the go-ahead, he has asked you to tell him what the equipment is likely to cost in each of the next 3 years. The cost of the equipment today is $300,000, and the MARR is 15% per year. (*a*) How much will it cost at the end of years 1, 2, and 3 if the cost increases only by the inflation rate of 4% per year? (*b*) What case is this from the descriptions in Section 14.3?

14.37 An engineer planning for her son's college education made deposits into a separate high-risk brokerage account every time she earned extra money from side consulting jobs. The deposits and their timings are as follows:

Year	Deposit, $
0	5,000
3	8,000
4	9,000
7	15,000
11	16,000
17	20,000

(*a*) If the account increased at 15% per year and inflation averaged 3% per year over the entire period, what was the *purchasing power* of the money in the account in terms of year zero CV dollars immediately after the last deposit in year 17?

(*b*) Use a spreadsheet to recalculate the purchasing power if the account actually earned 6% per year and inflation was higher than expected at 4% per year. Compare the purchasing power with the total amount deposited over the 17 years.

14.38 The cost of constructing a roundabout (R/A) in a low-traffic residential neighborhood 5 years ago was $625,000. A civil engineer designing another R/A that is almost the same design estimates the cost today will be $740,000. If the cost had increased only by the inflation rate over the 5 years, determine the inflation rate per year.

14.39 An engineer deposits $10,000 into an account when the market interest rate is 10% per year and the inflation rate is 5% per year. If the account is left undisturbed for 5 years,
(*a*) How much money will be in the account?
(*b*) What will be the purchasing power in terms of today's dollars?
(*c*) What is the real rate of return on the account?

14.40 Factors that increase costs and prices—especially for materials and manufacturing costs sensitive to market, technology, and labor availability—can be considered separately using the real interest rate, i, the inflation rate, f, and additional increases that grow at a geometric rate, g (commonly due to maintenance and repair cost increases as machinery ages). The future amount is calculated based on a current estimate by using the relation:

$$F = P(1 + i)^n (1 + f)^n (1 + g)^n = P[(1 + i)(1 + f)(1 + g)]^n$$

The current cost to manufacture an electronic subcomponent is $145,000 per year. Provided the average annual rates for i, f, and g are 8%, 4%, and 3%, respectively, determine the equivalent future cost (*a*) in 3 years, and (*b*) in 8 years.

14.41 The Nobel Prize is administered by The Nobel Foundation, a private institution that was founded in 1900 based on the will of Alfred Nobel, the inventor of dynamite. In part, his will stated: "The capital shall be invested by my executors in safe securities and shall constitute a fund, the interest on which shall be annually distributed in the form of prizes to those who, during the preceding year, shall have conferred the greatest benefit on mankind." The will further stated that the prizes were to be awarded in physics, chemistry, peace, physiology or medicine, and literature. In addition to a gold medal and a diploma, each recipient receives a substantial sum of money that depends on the Foundation's income that year. The first Nobel Prize was awarded in 1901 in the amount of $150,000. In 1996, the award was $653,000; it was $1.03 million in 2014.
(*a*) If the increase between 1996 and 2014 was strictly due to inflation, what was the average inflation rate per year during that 18-year period?

(*b*) If the Foundation expects to invest money with a return of 5% above the inflation rate, how much will a laureate receive in 2020, provided the inflation rate averages 3% per year between 2014 and 2020?

14.42 To retire at a decent age and move to Hawaii, an engineer plans to trust her account to an investment firm that promises to make a real rate of return of 10% per year when the inflation rate is 4% per year. If the account currently is valued at $422,000 and she wants to retire in 15 years, how much (in then-current dollars) will have to be in the account for the realized rate of return to be a real 10% per year? Also, write a single-cell spreadsheet function to display the answer.

14.43 An offshore services company is considering the purchase of equipment that has a cost today of $96,000. Inflation is a concern. The manufacturer plans to raise the price exactly in accordance with the inflation rate that may be somewhere between 1% and 8% per year. Develop a graph of how much the equipment will cost 3 years from now in terms of both (*a*) *CV*, and (*b*) *future* dollars.

14.44 You just made an investment in an insurance policy that is guaranteed to pay you $1.8 million 20 years from now provided you live that long. What will be the *purchasing power* of that amount with respect to today's dollars if the market interest rate is 8% per year and the inflation rate stays at 3.8% per year over the 20-year period?

14.45 A top-of-the-line 3D printer costs $40,000 today. The manufacturer plans to raise the price so that a real rate of return of 5% per year is realized. In a period of 4% per year inflation, how much will the printer cost 3 years from now in terms of *CV dollars*?

14.46 Well-managed companies set aside money to pay for emergencies that inevitably arise in the course of doing business. A commercial solid-waste recycling and disposal company in Mexico City puts 0.5% of its after-tax income into such an account. (*a*) How much will the company have after 7 years if after-tax income averages $15.2 million and inflation and market interest rates are 5% per year and 9% per year, respectively? (*b*) What will be the buying power of that amount in today's dollars?

14.47 A company has been invited to invest $1 million in a partnership and receive a guaranteed total of $2.5 million after 4 years. By corporate policy, the MARR is always established at 4% above the real cost of capital. The real interest rate paid on capital is currently 10% per year and the inflation rate during the 4-year period is expected to average 3% per year. (*a*) Is the investment economically justified? (*b*) At what real interest rate on capital will the decision made above change?

Capital Recovery with Inflation

14.48 Aquatech Microsystems spent $183,000 for a communications protocol to achieve interoperability among its utility systems. The company uses a real interest rate of 15% per year on such investments and a recovery period of 5 years. (*a*) What is the annual worth of the expenditure in future dollars at an inflation rate of 6% per year? (*b*) Write a single-cell spreadsheet function to display the correct AW value.

14.49 A DSL company has made an equipment investment of $40 million with the expectation that it will be recovered in 10 years. The company has a MARR based on a real rate of return of 12% per year. If inflation is 7% per year, how much must the company make each year (*a*) in constant-value dollars, and (*b*) in future dollars, to meet its expectation?

14.50 A European-based cattle genetics engineering research lab is planning for a major expenditure on research equipment. The lab needs $5 million of today's dollars so that it can make the acquisition 4 years from now. The inflation rate is steady at 5% per year. (*a*) How many future dollars will be needed when the equipment is purchased if purchasing power is maintained? (*b*) What is the required amount of the annual deposit into a fund that earns the market rate of 10% per year to ensure that the amount calculated in part (*a*) is accumulated? (*c*) Write a single-cell spreadsheet function that immediately displays the correct annual deposit required.

14.51 In wisely planning for your retirement, you invest $12,000 per year for 20 years into a 401-k investment account. (*a*) How much can you withdraw each year for 10 years, starting 1 year after your last deposit, if you want a real return of 10% per year when the inflation rate averages 2.8% per year? (*b*) For a spreadsheet function challenge, write a single-cell function that displays the correct 10-year annual withdrawal amount directly.

14.52 For a present sum of $750,000, what is the annual worth (in then-current dollars) in years 1 through 5 if the market interest rate is 10% per year and the inflation rate is 5% per year?

14.53 A recently graduated mechanical engineer wants to build a reserve fund as a safety net to pay his expenses in the unlikely event that an unexpected emergency arises. His aim is to have $45,000 developed over the next 3 years, with the proviso that the amount must have the same purchasing power as $45,000 today. If the expected market rate on investments is 8% per year and inflation is averaging 2% per year, find the annual amount necessary to meet his goal.

14.54 A multinational security software company is planning an overseas expansion that will cost $50 million of today's dollars 3 years from now. Due to a robust economy in Europe, the cost is expected to increase by 15% per year in each of the next 3 years. Assuming the inflation rate is 4% per year, determine the required annual deposit into a fund that earns the market rate of 10% per year to ensure that the amount needed in 3 years will be available. Also, write the spreadsheet function that displays the annual deposit directly.

14.55 (a) Calculate the perpetual equivalent annual worth in future dollars for years 1 through ∞ for income of $50,000 now and $5000 per year thereafter. Assume the market interest rate is 8% per year and inflation averages 4% per year. All amounts are quoted as future dollars. (b) If the

amounts had been quoted in CV dollars, what is the annual worth in future dollars?

14.56 The two machines shown are being considered for a chip manufacturing operation. Assume the MARR is a real return of 12% per year and that the inflation rate is 7% per year. Which machine should be selected on the basis of an annual worth analysis if the estimates are in (a) constant-value dollars, and (b) future dollars? Solve *by hand* and using a *spreadsheet*.

Machine	A	B
First cost, $	−150,000	−1,025,000
M&O, $ per year	−70,000	−5,000
Salvage value, $	40,000	200,000
Life, years	5	∞

ADDITIONAL PROBLEMS AND FE EXAM REVIEW QUESTIONS

14.57 For a real interest rate of 12% per year and an inflation rate of 7% per year, the market interest rate per year is closest to:
(a) 5.7%   (b) 7%   (c) 12%   (d) 19.8%

14.58 When all future cash flows are expressed in then-current dollars, the rate that should be used to find the present worth is the:
(a) Real MARR   (b) Inflation rate
(c) Inflated interest rate   (d) Real interest rate

14.59 In order to convert inflated dollars into constant-value dollars, it is necessary to:
(a) Divide by $(1 + i_f)^n$
(b) Divide by $(1 + f)^n$
(c) Divide by $(1 + i)^n$
(d) Multiply by $(1 + f)^n$

14.60 When the market interest rate is less than the real interest rate, then:
(a) The inflated interest rate is higher than the real interest rate
(b) The real interest rate is zero
(c) A deflationary condition exists
(d) An inflationary condition exists

14.61 The cost of a well-equipped F-150 truck was $29,350 three years ago. If the cost increased only by the inflation rate and the price today is $33,015, the inflation rate over the 3-year period was closest to:
(a) 3%   (b) 4%   (c) 5%   (d) 6%

14.62 If the market interest rate is 12% per year and the inflation rate is 5% per year, the number of future dollars in year 7 that will be equivalent to $2000 now is best represented by the equation:
(a) Future dollar amount = $2000(1 + 0.198)^7$
(b) Future dollar amount = $2000/(1.198)^7$

(c) Future dollar amount = $2000(1 + 0.12)^7$
(d) Future dollar amount = $2000/(1.07)^7$

14.63 An investment of $1000 was made 25 years ago. The amount available 10 years from now at the market interest rate of 5% per year and an inflation rate of 2% per year is closest to:
(a) $3085   (b) $5430
(c) $5515   (d) $35,000

14.64 You expect to receive an inheritance of $50,000 six years from now. Its present worth at a real interest rate of 4% per year and an inflation rate of 3% per year is closest to:
(a) $27,600   (b) $29,800
(c) $33,100   (d) $50,000

14.65 For a real interest rate of 1% per month and an inflation rate of 1% per month, the nominal inflated interest rate *per year* is closest to:
(a) 1%   (b) 2%   (c) 24%   (d) 24.12%

14.66 An industrial robot with a controller, teach pendants, and job-specific peripherals has an initial cost of $85,000 now, and the future cost will increase exactly by the inflation rate. The cost of a similar robot 3 years from now at a market interest rate of 10% per year and inflation rate of 4% per year is closest to:
(a) $95,610   (b) $102,414
(c) $125,931   (d) $127,261

14.67 Assume you save $6000 each year starting this year until your planned retirement 40 years from now. The buying power of the money in terms of today's dollars at the market interest rate of 10% per year and inflation rate of 5% per year is closest to:
(a) $377,200   (b) $605,350
(c) $1,318,150   (d) $2,655,550

CASE STUDY

INFLATION CONSIDERATIONS FOR STOCK AND BOND INVESTMENTS

Background

The savings and investments that an individual maintains should have some balance between equity (corporate stocks that rely on market growth and dividend income) and fixed-income investments (bonds that pay dividends to the purchaser and a guaranteed amount upon maturity). When inflation is moderately high, bonds offer a low return relative to stocks because the potential for market growth is not present with bonds. Additionally, the forces of inflation make the dividends worth less in future years because for most bonds there is no inflation adjustment made in the amount the dividend pays as time passes. However, bonds do offer a steady income that may be important to an individual, and they serve to preserve the principal invested in the bond because the face value is returned at maturity.

Information

Earl is an engineer who wants a predictable flow of money for travel and vacations. He has a collection of stocks in his retirement portfolio, but no bonds. He has accumulated a total of $50,000 of his own funds in low-yielding savings accounts and wants to improve his long-term return from this nonretirement program "nest egg." He can choose additional stocks or bonds, but has decided to not split the $50,000 between the two forms of investments. There are two choices he has outlined, with the best estimates he can make at this time. He assumes the effects of federal and state income taxes will be the same for both forms of investment.

Stock purchase: Stocks purchased through a mutual fund would pay an estimated 2% per year dividend and appreciate in value at 5% per year.

Bond purchase: If he purchased a bond, he would have a predictable income of 5% per year and the $50,000 face value after the 12-year maturity period.

Case Study Questions

The analysis that Earl has laid out has the following questions. Can you answer them for him for both choices?

1. What is the overall rate of return after 12 years?
2. If he decided to sell the stock or bond immediately after the fifth annual dividend, what is his minimum selling price to realize a 7% real return? Include an adjustment of 4% per year for inflation.
3. If Earl needed some money in the future, say, immediately after the fifth dividend payment, what would be the minimum selling price in future dollars, if he were only interested in recovering an amount that maintained the purchasing power of the original price?
4. As a follow-on to question 3, what happens to the selling price (in future dollars) 5 years after purchase, if Earl is willing to remove (net out) the future purchasing power of each of the dividends in the computation to determine the required selling price 5 years hence?
5. Earl plans to keep the stocks or bonds for 12 years, that is, until the bond matures. However, he wants to make the 7% per year real return and make up for the expected 4% per year inflation. For what amount must he sell the stocks after 12 years, or buy the bonds now, to ensure he realizes this return? Do these amounts seem reasonable to you, given your knowledge of the way that stocks and bonds are bought and sold?

CHAPTER 15

Cost Estimation and Indirect Cost Allocation

Shutterstock/Rawpixel.com

LEARNING OUTCOMES

Purpose: Make cost estimates using different methods; demonstrate the allocation of indirect costs using traditional and activity-based costing rates.

SECTION	TOPIC	LEARNING OUTCOME
15.1	Approach	• Explain the bottom-up and design-to-cost (top-down) approaches to cost estimation.
15.2	Unit method	• Use the unit method to make a preliminary cost estimate.
15.3	Cost index	• Use a cost index to estimate a present cost based on historical data.
15.4	Cost capacity	• Use a cost-capacity equation to estimate component, system, or plant costs.
15.5	Factor method	• Estimate total plant cost using the factor method.
15.6	Indirect cost rates	• Allocate indirect costs using traditional indirect cost rates.
15.7	ABC allocation	• Use the Activity-Based Costing (ABC) method to allocate indirect costs.
15.8	Ethics and profit	• Describe how biased estimation can become an ethical dilemma.

U p to this point, cost and revenue cash flow values have been stated or assumed as known. In reality, they are not; they must be estimated. This chapter explains what cost estimation involves and applies cost estimation techniques. **Cost estimation** is important in all aspects of a project, but especially in the stages of project conception, preliminary design, detailed design, and economic analysis. When a project is developed in the private or the public sector, questions about costs and revenues will be posed by individuals representing many different functions: management, engineering, construction, production, quality, finance, safety, environmental, legal, and marketing, to name some. In engineering practice, the estimation of costs receives much more attention than revenue estimation; costs are the topic of this chapter.

Unlike direct costs for labor and materials, indirect costs are not easily traced to a specific department, machine, or processing line. Therefore, **allocation of indirect costs** for functions such as utilities, safety, management and administration, purchasing, and quality is made using some rational basis. Both the traditional method of allocation and the Activity-Based Costing (ABC) method are covered in this chapter.

15.1 Understanding How Cost Estimation Is Accomplished ●●●

Cost estimation is a major activity performed in the initial stages of virtually every effort in industry, business, and government. In general, most cost estimates are developed for either a *project* or a *system;* however, combinations of these are very common. A **project** usually involves physical items, such as a building, bridge, manufacturing plant, or offshore drilling platform, to name just a few. A **system** is usually an operational design that involves processes, services, software, and other nonphysical items. Examples might be a purchase order system, a software package, an Internet-based remote-control system, or a health care delivery system. Of course, many projects will have major elements that are not physical, so estimates of both types must be developed. For example, consider a computer network system. There would be no operational system if only the costs of computer hardware plus wire and wireless connectors were estimated; it is equally important to estimate the software, personnel, and maintenance costs.

Thus far virtually all cash flow estimates in the examples, problems, progressive examples, and case studies were stated or assumed to be known. In real-world practice, the cash flows for costs and revenues must be estimated prior to the evaluation of a project or comparison of alternatives. We concentrate on cost estimation because costs are the primary values estimated for the economic analysis. Revenue estimates utilized by engineers are usually developed in marketing, sales, and other departments, and are often assumed to be equal for all alternatives being evaluated.

Costs are comprised of **direct costs** and **indirect costs.** Normally direct costs are estimated with some detail, then the indirect costs are added using standard rates and factors. However, direct costs in many industries, including manufacturing and assembly settings, have become a small percentage of overall product cost, while indirect costs have become much larger. Accordingly, many industrial settings require some estimating for indirect costs as well. Indirect cost allocation is discussed in detail in later sections of this chapter. Primarily, direct costs are discussed here.

Direct/Indirect costs

Because cost estimation is a complex activity, the following questions form a structure for the discussion that follows:

- What cost *components* must be estimated?
- What *strategy* to cost estimation will be applied?
- How *accurate* should the estimates be?
- What estimation *techniques* will be utilized?

Costs to Estimate If a project revolves around a single piece of equipment, for example, biofiltration equipment or an industrial robot, the *cost components* will be significantly simpler and fewer than the components for a complete system such as the manufacturing and testing line for a new product. Therefore, it is important to know up-front how much the cost estimation task will involve. Examples of cost components are the first cost P and the annual operating cost (AOC), also called the M&O costs (maintenance and operating) of equipment. Each component will have several *cost elements*. Listed below are sample elements of the first cost and AOC components.

> *First cost component P:*
>   Elements: Equipment cost
>   Delivery charges
>   Installation cost
>   Insurance coverage
>   Initial training of personnel for equipment use

Delivered-equipment cost is the sum of the first two elements in the list above; **installed-equipment cost** adds the third element.

> *AOC component* (part of the equivalent annual cost A):
>   Elements: Direct labor cost for operating personnel
>   Direct materials
>   Maintenance costs (daily, periodic, repairs, etc.)
>   Rework and rebuild

Some of these elements, such as equipment cost, can be determined with high accuracy; others, such as maintenance costs, are harder to estimate. When costs for an entire system must be estimated, the number of cost components and elements is likely to be in the hundreds. It is then necessary to prioritize the estimation tasks.

For familiar projects (houses, office buildings, highways, and some chemical plants) there are standard cost estimation software packages available. For example, state highway departments utilize software that prompts for the cost components (bridges, pavement, cut-and-fill profiles, etc.) and estimates costs with time-proven, built-in relations. Once these components are estimated, exceptions for the specific project are added. However, there are no "canned" software packages for a large percentage of industrial, business, service, and public sector projects.

Cost Estimation Strategies   In many industries, and in the public sector, a **bottom-up** strategy, also called the engineering estimation approach, to cost estimation has been historically applied. For a simple rendition of this approach, see Figure 15–1 (left side). The progression is as follows: cost components and their elements are identified, cost elements are estimated, and estimates are summed to obtain total direct cost. The price is then determined by adding indirect costs and the profit margin, which is usually a percentage of the total cost.

> The **bottom-up approach** treats the required price as an **output variable** and the cost estimates as input variables. This approach works well when competition is not a dominant factor in pricing the product or service and there is time and money available to make the estimates.

Figure 15–1 (right side) shows a simplistic progression for the **design-to-cost,** or top-down strategy. The competitive price establishes the target cost.

> The **design-to-cost,** or **top-down strategy,** treats the competitive price as an **input variable** and the cost estimates as output variables. This approach is useful in encouraging innovation, new design, manufacturing process improvement, and efficiency. These are some of the essentials of *value engineering* and value-added systems engineering.

This second strategy places greater emphasis on the accuracy of the price estimation activity. The target cost must be realistic, or else it can be a disincentive to design and engineering staff. The design-to-cost strategy is best applied in the early stages of a new or enhanced product design. The detailed design and specific equipment options are not yet known, but the price estimates assist in establishing target costs for different components.

Usually, the applied strategy is some combination of these two philosophies. However, it is helpful to understand from the beginning of the estimation effort which strategy will be emphasized. Historically, the bottom-up approach was more predominant in Western engineering cultures, especially in North America. The design-to-cost approach is considered routine in Eastern

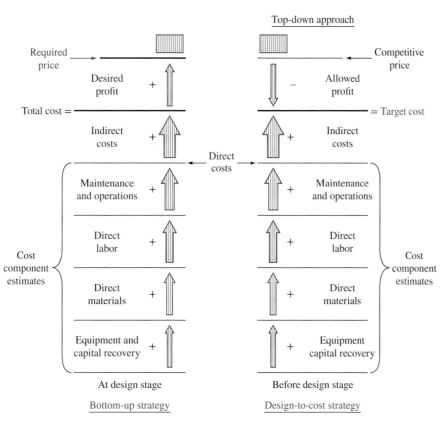

Figure 15–1
Simplified cost estimation processes for bottom-up and top-down approaches.

engineering cultures; however, globalization of engineering design and pursuit of international competitiveness have speeded the adoption of the design-to-cost strategy worldwide.

Accuracy of the Estimates   No cost estimates are expected to be exact; however, they are expected to be reasonable and accurate enough to support economic scrutiny. The accuracy required increases as the project progresses from preliminary design to detailed design and on to economic evaluation. Cost estimates made before and during the preliminary design stage are expected to be good "first-cut" estimates that serve as input to the project budget.

When utilized at early and conceptual design stages, estimates are referred to as **order-of-magnitude** estimates and generally range within ±20% of actual cost. At the detailed design stage, cost estimates are expected to be accurate enough to support economic evaluation for a go–no go decision. Every project setting has its own characteristics, but a range of ±5% of actual costs is expected at the detailed design stage. Figure 15–2 shows the general range of estimate accuracy for the construction cost of a building versus time spent in preparing the estimate. Obviously, the desire for better accuracy has to be balanced against the cost of obtaining it.

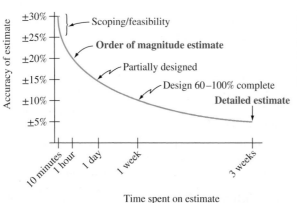

Figure 15–2
Characteristic curve of estimate accuracy versus time spent to estimate construction cost of a building.

Cost Estimation Techniques Methods such as expert opinion and comparison with comparable installations serve as excellent estimators. The use of the *unit method* and *cost indexes* base the present estimate on past cost experiences, with inflation considered. Models such as *cost-capacity equations* and the *factor method* are simple mathematical techniques applied at the preliminary design stage. They are called **cost estimating relationships (CERs).** There are many additional methods discussed in the handbooks and publications of different industries.

Most cost estimates made in a professional setting are accomplished in part or wholly using software packages linked to updated databases that contain cost indexes and rates for the locations, products, or processes being studied. There are a wide variety of estimators, cost trackers, and cost compliance software systems, most of them developed for specific industries. Corporations usually standardize on one or two packages to ensure consistency over time and projects.

15.2 Unit Method ● ● ●

The **unit method** is a popular preliminary estimation technique applicable to virtually all professions. The total estimated cost C_T is obtained by multiplying the number of units N by a per unit cost factor u.

$$C_T = u \times N \qquad\qquad [15.1]$$

Unit cost factors must be updated frequently to remain current with changing costs, areas, and inflation. Some sample unit cost factors (and values) are

    Total average cost of operating an automobile (55¢ per mile or 33¢ per kilometer)

    Cost to bury fiber cable in a suburban area ($30,000 per mile)

    Cost to construct a parking space in a parking garage ($4500 per space)

    Cost of constructing interstate highway ($6.2 million per mile)

    Cost of house construction per livable area ($225 per square foot)

Applications of the unit method to estimate costs are easily found. If house construction costs average $225 per square foot, a preliminary cost estimate for an 1800-square-foot house, using Equation [15.1], is $405,000. Similarly, a 200-km trip should cost about $66 for the car only at 33¢ per km.

When there are several components to a project or system, the unit cost factors for each component are multiplied by the amount of resources needed, and the results are **summed** to obtain the total cost C_T. This is illustrated in Example 15.1.

EXAMPLE 15.1

Justin, an ME with Dynamic Castings, has been asked to make a preliminary estimate of the total cost to manufacture 1500 sections of high-pressure gas pipe using an advanced centrifugal casting method. Since a ±20% estimate is acceptable at this preliminary stage, a unit method estimate is sufficient. Use the following resource and unit cost factor estimates to help Justin.

    Materials: 3000 tons at $45.90 per ton

    Machinery and tooling: 1500 hours at $120 per hour

    Direct labor in plant:

        Casting and treating: 3000 hours at $55 per hour

        Finishing and shipping: 1200 hours at $45 per hour

    Indirect labor: 400 hours at $75 per hour

Solution

Apply Equation [15.1] to each of the five areas and sum the results to obtain the total cost estimate of $566,700. Table 15–1 provides the details.

TABLE 15–1	Total Cost Estimate Using Unit Cost Factors for Several Resource Areas, Example 15.1		
Resource	**Amount *N***	**Unit Cost Factor *u*, $**	**Cost Estimate, *u* × *N*, $**
Materials	3000 tons	45.90 per ton	137,700
Machinery, tooling	1500 hours	120 per hour	180,000
Labor, casting	3000 hours	55 per hour	165,000
Labor, finishing	1200 hours	45 per hour	54,000
Labor, indirect	400 hours	75 per hour	30,000
Total cost estimate			**566,700**

15.3 Cost Indexes ● ● ●

This section explains indexes and their use in cost estimation. An *index* is a ratio or other number based on observation and used as an indicator or measure. A preliminary cost estimate is often based on a cost index.

A **cost index** is a ratio of the cost of something today to its cost sometime in the past. The index is **dimensionless** and **measures relative cost change over time.** Because these indexes are sensitive to technological change, the predefined quantity and quality of elements used to define the index may be hard to retain over time, thus causing "index creep." Timely updating of the index is very important.

One such index that most people are familiar with is the Consumer Price Index (CPI), which shows the relationship between present and past costs for many of the things that "typical" consumers must buy. This index includes such items as rent, food, transportation, and certain services. Other indexes track the costs of equipment, and goods and services that are more pertinent to the engineering disciplines. Table 15–2 is a listing of some of the more common indexes.

TABLE 15–2	Types and Sources of Various Cost Indexes
Type of Index	**Source**
Overall prices	
Consumer (CPI)	Bureau of Labor Statistics
Producer (wholesale)	U.S. Department of Labor
Construction	
Chemical plant overall	*Chemical Engineering*
Equipment, machinery, and supports	
Construction labor	
Buildings	
Engineering and supervision	
Engineering News Record overall	*Engineering News Record* (*ENR*)
Construction	(www.construction.com)
Building	
Common labor	
Skilled labor	
Materials	
FRED indexes	Federal Reserve Economic Data
Consumer price index (CPI)	(www.research.stlouisfed.org)
Producer price index (PPI)	
Construction	
House prices	
Import price indexes	
Equipment	
Marshall and Swift (M&S) overall	Marshall & Swift
M&S specific industries	
Labor	
Output per worker-hour by industry	U.S. Department of Labor

TABLE 15–3   Samples of Cost Index values

Year	CE Plant Cost Index (CEPCI) (1913 = 100)	Consumer Price Index (CPI) (1982–84 = 100)	Crude Oil Index (2005 = 100)	Pumps, Compressors, and Equipment Index (1982 = 100)
2008	575.4	211.4	105.1	210.0
2009	521.9	217.3	141.9	213.4
2010	550.8	220.5	160.6	217.0
2011	585.7	227.2	194.4	225.3
2012	584.6	231.3	192.3	230.3
2013	567.3	234.8	196.8	236.3
2014	579.8	236.5	140.2	242.5
2015	556.8	238.0	79.3	245.9

The general equation for updating costs through the use of a cost index over a period from time $t = 0$ (base) to another time t is

$$C_t = C_0 \left(\frac{I_t}{I_0} \right) \qquad\qquad [15.2]$$

where C_t = estimated cost at present time t
C_0 = cost at previous time t_0
I_t = index value at time t
I_0 = index value at time t_0

Generally, the indexes for equipment and materials are made up of a mix of components that are assigned certain weights, with the components sometimes further subdivided into more basic items. For example, the equipment, machinery, and support component of the chemical plant cost index is subdivided into process machinery, pipes, valves and fittings, pumps and compressors, and so forth. These subcomponents, in turn, are built up from even more basic items such as pressure pipe, black pipe, and galvanized pipe. Table 15–3 shows the *Chemical Engineering* plant cost index (CEPCI), and several indexes from FRED, including the consumer price index (CPI), crude oil index, and an equipment index from 2008 forward. The base period for these and most indexes varies considerably, as indicated in the table.

Current and past values of several of the indexes may be obtained from the Internet (usually for a fee). For example, the *CE* plant cost index is available at www.che.com/pci. The *ENR* construction cost index mentioned in Table 15–2 offers a comprehensive series of construction-related resources, including several *ENR* cost indexes and cost estimation systems. A website used by many engineering professionals in the form of a "technical chat room" for all types of topics, including estimation, is www.eng-tips.com.

EXAMPLE 15.2

In evaluating the feasibility of a major construction project, an engineer is interested in estimating the cost of skilled labor for the job. The engineer finds that a project of similar complexity and magnitude was completed 5 years ago at a skilled labor cost of $360,000. The *ENR* skilled labor index was 3496 then and is now 5127. What is the estimated skilled labor cost for the new project?

Solution

The base time t_0 is 5 years ago. Using Equation [15.2], the present cost estimate is

$$C_r = 360,000 \left(\frac{5127}{3496} \right)$$

$$= \$527,952$$

In the manufacturing and service industries, tabulated cost indexes are not readily available. The cost index will vary, perhaps with the region of the country, the type of product or service, and many other factors. When estimating costs for a manufacturing system, for example, it is often necessary to develop the cost index for high-priority items such as subcontracted components, selected materials, and labor costs. The development of the cost index requires the actual cost at different times for a prescribed quantity and quality of the item. The **base period** is a selected time when the index is defined with a basis value of 100 (or 1). The index each year (period) is determined as the cost divided by the base-year cost and multiplied by 100 (or 1). Future index values may be forecast using simple extrapolation or more refined mathematical techniques, such as time-series analysis.

EXAMPLE 15.3

Sean, owner of Alamo Pictures, makes fact-based documentaries about the Old West and sells them in a variety of outlets, by mail, and online. He has decided to expand into new areas and wants to make cost estimates for three of the more significant labor costs involved in making these types of films. The Director of Finance of Alamo Pictures generated the annual average hourly costs over the last 7 years (Table 15–4).

(a) Make 2014 the base year, and determine the cost indexes using a basis of 1.00. Comment on the trend of each index over the years.
(b) Sean expects to utilize a lot of graphics services in future years for planned documentaries, especially as 2020 approaches; however, it is the most rapidly increasing cost component. The cost in 2016 was \$78 per hour; assume a worst-case scenario such that the graphics index continues the same arithmetic trend it had from 2016 to 2017. Determine the hourly cost that he should budget for in 2020.

Solution

(a) For each type of service, calculate I_t/I_0 where $t = 2011, 2012, \ldots$ with 2014 as the base year 0. Table 15–5 presents the indexes. Observations about trend are as follows:

> Graphics labor cost: Constantly increasing over all years.
>
> Stuntmen labor cost: Rising until 2016, then stable.
>
> Actors labor cost: Higher in 2013 and 2014; comparatively lower and stable in other years.

TABLE 15–4 Average Hourly Costs for Three Services, Example 15.3

Type of Service	Cost of Service, Average \$ per Hour						
	2011	2012	2013	2014	2015	2016	2017
Graphics	50	56	65	67	70	78	90
Stuntmen	50	55	60	70	87	83	85
Actors	80	80	90	90	80	75	85

TABLE 15–5 Index Values with 2014 as Base Year, Example 15.3

Type of Service	Index I_t/I_0						
	2011	2012	2013	2014	2015	2016	2017
Graphics	0.75	0.84	0.97	1.00	1.04	1.16	1.34
Stuntmen	0.71	0.79	0.86	1.00	1.24	1.19	1.21
Actors	0.89	0.89	1.00	1.00	0.89	0.83	0.94

(b) The index in 2016 is 1.16. The increase to 2017 is 0.18; the index value in 2020 will be 1.34 + 3(0.18) = 1.88. Equation [15.2] finds the expected, worst-case cost in 2020.

$$C_{2020} = C_{2016}(I_{2020}/I_{2016}) = 78(1.88/1.16)$$
$$= 78(1.62)$$
$$= \$126 \text{ per hour}$$

15.4 Cost-Estimating Relationships: Cost-Capacity Equations ●●●

Design variables (speed, weight, thrust, physical size, etc.) for plants, equipment, and construction are determined in the early design stages. **Cost-estimating relationships (CERs)** use these design variables to predict costs. Thus, a CER is generically different from the cost index method because the index is based on the cost history of a defined quantity and quality of a variable.

One of the most widely used CER models is a **cost-capacity equation.** As the name implies, an equation relates the cost of a component, system, or plant to its capacity. This is also known as the *power law and sizing model.* Since many cost-capacity equations plot as a straight line on log-log paper, a common form is

$$C_2 = C_1\left(\frac{Q_2}{Q_1}\right)^x \qquad\qquad [15.3]$$

where C_1 = cost at capacity Q_1
C_2 = cost at capacity Q_2
x = correlating exponent

The value of the exponent for various components, systems, or entire plants can be obtained or derived from a number of sources, including *Plant Design and Economics for Chemical Engineers, Preliminary Plant Design in Chemical Engineering, Chemical Engineers' Handbook,* technical journals (especially *Chemical Engineering*), the U.S. Environmental Protection Agency, professional or trade organizations, consulting firms, handbooks, and equipment companies. Table 15–6 is a partial listing of typical values of the exponent for various units. When an exponent value for a particular unit is not known, it is common practice to use the value of $x = 0.6$. In fact, in the chemical processing industry, Equation [15.3] is referred to as the six-tenths model.

The exponent x in the cost-capacity equation is commonly in the range $0 < x \le 1$.

If $x < 1$, *economies of scale* provide a cost advantage for larger sizes.

If $x = 1$, a *linear* relationship is present.

If $x > 1$, there are *diseconomies of scale* present in that a larger size is more costly than that of a linear relation.

It is especially powerful to combine the time adjustment of the cost index (I_t/I_0) from Equation [15.2] with a cost-capacity equation to estimate costs that change over time. If the index is embedded into the cost-capacity computation in Equation [15.3], the cost at time t and capacity level 2 may be written as the product of two independent terms.

$C_{2,t}$ = (cost at time 0 of level 2) × (time adjustment cost index)

$$= \left[C_{1,0}\left(\frac{Q_2}{Q_1}\right)^x\right]\left(\frac{I_t}{I_0}\right)$$

This is commonly expressed without the time subscripts. Thus,

$$C_2 = C_1\left(\frac{Q_2}{Q_1}\right)^x\left(\frac{I_t}{I_0}\right) \qquad\qquad [15.4]$$

The following example illustrates the use of this relation.

TABLE 15–6	Sample Exponent Values for Cost-Capacity Equations	
Component/System/Plant	Size Range	Exponent
Activated sludge plant	1–100 MGD	0.84
Aerobic digester	0.2–40 MGD	0.14
Blower	1000–7000 ft/min	0.46
Centrifuge	40–60 in	0.71
Chlorine plant	3000–350,000 tons/year	0.44
Clarifier	0.1–100 MGD	0.98
Compressor, reciprocating (air service)	5–300 hp	0.90
Compressor	200–2100 hp	0.32
Cyclone separator	20–8000 ft^3/min	0.64
Dryer	15–400 ft^2	0.71
Filter, sand	0.5–200 MGD	0.82
Heat exchanger	500–3000 ft^2	0.55
Hydrogen plant	500–20,000 scfd	0.56
Laboratory	0.05–50 MGD	1.02
Lagoon, aerated	0.05–20 MGD	1.13
Pump, centrifugal	10–200 hp	0.69
Reactor	50–4000 gal	0.74
Sludge drying beds	0.04–5 MGD	1.35
Stabilization pond	0.01–0.2 MGD	0.14
Tank, stainless	100–2000 gal	0.67

Note: MGD = million gallons per day; hp = horsepower; scfd = standard cubic feet per day.

EXAMPLE 15.4

The total design and construction cost for a digester to handle a flow rate of 0.5 million gallons per day (MGD) was $1.7 million in 2010. Estimate the cost today for a flow rate of 2.0 MGD. The exponent from Table 15–6 for the MGD range of 0.2 to 40 is 0.14. The cost index in 2010 of 131 has been updated to 225 for this year.

Solution

Equation [15.3] can estimate the cost of the larger system in 2010, but it must be updated by the cost index to today's dollars. Equation [15.4] performs both operations at once. The estimated cost is

$$C_2 = 1,700,000 \left(\frac{2.0}{0.5} \right)^{0.14} \left(\frac{225}{131} \right)$$
$$= 1,700,000(1.214)(1.718) = \$3.546 \text{ million}$$

15.5 Cost-Estimating Relationships: Factor Method ●●●

Another widely used model for preliminary cost estimates of process plants is called the **factor method.** While the methods discussed above can be used to estimate the costs of major items of equipment, processes, and the total plant costs, the factor method was developed specifically for **total plant costs.** The method is based on the premise that fairly reliable total plant costs can be obtained by multiplying the cost of the major equipment by certain factors. Since major equipment costs are readily available, rapid plant estimates are possible if the appropriate factors are known. These factors are commonly referred to as *Lang factors* after Hans J. Lang, who first proposed the method.

In its simplest form, the factor method is expressed in the same form as the unit method

$$C_T = hC_E \qquad [15.5]$$

where C_T = total plant cost
h = overall cost factor or sum of individual cost factors
C_E = total cost of major equipment

The h may be one overall cost factor or, more realistically, the sum of individual cost components such as construction, maintenance, direct labor, materials, and indirect cost elements. This follows the cost estimation approaches presented in Figure 15–1.

Direct/Indirect costs

In his original work, Lang showed that *direct cost factors* and *indirect cost factors* can be combined into one overall factor for some types of plants as follows: solid process plants, 3.10; solid-fluid process plants, 3.63; and fluid process plants, 4.74. These factors reveal that the total installed-plant cost is many times the first cost of the major equipment.

EXAMPLE 15.5

An engineer with Valero Petroleum has learned that an expansion of the solid-fluid process plant is expected to have a delivered equipment cost of $2.08 million. If the overall cost factor for this type of plant is 3.63, estimate the plant's total cost.

Solution

The total plant cost is estimated by Equation [15.5].

$$C_T = 3.63(2,080,000)$$
$$= \$7,550,400$$

Subsequent refinements of the factor method have led to the development of separate factors for direct and indirect cost components. Direct costs as discussed in Section 15.1 are specifically identifiable with a product, function, or process. Indirect costs are not directly attributable to a single function, but are shared by several because they are necessary to perform the overall objective. Examples of indirect costs are general administration, computer services, quality, safety, taxes, security, and a variety of support functions. The factors for both direct and indirect costs are sometimes developed for use with *delivered-equipment costs* and other times for *installed-equipment costs,* as defined in Section 15.1. In this text, we assume that all factors apply to delivered-equipment costs, unless otherwise specified.

For indirect costs, some of the factors apply to equipment costs only, while others apply to the total direct cost. In the former case, the simplest procedure is to add the direct and indirect cost factors before multiplying by the delivered-equipment cost. The overall cost factor h can be written as

$$h = 1 + \sum_{i=1}^{n} f_i \qquad [15.6]$$

where f_i = factor for each cost component
i = 1 to n components, including indirect cost

If the indirect cost factor is applied to the total direct cost, only the direct cost factors are added to obtain h. Therefore, Equation [15.5] is rewritten as

$$C_T = \left[C_E \left(1 + \sum_{i=1}^{n} f_i \right) \right] (1 + f_I) \qquad [15.7]$$

where f_I = indirect cost factor
f_i = factors for direct cost components only

Examples 15.6 and 15.7 illustrate these equations.

EXAMPLE 15.6

The delivered-equipment cost for a small chemical process plant is expected to be $2 million. If the direct cost factor is 1.61 and the indirect cost factor is 0.25, determine the total plant cost.

Solution

Since all factors apply to the delivered-equipment cost, they are added to obtain h, the total cost factor in Equation [15.6].

$$h = 1 + 1.61 + 0.25 = 2.86$$

The total plant cost is

$$C_T = 2.86(2,000,000) = \$5,720,000$$

EXAMPLE 15.7

A new container-handling crane at the Port of Singapore is expected to have a delivered-equipment cost of $875,000. The cost factor for the installation of tracks, concrete, steel, noise abatement, supports, etc., is 0.49. The construction factor is 0.53, and the indirect cost factor is 0.21. Determine the total cost if (*a*) all cost factors are applied to the cost of the delivered equipment, and (*b*) the indirect cost factor is applied to the total direct cost.

Solution

(*a*) Total equipment cost is $875,000. Since both the direct and indirect cost factors are applied to only the equipment cost, the overall cost factor from Equation [15.6] is

$$h = 1 + 0.49 + 0.53 + 0.21 = 2.23$$

The total cost is

$$C_T = 2.23(875,000) = \$1,951,250$$

(*b*) Now the total direct cost is calculated first, and Equation [15.7] is used to estimate total cost.

$$h = 1 + \sum_{i=1}^{n} f_i = 1 + 0.49 + 0.53 = 2.02$$

$$C_T = [875,000(2.02)](1.21) = \$2,138,675$$

Comment

Note the lower estimated cost when the indirect cost is applied to the equipment cost only in part (*a*). This illustrates the importance of determining exactly what the factors apply to before they are used.

15.6 Indirect Cost Rates and Allocation: The Traditional Method ● ● ●

Costs incurred in the production of an item or delivery of a service are tracked and assigned by a **cost accounting system.** For the manufacturing environment, it can be stated generally that the *statement of cost of goods sold* (discussed in Appendix B) is one end product of this system. The cost accounting system accumulates material costs, labor costs, and indirect costs (also called overhead costs or factory expenses) by using **cost centers.** All costs incurred in one department or process line are collected under a cost center title, for example, Department 3X. Since direct materials and direct labor are usually directly assignable to a cost center, the system need only identify and track these costs. Of course, this in itself is no easy chore, and the cost of the tracking system may prohibit collection of all direct cost data to the level of detail desired.

One of the primary and more difficult tasks of cost accounting is the allocation of **indirect costs** when it is necessary to allocate them separately to departments, processes, and product lines.

TABLE 15–7	Sample Indirect Cost Allocation Bases
Cost Category	**Possible Allocation Basis**
Taxes	Space occupied
Heat, light	Space, usage, number of outlets
Power	Space, direct labor hours, horsepower-hours, machine hours
Receiving, purchasing	Cost of materials, number of orders, number of items
Personnel, machine shop	Direct labor hours, direct labor cost
Building maintenance	Space occupied, direct labor cost
Software	Number of accesses
Quality control	Number of inspections

Indirect costs

Indirect costs (IDC) are costs associated with property taxes, service and maintenance departments, personnel, legal, quality, supervision, purchasing, utilities, software development, etc. They must all be allocated to the using cost center. Detailed collection of these data is cost-prohibitive and often impossible; thus, allocation schemes are utilized to distribute the expenses on a reasonable basis.

A listing of possible allocation bases is included in Table 15–7. Historically, common bases have been direct labor cost, direct labor hours, machine-hours, number of employees, space, and direct materials.

Traditionally, most allocation is accomplished utilizing a predetermined *indirect cost rate,* computed by using the following relation. The rate is also referred to as the IDC rate or overhead rate.

$$\text{Indirect cost rate} = \frac{\text{estimated total indirect costs}}{\text{estimated basis level}} \qquad [15.8]$$

The estimated indirect cost is the amount allocated to a cost center. For example, if a division has two producing departments, the total indirect cost allocated to a department is used as the numerator in Equation [15.8] to determine the department rate. Example 15.8 illustrates allocation when the cost center is a machine.

EXAMPLE 15.8

The manager of beauty products at BestWay wants to determine allocation rates for $150,000 per year of indirect costs for the three machines used to process beauty lotions. The following information was obtained from last year's budget for the three machines. Determine rates for each machine if the amount is equally distributed to each machine.

Cost Source	Allocation Basis	Estimated Activity Level
Machine 1	Direct labor cost	$100,000 per year
Machine 2	Direct labor hours	2000 hours per year
Machine 3	Direct material cost	$250,000 per year

Solution
Applying Equation [15.8] for $50,000 for each machine, the annual rates are

$$\text{Machine 1 rate} = \frac{\text{indirect budget}}{\text{direct labor cost}} = \frac{50,000}{100,000}$$

$$= \$0.50 \text{ per direct labor dollar}$$

$$\text{Machine 2 rate} = \frac{\text{indirect budget}}{\text{direct labor hours}} = \frac{50,000}{2000}$$

$$= \$25 \text{ per direct labor hour}$$

$$\text{Machine 3 rate} = \frac{\text{indirect budget}}{\text{material cost}} = \frac{50,000}{250,000}$$

$$= \$0.20 \text{ per direct material dollar}$$

> Now the actual direct labor costs and hours and material costs are determined for this year, and each dollar of direct labor cost spent on machine 1 implies that $0.50 in indirect cost will be added to the cost of the product. Similarly, indirect costs are added for machines 2 and 3.

When the same allocation basis is used to distribute indirect costs to several cost centers, a **blanket rate** may be determined. For example, if direct materials are the basis for allocation to four separate processing lines, the blanket rate is

$$\text{Indirect cost rate} = \frac{\text{total indirect costs}}{\text{total direct materials cost}}$$

If $500,000 in indirect costs and $3 million in materials are estimated for the four lines, the blanket indirect rate is $500,000/3,000,000 = \$0.167$ per materials cost dollar. Blanket rates are easier to calculate and apply, but they do not account for differences in the type of activities accomplished in each cost center.

In most cases, machinery or processes add value to the end product at different rates per unit or hour of use. For example, light machinery may contribute less per hour than heavy, more expensive machinery. This is especially true when advanced technology processing, such as integrated manufacturing systems are used along with classic methods, like nonautomated finishing equipment. The use of blanket rates in these cases is not recommended, as the indirect cost will be incorrectly allocated. The lower-value-contribution machinery will accumulate too much of the indirect cost. The approach should be the application of different bases for different machines, activities, etc., as discussed earlier and illustrated in Example 15.8. The use of different, appropriate bases is often called the **productive hour rate method** since the cost rate is determined based on the value added, not a uniform or blanket rate. Realization that more than one basis should be normally used in allocating indirect costs has led to the use of activity-based costing methods, as discussed in the next section.

Once a period of time (month, quarter, or year) has passed, the indirect cost rates are applied to determine the indirect cost *charge,* which is then added to direct costs. This results in the total cost of production, which is called the cost of goods sold, or **factory cost.** These costs are all accumulated by *cost center.*

If the total indirect cost budget is correct, the indirect costs charged to all cost centers for the period of time should equal this budget amount. However, since some error in budgeting always exists, there will be overallocation or underallocation relative to actual charges, which is termed *allocation variance.* Experience in indirect cost estimation assists in reducing the variance at the end of the accounting period.

EXAMPLE 15.9

Since the manager determined indirect cost rates for BestWay (Example 15.8), she wants to know the variance for indirect cost allocation for this month. Calculate it for her using the actual cost and hour data in Table 15-8.

Solution

To determine indirect cost, the rates from Example 15.8 are applied:

$$\text{Machine 1 indirect} = (\text{labor cost})(\text{rate}) = 2500(0.50) = \$1250$$
$$\text{Machine 2 indirect} = (\text{labor hours})(\text{rate}) = 450(25.00) = \$11,250$$
$$\text{Machine 3 indirect} = (\text{material cost})(\text{rate}) = 10,550(0.20) = \$2110$$

Total charged indirect cost $= \$14,610$

TABLE 15–8   Actual Monthly Data Used for Indirect Cost Allocation

Cost Source	Machine Number	Actual Cost, $	Actual Hours
Material	1	3,800	
	3	10,550	
Labor	1	2,500	650
	2	3,200	450
	3	2,800	720

Based on the annual indirect cost budget of $150,000, one month represents 1/12 of the total or

$$\text{Monthly budget} = \frac{150{,}000}{12}$$
$$= \$12{,}500$$

The allocation variance for total indirect cost is

$$\text{Variance} = 12{,}500 - 14{,}610 = \$-2110$$

This is a budget underallocation, since more was actually charged than allocated. This analysis for only one month of a year may prompt a review of the rates, an increase in the annual indirect cost budget, or a redistribution of the annual indirect budget between the three machines.

Once estimates of indirect costs are determined, it is possible to perform an economic analysis of the present operation versus a proposed operation. Such a study is described in Example 15.10.

EXAMPLE 15.10

For several years, Cuisinart Corporation has purchased the carafe assembly of its major coffeemaker line at an annual cost of $2.2 million. The suggestion to make the component in-house has been made. For the three departments involved, the annual indirect cost rates, estimated material, labor, and hours are found in Table 15–9. The allocated hours column is the time necessary to produce the carafes for a year.

Equipment that must be purchased has the following estimates: first cost of $2 million, salvage value of $50,000, and life of 10 years. Perform an economic analysis for the make alternative, assuming that a market rate of 15% per year is the MARR.

Solution

For making the components in-house, the AOC is comprised of direct labor, direct material, and indirect costs. Use the data of Table 15–9 to calculate the IDC allocation.

Department A: 25,000(10) = $250,000
Department B: 25,000(5) = 125,000
Department C: 10,000(15) = 150,000
                             $525,000

$$\text{AOC} = 500{,}000 + 300{,}000 + 525{,}000 = \$1{,}325{,}000$$

TABLE 15–9	Production Cost Estimates for Example 15.10				
		Indirect Costs (IDC)			**Direct Labor Cost, $**
Department	**Basis, Hours**	**Rate per Hour, $**	**Allocated Hours**	**Material Cost, $**	**Direct Labor Cost, $**
A	Labor	10	25,000	200,000	200,000
B	Machine	5	25,000	50,000	200,000
C	Labor	15	10,000	50,000	100,000
				300,000	500,000

The make alternative annual worth is

$$\begin{aligned} AW_{make} &= -P(A/P,i,n) + S(A/F,i,n) - AOC \\ &= -2,000,000(A/P,15\%,10) + 50,000(A/F,15\%,10) - 1,325,000 \\ &= \$-1,721,037 \end{aligned}$$

Currently, the carafes are purchased with an AW of

$$AW_{buy} = \$-2,200,000$$

It is cheaper to make because the AW of costs is less.

15.7 Activity-Based Costing (ABC) for Indirect Costs ● ● ○

As automation, software, information, and manufacturing technologies have advanced over the years, the number of direct labor hours necessary to manufacture a product or provide a service has decreased substantially. Where once as much as 35% to 45% of the final product cost was represented in labor, now the labor component is commonly 5% to 15% of total cost. However, the indirect cost may represent as much as 35% to 65% of the total manufacturing cost. The use of bases, such as direct labor hours, to allocate indirect cost is not accurate enough for automated and technologically advanced environments. This transformation has led to the development of methods that replace or supplement traditional cost allocations that rely upon one form or another of Equation [15.8]. Also, allocation bases different from traditional ones are commonly utilized.

A product that by traditional IDC rate allocation may have contributed a large portion to profit *may actually be a loser* when indirect costs are allocated more correctly. Companies that have a wide variety of products, some produced in small lots, may find that traditional allocation methods have a tendency to underallocate the indirect cost to small-lot products. This may indicate that they are profitable, when in actuality they are losing money.

A better allocation method for high-overhead industries is **Activity-Based Costing (ABC).** It is designed to identify *cost centers, activities,* and *cost drivers.* Descriptions of each follow.

Cost centers: The final products or services of the corporation are called cost centers or cost pools. They *receive* the allocated indirect costs.

Activities: These are usually support departments (purchasing, quality, IT, maintenance, engineering, supervision) that *generate* the indirect costs which are then distributed to the cost centers.

Cost drivers: Commonly expressed in volumes, these *drive* the consumption of a shared resource. Examples are the number of purchase orders, cost of engineering change orders, number of machine setups, number of safety violations, and the like.

Implementing ABC is not a simple task for a corporation or business. It involves several steps.

1. Identify each *activity* and its total cost.
2. Identify the *cost drivers* and their usage volumes.
3. Calculate the indirect cost *rate* for each activity.

$$\text{ABC indirect cost rate} = \frac{\text{total cost of activity}}{\text{total volume of cost driver}} \qquad [15.9]$$

4. Use the rate to *allocate* indirect cost to cost centers for each activity.

As an illustration, assume a company that produces two types of industrial lasers (cost centers) has three primary support departments (activities; step 1 in the procedure). The allocation for costs generated by the purchasing department, for example, is based on the number of purchase orders (step 2) to support laser production. The ABC rate (step 3) in dollars per purchase order is used to allocate indirect costs to the two laser products (step 4).

EXAMPLE 15.11

A multinational aerospace firm uses traditional methods to allocate manufacturing and management support costs for its European division. However, accounts such as business travel have historically been allocated on the basis of the number of employees at the plants in France, Italy, Germany, and Spain.

The president recently observed that some product lines are likely generating much more travel than others. The ABC system is chosen to augment the traditional method to more precisely allocate travel costs to major product lines at each plant.

(*a*) First, assume that allocation of total observed travel expenses of $500,000 to the plants using a traditional basis of workforce size is sufficient. If total employment of 29,100 is distributed as follows, allocate the $500,000.

Paris, France plant	12,500 employees
Florence, Italy plant	8,600 employees
Hamburg, Germany plant	4,200 employees
Barcelona, Spain plant	3,800 employees

(*b*) Now, assume that corporate management wants to know more about travel expenses based on product line, not merely plant location and workforce size. The ABC method will be applied to allocate travel costs to major product lines. Annual plant support budgets indicate that the following percentages are expended for travel:

Paris	5% of $2 million
Florence	15% of $500,000
Hamburg	17.5% of $1 million
Barcelona	30% of $500,000

Further, the study indicates that in 1 year, a total of 500 travel vouchers were processed by the management of the major five product lines produced at the four plants. The distribution is as follows:

Paris	Product lines—1 and 2; number of vouchers—50 for line 1, 25 for 2.
Florence	Product lines—1, 3, and 5; vouchers—80 for line 1, 30 for 3, 30 for 5.
Hamburg	Product lines—1, 2, and 4; vouchers—100 for line 1, 25 for 2, 20 for 4.
Barcelona	Product line—5; vouchers—140 for line 5.

Use the ABC method to determine how the product lines drive travel costs at the plants.

Solution

(*a*) In this case, Equation [15.8] takes the form of a blanket rate per employee.

$$\text{Indirect cost rate} = \frac{\text{travel budget}}{\text{total workforce}}$$

$$= \frac{\$500,000}{29,100} = \$17.1821 \text{ per employee}$$

Using this traditional basis of rate times workforce size results in a plant-by-plant allocation.

Paris	$17.1821(12,500) = $214,777
Florence	$147,766
Hamburg	$72,165
Barcelona	$65,292

(*b*) The ABC method is more involved to apply, and the by-plant amounts will be different from those in part (*a*) since completely different bases are applied. Use the four-step procedure to allocate travel costs to the five products.

> **Step 1.** The total amount to be allocated is determined from the percentages of each plant's support budget devoted to travel. The number is determined from the percent-of-budget data as follows:
>
> $$0.05(2,000,000) + \cdots + 0.30(500,000) = \$500,000$$

TABLE 15–10	ABC Allocation of Travel Cost ($ in Thousands), Example 15.11						
	Vouchers by Product Line					**Totals**	
	1	**2**	**3**	**4**	**5**	**Vouchers**	**Allocated, $1000**
Paris	50	25				75	75
Florence	80		30		30	140	140
Hamburg	100	25		20		145	145
Barcelona					140	140	140
Total Allocated	$230	$50	$30	$20	$170	500	500

Step 2. The cost driver is the number of travel vouchers submitted by the management unit responsible for each product line at each plant. The allocation will be to the products directly, not to the plants. However, the travel allocation to the plants can be determined afterward since we know what product lines are produced at each plant.

Step 3. Equation [15.9] determines an ABC allocation rate.

$$\text{ABC allocation rate} = \frac{\text{total cost of travel}}{\text{total number of vouchers}}$$

$$= \frac{\$500,000}{500}$$

$$= \$1000 \text{ per voucher}$$

Step 4. Table 15–10 summarizes the vouchers and allocation by product line and by city. Product 1 ($230,000) and product 5 ($170,000) drive the travel costs based on the ABC analysis. Comparison of the by-plant totals in Table 15–10 (far right column) with the respective totals in part (a) indicates a substantial difference in the amounts allocated, especially to Paris ($75,000 versus $214,777), Hamburg, and Barcelona. This comparison verifies the president's statement that product lines, not plants, drive travel requirements.

Comment

Let's assume that product 1 has been produced in small lots at the Hamburg plant for a number of years. The ABC analysis, when compared to the traditional cost allocation method in part (a), reveals a very interesting fact. In the ABC analysis, Hamburg has a total of $145,000 travel dollars allocated, $100,000 from product 1. In the traditional analysis based on workforce size, Hamburg was allocated only $72,165—about 50% of the more precise ABC analysis amount. This indicates to management the need to examine the manufacturing lot size practices at Hamburg and possibly other plants, especially when a product is currently manufactured at more than one plant.

Some proponents of the ABC method recommend discarding the traditional IDC rates and utilizing ABC exclusively. This is not a good approach, since ABC is not a complete cost system. The ABC method provides information that assists in *cost control,* while the traditional method emphasizes cost allocation and cost estimation. The two systems work well together with the traditional methods allocating costs that have identifiable direct bases, for example, direct labor. ABC analysis is usually more expensive and time-consuming than a traditional cost allocation system, but in many cases it can assist in understanding the economic impact of management decisions and in controlling certain types of indirect costs.

15.8 Making Estimates and Maintaining Ethical Practices ●●●

Making estimates about the future of costs, revenues, cash flows, rates of return, and many other parameters is routine when one is engaged in any type of economic analysis. Public agencies, private corporations, and not-for-profit businesses all make economic decisions based on these estimates, most of which are made by employees of the organizations or by outside consultants hired to perform specific activities under contract. The opportunities for bias, poor accuracy, deception, and profit-driven or other motives are always present. The personal morals and adherence to codes of professional ethics discussed in Section 1.3 guide individuals in their work to make fair and believable estimates for the analyses and decisions that follow.

The NSPE Code of Ethics for Engineers (Appendix C) referenced previously starts with a list of six Fundamental Canons. One very relevant to estimation integrity is "Avoid deceptive acts." Acts that bias the results from experimental samples, previous cost data, or survey results for the purposes of personal gain, increased profits, or favoritism are examples of unethical behavior. Estimates of all types should be founded on practices such as the following:

Base estimates on sound information gathered over a range of situations representative of the current one.

Use accepted theory and techniques in taking statistical samples, building budget elements, and drawing conclusions that are included in proposals, applications, and recommendations.

As a consultant or contractor, keep personal and working relationships separate when making estimates and delivering the final documentation to a client or sponsor.

The second case study at the end of this chapter presents an example of some ethical challenges present when preparing estimates and proposals for contract work.

CHAPTER SUMMARY

Cost estimates are not expected to be exact, but they should be accurate enough to support a thorough economic analysis using an engineering economy approach. There are bottom-up and top-down strategies; each treats price and cost estimates differently.

Costs can be updated via a cost index, which is a ratio of costs for the same item at two separate times. The Consumer Price Index (CPI) is an often-quoted example of cost indexing. Cost estimating may also be accomplished with a variety of models called cost-estimating relationships. Two of them are

Cost-capacity equation—good for estimating costs from design variables for equipment, materials, and construction

Factor method—good for estimating total plant cost

Traditional indirect cost allocation uses a rate determined for a machine, department, product line, etc. Bases such as direct labor cost, direct material cost, and direct labor hours are common. With increased automation and information technology, different techniques of indirect cost allocation have been developed. The Activity-Based Costing (ABC) method is an excellent technique to augment traditional allocation methods. The ABC method allocates indirect costs on the rationale that purchase orders, inspections, machine setups, reworks, etc. *drives* cost accumulation in departments and functions, such as quality, purchasing, accounting, and maintenance. Improved understanding of how the company or plant accumulates indirect costs is a major by-product of implementing the ABC method.

PROBLEMS

Understanding Cost Estimation

15.1   Rank the following estimate types in terms of accuracy of the estimate from least accurate (1) to most accurate (5): partially designed, design 60% to 100% complete, order of magnitude, scoping/feasibility, detailed estimate.

15.2   Identify the following cost elements as first-cost (FC) components or annual operating cost (AOC)

components for a piece of equipment in a chemical processing plant: rent, direct labor, equipment overhaul, overhead, supplies, insurance, equipment cost, utility cost, installation, delivery charges, salaried personnel, and periodic safety training.

15.3   When competition is the dominant factor in pricing a product or service, what is the best cost estimation strategy to adopt?

15.4 State whether actual (A) or estimated (E) costs are more likely to be used to carry out the following activities: calculate taxes, make bids, pay bonuses, determine profit or loss, predict sales, set prices, evaluate proposals, distribute resources, plan production, and set goals.

15.5 Identify the output and input variables in both the bottom-up and top-down strategies to cost estimation.

15.6 Classify the following costs as typically direct (D) or indirect (I):

Project staff	Audit and legal
Utilities	Rent
Raw materials	Training on equipment
Project supplies	Labor
Administrative staff	Miscellaneous office supplies
Quality assurance	IT department
Scheduled maintenance	Shared software packages

15.7 Identify each of the following costs associated with owning an automobile as direct (D) or indirect (I). Assume a direct cost is one that is directly attributable to the number of miles driven. (*a*) License plate, (*b*) Drivers license, (*c*) Gasoline, (*d*) Highway toll fee, (*e*) Oil change, (*f*) Repairs after collision, (*g*) Gasoline tax, (*h*) Monthly loan payment, (*i*) Annual inspection fee, and (*j*) Garage rental.

15.8 In the early and conceptual design stages of a project, what are the cost estimates called? Approximately how close should they be to the actual cost?

Unit Costs

15.9 The preliminary cost for a 50,000 ft^2 aircraft hangar must be estimated. (*a*) The first unit estimate was $98.23 per square foot. Estimate the cost of the hangar. (*b*) The budget limit came in at $4 million. Perform two analyses: determine the maximum allowed size to be within budget at the first estimated unit cost, and determine the maximum unit cost to stay within budget and maintain the 50,000 ft^2 size.

15.10 The cost for 4″ diameter auger holes in earth is $18.85 per linear foot, but if they are cased, the cost is $69.18 per linear foot. What is the estimated cost for a 570-foot cased 4″ boring?

15.11 Low, medium, and high cost estimates for a 30,000 ft^2 apartment building in zip code 79912 are $3,111,750, $3,457,500, and 4,321,875, respectively. (*a*) What is the cost per square foot for the medium cost estimate, and (*b*) what is the percent increase between low and high cost per square foot estimates?

15.12 Estimate the total cost of a house (from purchasing a lot to furnishing it) using the following estimates:

Cost Data	House Data
Purchasing a lot = $2.50/ft^2	Lot size = 100 × 150 ft
Construction = $125/ ft^2 of livable space	House size = 50 × 46, with 75% livable
Furnishings = $3000 per room	Number of rooms = 6

15.13 A labor crew for placing concrete consists of one labor foreman at $26.70 per hour, one cement finisher at $29.30 per hour, five laborers at $24.45 per hour each, and one equipment operator at $33.95 per hour. Such a crew, called a C20 crew, can place 165 cubic yards of concrete per 8-hour day. Determine (*a*) the cost per day of labor for the C20 crew, (*b*) the cost of the C20 crew per cubic yard of concrete, and (*c*) the cost to place 550 cubic yards of concrete.

15.14 The Department of Defense uses area cost factors (ACFs) to compensate for differences in construction costs in different parts of the country (and world). The area cost factor for Waianae, Hawaii is 2.19 and for Moody, Alabama is 0.83. If a cold-storage processing warehouse will cost $1,350,000 in Moody, Alabama, what would be the monetary difference in cost for a similar warehouse constructed in Waianae, Hawaii.

15.15 Oil spill cleanup costs for a nation or region can be estimated by taking the average cost for the specified region and multiplying it by various "modification factors" that are a function of oil type (light crude, heavy crude, No. 4 fuel, etc.), spill size (<34 tons, 34–340 tons, etc.), location type (offshore, nearshore, in-port), cleanup method (dispersants, burning, manual), and extent of shoreline oiling (0–1 km, 2–5 km, etc.). For an average cost of $23.02 per liter and modifiers for a region in the United States of 1.32, 0.65, 1.28, 0.25, and 1.53 for oil type, spill size, location type, cleanup method, and extent of shoreline oiling, respectively, determine the estimated cleanup cost per liter of an oil spill in the region of the United States.

15.16 Two people developed first-cut cost estimates to construct a new 130,000 square foot building on a university campus. Person A applied a general-purpose per unit cost estimate of $180 per square foot for the estimate. Individual B was more specific: she used the area estimates and per unit cost factors shown below. (*a*) What are the cost estimates developed by the two people? (*b*) What is the percent increase between the lower and higher cost estimates?

Type of Usage	Area, %	Cost per ft², $
Classroom	30	125
Laboratory	40	185
Offices	30	110
Furnishings—labs	25	150
Furnishings—all others	75	25

Cost Indexes

15.17 Equipment necessary to manufacture adjustable ball-lock pins for pulling two components together had a cost of $185,000 when the FRED Equipment Cost Index was 210.0. What was the estimated cost of a similar system in 2015 when the index value was 245.9?

15.18 The FRED CPI Index (Table 15–3) can be used to predict the price of household items over the years. In 2011, Sandy and Malcolm purchased a combo clothes washing machine and dryer for $800. Assume it is now 2018 and the combo system has experienced a serious failure and must be replaced immediately. If the CPI index continued to change at the same linear rate as it did between 2014 and 2015, estimate the cost of a similar combo washer and dryer in 2018.

15.19 AB Jackson Industries installed a compressor and pump system in 2010 at a cost of $87,200. If the FRED equipment cost index applies, what will be the estimated cost now for a similar system when the index value is 273?

15.20 In 2009, the Office of the Under Secretary of Defense projected that a critical International Defense Construction Cost Index would increase from its value then of 3423 to 4098 in 2016. If the actual value in 2016 was 5167, (a) what was the actual building cost inflation rate between 2009 and 2016, and (b) what was the increase in the inflation rates between projected and actual?

15.21 The cost of a piece of equipment was $67,900 when the relevant cost index was 1457.4. Determine the index value when the same equipment was estimated to cost $83,400?

15.22 If the PCI editor at *Chemical Engineering (CE)* decides to update the Plant Cost Index so that the year 2010 has a base value of 100 instead of the current value of 550.8, determine the PCI values for 2014 and for 2015. (Hint: Use Table 15–3 as a reference point.)

15.23 A consulting engineering firm is preparing a preliminary cost estimate for a 40-MW advanced combined cycle natural gas-fired power plant. The firm, which completed a similar project in 2012 with a total cost (construction and equipment) of $809.2 million, wants to use a construction cost index to update the construction cost. If the construction cost was assumed to be 25% of the total cost and the index value in 2012 was 8570, estimate the cost of construction for a similar-size plant in 2017 when the index is 9324.

15.24 The Marshall and Swift (M&S) equipment cost index started with a base value of 100 in 1926. Determine the average (compound) percentage increase per year between 1926 and 2011 when the index value was 1490.2.

15.25 An engineer who owns a construction company that specializes in large commercial projects noticed that material costs increased at a rate of 2% *per quarter* over the past 2 years. If a material cost index were created for that 2-year period with the value of the index set at 100 at the beginning of the period, what would be the value of the index at the end? Express your answer to two decimal places.

15.26 A survey of total project construction costs for a wide range of buildings (hospitals, schools, banks, nursing homes, etc.) in 2006 found that the average cost was $13,136,431. The mechanical and electrical portions of that cost were $2,511,893 and $1,585,384, respectively. If the total project cost in 2017 increases to $15,700,000, what would the mechanical portion of the total cost equal, provided the percentage (a) remained the same as in 2006, and (b) increased by 20% from the 2006 portion?

15.27 A quadrupole mass spectrometer can be purchased today for $85,000. The owner of a mineral analysis laboratory expects the cost to increase exactly by the specialty laboratory equipment cost index over the next 10 years. (a) If the specialty equipment inflation rate is estimated to be 2% per year for the next 3 years and 5% per year thereafter, how much will the spectrometer cost in 10 years? (b) If the applicable cost index is 1203 now, what will the index be 10 years from now at the expected inflation rates?

15.28 In 2015, Mariam, an engineer with Total Petroleum, decided to invest in stocks of a corporation that concentrates on crude oil production and refined products, though the price of oil is a very volatile commodity during recent years. She found the Global Crude Oil Index (GCO Index as shown in Table 15–3) on the web, plus the recorded annual prices (in dollars per barrel, $/bbl, as of December of each year) for two very well-known oil categories—Brent-Europe (B-E) and West Texas Intermediate (WTI). The data for all three are presented below for the period 2008 through 2015. The GCO Index has a base year of 2005 of 100 points.

Year	GCO Index	B-E Recorded, $/bbl	WTI Recorded, $/bbl
2008	105.1	35.82	44.60
2009	141.9	77.91	79.39
2010	160.6	93.23	91.38
2011	194.4	108.09	98.83
2012	192.3	110.80	91.83
2013	196.8	109.95	98.17
2014	140.2	62.16	53.45
2015	79.3	36.61	37.13

Use the cost index equation to do the following:

(a) Predict the price for years 2008 through 2015 for both B-E and WTI crude.

(b) Develop a spreadsheet line graph for the Brent-Europe recorded and predicted price from 2008 to 2015.

(c) Develop a spreadsheet line graph for the West Texas Intermediate recorded and predicted price from 2008 to 2015.

(d) Which of the two categories of oil does the index appear to track more closely?

Cost-capacity and Factor Methods

15.29 What is the fundamental difference between estimating costs using a CER and a cost index?

15.30 Estimate the cost of a 0.75 million gallon per day (MGD) induced-draft packed tower for air-stripping trihalomethanes from drinking water if the cost for a 2-MGD tower is $153,200. The exponent in the cost capacity equation is 0.58.

15.31 What is the value of the exponent in the cost-capacity equation if a pump half the size of a similar, but larger one, costs exactly half as much as the larger one?

15.32 The cost of a high-quality 250-horsepower compressor was $14,000 when recently purchased. What would a 600-horsepower compressor be expected to cost?

15.33 The cost for implementing a manufacturing process that has a capacity of 6000 units per day was $550,000. If the cost for a plant with a capacity of 100,000 units per day was $3 million, what is the value of the exponent in the cost-capacity equation?

15.34 The estimated cost for a multitube cyclone system with a capacity of 60,000 cubic feet per minute is $450,000. (a) If the $200,000 actual cost for a 35,000 cubic feet per minute system was used in the cost-capacity equation, what was the value of the exponent? (b) What can be concluded about the economy of scale of the costs between the two systems?

15.35 The cost for construction of a desulfurization system for flue gas from boilers at a 600 MW power plant was estimated to be $250 million. If a similar system for a smaller plant had a cost of $55 million and the exponent in the cost-capacity equation is 0.67, what is the size of the smaller plant that served as the basis for the cost projection?

15.36 The net annual operating cost (AOC) for a water treatment filtration plant for a semi-conductor fabrication line was estimated to be $1.5 million per year. The estimate was based on the $200,000 per year cost of a 1-MGD plant. The exponent in the cost-capacity equation is 0.80. (a) What was the size of the larger plant? (b) Build a spreadsheet to answer the same question. (c) If the plant manager has limited the AOC to $1 million per year, use your spreadsheet to determine the maximum MGD size plant that can be planned.

15.37 Estimate the cost in 2011 of chemical processing equipment if the cost of a unit one-fourth its size was $60,000 in 2005. The exponent in the cost capacity equation is 0.24. Use the M&S equipment cost index to update the cost.

15.38 Estimate the cost in 2018 of a 1000-horsepower steam turbine air compressor if a 200-horsepower unit costs $160,000 in 1999. The exponent in the cost capacity equation is 0.35 and the equipment cost index increased by 35% over that time period.

15.39 A mini-wind tunnel for calibrating vane or hotwire anemometers costs $3750 in 2011 when the Marshall and Swift (M&S) equipment cost index value was 1490.2. If True-Tech Instruments estimated the cost of a wind tunnel twice as large to be $10,200 in 2016, what did it expect the equipment cost index value to be, provided the equipment cost increased accordingly? The cost-capacity equation exponent is 0.89.

15.40 The cost of equipment for manufacturing mineral products in a harsh environment is $2.3 million. If the overall cost factor for this type of facility is 2.25, what is the estimated total plant cost?

15.41 The total cost for a production plant that supplies hydrogen to the refining industry is estimated at $55.4 million. The equipment is expected to cost

$17.8 million. (*a*) What is the overall cost factor for this type of plant? (*b*) What is included in the cost factor that you determined?

15.42 A chemical engineer at Western Refining estimated the total cost for a diesel fuel desulfurization system at $2.3 million. If the direct cost factor is 1.55 and the indirect cost factor is 0.43, what is the total equipment cost? Both factors apply to delivered equipment cost.

15.43 The delivered equipment cost for a fully equipped CNC machining system is $1.6 million. The direct cost factor is 1.52 and the indirect cost factor is 0.31. Estimate the total plant cost if the indirect cost factor applies to (*a*) the delivered equipment cost only, and (*b*) the total direct cost.

15.44 During a major expansion in 2004, Douwalla's Import Company developed a new processing line for which the delivered equipment cost was $1.75 million. This year, the board of directors decided to expand into new markets and expects to build the current version of the same line. Estimate the cost if the following factors are applicable: construction cost factor is 0.20, installation cost factor is 0.50, indirect cost factor applied against equipment is 0.25, and the total plant cost index has risen from 2509 to 3713 over the years.

15.45 Nicole is an engineer on temporary assignment at a refinery operation in Seaside. She has reviewed a cost estimate for $430,000, which covers some new processing equipment for the ethylene line. The equipment itself is estimated at $250,000 with a construction cost factor of 0.30 and an installation cost factor of 0.30. No indirect cost factor is listed, but she knows from other sites that indirect cost is a sizable amount that increases the cost of the line's equipment. (*a*) If the indirect cost factor should be 0.30, determine whether the current estimate includes a factor comparable to this value, and (*b*) determine the cost estimate if the 0.30 indirect cost factor is used.

Indirect Cost (IDC) Rates and Allocation

15.46 The company you work for currently allocates insurance costs on the basis of cost per direct labor hour. This indirect cost component for the year is budgeted at $36,000. If the annual direct labor hours for departments A, B, and C are expected to be 2000, 8000, and 5000, respectively, determine the allocation to each department.

15.47 The Director of Street Maintenance wants to allocate his annual IDC budget of $1.2 million to the three divisions around the city. The recorded amounts for this year are as follows:

Division	Recorded for This Year	
	Miles Driven	Direct Labor Hours
North	275,000	38,000
South	247,000	31,000
Midtown	395,000	55,500

The Director plans to use the allocation and information from last year to determine the rates for this year. This information follows:

Division	Miles Driven	Direct Labor Hours	Basis	IDC Allocation Last Year, $
North	350,000	40,000	Miles	300,000
South	200,000	20,000	Labor	200,000
Midtown	500,000	64,000	Labor	450,000

(*a*) Determine the rates for this year for each Division.

(*b*) Use the rates to allocate this year's total budget. What percentage of this year's IDC budget did he allocate?

15.48 A company has a processing department with 10 stations. Because of the nature and use of three of these stations, each is considered a separate cost center for IDC allocation. The remaining seven are grouped as one cost center, CC190. Operating hours are used as the allocation basis for all stations. A total of $250,000 is allocated to the department for next year. Use the data collected this year to determine the IDC rate for each center.

Cost Center	IDC Allocated	Operating Hours
CC100	$25,000	800
CC110	$50,000	200
CC120	$75,000	1200
CC190	$100,000	1600

15.49 At Williams Corp., all indirect costs are allocated by accounting. One department manager obtained records for her department of allocation rates and actual charges for the prior 3 months and estimates for this and next month (May and June; see the table). The basis of allocation is not indicated, and the company accountant would not share the basis used. However, the accountant advised the manager to not be concerned, since it is a good sign when the allocation rates decrease each month.

	Indirect Cost		
Month	**Rate**	**Allocated**	**Charged**
February	$1.40	$2800	$2600
March	$1.33	$3400	$3800
April	$1.37	$3500	$3500
May	$1.03	$3600	
June	$0.92	$6000	

During the evaluation, the following additional information from departmental and accounting records was obtained.

	Direct Labor		Material	Departmental
Month	**Hours**	**Cost, $**	**Cost, $**	**Space, ft²**
February	640	2560	5400	2000
March	640	2560	4600	2000
April	640	2560	5700	3500
May	640	2720	6300	3500
June	800	3320	6500	3500

(*a*) Determine the allocation basis used each month with this information.

(*b*) Comment on the accountant's statement about decreasing allocation rates.

15.50 The Mechanical Components Division manager asks you to recommend a make/buy decision on a major automotive subassembly that is currently purchased externally for a total of $3.9 million this year. This cost is expected to continue rising at a rate of $300,000 per year. Your manager asks that both direct and indirect costs be included when in-house manufacturing (make alternative) is evaluated. New equipment will cost $3 million, have a salvage of $0.5 million and a life of 6 years. Estimates of materials, labor costs, and other direct costs are $1.5 million, per year. Typical indirect rates, bases, and expected usage are shown. Perform the AW evaluation at MARR = 12% per year over a 6-year study period. Show both *hand* and *spreadsheet* solutions.

Department	Basis	Rate	Expected Usage
X	Direct labor cost	$2.40 per $	$450,000
Y	Materials cost	$0.50 per $	$850,000
Z	Number of inspections	$20 per inspection	$4,500

ABC Method of IDC Allocation

15.51 CarryALL, Inc., makes and sells small cargo trailers to individuals and small businesses. Since its opening in 1990, it has allocated indirect costs (IDC) to its three manufacturing plants based on *direct materials cost per unit*. Each plant builds different models and sizes. Because of advances in automation and materials, Judy, the CFO, plans to use *build-time per unit* as the new basis. Build-time is the average number of work-hours to complete a trailer. However, she initially wants to determine what the allocation would have been this year had the build-time basis been used prior to the incorporation of new technology and materials. The data shown below represents average costs and times. Use this data and the bases indicated to determine the allocation rates and IDC allocation of $1,000,000 for this year for the three bases.

Plant	New York	Virginia	Tennessee	Total
Direct material cost, $ per unit	20,000	12,700	18,600	51,300
Previous build-time per unit, work-hours	400	415	580	1395
New build-time per unit, work-hours	425	355	480	1260

15.52 The municipal water and desalinization utility in a California city currently allocates some costs for maintenance shop workers to pumping stations based on the number of pumps at each station. At the last Director's Semiannual Meeting, a suggestion was made to change the allocation basis to the number of trips that pump service personnel make to each station, because some stations have old pumps that require more maintenance. Information about the stations is below. The indirect cost budget is $20,000 per pump. (*a*) Allocate the budget to each station based on the number of service trips. (*b*) Determine the old allocation on the basis of number of pumps and comment on any significant differences in the amounts allocated to the stations.

Station ID	No. of Pumps	Service Trips/Year
Sylvester	5	190
Laurel	7	55
7th St	3	38
Spicewood	4	104

Problems 15.53 through 15.55 use the following information.

Jet Green Airways historically distributes the indirect costs of lost and damaged baggage to its three major hubs using a basis of annual number of flights in and out of each hub. Last year, $667,500 was distributed as follows:

Hub Airport	Flights	Rate	Allocation
HUA	55,000	$ 6/flight	$330,000
DFW	20,833	$ 9/flight	187,500
SAT	15,000	$10/flight	150,000

The airline's Baggage Management Director suggests that an allocation on the basis of baggage traffic, not flights, will be better at representing the distribution, primarily based on the fact that the high fees now charged to passengers to check luggage have significantly changed the number of bags handled at the major hubs. Total number of bags handled during the year are 2,490,000 at HUA, 1,582,400 at DFW, and 763,500 at SAT.

15.53 What is (a) the activity, and (b) the cost driver for the suggested baggage traffic basis?

15.54 Using the baggage traffic basis, determine the IDC rate (to the nearest 0.1¢) using last year's total of $667,500 and allocate this amount to the hubs this year.

15.55 What are the percentage changes in allocation at each hub using the two different bases?

15.56 Historically, The Travel Club has distributed advertising costs to its resort sites in Europe on the basis of the size of the resort budget. For this year, in round numbers, the budgets and allocation of $1 million advertising indirect costs are as follows to the four sites:

	Site			
	1	2	3	4
Budget, $	2 million	3 million	4 million	1 million
Allocation, $	200,000	300,000	400,000	100,000

(a) Determine the allocation if the ABC method is used with a new basis. Define the activity as the advertising department at each resort. The cost driver is the number of guests during the year.

	Site			
	1	2	3	4
Guests	3500	4000	8000	1000

(b) Again use the ABC method, but now make the cost driver the total number of guest-nights at each resort. The average number of lodging-nights for guests at each site is:

	Site			
	1	2	3	4
Length of stay, nights	3.0	2.5	1.25	4.75

(c) Comment on the distribution of advertising costs using the two methods.
(d) Identify one additional cost driver that might be considered for the ABC approach that could reflect a realistic allocation of the costs.

ADDITIONAL PROBLEMS AND FE EXAM REVIEW QUESTIONS

15.57 The total mechanical and electrical unit costs for a library are listed as $40 per square foot, while total project cost is estimated at $119 per square foot. Based on these values, the percentage of total costs represented by mechanical and electrical costs is closest to:
(a) 21% (b) 29% (c) 34% (d) 38%

15.58 The cost for constructing a small building in 1999 was $700,000. The construction cost index was 6059 at that time. The estimated cost of a similar building when the index value is 10,000 is closest to:
(a) $424,130   (b) $1,155,300
(c) $1,347,200   (d) $1,384,500

15.59 An assembly line robot arm with a first cost of $75,000 in 2001 had a cost of $89,750 in 2017. If the equipment cost index for robots was 307 in 1995 and the robot cost increased exactly in proportion to the index, the value of the index in 2017 was closest to:
(a) 367 (b) 395 (c) 437 (d) 4

15.60 The cost for a 200-horsepower pump with a VFD controller is $42,000. The cost estimate of a similar pump of 300 horsepower capacity, provided the exponent in the cost-capacity equation is 0.64, is closest to:
(a) $47,240   (b) $48,520
(c) $53,780   (d) $54,440

15.61 The cost of an electrode position control system in an electric arc furnace was $16,000 when the applicable equipment cost index had a value of 1192. If the current index value is 1364 and the exponent in the cost-capacity equation has a value of 0.65, the estimated cost now of a similar system twice as large is closest to:
(a) $17,465   (b) $21,545
(c) $28,730   (d) $36,130

15.62 If the cost of an imported high-speed assembly-line robot is $80,000 and the in-country cost for one with twice the capacity is $120,000, the value of the exponent in the cost-capacity equation is closest to:
(a) 0.51 (b) 0.58 (c) 0.62 (d) 0.69

15.63 The delivered equipment cost for setting up a production and assembly line for high-sensitivity,

gas-damped accelerometers is $650,000. If the direct cost and indirect cost factors are 1.36 and 0.31, respectively, and both factors apply to delivered equipment cost, the total plant cost estimate is approximately:
(a) $2,034,500   (b) $1,815,500
(c) $1,735,500   (d) $1,183,000

15.64 The equipment for applying specialty coatings that provide a high angle of skid for the paperboard and corrugated box industries has a delivered cost of $390,000. If the *overall* cost factor for the complete system is 3.96, the total plant cost is approximately:
(a) $955,500   (b) $1,054,400
(c) $1,154,400   (d) $1,544,400

15.65 The IT department allocates indirect costs to user departments on the basis of CPU time at the rate of $2000 per second. For the first quarter, the two heaviest use departments logged 900 and 1300 seconds, respectively. If the IT indirect budget for the year is $19.56 million, the percentage of this year's allocation consumed by these departments is closest to:
(a) 22.5%   (b) 32%
(c) 55%   (d) 67%

15.66 If a hospital cafeteria is the activity to receive indirect cost allocation for the year, cost driver(s) for the ABC method that seem reasonable may be:
1 – Number of cafeteria employees
2 – Number of meals
3 – Hospital patient volume
(a) 1   (b) 2
(c) 3   (d) 1 and 2

15.67 The delivered equipment cost for setting up a production and assembly line for two-way floating ball valves is $650,000. If the direct cost and indirect cost factors are 1.82 and 0.31, respectively, and only the direct cost factor applies to delivered equipment cost, the total plant cost estimate is approximately:
(a) $2,034,500   (b) $2,401,230
(c) $2,684,500   (d) $2,983,000

15.68 A possible disadvantage of the ABC method is:
(a) It does not allocate costs to the way production work occurs
(b) It is more expensive to operate than a traditional allocation system
(c) It does not help understand the economic impact of management decisions
(d) It cannot identify wasteful products and unnecessary costs

CASE STUDY

INDIRECT COST ANALYSIS OF MEDICAL EQUIPMENT MANUFACTURING COSTS

Background

A portable sterilizing unit, PS6, was introduced recently to the market. It utilizes nitrogen dioxide (NO_2) gas as the sterilant. This unit sterilizes and makes available at the bedside some of the reusable instruments that nurses and doctors usually obtain by walking to or receiving delivery from a centralized area. The PS6 unit makes the instruments available at the point and time of use for burn and severe wound patients who are in a regular patient room.

There are two models of PS6 available. The standard version sells for $10.75, and a premium version with customized trays and a battery backup system sells for $29.75. The product has sold well to hospitals, convalescent units, and nursing homes at the level of about 1 million units per year.

Information

The manufacturer, Health Care Services, has historically used an indirect cost allocation system based upon direct hours to manufacture for all its other product lines. The same was applied when PS6 was priced. However, Arnie, the person who performed the indirect cost analysis and set the sales price, is no longer at the company, and the detailed analysis is no longer available. Through e-mail and telephone conversations, Arnie said the current price was set at about 10% above the total manufacturing cost determined 2 years ago, and that some records were available in the design department files. A search of these files revealed the manufacturing and cost information in Table 15–11. It is clear from these and other records that Arnie used traditional indirect cost analysis based on direct labor hours to estimate the total manufacturing costs of $9.73 per unit for the standard model and $27.07 per unit for the premium model.

Last year management decided to place the entire plant on the ABC system of indirect cost allocation. The costs and sales figures collected for the PS6 line the year before were still accurate. Five activities and their cost drivers were identified for the Health Care Services manufacturing operations (Table 15–12). Also, the volumes for each model are summarized in this table.

The ABC method will be used henceforth, with the intention of determining the total cost and price based on its results.

TABLE 15–11 Historical Records of Direct and Indirect Cost Analyses for PS6

PS6 Direct Cost (DC) Evaluation

Model	Direct Labor, $/Unit[1]	Direct Material, $/Unit	Direct Labor, Hours/Unit	Total Direct Labor Hours
Standard	5.00	2.50	0.25	187,500
Premium	10.00	3.75	0.50	125,000

PS6 Indirect Cost (IDC) Evaluation

Model	Direct Labor, Hours/Unit	Fraction IDC Allocated	Allocated IDC, $	Sales, Units/Year
Standard	0.25	$\frac{1}{3}$	1.67 million	750,000
Premium	0.50	$\frac{2}{3}$	3.33 million	250,000

[1]Average direct labor rate is $20 per hour.

TABLE 15–12 PS6 Activities, Cost Drivers, and Volume Levels for ABC-Based Indirect Cost Allocation

Activity	Cost Driver	Volume/Year	Actual Cost, $/Year
Quality	Inspections	20,000 inspections	800,000
Purchasing	Purchase orders	40,000 orders	1,200,000
Scheduling	Change orders	1,000 orders	800,000
Production setup	Setups	5,000 setups	1,000,000
Machine operations	Machine hours	10,000 hours	1,200,000

Cost Driver	Volume Level for the Year	
	Standard	Premium
Quality inspections	8,000	12,000
Purchase orders	30,000	10,000
Number of change orders	400	600
Production setup	1,500	3,500
Machine hours	7,000	3,000

The first impression of the production manager is that the new system will show that indirect costs for PS6 are about the same as they have been for other products over the last several years when a standard model and an upgrade (premium) model were sold. Predictably, they state, the standard model will receive about 1/3 of the indirect cost, and the premium will receive the remaining 2/3. Fundamentally, there are two reasons why the production manager does not like to produce premium versions: They are less profitable for the company, and they require significantly more time and operations to manufacture.

Case Study Exercises

1. Use traditional indirect cost allocation to verify Arnie's cost and price estimates.
2. Use the ABC method to estimate the indirect cost allocation and total cost for each model.
3. If the prices and number of units sold are the same next year (750,000 standard and 250,000 premium), and all other costs remain constant, compare the profit from PS6 under the ABC method with the profit using the traditional indirect cost allocation method.
4. What prices should Health Care Services charge next year based on the ABC method and a 10% markup over cost? What is the total profit from PS6 predicted to be if sales hold steady?
5. Using the results above, comment on the production manager's prediction of indirect costs using ABC (1/3 standard; 2/3 premium) and the two reasons given to not produce the premium version of PS6.

DECEPTIVE ACTS CAN GET YOU IN TROUBLE

Contributed by Dr. Paul Askenasy, Agronomist, Texas Commission on Environmental Quality

Background

Surface mining of coal is the removal of soil and sediments from underlying strata that lie above the material to be mined. The law requires that land disturbed by these types of mining activities be returned to a productive capacity that is as good as or better than its productive capacity before mining.

The productive capacity of soils is directly correlated to the textural (sand, silt, and clay content) and chemical characteristics of the soil (e.g., pH). To this end, mining companies must sample the different soils found in the areas to be disturbed by mining activities. The purpose of the sampling is to establish a baseline characterizing the textural and chemical makeup of the soils prior to mining. Soils in a low pH range (pH values < 5) are indicative of low fertility. Once the natural resource, such as coal, is removed, the pit is backfilled with sediments and the terrain surface is contoured to reestablish the premine drainages. To meet the baseline for pH, the acreage of the mine soils with low pH should not exceed the acreage of the unmined soils with low pH.

Information

Yucatan Mining Company (not the actual name) planned to disturb 600 acres due to mining activities. The different soils within the 600 acres were depicted in the County Soil Survey where the mining activities were to take place. Prior to mining, the company obtained soil samples from 10 different locations within each soil type and had them analyzed for a number of parameters including pH. Assessment of the data indicated that 30% of the area (180 acres) occupied by the soils in the area to be disturbed had pH values between 4.0 and 4.9. The application for mining was approved by the State Department of Mining and Reclamation.

Six years later, 450 acres had been mined and the terrain surface had been leveled to reestablish premine slopes. Of the 450 acres leveled, 175 acres had pH values between 4.0 and 4.9. The president of Yucatan indicated that the company would submit a revised soil baseline based on new sampling in the remaining 150 acres of unmined soils because, in his opinion, the first soil baseline was biased.

The request to do more soil analyses to augment their existing soil baseline was approved. The company quickly hired a consultant to develop the new baseline, and subsequently Yucatan submitted the final report from its consultant to the State Department of Mining and Reclamation. This revised premine soil baseline indicated that 45% of the premine soils had pH values between 4.0 and 4.9. Comparative results between the old and new samples can be expressed as follows:

Soil Baselines	Percent and Acreage of Area	
	Old Soil Baseline	Revised Soil Baseline
pH: 4.0–4.9	30%	45%
	180 acres	270 acres

A rough statistical check between the old and revised soil baselines indicated that the results were mixed. Based on this preliminary result and the fact that there was a significant increase in the percent of area with low-productivity soil, an in-depth analysis of the revised baseline sample study was performed. Contained in the submitted new-sample package was a letter from the Yucatan consultant. It indicated to Yucatan's management that 100 separate soil samples had been obtained and analyzed and that the revised premine soil baseline had been developed using the data from the 30 samples *with the lowest pH values.*

The State Department of Mining and Reclamation staff concluded that the revised soil pH sample data had been carefully "screened" to reduce the amount of remediation work that Yucatan Mining would have to complete. Within a week, Yucatan was notified that further review of the revised soil baseline could not be pursued, because it appeared the revised soil baseline was developed using a technique that skewed results in favor of lower pH values. It was also noted that should Yucatan Mining disagree with this response, the case would be filed with the legal staff as a contested case.

Within several days, the Yucatan president responded indicating that the company was withdrawing the new application from consideration by the department.

Case Study Questions

1. Assume you are the director of the State Department of Mining and Reclamation and were informed of the findings on the new samples versus the old samples. What actions would you direct your staff to take concerning this situation?

2. Suppose there had been several cases of deceptive acts similar to this one over the last few years. What type of "audit" procedures might you want implemented to identify these possibly unethical activities?

3. Yucatan clearly would state that both the old and new samples were randomly located about the entire mining area. When the 30 lowest pH samples were used to establish the new baseline, were the samples still random, according to experimental design standards? If so, why? If not, why not?

4. You and the president of Yucatan Mining have been acquaintances for some years. You have golfed together several times, your and his children are on the same soccer team at school, and your families are members of the same community swimming pool club. What effect would this event have upon your and your family's relationships with the family of the Yucatan president? How would you handle this situation?

5. As a matter of principle and practice, do you believe there is some amount of data-altering or bias-making that is allowed before an application (such as the one described here) should be considered the result of professionally unethical acts? How would you define such a threshold limit?

Depreciation Methods

Daniel Jensen/Getty Images

LEARNING OUTCOMES

Purpose: Use depreciation or depletion methods to reduce the book value of a capital investment in an asset or natural resource.

SECTION	TOPIC	LEARNING OUTCOME
16.1	Terminology	• Define and use the basic terms of asset depreciation.
16.2	Straight line	• Apply the straight line (SL) method of depreciation.
16.3	Declining balance	• Apply the declining balance (DB) and double declining balance (DDB) methods of depreciation.
16.4	MACRS	• Apply the modified accelerated cost recovery system for tax depreciation purposes for U.S.-based corporations.
16.5	Recovery period	• Select the asset recovery period for MACRS depreciation.
16.6	Depletion	• Explain depletion; apply cost depletion and percentage depletion methods.
Chapter 16 Appendix		
16A.1	Historical methods	• Apply the sum-of-years-digits (SYD) and unit-of-production (UOP) methods of depreciation.
16A.2	Switching	• Switch between classical depreciation methods; explain how MACRS provides for switching.
16A.3	MACRS and switching	• Calculate MACRS rates using switching between classical methods and MACRS rules.

The capital investments of a corporation in tangible assets—equipment, computers, vehicles, buildings, and machinery—are commonly recovered on the books of the corporation through *depreciation*. Although the depreciation amount is **not an actual cash flow,** the process of depreciating an asset on the books of the corporation accounts for the decrease in an asset's value because of age, wear, and obsolescence. Even though an asset may be in excellent working condition, the fact that it is worth less through time is taken into account in after-tax economic evaluation studies. An introduction to depreciation types, terminology, and classical methods is followed by a discussion of the *Modified Accelerated Cost Recovery System (MACRS)*, which is the standard in the United States for tax purposes. Other countries commonly use the classical methods for tax computations.

Why is depreciation important to engineering economy? Depreciation is a **tax-allowed deduction** included in tax calculations in virtually all industrialized countries. Depreciation lowers income taxes via the general relation

$$\text{Taxes} = (\text{income} - \text{deductions})(\text{tax rate})$$

Income taxes are discussed further in Chapter 17.

This chapter concludes with an introduction to two methods of **depletion,** which are used to recover capital investments in deposits of natural resources such as oil, gas, minerals, ores, and timber.

The chapter appendix describes two historically useful methods of depreciation—sum-of-years-digits and unit-of-production. Additionally, the appendix includes an in-depth derivation of the MACRS depreciation rates from the straight line and declining balance rates. This is accomplished using a procedure called **switching** between classical depreciation methods.

16.1 Depreciation Terminology ●●●

The concept and types of depreciation are defined here. Most descriptions are applicable to corporations as well as individuals who own depreciable assets.

> **Depreciation** is a book method (noncash) to represent the reduction in value of a tangible asset. The method used to depreciate an asset is a way to account for the decreasing value of the asset to the owner *and* to represent the diminishing value (amount) of the capital funds invested in it. The annual depreciation amount is **not an actual cash flow,** nor does it necessarily reflect the actual usage pattern of the asset during ownership.

Though the term **amortization** is sometimes used interchangeably with the term *depreciation,* they are different. Depreciation is applied to tangible assets, while amortization is used to reflect the decreasing value of intangibles, such as loans, mortgages, patents, trademarks, and goodwill. In addition, the term *capital recovery* is sometimes used to identify depreciation. This is clearly a different use of the term than what we learned in Chapter 5. The term *depreciation* is used throughout this book.

There are two different purposes for using the depreciation methods we will cover in this chapter:

Book depreciation Used by a corporation or business for *internal* financial accounting to *track the value* of an asset or property over its life.

Tax depreciation Used by a corporation or business to *determine taxes due* based on current tax laws of the government entity (country, state, province, etc.). Even though depreciation itself is not a cash flow, it can result in actual cash flow changes because the amount of tax depreciation is a deductible item when calculating annual income taxes for the corporation or business.

The methods applied for these two purposes may or may not utilize the same formulas. Book depreciation indicates the reduced investment in an asset based upon the usage pattern and expected useful life of the asset. There are classical, internationally accepted depreciation methods used to determine book depreciation: straight line, declining balance, and the historical sum-of-years-digits method. The amount of tax depreciation is important in an after-tax engineering economy study and will vary among nations.

In most industrialized countries, the annual **tax depreciation is tax deductible;** that is, it is subtracted from income when calculating the amount of taxes due each year. However, the tax depreciation amount must be calculated using a government-approved method.

Tax depreciation may be calculated and referred to differently in countries outside the United States. For example, in Canada the equivalent is CCA (capital cost allowance), which is calculated based on the undepreciated value of all corporate properties that form a particular class of assets, whereas in the United States depreciation may be determined for each asset separately.

Where allowed, tax depreciation is usually based on an **accelerated method,** whereby the depreciation for the first years of use is larger than that for later years. In the United States, this method is called MACRS, as covered in later sections. In effect, accelerated methods defer some of the income tax burden to later in the asset's life; they do not reduce the total tax burden.

Common terms used in depreciation are explained here.

First cost P or **unadjusted basis B** is the delivered and installed cost of the asset including purchase price, delivery and installation fees, and other depreciable direct costs incurred to prepare the asset for use. The term *unadjusted basis,* or simply *basis,* is used when the asset is new, with the term *adjusted basis* used after some depreciation has been charged. When the first cost has no added, depreciable costs, the basis is the first cost, that is, $P = B$.

Book value BV_t represents the remaining, undepreciated capital investment on the books after the total amount of depreciation charges to date has been subtracted from the basis. The book value is determined at the end of each year t ($t = 1, 2, \ldots, n$), which is consistent with the end-of-year convention.

Recovery period n is the depreciable life of the asset in years. Often there are different n values for book and tax depreciation. Both of these values may be different from the asset's estimated productive life.

Market value MV, a term also used in replacement analysis, is the estimated amount realizable if the asset were sold on the open market. Because of the structure of depreciation laws, the book value and market value may be substantially different. For example, a commercial building tends to increase in market value, but the book value will decrease as depreciation charges are taken. However, a computer workstation may have a market value much lower than its book value due to rapidly changing technology.

Salvage value S is the estimated trade-in or market value at the end of the asset's useful life. The salvage value, expressed as an estimated dollar amount or as a percentage of the first cost, may be positive, zero, or negative due to dismantling and carry-away costs.

Depreciation rate or **recovery rate d_t** is the fraction of the first cost removed by depreciation each year t. This rate may be the same each year, which is called the straight line rate d, or different for each year of the recovery period.

Personal property, one of the two types of property for which depreciation is allowed, is the income-producing, tangible possessions of a corporation used to conduct business. Included is most manufacturing and service industry property—vehicles, manufacturing equipment, materials handling devices, computers and networking equipment, communications equipment, office furniture, refining process equipment, construction assets, and much more.

Real property includes real estate and all improvements—office buildings, manufacturing structures, test facilities, warehouses, apartments, and other structures. *Land itself is considered real property, but it is not depreciable.*

Half-year convention assumes that assets are placed in service or disposed of in midyear, regardless of when these events actually occur during the year. This convention is utilized in this text and in most U.S.-approved tax depreciation methods. There are also midquarter and midmonth conventions.

As mentioned before, there are several models for depreciating assets. The straight line (SL) method is used historically and internationally. Accelerated models, such as the declining

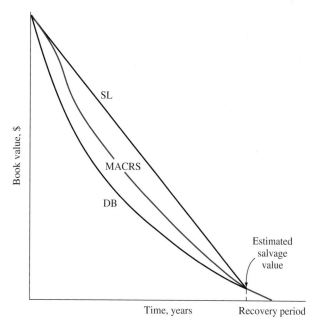

Figure 16–1
General shape of book
value curves for different
depreciation methods.

balance (DB) method, decrease the book value to zero (or to the salvage value) more rapidly than the straight line method, as shown by the general book value curves in Figure 16–1.

For each of the methods—straight line, declining balance, MACRS, and sum-of-years-digits—there are spreadsheet functions available to determine annual depreciation. Each function is introduced and illustrated as the method is explained.

As expected, there are many rules and exceptions to the depreciation laws of a country. One that may be of interest to a U.S.-based *small or medium-sized business* performing an economic analysis is the *Section 179 Deduction*. This is an economic incentive that changes over the years and encourages businesses to invest capital in equipment directly used in the company. Up to a specified amount, the entire basis of an asset is treated as a business expense in the year of purchase. This tax treatment reduces federal income taxes, just as depreciation does, but it is allowed in lieu of depreciating the first cost over several years. The limit changes with time; it was $250,000 in 2008–09 and increased to $500,000 for and after 2010. (An expected $10,000 per year inflation-index addition was to start with 2016.) The economic stimulus efforts in the United States and around the world during the latter part of the decade made many attempts to put investment capital to work in small and medium-sized businesses. Investments above these limits must be depreciated using MACRS.

In the 1980s, the U.S. government standardized accelerated methods for *federal tax depreciation* purposes. In 1981, all classical methods, including straight line, declining balance, and sum-of-years-digits depreciation, were disallowed as tax deductible and replaced by the Accelerated Cost Recovery System (ACRS). In a second round of standardization, MACRS (Modified ACRS) was made the required tax depreciation method in 1986. To this date, the following is the law in the United States.

Tax depreciation must be calculated using MACRS; **book depreciation** may be calculated using any classical method or MACRS.

MACRS has the DB and SL methods, in slightly different forms, embedded in it, but these two methods cannot be used directly if the annual depreciation is to be tax deductible. Many U.S. companies still apply the classical methods for keeping their own books because these methods are more representative of how the usage patterns of the asset reflect the remaining capital invested in it. Most other countries still recognize the classical methods of straight line and declining balance for tax or book purposes. Because of the continuing importance of the SL and DB methods, they are explained in the next two sections prior to MACRS. Appendix Section 16A.1 discusses two other historical methods of depreciation.

Tax law revisions occur often, and depreciation rules are changed from time to time in the United States and other countries. For more depreciation and tax law information, consult the U.S. Department of the Treasury, Internal Revenue Service (IRS), website at www.irs.gov. Pertinent publications can be downloaded. Publication 946, *How to Depreciate Property,* is especially applicable to this chapter. MACRS and most corporate tax depreciation laws are discussed in it.

16.2 Straight Line (SL) Depreciation ● ● ●

Straight line depreciation derives its name from the fact that the book value decreases **linearly with time.** The depreciation rate d_t is the same $(1/n)$ each year t of the recovery period n.

Straight line depreciation is considered the standard against which any depreciation model is compared. For *book depreciation* purposes, it offers an excellent representation of book value for any asset that is used regularly over an estimated number of years. For *tax depreciation,* as mentioned earlier, it is not used directly in the United States, but it is commonly used in most other nations for tax purposes. However, the U.S. MACRS method includes a version of SL depreciation with a larger n value than that prescribed by regular MACRS (see Section 16.5).

The annual SL depreciation is determined by multiplying the first cost minus the salvage value by d_t. In equation form,

$$D_t = (B - S)d_t$$
$$= \frac{B - S}{n} \qquad\qquad [16.1]$$

where t = year $(t = 1, 2, \ldots, n)$
$\quad D_t$ = annual depreciation charge
$\quad B$ = first cost or unadjusted basis
$\quad S$ = estimated salvage value
$\quad n$ = recovery period
$\quad d_t$ = depreciation rate = $1/n$

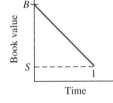

Since the asset is depreciated by the same amount each year, the book value after t years of service, denoted by BV_t, will be equal to the first cost B minus the annual depreciation times t.

$$BV_t = B - tD_t \qquad\qquad [16.2]$$

Earlier we defined d_t as a depreciation rate for a specific year t. However, the SL model has the same rate for all years, that is,

$$d = d_t = \frac{1}{n} \qquad\qquad [16.3]$$

The format for the spreadsheet function to display the annual depreciation D_t in a single-cell operation is

$$= SLN(B,S,n) \qquad\qquad [16.4]$$

EXAMPLE 16.1

If an asset has a first cost of $50,000 with a $10,000 estimated salvage value after 5 years, (*a*) calculate the annual depreciation, and (*b*) calculate and plot the book value of the asset after each year, using straight line depreciation.

Solution
(*a*) The depreciation each year for 5 years can be found by Equation [16.1].

$$D_t = \frac{B - S}{n} = \frac{50,000 - 10,000}{5} = \$8000$$

Enter the function = SLN(50000,10000,5) in any cell to display the D_t of $8000.

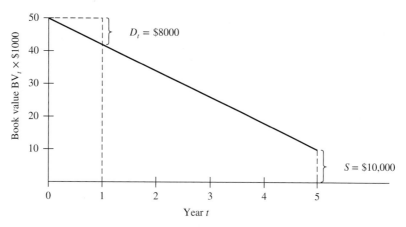

(b) The book value after each year t is computed using Equation [16.2]. The BV_t values are plotted in Figure 16–2. For years 1 and 5, for example,

$$BV_1 = 50,000 - 1(8000) = \$42,000$$
$$BV_5 = 50,000 - 5(8000) = \$10,000 = S$$

16.3 Declining Balance (DB) and Double Declining Balance (DDB) Depreciation ●●●

The **declining balance method** is commonly applied as the book depreciation method. Like the SL method, DB is embedded in the MACRS method, but the DB method itself cannot be used to determine the annual tax-deductible depreciation in the United States. This method is used routinely in most other countries for tax and book depreciation purposes.

Declining balance is also known as the *fixed percentage* or *uniform percentage* method. DB depreciation accelerates the write-off of asset value because the annual depreciation is determined by multiplying the *book value at the beginning of a year* by a fixed (uniform) percentage d, expressed in decimal form. If $d = 0.1$, then 10% of the book value is removed each year. Therefore, the depreciation amount decreases each year.

The maximum annual depreciation rate for the DB method is twice the straight line rate, that is,

$$d_{max} = 2/n \qquad \text{[16.5]}$$

In this case the method is called *double declining balance (DDB)*. If $n = 10$ years, the DDB rate is $2/10 = 0.2$; so 20% of the book value is removed annually. Another commonly used percentage for the DB method is 150% of the SL rate, where $d = 1.5/n$.

The depreciation for year t is the fixed rate d times the book value at the end of the previous year.

$$D_t = (d)BV_{t-1} \qquad \text{[16.6]}$$

The actual depreciation rate for each year t, relative to the basis B, is

$$d_t = d(1 - d)^{t-1} \qquad \text{[16.7]}$$

If BV_{t-1} is not known, the depreciation in year t can be calculated using B and d.

$$D_t = dB(1 - d)^{t-1} \qquad \text{[16.8]}$$

Book value in year t is determined in one of two ways: by using the rate d and basis B or by subtracting the current depreciation charge from the previous book value. The equations are

$$BV_t = B(1 - d)^t \qquad \text{[16.9]}$$
$$BV_t = BV_{t-1} - D_t \qquad \text{[16.10]}$$

It is important to understand that the book value for the DB method never goes to zero because the book value is always decreased by a fixed percentage. The **implied salvage value** after n years is the BV_n amount, that is,

$$\text{Implied } S = BV_n = B(1 - d)^n \qquad [16.11]$$

If a salvage value is estimated for the asset, this *estimated S value is not used in the DB or DDB method* to calculate annual depreciation. However, if the implied $S <$ estimated S, it is necessary to stop charging further depreciation when the book value is at or below the estimated salvage value. In most cases, the estimated S is in the range of zero to the implied S value. (This guideline is important when the DB method can be used directly for tax depreciation purposes.)

If the fixed percentage d is not stated, it is possible to determine an implied fixed rate using the estimated S value, if $S > 0$. The range for d is $0 < d < 2/n$.

$$\text{Implied } d = 1 - \left(\frac{S}{B}\right)^{1/n} \qquad [16.12]$$

The spreadsheet functions DDB and DB are used to display depreciation amounts for specific years. The function is repeated in consecutive spreadsheet cells because the depreciation amount D_t changes with t. For the double declining balance method, the format is

$$= \mathbf{DDB}(B,S,n,t,d) \qquad [16.13]$$

The entry d is the fixed rate expressed as a number between 1 and 2. If omitted, this optional entry is assumed to be 2 for DDB. An entry of $d = 1.5$ makes the DDB function display 150% declining balance method amounts. The DDB function automatically checks to determine when the book value equals the estimated S value. No further depreciation is charged when this occurs. (To allow *full* depreciation charges to be made, ensure that the S entered is between zero and the implied S from Equation [16.11].) Note that $d = 1$ is the same as the straight line rate $1/n$, but D_t *will not be the SL amount* because declining balance depreciation is determined as a fixed percentage of the previous year's book value, which is completely different from the SL calculation in Equation [16.1].

The format for the spreadsheet DB function is $= \text{DB}(B,S,n,t)$. Caution is needed when using this function. The fixed rate d is not entered in the DB function; d is an embedded calculation using a spreadsheet equivalent of Equation [16.12]. Also, only three significant digits are maintained for d, so the book value may go below the estimated salvage value due to round-off errors. Therefore, if the depreciation rate is known, always use the DDB function to ensure correct results. Examples 16.2 and 16.3 illustrate DB and DDB depreciation and their spreadsheet functions.

EXAMPLE 16.2

Underwater electroacoustic transducers were purchased for use in SONAR applications. The equipment will be DDB depreciated over an expected life of 12 years. There is a first cost of $25,000 and an estimated salvage of $2500. (*a*) Calculate the depreciation and book value for years 1 and 4. Write the spreadsheet functions to display depreciation for years 1 and 4. (*b*) Calculate the implied salvage value after 12 years.

Solution

(*a*) The DDB fixed depreciation rate is $d = 2/n = 2/12 = 0.1667$ per year. Use Equations [16.8] and [16.9].

Year 1: $D_1 = (0.1667)(25,000)(1 - 0.1667)^{1-1} = \$4167$
$BV_1 = 25,000(1 - 0.1667)^1 = \$20,833$

Year 4: $D_4 = (0.1667)(25,000)(1 - 0.1667)^{4-1} = \$2411$
$BV_4 = 25,000(1 - 0.1667)^4 = \$12,054$

The DDB functions for D_1 and D_4 are, respectively, $= \text{DDB}(25000,2500,12,1)$ and $= \text{DDB}(25000,2500,12,4)$.

(b) From Equation [16.11], the implied salvage value after 12 years is

$$\text{Implied } S = 25{,}000(1 - 0.1667)^{12} = \$2803$$

Since the estimated $S = \$2500$ is less than $\$2803$, the asset is not fully depreciated when its 12-year expected life is reached.

EXAMPLE 16.3

Freeport-McMoRan Copper and Gold has purchased a new ore grading unit for $80,000. The unit has an anticipated life of 10 years and a salvage value of $10,000. Use the DB and DDB methods to compare the schedule of depreciation and book values for each year. Solve by hand and by spreadsheet.

Solution by Hand

An implied DB depreciation rate is determined by Equation [16.12].

$$d = 1 - \left(\frac{10{,}000}{80{,}000}\right)^{1/10} = 0.1877$$

Note that $0.1877 < 2/n = 0.2$, so this DB model does not exceed twice the straight line rate. Table 16–1 presents the D_t values using Equation [16.6] and the BV_t values from Equation [16.10] rounded to the nearest dollar. For example, in year $t = 2$, the DB results are

$$D_2 = d(BV_1) = 0.1877(64{,}984) = \$12{,}197$$
$$BV_2 = 64{,}984 - 12{,}197 = \$52{,}787$$

Because we round off to even dollars, $2312 is calculated for depreciation in year 10, but $2318 is deducted to make $BV_{10} = S = \$10{,}000$ exactly. Similar calculations for DDB with $d = 0.2$ result in the depreciation and book value series in Table 16–1.

TABLE 16–1 D_t and BV_t Values for DB and DDB Depreciation, Example 16.3

Year t	Declining Balance, $		Double Declining Balance, $	
	D_t	BV_t	D_t	BV_t
0	—	80,000	—	80,000
1	15,016	64,984	16,000	64,000
2	12,197	52,787	12,800	51,200
3	9,908	42,879	10,240	40,960
4	8,048	34,831	8,192	32,768
5	6,538	28,293	6,554	26,214
6	5,311	22,982	5,243	20,972
7	4,314	18,668	4,194	16,777
8	3,504	15,164	3,355	13,422
9	2,846	12,318	2,684	10,737
10	2,318	10,000	737	10,000

Solution by Spreadsheet

The spreadsheet in Figure 16–3 displays the results for the DB and DDB methods. The chart plots book values for each year. Since the fixed rates are close—0.1877 for DB and 0.2 for DDB—the annual depreciation and book value series are approximately the same for the two methods.

The depreciation rate (cell B5) is calculated by Equation [16.12], but note in the cell tags that the DDB function is used in both columns B and D to determine annual depreciation. As mentioned earlier, the DB function automatically calculates the implied rate by Equation [16.12] and maintains it to only three significant digits. Therefore, if the DB function

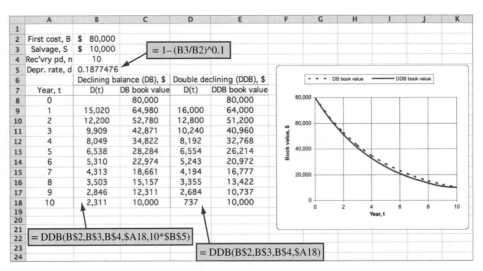

Figure 16–3
Annual depreciation and book value using DB and DDB methods, Example 16.3.

were used in column B (Figure 16–3), the fixed rate applied would be 0.188. The resulting D_t and BV_t values for years 8, 9, and 10 would be as follows:

t	D_t, $	BV_t, $
8	3,501	15,120
9	2,842	12,277
10	2,308	9,969

Also noteworthy is the fact that the DB function uses the implied rate without a check to halt the book value at the estimated salvage value. Thus, BV_{10} will go slightly below $S = \$10,000$, as shown above. However, the DDB function uses a relation different from that of the DB function to determine annual depreciation—one that correctly stops depreciating at the estimated salvage value, as shown in Figure 16–3, cells E17–E18.

16.4 Modified Accelerated Cost Recovery System (MACRS) ● ● ○

MACRS has been the **U.S. required tax depreciation method** for all depreciable assets since the 1980s. It defines statutory depreciation rates that take advantage of the accelerated DB and DDB methods. Corporations are free to apply any of the classical methods for book depreciation. When developed, MACRS and its predecessor ACRS were intended to create economic growth through the investment of new capital and the tax advantages that accelerated depreciation methods offer corporations and businesses.

Many aspects of MACRS deal with the specific depreciation accounting aspects of tax law. This section covers only the elements that materially affect after-tax economic analysis. Additional information on how the DDB, DB, and SL methods are embedded into MACRS and how to derive the MACRS depreciation rates is presented and illustrated in the chapter appendix, Sections 16A.2 and 16A.3.

MACRS determines annual depreciation amounts using the relation

$$D_t = d_t B \qquad\qquad [16.14]$$

where the depreciation rate d_t is provided in tabulated form. As for other methods, the book value in year t is determined by subtracting the depreciation amount from the previous year's book value

$$BV_t = BV_{t-1} - D_t \qquad [16.15]$$

or by subtracting the total depreciation from the first cost or unadjusted basis.

$$BV_t = \text{unadjusted basis} - \text{sum of accumulated depreciation}$$

$$= B - \sum_{j=1}^{j=t} D_j \qquad [16.16]$$

MACRS has standardized and simplified many of the decisions and calculations of depreciation.

The basis B (or first cost P) is completely depreciated; salvage is always assumed to be zero, or $S = \$0$.

Recovery periods are standardized to specific values:

$n = 3, 5, 7, 10, 15$, or 20 years   for personal property (e.g., equipment or vehicles)
$n = 27.5$ or 39 years   for real property (e.g., rental property or structures)

Depreciation rates provide accelerated write-off by incorporating switching between classical methods.

Section 16.5 explains how to determine an allowable MACRS recovery period. The MACRS personal property depreciation rates (d_t values) for $n = 3, 5, 7, 10, 15$, and 20 for use in Equation [16.14] are included in Table 16–2.

MACRS depreciation rates incorporate the DDB method ($d = 2/n$) and switch to SL depreciation during the recovery period as an inherent component for *personal property* depreciation. The MACRS rates start with the DDB rate or the 150% DB rate and switch when the SL method offers faster write-off.

TABLE 16–2   Depreciation Rates d_t Applied to the Basis B for the MACRS Method

	Depreciation Rate (%) for Each MACRS Recovery Period in Years					
Year	$n = 3$	$n = 5$	$n = 7$	$n = 10$	$n = 15$	$n = 20$
1	33.33	20.00	14.29	10.00	5.00	3.75
2	44.45	32.00	24.49	18.00	9.50	7.22
3	14.81	19.20	17.49	14.40	8.55	6.68
4	7.41	11.52	12.49	11.52	7.70	6.18
5		11.52	8.93	9.22	6.93	5.71
6		5.76	8.92	7.37	6.23	5.29
7			8.93	6.55	5.90	4.89
8			4.46	6.55	5.90	4.52
9				6.56	5.91	4.46
10				6.55	5.90	4.46
11				3.28	5.91	4.46
12					5.90	4.46
13					5.91	4.46
14					5.90	4.46
15					5.91	4.46
16					2.95	4.46
17–20						4.46
21						2.23

For *real property,* MACRS utilizes the SL method for $n = 39$ throughout the recovery period. The annual percentage depreciation rate is $d = 1/39 = 0.02564$. However, MACRS forces partial-year recovery in years 1 and 40. The MACRS real property rates in percentage amounts are

Year 1	$100d_1 = 1.391\%$
Years 2–39	$100d_t = 2.564\%$
Year 40	$100d_{40} = 1.177\%$

The real property recovery period of 27.5 years, which applies only to residential rental property, uses the SL method in a similar fashion.

Note that all MACRS depreciation rates in Table 16–2 are presented for 1 year longer than the stated recovery period. Also note that the extra-year rate is one-half of the previous year's rate. This is so because a built-in **half-year convention** is imposed by MACRS. This convention assumes that all property is placed in service at the midpoint of the tax year of installation. Therefore, only 50% of the first-year DB depreciation applies for tax purposes. This removes some of the accelerated depreciation advantage and requires that one-half year of depreciation be taken in year $n + 1$.

No specially designed spreadsheet function is present for MACRS depreciation. However, the variable declining balance (VDB) function, which is used to determine when to switch between classical methods, can be adapted to display MACRS deprecation for each year. (The VDB function is explained in detail in Section 16A.2 of this chapter and Appendix A of the text.) The MACRS depreciation format of the VDB function requires embedded MAX and MIN functions, as follows:

$$= \text{VDB}(B, 0, n, \text{MAX}(0, t-1.5), \text{MIN}(n, t-0.5), d) \qquad [16.17]$$

where
B = unadjusted basis
0 = salvage value of $S = 0$
n = recovery period
$d = \begin{cases} 2 & \text{if MACRS } n = 3, 5, 7, \text{ or } 10 \\ 1.5 & \text{if MACRS } n = 15 \text{ or } 20 \end{cases}$

The MAX and MIN functions ensure that the MACRS half-year conventions are followed; that is, only one-half of the first year's depreciation is charged in year 1, and one-half of the last year's charge is carried over to year $n + 1$.

EXAMPLE 16.4

Chevron Phillips Chemical Company in Baytown, Texas, acquired new equipment for its polyethylene processing line. This chemical is a resin used in plastic pipe, retail bags, blow molding, and injection molding. The equipment has an unadjusted basis of $B = \$400{,}000$, a life of only 3 years, and a salvage value of 5% of B. The chief engineer asked the finance director to provide an analysis of the difference between (1) the DDB method, which is the internal book depreciation and book value method used at the plant, and (2) the required MACRS tax depreciation and its book value. He is especially curious about the differences after 2 years of service for this short-lived, but expensive asset. Use hand and spreadsheet solutions to do the following:

(a) Determine which method offers the larger total depreciation after 2 years.
(b) Determine the book value for each method after 2 years and at the end of the recovery period.

Solution by Hand
The basis is $B = \$400{,}000$ and the estimated $S = 0.05(400{,}000) = \$20{,}000$. The MACRS rates for $n = 3$ are taken from Table 16–2, and the depreciation rate for DDB is $d_{max} = 2/3 = 0.6667$. Table 16–3 presents the depreciation and book values. Year 3 depreciation for DDB would be $\$44{,}444(0.6667) = \$29{,}629$, except this would make $BV_3 < \$20{,}000$. Only the remaining amount of $\$24{,}444$ is removed.

		MACRS		**DDB**	
Year	Rate	Tax Depreciation, $	Book Value, $	Book Depreciation, $	Book Value, $
0			400,000		400,000
1	0.3333	133,320	266,680	266,667	133,333
2	0.4445	177,800	88,880	88,889	44,444
3	0.1481	59,240	29,640	24,444	20,000
4	0.0741	29,640	0		

TABLE 16–3   Comparing MACRS and DDB Depreciation, Example 16.4

(*a*) The 2-year accumulated depreciation values from Table 16–3 are

MACRS: $D_1 + D_2 = \$133{,}320 + 177{,}800 = \$311{,}120$

DDB: $D_1 + D_2 = \$266{,}667 + 88{,}889 = \$355{,}556$

The DDB depreciation is larger. (Remember that for tax purposes, the company does not have the choice in the United States of DDB as applied here.)

(*b*) After 2 years the book value for DDB at $44,444 is 50% of the MACRS book value of $88,880. At the end of recovery (4 years for MACRS due to the built-in half-year convention, and 3 years for DDB), the MACRS book value is $BV_4 = 0$ and for DDB, $BV_3 = \$20{,}000$. This occurs because MACRS always removes the entire first cost, regardless of the estimated salvage value. This is a tax depreciation advantage of the MACRS method (unless the asset is disposed of for more than the MACRS-depreciated book value, as discussed in Section 17.4).

Solution by Spreadsheet

Figure 16–4 presents the spreadsheet solution using the VDB function (column B) for MACRS depreciation (in lieu of the MACRS rates) and applying the DDB function in column D.

(*a*) The 2-year accumulated depreciation values are

MACRS (add cells B6 + B7): $\$133{,}333 + 177{,}778 = \$311{,}111$

DDB (add cells D6 + D7): $\$266{,}667 + 88{,}889 = \$355{,}556$

(*b*) Book values after 2 years are

MACRS (cell C7): $88,889

DDB (cell E7): $44,444

The book values are plotted in Figure 16–4. Observe that MACRS goes to zero in year 4, while DDB stops at $20,000 in year 3.

Comment

It is advisable to set up a *spreadsheet template* for use with depreciation problems in this and future chapters. The format and functions of Figure 16–4 are a good template for MACRS and DDB methods.

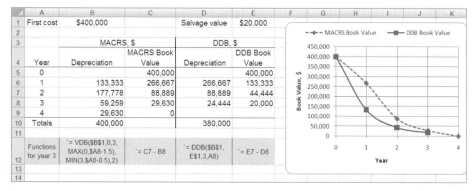

Figure 16–4
Spreadsheet screen shot of MACRS and DDB depreciation and book value, Example 16.4.

MACRS simplifies depreciation computations, but it removes much of the flexibility of method selection for a business or corporation. In general, an economic comparison that includes depreciation may be performed more rapidly and usually without altering the final decision by applying the *classical straight line method in lieu of MACRS*.

16.5 Determining the MACRS Recovery Period ● ● ●

The expected useful life of property is estimated in years and used as the *n* value in alternative evaluation and in depreciation computations. For book depreciation the *n* value should be the expected useful life. However, when the depreciation will be claimed as tax deductible, the *n* value should be lower. There are tables that assist in determining the life and recovery period for tax purposes.

> The advantage of a recovery period **shorter** than the anticipated useful life is leveraged by the accelerated depreciation methods that write off more of the basis *B* in the initial years.

The U.S. government requires that all depreciable property be classified into a **property class** which identifies its MACRS-allowed recovery period. Table 16–4, a selection of material from IRS Publication 946, gives examples of assets and the MACRS *n* values. Virtually any property considered in an economic analysis has a MACRS *n* value of 3, 5, 7, 10, 15, or 20 years.

Table 16–4 provides two MACRS *n* values for each property. The first is the **general depreciation system (GDS)** value, which we use in examples and problems. The depreciation rates in Table 16–2 correspond to the *n* values for the GDS column and provide the fastest write-off allowed. The rates utilize the DDB method or the 150% DB method with a switch to SL depreciation. Note that any asset not in a stated class is automatically assigned a 7-year recovery period under GDS.

The far right column of Table 16–4 lists the **alternative depreciation system (ADS)** recovery period range. This alternative method allows the use of *SL depreciation over a longer*

TABLE 16–4	Example MACRS Recovery Periods for Various Asset Descriptions	
	MACRS **_n_ Value, Years**	
Asset Description (Personal and Real Property)	**GDS**	**ADS Range**
Special manufacturing and handling devices, tractors, racehorses	3	3–5
Computers and peripherals, oil and gas drilling equipment, construction assets, autos, trucks, buses, cargo containers, some manufacturing equipment	5	6–9.5
Office furniture; some manufacturing equipment; railroad cars, engines, tracks; agricultural machinery; petroleum and natural gas equipment; **all property not in another class**	7	10–15
Equipment for water transportation, petroleum refining, agriculture product processing, durable-goods manufacturing, shipbuilding	10	15–19
Land improvements, docks, roads, drainage, bridges, landscaping, pipelines, nuclear power production equipment, telephone distribution	15	20–24
Municipal sewers, farm buildings, telephone switching buildings, power production equipment (steam and hydraulic), water utilities	20	25–50
Residential rental property (house, mobile home)	27.5	40
Nonresidential real property attached to the land, but not the land itself	39	40

recovery period than the GDS. The half-year convention applies, and any salvage value is neglected, as it is in regular MACRS. The use of ADS is generally a choice left to a company, but it is required for some special asset situations. Since it takes longer to depreciate the asset, and since the SL model is required (thus removing the advantage of accelerated depreciation), ADS is usually not considered an option for the economic analysis. This electable SL option is, however, sometimes chosen by businesses that are young and do not need the tax benefit of accelerated depreciation during the first years of operation and asset ownership. If ADS is selected, tables of d_t rates are available.

16.6 Depletion Methods ●●●

Previously, for all assets, facilities, and equipment that can be replaced, we have applied depreciation. We now turn to irreplaceable natural resources and the equivalent of depreciation, which is called *depletion.*

> **Depletion** is a book method (noncash) to represent the decreasing value of a **natural resource** as it is recovered, removed, or felled. The two methods of depletion for book or tax purposes are used to write off the first cost, or value of the estimated quantity, of resources in mines, wells, quarries, geothermal deposits, forests, and the like.

The two methods of depletion are *cost* and *percentage depletion,* as described below. Details for U.S. taxes on depletion are found in IRS Publication 535, *Business Expenses.*

Cost depletion Sometimes referred to as factor depletion, cost depletion is based on the level of activity or usage, not time, as in depreciation. Cost depletion may be applied to most types of natural resources and *must* be applied to timber production. The cost depletion factor for year *t*, denoted by CD_t, is the ratio of the first cost of the resource to the estimated number of units recoverable.

$$CD_t = \frac{\text{first cost}}{\text{resource capacity}} \qquad [16.18]$$

The annual depletion charge is CD_t times the year's usage or volume. *The total cost depletion cannot exceed the first cost of the resource.* If the capacity of the property is reestimated some year in the future, a new cost depletion factor is determined based upon the undepleted amount and the new capacity estimate.

Percentage depletion This is a special consideration given for natural resources. A constant, stated percentage of the resource's **gross income** may be depleted each year *provided it does not exceed 50% of the company's taxable income.* The depletion amount for year *t* is calculated as

$$\text{Percentage depletion}_t = \text{percentage depletion rate} \\ \times \text{ gross income from property} \\ = PD \times GI_t \qquad [16.19]$$

Using percentage depletion, total depletion charges may exceed first cost with no limitation. The U.S. government does not generally allow percentage depletion to be applied to oil and gas wells (except small independent producers).

The annual percentage depletion rates for some common natural deposits are listed below per U.S. tax law.

Deposit	Percentage of Gross Income, PD
Sulfur, uranium, lead, nickel, zinc, and some other ores and minerals	22
Gold, silver, copper, iron ore, and some oil shale	15
Oil and natural gas wells (varies)	15–22
Coal, lignite, sodium chloride	10
Gravel, sand, clay, some stones	5
Most other minerals, metallic ores	14

EXAMPLE 16.5

Weyerhaeuser has negotiated the rights to cut timber on privately held forest acreage for $700,000. An estimated 350 million board feet of lumber is harvestable.

(a) Determine the depletion amount for the first 2 years if 15 million and 22 million board feet are removed.
(b) After 2 years the total recoverable board feet was reestimated upward to be 450 million from the time the rights were purchased. Compute the new cost depletion factor for years 3 and later.

Solution

(a) Use Equation [16.18] for CD_t in dollars per million board feet.

$$CD_t = \frac{700,000}{350} = \$2000 \text{ per million board feet}$$

Multiply CD_t by the annual harvest to obtain depletion of $30,000 in year 1 and $44,000 in year 2. Continue until a total of $700,000 is written off.

(b) After 2 years, a total of $74,000 has been depleted. A new CD_t value must be calculated based on the remaining $700,000 - 74,000 = \$626,000$ investment. Additionally, with the new estimate of 450 million board feet, a total of $450 - 15 - 22 = 413$ million board feet remains. For years $t = 3, 4, \ldots,$ the cost depletion factor is

$$CD_t = \frac{626,000}{413} = \$1516 \text{ per million board feet}$$

EXAMPLE 16.6

A gold mine was purchased for $10 million. It has an anticipated gross income of $5.0 million per year for years 1 to 5 and $3.0 million per year after year 5. Assume that depletion charges do not exceed 50% of taxable income. Compute annual depletion amounts for the mine. How long will it take to recover the initial investment at $i = 0\%$?

Solution

The rate for gold is PD = 0.15. Depletion amounts are

Years 1 to 5:   0.15(5.0 million) = $750,000

Years thereafter: 0.15(3.0 million) = $450,000

A total of $3.75 million is written off in 5 years, and the remaining $6.25 million is written off at $450,000 per year. The total number of years is

$$5 + \frac{\$6.25 \text{ million}}{\$450,000} = 5 + 13.9 = 18.9$$

In 19 years, the initial investment could be fully depleted.

In many of the natural resource depletion situations, the tax law allows the larger of the two depletion amounts to be claimed each year. This is allowed provided the percentage depletion amount does not exceed 50% of taxable income. Therefore, it is wise to calculate both depletion amounts and select the larger. Use the following terminology for year t ($t = 1, 2, \ldots$).

$$CDA_t = \text{cost depletion amount}$$
$$PDA_t = \text{percentage depletion amount}$$
$$TI_t = \text{taxable income}$$

The guideline for the tax-allowed depletion amount for year t is

$$\text{Depletion} = \begin{cases} \max[CDA_t, PDA_t] & \text{if } PDA_t \leq 50\% \text{ of } TI_t \\ \max[CDA_t, 50\% \text{ of } TI_t] & \text{if } PDA_t < 50\% \text{ of } TI_t \end{cases}$$

For example, assume a medium-sized quarry owner calculates the following for 1 year.

$$TI = \$500{,}000 \qquad CDA = \$275{,}000 \qquad PDA = \$280{,}000$$

Since 50% of TI is $250,000, the PDA is too large and, therefore, is not allowed. For tax purposes, apply the guideline above and use the cost depletion of $275,000, since it is larger than 50% of TI.

CHAPTER SUMMARY

Depreciation may be determined for internal company records (book depreciation) or for income tax purposes (tax depreciation). In the United States, the MACRS method is the only one allowed for tax depreciation. In many other countries, straight line and declining balance methods are applied for both tax and book depreciation. Depreciation does not result in cash flow directly. It is a book method by which the capital investment in tangible property is recovered. The annual depreciation amount is tax deductible, which can result in actual cash flow changes.

Some important points about the straight line, declining balance, and MACRS methods are presented below. Common relations for each method are summarized in Table 16–5.

Straight Line (SL)

- It writes off capital investment linearly over n years.
- The estimated salvage value is always considered.
- This is the classical, nonaccelerated depreciation model.

Declining Balance (DB)

- The method accelerates depreciation compared to the straight line method.
- The book value is reduced each year by a fixed percentage.
- The most used rate is twice the SL rate, which is called double declining balance (DDB).
- It has an implied salvage that may be lower than the estimated salvage.
- It is not an approved tax depreciation method in the United States. It is frequently used for book depreciation purposes.

Modified Accelerated Cost Recovery System (MACRS)

- It is the only approved tax depreciation system in the United States.
- It automatically switches from DDB or DB to SL depreciation.
- It always depreciates to zero; that is, it assumes $S = 0$.
- Recovery periods are specified by property classes.
- Depreciation rates are tabulated.
- The actual recovery period is 1 year longer due to the imposed half-year convention.
- MACRS straight line depreciation is an option, but recovery periods are longer than those for regular MACRS.

Cost and *percentage depletion methods* recover investment in natural resources. The annual cost depletion factor is applied to the amount of resource removed. No more than the initial investment can be recovered with cost depletion. Percentage depletion, which can recover more than the initial investment, reduces the investment value by a constant percentage of gross income each year.

Method	MACRS	SL	DDB
TABLE 16–5 Summary of Common Depreciation Method Relations			
Fixed depreciation rate d	Not defined	$\dfrac{1}{n}$	$\dfrac{2}{n}$
Annual rate d_t	Table 16–2	$\dfrac{1}{n}$	$d(1-d)^{t-1}$
Annual depreciation D_t	$d_t B$	$\dfrac{B-S}{n}$	$d(\text{BV}_{t-1})$
Book value BV_t	$\text{BV}_{t-1} - D_t$	$B - tD_t$	$B(1-d)^t$

CHAPTER 16   APPENDIX

16A.1 Sum-of-Years-Digits (SYD) and Unit-of-Production (UOP) Depreciation ● ● ◐

The **SYD method** is the first historical accelerated depreciation technique that can remove over 40% of the first cost in the first 25% of a 20-year recovery period; however, write-off is not as rapid as for DDB or MACRS. This technique may be used in an engineering economy analysis in the book depreciation of multiple-asset accounts (group and composite depreciation).

The mechanics of the method involve the sum of the year's digits from 1 through the recovery period n. The depreciation charge for any given year is obtained by multiplying the basis of the asset, less any salvage value, by the ratio of the number of years remaining in the recovery period to the sum of the year's digits, SUM.

$$D_t = \frac{\text{depreciable years remaining}}{\text{sum of years digits}}(\text{basis} - \text{salvage value})$$

$$D_t = \frac{n - t + 1}{\text{SUM}}(B - S) \qquad \text{[16A.1]}$$

where SUM is the sum of the digits 1 through n.

$$\text{SUM} = \sum_{j=1}^{j=n} j = \frac{n(n+1)}{2}$$

The book value for any year t is calculated as

$$BV_t = B - \frac{t(n - t/2 + 0.5)}{\text{SUM}}(B-S) \qquad \text{[16A.2]}$$

The rate of depreciation decreases each year and equals the multiplier in Equation [16A.1].

$$d_t = \frac{n - t + 1}{\text{SUM}} \qquad \text{[16A.3]}$$

The SYD spreadsheet function displays the depreciation for the year t. The function format is

$$= \text{SYD}(B,S,n,t)$$

EXAMPLE 16A.1

Calculate the SYD depreciation charges for year 2 for electro-optics equipment with $B = \$25,000$, $S = \$4000$, and an 8-year recovery period.

Solution

The sum of the year's digits is 36, and the depreciation amount for the second year by Equation [16A.1] is

$$D_2 = \frac{7}{36}(21,000) = \$4083$$

The SYD function is $= \text{SYD}(25000,4000,8,2)$.

Figure 16A–1 is a plot of the book values for an \$80,000 asset with $S = \$10,000$ and $n = 10$ years using the four depreciation methods that we have learned. The MACRS, DDB, and SYD curves track closely except for year 1 and years 9 through 11.

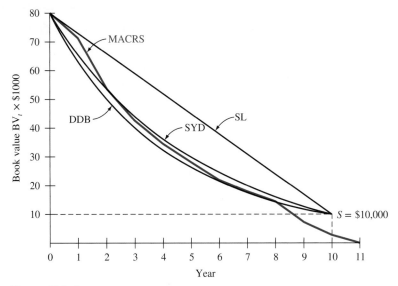

Figure 16A–1
Comparison of book values using SL, SYD, DDB, and MACRS depreciation.

A second depreciation method that is useful when the total number of units to be produced or hours of usage over the useful lifetime of an asset is estimated is the **unit-of-production (UOP) method.** When the decreasing value of equipment is **based on usage, not time,** the UOP method is quite applicable. Suppose a highway contractor has a series of state highway department contracts that will last several years and that earth moving equipment is purchased for use on all contracts. If the equipment usage (e.g., in hours) goes up and down significantly over the years, the UOP method is ideal for book depreciation. For year t, UOP deprecation is calculated as

$$D_t = \frac{\text{actual usage for year } t}{\text{total lifetime usage}} \text{(basis – salvage)} \qquad \textbf{[16A.4]}$$

EXAMPLE 16A.2

Zachry Contractors purchased an $80,000 mixer for use during the next 10 years for contract work on IH-10 in San Antonio. The mixer will have a negligible salvage value after 10 years, and the total amount of material to process is estimated at 2 million m^3. Use the actual usage per year shown in Table 16A–1 and the unit-of-production method to determine annual depreciation.

Solution

The actual usage each year is placed in the numerator of Equation [16A.4] to determine the annual depreciation based on the estimated total lifetime amount of material, 2 million m^3 in this case. Table 16A–1 shows the annual and cumulative depreciation over the 10 years. If the mixer is continued in service after the 2 million m^3 is processed, no further depreciation is allowed.

TABLE 16A–1	Unit-of-Production Method of Depreciation, Example 16A.2		
Year t	Actual Usage, 1000 m^3	Annual Depreciation D_t, $	Cumulative Depreciation, $
1	400	16,000	16,000
2–8	200	8,000	72,000
9–10	100	4,000	80,000
Total	2,000	80,000	

16A.2 Switching between Depreciation Methods ● ● ○

Switching between depreciation methods may assist in accelerated reduction of the book value. It also maximizes the present value of accumulated and total depreciation over the recovery period. Therefore, switching usually increases the tax advantage in years where the depreciation is larger. The approach below is an inherent part of MACRS.

Switching from a DB method to the SL method is the most common switch because it usually offers a real advantage, especially if the DB method is DDB. General rules of switching are summarized here.

1. Switching is recommended when the depreciation for year t by the currently used method is less than that for a new method. The selected depreciation D_t is the larger amount.
2. Only one switch can take place during the recovery period.
3. Regardless of the (classical) depreciation methods, the book value cannot go below the estimated salvage value. When switching from a DB method, the estimated salvage value, not the DB-implied salvage value, is used to compute the depreciation for the new method; we assume $S = 0$ in all cases. (This does not apply to MACRS, since it already includes switching.)
4. The undepreciated amount, that is, BV_t, is used as the new adjusted basis to select the larger D_t for the next switching decision.

In all situations, the criterion is to **maximize the present worth of the total depreciation PW_D.** The combination of depreciation methods that results in the largest present worth is the best switching strategy.

$$PW_D = \sum_{t=1}^{t=n} D_t(P/F,i,t) \qquad [16A.5]$$

This logic minimizes tax liability in the early part of an asset's recovery period.

Switching is most advantageous from a rapid write-off method such as DDB to the SL model. This switch is predictably advantageous if the implied salvage value computed by Equation [16.11] exceeds the salvage value estimated at purchase time; that is, switch if

$$BV_n = B(1 - d)^n > \text{estimated } S \qquad [16A.6]$$

Since we assume that S will be zero per rule 3 above, and since BV_n *will be greater than zero,* for a DB method a switch to SL is always advantageous. Depending upon the values of d and n, the switch may be best in the later years or last year of the recovery period, which removes the implied S inherent to the DDB model.

The procedure to switch from DDB to SL depreciation is as follows:

1. For each year t, compute the two depreciation charges.

For DDB: $$D_{DDB} = d(BV_{t-1}) \qquad [16A.7]$$

For SL: $$D_{SL} = \frac{BV_{t-1}}{n - t + 1} \qquad [16A.8]$$

2. Select the larger depreciation value. The depreciation for each year is

$$D_t = \max[D_{DDB}, D_{SL}] \qquad [16A.9]$$

3. If needed, determine the present worth of total depreciation, using Equation [16A.5].

It is acceptable, though not usually financially advantageous, to state that a switch will take place in a particular year, for example, a mandated switch from DDB to SL in year 7 of a 10-year recovery period. This approach is usually not taken, but the switching technique will work correctly for all depreciation methods.

To use a spreadsheet for switching, first understand the depreciation model switching rules and practice the switching procedure from declining balance to straight line. Once these are understood, the mechanics of the switching can be speeded up by applying the spreadsheet function

VDB (variable declining balance). This is a quite powerful function that determines the depreciation for 1 year or the total over several years for the DB-to-SL switch. The function format is

$$= \text{VDB}(B,S,n,\text{start\_}t,\text{end\_}t,d,\text{no\_switch}) \qquad [16A.10]$$

Appendix A explains all the fields in detail, but for simple applications, where the DDB and SL annual D_t values are needed, the following are correct entries:

start_t is the year $(t-1)$

end_t is year t

d is optional; 2 for DDB is assumed, the same as in the DDB function

no_switch is an optional logical value:

FALSE or omitted—switch to SL occurs, if advantageous

TRUE—DDB or DB method is applied with no switching to SL depreciation considered.

Entering TRUE for the no_switch option obviously causes the VDB function to display the same depreciation amounts as the DDB function. This is discussed in Example 16A.3d. You may notice that the VDB function is the same one used to calculate annual MACRS depreciation.

EXAMPLE 16A.3

The Outback Steakhouse main office has purchased a $100,000 online document imaging system with an estimated useful life of 8 years and a tax depreciation recovery period of 5 years. Compare the present worth of total depreciation for (*a*) the SL method, (*b*) the DDB method, and (*c*) DDB-to-SL switching. (*d*) Perform the DDB-to-SL switch using a spreadsheet and plot the book values. Use an interest rate of $i = 15\%$ per year.

Solution by Hand
The MACRS method is not involved in this solution.

(*a*) Equation [16.1] determines the annual SL depreciation.

$$D_t = \frac{100,000 - 0}{5} = \$20,000$$

Since D_t is the same for all years, the P/A factor replaces P/F to compute PW_D.

$$\text{PW}_D = 20,000(P/A,15\%,5) = 20,000(3.3522) = \$67,044$$

(*b*) For DDB, $d = 2/5 = 0.40$. The results are shown in Table 16A–2. The value $\text{PW}_D = \$69,915$ exceeds $67,044 for SL depreciation. As is predictable, the accelerated depreciation of DDB increases PW_D.

(*c*) Use the DDB-to-SL switching procedure.

1. The DDB values for D_t in Table 16A–2 are repeated in Table 16A–3 for comparison with the D_{SL} values from Equation [16A.8]. The D_{SL} values change each year because BV_{t-1} is different. Only in year 1 is $D_{\text{SL}} = \$20,000$, the same as computed in part (*a*). For illustration, compute D_{SL} values for years 2 and 4. For $t = 2$, $\text{BV}_1 = \$60,000$ by the DDB method and

$$D_{\text{SL}} = \frac{60,000 - 0}{5 - 2 + 1} = \$15,000$$

For $t = 4$, $\text{BV}_3 = \$21,600$ by the DDB method and

$$D_{\text{SL}} = \frac{21,600 - 0}{5 - 4 + 1} = \$10,800$$

2. The column "Larger D_t" indicates a switch in year 4 with $D_4 = \$10,800$. The $D_{\text{SL}} = \$12,960$ in year 5 would apply *only* if the switch occurred in year 5. Total depreciation with switching is $100,000 compared to the DDB amount of $92,224.

3. With switching, $\text{PW}_D = \$73,943$, which is an increase over both the SL and DDB methods.

TABLE 16A–2 DDB Model Depreciation and Present Worth Computations, Example 16A.3*b*

Year t	D_t, $	BV_t, $	$(P/F, 15\%, t)$	Present Worth of D_t, $
0		100,000		
1	40,000	60,000	0.8696	34,784
2	24,000	36,000	0.7561	18,146
3	14,400	21,600	0.6575	9,468
4	8,640	12,960	0.5718	4,940
5	5,184	7,776	0.4972	2,577
Totals	92,224			69,915

TABLE 16A–3 Depreciation and Present Worth for DDB-to-SL Switching, Example 16A.3*c*

Year t	DDB Method, $ D_{DDB}	DDB Method, $ BV_t	SL Method D_{SL}, $	Larger D_t, $	P/F Factor	Present Worth of D_t, $
0	—	100,000				
1	40,000	60,000	20,000	40,000	0.8696	34,784
2	24,000	36,000	15,000	24,000	0.7561	18,146
3	14,400	21,600	12,000	14,400	0.6575	9,468
4*	8,640	12,960	10,800	10,800	0.5718	6,175
5	5,184	7,776	12,960	10,800	0.4972	5,370
Totals	92,224			100,000		73,943

*Indicates year of switch from DDB to SL depreciation.

Solution by Spreadsheet

(*d*) In Figure 16A–2, column D entries are the VDB functions to determine that the DDB-to-SL switch should take place in year 4. The entries "2,FALSE" at the end of the VDB function are optional (see the VDB function description). If TRUE were entered, the declining balance model would be maintained throughout the recovery period, and the annual depreciation amounts would be equal to those in column B. The plot in Figure 16A–2 indicates another difference in depreciation methods. The terminal book value in year 5 for the DDB method is $BV_5 = \$7776$, while the DDB-to-SL switch reduces the book value to zero.

The NPV function determines the PW of depreciation (row 9). The results here are the same as in parts (*b*) and (*c*) above. The DDB-to-SL switch has the larger PW_D value.

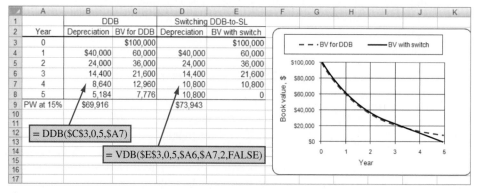

Figure 16A–2
Depreciation for DDB-to-SL switch using the VDB function, Example 16A.3.

In MACRS, recovery periods of 3, 5, 7, and 10 years apply DDB depreciation with half-year convention switching to SL. When the switch to SL takes place, which is usually in the last 1 to 3 years of the recovery period, any remaining basis is charged off in year $n + 1$ so that the book value reaches zero. Usually 50% of the applicable SL amount remains after the switch has occurred. For recovery periods of 15 and 20 years, 150% DB with the half-year convention and the switch to SL apply.

The present worth of depreciation PW_D will always indicate which method is the most advantageous. Only the MACRS rates for the GDS recovery periods (Table 16–4) utilize the DDB-to-SL switch. The MACRS rates for the alternative depreciation system (ADS) have longer recovery periods and impose the SL model for the entire recovery period.

EXAMPLE 16A.4

In Example 16A.3, parts (c) and (d), the DDB-to-SL switching method was applied to a $100,000, $n = 5$ year asset resulting in $PW_D = \$73,943$ at $i = 15\%$. Use MACRS to depreciate the same asset for a 5-year recovery period, and compare PW_D values.

Solution

Table 16A–4 summarizes the computations for depreciation (using Table 16–2 rates), book value, and present worth of depreciation. The PW_D values for all four methods are as follows:

DDB-to-SL switching	$73,943
Double declining balance	$69,916
MACRS	$69,016
Straight line	$67,044

MACRS provides a slightly less accelerated write-off. This is so, in part, because the half-year convention disallows 50% of the first-year DDB depreciation (which amounts to 20% of the basis). Also the MACRS recovery period extends to year 6, further reducing PW_D.

TABLE 16A–4	Depreciation and Book Value Using MACRS, Example 16A.4		
t	d_t	D_t, $	BV_t, $
0	—	—	100,000
1	0.20	20,000	80,000
2	0.32	32,000	48,000
3	0.192	19,200	28,800
4	0.1152	11,520	17,280
5	0.1152	11,520	5,760
6	0.0576	5,760	0
	1.000	100,000	

$$PW_D = \sum_{t=1}^{t=6} D_t(P/F,15\%,t) = \$69,016$$

16A.3 Determination of MACRS Rates ● ● ●

The depreciation rates for MACRS incorporate the DB-to-SL switching for all GDS recovery periods from 3 to 20 years. In the first year, some adjustments have been made to compute the MACRS rate. The adjustments vary and are not usually considered in detail in economic analyses. The half-year convention is always imposed, and any remaining book value in year n is removed in year $n + 1$. The value $S = 0$ is assumed for all MACRS schedules.

Since different DB depreciation rates apply for different n values, the following summary may be used to determine D_t and BV_t values. The symbols D_{DB} and D_{SL} are used to identify DB and SL depreciation, respectively.

For n = 3, 5, 7, and 10   Use DDB depreciation with the half-year convention, switching to SL depreciation in year t when $D_{SL} \geq D_{DB}$. Use the switching rules of Section 16A.2, and add one-half year when computing D_{SL} to account for the half-year convention. The yearly depreciation rates are

$$d_t = \begin{cases} \dfrac{1}{n} & t = 1 \\ \dfrac{2}{n} & t = 2, 3, \ldots \end{cases} \qquad [16A.11]$$

Annual depreciation values for each year t applied to the adjusted basis, allowing for the half-year convention, are

$$D_{DB} = d_t(BV_{t-1}) \qquad [16A.12]$$

$$D_{SL} = \begin{cases} \dfrac{1}{2}\left(\dfrac{1}{n}\right)B & t = 1 \\ \dfrac{BV_{t-1}}{n - t + 1.5} & t = 2, 3, \ldots, n \end{cases} \qquad [16A.13]$$

After the switch to SL depreciation takes place—usually in the last 1 to 3 years of the recovery period—any remaining book value in year n is removed in year $n + 1$.

For n = 15 and 20   Use 150% DB with the half-year convention and the switch to SL when $D_{SL} \geq D_{DB}$. Until SL depreciation is more advantageous, the annual DB depreciation is computed using a form of Equation [16A.7]

$$D_{DB} = d_t(BV_{t-1})$$

where

$$d_t = \begin{cases} \dfrac{0.75}{n} & t = 1 \\ \dfrac{1.50}{n} & t = 2, 3, \ldots \end{cases} \qquad [16A.14]$$

EXAMPLE 16A.5

A wireless tracking system for shop floor control with a MACRS 5-year recovery period has been purchased for $10,000. (*a*) Use Equations [16A.11] through [16A.13] to obtain the annual depreciation and book value. (*b*) Determine the resulting annual depreciation rates and compare them with the MACRS rates in Table 16–2 for $n = 5$.

Solution

(*a*) With $n = 5$ and the half-year convention, use the DDB-to-SL switching procedure to obtain the results in Table 16A–5. The switch to SL depreciation, which occurs in year 4 when both depreciation values are equal, is indicated by

$$D_{DB} = 0.4(2880) = \$1152$$

$$D_{SL} = \frac{2880}{5 - 4 + 1.5} = \$1152$$

The SL depreciation of $1000 in year 1 results from applying the half-year convention included in the first relation of Equation [16A.13]. Also, the SL depreciation of $576 in year 6 is the result of the half-year convention.

(*b*) The actual rates are computed by dividing the "Larger D_t" column values by the first cost of $10,000. The rates below are the same as the Table 16–2 rates.

t	1	2	3	4	5	6
d_t	0.20	0.32	0.192	0.1152	0.1152	0.0576

TABLE 16A–5	Depreciation Amounts Used to Determine MACRS Rates for $n = 5$, Example 16A.5				
Years	**DDB**		**SL Depreciation**	**Larger**	
t	d_t	D_{DB}, $	D_{SL}, $	D_t, $	BV$_t$, $
0	—	—	—	—	10,000
1	0.2	2,000	1,000	2,000	8,000
2	0.4	3,200	1,777	3,200	4,800
3	0.4	1,920	1,371	1,920	2,880
4	0.4	1,152	1,152	1,152	1,728
5	0.4	691	1,152	1,152	576
6	—	—	576	576	0
				10,000	

It is clearly easier to use the rates in Table 16–2 or the VDB spreadsheet function than to determine each MACRS rate using the switching logic above. But the logic behind the MACRS rates is described here for those interested. The annual MACRS rates may be derived by using the applicable rate for the DB method. The subscripts DB and SL have been inserted along with the year t. For the first year $t = 1$,

$$d_{DB,1} = \frac{1}{n} \quad \text{or} \quad d_{SL,1} = \frac{1}{2}\left(\frac{1}{n}\right)$$

For summation purposes only, we introduce the subscript i ($i = 1, 2, \ldots, t$) on d. Then the depreciation rates for years $t = 2, 3, \ldots, n$ are

$$d_{DB,t} = d\left(1 - \sum_{i=1}^{i=t-1} d_i\right) \qquad [16A.15]$$

$$d_{SL,t} = \frac{\left(1 - \sum_{i=1}^{i=t-1} d_i\right)}{n - t + 1.5} \qquad [16A.16]$$

Also, for year $n + 1$, the MACRS rate is one-half the SL rate of the previous year n.

$$d_{SL,n+1} = 0.5(d_{SL,n}) \qquad [16A.17]$$

The DB and SL rates are compared each year to determine which is larger and when the switch to SL depreciation should occur.

EXAMPLE 16A.6

Verify the MACRS rates in Table 16–2 for a 3-year recovery period. The rates in percent are 33.33, 44.45, 14.81, and 7.41.

Solution

The fixed rate for DDB with $n = 3$ is $d = 2/3 = 0.6667$. Using the half-year convention in year 1 and Equations [16A.15] through [16A.17], the results are as follows:

d_1: $\qquad\qquad\qquad d_{DB,1} = 0.5d = 0.5(0.6667) = 0.3333$

d_2: Cumulative depreciation rate is 0.3333.

$$d_{DB,2} = 0.6667(1 - 0.3333) = 0.4445 \qquad \text{(larger value)}$$

$$d_{SL,2} = \frac{1 - 0.3333}{3 - 2 + 1.5} = 0.2267$$

d_3: Cumulative depreciation rate is $0.3333 + 0.4445 = 0.7778$.

$$d_{DB,3} = 0.6667(1 - 0.7778) = 0.1481$$

$$d_{SL,2} = \frac{1 - 0.7778}{3 - 3 + 1.5} = 0.1481$$

Both values are the same; switch to straight line depreciation.
d_4: This rate is 50% of the last SL rate.

$$d_4 = 0.5(d_{SL,3}) = 0.5(0.1481) = 0.0741$$

PROBLEMS

Fundamentals of Depreciation

16.1 How does tax depreciation affect income taxes?

16.2 What is meant by an asset's unadjusted basis? Adjusted basis?

16.3 (a) State the difference between book value and market value. (b) Describe a condition under which they will have significantly different values.

16.4 There are three different life (recovery period) values associated with a depreciable asset. Identify each by name and explain how it is correctly used.

16.5 Cyber Manufacturing purchased a complete video borescope system for applications that require work in places where eyes cannot see. The purchase price was $18,000, shipping and delivery was $300, installation cost was $1200, tax recovery period was 7 years, estimated useful life was 10 years, salvage value was nil, and annual operating cost was $45,000 per year. For tax depreciation purposes, what are the values of B, S, and n that should be used in depreciating the asset?

16.6 An energy production company has the following information regarding the acquisition of new gas-turbine equipment.

Purchase price = $780,000
Trans-oceanic shipping and delivery cost = $4300
Installation cost (1 technician at $1600 per day for 4 days) = $6400
Tax recovery period = 15 years
Book depreciation recovery period = 10 years
Salvage value = 10% of purchase price
Operating cost (with technician) = $185,000 per year

The manager of the department asked your friend in Accounting to enter the appropriate data into the tax-accounting program. What are the values of B, n, and S in depreciating the asset for tax purposes that he should enter?

16.7 Jobe Concrete Products placed a new sand sifter into production 3 years ago. It had an installed cost of $100,000, a life of 5 years, and an anticipated salvage of $20,000. Book depreciation charges for the 3 years are $40,000, $24,000, and $14,000, respectively. (a) Determine the book value after 2 years. (b) If the sifter's market value today is $20,000, determine the difference between its current book value and its market value. (c) Determine the total percentage of the unadjusted basis written off through year 3.

16.8 Quantum Electronic Services paid $P = \$40,000$ for its networked computer system. Both tax and book depreciation accounts are maintained. The annual tax recovery rate is based on the previous year's book value (BV), while the book depreciation rate is based on the original first cost (P). Use the rates listed below to calculate (a) annual depreciation, and (b) book values for each method.

Year of Ownership	1	2	3	4
Tax rate, % of BV	40	40	40	40
Book rate, % of P	25	25	25	25

Tax depreciation: $D_t = \text{Rate} \times BV_{t-1}$
Book depreciation: $D_t = \text{Rate} \times P$

16.9 Visit the U.S. Internal Revenue Service website at www.irs.gov and answer the following questions about depreciation and MACRS by consulting Publication 946, *How to Depreciate Property*.
(a) What is the definition of depreciation according to the IRS?
(b) What is the description of the term salvage value?
(c) What are the two depreciation systems within MACRS, and what are the major differences between them?
(d) What are the properties listed that cannot be depreciated under MACRS?
(e) When does depreciation begin and end?
(f) What is a Section 179 Deduction?

Straight Line Depreciation

16.10 What is the depreciation rate d_t each year t for an asset that has a 5-year recovery period and is straight line depreciated?

16.11 Goodson Healthcare purchased a new sonogram imaging unit for $300,000 and a truck body and chassis for an additional $100,000 to make the unit mobile. The unit-truck system will be depreciated as one asset. The functional life is 8 years, and the salvage is estimated to be 10% of the purchase price of the imaging unit regardless of the number of years of service. Use classical straight line depreciation to determine the salvage value, annual depreciation, and book value after 4 years of service.

16.12 Air handling equipment that costs $12,000 has a life of 7 years with a $2000 salvage value.
(a) Calculate the straight line depreciation charge for each year.
(b) Write the spreadsheet function to determine the annual depreciation.
(c) Determine the book value after 3 years.
(d) Determine the rate of depreciation.
(e) Use a *spreadsheet* to plot the accumulated depreciation and book value for each year.

16.13 A company that manufactures pulse doppler insertion flow meters uses the straight line method for book depreciation purposes. Newly-acquired equipment has a first cost of $170,000 with a 3-year life and $20,000 salvage value. Determine the depreciation charge and book value for year 2.

16.14 Kobi Technologies book-depreciated an asset at $27,500 per year for 4 years using the straight line method. If the book value at the end of year 2 was $65,000, determine the asset's (a) salvage value, and (b) first cost.

16.15 An asset owned by Photon Environmental was book-depreciated by the straight line method over a 5-year period with book values of $296,000 and $224,000 in years 2 and 3, respectively. Determine (a) the salvage value S used in the calculation, and (b) the unadjusted basis B.

16.16 A special-purpose digitizing system just acquired by Bzybee Consultants has $B = \$50,000$ and a 4-year recovery period. (a) What is the accumulated depreciation after year 3 if the salvage value of the system is $5000? (b) What is the total amount of depreciation remaining?

16.17 Bristol-Myers-Squibb purchased a tablet-forming machine in 2010 for $750,000. The company planned to use the machine for 10 years and then sell it for $50,000; however, due to rapid obsolescence, it will be retired after only 6 years in 2016.
(a) Determine the capital investment remaining when the asset was prematurely retired.
(b) If the asset is sold at the end of 6 years for $175,000, determine the capital investment loss based on straight line depreciation.
(c) If the new-technology machine has an estimated cost of $260,000, how many more years would the company have had to depreciate the currently-owned machine to make its book value and the first cost of the new machine equal each other?

Declining Balance Depreciation

16.18 Halcrow Yolles purchased equipment for new highway construction in Manitoba, Canada, costing $500,000 Canadian. Estimated salvage at the end of the expected life of 5 years is $50,000. Various acceptable depreciation methods are being studied currently. Determine the depreciation for year 2 using the DDB, 150% DB and SL methods. Solve (a) by hand, and (b) by spreadsheet function.

16.19 Software and hardware for optimizing cell design of robotic picking lines have an installed cost of $78,000 with no residual value after 5 years. For years 2 and 4, use DDB book depreciation to determine (a) the depreciation charge, and (b) the book value.

16.20 Determine the first cost of a machine that is used for making spill-containment pallets if its book value in year 3 is $25,000. The machine has a 5-year life and the double declining balance method is applied.

16.21 If an asset is book-depreciated by the DDB method over a 5-year period, how long will it take to reach its salvage value if the estimated salvage is 25% of the first cost?

16.22 If the salvage value of an asset is nil and it is depreciated by the double declining balance method, what percentage of the asset's first cost will remain after its 5-year life?

16.23 A video recording system was purchased 3 years ago at a cost of $40,000. A 5-year recovery period and DDB depreciation have been used to write off the basis. The system is to be replaced this year with a trade-in value of $4000. What is the difference between the book value and the trade-in value?

16.24 New equipment to read 96-bit product codes that are replacing old bar codes has just been purchased by General Food Stores. As a trial, 1000 of the items will be initially purchased. For book depreciation purposes (not tax), the total investment of $50,000 will be written off over a 4-year period, with no salvage value, by applying one of three methods—SL, 1.75% DB, or DDB. The objective is to have the largest amount of *accumulated depreciation after 2 years*, in order to minimize the lost depreciation if it is again necessary to purchase new-technology readers prior to the end of the 4-year useful life of the readers just purchased. You know that the answer is the DDB method, but you must graphically demonstrate this result to your supervisor. Use a spreadsheet to do so.

MACRS Depreciation and Recovery Periods

16.25 For the MACRS method of depreciation, identify (*a*) the six standardized recovery periods for personal property, (*b*) the two recovery periods for real property, (*c*) the assumed salvage value for personal property, and (*d*) the default recovery period for an asset not in a stated GDS class.

16.26 (*a*) What is meant by half-year convention? (*b*) Where does it show up in the MACRS depreciation rate table? What is its effect?

16.27 Del Norte Brick Co. is located near the intersection of Texas, New Mexico, and Mexico. Improved access to the company's property is via a small bridge across the Rio Grande. The cost of the bridge was $770,000. Determine the depreciation and book value for year 3 according to the MACRS method.

16.28 An automated assembly robot that cost $400,000 has a depreciable life of 5 years with a $100,000 salvage value. The MACRS depreciation rates for years 1, 2, and 3 are 20.00%, 32.00%, and 19.20%, respectively. What is the book value at the end of year 3? Year 5? Year 6?

16.29 A company report stated that a $140,000 asset purchased 3 years ago has a current MACRS book value that is 57.6% of the asset's basis. (*a*) Determine the recovery period used. (*b*) Determine the depreciation next year using the VDB spreadsheet function.

16.30 The manager of a plant that manufactures stepper drives knew that MACRS and DDB were both accelerated depreciation methods, but out of curiosity, he wanted to know which one would provide the faster write-off in the first 3 years for equipment that has a first cost of $300,000, a 5-year life, and a $60,000 salvage value. (*a*) Use *hand* calculations to determine which method yields the lower book value and by how much. (*b*) Develop the complete depreciation schedules on a *spreadsheet* and use them to answer the questions above. (*c*) Using your spreadsheet results, explain what happened to the DDB depreciation amounts for years 4 and 5.

16.31 Bison Gear and Engineering of St Charles, Illinois, makes sensorless and brushless dc gear motors suited for foodservice equipment, factory automation, alternative energy systems, and other specialty machinery applications. The company purchased an asset 2 years ago that has a 5-year recovery period. The MACRS depreciation charge for year 3 is $14,592. (*a*) What was the first cost of the asset? (*b*) Develop the entire MACRS schedule for the asset and determine the depreciation charge in year 1 and next year.

16.32 Fairfield Properties owns real property that is MACRS depreciated with $n = 39$ years. They paid $3.4 million for an apartment complex and hope to sell it after 10 years of ownership for 50% more than the book value at that time. Compare the expected selling price with the amount that Fairfield paid for the property.

16.33 Blackwater Spring and Metal utilizes the same computerized spring forming machinery in its U.S. and Malaysian plants. The first cost was $750,000 with $S = \$150,000$ after $n = 10$ years. MACRS depreciation with $n = 5$ years is applied in the United States and standard SL depreciation with $n = 10$ years is used by the Malaysian facility. (*a*) If the equipment is sold after 6 years for $100,000, calculate the over- or under-depreciation amounts for each method. (*b*) Use a *spreadsheet* to plot the book values for both methods on a single graph.

16.34 A plant manager for a fiber optics cable manufacturing company knows that the remaining capital investment in several types of equipment is more closely approximated when the equipment is depreciated linearly by the SL method compared to a rapid write-off method like MACRS. Therefore, he keeps two sets of books, one for tax purposes (MACRS) and one for equipment-management purposes (SL). For an asset that has a first cost of $80,000, a depreciable life of 5 years, and a salvage value equal to 25% of the first cost, determine the difference in the book values shown in the two sets of books at the end of year 3. Which method has the lower book value and by how much?

16.35 Youngblood Shipbuilding Yard just purchased $1 million in capital equipment for ship repairing functions on dry-docked ships. Estimated salvage is $150,000 for any year after 5 years of use. Compare the depreciation and book value for year 3 for each of the following depreciation methods:
   (*a*) GDS MACRS, where a recovery period of 10 years is allowed.

(b) Double declining balance, with a recovery period of 15 years.

(c) ADS straight line as an alternative to MACRS, with a recovery period of 15 years.

Depletion

16.36 What is the difference between depreciation and depletion?

16.37 When WTA, Inc. purchased rights to extract silver from a mine for a total price of $2.1 million 3 years ago, the estimated 350,000 ounces of silver was to be removed over the next 10 years. A total of 175,000 ounces has been removed and sold thus far. (a) What is the total cost depletion allowed over the 3 years? (b) New exploratory tests indicate that only an estimated 100,000 ounces remain in the veins of the mine. What is the cost depletion factor applicable for the next year?

16.38 A relatively small privately-owned coal-mining company has the sales results summarized below. Determine the annual percentage depletion for the coal mine. Assume the company's taxable income is $140,000 each year.

Year	Sales, Tons	Spot Sales Price, $/Ton
1	34,300	9.82
2	50,100	10.50
3	71,900	11.23

16.39 SA Forest Resources purchased forest acreage for $500,000 from which an estimated 200 million board feet of lumber are recoverable. The company will sell the lumber for $0.10 per board foot. No lumber will be sold next year because an environmental impact statement must be completed before harvesting can

begin. In years 2 through 10, however, the company expects to remove 20 million board feet per year. The inflation rate is 8%, and the company's MARR is 10%. Determine the depletion amount in years 1 and 2 by the cost depletion method.

16.40 Carrolton Oil and Gas, an independent oil and gas producer, is approved to use a 20% of gross income depletion allowance. The write-off last year was $700,000 on its horizontal directional drill wells. Determine the estimated total reserves in barrels if the volume pumped last year amounted to 1% of the total and the delivered product price averaged $40 per barrel.

16.41 Ederly Quarry sells a wide variety of cut limestone for residential and commercial building construction. A recent quarry expansion cost $2.9 million and added an estimated 100,000 tons of reserves. (a) Estimate the cost depletion allowance for the next 5 years, using the projections made by the owner, John Ederly. (b) Will the cost depletion in any of the 5 years be limited by law? Why or why not?

Year	1	2	3	4	5
Volume, 1000 tons	10	9	15	15	18
Price, $ per ton	75	70	70	75	85

16.42 For the last 10 years, Am-Mex Coal has used the cost depletion factor of $2500 per 100 tons to write off the investment of $35 million in its Pennsylvania anthracite coal mine. Depletion thus far totals $24.8 million. A new study to appraise mine reserves indicates that no more than 800,000 tons of salable coal remains. Determine next year's cost and percentage depletion amounts, if estimated gross income is expected to be between $6.125 and $8.50 million on a production level of 72,000 tons.

ADDITIONAL PROBLEMS AND FE EXAM REVIEW QUESTIONS

16.43 All of the following assets can be depreciated, except:
(a) A bulldozer
(b) A copper mine
(c) A surgical robot
(d) A conveyor belt

16.44 Classic straight line depreciation of a $100,000 asset takes place over a 5-year recovery period. If the salvage value is 15% of first cost, the depreciation charge for year 3 is closest to:
(a) $17,000   (b) $20,000
(c) $24,000   (d) $28,000

16.45 A commercial generator with an unadjusted basis of $50,000 is straight-line depreciated over a

5-year period. The asset will have an AOC of $35,000 and a salvage value of $10,000. The book value at the end of year 3 will be closest to:
(a) $8000   (b) $20,000
(c) $24,000   (d) $26,000

16.46 A motorized cultivator with a first cost of $28,000 and salvage value of 25% of the first cost is depreciated by the DDB method over a 5-year period. If the operating cost is $43,000 per year, the depreciation charge for year 2 is closest to:
(a) $18,000   (b) $11,200
(c) $6720   (d) $4350

16.47 An 8-year recovery-period conveyor system for grading eggs is planned for DDB depreciation.

The system had a first cost of $30,000 with a $9000 salvage value. The annual operating cost allocated to the conveyor is $7000 per year. The book value at the end of year 2 is nearest to:

(a) $7560   (b) $11,812
(c) $16,875   (d) $14,300

16.48   Gisele is performing a make/buy study involving the retention or disposal of a 4-year-old machine that was to be in production for 8 years. It cost $500,000 originally. Without a market value estimate, she decided to use the current book value plus 20% for the market value. If DDB depreciation is applied, the market value estimate is closest to:

(a) $158,200   (b) $253,125
(c) $217,900   (d) $189,840

16.49   A currently-owned asset has $B = \$150,000$, $S = \$95,000$, and a 10-year depreciable life. The book value at the end of year 3 according to the MACRS method would be closest to: (Note: d_t values for years 1, 2, 3, and 4 are 10.00%, 18.00%, 14.40%, and 11.52%, respectively.)

(a) $86,400   (b) $62,400
(c) $43,900   (d) $23,320

16.50   An industrial robot that is depreciated by the MACRS method has $B = \$60,000$ and a 5-year depreciable life. If the depreciation charge in year 3 is $8640, the salvage value that was used in the depreciation calculation is closest to: (d_t values for years 1, 2, 3, 4, and 5 are 10.00%, 18.00%, 14.40%, 11.52%, and 9.22%, respectively.)

(a) $0   (b) $10,000
(c) $20,000   (d) $30,000

16.51   Under the General Depreciation System (GDS) of asset classification, any asset that is not in a stated class is automatically assigned a recovery period of:

(a)   5 years   (b)   7 years
(c)   10 years   (d)   15 years

16.52   All of the following statements about the Alternative Depreciation System (ADS) are true, except:

(a)   The half-year convention applies
(b)   Salvage value is neglected
(c)   The recovery periods are shorter than in GDS
(d)   The straight line method is required

16.53   South African Gold Mines, Inc. is writing off its $210 million investment using the cost depletion method. An estimated 700,000 ounces of gold are available in its developed mines. This year 35,000 ounces were produced at an average price of $400 per ounce. The depletion for the year is closest to:

(a) $2.1 million, which is 15% of the gross income of $14.0 million
(b) $2.4 million
(c) $10.5 million
(d) $42 million, which is 15% of the gross income of $280 million

16.54   Rayonier Forest Resources purchased a small tract of timberland for $70,000 that contained 25,000 trees. The value of the bare land was estimated to be $20,000. The annual percentage depletion factor for timber is 10%. In the first year of operation, the lumber company cut down 7000 trees. According to the cost depletion method, the depletion deduction for year 1 would be closest to:

(a) $2000   (b) $7000
(c) $10,000   (d) $14,000

16.55   A stone and gravel quarry in Texas can use a percentage depletion rate of 5% of gross income, or a cost depletion rate of $1.28 per ton. Quarry first cost = $3.2 million; estimated total tonnage = 2.5 million tons; tonnage this year = 65,000; gross income = $40 per ton. Of the two depletion charges, the method and larger amount are:

(a)   Percentage at $83,200
(b)   Percentage at $130,000
(c)   Cost at $80,000
(d)   Cost at $130,000

APPENDIX PROBLEMS

Sum-of-Years-Digits Depreciation

16A.1   A European manufacturing company has new equipment with a first cost of 12,000 euros, an estimated salvage value of 2000 euros, and a recovery period of 8 years. Use the SYD method to tabulate annual depreciation and book value. Solve by hand and spreadsheet.

16A.2   Earthmoving equipment with a first cost of $150,000 is expected to have a life of 10 years.

The salvage value is expected to be 10% of the first cost. Calculate (a) by *hand*, and (b) by *spreadsheet* the depreciation charge and book value for years 2 and 7 using the SYD method.

16A.3   If $B = \$400,000$, $n = 6$ years, and S is estimated at 15% of B for a new pavement recycling machine, use the SYD method to determine (a) the book value after 3 years, and (b) the rate of depreciation and the depreciation amount in year 4.

Unit-of-Production Depreciation

16A.4 A robot used in simulated car crashes cost $70,000, has no salvage value, and has an expected capacity of tests not to exceed 10,000 according to the manufacturer. Volvo Motors decided to use the unit-of-production depreciation method because the number of test crashes per year in which the robot would be involved was not estimable. Determine the annual depreciation and book value for the first 3 years if the number of tests were 3810, 2720, and 5390 per year.

16A.5 A new hybrid car was purchased by Pedernales Electric Cooperative as a courier vehicle to transport items between its 12 city offices. The car cost $35,000 and was retained for 5 years. Alternatively, it could have been retained for 100,000 miles. Salvage value is nil. Five-year DDB depreciation was applied. The car pool manager stated that he prefers UOP depreciation on vehicles because it writes off the first cost faster. Use the actual annual miles driven, listed below, to plot the book values for both methods. Determine which method would have removed the $35,000 faster. Show hand or spreadsheet solution, as instructed.

Year	1	2	3	4	5
Miles driven, 1000	15	22	16	18	25

Switching Methods

16A.6 An asset has a first cost of $45,000, a recovery period of 5 years, and a $3000 salvage value. Use the switching procedure from DDB to SL depreciation, and calculate the present worth of depreciation at $i = 18\%$ per year.

16A.7 If $B = \$45,000$, $S = \$3000$, and $n = 5$-year recovery period, use a spreadsheet and $i = 18\%$ per year to maximize the present worth of depreciation, using the following methods: DDB-to-SL switching (this was determined in Problem 16A.6) and MACRS. Given that MACRS is the required depreciation system in the United States, comment on the results.

16A.8 Hempstead Industries has a new milling machine with $B = \$110,000$, $n = 10$ years, and $S = \$10,000$. Determine the depreciation schedule and present worth of depreciation at $i = 12\%$ per year, using the 175% DB method for the first 5 years and switching to the classical SL method for the last 5 years. Use a spreadsheet to solve this problem and plot the book values.

16A.9 Reliant Electric Company has erected a large portable building with a first cost of $255,000 and an anticipated salvage of $50,000 after 25 years. (*a*) Should the switch from DDB to SL depreciation be made? (*b*) For what values of the uniform depreciation rate in the DB method would it be advantageous to switch from DB to SL depreciation at some point in the life of the building? How does this rate compare with the DDB rate of $2/n$?

MACRS Rates

16A.10 Verify the 5-year recovery period rates for MACRS given in Table 16–2. Start with the DDB method in year 1, and switch to SL depreciation when it offers a larger recovery rate. Solve *by hand* and *spreadsheet* using $B = \$1$ and $S = 0$.

16A.11 A video recording system was purchased 3 years ago at a cost of $30,000. A 5-year recovery period and MACRS depreciation have been used to write off the basis. The system is to be prematurely replaced with a trade-in value of $5000. Determine the MACRS depreciation, using the switching rules to find the difference between the book value and the trade-in value after 3 years.

16A.12 Use the computations in Equations [16A.11] through [16A.13] to determine the MACRS annual depreciation for the following asset data: $B = \$50,000$ and a recovery period of 7 years.

16A.13 The 3-year MACRS recovery rates are 33.33%, 44.45%, 14.81%, and 7.41%, respectively. (*a*) What are the corresponding rates for the alternative MACRS straight line ADS method with the half-year convention imposed? (*b*) Compare the PW_D values for these two methods if $B = \$80,000$ and $i = 15\%$ per year.

CHAPTER 17

After-Tax Economic Analysis

Photodisc/Getty Images

LEARNING OUTCOMES

Purpose: Perform an after-tax economic evaluation considering the impact of pertinent tax regulations, income taxes, and depreciation.

SECTION	TOPIC	LEARNING OUTCOME
17.1	Terminology and rates	• Know the fundamental terms and relations of after-tax analysis; use a marginal tax rate table.
17.2	CFBT and CFAT	• Determine cash flow series before taxes and after taxes.
17.3	Taxes and depreciation	• Demonstrate the tax advantage of accelerated depreciation and shortened recovery periods.
17.4	Depreciation recapture (DR)	• Calculate the tax impact of DR; explain the consideration of capital gains and capital losses.
17.5	After-tax analysis	• Evaluate one project or multiple alternatives using after-tax PW, AW, and ROR analysis.
17.6	After-tax replacement	• Evaluate a defender and challenger in an after-tax replacement study.
17.7	EVA analysis	• Evaluate an alternative using after-tax economic value-added analysis; compare to CFAT analysis.
17.8	Taxes outside the United States	• Understand the fundamental practices for depreciation and tax rates in international settings.
17.9	VAT	• Demonstrate the use and computation of a value-added tax on manufactured products.

his chapter provides an overview of tax terminology, income tax rates, and tax equations pertinent to an after-tax economic analysis. The change from estimating cash flow before taxes (CFBT) to **cash flow after taxes (CFAT)** involves a consideration of significant tax effects that may alter the final decision, and estimates the magnitude of the effect on cash flow over the life of the alternative that taxes may have.

Mutually exclusive alternative comparisons using after-tax PW, AW, and ROR methods are explained with major tax implications considered. **Replacement studies** are discussed with tax effects that occur at the time that a defender is replaced. Also, the after-tax **economic value added** by an alternative is discussed in the context of annual worth analysis. All these methods use the procedures learned in earlier chapters, except now with tax effects considered.

An after-tax evaluation using any method requires more computations than in previous chapters. Templates for tabulation of cash flow after taxes by hand and by spreadsheet are developed. Additional information on U.S. federal taxes—tax law and annually updated tax rates—is available through Internal Revenue Service publications and, more readily, on the IRS website www.irs.gov. Publications 542, *Corporations*, and 544, *Sales and Other Dispositions of Assets*, are especially applicable to this chapter. Some differences in tax considerations outside the United States are summarized.

17.1 Income Tax Terminology and Basic Relations ● ● ●

The perspective taken in engineering economy when performing an after-tax evaluation is that of the *project* and how relevant tax rules and allowances influence the economic decision. The perspective of a financial study is that of the *corporation* and how the tax structure and laws affect profitability. We take the engineering economy viewpoint in the sections that follow.

There are many types of taxes levied upon corporations and individuals in all countries, including the United States. Some are sales tax, value-added tax, import tax, income tax, highway tax, gasoline tax, and property (real estate) tax. Federal governments rely on income taxes for a significant portion of their annual revenue. States, provinces, and municipal-level governments rely on sales, value-added, and property taxes to maintain services, schools, etc. for the citizenry. For a starting point on *corporate income taxes* and how they are used when performing an after-tax economic evaluation of a project or multiple alternatives, this section covers basic definitions, terms, and relations.

> **Income tax** is the amount of the payment (taxes) on income or profit that must be delivered to a federal (or lower-level) government unit. Taxes are **real cash flows;** however, for corporations, tax computation requires some noncash elements, such as depreciation. Corporate income taxes are usually submitted quarterly, and the last payment of the year is submitted with the annual tax return.

In the United States, the Internal Revenue Service (IRS), a part of the Department of the Treasury, collects the taxes and enforces tax laws. The website www.irs.gov provides information on tax laws, rates, publications, etc. that are referenced in this chapter. Every country has a comparable tax-collection and enforcement unit of government; some states and provinces have equivalent units.

Though the formulas are much more complex when applied to a specific situation, two fundamental relations form the basis for income tax computations. The first involves only actual cash flows:

$$\text{Net operating income} = \text{revenue} - \text{operating expenses}$$

The second involves actual cash flows and noncash deductibles, such as depreciation.

$$\text{Taxable income} = \text{revenue} - \text{operating expenses} - \text{depreciation}$$

These terms and relations for corporations are now described. Since each term is calculated for 1 year, there can be a subscript t ($t = 1, 2, \ldots$) added; t is omitted here for simplicity.

> **Operating revenue R,** also commonly called **gross income GI,** is the total income realized from all revenue-producing sources. These incomes are listed in the income statement. (See Appendix B on accounting reports.) Other, nonoperating revenues such as sale of assets, license fee income, and royalties are considered separately for tax purposes.

Operating expenses OE include all costs incurred in the transaction of business. These expenses are tax-deductible for corporations. For after-tax economic evaluations, the AOC (annual operating costs) and M&O (maintenance and operating) costs are applicable here. Depreciation is not included here since it is *not an operating expense*.

Net operating income NOI, often called **EBIT** (earnings before interest and income taxes), is the difference between gross income and operating expenses.

$$\text{NOI} = \text{EBIT} = \text{GI} - \text{OE} \qquad [17.1]$$

Taxable income TI is the amount of income upon which taxes are based. A corporation is allowed to **remove depreciation,** depletion and amortization, and some other deductibles from net operating income in determining the taxable income for a year. For our evaluations, we define taxable income as

$$\text{TI} = \text{gross income} - \text{operating expenses} - \text{depreciation}$$
$$= \text{GI} - \text{OE} - \text{D} \qquad [17.2]$$

Though there may be subtleties and varying interpretations over time, in essence, the differences between NOI and TI are tax-law-allowed deductibles, such as depreciation. (In keeping with the project view of engineering economics, we will primarily use the TI relation when conducting an after-tax evaluation.)

Tax rate T is a percentage, or decimal equivalent, of TI that is owed in taxes. The tax rates in many countries (including the United States) are **graduated** (or **progressive**) by level of TI; that is, higher rates apply as the TI increases. The **marginal tax rate** is the percentage paid on the *last dollar of income*. The average tax rate paid is calculated separately from the highest marginal rate used, as shown later. The general tax computation relation is

$$\text{Income taxes} = \text{applicable tax rate} \times \text{taxable income}$$
$$= (T)(\text{TI}) \qquad [17.3]$$

Net operating profit after taxes NOPAT is the amount remaining each year after taxes are subtracted from taxable income.

$$\text{NOPAT} = \text{TI} - \text{taxes} = \text{TI} - (T)(\text{TI})$$
$$= \text{TI}(1 - T) \qquad [17.4]$$

Basically, NOPAT represents the money remaining in the corporation as a result of the capital invested during the year. It is also called *net profit after taxes* (NPAT).

The graduated tax rate schedule for corporations is presented in Table 17–1 as taken from IRS Publication 542, *Corporations*. These are rates for the entire corporation, not for an individual project, though they are often applied in the after-tax analysis of a single project. The rates can change based upon government legislation; however, the U.S. corporate tax rate schedule has remained the same for some years. To illustrate the use of the graduated tax rate, assume a company is expected to generate a taxable income of $500,000 in 1 year. From Table 17–1, the marginal tax rate for the last dollar of TI is 34%, but the graduated rates become progressively larger as TI increases. In this case, for TI = $500,000,

$$\text{Taxes} = 113,900 + 0.34(500,000 - 335,000)$$
$$= 113,900 + 56,100$$
$$= \$170,000$$

Alternatively, the rates for each TI level can be used to calculate taxes the longer way.

$$\text{Taxes} = 0.15(50,000) + 0.25(75,000 - 50,000) + 0.34(100,000 - 75,000)$$
$$+ 0.39(335,000 - 100,000) + 0.34(500,000 - 335,000)$$
$$= 7500 + 6250 + 8500 + 91,650 + 56,100$$
$$= \$170,000$$

TABLE 17–1	U.S. Corporate Income Tax Rate Schedule		
	If Taxable Income ($) Is:		
Over	**But Not Over**	**Tax Is**	**Of the Amount Over**
0	50,000	15%	0
50,000	75,000	7,500 + 25%	50,000
75,000	100,000	13,750 + 34%	75,000
100,000	335,000	22,250 + 39%	100,000
335,000	10,000,000	113,900 + 34%	335,000
10,000,000	15,000,000	3,400,000 + 35%	10,000,000
15,000,000	18,333,333	5,150,000 + 38%	15,000,000
18,333,333	—	35%	0

Smaller businesses (with TI < $335,000) receive a slight tax advantage compared to large corporations. Once the TI exceeds $335,000, an effective federal tax rate of 34% applies, and when TI > $18.33 million, there is a flat tax rate of 35%.

As we move forward with after-tax analysis, it is important to keep the following in mind:

The corporate tax rates apply to a corporation as a whole, not to a specific project, unless the project *is* the company. Tax rates are usually graduated by level of taxable income. Therefore, the last dollar of TI is taxed at a marginal rate. The average federal tax rate actually paid is lower than the highest marginal rate paid because the rates, in general, graduate to higher percentages as TI increases.

Because the marginal tax rates change with TI, it is not possible to quote directly the percentage of TI paid in income taxes. Alternatively, a single-value number, the *average tax rate,* is calculated as

$$\text{Average tax rate} = \frac{\text{total taxes paid}}{\text{taxable income}} = \frac{\text{taxes}}{\text{TI}} \qquad [17.5]$$

Referring to Table 17–1, for a small business with TI = $100,000, the federal income tax burden averages $22,250/100,000 = 22.25%. If TI = $15 million, the average tax rate is $5.15 million/15 million = 34.33%.

As mentioned earlier, there are federal, state, and local taxes imposed. For the sake of simplicity, the tax rate used in an economy study is often a single-figure **effective tax rate T_e,** which accounts for all taxes. Effective tax rates are in the range of 35% to 50%. One reason to use the effective tax rate is that state taxes are deductible for federal tax computation. The effective tax rate and taxes are calculated as

$$T_e = \text{state rate} + (1 - \text{state rate})(\text{federal rate}) \qquad [17.6]$$
$$\text{Taxes} = (T_e)(\text{TI}) \qquad [17.7]$$

EXAMPLE 17.1

REI (Recreational Equipment Incorporated) sells outdoor equipment and sporting goods through retail outlets, the Internet, and catalogs. Assume that for 1 year REI has the following financial results in the state of Kentucky, which has a flat tax rate of 6% on corporate taxable income:

Total revenue	$19.9 million
Operating expenses	$8.6 million
Depreciation and other allowed deductions	$1.8 million

(*a*) Determine the state taxes and federal taxes due using Table 17–1 rates.
(*b*) Find the average federal tax rate paid for the year.
(*c*) Determine a single-value tax rate useful in economic evaluations using the average federal tax rate determined in part (*b*).
(*d*) Estimate federal and state taxes using the single-value rate, and compare their total with the total in part (*a*).

Solution

(*a*) Calculate TI by Equation [17.2] and use Table 17–1 rates for federal taxes due.

$$\text{Kentucky state TI} = \text{GI} - \text{OE} - D = 19.9 \text{ million} - 8.6 \text{ million} - 1.8 \text{ million}$$
$$= \$9.5 \text{ million}$$

$$\text{Kentucky state taxes} = 0.06(\text{TI}) = 0.06(9,500,000) = \$570,000$$

$$\text{Federal TI} = \text{GI} - \text{OE} - D - \text{state taxes} = 9,500,000 - 570,000$$
$$= \$8,930,000$$

$$\text{Federal taxes} = 113,900 + 0.34(8,930,000 - 335,000) = \$3,036,200$$

$$\text{Total federal and state taxes} = 3,036,200 + 570,000 = \$3,606,200 \qquad [17.8]$$

(*b*) From Equation [17.5], the average tax rate paid is approximately 32% of TI.

$$\text{Average federal tax rate} = 3,036,200/9,500,000 = 0.3196$$

(*c*) By Equation [17.6], T_e is slightly over 36% per year for combined state and federal taxes.

$$T_e = 0.06 + (1 - 0.06)(0.3196) = 0.3604 \quad (36.04\%)$$

(*d*) Use the effective tax rate and TI = \$9.5 million from part (*a*) in Equation [17.7] to approximate total taxes.

$$\text{Taxes} = 0.3604(9,500,000) = \$3,423,800$$

Compared to Equation [17.8], this approximation is \$182,400 low, a 5.06% underestimate.

We should understand how our *individual* federal-level income tax liabilities and associated rates are imposed on salaries and other income. First, we can compare them with the corporate tax structure we have just discussed. Gross income for an individual taxpayer is comparable if corporate revenue is replaced by salaries and wages. However, for an individual's taxable income, most of the expenses for living and working are not tax deductible to the same degree as operating expenses (OE) are for corporations. For individual taxpayers,

$$\text{GI} = \text{salaries} + \text{wages} + \text{interest and dividends} + \text{other income}$$

$$\text{TI} = \text{GI} - \text{personal exemption} - \text{standard or itemized deductions}$$

$$\text{Taxes} = (T)(\text{TI})$$

For TI, corporate operating expenses are replaced by individual exemptions and specific deductions. Exemptions are yourself, your spouse, your children, and your other dependents. Each exemption reduces TI by \$4000 to \$4500 per year, depending upon current exemption allowances.

In the United States, the tax rates for individuals, like those for corporations, are graduated by level of TI. In 2015, the marginal rates ranged from 10% to 39.6%. The top marginal rates have increased for individuals with larger TI amounts in recent times and are likely to continue this trend. Once the marginal rates are set, the TI levels are adjusted each year to account for inflation and other factors. This process is called **indexing,** also referred to as "bracket creep." Clearly, tax rates for individuals change much more frequently than the rates for corporations change. Current information is available on the IRS website www.irs.gov through Publication 17, *Your Federal Income Tax,* which you will be asked to consult when solving some problems at the end of this chapter. The current rate schedule is published in the back of this document for four filing status categories:

Unmarried individuals (single)
Married filing jointly
Married filing separately
Head of household

17.2 Calculation of Cash Flow after Taxes ●●●

Early in the text, the term *net cash flow* (*NCF*) was identified as the best estimate of actual cash flow each year. The NCF is calculated as cash inflows minus cash outflows. Since then, the annual NCF amounts have been used many times to perform alternative evaluations via the PW, AW, ROR, and B/C methods. Now that the impact on cash flow of depreciation and related taxes will be considered, it is time to expand our terminology. NCF is replaced by the term **cash flow before taxes (CFBT),** and we introduce the new term **cash flow after taxes (CFAT).**

CFBT and CFAT are **actual cash flows;** that is, they represent the estimated actual flow of money into and out of the corporation that will result from the alternative. The remainder of this section explains how to transition from before-tax to after-tax cash flows for solutions by hand and by spreadsheet, using tax regulations described in the next few sections. Once the CFAT estimates are developed, the economic evaluation is performed using the same methods and selection guidelines applied previously. However, the analysis is performed on the CFAT estimates.

We learned that net operating income (NOI) does not include the purchase or sale of capital assets. However, the annual CFBT estimate *must include* the initial capital investment and salvage value for the years in which they occur. Incorporating the definitions of gross income and operating expenses from NOI, CFBT for any year is defined as

$$\text{CFBT} = \text{gross income} - \text{operating expenses} - \text{initial investment} + \text{salvage value}$$
$$= \text{GI} - \text{OE} - P + S \qquad [17.9]$$

As in previous chapters, P is the initial investment (year 0) and S is the estimated salvage value in year n. Therefore, only in year 0 will the CFBT include P, and only in year n will an S value be present. Once all taxes are estimated, the annual after-tax cash flow is simply

$$\text{CFAT} = \text{CFBT} - \text{taxes} \qquad [17.10]$$

where taxes are estimated using the relation $(T)(\text{TI})$ or $(T_e)(\text{TI})$.

We know from Equation [17.2] that depreciation D is considered when calculating TI. It is very important to understand the different roles of depreciation for income tax computations and in CFAT estimation.

Depreciation is not an operating expense and is a *non*cash flow. Depreciation is tax deductible for determining the amount of income taxes only, but it does not represent a direct, after-tax cash flow. Therefore, the after-tax engineering economy study must be based on actual cash flow estimates, that is, annual CFAT estimates that do not include depreciation as an expense (negative cash flow).

Accordingly, if the CFAT expression is determined using the TI relation, depreciation must not be included outside of the TI component. Equations [17.2], [17.7], [17.9], and [17.10] can now be combined.

$$\text{CFAT} = \text{GI} - \text{OE} - P + S - (\text{GI} - \text{OE} - D)(T_e) \qquad [17.11]$$

Suggested table column headings for CFBT and CFAT calculations by hand or by spreadsheet are shown in Table 17–2. The equations are shown in column numbers, with the effective tax rate T_e used for income tax estimation. Operating expenses OE and initial investment P carry negative signs in all tables and spreadsheets.

A **negative TI value** may occur in some years due to a depreciation amount that is larger than $(\text{GI} - \text{OE})$. It is possible to account for this in a detailed after-tax analysis using carry-forward and carry-back rules for operating losses. It is the exception that the engineering economy study will consider this level of detail. Rather, the associated negative income tax is considered as a **tax savings for the year.** The assumption is that the negative tax will offset taxes for the same year in other income-producing areas of the corporation.

			Investment						
	Gross	Operating	and				Taxable		
	Income	Expenses	Salvage			Depreciation	Income		
Year	GI	OE	P and S	CFBT		D	TI	Taxes	CFAT
	(1)	(2)	(3)	(4) = (1) + (2) + (3)		(5)	(6) = (1) + (2) − (5)	(7) = T_e(6)	(8) = (4) − (7)

TABLE 17–2   Suggested Column Headings for Calculation of CFAT

EXAMPLE 17.2

Wilson Security has received a contract to provide additional security for corporate and government personnel along the international border between two countries in South America. Wilson plans to purchase listening and detection equipment for use in the 6-year contract. The equipment is expected to cost $550,000 and have a resale value of $150,000 after 6 years. Based on the incentive clause in the contract, Wilson estimates that the equipment will increase contract revenue by $200,000 per year and require an additional M&O expense of $90,000 per year. MACRS depreciation allows recovery in 5 years, and the effective corporate tax rate is 35% per year. Tabulate and plot the CFBT and CFAT series.

Solution

The spreadsheet in Figure 17–1 presents before-tax and after-tax cash flows using the format of Table 17–2. The functions for year 6 are detailed in row 11. Discussion and sample calculations follow.

CFBT: The operating expenses OE and initial investment P are shown as negative cash flows. The $150,000 salvage (resale) is a positive cash flow in year 6. CFBT is calculated by Equation [17.9]. In year 6, for example, when the equipment is sold, the function in row 11 indicates that

$$CFBT_6 = 200,000 - 90,000 + 150,000 = \$260,000$$

CFAT: Column F for MACRS depreciation, which is determined using the VDB function over the 6-year period, writes off the entire $550,000 investment. Using year 4 as an example, taxable income, taxes, and CFAT are calculated as follows:

$$TI_4 = GI - OE - D = 200,000 - 90,000 - 63,360 = \$46,640$$

$$Taxes_4 = (0.35)(TI) = (0.35)(46,640) = \$16,324$$

$$CFAT_4 = GI - OE - taxes = 200,000 - 90,000 - 16,324 = \$93,676$$

In year 2, MACRS depreciation is large enough to cause TI to be negative ($-66,000$). As mentioned above, the negative tax ($-23,100$) is considered a *tax savings* in year 2, thus increasing CFAT.

Comment

MACRS depreciates to a salvage value of $S = 0$. Later we will learn about a tax implication due to "recapturing of depreciation" when an asset is sold for an amount larger than zero and MACRS was applied to fully depreciate the asset to zero.

Figure 17–1

Computation of CFBT and CFAT using MACRS depreciation and $T_e = 35\%$, Example 17.2.

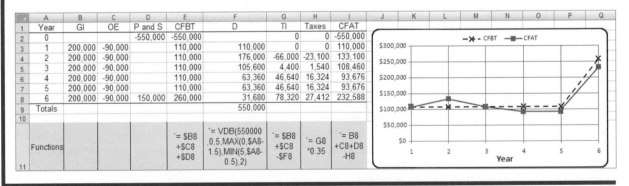

	A	B	C	D	E	F	G	H	I
1	Year	GI	OE	P and S	CFBT	D	TI	Taxes	CFAT
2	0			-550,000	-550,000		0	0	-550,000
3	1	200,000	-90,000		110,000	110,000	0	0	110,000
4	2	200,000	-90,000		110,000	176,000	-66,000	-23,100	133,100
5	3	200,000	-90,000		110,000	105,600	4,400	1,540	108,460
6	4	200,000	-90,000		110,000	63,360	46,640	16,324	93,676
7	5	200,000	-90,000		110,000	63,360	46,640	16,324	93,676
8	6	200,000	-90,000	150,000	260,000	31,680	78,320	27,412	232,588
9	Totals					550,000			
10									
11	Functions				'= $B8 +$C8 +$D8	'= VDB(550000 ,0,5,MAX(0,$A8-1.5),MIN(5,$A8-0.5),2)	'= $B8 +$C8 -$F8	'= G8 *0.35	'= B8 +C8+D8 -H8

17.3 Effect on Taxes of Different Depreciation Methods and Recovery Periods ●●●

It is important to understand why *accelerated depreciation rates* give the corporation a tax advantage relative to the straight line method with the same recovery period. Larger rates in earlier years of the recovery period require less taxes due to the larger reductions in taxable income. The criterion of **minimizing the present worth of taxes** is used to demonstrate the tax effect. For the recovery period n, choose the depreciation rates that result in the **minimum** present worth value for taxes.

$$\text{PW}_{\text{tax}} = \sum_{t=1}^{t=n} (\text{taxes in year } t)(P/F,i,t) \qquad [17.12]$$

This is equivalent to maximizing the present worth of total depreciation PW_D.

Compare any two depreciation methods. Assume the following: (1) There is a constant single-value tax rate, (2) CFBT exceeds the annual depreciation amount, (3) both methods reduce book value to the same salvage value, and (4) the same recovery period is used. On the basis of these assumptions, the following statements are correct:

The total taxes paid are **equal** for all depreciation methods.
The present worth of taxes is **less** for accelerated depreciation methods.

As we learned in Chapter 16, MACRS is the prescribed tax depreciation method in the United States, and the only alternative is MACRS straight line depreciation with an extended recovery period. The accelerated write-off of MACRS always provides a smaller PW_{tax} compared to less accelerated methods. If the DDB method were still allowed directly as it is in most countries of the world, rather than embedded in MACRS, DDB would not fare as well as MACRS. This is correct because DDB does not reduce the book value to zero. This is illustrated in Example 17.3.

EXAMPLE 17.3

An after-tax analysis for a new $50,000 machine proposed for a fiber optics manufacturing line is in process. The CFBT for the machine is estimated at $20,000. If a recovery period of 5 years applies, use the present worth of taxes criterion, an effective tax rate of 35%, and a return of 8% per year to compare the following: classical straight line, classical DDB, and MACRS depreciation. Use a 6-year period for the comparison to accommodate the half-year convention imposed by MACRS.

Solution

Table 17–3 presents a summary of annual depreciation, taxable income, and taxes for each method. For classical straight line depreciation, the numbers are $n = 5$, $D_t = \$10,000$ for 5 years, and $D_6 = 0$ (column 3). The CFBT of $20,000 is fully taxed at 35% in year 6.

The classical DDB percentage of $d = 2/n = 0.40$ is applied for 5 years. The implied salvage value is $\$50,000 - 46,112 = \$3888$, so not all $50,000 is tax deductible. The taxes using classical DDB will be $\$3888(0.35) = \$1361$ larger than for the classical SL method.

TABLE 17–3		Comparison of Taxes and Present Worth of Taxes for Different Depreciation Methods								
		Classical Straight Line			Classical Double Declining Balance			MACRS		
(1) Year t	(2) CFBT, $	(3) D_t, $	(4) TI, $	(5) = 0.35(4) Taxes, $	(6) D_t, $	(7) TI, $	(8) = 0.35(7) Taxes, $	(9) D_t, $	(10) TI, $	(11) = 0.35(10) Taxes, $
1	+20,000	10,000	10,000	3,500	20,000	0	0	10,000	10,000	3,500
2	+20,000	10,000	10,000	3,500	12,000	8,000	2,800	16,000	4,000	1,400
3	+20,000	10,000	10,000	3,500	7,200	12,800	4,480	9,600	10,400	3,640
4	+20,000	10,000	10,000	3,500	4,320	15,680	5,488	5,760	14,240	4,984
5	+20,000	10,000	10,000	3,500	2,592	17,408	6,093	5,760	14,240	4,984
6	+20,000	0	20,000	7,000	0	20,000	7,000	2,880	17,120	5,992
Totals		50,000		24,500	46,112		25,861*	50,000		24,500
PW$_{\text{tax}}$				18,386			18,549			18,162

*Larger than other values since there is an implied salvage value of $3888 not recovered.

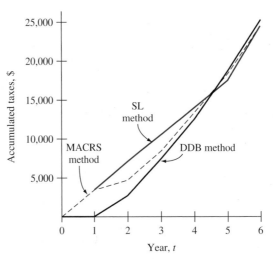

Figure 17–2
Cumulative taxes incurred by different depreciation rates for a
6-year comparison period, Example 17.3.

MACRS writes off $50,000 in 6 years using the rates of Table 16–2. Total taxes are $24,500, the same as for classical SL depreciation.

The annual taxes (columns 5, 8, and 11) are accumulated year by year in Figure 17–2. Note the pattern of the curves, especially the lower total taxes relative to the SL model after year 1 for MACRS and in years 1 through 4 for DDB. These higher tax values for SL cause PW_{tax} for SL depreciation to be larger. The PW_{tax} values at the bottom of Table 17–3 are calculated using Equation [17.12]. The MACRS PW_{tax} value is the smallest at $18,162.

To compare taxes for different *recovery periods*, change only assumption 4 at the beginning of this section to read: The same depreciation method is applied. It can be shown that a shorter recovery period will offer a tax advantage over a longer period using the criterion to minimize PW_{tax}. Comparison will indicate that

> The total taxes paid are **equal** for all n values.
> The present worth of taxes is **less** for smaller n values.

This is why corporations want to use the shortest MACRS recovery period allowed for income tax purposes. Example 17.4 demonstrates these conclusions for classical straight line depreciation, but the conclusions are correct for MACRS or any other tax depreciation method.

EXAMPLE 17.4

Grupo Grande Maquinaría, a diversified manufacturing corporation based in Mexico, maintains parallel records for depreciable assets in its operations in Berlin and Dubai. This is common for multinational corporations. One set is for corporate use that reflects the estimated useful life of assets. The second set is for foreign government purposes, such as depreciation and taxes.

The company just purchased an asset for $90,000 with an estimated useful life of 9 years; however, a shorter recovery period of 5 years is allowed by German and UAE tax laws. Demonstrate the tax advantage for the smaller n if net operating income (NOI) is $30,000 per year, an effective tax rate of 35% applies, invested money is returning 5% per year after taxes, and classical SL depreciation is allowed. Neglect the effect of any salvage value.

Solution

Determine the annual TI and taxes by Equations [17.2] through [17.3] and the present worth of taxes using Equation [17.12] for both n values.

Useful life $n = 9$ years:

$$D = \frac{90,000}{9} = 10,000$$

$$TI = 30,000 - 10,000 = \$20,000 \text{ per year}$$

$$\text{Taxes} = (0.35)(20,000) = \$7000 \text{ per year}$$

$$PW_{tax} = 7000(P/A,5\%,9) = \$49,755$$

$$\text{Total taxes} = (7000)(9) = \$63,000$$

Recovery period $n = 5$ years:

Use the same comparison period of 9 years, but depreciation occurs only during the first 5 years.

$$D_t = \begin{cases} \dfrac{90,000}{5} = \$18,000 & t = 1 \text{ to } 5 \\ 0 & t = 6 \text{ to } 9 \end{cases}$$

$$\text{Taxes} = \begin{cases} (0.35)(30,000 - 18,000) = \$4200 & t = 1 \text{ to } 5 \\ (0.35)(30,000) = \$10,500 & t = 6 \text{ to } 9 \end{cases}$$

$$\begin{aligned} PW_{tax} &= 4200(P/A,5\%,5) + 10,500(P/A,5\%,4)(P/F,5\%,5) \\ &= \$47,356 \end{aligned}$$

$$\text{Total taxes} = 4200(5) + 10,500(4) = \$63,000$$

A total of \$63,000 in taxes is paid in both cases. However, the more rapid write-off for $n = 5$ results in a present worth of tax savings of nearly \$2400 (49,755 − 47,356).

17.4 Depreciation Recapture and Capital Gains (Losses) ●●●

When the *tax implications* of the disposal of a large-investment, depreciable asset before, at, or after its recovery period are of economic interest, the material in this section is of great value. Otherwise, it is usually not considered in the economic analysis of an in-place or future purchase of a depreciable asset. In an after-tax economic analysis of large investment assets, the tax effects should be considered. The key is the size of the selling price (or salvage or market value) relative to the current book value at disposal (selling or trade-in) time and relative to the first cost, which is called the *unadjusted basis B* in depreciation terminology. There are three relevant tax-related terms.

Depreciation recapture DR, also called **ordinary gain,** occurs when a depreciable asset is sold for more than the current book value BV_t. As shown in Figure 17–3,

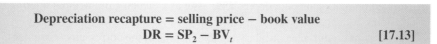

Depreciation recapture = selling price − book value

$$DR = SP_2 - BV_t \qquad\qquad [17.13]$$

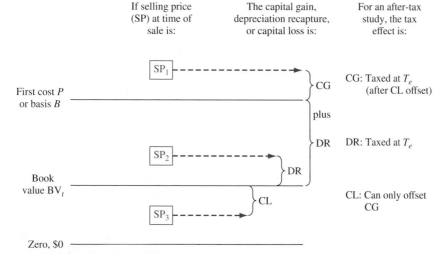

Figure 17–3

Summary of calculations and tax treatment for depreciation recapture (DR) and capital gains (losses).

Depreciation recapture is often present in the after-tax study. In the United States, an amount equal to the estimated salvage value can always be anticipated as DR when the asset is disposed of after the MACRS recovery period. This is correct simply because MACRS depreciates every asset to zero in $n + 1$ years. The amount of DR is treated as ordinary taxable income in the year of asset disposal.

Capital gain CG is an amount incurred when the selling price exceeds its (unadjusted) basis *B*. See Figure 17–3.

$$\text{Capital gain} = \text{selling price} - \text{basis}$$
$$\text{CG} = \text{SP}_1 - B \qquad [17.14]$$

Since future capital gains are difficult to predict, they are usually not detailed in an after-tax economy study. An exception is for assets that historically increase in value, such as buildings and land.

When the selling price exceeds *B*, the TI due to the sale is the capital gain *plus* the depreciation recapture, as shown in Figure 17–3. The DR is now the total amount of depreciation taken thus far, that is, $B - \text{BV}$.

Capital loss CL occurs when a depreciable asset is disposed of for less than its current book value. In Figure 17–3,

$$\text{Capital loss} = \text{book value} - \text{selling price}$$
$$\text{CL} = \text{BV}_t - \text{SP}_3 \qquad [17.15]$$

An economic analysis does not commonly account for capital loss, simply because it is not estimable for a specific alternative. However, an *after-tax replacement study* should account for any capital loss if the defender must be traded at a "sacrifice" price. For the purposes of the economic study, this provides a tax savings in the year of replacement. Use the effective tax rate to estimate the tax savings. These savings are assumed to be offset elsewhere in the corporation by other income-producing assets that generate taxes.

There are several additional points worth mentioning about capital gains and capital losses for a corporation, apart from their presence in an economic evaluation.

- U.S. tax law defines capital gains as long-term (items retained for more than 1 year) or short-term.
- Capital gains actually take place for property that is not depreciated or amortized. The term *capital gain* correctly applies at sale time to property such as investments (stocks and bonds), art, jewelry, land, and the like. When a depreciable asset's selling price is higher than the original cost (its basis), it is correctly termed an **ordinary gain.** Corporate tax treatment at this time is the same for both: taxed as ordinary income. All said, it is common to classify an expected ordinary gain on a depreciable asset as a capital gain, without altering the economic decision.
- Capital gains are taxed as ordinary taxable income at the corporation's regular tax rates.
- Capital losses do not directly reduce annual income taxes because they can only be *netted* against capital gains to the maximum extent of the capital gains for the year. The terms used then are *net* capital gains (losses).
- When capital losses exceed capital gains, the corporation can take advantage of carry-back and carry-forward tax laws for the excess, something of value to a finance or tax officer, not an engineer doing an economic analysis.
- When an asset is disposed of, the tax treatment is referred to as a *Section 1231 transaction*, which is the IRS rule section of the same number.
- IRS Publication 544, *Sales and Other Dispositions of Assets*, may be helpful if gains and losses are present in a study.
- All these rules apply to corporations. Individual tax rules and rates are different concerning asset disposition.

If the three additional income and tax elements covered here are incorporated into Equation [17.2], *taxable income* is defined as

$$
\begin{aligned}
\text{TI} &= \text{gross income} - \text{operating expenses} - \text{depreciation} \\
&\quad + \text{depreciation recapture} + \text{net capital gain} - \text{net capital loss} \\
&= \text{GI} - \text{OE} - D + \text{DR} + \text{CG} - \text{CL} \qquad [17.16]
\end{aligned}
$$

In keeping with our perspective of an engineering economy study rather than that of a financial study, deprecation recapture (i.e., ordinary gain) will be the primary element considered in after-tax evaluations. Only when a capital gain or loss must be included due to the nature of the problem will the calculations involve it.

EXAMPLE 17.5

Biotech, a medical imaging and modeling company, must purchase a bone cell analysis system for use by a team of bioengineers and mechanical engineers studying bone density in athletes. This particular part of a 3-year contract with the NBA will provide additional gross income of $100,000 per year. The effective tax rate is 35%. Estimates for two alternatives are summarized below.

	Analyzer 1	Analyzer 2
Basis B, $	150,000	225,000
Operating expenses, $ per year	30,000	10,000
MACRS recovery, years	5	5

Answer the following questions, solving by hand and spreadsheet:

(a) The Biotech president, who is very tax conscious, wishes to use a criterion of minimizing total taxes incurred over the 3 years of the contract. Which analyzer should be purchased?

(b) Assume that 3 years have now passed, and the company is about to sell the analyzer. Using the same total tax criterion, did either analyzer have an advantage? Assume the selling price is $130,000 for analyzer 1, or $225,000 for analyzer 2.

Solution by Hand

(a) Table 17–4 details the tax computations. First, the yearly MACRS depreciation is determined. Equation [17.2], $TI = GI - OE - D$, is used to calculate TI, after which the 35% tax rate is applied each year. Taxes for the 3-year period are summed, with no consideration of the time value of money.

Analyzer 1 tax total: $36,120   Analyzer 2 tax total: $38,430

The two analyzers are very close, but analyzer 1 wins with $2310 less in total taxes.

(b) When the analyzer is sold after 3 years of service, there is a depreciation recapture (DR) that is taxed at the 35% rate. This tax is in addition to the third-year tax. For each analyzer, account for the DR by Equation [17.13], $SP - BV_3$; then determine the TI, using Equation [17.16], $TI = GI - OE - D + DR$. Again, find the total taxes over 3 years, and select the analyzer with the smaller total.

	Gross Income GI, $	Operating Expenses OE, $	Basis B, $	MACRS Depreciation D, $	Book Value BV, $	Taxable Income TI, $	Taxes at 0.35TI, $
TABLE 17–4	Comparison of Total Taxes for Two Alternatives, Example 17.5*a*						
Year							
				Analyzer 1			
0			150,000		150,000		
1	100,000	30,000		30,000	120,000	40,000	14,000
2	100,000	30,000		48,000	72,000	22,000	7,700
3	100,000	30,000		28,800	43,200	41,200	14,420
							36,120
				Analyzer 2			
0			225,000		225,000		
1	100,000	10,000		45,000	180,000	45,000	15,750
2	100,000	10,000		72,000	108,000	18,000	6,300
3	100,000	10,000		43,200	64,800	46,800	16,380
							38,430

$$\text{Analyzer 1:} \qquad DR = 130{,}000 - 43{,}200 = \$86{,}800$$
$$\text{Year 3 TI} = 100{,}000 - 30{,}000 - 28{,}800 + 86{,}800 = \$128{,}000$$
$$\text{Year 3 taxes} = (0.35)(128{,}000) = \$44{,}800$$
$$\text{Total taxes} = 14{,}000 + 7700 + 44{,}800 = \$66{,}500$$

$$\text{Analyzer 2:} \qquad DR = 225{,}000 - 64{,}800 = \$160{,}200$$
$$\text{Year 3 TI} = 100{,}000 - 10{,}000 - 43{,}200 + 160{,}200 = \$207{,}000$$
$$\text{Year 3 taxes} = (0.35)(207{,}000) = \$72{,}450$$
$$\text{Total taxes} = 15{,}750 + 6300 + 72{,}450 = \$94{,}500$$

Now, analyzer 1 has a considerable advantage in total taxes ($94,500 versus $66,500).

Solution by Spreadsheet

(*a*) Rows 5 through 9 of Figure 17–4 perform the same computations as the hand solution for analyzer 1 with total taxes of $36,120. Similar analysis in rows 14 to 18 results in total taxes of $38,430 for analyzer 2, indicating that the company should select analyzer 1, based on taxes only.

(*b*) Revised year 3 entries for analyzer 1 in row 10 show the sales price of $130,000, the updated TI of $128,000, and a 3-year tax total of $66,500. The new TI in year 3 has the depreciation recapture incorporated as DR = selling price − book value = $SP - BV_3$, which is shown in the cell tag as the last term (D10 − F10). With a similar updating for analyzer 2 (row 19), total taxes of $94,500 now show a significantly larger tax advantage for analyzer 1 over 3 years.

Comment

Note that no time value of money is considered in these analyses, as we have used in previous alternative evaluations. In Section 17.5 below we will rely upon PW, AW, and ROR analyses at an established MARR to make an after-tax decision based upon CFAT values.

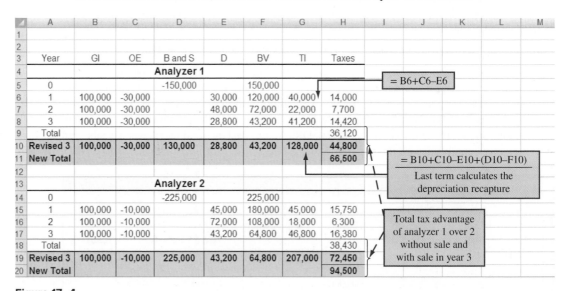

Figure 17–4
Impact of depreciation recapture on total taxes, Example 17.5.

17.5 After-Tax Economic Evaluation ●●●

The required after-tax MARR is established using the market interest rate, the corporation's effective tax rate, and its weighted average cost of capital. The CFAT estimates are used to compute the PW or AW at the after-tax MARR. When *positive and negative CFAT values* are present, a PW or AW < 0 indicates the MARR is not met. For a single project or mutually exclusive alternatives, apply the same logic as in Chapters 5 and 6. The guidelines are:

One project. PW or AW $\geq$ 0, the project is financially viable because the after-tax MARR is met or exceeded.

Two or more alternatives. Select the alternative with the best (numerically largest) PW or AW value.

☑ ME alternative selection

☑ Equal service

If *only cost CFAT amounts* are estimated, calculate the after-tax savings generated by the operating expenses and depreciation. Assign a plus sign to each saving and apply the guidelines above.

Remember, the **equal-service** assumption requires that the PW analysis be performed over the least common multiple (LCM) of alternative lives. This requirement must be met for every analysis—before or after taxes.

Since the CFAT estimates usually vary from year to year in an after-tax evaluation, the spreadsheet offers a much speedier analysis than solution by hand.

For AW analysis: Use the PMT function with an embedded NPV function *over one life cycle.* The general format is as follows, with the NPV function in italics for the CFAT series.

$$= -PMT(MARR, n, NPV(MARR, year\_1: year\_n) + year\_0) \qquad [17.17]$$

For PW analysis: Obtain the PMT function results first, followed by the PV function taken over the LCM. (There is an LCM function in Excel.) The cell containing the PMT function result is entered as the *A* value. The general format is

$$= -PV(MARR, LCM\_years, PMT\_result\_cell) \qquad [17.18]$$

EXAMPLE 17.6

Paul is designing the interior walls of an industrial building. In some places, it is important to reduce noise transmission across the wall. Two construction options—stucco on metal lath (S) and bricks (B)—each have about the same transmission loss, approximately 33 decibels. This will reduce noise attenuation costs in adjacent office areas. Paul has estimated the first costs and after-tax savings each year for both designs. (*a*) Use the CFAT values and an after-tax MARR of 7% per year to determine which is economically better. (*b*) Use a spreadsheet to select the alternative and determine the required first cost for the plans to break even.

Plan S		Plan B	
Year	CFAT, $	Year	CFAT, $
0	−28,800	0	−50,000
1–6	5,400	1	14,200
7–10	2,040	2	13,300
10	2,792	3	12,400
		4	11,500
		5	10,600

Solution by Hand

(*a*) In this example, both AW and PW analyses are shown. Develop the AW relations using the CFAT values over each plan's life. Select the larger value.

$$\begin{aligned} AW_S &= [-28,800 + 5400(P/A,7\%,6) + 2040(P/A,7\%,4)(P/F,7\%,6) \\ &\quad + 2792(P/F,7\%,10)](A/P,7\%,10) \\ &= \$422 \end{aligned}$$

$$\begin{aligned} AW_B &= [-50,000 + 14,200(P/F,7\%,1) + \cdots + 10,600(P/F,7\%,5)](A/P,7\%,5) \\ &= \$327 \end{aligned}$$

Both plans are financially viable; select plan S because AW_S is larger.

For the PW analysis, the LCM is 10 years. Use the AW values and the *P/A* factor for the LCM of 10 years to select stucco on metal lath, plan S.

$$PW_S = AW_S(P/A,7\%,10) = 422(7.0236) = \$2964$$
$$PW_B = AW_B(P/A,7\%,10) = 327(7.0236) = \$2297$$

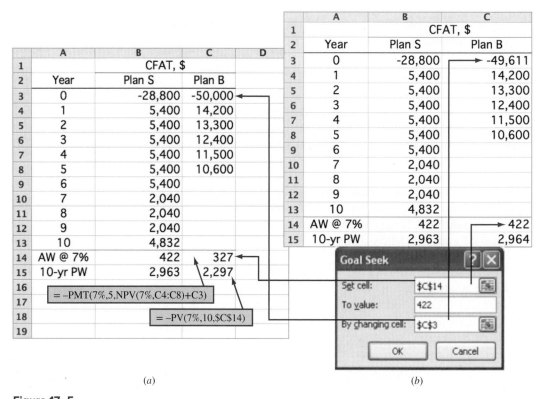

Figure 17–5
(a) After-tax AW and PW analysis and (b) breakeven first cost using Goal Seek, Example 17.6.

Solution by Spreadsheet

(b) Figure 17–5a, row 14, displays the AW value calculated using the PMT function defined by Equation [17.17], and row 15 shows the 10-year PW that results from the PV function in Equation [17.18]. Plan S is chosen by a relatively small margin.

Figure 17–5b shows the Goal Seek template used to equate the AW values and determine the plan B first cost of $−49,611 that causes the plans to break even. This is a small reduction from the $−50,000 first cost initially estimated.

Comment

It is important to remember the minus signs in PMT and PV functions when utilizing them to obtain the corresponding PW and AW values. If the minus is omitted, the AW and PW values have the wrong sign and it appears that the plans are not financially viable in that they do not return at least the after-tax MARR. That would happen in this example.

To utilize the **ROR method,** apply exactly the same procedures as in Chapter 7 (single project) and Chapter 8 (two or more alternatives) to the CFAT series. A PW or AW relation is developed to estimate the rate of return i^* for a project, or Δi^* for the incremental CFAT between two alternatives. Multiple roots may exist in the CFAT series, as they can for any cash flow series. For a single project, set the PW or AW equal to zero and solve for i^*.

Project evaluation

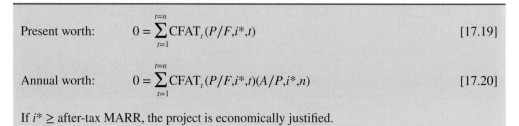

Present worth: $$0 = \sum_{t=1}^{t=n} \text{CFAT}_t (P/F,i^*,t)$$ [17.19]

Annual worth: $$0 = \sum_{t=1}^{t=n} \text{CFAT}_t (P/F,i^*,t)(A/P,i^*,n)$$ [17.20]

If $i^* \geq$ after-tax MARR, the project is economically justified.

Spreadsheet solution for i^* is faster for most CFAT series. It is performed using the IRR function with the general format

$$= \text{IRR(year\_0\_CFAT:year\_n\_CFAT)} \qquad [17.21]$$

If the after-tax ROR is important to the analysis, but the details of an after-tax study are not of interest, the before-tax ROR (or MARR) can be adjusted with the effective tax rate T_e by using the *approximating* relation

$$\textbf{Before-tax ROR} = \frac{\textbf{after-tax ROR}}{\textbf{1} - T_e} \qquad [17.22]$$

For example, assume a company has an effective tax rate of 40% and normally uses an after-tax MARR of 12% per year for economic analyses that consider taxes explicitly. To *approximate* the effect of taxes without performing the details of an after-tax study, the before-tax MARR can be estimated as

$$\text{Before-tax MARR} = \frac{0.12}{1 - 0.40} = 20\% \text{ per year}$$

If the decision concerns the economic viability of a project and the resulting PW or AW value is close to zero, the details of an after-tax analysis should be developed.

EXAMPLE 17.7

A fiber optics manufacturing company operating in Hong Kong has spent $50,000 for a 5-year-life machine that has a projected $20,000 annual NOI and annual depreciation of $10,000 for years 1 through 5. The company has a T_e of 40%. (*a*) Determine the after-tax rate of return. (*b*) Approximate the before-tax return and compare it with the actual before-tax ROR.

Solution

(*a*) The CFAT in year 0 is $-50,000. For years 1 through 5, there is no capital purchase or sale, so NOI = CFBT. (See Equations [17.1] and [17.9].) Determine CFAT.

$$\text{TI} = \text{NOI} - D = 20{,}000 - 10{,}000 = \$10{,}000$$

$$\text{Taxes} = T_e\,(\text{TI}) = 0.4(10{,}000) = \$4000$$

$$\text{CFAT} = \text{CFBT} - \text{taxes} = 20{,}000 - 4000 = \$16{,}000$$

Since the CFAT for years 1 through 5 has the same value, use the P/A factor in Equation [17.19].

$$0 = -50{,}000 + 16{,}000(P/A,i^*,5)$$
$$(P/A,i^*,5) = 3.125$$

Solution gives $i^* = 18.03\%$ as the after-tax rate of return.

(*b*) Use Equation [17.22] for the before-tax return estimate.

$$\text{Before-tax ROR} = \frac{0.1803}{1 - 0.40} = 0.3005 \quad (30.05\%)$$

The actual before-tax i^* using CFBT = $20,000 for 5 years is 28.65% from the relation

$$0 = -50{,}000 + 20{,}000(P/A,i^*,5)$$

The tax effect will be slightly overestimated if a MARR of 30.05% is used in a before-tax analysis.

A rate of return evaluation performed by hand on two or more alternatives utilizes a PW or AW relation to determine the incremental return Δi^* of the incremental CFAT series between two alternatives. Solution by spreadsheet is accomplished using the incremental CFAT values and the IRR function. The equations and procedures applied are the same as in Chapter 8 (Sections 8.4 through 8.6) for selection from mutually exclusive alternatives using the ROR method. You should review and understand these sections before proceeding with this section.

From this review, several important facts can be recalled:

> **Selection guideline:** The fundamental rule of incremental ROR evaluation at a stated MARR is as follows:

> Select the one alternative that requires the largest initial investment, provided the extra investment is justified relative to another justified alternative.

> **Incremental ROR:** Incremental analysis must be performed. Overall i^* values cannot be depended upon to select the correct alternative, unlike the PW or AW method at the MARR, which will always indicate the correct alternative.

> **Equal-service requirement:** Incremental ROR analysis requires that the alternatives be evaluated over equal time periods. The LCM of the two alternative lives must be used to find the PW or AW of incremental cash flows. (The only exception, mentioned in Section 8.5, occurs when the AW analysis is performed on *actual cash flows, not the increments;* then one-life-cycle analysis is acceptable over the respective alternative lives.)

> **Revenue and cost alternatives:** Revenue alternatives (positive and negative cash flows) may be treated differently from cost alternatives (cost-only cash flow estimates). For revenue alternatives, the overall i^* may be used to perform an initial screening. Alternatives with $i^* <$ MARR can be removed from further evaluation. An i^* for cost-only alternatives cannot be determined, so incremental analysis is required with all alternatives included.

Breakeven ROR

Once the CFAT series are developed, the **breakeven ROR** can be obtained using a plot of PW versus i^* by solving the PW relation for each alternative over the LCM at several interest rates. For any after-tax MARR greater than the breakeven ROR, the extra investment is not justified.

The next examples solve CFAT problems using incremental ROR analysis and the breakeven ROR plot of PW versus i.

EXAMPLE 17.8

In Example 17.6, Paul estimated the CFAT for interior wall materials to reduce sound transmission; plan S is to construct with stucco on metal lath, and plan B is to construct using brick. Figure 17–5a presented both a PW analysis over 10 years and an AW analysis over the respective lives. Plan S was selected. After reviewing this earlier solution, (a) perform an ROR evaluation at the after-tax MARR of 7% per year, and (b) plot the PW versus Δi graph to determine the breakeven ROR.

Solution by Spreadsheet

(a) The LCM is 10 years for the incremental ROR analysis, and plan B requires the extra investment that must be justified. Apply the procedure in Section 8.6 for incremental ROR analysis. Figure 17–6 shows the estimated CFAT for each alternative and the incremental CFAT series. Since these are revenue alternatives, the overall Δi^* is calculated first to ensure that they both make at least the MARR of 7%. Row 14 indicates they do. The IRR function (cell E14) is applied to the incremental CFAT, indicating that $\Delta i^* = 6.35\%$. Since this is lower than the MARR, the extra investment in brick walls is not justified. Plan S is selected, the same as with the PW and AW methods.

(b) The NPV function is used to find the PW of the incremental CFAT series at various i values. The graph indicates that the breakeven Δi^* occurs at 6.35%—the same that the IRR function found. Whenever the after-tax MARR is above 6.35%, as is the case here with MARR = 7%, the extra investment in plan B is not justified.

Comment

Note that the incremental CFAT series has three sign changes. The cumulative series also has three sign changes (Norstrom's criterion). Accordingly, there may be multiple Δi^* values. The application of the IRR function using the "guess" option finds no other real number roots in the normal rate of return range.

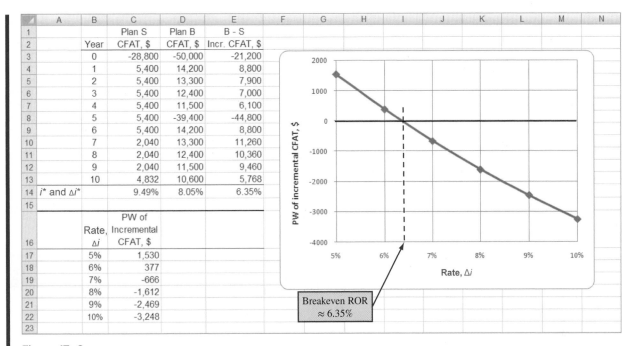

Figure 17–6
Incremental evaluation of CFAT and determination of breakeven ROR, Example 17.8.

EXAMPLE 17.9

In Example 17.5 an after-tax analysis of two bone cell analyzers was initiated due to a new 3-year NBA contract. The criterion used to select analyzer 1 was the total taxes for the 3 years. The complete solution is in Table 17–4 (hand) and Figure 17–4 (spreadsheet).

Continue the spreadsheet analysis by performing an after-tax ROR evaluation, assuming the analyzers are sold after 3 years for the amounts estimated in Example 17.5: $130,000 for analyzer 1 and $225,000 for analyzer 2. The after-tax MARR is 10% per year.

Solution

A spreadsheet solution is presented here, but a hand solution is equivalent, just slower. Figure 17–7 is an updated version of the spreadsheet in Figure 17–4 to include the sale of the analyzers in year 3. The CFAT series (column I) are determined by the relation CFAT = CFBT − taxes, with the taxable income determined using Equation [17.16], where DR is included. For example, in year 3 when analyzer 2 is sold for $S = \$225,000$, the CFAT calculation is

$$CFAT_3 = CFBT - (TI)(T_e) = GI - OE - P + S - (GI - OE - D + DR)(T_e)$$

The depreciation recapture DR is the amount above the year 3 book value received at sale time. Using the book value after 3 years (F14),

$$DR = \text{selling price} - BV_3 = 225,000 - 64,800 = \$160,200$$

Now the CFAT in year 3 for analyzer 2 can be determined.

$$\begin{aligned}
CFAT_3 &= 100,000 - 10,000 + 0 + 225,000 \\
&\quad -(100,000 - 10,000 - 43,200 + 160,200)(0.35) \\
&= 315,000 - 207,000(0.35) = \$242,550
\end{aligned}$$

The cell tags in row 14 of Figure 17–7 follow this same progression. The incremental CFAT is calculated in column J, ready for the after-tax incremental ROR analysis.

These are revenue alternatives, so the overall i^* values indicate that both CFAT series are acceptable. The value $\Delta i^* = 23.6\%$ (cell J17) also exceeds MARR = 10%, so **analyzer 2 is** *selected.* This decision applies the ROR method guideline: Select the alternative that requires the largest, *incrementally* justified investment.

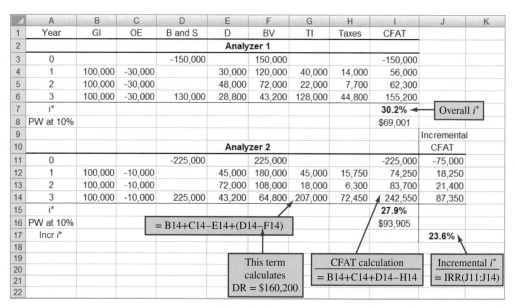

Figure 17–7
Incremental ROR analysis of CFAT with depreciation recapture, Example 17.9.

Comment

In Section 8.4, we demonstrated the fallacy of selecting an alternative based solely on the overall $i*$, because of the ranking inconsistency problem of the ROR method. The incremental ROR must be used. The same fact is demonstrated in this example. If the larger $i*$ alternative is chosen, analyzer 1 is incorrectly selected. When $\Delta i*$ exceeds the MARR, the larger investment is correctly chosen—analyzer 2 in this case. For verification, the PW at 10% is calculated for each analyzer (column I). Again, analyzer 2 is the winner, based on its larger PW of $93,905.

17.6 After-Tax Replacement Study ● ● ●

When a currently-installed asset (the defender) is challenged with possible replacement, the effect of taxes can have an impact upon the decision of the replacement study. The final decision may not be reversed by taxes, but the difference between before-tax AW values of the defender and challenger may be significantly different from the after-tax difference. Tax considerations in the year of the replacement are as follows:

Depreciation recapture or tax savings due to a sizable **capital loss** are possible if it is necessary to trade the defender at a sacrifice price. Additionally, the after-tax replacement study considers tax-deductible **depreciation** and **operating expenses** not accounted for in a before-tax analysis.

The effective tax rate T_e is used to estimate the amount of annual taxes (or tax savings) from TI. The same procedure as the before-tax replacement study in Chapter 11 is applied here, but for CFAT estimates. The procedure should be thoroughly understood before proceeding. Special attention to Sections 11.3 and 11.5 is recommended.

Example 17.10 presents a solution by hand of an after-tax replacement study using a simplifying assumption of classical SL (straight line) depreciation. Example 17.11 solves the same problem by spreadsheet, but includes the detail of MACRS depreciation. This provides an opportunity to observe the difference in the AW values between the two depreciation methods.

EXAMPLE 17.10

Midcontinent Power Authority purchased emission control equipment 3 years ago for $600,000. Management has discovered that it is technologically and legally outdated now. New equipment has been identified. If a market value of $400,000 is offered as the trade-in for the current equipment, perform a replacement study using (*a*) a before-tax MARR of 10% per year, and

(b) a 7% per year after-tax MARR. Assume an effective tax rate of 34%. As a simplifying assumption, use classical straight line depreciation with $S = 0$ for both alternatives.

	Defender	Challenger
Market value, $	400,000	
First cost, $		−1,000,000
Annual cost, $/year	−100,000	−15,000
Recovery period, years	8 (originally)	5

Solution

Assume that an ESL (economic service life) analysis has determined the best life values to be 5 more years for the defender and 5 years total for the challenger.

(a) For the *before-tax replacement study*, find the AW values. The defender AW uses the market value as the first cost, $P_D = \$-400,000$.

$$AW_D = -400,000(A/P,10\%,5) - 100,000 = \$-205,520$$
$$AW_C = -1,000,000(A/P,10\%,5) - 15,000 = \$-278,800$$

Applying step 1 of the replacement study procedure (Section 11.3), we select the better AW value. The defender is retained now with a plan to keep it for the five remaining years. The defender has a $73,280 lower equivalent annual cost compared to the challenger. This complete solution is included in Table 17–5 (left half) for comparison with the after-tax study.

(b) For the *after-tax replacement study*, there are no tax effects other than income tax for the defender. The annual SL depreciation is $75,000, determined when the equipment was purchased 3 years ago.

$$D_t = 600,000/8 = \$75,000 \qquad t = 1 \text{ to } 8 \text{ years}$$

Table 17–5 shows the TI and taxes at 34%. The taxes are actually tax savings of $59,500 per year, as indicated by the minus sign. (Remember that for tax savings in an economic

TABLE 17–5   Before-Tax and After-Tax Replacement Analyses, Example 17.10

		Before Taxes			After Taxes			
Defender Age	Year	Expenses OE, $	P and S, $	CFBT, $	Depreciation D, $	Taxable Income TI, $	Taxes* at 0.34TI, $	CFAT, $
				Defender				
3	0		−400,000	−400,000				−400,000
4	1	−100,000		−100,000	75,000	−175,000	−59,500	−40,500
5	2	−100,000		−100,000	75,000	−175,000	−59,500	−40,500
6	3	−100,000		−100,000	75,000	−175,000	−59,500	−40,500
7	4	−100,000		−100,000	75,000	−175,000	−59,500	−40,500
8	5	−100,000	0	−100,000	75,000	−175,000	−59,500	−40,500
AW at 10%				−205,520	AW at 7%			−138,056
				Challenger				
	0		−1,000,000	−1,000,000		+25,000†	8,500	−1,008,500
	1	−15,000		−15,000	200,000	−215,000	−73,100	+58,100
	2	−15,000		−15,000	200,000	−215,000	−73,100	+58,100
	3	−15,000		−15,000	200,000	−215,000	−73,100	+58,100
	4	−15,000		−15,000	200,000	−215,000	−73,100	+58,100
	5	−15,000	0	−15,000	200,000	−215,000‡	−73,100	+58,100
AW at 10%				−278,800	AW at 7%			−187,863

* Minus sign indicates a tax savings for the year.
† Depreciation recapture on defender trade-in.
‡ Assumes challenger's salvage actually realized is $S = 0$; no tax.

analysis, it is assumed that there is positive taxable income elsewhere in the corporation to offset the saving.) Since only costs are estimated, the annual CFAT is negative, but the $59,500 tax savings has reduced it. The CFAT and AW at 7% per year are

$$\text{CFAT} = \text{CFBT} - \text{taxes} = -100,000 - (-59,500) = \$-40,500$$

$$\text{AW}_D = -400,000(A/P,7\%,5) - 40,500 = \$-138,056$$

For the challenger, depreciation recapture on the defender occurs when it is replaced because the trade-in amount of $400,000 is larger than the current book value. In year 0 for the challenger, Table 17–5 includes the following computations to arrive at a tax of $8500:

Defender book value, year 3: $BV_3 = 600,000 - 3(75,000) = \$375,000$
Depreciation recapture: $DR_3 = TI = 400,000 - 375,000 = \$25,000$
Taxes on trade-in, year 0: $\text{Taxes} = 0.34(25,000) = \$8500$

The SL depreciation is $1,000,000/5 = \$200,000$ per year. This results in tax saving and CFAT as follows:

$$\text{Taxes} = (-15,000 - 200,000)(0.34) = \$-73,100$$

$$\text{CFAT} = \text{CFBT} - \text{taxes} = -15,000 - (-73,100) = \$+58,100$$

In year 5, it is assumed the challenger is sold for $0; there is no depreciation recapture. The AW for the challenger at the 7% after-tax MARR is

$$\text{AW}_C = -1,008,500(A/P,7\%,5) + 58,100 = \$-187,863$$

The defender is again selected; however, the equivalent annual advantage has decreased from $73,280 before taxes to $49,807 after taxes.

Conclusion: By either analysis, retain the defender now and plan to keep it for 5 more years. Additionally, plan to evaluate the estimates for both alternatives 1 year hence or when another challenger is identified. If and when cash flow estimates change significantly, perform another replacement analysis.

Comment

If the market value (trade-in) had been less than the current defender book value of $375,000, a capital loss, rather than depreciation recapture, would occur in year 0. The resulting tax savings would decrease the CFAT (which is to reduce costs if CFAT is negative) of the challenger. For example, a trade-in amount of $350,000 would result in a TI of $350,000 - 375,000 = \$-25,000$ and a *tax savings* of $-8500 in year 0. The CFAT is then $-1,000,000 - (-8500) = \$-991,500$.

EXAMPLE 17.11

Repeat the after-tax replacement study of Example 17.10b using 7-year MACRS depreciation for the defender and 5-year MACRS depreciation for the challenger. Assume either asset is sold after 5 years for exactly its book value. Determine if the answers are significantly different from those obtained when the simplifying assumption of classical SL depreciation was made.

Solution

Figure 17–8 shows the complete analysis. MACRS requires more computation than SL depreciation, but this effort is easily reduced by the use of a spreadsheet. Again the *defender is selected* for retention, but now by an advantage of $44,142 annually. This compares to the $49,807 advantage using classical SL depreciation and the $73,280 before-tax advantage of the defender. Therefore, taxes and MACRS have reduced the defender's economic advantage, but not enough to reverse the decision to retain it.

Several other differences in the results between SL and MACRS depreciation are worth noting. There is depreciation recapture in year 0 of the challenger due to trade-in of the defender at $400,000, a value larger than the book value of the 3-year-old defender. This amount,

	A	B	C	D	E	F	G	H	I	J
1	MARR =	7%								
2	Purchase =	$ 600,000			**Defender after-tax MACRS analysis**					
3	Defender		Basis B &	(Expenses)	MACRS					
4	age	Year	salvage S$^{(1)}$	CFBT	rates	Depr	TI	Tax savings	CFAT	
5	3	0	-400,000						-400,000	
6	4	1		-100,000	0.1249	74,940	-174,940	-59,480	-40,520	
7	5	2		-100,000	0.0893	53,580	-153,580	-52,217	-47,783	
8	6	3		-100,000	0.0892	53,520	-153,520	-52,197	-47,803	
9	7	4		-100,000	0.0893	53,580	-153,580	-52,217	-47,783	
10	8	5	0	-100,000	0.0446	26,760	-126,760	-43,098	-56,902	
11	Total					262,380				
12	$^{(1)}$Defender assumed to be sold in year 5 (year 8 of its life) for exactly BV = 0.							AW at 7%	**-$145,273**	
13	All the original B = $600,000 is depreciated over the 8 years; no tax effect.									
14										
15	Purchase =	$1,000,000			**Challenger after-tax MACRS analysis**			Taxes or		
16	Challenger		Basis B &	(Expenses)	MACRS					
17	age	Year	salvage S$^{(1)}$	CFBT	rates	Depr	TI $^{(2)}$	Tax savings	CFAT	
18	0	0	-1,000,000				137,620	46,791	-1,046,791	
19	1	1		-15,000	0.2000	200,000	-215,000	-73,100	58,100	
20	2	2		-15,000	0.3200	320,000	-335,000	-113,900	98,900	
21	3	3		-15,000	0.1920	192,000	-207,000	-70,380	55,380	
22	4	4		-15,000	0.1152	115,200	-130,200	-44,268	29,268	
23	5	5	57,600	-15,000	0.1152	115,200	-130,200	-44,268	86,868	
24	Total					942,400				
25	$^{(1)}$Challenger assumed to be sold in year 5 for exactly							AW at 7%	**-$189,415**	
26	BV = 1,000,000-942,400 = $57,600.									
27	No tax effect, but CFAT increases in year 5.									
28	$^{(2)}$ TI of $137,620 in year 0 is depreciation recapture from									
29	trade of defender. DR = B - current BV = 400,000 - 262,380.									

Depreciation recapture
= −C5 − F11

Challenger sales price
= B15 − F24

Figure 17–8
After-tax replacement study with MACRS depreciation and depreciation recapture, Example 17.11.

$137,620 (cell G18), is treated as ordinary taxable income. The calculations for the DR and associated tax, by hand, are as follows:

$$BV_3 = \text{first cost} - \text{MACRS depreciation for 3 years}$$
$$= \text{total MACRS depreciation for years 4 through 8}$$
$$= \$262,380 \qquad\qquad\qquad\qquad \text{(cell F11)}$$

$$DR = TI_0 = \text{trade-in} - BV_3$$
$$= 400,000 - 262,380 = \$137,620 \quad \text{(cell G18)}$$

$$\text{Taxes} = (0.34)(137,620) = \$46,791 \qquad\qquad\qquad \text{(cell H18)}$$

See the cell tags and table notes that duplicate this logic.

The assumption that the challenger is sold after 5 years at its book value implies a positive cash flow in year 5. The entry $57,600 (C23) reflects this assumption since the forgone MACRS depreciation in year 6 would be 1,000,000(0.0576) = $57,600. The spreadsheet relation = B15 − F24 determines this value using the accumulated depreciation in F24. [*Note:* If the salvage S = 0 is anticipated after 5 years, then a capital loss of $57,600 will be incurred. This implies an additional tax saving of 57,600(0.34) = $19,584 in year 5. Conversely, if the salvage value exceeds the book value, a depreciation recapture and associated tax should be estimated.]

17.7 After-Tax Value-Added Analysis ● ● ●

When a person or company is willing to pay more for an item, it is likely that some processing has been performed on an earlier version of the item to make it more valuable now to the purchaser. This is value added.

Value added is a term used to indicate that a product or service has **added worth** from the perspective of a consumer, owner, investor, or purchaser. It is common to leverage value-adding activities on a product or service.

Value added

For an example of highly leveraged value-added activities, consider onions that are grown and sold at the farm level for cents per pound. They may be purchased by the shopper in a store at 50 cents to $1.25 per pound. But when onions are cut and coated with a special batter, they may be fried in hot oil and sold as onion rings for several dollars per pound. Thus, from the perspective of the consumer, there has been a large amount of value added by the processing from raw onions in the ground into onion rings sold at a restaurant or fast-food shop.

The value-added measure was briefly introduced in conjunction with AW analysis before taxes. When value-added analysis is performed after taxes, the approach is somewhat different from that of CFAT analysis developed previously in this chapter. However, as shown below,

> The decision about an alternative will be the same for both the value-added and CFAT methods, because the AW of economic value-added estimates is the same as the AW of CFAT estimates.

Value-added analysis starts with Equation [17.4], net operating profit after taxes (NOPAT), which includes the depreciation for *year 1* through *year n*. Depreciation D is included in that TI = GI − OE − D. This is different from CFAT, where the depreciation has been specifically removed so that only *actual* cash flow estimates are used for *years 0* through *n*.

The term **economic value added (EVA)** indicates the monetary worth added by an alternative to the corporation's bottom line. (The term *EVA* is a service mark of Stern Value Management.) The technique, introduced in the 1990s, has become popular as a means to evaluate the ability of a corporation to increase its economic worth, especially from the shareholders' viewpoint.

> The annual EVA is the amount of NOPAT remaining on corporate books after removing the **cost of invested capital** during the year. That is, EVA indicates the project's **contribution to the net profit** of the corporation after taxes.

The **cost of invested capital** is the after-tax rate of return (usually the MARR value) multiplied by the book value of the asset during the year. This is the interest incurred by the current level of capital invested in the asset. (If different tax and book depreciation methods are used, the *book depreciation value is used* here because it more closely represents the remaining capital invested in the asset from the corporation's perspective.) Computationally,

$$\text{EVA} = \text{NOPAT} - \text{cost of invested capital}$$
$$= \text{NOPAT} - (\text{after-tax interest rate})(\text{book value in year } t-1)$$
$$= \text{TI}(1 - T_e) - (i)(\text{BV}_{t-1}) \qquad [17.23]$$

Since both TI and the book value consider depreciation, EVA is a measure of worth that mingles actual cash flow with noncash flows to determine the estimated financial worth contribution to the corporation. This financial worth is the amount used in public documents of the corporation (balance sheet, income statement, stock reports, etc.). Because corporations want to present the largest value possible to the stockholders and other owners, the EVA method is often more appealing than the AW method from the financial perspective.

The result of an EVA analysis is a series of annual EVA estimates. Two or more alternatives are compared by calculating the AW of EVA estimates and selecting the alternative with the larger AW value. If only one project is evaluated, AW > 0 means the after-tax MARR is exceeded, thus making the project value-adding.

Sullivan and Needy[1] demonstrated that the AW of EVA and the AW of CFAT are identical in amount. Thus, either method can be used to make a decision. The annual EVA estimates indicate added worth to the corporation generated by the alternative, while the annual CFAT estimates describe how cash will flow. This comparison is made in Example 17.12.

---

[1] W. G. Sullivan and K. L. Needy, "Determination of Economic Value Added for a Proposed Investment in New Manufacturing." *The Engineering Economist,* vol. 45, no. 2 (2000), pp. 166–181.

EXAMPLE 17.12

Biotechnics Engineering has developed two mutually exclusive plans for investing in new capital equipment with the expectation of increased revenue from its medical diagnostic services to cancer patients. The estimates are summarized below. (*a*) Use classical straight line depreciation, an after-tax MARR of 12%, and an effective tax rate of 40% to perform two annual worth after-tax analyses: EVA and CFAT. (*b*) Explain the fundamental difference between the results of the two analyses.

	Plan A	Plan B
Initial investment, $	−500,000	−1,200,000
Gross income − expenses, $	170,000 per year	600,000 in year 1, decreasing by 100,000 per year thereafter
Estimated life, years	4	4
Salvage value	None	None

Solution by Spreadsheet

(*a*) Refer to the spreadsheet and function cells (row 22) in Figure 17–9.

EVA evaluation: All the necessary information for EVA estimation is determined in columns B through G. The net operating profit after taxes (NOPAT) in column H is calculated by Equation [17.4], TI − taxes. The book values (column E) are used to determine the cost of invested capital in column I, using the second term in Equation [17.23], that is, $i(BV_{t-1})$, where i is the 12% after-tax MARR. This represents the amount of interest at 12% per year, after taxes, for the currently invested capital as reflected by the book value at the beginning of the year. The EVA estimate is the sum of columns H and I for years 1 through 4. *Notice there is no EVA estimate for year 0*, since NOPAT and the cost of invested capital are estimated for years 1 through *n*. Finally, the larger AW of the EVA value is selected, which indicates that plan B is better and that plan A does not make the 12% return.

CFAT evaluation: As shown in function row 22 (plan B for year 3), CFAT estimates (column K) are calculated as (GI − OE) − P − taxes. The AW of CFAT again concludes that plan B is better and that plan A does not return the after-tax MARR of 12% (K10).

(*b*) What is the fundamental difference between the EVA and CFAT series in columns J and K? They are clearly equivalent from the time value of money perspective since the AW values are numerically the same. To answer the question, consider plan A, which has a constant CFAT estimate of $152,000 per year. To obtain the AW of EVA estimate of $−12,617 for

	A	B	C	D	E	F	G	H	I	J	K
1						**PLAN A**					
2									EVA analysis		CFAT analysis
3			Investment P	SL	Book value	Taxable			Cost of		
4	Year	GI - OE	(Basis B)	Depreciation	BV	income, TI	Taxes	NOPAT	inv. capital	EVA	CFAT
5	0		-500,000		500,000						-500,000
6	1	170,000		125,000	375,000	45,000	18,000	27,000	-60,000	-33,000	152,000
7	2	170,000		125,000	250,000	45,000	18,000	27,000	-45,000	-18,000	152,000
8	3	170,000		125,000	125,000	45,000	18,000	27,000	-30,000	-3,000	152,000
9	4	170,000		125,000	0	45,000	18,000	27,000	-15,000	12,000	152,000
10	AW values									-$12,617	-$12,617
11						**PLAN B**					
12									EVA analysis		CFAT analysis
13			Investment P	SL	Book value	Taxable			Cost of		
14	Year	GI - OE	(Basis B)	Depreciation	BV	income, TI	Taxes	NOPAT	inv. capital	EVA	CFAT
15	0		-1,200,000		1,200,000						-1,200,000
16	1	600,000		300,000	900,000	300,000	120,000	180,000	-144,000	36,000	480,000
17	2	500,000		300,000	600,000	200,000	80,000	120,000	-108,000	12,000	420,000
18	3	400,000		300,000	300,000	100,000	40,000	60,000	-72,000	-12,000	360,000
19	4	300,000		300,000	0	0	0	0	-36,000	-36,000	300,000
20	AW values									$3,388	$3,388
21											
22	Functions for Plan B, year 3			` = -$C$15/4`	` = E17-D18`	` = B18-D18`	` = F18*0.4`	` = F18-G18`	` =-0.12*E17`	` = H18+I18`	` = B18+C18-G18`

Figure 17–9
Comparison of two plans using EVA and CFAT analyses, Example 17.12.

years 1 through 4, the initial investment of \$500,000 is distributed over the 4-year life using the A/P factor at 12%. That is, an equivalent amount of $\$500,000(A/P,12\%,4) = \$164,617$ is "charged" against the cash inflows in each of years 1 through 4. In effect, the yearly CFAT is reduced by this charge.

$$\text{CFAT} - (\text{initial investment})(A/P,12\%,4) = 152,000 - 500,000(A/P,12\%,4)$$
$$152,000 - 164,617 = \$-12,617$$
$$= \text{AW of EVA}$$

This is the AW value for both series, demonstrating that the two methods are economically equivalent. However, the EVA method indicates an alternative's yearly estimated contribution to the *value of the corporation,* whereas the CFAT method estimates the actual cash flows to the corporation. This is why the EVA method is often more popular than the cash flow method with corporate executives.

Comment

The calculation $P(A/P,i,n) = \$500,000(A/P,12\%,4)$ is exactly the same as the capital recovery in Equation [6.3], assuming an estimated salvage value of zero. Thus, the cost of invested capital for EVA is the same as the capital recovery discussed in Chapter 6. This further demonstrates why the AW method is economically equivalent to the EVA evaluation.

17.8 After-Tax Analysis for International Projects ● ● ●

Primary questions to be answered prior to performing a corporate-based after-tax analysis for international settings revolve around tax-deductible allowances—depreciation, business expenses, capital asset evaluation—and the effective tax rate needed for Equation [17.7], taxes $= (T_e)(\text{TI})$. As discussed in Chapter 16, most governments of the world recognize and use the straight line (SL) and declining balance (DB) methods of depreciation with some variations to determine the annual tax-deductible allowance. Expense deductions vary widely from country to country. By way of example, some of these are summarized here.

Canada

Depreciation: This is deductible and is normally based on DB calculations, although SL may be used. An equivalent of the half-year convention is applied in the first year of ownership. The annual tax-deductible allowance is termed *capital cost allowance (CCA).* As in the U.S. system, recovery rates are standardized, so the depreciation amount does not necessarily reflect the useful life of an asset.

Class and CCA rate: Asset classes are defined and annual depreciation rates are specified by class. No specific recovery period (life) is identified, in part because assets of a particular class are grouped together and the annual CCA is determined for the entire class, not individual assets. There are some 44 classes, and CCA rates vary from 4% per year (the equivalent of a 25-year-life asset) for buildings (class 1) to 100% (1-year life) for applications software, chinaware, dies, etc. (class 12). Most rates are in the range of 10% to 30% per year.

Expenses: Business expenses are deductible in calculating TI. Expenses related to capital investments are not deductible, since they are accommodated through the CCA.

Internet: Further details are available on the Canada Revenue Agency website at www.cra.gc.ca in the Forms and Publications section.

China (PRC)

Depreciation: Officially, SL is the primary method of tax-deductible depreciation; however, assets employed in selected industries or types of assets can utilize accelerated DB or SYD (sum-of-years-digits) depreciation, when approved by the government. The selected industries and assets can change over time; currently favored industries serve areas such as technology and oil exploration, and equipment subjected to large vibrations during normal usage is allowed accelerated depreciation.

Recovery period: Standardized recovery periods are published that vary from 3 years (electronic equipment) to 10 years (aircraft, machinery, and other production equipment) to 20 years (buildings). Shortened periods can be approved, but the minimum recovery period cannot be less than 60% for the normal period defined by current tax law.

Expenses: Business expenses are deductible with some limitations and some special incentives. Limitations are placed, for example, on advertising expense deductions (15% of sales for the year). Incentives are generous in some cases; for example, 150% of actual expenses is deductible for new technology and new product R&D activities.

Internet: Summary information for China and several other countries is available at www.worldwide-tax.com.

Mexico

Depreciation: This is a fully deductible allowance for calculating TI. The SL method is applied with an index for inflation considered each year. For some asset types, an immediate deduction of a percentage of the first cost is allowed. (This is a close equivalent to the Section 179 Deduction in the United States.)

Class and rates: Asset types are identified, though not as specifically defined as in some countries. Major classes are identified, and annual recovery rates vary from 5% for buildings (the equivalent of a 20-year life) to 100% for environmental machinery. Most rates range from 10% to 30% per year.

Profit tax: The income tax is levied on profits on income earned from carrying on business in Mexico. Most business expenses are deductible. Corporate income is taxed only once, at the federal level; no state-level taxes are imposed.

Tax on Net Assets (TNA): Under some conditions, a tax of 1.8% of the average value of assets located in Mexico is paid annually in addition to income taxes.

Internet: The best information is via websites for companies that assist international corporations located in Mexico. One example is PriceWaterhouseCoopers at www.pwcglobal .com/mx/eng.

The effective tax rate varies considerably among countries. Some countries levy taxes only at the federal level, while others impose taxes at several levels of government (federal, state or provincial, prefecture, county, and city). A summary of international corporate average tax rates is presented in Table 17–6 for a wide range of industrialized countries. These include income taxes at all reported levels of government within each country; however, other types of taxes may be imposed by a particular government. Although these average rates of taxation will vary

TABLE 17–6	Summary of International Corporate Average Tax Rates (2015)
Tax Rate Levied on Taxable Income, %	**For These Countries**
≥40	United States
35 to <40	South Africa
32 to <35	France, India, Japan, Pakistan
28 to <32	Australia, New Zealand, Spain, Germany, Mexico
24 to <28	China, Republic of Korea, Canada
20 to <24	Russia, Turkey, Saudi Arabia, Chile, United Kingdom
<20	Singapore, Hong Kong, Ireland, Iceland, Hungary, Switzerland, Taiwan

Sources: Extracted from KPMG's *Corporate and Indirect Tax Survey 2015* (www.kpmg.com/taxrates) and from country websites on corporate taxation.

from year to year, especially as tax reform is enacted, it can be surmised that most corporations face effective rates of about 20% to 40% of taxable income. A close examination of international rates shows that they have decreased significantly over the last decade. In fact, the KPMG report noted in Table 17–6 indicates that the global average corporate tax rate on TI has decreased from 32.7% (1999) to 25.5% (2009) to 23.7% (2015). This has encouraged corporate investment and business expansion within country borders and helped soften the economic downturns experienced over the years.

One of the prime ways that governments have been able to reduce corporate tax rates is by shifting to **indirect taxes** on goods and services for additional tax revenue. These taxes are usually in the form of a **value-added tax (VAT),** goods and services tax (GST), and taxes on products imported from outside its borders. As corporate tax rates have declined, in general the indirect tax rates have increased. This has been especially true during the first 10 to 15 years of the 21st century. However, as worldwide economic slumps are experienced, governments around the world are more cautious of how they tax corporations and maintain a reasonable balance between regular tax rates (as listed in Table 17–6) and indirect tax rates. The primary indirect-tax system called the VAT system is explained now.

17.9 Value-Added Tax ●●●

A value-added tax (VAT) has facetiously been called a sales tax on steroids because the VAT rates on some items in some countries that impose a VAT can be as high as 90%. It is also called a GST (goods and service tax).

> A **value-added tax** is an indirect tax; that is, it is a tax on goods and services rather than on people or corporations. It differs from a sales tax in two ways: (1) when it is charged, and (2) who pays (as explained below). A specific percentage, say 10%, is a charge added to the price of the item and paid by the buyer. The seller then sends this 10% VAT to the taxing entity, usually a government unit. This process of 10% VAT continues every time the item is resold—as purchased or in a modified form—thus the term *value-added*.

A value-added tax is commonly used throughout the world. In some countries, VAT is used in lieu of business or individual income taxes. There is no VAT system in the United States yet. In fact, the United States is the only major industrialized country in the world that does not have a VAT system, though other forms of indirect taxation are used liberally. There is mounting evidence, however, that a VAT/GST system will be necessary in the United States in the near future, but not without a great amount of political discord.

A **sales tax** is used by the U.S. government, by nearly all of the states, and by many local entities. A sales tax is charged on goods and services at the time the goods and services reach *the end user or consumer.* That is, businesses do not pay a sales tax on raw material, unfinished goods, or items they purchase that will ultimately be sold to an end user; only the end user pays the sales tax. Businesses *do pay* a sales tax on items for which they are the end user. Total sales tax percentages imposed by multiple government levels can range from 5% to 11%, sometimes larger on specific items. For example, when Home Depot (HD) purchases microwave ovens from Jenn-Air, Inc., Home Depot does not pay a sales tax on the microwave ovens because they will be sold to HD customers who *will* pay the sales tax. On the other hand, if Home Depot purchases a forklift from Caterpillar for loading and unloading merchandise in one of its stores, HD will pay the sales tax on the forklift, since HD is the end user. Thus, a sales tax is **paid only one time,** and that time is when the goods or services are purchased **by the end user.** The sales tax is the responsibility of the merchant to collect and remand to the taxing entity.

A **value-added tax (VAT),** on the other hand, is charged to the buyer at purchase time, whether the buyer is a business or end user. The seller sends the collected VAT to the taxing entity. If the buyer subsequently resells the goods to another buyer (as is or a modification of it), another VAT is collected by the seller. Now, this second seller will send to the taxing entity an amount that is equal to the total tax collected *minus* the amount of VAT already paid.

As an illustration, assume the U.S. government charged a 10% VAT. Here is how the VAT might work.

> Northshore Mining Corporation of Babbitt, Minnesota, sells $100,000 worth of iron ore to Westfall Steel. As part of the price, Northshore collects $110,000, that is, $100,000 + 0.1 × 100,000, from Westfall Steel. Northshore remits the $10,000 VAT to the U.S. Treasury.
>
> Westfall Steel sells all of the steel it made from the iron ore at a price of $300,000 to General Electric (GE). Westfall collects $330,000 from GE and then sends $20,000 to the U.S. Treasury, that is, $30,000 it collected in VAT from GE minus $10,000 it paid in taxes to Northshore Mining.
>
> GE uses the steel to make refrigerators that it sells for $700,000 to retailers, such as Home Depot, Lowe's, and others. GE collects $770,000 and then remits $40,000 to the U.S. Treasury, that is, $70,000 it collected in taxes from retailers minus $30,000 it paid in VAT to Westfall Steel.
>
> If GE purchased machines, tools, or other items to make the refrigerators during this accounting period, and it paid taxes on those items, the taxes paid would also be deducted from the VAT that GE collected before sending the money to the U.S. Treasury. For example, if GE paid $5000 in taxes on motors it purchased for the refrigerators, the amount GE would remit to the U.S. Treasury would be $35,000 (that is, $70,000 it collected from retailers minus $30,000 it paid in taxes to Westfall Steel minus $5000 it paid in taxes on the motors).
>
> The retailers sell the refrigerators for $950,000 and collect $95,000 in taxes from end users—consumers. The retailers remit $25,000 to the U.S. Treasury (that is, $95,000 they collected minus $70,000 they paid previously).

Through this process, the U.S Treasury has received $10,000 from Northshore, $20,000 from Westfall Steel, $35,000 from GE, $5000 from the supplier of the motors, and $25,000 from the retailers, for a total of $95,000. This is 10% of the final sales price of $950,000. The VAT money was deposited into the Treasury at several different times from several different companies.

The taxes that a company *pays* for materials or items it purchases in order to produce goods or services that will subsequently be sold to another business or end user are called *input taxes*, and the company is able to recover them when it collects the VAT from the sale of its products. The taxes that a company *collects* are called *output taxes*, and these are forwarded to the taxing entity, less the amount of input taxes the company paid. Hence, the businesses incur no taxes themselves, the same as with a sales tax.

Several dimensions of a VAT distinguish it from a sales tax or corporate income taxes. Some are as follow:

- Value-added taxes are taxes on consumption, not production or taxable income.
- The end user pays all of the value-added taxes, but VATs are not as obvious as a sales tax that is added to the price of the item at the time of purchase (and displayed on the receipt). Therefore, VAT taxing entities encounter less resistance from consumers.
- Value-added taxes are generally considerably higher than sales taxes, with the average European VAT rate at 20% and the worldwide average at 15.79%.
- When a VAT system is initiated in a country, the rate starts low, but creeps up over time to settle in the range of 13% to 25%.
- The VAT is essentially a "sales tax," but it is charged at each stage of the product development process instead of when the product is sold.
- With VAT, there is less evasion of taxes because it is harder for multiple entities to evade collecting and paying the taxes than it is for one entity to do so.
- VAT rates vary from country to country and from category to category. For example, in some countries, food has a 0% VAT rate, while aviation fuel is taxed at 32%.

EXAMPLE 17.13

Tata Motors is a major automobile manufacturer in India. It has three different manufacturing units that specialize in manufacturing different transportation-related products, such as trucks, engines and axles, commercial vehicles, utility vehicles, and passenger cars. The company buys products that fall under different sections of the Indian government VAT tax code, and therefore the products have different VAT rates. In one particular accounting period, Tata had invoices from four different suppliers (vendors A, B, C, and D) in the respective amounts of $1.5 million, $3.8 million, $1.1 million, and $900,000. The products Tata purchased were subject to VAT rates of 4%, 4%, 12.5%, and 22%, respectively.

(*a*) How much total VAT did Tata pay to its vendors?

(*b*) Assume that Tata's products have a VAT rate of 12.5%. If Tata's sales during the period were $9.2 million, how much VAT did the Indian Treasury receive from Tata?

Solution

(*a*) Let X equal the product price before the VAT is added. Solve for X and then subtract it from the purchase amount to determine the VAT charged by each vendor. Table 17–7 shows the VAT that Tata paid its four vendors. An example computation for vendor A is as follows:

$$X + 0.04X = 1,500,000$$
$$1.04X = 1,500,000$$
$$X = \$1,442,308$$
$$\text{VAT}_A = 1,500,000 - 1,442,308$$
$$= \$57,692$$
$$\text{Total VAT paid} = 57,692 + 146,154 + 122,222 + 162,295$$
$$= \$488,363$$

(*b*) Total from Tata = total VAT − VAT paid by vendors
$$= 9,200,000(0.125) - 488,363$$
$$= \$661,637$$

TABLE 17–7	VAT Computation, Example 17.13			
Vendor	**Purchases, $**	**VAT Rate, %**	**Price before VAT, *X*, $**	**VAT, $**
A	1,500,000	4.0	1,442,308	57,692
B	3,800,000	4.0	3,653,846	146,154
C	1,100,000	12.5	977,778	122,222
D	900,000	22.0	737,705	162,295
Total				488,363

CHAPTER SUMMARY

After-tax analysis does not usually change the decision to select one alternative over another; however, it does offer a much clearer estimate of the monetary impact of taxes. After-tax PW, AW, and ROR evaluations of one or more alternatives are performed on the CFAT series using exactly the same procedures as in previous chapters.

Income tax rates for U.S. corporations and individual taxpayers are graduated or progressive— higher taxable incomes pay higher income taxes. A single-value, effective tax rate T_e is usually applied in an after-tax economic analysis. Taxes are reduced because of tax deductible items, such as depreciation and operating expenses. Because depreciation is a noncash flow, it is important to consider depreciation only in TI computations, and not directly in the CFBT and

CFAT calculations. Accordingly, key general cash flow after-tax relations for each year are as follows:

$$NOI = \text{gross income} - \text{operating expenses}$$
$$TI = \text{gross income} - \text{operating expenses} - \text{depreciation} + \text{depreciation recapture}$$
$$CFBT = \text{gross income} - \text{operating expenses} - \text{initial investment} + \text{salvage value}$$
$$CFAT = CFBT - \text{taxes} = CFBT - (T_e)(TI)$$

If an alternative's estimated contribution to corporate financial worth is the economic measure, the economic value added (EVA) should be determined. Unlike CFAT, the EVA includes the effect of depreciation. The equivalent annual worths of CFAT and EVA estimates are the same numerically, because they interpret the annual cost of the capital investment in different, but equivalent manners when the time value of money is taken into account.

In a replacement study, the tax impact of depreciation recapture, which may occur when the defender is traded for the challenger, is accounted for in an after-tax analysis. The replacement study procedure of Chapter 11 is applied. The tax analysis may not reverse the decision to replace or retain the defender, but the effect of taxes will likely reduce (possibly by a significant amount) the economic advantage of one alternative over the other.

International corporate tax rates have remained steady, but indirect taxes, such as value-added tax (VAT), have increased. The mechanism of a VAT is explained and compared to a sales tax. The United States is an exception; in that it currently has no VAT system.

PROBLEMS

Terminology and Basic Tax Computations

17.1 State what the following abbreviations stand for: NOI, GI, T_e, NOPAT, TI, R, OE, EBIT.

17.2 For a corporation that has a taxable income of $250,000, determine (a) the marginal tax rate, (b) the total taxes, and (c) the average tax rate.

17.3 For the events described below, select the tax-related term from the following list that best applies: gross income, depreciation, operating expense, taxable income, income tax, or net operating profit after taxes.

(a) A corporation reports that it had a negative $1,750,000 net profit on its annual income statement.

(b) An asset with a current book value of $120,000 was utilized on a new processing line to increase sales by $200,000 this year.

(c) A machine has an annual write-off of $21,000.

(d) The cost to maintain quality assurance equipment during the past year was $75,000.

(e) A supermarket collected $24,000 in lottery ticket sales last year. Based on the winnings paid to individuals holding these tickets, a rebate of $250 was sent to the store manager.

(f) An asset with a book value of $8000 was retired and sold for $8450.

(g) The cost of goods sold in the past year was $3,680,200.

(h) An over-the-counter software system will generate $420,000 in revenue this quarter.

17.4 For a company that had net operating income of $51.3 million and operating expenses of $23.6 million, what was the (a) gross income, and (b) earnings before interest and income taxes?

17.5 Determine the taxable income for a company that had gross income of $36.7 million, earnings before interest and income taxes of $21.4 million, and depreciation of $9.5 million.

17.6 Determine the single-value effective tax rate for a corporation that has a federal tax rate of 35% and state tax rate of 7%.

17.7 Knorr & Associates paid total taxes of $72,000 in the first year of their consulting business. (a) What was the company's taxable income? (b) What was the average tax rate? (c) What was the estimated NOPAT for the year?

17.8 Helical Products makes machined springs with elastic redundant elements so that a broken spring will continue to function. The company had GI of $450,000 with OE of $230,000 and depreciation of $48,000. (a) How much did the company owe in taxes if its effective tax rate was 38%? (b) Use a *spreadsheet* to plot the projected taxes for next year if GI and depreciation are the same as this year, but expenses may vary, due to cost cutting and increased prices, between $180,000 and $300,000. Is this a linear or nonlinear relation?

17.9 The last annual report of Harrison Engineering's 3-D Imaging Division showed GI = $4.9 million, OE = $2.1 million, and D = $1.4 million. If the average federal tax rate is 31% and state/local tax rates total 9.8%, estimate (a) federal income taxes, and (b) the percentage of GI that the federal government took in income taxes.

17.10 Two companies, ABC and XYZ, have the following values on their annual tax returns:

Company	ABC	XYZ
Sales revenue, $	1,500,000	820,000
Interest revenue, $	31,000	25,000
Operating expenses, $	−754,000	−591,000
Depreciation, $	148,000	18,000

(a) Calculate the exact federal income taxes for the year.

(b) Determine the percentage of sales revenue each company will pay in federal income taxes.

(c) Estimate the taxes using an effective rate on the entire TI of the marginal percentages in Table 17–1, that is, 34% for ABC and 39% for XYZ. Determine the percentage errors made relative to the exact taxes in (a).

17.11 Borsberry Medical has a gross income of $6.5 million for the year. Depreciation and operating expenses total $4.1 million. The combined state and local tax rate is 7.6%. (a) Use an effective federal rate of 34% to estimate the income taxes. (b) Borsberry's president hopes to have a total of $2 million left after taxes. What reduction in OE is necessary to realize this goal if the effective tax rate and depreciation are constant?

17.12 Aquatech Microsystems reported a TI of $80,000 last year. If the state income tax rate is 6%, determine the (a) average federal tax rate, (b) overall effective tax rate, (c) total taxes to be paid based on the effective tax rate, and (d) total taxes paid to the state and paid to the federal government.

17.13 C. F. Jordon Management Services has operated for the last 26 years in a northern province where the provincial income tax on corporate revenue is 6% per year. C. F. Jordon pays an average federal tax of 23% and reports taxable income of $7 million. Because of excessive labor cost increases, the president wants to move to another province to reduce the total tax burden. The new province may have to be willing to offer tax allowances or an interest-free grant for the first couple of years to attract the company. You are an engineer with the company and you are asked to do the following:

(a) Determine the effective tax rate for C. F. Jordon.

(b) Estimate the provincial tax rate that would be necessary to reduce the overall effective tax rate by 10% per year.

(c) Determine what the new province would have to do financially for C. F. Jordon to move and to reduce its effective tax rate to 22% per year.

17.14 The Johnson's had two children, both of them are married and now have their own families, call them Family A and Family B. They both file their U.S. income taxes as married, filing jointly. Information collected for a year for each family is shown below. Neglecting any effect of state taxes, use a *spreadsheet* and the latest tax rates from the IRS publication 17, *Your Federal Income Tax* (www.irs.gov), to determine the following for each family:

(a) Percentage of TI paid in federal taxes.

(b) Percentage of total income (salaries, dividends, and other) paid in federal taxes.

Family	A	B
Salaries, $1000	65	290
Dividends, $1000	8	58
Other income	0	14
Exemptions	5	3
Deductions, $1000	12	25

Exemptions are $4000 per individual (adult or child). Deductions for Family A are standard and itemized for B.

17.15 Joyce and Vincent, both engineers, got married and raised three children over an 18-year period and the first one is now ready for college. There have been good and bad years financially; Joyce quit work for some years to raise small children, Vincent lost his job in year 10 and had to start a new career.

The two have summarized the basic information from their tax returns for the 18 years. They wonder what percentage of their gross income has gone to federal taxes over the years. Apply the tax rate in the latest IRS Publication 17 (www.irs.gov) for married, filing jointly to calculate their taxes each year using a spreadsheet and plot the percentage of GI.

Assume exemptions are deducted as follows for each person (adult or child): years 1 to 8, $3500; years 9 to 14, $4000; years 15 to 18, $4500. (All monetary amounts are in $1000 units.)

	Salaries					
Year	Joyce	Vincent	Dividends	Other Income	Personal Exemptions	Itemized Deductions
1	69	61	5	—	2	8
2	71	65	6	—	2	10
3	75	72	6	5	2	10
4	80	78	7	10	2	11
5	25	79	7	12	4	11
6	25	83	7	10	4	11
7	25	85	8	8	4	12
8	27	90	8	5	4	12
9	28	92	8	—	4	14
10	30	0	4	—	5	10
11	70	20	4	—	5	10
12	80	20	5	—	5	8
13	90	20	5	—	5	8
14	95	60	6	—	5	10
15	100	62	10	—	5	12
16	105	65	15	—	5	14
17	107	70	20	—	5	16
18	110	75	15	—	5	20

CFBT and CFAT

17.16 Identify which of the following items are *not* included in the calculation of cash flow before taxes, CFBT: life of asset, operating expenses, salvage value, depreciation, initial investment, gross income, tax rate.

17.17 What is the basic difference between cash flow after taxes (CFAT) and net operating profit after taxes (NOPAT)?

17.18 Estimate the CFAT for a company that has taxable income of $120,000, depreciation of $133,350, and an effective tax rate of 35%.

17.19 If PSK Engineering had a CFAT of $750,000, charged depreciation of $400,000, and had a T_e of 36%, what was PSK's CFBT?

17.20 Estimate the gross income for Lopez Enterprises when it had a CFAT of $2.5 million, $900,000 in expenses, $900,000 in depreciation charges, and an effective tax rate of 26.4%.

17.21 Fill in the missing values for CFBT, D, TI, taxes, and CFAT in the following table. Depreciation amounts are based on the 3-year MACRS method and T_e is 35%. Solve (*a*) *by hand*, and (*b*) *by spreadsheet*.

Year	GI	OE	P and S	CFBT	D	TI	Taxes	CFAT
0			−1900	−1900	—	—	—	−1900
1	800	−100	0	700	633	67	23	677
2	950	−150	0	—		−45	—	816
3	600	−200	0	400	281		42	—
4	300	−250	700	750	—	−91	−32	782

17.22 Four years ago, Sierra Instruments of Monterey, California, spent $200,000 for equipment for manufacturing standard gas flow calibrators. The equipment was depreciated by MACRS using a 3-year recovery period. For year 4, GI was $100,000, OE was $50,000, and T_e was 40%. Develop *hand* and *spreadsheet* solutions for year 4 only that determine the CFAT if the asset was disposed of as follows (Neglect any tax effects caused by the sale.):

(*a*) Discarded with no salvage value at the end of year 4.

(*b*) Sold for $20,000 at the end of year 4.

(*c*) Sold for $20,000 at the end of year 4; however, SL depreciation was applied throughout the 3-year recovery period.

17.23 Four years ago, a division of Harcourt-Banks purchased an asset that was depreciated by the MACRS method using a 3-year recovery period. If the total revenue for year 2 was $48 million with depreciation of $8.2 million and operating expenses of $28 million, use a federal tax rate of 35% and a state tax rate of 6.5% to determine (*a*) CFAT, (*b*) percentage of total revenue expended on taxes, and (*c*) net profit after taxes for the year.

17.24 Advanced Anatomists, Inc., researchers in medical science, is contemplating a commercial venture concentrating on proteins based on the new X-ray technology of free-electron lasers. To recover the huge investment needed, an annual $2.5 million CFAT is needed. A favored average federal tax rate of 20% is expected; however, state taxing authorities will levy an 8% tax on TI. Over a 3-year period, the deductible expenses and depreciation are estimated to total $1.3 million the first year, increasing by $500,000 per year thereafter. Of this, 50% is operating expenses and 50% is depreciation. What is the required gross income each year?

17.25 Elias wants to perform an after-tax evaluation of equivalent methods to electrostatically remove airborne particulate matter from clean rooms used to package liquid pharmaceutical products. Using the information shown, MACRS depreciation with $n = 3$ years, a 5-year study period, after-tax MARR = 7% per year, and $T_e = 34\%$ and a *spreadsheet*, he obtained the results $AW_A = \$ -2176$ and $AW_B = \$3545$. Any tax effects when the equipment is salvaged were neglected. Method B is the better method.

Now, use classical SL depreciation with $n = 5$ years to select the better method. Is the decision different from that reached using MACRS? *After solving by hand, verify your answer using a spreadsheet.*

Method	A	B
First cost, $	−100,000	−150,000
Salvage value, $	10,000	20,000
Savings, $ per year	35,000	45,000
AOC, $ per year	−15,000	−6,000
Expected life, years	5	5

Depreciation Effects on Taxes

17.26 Explain why minimizing PW of taxes is equivalent to maximizing PW of depreciation when corporate taxes are calculated.

17.27 Equipment associated with manufacturing small railcars had a first cost of $180,000 with an expected salvage value of $30,000 at the end of its 5-year life. The revenue was $620,000 in year 2, with operating expenses of $98,000. If the company's effective tax rate was 36%, what would be the difference in taxes paid in year 2 if the depreciation method were straight line instead of MACRS? The MACRS depreciation rate for year 2 is 32%.

17.28 Vibrations Dynamics uses vibrations sensitive equipment based on lithography to measure vibrations in building foundations caused by subway, rail, and automobile traffic in major cities of the world. Use a spreadsheet to (a) plot the annual tax curves, and (b) calculate total taxes and the PW_{tax} values at $i = 8\%$ per year over an 8-year study period for two depreciation methods: SL with $n = 6$ years and DDB with $n = 8$ years. Comment on your total tax and PW_{tax} results. (Note: Neglect any tax effects connected with the salvage value.) The estimates are:

$P = \$200,000, B = \$280,000, S = 20\%$ of P, CFBT $= \$100,000$ per year, $T_e = 30\%$.

17.29 Guess which depreciation method will have the lower 10%-per-year PW of taxes between the following methods over a 6-year study period: $P = \$100, S = 0$, and GI-OE $= \$50$, and $T_e = 30\%$. Verify your answer via spreadsheet.
    I. Straight line with $n = 4$ years
    II. MACRS with $n = 5$ years
    III. DDB with $n = 6$ years

17.30 Hibra, an EE student who is also taking a business minor, is studying depreciation in her engineering management and finance courses. The assignment in both classes is to demonstrate that shorter recovery periods require the same total taxes, but they offer a time-value-of-taxes advantage for depreciable assets. Help her using asset estimates made for a 6-year study period: $P = \$65,000$, $S = \$5000$, GI = $32,000 per year, AOC = $10,000 per year, SL depreciation, $i = 12\%$ per year, and $T_e = 31\%$. (a) Perform the comparison using recovery periods of 3 and 6 years. (b) Develop the *spreadsheet* solution. (c) For a *spreadsheet function* challenge, construct a single-cell function to determine the PW of taxes for the 3-year and 6-year analyses.

17.31 Complete the last four columns of the table below using an effective tax rate of 40% for an asset that has a first cost of $20,000, no salvage value, and a 3-year recovery period. Use (a) straight line depreciation, and (b) MACRS depreciation. (All cash flows are in $1000 units.)

			Estimates, $				
Year	GI	P	OE	D	TI	Taxes	CFAT
0	—	−20	—	—	—	—	−20
1	8	—	−2	—	—	—	—
2	15	—	−4	—	—	—	—
3	12	0	−3	—	—	—	—
4	10	0	−5	—	—	—	—

17.32 An asset with a first cost of $9000 is depreciated using 5-year MACRS recovery. The CFBT is estimated at $10,000 for the first 4 years and $5000 thereafter as long as the asset is retained. The effective tax rate is 40%, and money is worth 10% per year. In present worth dollars, how much of the cash flow generated by the asset over its recovery period is lost to taxes?

17.33 Use an effective tax rate of 32% to determine three parameters: CFAT, NOPAT, and PW of taxes ($i = 6\%$ per year) associated with a new-technology MRI machine (first cost of $30,000) recently located at the county-government-funded Zendra Hospital in San Francisco. Analyze two different depreciation scenarios: (1) uniform write-off of $6000 per year, and (2) accelerated depreciation of $6000, $9600, $5760, and $3456 in years 1 through 4, respectively. Which one is the better depreciation schedule using the criteria (a) total NOPAT, and (b) PW of taxes? (All cash amounts are in $1000 units.)

			Estimates, $				
Year	GI	OE	P	D	TI	Taxes	CFAT	NOPAT
0	—	—	−30	—	—	—	−30	—
1	8	−2	—	—	—	—	—	—
2	15	−4	—	—	—	—	—	—
3	12	−3	—	—	—	—	—	—
4	10	−5	—	—	—	—	—	—

Depreciation Recapture and Capital Gains (Losses)

17.34 In what alternative evaluation is it considered important that an anticipated capital loss (CL) on an in-place asset should be included in an after-tax analysis? Why?

17.35 An in-place machine with $B = \$120,000$ was depreciated by MACRS over a 3-year period. The machine was sold for $60,000 at the end of year 2 when the company decided to import the item that required the use of the machine. In year 2, GI = $1.4 million and OE = $500,000. Determine the tax liability in year 2 if $T_e = 35\%$.

17.36 Last month, HighPower, which specializes in wind power plant design and engineering, made a capital investment of $400,000 in physical simulation equipment that will be used for at least 5 years, then sold for approximately 25% of the first cost. By law, the assets are MACRS depreciated using a 3-year recovery period. By how much will the sale cause HighPower's TI and taxes to change in year 5 if the federal tax rate is 35% and the state tax rate is 6.5%?

17.37 Determine the amount of any DR, CG, or CL generated by each event described below. Use the result to determine the income tax effect, if $T_e = 30\%$.
   (a) A strip of land zoned as "Commercial A" purchased 8 years ago for $2.6 million was just sold at a 15% profit.
   (b) Earthmoving equipment purchased for $155,000 was depreciated using MACRS over a 5-year recovery period. It was sold at the end of the fifth year of ownership for $10,000.
   (c) A MACRS-depreciated asset with a 7-year recovery period has been sold after 8 years at an amount equal to 20% of its first cost of $150,000.

17.38 A touch-sensitive assembly robot that costs $300,000 had a depreciable life of 5 years with an expected $50,000 salvage value when purchased 3 years ago. Using MACRS depreciation, determine any DR, CG, or CL if the company sold the robot after 3 years for $80,000.

17.39 A nanotube forming asset was purchased 3 years ago for $240,000. It was just sold for $285,000. The asset was depreciated by the MACRS method with $n = 5$ years and has a current book value of $69,120. Determine the amount of (a) CG and DR, and (b) prepare a spreadsheet showing the taxes each year for $T_e = 28\%$. Assume the asset's

GI = $100,500 and OE = $50,000 for each of the 3 years.

17.40 A couple of years ago, the company Health4All purchased land, a building, and two depreciable assets from another corporation. These have all recently been disposed of. Use the information shown to determine the presence and amount of any capital gain, capital loss, or depreciation recapture.

Asset	Purchase Price, $	Recovery Period, Years	Current Book Value, $	Sales Price, $
Land	−200,000	—		245,000
Building	−800,000	27.5	300,000	255,000
Asset 1	−50,500	3	15,500	18,500
Asset 2	−10,000	3	5,000	10,000

17.41 Freeman Engineering paid $28,500 for specialized equipment for use with their new GPS/GIS system. The equipment was depreciated over a 3-year recovery period using MACRS depreciation. The company sold the equipment after 2 years for $5000 when it purchased an upgraded system. (a) Determine the amount of the depreciation recapture or capital loss involved in selling the asset. (b) What tax effect will this amount have?

17.42 Thomas completed a study of a $1 million 3-year-old DNA analysis and modeling system that DynaScope Enterprises wants to keep for 1 more year or dispose of now. His table (in $1000 units) details the analysis, including an anticipated $100,000 selling price (SP) at the end of next year, SL depreciation, taxes at the all inclusive rate of $T_e = 52\%$, and PW at the after-tax MARR of 5% per year. Thomas recommends retention since PW > 0. Critique the analysis to determine if he made the correct recommendation.

Year	CFBT	SP	SL Depr.	TI	Taxes	CFAT
0	−1000					−1000
1	275		250	25	13	262
2	275		250	25	13	262
3	275		250	25	13	262
4	275	100	250	25	13	362
PW at 5%						11.3

After-Tax Economic Evaluation

17.43 Determine (a) the before-tax ROR, and (b) approximate after-tax ROR for a project that has a first cost of $750,000, a salvage value of 25% of the first cost after 3 years, and annual (GI-OE) of $260,000. Assume the company has a T_e of 37%.

17.44 Determine the required before-tax rate of return if an after-tax return of 9% per year is expected and the state and local tax rates total 6%. The effective federal tax rate is 35%.

17.45 An engineer who made an annual return of 8% after taxes on a stock investment was told by his accountant that this is equivalent to a 12% per year before-tax return. What percentage of taxable income is the accountant assuming will be taken by taxes?

17.46 Estimate the after-tax ROR for a project that has a before-tax ROR of 24%. Assume the company is in the 35% tax bracket and it used MACRS depreciation for an asset that has a $27,000 salvage value.

17.47 A division of Midland Oil & Gas has a TI of $8.95 million for a tax year. If the state tax rate averages 5% for all states in which the corporation operates, find the equivalent after-tax ROR required of projects that are justified only if they can demonstrate a before-tax return of 22% per year.

17.48 When JJ and Sons was awarded a contract to pour a skyscraper's foundation, the father had to choose between two pieces of equipment needed to supplement pumping of concrete into foundation settings. Estimates are below. Both machines have an estimated 7-year useful life; however, MACRS depreciation is over a 5-year recovery period. The effective tax rate is 40%, and the after-tax MARR is 8% per year. Your father asked you, one of the sons, to recommend one machine. (a) Perform the analysis by *spreadsheet* using after-tax PW analysis, and (b) plot the CFAT curves.

Machine	CreteHelper (CH)	Hoister (H)
First cost, $	−50,000	−40,000
Salvage, $	4,000	3,000
CFBT, $ per year	10,000	8,500
Life, years	7	7

17.49 A European candy manufacturing plant manager must select a new irradiation system to ensure the safety of specific ingredients, while being economical. The two alternatives available have the following estimates:

System	A	B
First cost, $	−150,000	−85,000
CFBT, $ per year	60,000	20,000
Life, years	3	5

The company is in the 35% tax bracket and assumes classical straight line depreciation for alternative comparisons performed at an after-tax MARR of 6% per year. A salvage value of zero is used when depreciation is calculated; however, System B can be sold after 5 years for an estimated 10% of its first cost. System A has no anticipated salvage value. Determine which is more economical using an AW analysis worked *by hand*.

17.50 Process control equipment purchased for $78,000 by Debco, Incorporated generated a CFBT of $26,080 during the first year of its 10-year estimated life. This would represent a return of 31.2% per year if maintained throughout the 10 years. However, the corporate finance officer determined that the CFAT was only $18,000 for the first year, and it is expected to decrease by $1000 per year thereafter. If the president wants to realize an after-tax return of 12% per year, for how many years must the equipment remain in service?

17.51 Perform a PW-based evaluation of the two alternatives below using (a) by *hand*, and (b) by *spreadsheet* solutions. The after-tax MARR is 8% per year, MACRS depreciation applies, and $T_e = 40\%$. The (GI − OE) estimate is made for the first 3 years; it is zero in year 4 when each asset is sold.

Alternative	X	Y
First cost, $	−8,000	−13,000
Salvage value, year 4, $	0	2,000
GI − OE, $ per year	3,500	5,000
Recovery period, years	3	3

17.52 Elias wants to perform an after-tax evaluation of equivalent methods to electrostatically remove airborne particulate matter from clean rooms used to package liquid pharmaceutical products. Two alternatives are available, but others may be identified if these are not acceptable. Using the information shown, MACRS depreciation with $n = 3$ years, a 5-year study period, after-tax MARR = 7% per year, $T_e = 34\%$, and a spreadsheet, he obtained the results $AW_G = \$ - 2176$ and $AW_H = \$3545$. Any tax effects when the equipment is salvaged were neglected. Thus, with MACRS depreciation, method H is the better method.
Use classical SL depreciation with $n = 5$ years to evaluate the same alternatives. Is the decision different from that reached using MACRS?

Method	G	H
First cost, $	−100,000	−150,000
Salvage value, $	10,000	20,000
Savings, $ per year	35,000	45,000
AOC, $/year	−15,000	−6,000
Expected life, years	5	5

17.53 AAA Pest Control uses the following: before-tax MARR = 14% per year, after-tax MARR = 7% per year, and T_e = 50%. Two new spray machine options have the following estimates and will generate the same GI each year.

Machine	A	B
First cost, $	−15,000	−22,000
Salvage value, $	3,000	5,000
AOC, $ per year	−3,000	−1,500
Life, years	10	10

Select A or B under the following conditions using the method described or as instructed:
(a) Before-tax PW analysis using *spreadsheet functions.*
(b) After-tax PW analysis using classical SL depreciation over the 10-year life using *hand* solution.
(c) After-tax PW analysis using MACRS depreciation with a 5-year recovery period using a *spreadsheet.* Assume the machines will be retained for 10 years, then sold at the estimated salvage values.

After-Tax Replacement

17.54 In an after-tax replacement study between a defender and a challenger, there may be a capital gain (CG) or loss (CL) when the defender is sold. (a) How is the gain or loss calculated, and (b) how does it affect the AW values in the study?

17.55 The defender in a multiple-effect solar cell manufacturing plant has a market value of $130,000 and expected annual operating costs of $70,000 with no salvage value after its remaining life of 3 years. The depreciation charges for the next 3 years will be $69,960, $49,960, and $35,720. Using an effective tax rate of 35% and an after-tax MARR of 12% per year, determine the cash flow after taxes (CFAT) for *year 2 only* that can be used in a PW equation for comparing the defender against a challenger that also has a 3-year life.

17.56 A 2-year-old injection molding machine was expected to be kept in service for its projected life of 5 years, but a new challenger promises to be more efficient and have lower operating costs. You have been asked to determine if it would be economically attractive to replace the defender now or keep it for 3 more years as originally planned. The defender had a first cost of $300,000, but its market value now is only $100,000. It has chargeable expenses of $120,000 per year and no expected salvage value. To simplify calculations

for this problem only, assume that SL depreciation was charged at $60,000 per year, and that it will continue at that rate for the next 3 years.

The challenger will cost $420,000, have a 3-year life, and no salvage value. It will have chargeable expenses of $30,000 per year, and it will be depreciated at $140,000 per year (again, using SL depreciation for simplicity for this problem only). Assume a T_e of 35%, and an after-tax MARR of 15% per year. Determine (a) through (c) *by hand.*
(a) Determine the CFAT in year 0 for the challenger and defender. (Hint: There may be a DR, CG, or CL to consider.)
(b) Determine the CFAT in years 1 through 3 for the challenger and defender.
(c) Conduct an AW-based evaluation to determine if the defender should be kept for 3 more years or replaced now.
(d) Use a *spreadsheet* to perform the AW-based evaluation.

17.57 Perform a PW-based after-tax replacement study from the information shown below using an after-tax MARR of 12% per year, T_e of 35%, and a study period of 4 years. All monetary values are in $1000 units. Assume the assets will be traded at their original salvage estimates. Since no revenues are estimated, all taxes are negative and considered savings to the alternative. Solve *by hand* and *by spreadsheet.*

	Defender	Challenger
First cost P, $	−45	−24
Estimated S at purchase, $	5	0
Market value now, $	35	—
OE, $ per year	−7	−8
Depreciation method	SL	MACRS
Recovery period, years	8	3
Useful life, years	8	5
Years owned	3	—

17.58 After 8 years of use, the heavy-truck engine-overhaul equipment at Pete's Truck Repair was evaluated for replacement. Pete's accountant used an after-tax MARR of 8% per year, T_e of 30%, and a current market value of $25,000 to determine AW = $2100. The new equipment costs $75,000, uses SL depreciation over a 10-year recovery period, and has a $15,000 salvage estimate. Estimated CFBT is $15,000 per year. Pete asked his engineer son Ramon to determine if the new equipment should replace what is owned currently. From the accountant, Ramon learned the current equipment cost $20,000 when purchased and reached a zero book value several years ago. Help Ramon answer his father's question.

17.59 Needco Supplies-Canada employee Stella Needleson was asked to determine if the current process of dying writing paper should be retained or a new, environment friendly process should be implemented. Estimates or actual values for the two processes are summarized below. She performed an after-tax replacement analysis at 10% per year and the corporation's effective tax rate of 32% to determine that economically, the new process should be chosen. Was she correct? Why or why not? (*Note:* Canadian tax law does not impose the half-year convention requirement. Monetary units are in Canadian dollars.)

	Current Process	New Process
First cost 7 years ago, $	−450,000	—
First cost, $	—	−700,000
Remaining life, years	5	10
Current market value, $	50,000	—
OE, $/year	−160,000	−150,000
Future salvage, $	0	50,000
Depreciation method	SL	SL

17.60 The Los Angeles, California, city engineer is analyzing a for-profit public works project at the port authority using an after-tax replacement analysis of the system installed 5 years ago (defender) and a challenger as detailed below. All values are in $1000 units. The effective state tax rate of 6% is applicable, but no federal taxes are imposed. The municipal after-tax return of 6% per year is required. Assume salvages in the future occur at the estimated amounts and use classical SL depreciation.
    (*a*) Perform the AW analysis *by hand.*
    (*b*) Perform the evaluation using a *spreadsheet.*
    (*c*) Would the decision be different if a before-tax replacement analysis were performed at $i = 12\%$ per year? Also, write the spreadsheet functions to display the AW values.

	Defender	Challenger
First cost, $	−28,000	−15,000
OE, $ per year	−1200	−1500
Salvage estimate, $	2000	3000
Market value, $	15,000	—
Life, years	10	8

17.61 Nuclear safety devices installed several years ago have been depreciated from a first cost of $200,000 to zero using MACRS. The devices can be sold on the used equipment market for an estimated $15,000, or they can be retained in service for 5 more years with a $9000 upgrade now and an OE of $6000 per year. The upgrade investment will be depreciated over 3 years with no salvage value. The challenger is a replacement with newer technology at a first cost of $40,000, $n = 5$ years,

and $S = 0$. The new units will have operating expenses of $7000 per year.
    (*a*) Use a 5-year study period, an effective tax rate of 40%, an after-tax MARR of 12% per year, and an assumption of classical straight line depreciation (no half-year convention) to perform an after-tax AW-based replacement study.
    (*b*) If the challenger is known to be salable after 5 years for an amount between $2000 and $4000, will the challenger AW value become more or less costly? Why?

Economic Value Added

17.62 (*a*) What does the term economic value added (EVA) mean relative to the bottom line of a corporation? (*b*) Why might an investor in a public corporation prefer to use the EVA estimates over the CFAT estimates for a project?

17.63 An asset with a first cost of $300,000 is depreciated by the MACRS method using a 5-year recovery period. Determine the economic value added in year 2, if the net operating profit after taxes is $70,000 and the company uses an after-tax MARR of 15% per year. The MACRS rates for years 1 and 2 are 20% and 32%, respectively.

17.64 In conducting an EVA analysis for year 2 for new equipment on its primary product line, Analogy, Inc., manufacturers of preassembled blower packages and other water treatment components, determined the EVA to be $28,000. Analogy uses an after-tax interest rate of 14% per year and its T_e is 35%. The new equipment had a first cost of $550,000 and was MACRS depreciated using a 3-year recovery period. Since the company CEO knew that the GI was $500,000, he asked you to determine the operating expenses (OE) associated with the equipment in year 2.

17.65 Cardenas and Moreno Engineering is evaluating a large flood control program for several southern cities. One component is a 4-year project for a special-purpose transport ship-crane for use in building permanent storm surge protection against hurricanes on the New Orleans, Louisiana coastline. The estimates are $P = \$300,000$, $S = 0$, and $n = 3$ years. MACRS depreciation with a 3-year recovery is indicated. GI and OE are estimated at $200,000 and $80,000, respectively, for each of 4 years, $T_e = 35\%$, and the after-tax MARR is 5% per year. The CFAT has been calculated. Determine the AW values of the CFAT and EVA series. They should have the same value.

Year	GI, $	OE, $	P, $	Depr., $	TI, $	Taxes, $	CFAT, $
0	—	—	−300,000	—	—	—	−300,000
1	200,000	−80,000		99,990	20,010	7,003	112,997
2	200,000	−80,000		133,350	−13,350	−4,673	124,673
3	200,000	−80,000		44,430	75,570	26,450	93,551
4	200,000	−80,000		22,230	97,770	34,220	85,781

17.66 Triple Play Innovators Corporation (TPIC) plans to offer IPTV (Internet Protocol TV) service to North American customers starting soon. Perform an AW analysis of the EVA series for the two alternative suppliers available for the hardware and software. Use $T_e = 30\%$, after-tax MARR = 8%, a study period of 8 years, and SL depreciation (for simplicity, omit the half-year convention and MACRS). Use a *spreadsheet*.

Vendor	Hong Kong	Vietnam
First cost, $	4.2 million	3.6 million
Recovery period, years	8	5
Salvage value, $	0	0
GI − OE, $ per year	1,500,000 in year 1; increasing by 300,000 per year up to 8 years	

17.67 FreeBird Software has developed partnerships with several large manufacturing corporations to use a Java-derivative software in their consumer and industrial products. A new corporation will be formed to manage these applications. One major project involves using Java in commercial and industrial appliances that store and cook food. The gross income and expenses are expected to follow the relations below for the estimated life of 6 years. For $t = 1$ to 6 years:

$$\text{Annual GI} = 2,800,000 - 100,000t$$

$$\text{Annual OE} = 950,000 + 50,000t$$

For the new company, $T_e = 35\%$, after-tax MARR = 12% per year, and the depreciation method chosen for the $3,000,000 in capital investment is the 5-year MACRS alternative that allows straight line write-off with the half-year convention in years 1 and 6. Using a *spreadsheet*, estimate (*a*) the annual economic contribution of the project to the new corporation, and (*b*) the equivalent annual worth of these contributions. (*c*) At what after-tax MARR will the AW of contributions exceed $400,000 per year?

Value-Added Tax (VAT)

17.68 What is the primary difference between a sales tax and a value-added tax?

17.69 In Denmark, VAT is applied at a rate of 25%, with few exceptions. If vendor A sells raw materials to vendor B for $60,000 plus VAT, and vendor B sells a product to vendor C for $130,000 plus VAT, and vendor C sells an improved product to an enduser for $250,000 plus VAT, (*a*) what is the amount of tax collected by vendor B? (*b*) How much tax does vendor B send to Denmark's Treasury? and (*c*) What is the total amount of tax collected by the Treasury department?

The following information is used in Problems 17.70 through 17.75.

Ajinkya Electronic Systems, a company in India that manufactures many different electronic products, has to purchase goods and services from a variety of suppliers (wire, diodes, LED displays, plastic components, etc.). The table below shows several suppliers and the VAT tax rates associated with each. It also shows the purchases (in $1000 units) that Ajinkya made (before tax) from each supplier in the previous accounting period. Assume Ajinkya's sales to end users were $9.2 million and Ajinkya's products carry a 15% VAT.

Supplier	VAT Rate, %	Purchases by Ajinkya, $1000
A	4.0	350
B	12.5	870
C	12.5	620
D	21.3	90
E	32.6	50

17.70 How much VAT did supplier C collect?

17.71 How much tax did Ajinkya keep from the tax it collected based on the purchases it made from supplier A?

17.72 What was the total amount of VAT paid by Ajinkya to the suppliers?

17.73 What was the average VAT rate paid by Ajinkya in purchasing goods and services?

17.74 What was the amount of VAT Ajinkya sent to the Treasury of India?

17.75 What was the total amount of VAT collected by the Treasury of India from Ajinkya and Ajinkya's suppliers?

ADDITIONAL PROBLEMS AND FE EXAM REVIEW QUESTIONS

17.76 If the after-tax ROR is 11.4% and the corporate T_e is 39%, the approximate before-tax rate of return is closest to:
  (a) 6.8%
  (b) 15.4%
  (c) 18.7%
  (d) 19.7%

17.77 If the federal tax rate is 36% and the state tax rate is 7%, the effective tax rate is closest to:
  (a) 40.5%
  (b) 37.3%
  (c) 35.4%
  (d) 31.8%

17.78 If all values carry a + sign, cash flow before taxes (CFBT) is represented by the equation:
  (a) Gross income – operating expenses – depreciation – initial investment + salvage value
  (b) Gross income – operating expenses – depreciation + salvage value
  (c) Gross income – operating expenses – initial investment + salvage value
  (d) Gross income – operating expenses + initial investment + salvage value

17.79 A graduated income tax system means:
  (a) Only taxable incomes above a certain level pay any taxes
  (b) A higher flat rate goes with all of the taxable income
  (c) Higher tax rates go with higher taxable incomes
  (d) Rates are indexed each year to keep up with inflation

17.80 All of the following are characteristics of a value-added tax system, except:
  (a) Value-added taxes are taxes on consumption
  (b) The end user pays value-added taxes
  (c) Value-added taxes are charged at each stage of product development
  (d) Value-added taxes are charged only on the raw materials for product development

17.81 Depreciation recapture occurs when a depreciable asset is sold for:
  (a) More than the current book value
  (b) More than the current market value
  (c) More than the estimated salvage value
  (d) More than the first cost

17.82 A capital gain is calculated by the equation:
  (a) Capital gain = book value – selling price
  (b) Capital gain = book value – first cost
  (c) Capital gain = market value – selling price
  (d) Capital gain = selling price – first cost

17.83 Pennington Oil is in the 50% effective tax bracket. It had gross income of $470 million in each of the last 2 years. In the first year, deductions were $160 million. In the second year, deductions were lower at $120 million. The *difference* in income taxes paid by the company in the 2 years was closest to:
  (a) $10 million
  (b) $20 million
  (c) $50 million
  (d) $40 million

17.84 A small manufacturing company with a gross income of $360,000 has the following operating expenses: M&O = $76,000, insurance = $7000, labor = $110,000, utilities = $29,000, debt service = $37,000, and taxes = $9000. The net operating income (NOI) is closest to:
  (a) $92,000
  (b) $101,000
  (c) $138,000
  (d) $174,000

17.85 An after-market auto parts company sells a machining robot that had been depreciated to zero for $16,000. If the company's effective tax rate is 36%, the sale will:
  (a) Increase the company's taxes by $16,000
  (b) Increase the company's taxes by $5760
  (c) Reduce the company's taxes by $16,000
  (d) Reduce the company's taxes by $5760

17.86 A contractor who files as unmarried (single) to the IRS has an effective tax rate of 28%. His gross income is $155,000, other income is $4000, personal expenses are $45,000, and deductions and exemptions are $12,000. His total income tax due is closest to:
  (a) $28,550
  (b) $41,160
  (c) $43,400
  (d) $55,750

17.87 The after-tax analysis for a $60,000 investment with associated gross income minus expenses (GI – OE) is shown below for the first 2 years only. If the effective tax rate is 40%, the values for depreciation D, taxable income TI, and taxes for year 1 are closest to:

Year	Investment, $	GI – OE, $	D, $	TI, $	Taxes, $	CFAT, $
0	–60,000					–60,000
1		30,000				26,000
2		35,000	15,000	6,000		29,000

(a) $D = \$5,000$, TI $= \$25,000$, taxes $= \$10,000$
(b) $D = \$30,000$, TI $= \$30,000$, taxes $= \$4,000$
(c) $D = \$20,000$, TI $= \$50,000$, taxes $= \$20,000$
(d) $D = \$20,000$, TI $= \$10,000$, taxes $= \$4,000$

17.88 An asset purchased for $100,000 with $S = \$20,000$ after 5 years was depreciated using the 5-year MACRS rates. Expenses averaged $18,000 per year and the effective tax rate is 30%. The asset was actually sold after 5 years of service for $22,000. MACRS rates in years 5 and 6 are 11.52% and 5.76%, respectively. The after-tax cash flow from the sale is closest to:

(a) $27,760
(b) $17,130
(c) $26,870
(d) $20,585

CASE STUDY

AFTER-TAX ANALYSIS FOR BUSINESS EXPANSION

Background

Charles was always a hands-on type of person. Within a couple of years of graduating from college, he started his own business. After some 20 years, it has grown significantly. He owns and operates Pro-Fence, Inc. in the Metroplex, specializing in custom-made metal and stone fencing for commercial and residential sites. For some time, Charles has thought he should expand into a new geographic region, with the target area being another large metropolitan area about 500 miles north, called Victoria.

Pro-Fence is privately owned by Charles; therefore, the question of how to finance such an expansion has been, and still is, the major challenge. Debt financing would not be a problem in that the Victoria Bank has already offered a loan of up to $2 million. Taking capital from the retained earnings of Pro-Fence is a second possibility, but taking too much will jeopardize the current business, especially if the expansion were not an economic success and Pro-Fence were stuck with a large loan to repay.

This is where you come in as a long-time friend of Charles. He knows you are quite economically oriented and that you understand the rudiments of debt and equity financing and economic analysis. He wants you to advise him on the balance between using Pro-Fence funds and borrowed funds. You have agreed to help him, as much as you can.

Information

Charles has collected some information that he shares with you. Between his accountant and a small market survey of the business opportunities in Victoria, the following generalized estimates seem reasonable:

Initial capital investment = $1.5 million

Annual gross income = $700,000

Annual operating expenses = $100,000

Effective income tax rate for Pro-Fence = 35%

Five-year MACRS depreciation for all $1.5 million investment

The terms of the Victoria Bank loan would be 6% per year simple interest based on the initial loan principal. Repayment would be in five equal payments of interest and principal. Charles comments that this is not the best loan arrangement he hopes to get, but it is a good worst-case scenario upon which to base the debt portion of the analysis. A range of D-E mixes should be analyzed. Between Charles and yourself, you have developed the following viable options:

Debt		Equity	
Percentage	Loan Amount, $	Percentage	Investment Amount, $
0		100	1,500,000
50	750,000	50	750,000
70	1,050,000	30	450,000
90	1,350,000	10	150,000

Case Study Exercises

1. For each funding option, perform a spreadsheet analysis that shows the total CFAT and its present worth over a 6-year period, the time it will take to realize the full advantage of MACRS depreciation. An after-tax return of 10% is expected. Which funding option is best for Pro-Fence? (*Hint:* For the spreadsheet, sample column headings are: year, GI − OE, loan interest, loan principal, equity investment, depreciation rate, depreciation, book value, TI, taxes, and CFAT.)

2. Observe the changes in the total 6-year CFAT as the D-E percentages change. If the time value of money is neglected, what is the constant amount by which this sum changes for every 10% increase in equity funding?

3. Charles noticed that the CFAT total and PW values go in opposite directions as the equity percentage increases. He wants to know why this phenomenon occurs. How should you explain this to Charles?

4. After deciding on the 50-50 split of debt and equity financing, Charles wants to know what additional bottom-line contributions to the economic worth of the company may be added by the new Victoria site. What are the best estimates at this time?

baona/Getty Images

CHAPTER 18

Sensitivity Analysis and Staged Decisions

This chapter includes several related topics about alternative evaluation. Initially, we expand our capability to perform a **sensitivity analysis** of one or more parameters and of an entire alternative. Then the determination and use of the **expected value** of a cash flow series are treated. The techniques of **decision trees** help make a series of economic decisions for alternatives that have different, but closely connected stages.

Economic decisions that involve **staged funding** are very common in professional and everyday life. The last topic of **real options analysis** introduces a method useful in these circumstances.

18.1 Determining Sensitivity to Parameter Variation ● ● ●

The term *parameter* is used in this chapter to represent any variable or factor for which an estimate or stated value is necessary. Example parameters are first cost, salvage value, AOC, estimated life, production rate, and materials costs. Estimates such as the loan interest rate and the inflation rate can also be parameters of the analysis.

Economic analysis uses estimates of a parameter's future value to assist decision makers. Since future estimates are always incorrect to some degree, inaccuracy is present in the economic projections. The effect of variation may be determined by using sensitivity analysis.

> **Sensitivity analysis** determines how a measure of worth—PW, AW, FW, ROR, B/C, or CER—is altered when one or more parameters vary over a **selected range of values.** Usually one parameter at a time is varied, and independence with other parameters is assumed. Though this approach is an oversimplification in real-world situations, since the dependencies are difficult to accurately model, the end results are usually correct.

Sensitivity analysis

In reality, we have applied this approach (informally) throughout previous chapters to determine the response to variation in a variety of parameters. Variation in a parameter such as MARR will not alter the decision to select an alternative when all compared alternatives return considerably more than the MARR; thus, the decision is relatively insensitive to the MARR. However, variation in the *n* or AOC value may indicate that the alternative's measure of worth is very sensitive to the estimated life or annual operating costs.

Usually the variations in life, annual costs, and revenues result from variations in selling price, operation at different levels of capacity, inflation, etc. For example, if an operating level of 90% of airline seating capacity for a domestic route is compared with 70% for a proposed international route, the operating cost and revenue per passenger-mile will increase, but anticipated aircraft life will probably decrease only slightly. Usually several important parameters are studied to learn how the uncertainty of estimates affects the economic analysis.

Sensitivity analysis routinely concentrates on the variation expected in estimates of *P*, AOC, *S*, *n*, unit costs, unit revenues, and similar parameters. These parameters are often the result of design questions and their answers, as discussed in Chapter 15. Parameters that are interest rate-based are not treated in the same manner.

> Parameters such as MARR and other interest rates (loan rates, inflation rate) are more stable from project to project. If performed, sensitivity analysis on them is for specific values or over a **narrow range of values.** This point is important to remember if simulation is used for decision making under risk (Chapter 19).

Plotting the sensitivity of PW, AW, or ROR versus the parameter(s) studied is very helpful. Two alternatives can be compared with respect to a given parameter and the breakeven point. This is the value at which the two alternatives are economically equivalent. However, the breakeven chart commonly represents only one parameter per chart. Thus, several charts are constructed, and independence of each parameter is assumed. In previous uses of breakeven analysis, we often computed the measure of worth at only two values of a parameter and connected the points with a straight line. However, if the results are sensitive to the parameter value, several intermediate points should be used to better evaluate the sensitivity, especially if the relationships are not linear.

When several parameters are studied, sensitivity analysis can become quite complex. It may be performed one parameter at a time using a spreadsheet or computations by hand or calculator.

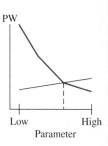

The computer facilitates comparison of multiple parameters and multiple measures of worth, and the software can rapidly plot the results.

Here is a general procedure to follow when conducting a thorough sensitivity analysis.

1. Determine which parameter(s) of interest might vary from the most likely estimated value.
2. Select the probable range and an increment of variation for each parameter.
3. Select the measure of worth.
4. Compute the results for each parameter, using the measure of worth as a basis.
5. To better interpret the sensitivity, graphically display the parameter versus the measure of worth.

This sensitivity analysis procedure should indicate the parameters that warrant closer study or require additional information. When there are two or more alternatives, it is better to use the PW or AW measure of worth in step 3. If ROR is used, it requires the extra efforts of incremental analysis between alternatives. Example 18.1 illustrates sensitivity analysis for one project.

EXAMPLE 18.1

Wild Rice, Inc. expects to purchase a new asset for automated rice handling. The most likely estimates are a first cost of $80,000, zero salvage value, and a cash flow before taxes (CFBT) per year t that follows the relation $27,000 - 2000t$. The MARR for the company varies over a wide range from 10% to 25% per year for different types of investments. The economic life of similar machinery varies from 8 to 12 years. Evaluate the sensitivity of PW by varying (a) MARR, while assuming a constant n value of 10 years, and (b) n, while MARR is constant at 15% per year. Perform the analysis by hand and by spreadsheet.

Solution by Hand

(a) Follow the procedure above to understand the sensitivity of PW to MARR variation.
1. MARR is the parameter of interest.
2. Select 5% increments to evaluate sensitivity to MARR; the range is 10% to 25%.
3. The measure of worth is PW.
4. Set up the PW relation for 10 years. When MARR = 10%,

$$PW = -80,000 + 25,000(P/A,10\%,10) - 2000(P/G,10\%,10)$$
$$= \$27,830$$

The PW for all MARR values at 5% intervals is as follows:

MARR, %	PW, $
10	27,830
15	11,512
20	−962
25	−10,711

5. A plot of MARR versus PW is shown in Figure 18–1. The steep negative slope indicates that the decision to accept the proposal based on PW is quite sensitive to variations in the MARR. If the MARR is established at the upper end of the range, the investment is not attractive.

(b) 1. Asset life n is the parameter.
2. Select 2-year increments to evaluate PW sensitivity over the range 8 to 12 years.
3. The measure of worth is PW.
4. Set up the same PW relation as in part (a) at $i = 15\%$. The PW results are

n	PW, $
8	7,221
10	11,511
12	13,145

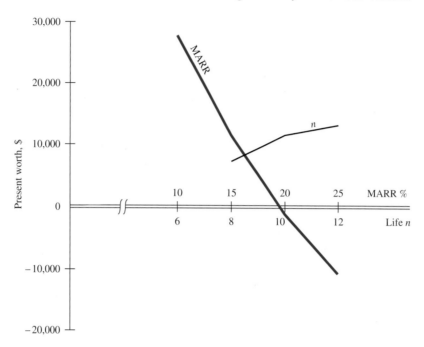

Figure 18–1
Plot of PW versus MARR
and *n* for sensitivity
analysis, Example 18.1.

5. Figure 18–1 presents the plot of PW versus *n*. Since the PW measure is positive for all values of *n*, the decision to invest is not materially affected by the estimated life. The PW curve levels out above *n* = 10. This insensitivity to changes in cash flow in the distant future is a predictable observation, because the P/F factor gets smaller as *n* increases.

Solution by Spreadsheet

Figure 18–2 presents two spreadsheets and accompanying plots of PW versus MARR (fixed *n*) and PW versus *n* (fixed MARR). The NPV function calculates PW for *i* values from 10% to 25% and *n* values from 8 to 12 years. As the solution by hand indicated, so do the charts; PW is sensitive to changes in MARR values, but not very sensitive to variations in *n*.

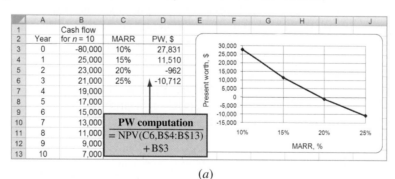

Figure 18–2
Sensitivity analysis of
PW to variation in
(*a*) MARR values and
(*b*) life estimates,
Example 18.1.

Figure 18–3
Sensitivity analysis graph
of percent variation from
the most likely estimate.

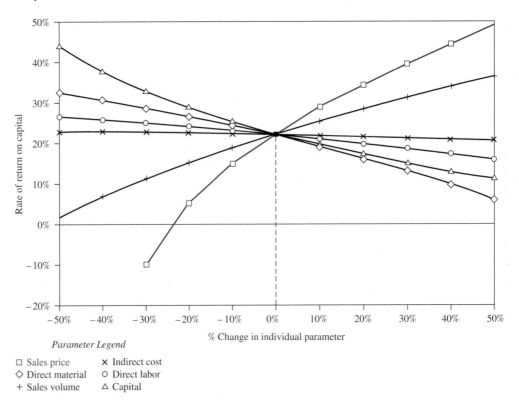

Parameter Legend

☐ Sales price   ✕ Indirect cost
◇ Direct material   ○ Direct labor
+ Sales volume   △ Capital

When the sensitivity of *several parameters* is considered for *one alternative* using a *single measure of worth,* it is helpful to **graph percentage change** for each parameter versus the measure of worth. This is sometimes called a **spider graph.** Figure 18–3 illustrates ROR versus six different parameters for one alternative. The variation in each parameter is indicated as a percentage deviation from the most likely estimate on the horizontal axis. If the ROR response curve is flat and approaches horizontal over the range of total variation graphed for a parameter, there is little sensitivity of ROR to changes in the parameter's value. This is the conclusion for indirect cost in Figure 18–3. On the other hand, ROR is very sensitive to sales price. A reduction of 30% from the expected sales price reduces the ROR from approximately 20% to −10%, whereas a 10% increase in price raises the ROR to about 30%.

If *two alternatives* are compared and the sensitivity to *one parameter* is sought, the graph may show quite nonlinear results. Observe the general shape of the sample sensitivity graphs in Figure 18–4. The plots are shown as linear segments between specific computation points. The graph indicates that the PW of each plan is a nonlinear function of hours of operation. Plan A is

Figure 18–4
Sample PW sensitivity to
hours of operation for two
alternatives.

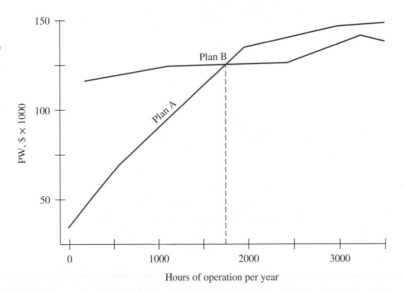

very sensitive in the range of 0 to 2000 hours, but it is comparatively insensitive above 2000 hours. Plan B is more attractive due to its relative insensitivity. The breakeven point is at about 1750 hours per year. It may be necessary to plot the measure of worth at intermediate points to better understand the nature of the sensitivity.

EXAMPLE 18.2

Columbus, Ohio, needs to resurface a 3-kilometer stretch of highway. Knobel Construction has proposed two methods of resurfacing. The first method is a concrete surface for a cost of $1.5 million and an annual maintenance cost of $10,000. The second method is an asphalt covering with a first cost of $1 million and a yearly maintenance of $50,000. However, Knobel requests that every third year the asphalt highway be touched up at a cost of $75,000.

The city uses the interest rate on bonds, 6% on its last bond issue, as the discount rate.

(a) Determine the breakeven number of years of the two methods. If the city expects an inter-state to replace this stretch of highway in 10 years, which method should be selected?
(b) If the touch-up cost increases by $15,000 per kilometer every 3 years, is the decision sensitive to this increase?

Solution

(a) Use PW analysis to determine the breakeven n value.

$$PW \text{ of concrete} = PW \text{ of asphalt}$$

$$-1,500,000 - 10,000(P/A,6\%,n) = -1,000,000 - 50,000(P/A,6\%,n)$$
$$-75,000\left[\sum_j (P/F,6\%,j)\right]$$

where $j = 3, 6, 9, \ldots, n$. The relation can be rewritten to reflect the incremental cash flows.

$$-500,000 + 40,000(P/A,6\%,n) + 75,000\left[\sum_j (P/F,6\%,j)\right] = 0 \qquad [18.1]$$

The breakeven n value can be determined by hand solution by increasing n until Equation [18.1] switches from negative to positive PW values. Alternatively, a spreadsheet solution using the NPV function can find the breakeven n value (Figure 18–5). The NPV functions in column C are the same each year, except that the cash flows are extended

	A	B	C	D	E
1		Part (a)		Part (b)	
2		Incremental	PW for	Incremental	PW for
3	Year, n	cash flow, $	n years, $	cash flow, $	n years, $
4	0	-500,000		-500,000	
5	1	40,000	-462,264	40,000	-462,264
6	2	40,000	-426,664	40,000	-426,664
7	3	115,000	-330,108	115,000	-330,108
8	4	40,000	-298,424	40,000	-298,424
9	5	40,000	-268,534	40,000	-268,534
10	6	115,000	-187,464	130,000	-176,889
11	7	40,000	-160,861	40,000	-150,287
12	8	40,000	-135,765	40,000	-125,190
13	9	115,000	-67,696	145,000	-39,365
14	10	40,000	-45,361	40,000	**-17,029**
15	11	40,000	**-24,289**	40,000	**4,042**
16	12	115,000	**32,862**	160,000	83,557
17	13	40,000	51,616	40,000	102,311
18	14	40,000	69,308	40,000	120,003
19	15	115,000	117,293	175,000	193,024
20	16	40,000	133,039	40,000	208,770
21					
22				PW for 12 years	
23		PW for 11 years		= NPV(6%,$B$5:$B16)+$B$4	
24		= NPV(6%,$B$5:$B15)+$B$4			
25					
26					
27			Year counter advances by 1		

Figure 18–5
Sensitivity of the break-even life between two alternatives, Example 18.2.

1 year for each present worth calculation. At approximately $n = 11.4$ years, concrete and asphalt resurfacing break even economically. Since the road is needed for 10 more years, the extra cost of concrete is not justified; select the asphalt alternative.

(b) The total touch-up cost will increase by \$15,000 every 3 years. Equation [18.1] is now

$$-500,000 + 40,000(P/A,6\%,n) + \left[75,000 + 15,000\left(\frac{j-3}{3}\right)\right]\left[\sum_j (P/F,6\%,j)\right] = 0$$

Now the breakeven n value is between 10 and 11 years—10.8 years using linear interpolation (Figure 18–5, column E). The decision has become marginal for asphalt since the interstate is planned for 10 years hence.

Noneconomic considerations may be used to determine if asphalt is still the better alternative. One conclusion is that the asphalt decision becomes more questionable as the asphalt maintenance costs increase; that is, the PW value is sensitive to increasing touch-up costs.

18.2 Sensitivity Analysis Using Three Estimates ●●●

We can thoroughly examine the economic advantages and disadvantages among two or more alternatives by borrowing from the field of project scheduling the concept of making three estimates for each parameter: **a pessimistic, a most likely, and an optimistic estimate.** Depending upon the nature of a parameter, the pessimistic estimate may be the lowest value (alternative life is an example) or the largest value (such as asset first cost).

This approach allows us to study measure of worth and alternative selection sensitivity within a predicted range of variation for each parameter. Usually the most likely estimate is used for all other parameters when the measure of worth is calculated for one particular parameter or one alternative.

EXAMPLE 18.3

An engineer is evaluating three alternatives for new equipment at Emerson Electronics. She has made three estimates for the salvage value, annual operating cost, and life. The estimates are presented on an alternative-by-alternative basis in Table 18–1. For example, alternative B has pessimistic estimates of $S = \$500$, AOC $= \$-4000$, and $n = 2$ years. The first costs are known, so they have the same value. Perform a sensitivity analysis and determine the most economical alternative, using AW analysis at a MARR of 12% per year.

TABLE 18–1	Competing Alternatives with Three Estimates Made for Salvage Value, AOC, and Life Parameters			
Strategy	**First Cost, $**	**Salvage Value S, $**	**AOC, $ per Year**	**Life n, Years**
Alternative A				
P	−20,000	0	−11,000	3
Estimates ML	−20,000	0	−9,000	5
O	−20,000	0	−5,000	8
Alternative B				
P	−15,000	500	−4,000	2
Estimates ML	−15,000	1,000	−3,500	4
O	−15,000	2,000	−2,000	7
Alternative C				
P	−30,000	3,000	−8,000	3
Estimates ML	−30,000	3,000	−7,000	7
O	−30,000	3,000	−3,500	9

P = pessimistic; ML = most likely; O = optimistic.

TABLE 18–2	Annual Worth Values, Example 18.3		
	Alternative AW Values, $		
Estimates	**A**	**B**	**C**
P	−19,327	−12,640	−19,601
ML	−14,548	−8,229	−13,276
O	−9,026	−5,089	−8,927

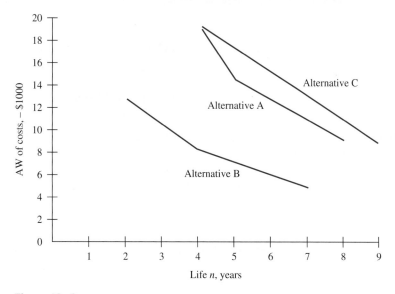

Figure 18–6
Plot of AW of costs for different-life estimates, Example 18.3.

Solution

For each alternative in Table 18–1, calculate the AW value of costs. For example, the AW relation for alternative A, pessimistic estimates, is

$$AW = -20,000(A/P,12\%,3) - 11,000 = \$-19,327$$

Table 18–2 presents all AW values. Figure 18–6 is a plot of AW versus the three estimates of *life* for each alternative. Since the AW calculated using the ML estimates for alternative B ($−8229) is economically better than even the optimistic AW value for alternatives A and C, alternative B is clearly favored.

Comment

While the alternative that should be selected here is quite obvious, this is not normally the case. For example, in Table 18–2, if the pessimistic alternative B equivalent AW value was much higher, say, $−21,000 per year (rather than $−12,640), and the optimistic AW values for alternatives A and C were less than that for B ($−5089), the choice of B would not be apparent or correct. In this case, it would be necessary to select one set of estimates (P, ML, or O) upon which to base the decision. Alternatively, the different estimates can be used in an expected value analysis, which is introduced next.

18.3 Estimate Variability and the Expected Value ● ● ●

Engineers and economic analysts usually deal with estimate variation and risk about an uncertain future by placing appropriate reliance on past data, if any exist. This means that *probability* and *samples* are used. Actually the use of probabilistic analysis is not as common as might be expected. The reason is not that the computations are difficult to perform or understand, but that realistic probabilities associated with cash flow estimates are difficult to assign. Experience and judgment can often be used in conjunction with probabilities and expected values to evaluate the desirability of an alternative.

> The **expected value** can be interpreted as a long-run average observable if the project is repeated many times. Since a particular alternative is evaluated or implemented only once, the expected value results in a **point estimate.** However, even for a single occurrence, the expected value is a meaningful number.

The expected value $E(X)$ is computed using the relation

$$E(X) = \sum_{i=1}^{i=m} X_i P(X_i) \qquad\qquad [18.2]$$

where
$$X_i = \text{value of the variable } X \text{ for } i \text{ from 1 to } m \text{ different values}$$
$$P(X_i) = \text{probability that a specific value of } X \text{ will occur}$$

Probabilities are always correctly stated in decimal form, but they are routinely spoken of in percentages and often referred to as **chance,** such as *the chances are about 10%.* When placing the probability value in Equation [18.2] or any other relation, use the decimal equivalent, for example, 0.1 for a 10% chance. In all probability statements, the $P(X_i)$ values for a variable X must total to 1.0.

$$\sum_{i=1}^{i=m} P(X_i) = 1.0$$

We may frequently omit the subscript i on X for simplicity.

If X represents the estimated cash flows, some will be positive and others will be negative. If a cash flow sequence includes revenues and costs, and the measure of worth is present worth calculated at the MARR, the result is the expected value of the discounted cash flows $E(PW)$. If the expected value is negative, the overall outcome is expected to be a cash outflow. For example, if $E(PW) = \$-1500$, this indicates that the proposal is not *expected* to return the MARR.

EXAMPLE 18.4

ANA airlines plans to offer several new electronic services on flights between Tokyo and selected European destinations. The marketing director estimates that for a typical 24-hour period there is a 50% chance of having a net cash flow of $5000 and a 35% chance of $10,000. He also estimates there is a small 5% chance of no cash flow and a 10% chance of a loss of $1000, which is the estimated extra personnel and utility costs to offer the services. Determine the expected net cash flow.

Solution

Let NCF be the net cash flow in dollars, and let $P(\text{NCF})$ represent the associated probabilities. Using Equation [18.2],

$$E(\text{NCF}) = 5000(0.5) + 10,000(0.35) + 0(0.05) - 1000(0.1) = \$5900$$

Although the "no cash flow" possibility does not increase or decrease $E(\text{NCF})$, it is included because it makes the probability values sum to 1.0 and it makes the computation complete.

18.4 Expected Value Computations for Alternatives ● ● ●

The expected value computation $E(X)$ is utilized in a variety of ways. Two prime ways are to:

- Prepare information for use in an economic analysis.
- Evaluate the expected value of the measure of worth of an alternative.

Example 18.5 illustrates the first situation, and Example 18.6 determines the expected PW when the entire cash flow series and its probabilities are estimated.

EXAMPLE 18.5

There are many government incentives to become more energy efficient. Installing solar panels on homes, business buildings, and multiple-family dwellings is one of them. The owner pays a portion of the total installation costs, and the government agency pays the rest. Nichole works for the Department of Energy and is responsible for approving solar panel incentive payouts. She has exceeded the annual budgeted amount of $50 million per year in each of the previous 2 years. Disappointed with this situation, Nichole and her boss decided to collect data to determine what size increase in annual budget the incentive program needs in the future. Over the last 36 months, the amount of average monthly payout and number of months are shown in Table 18–3. She categorized by level the monthly averages according to her experience with the program. Provided the same pattern continues, what is the expected value of the dollar increase in annual budget that is needed to meet the requests?

TABLE 18–3	Solar Panel Incentive Payouts, Example 18.5	
Level	Average Payout, $ Million per Month	Months over Past 3 Years
Very high	6.5	15
High	4.7	10
Moderate	3.2	7
Low	2.9	4

Solution

Use the 36 months of payouts PO_j (j = low, . . . ,very high) to estimate the probability $P(PO_j)$ for each level, and make sure the total is 1.0.

Level, j	Probability of Payout Level, $P(PO_j)$
Very high	$P(PO_1) = 15/36 = 0.417$
High	$P(PO_2) = 10/36 = 0.278$
Moderate	$P(PO_3) = 7/36 = 0.194$
Low	$P(PO_4) = 4/36 = \underline{0.111}$
	1.000

The expected monthly payout is calculated using Equation [18.2]. In $ million units,

$$E[PO] = 6.5(0.417) + 4.7(0.278) + 3.2(0.194) + 2.9(0.111)$$
$$= 2.711 + 1.307 + 0.621 + 0.322$$
$$= \$4.961 \quad (\$4,961,000)$$

The annual expected budget need is 12×4.961 million = $59.532 million. The current budget of $50 million should be increased by an average of $9.532 million per year.

EXAMPLE 18.6

Lite-Weight Wheelchair Company has a substantial investment in tubular steel bending equipment. A new piece of equipment costs $5000 and has a life of 3 years. Estimated cash flows (Table 18–4) depend on economic conditions classified as receding, stable, or expanding. A probability is estimated that each of the economic conditions will prevail during the 3-year period. Apply expected value and PW analysis to determine if the equipment should be purchased. Use a MARR of 15% per year.

TABLE 18–4	Equipment Cash Flow and Probabilities, Example 18.6		
	Economic Condition		
	Receding (Prob. = 0.4)	**Stable** (Prob. = 0.4)	**Expanding** (Prob. = 0.2)
Year	**Annual Cash Flow Estimates, $**		
0	−5000	−5000	−5000
1	+2500	+2500	+2000
2	+2000	+2500	+3000
3	+1000	+2500	+3500

Solution

First determine the PW of the cash flows in Table 18–4 for each economic condition, and then calculate $E(\text{PW})$ using Equation [18.2]. Define subscripts R for receding economy, S for stable, and E for expanding. The PW values for the three scenarios are

$$\text{PW}_R = -5000 + 2500(P/F,15\%,1) + 2000(P/F,15\%,2) + 1000(P/F,15\%,3)$$
$$= -5000 + 4344 = \${-656}$$
$$\text{PW}_S = -5000 + 5708 = \${+708}$$
$$\text{PW}_E = -5000 + 6309 = \${+1309}$$

Only in a receding economy will the cash flows not return the 15% to justify the investment. The expected present worth is

$$E(\text{PW}) = \sum_{j=R,S,E} \text{PW}_j[P(j)]$$
$$= -656(0.4) + 708(0.4) + 1309(0.2)$$
$$= \$283$$

At 15%, $E(\text{PW}) > 0$; the equipment is justified, using an expected value analysis.

Comment

It is also correct to calculate the $E(\text{cash flow})$ for each year and then determine PW of the $E(\text{cash flow})$ series because the PW computation is a linear function of cash flows. Computing $E(\text{cash flow})$ first may be easier in that it reduces the number of PW computations. In this example, calculate $E(\text{CF}_t)$ for each year, then determine $E(\text{PW})$.

$$E(\text{CF}_0) = \${-5000}$$
$$E(\text{CF}_1) = 2500(0.4) + 2500(0.4) + 2000(0.2) = \$2400$$
$$E(\text{CF}_2) = \$2400$$
$$E(\text{CF}_3) = \$2100$$
$$E(\text{PW}) = -5000 + 2400(P/F,15\%,1) + 2400(P/F,15\%,2) + 2100(P/F,15\%,3)$$
$$= \$283$$

18.5 Staged Evaluation of Alternatives Using a Decision Tree ●●●

Alternative evaluation may require a series of decisions in which the outcome from one stage is important to the next stage of decision making. When each alternative is clearly defined and probability estimates can be made to **account for risk,** it is helpful to perform the evaluation using a **decision tree.**

A decision tree includes:

- More than one stage of alternative selection.
- Selection of an alternative at one stage that leads to another stage.
- Expected results from a decision at each stage.
- Probability estimates for each outcome.
- Estimates of economic value (cost or revenue) for each outcome.
- Measure of worth as the selection criterion, such as $E(\text{PW})$.

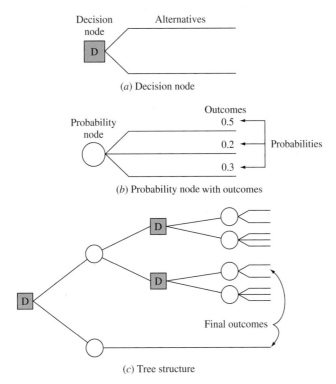

Figure 18–7
Decision and probability nodes used to construct a decision tree.

The decision tree is constructed left to right and includes each possible decision and outcome.

- A square represents a **decision node** with the possible alternatives indicated on the **branches** from the decision node (Figure 18–7a).
- A circle represents a **probability node** with the possible **outcomes** and estimated **probabilities** on the branches (Figure 18–7b).
- The treelike structure in Figure 18–7c results, with outcomes following a decision.

Usually each branch of a decision tree has some estimated economic value (often referred to as *payoff*) in cost, revenue, saving, or benefit. These cash flows are expressed in terms of PW, AW, or FW values and are shown to the right of each final outcome branch. The cash flow and probability estimates on each outcome branch are used in calculating the expected economic value of each decision branch. This process, called **solving the tree** or **rollback,** is explained after Example 18.7, which illustrates the construction of a decision tree.

EXAMPLE 18.7

Jerry Hill is president and CEO of a U.S.-based food processing company, Hill Products and Services. He was recently approached by an international supermarket chain that wants to market in-country its own brand of frozen microwaveable dinners. The offer made to Jerry by the supermarket corporation requires that a series of two decisions be made, now and 2 years hence. The current decision involves two alternatives: (1) *Lease* a facility in Germany from the supermarket chain, which has agreed to convert a current processing facility for immediate use by Jerry's company; or (2) *build and own* a processing and packaging facility in Germany. Possible outcomes of this first decision stage are good market or poor market depending upon the public's response.

The decision choices 2 years hence are dependent upon the lease-or-own decision made now. If Hill *decides to lease,* good market response means that the future decision alternatives are to produce at twice, equal to, or one-half of the original volume. This will be a mutual decision between the supermarket chain and Jerry's company. A poor market response will indicate a one-half level of production, or complete removal from the German market. Outcomes for the future decisions are, again, good and poor market responses.

As agreed by the supermarket company, the current decision for Jerry *to own* the facility will allow him to set the production level 2 years hence. If market response is good, the decision alternatives are four or two times original levels. The reaction to poor market response will be production at the same level or no production at all.

Construct the tree of decisions and outcomes for Hill Products and Services.

Solution

This is a two-stage decision tree that has alternatives now and 2 years hence. Identify the decision nodes and branches, and then develop the tree using the branches and the outcomes of good and poor market for each decision. Figure 18–8 details the decision stages and outcome branches.

Stage 1 (decision now):
  Label it D1.
  Alternatives: lease (L) and own (O).
  Outcomes: good and poor markets.

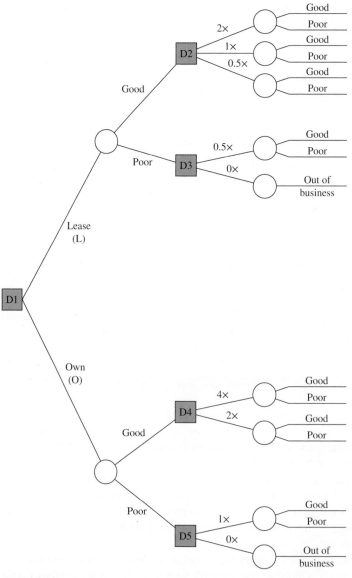

Figure 18–8
A two-stage decision tree identifying alternatives and possible outcomes.

Stage 2 (decisions 2 years hence):
Label them D2 through D5.
Outcomes: good market, poor market, and out of business.

Choice of production levels for D2 through D5:
Quadruple production (4×); double production (2×); level production (1×); one-half production (0.5×); stop production (0×)

The alternatives for future production levels (D2 through D5) are added to the tree and followed by the market responses of good and poor. If the stop-production (0×) decision is made at D3 or D5, the only outcome is 'out of business'.

To utilize the decision tree for alternative evaluation and selection, the following additional information is necessary for each branch:

- The **estimated probability** that each outcome may occur. These probabilities must sum to 1.0 for each set of outcomes (branches) that result from a decision.
- **Economic information** for each decision alternative and possible outcome, such as initial investment and estimated cash flows.

Decisions are made using the probability estimate and economic value estimate for each outcome branch. Commonly the present worth at the MARR is used in an expected value computation of the type in Equation [18.2]. This is the general procedure to solve the tree using PW analysis:

1. Start at the top right of the tree. Determine the PW value for each outcome branch considering the time value of money.
2. Calculate the expected value for each decision alternative.

$$E(\text{decision}) = \Sigma(\text{outcome estimate})P(\text{outcome}) \qquad [18.3]$$

where the summation is taken over all possible outcomes for each decision alternative.

3. At each decision node, select the best E(decision) value—minimum cost or maximum value (if both costs and revenues are estimated).

4. Continue moving to the left of the tree to the root decision in order to select the best alternative.

5. Trace the best decision path through the tree.

EXAMPLE 18.8

A decision is needed to either market or sell a new invention. If the product is marketed, the next decision is to take it international or national. Assume the details of the outcome branches result in the decision tree of Figure 18–9. The probabilities for each outcome and PW of CFBT (cash flow before taxes) are indicated. These payoffs are in millions of dollars. Determine the best decision at the decision node D1.

Solution

Use the procedure above to determine that the D1 decision alternative to sell the invention should maximize E(PW of CFBT).

1. Present worth of CFBT is supplied.
2. Calculate the expected PW for alternatives from nodes D2 and D3, using Equation [18.3]. In Figure 18–9, to the right of decision node D2, the expected values of 14 and 0.2 in ovals are determined as

$$E(\text{international decision}) = 12(0.5) + 16(0.5) = 14$$
$$E(\text{national decision}) = 4(0.4) - 3(0.4) - 1(0.2) = 0.2$$

The expected PW values of 4.2 and 2 for D3 are calculated in a similar fashion.

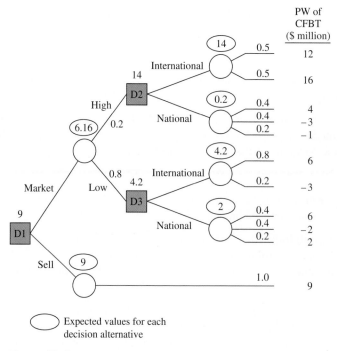

Figure 18–9
Solution of a decision tree with present worth of estimated CFBT values, Example 18.8.

3. Select the larger expected value at each decision node. These are 14 (international) at D2 and 4.2 (international) at D3.
4. Calculate the expected PW for the two D1 branches.

$$E(\text{market decision}) = 14(0.2) + 4.2(0.8) = 6.16$$
$$E(\text{sell decision}) = 9(1.0) = 9$$

The expected value for the sell decision is simple since the one outcome has a payoff of 9. The sell decision yields the larger expected PW of 9.
5. The largest expected PW of CFBT path is to select the sell branch at D1 for a guaranteed $9,000,000.

18.6 Real Options in Engineering Economics ● ● ●

As we learned with decision trees, many of the problems in engineering economy can be viewed as staged decisions. When the decision to invest more or less can be delayed into the future, the problem is called **staged funding.** As an illustration, assume a large company expects to sell an energy-saving, window-mounted residential air conditioning unit at the rate of 100,000 per month by the end of 2 years on the market. Decision makers may opt to (1) build the capacity to supply 100,000 per month to market immediately, or (2) build capacity to supply 25,000 per month now and test the market's receptivity. If positive, they can stage the increase by 25,000 additional units each 6 months to meet current demand. Of course, if an aggressive competitor enters the scene, or the economy falters, the staged funding decision will change as warranted. These alternatives provide time-based options to the company. Before we go further, some definitions are needed.

An **option** is a purchase or investment that contractually provides the privilege to take a stated action by some stated time in the future, or the right to not accept the offer and forfeit the option.

A **real option,** in engineering economy terms, is the investment (cost) in a project, process, or system. The options usually involve **physical (real)** assets, buildings, equipment, materials, and the like, thus, the word *real*. Options may also be leases, subcontracts, or franchises. The investment alternatives present varying amounts of **risk,** which is estimated by probabilities of occurrence for predictable future events.

Real options analysis is the application of techniques to determine the economic consequences of **delaying** the funding decisions as allowed by the option. The estimated cash flows and other consequences of these delays are analyzed with risk taken into account to the degree possible. A measure of worth, for example, PW or AW, is the criterion used to make the staged funding decisions. A decision may be to expand, continue as is, contract, abandon, or replicate the alternative at the time the option must be exercised.

An inherent part of real options analysis is the uncertainty of future estimates, as it is for most economic analyses. After some illustrations of real options, we will discuss the probabilistic dimensions. Samples from industry and everyday personal life that can be formulated as real options follow:

Industrial or Business Setting

New markets—Purchase equipment and staff to enter an expanding international market over the next 5 years.

New planes—Purchase commercial airplanes now with an option to buy an additional five planes over the next 3 years at the same price as that paid for the current order.

Removing car models—Ford Motor Company can decide to maintain production on an established car model with dwindling sales for the next 3 years or can opt to discontinue the model in stages over a 1- or 2-year period.

Drilling lease—Buy a drilling option contract from landowners to drill for oil and gas at some time in the next 10 years. The drilling may not be justified at this time, but the contract offers the option to drill were it to become economically advantageous based on events such as increased oil prices or improved recovery technology.

Personal Decision Making

Extended car warranty—When purchasing a new car, the option to buy an extended-coverage warranty beyond the manufacturer's warranty is always an option. The price of the option is the cost of the extended warranty. The uncertainties and risks are the future unknown costs for repairs and failed components.

House insurance—When a homeowner has no mortgage to pay, maintaining house insurance is an option. Deductibles are high enough, for example, 1% to 5% of the fully appraised value, that insurance primarily covers only catastrophic damage to the structure. Self-insurance, where money is set aside for potential damages while accepting the risk that a major event will take place, is an option for the homeowner.

Some of the primary characteristics (with an example) of a real options analysis performed within the context of engineering economics are as follows:

- **Cost** to obtain the option to delay a decision (PW of initial investment, lease cost, or future investment amount).
- Anticipated future **options** and **cash flow** estimates (double production with annual net cash flows estimated).
- **Time period** for follow-on decisions (staged decision time, such as 1 year or a 3-year test period).
- Market and risk-free **interest rates** (expected market MARR of 12% per year and inflation rate estimate of 4% per year).
- Estimates of risk and future **uncertainty** for each option (probability that an estimated cash flow series will actually occur, if a specific option is selected).
- **Economic criterion** used to make a decision (PW, ROR, or other measure of worth).

It is common to use a decision tree to record and understand the options prior to performing a real options analysis with risk included. Example 18.9 demonstrates the use of a decision tree and PW analysis.

EXAMPLE 18.9

A start-up company in the solar energy production business, SolarScale Energy, Inc. has developed and field-tested a modularized, scalar solar thermal electric (STE) generation system that is relatively inexpensive to purchase and has an efficiency considerably better than traditional photovoltaic (PV) panels. The technology is promising enough that Capital Investor Funds (CIF) has provided $10 million for manufacturing. Additionally, a contract with a consortium of sunbelt states has been offered, but not accepted thus far, for a total of $1.5 million per year for a 2-year test period. By contract, the units will be marketed through the state energy departments with all revenue going to the state treasuries. The lead engineer at SolarScale, the manager of CIF, and a conservation representative for the state consortium have developed the following staged-funding options, based on the delayed decision to increase manufacturing production level until preliminary results of the 2-year contract are in hand.

Condition	Option	CIF Funding	Consortium Contract
Sales are excellent (> 5000 units/year)	2 × production level	Additional $10 million in year 2	Additional 8 years; $4 million in years 3–10
Sales are excellent (3000–5000 units/year)	1 × production level	Nothing; no salvage after 10 years	Additional 8 years; $1.5 million in years 3–10
Sales are poor (2000–3000 units/year)	Reduce to ½ × production	Nothing; sell for $2.5 million after 5 years	Additional 3 years; $1.5 million in years 3–5
Sales are poor (< 2000 units/year)	Stop after 2 years	Nothing; sell for $5 million after 2 years	Nothing

(a) Develop the two-stage decision tree for the options described.
(b) The base case is the 1 × production level with the 8-year follow-on contract from the consortium. If the estimates for this option are considered the most likely (expected value) estimates, determine the present worth at a MARR of 10% per year.
(c) Determine the PW values for each possible final outcome at 10% per year, and identify the best economic option when the stage 2 funding decision must be made.

Solution

(a) Figure 18–10 details the options with the year shown at the bottom. There are two outcome branches initially (accept option; decline option) and four final branches for the accept decision at D1, based upon sales level. The decline option has a $0 outcome. (SolarScale has other ways to pursue revenue that are not represented in this abbreviated example.)

(b) Perform a PW evaluation as we have in all previous chapters, assuming the estimates are point estimates over the 10-year life of the project. The resulting $PW_{1x} < 0$ as shown below indicates the contract is not justified economically. In $ millions,

$$PW_{1x} = -10 + 1.5(P/A, 10\%, 10)$$
$$= \$-0.78 \quad (\$-780,000)$$

(c) Figure 18–11 is a spreadsheet image that calculates the PW for each option using the NPV function. The i^* values are also shown using the IRR function. Note that the sale of production assets after 5 years for $2.5 million (½ × level) or after 2 years for $5 million (stop) is included. Also, the extra $10 million investment in year 2 for the 2 × level option is included.

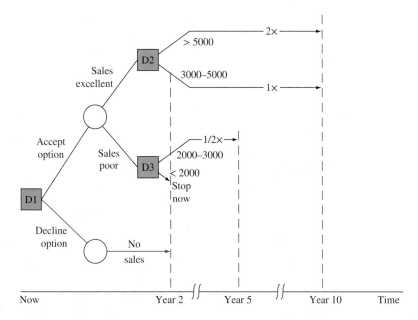

Figure 18–10
Decision tree showing real options over 10-year period, Example 18.9.

	A	B	C	D	E	F
1		Net cash flow estimates, $ million per year				
2	Year	2X level	1X level	1/2X level	Stop	
3	0	-10	-10	-10	-10	
4	1	1.5	1.5	1.5	1.5	
5	2	-8.5	1.5	1.5	6.5	
6	3	4.0	1.5	1.5		
7	4	4.0	1.5	1.5		
8	5	4.0	1.5	4.0		
9	6	4.0	1.5			
10	7	4.0	1.5	Sale of assets in this year		
11	8	4.0	1.5			
12	9	4.0	1.5			
13	10	4.0	1.5			
14	i^*	12.6%	8.1%	0.0%	-11.5%	
15	PW @ 10%	1.97	-0.78	-2.76	-3.26	

Figure 18–11
PW analysis of real options without risk considered, Example 18.9.

Only the 2 × production-level option is justified at MARR = 10% per year. If SolarScale and CIF, the financial backers, are not convinced that the sales level will exceed 5000 units per year, the contract option should be declined. Some marketing survey information and risk analysis may be very helpful before this important decision is made.

Of the characteristics listed above for real options situations, the primary one absent in Example 18.9 is that of estimate variation and some measure of **risk.** Decision making under risk is covered more extensively in the next chapter; however, we can use the following definition for this discussion of real options analysis.

When a parameter can take on more than one value and there is any estimate of **chance** or **probability** about the opportunity that each value may be observed, risk is present.

Risk

A coin has two sides. If it is a perfectly balanced coin, a flip of the coin should result in heads 50% of the time and tails 50% of the time. If the coin is intentionally biased in weight, such that 58% of the time it lands heads up, then the long-run probability of heads is $P(\text{heads}) = 0.58$. Since the sum of probabilities across all possible values must add to 1, the biased coin has $P(\text{tails}) = 0.42$.

There are a couple of points worth mentioning about risk and the calculated PW values for real options analysis.

- When the risk is higher and the stakes are larger, the real options analysis is often more valuable. (We shall see this in Example 18.10.)
- Since one of the objectives of real options analysis is to evaluate the economic consequences of delaying a decision, a PW value of the base case that is *moderately positive* means that the project is justified and should be accepted immediately, without the decision delay. On the other hand, if the PW is *largely negative,* the delay is likely not worthwhile, as it would take a very large positive PW to result in $E(\text{PW}) > 0$. Thus, the project should be rejected now, not delayed for a future decision.

We return to the previous example with some risk assessment added to determine if the contract option should be declined, as indicated by the base case.

EXAMPLE 18.10

Before the stage 1 decision is made in Example 18.9 about SolarScale's state-consortium contract offer, some expected sales information was collected. Once the results were reviewed by the three individuals in attendance at the final decision-making meeting—one representative each for SolarScale, CIF, and the consortium—each person recorded her or his estimated probability as a measure of risk that the sales would be excellent or poor, which represents the two outcome possibilities were the option accepted. The results are as follows:

	Probability of Outcome	
	Excellent	Poor
SolarScale	0.5	0.5
CIF	0.8	0.2
Consortium	0.6	0.4

Use these probability estimates to determine the expected PW value, provided equal weighting is given to each representative's input.

Solution

For each outcome (excellent and poor), select the best PW value from Figure 18–11, and then find $E(\text{PW})$ for each representative.

Excellent: From $2 \times$ level and $1 \times$ level, select $2 \times$ with PW = $1.97 million.

Poor: From $\frac{1}{2} \times$ level and stop now, select $\frac{1}{2} \times$ with PW = $-2.76 million.

In $ million, $E(\text{PW})$ for each organization is

$$E(\text{PW for SolarScale}) = 1.97(0.5) - 2.76(0.5) = \$-0.40$$

$$E(\text{PW for CIF}) = 1.97(0.8) - 2.76(0.2) = \$1.02$$

$$E(\text{PW for consortium}) = 1.97(0.6) - 2.76(0.4) = \$0.08$$

With a 1/3 chance assigned to each representative, the overall $E(\text{PW of stage 2 decision})$ is

$$E(\text{PW of stage 2 decision}) = 0.33(-0.40 + 1.02 + 0.08)$$
$$= \$0.23 \quad (\$230,000)$$

The base case of $1 \times$ level production in part (*b*) of Example 18.9 resulted in $E(\text{PW}) = \$-780,000$. When compared with the positive $E(\text{PW})$ result here, we see that with consideration of the different options of production level and probabilities for sales level, the expected PW has increased to a positive value. All other things being equal, the state consortium offer should be accepted; that is, accept the real option of the contract.

There are many other examples and dimensions of real options analysis in engineering economics and in the area of financial analysis, where options analysis got its start some years ago. If you are interested in this area of analysis, consult more advanced texts and journal articles on the topic of real options.

CHAPTER SUMMARY

In this chapter, the emphasis is on sensitivity to variation in one or more parameters using a specific measure of worth. When two alternatives are compared, compute and graph the measure of worth for different values of the parameter to determine when each alternative is better.

When several parameters are expected to vary over a predictable range, the measure of worth is plotted and calculated using three estimates for a parameter—most likely, pessimistic, and optimistic. This approach can help determine which alternative is best among several. Independence between parameters is assumed in all these analyses.

The combination of parameter and probability estimates results in the expected value relation

$$E(X) = \Sigma X P(X)$$

This expression is also used to calculate E(revenue), E(cost), E(cash flow), and E(PW) for the entire cash flow sequence of an alternative.

Decision trees are used to make a series of alternative selections. This is a way to explicitly take risk into account. It is necessary to make several types of estimates for a decision tree: outcomes for each possible decision, cash flows, and probabilities. Expected value computations are coupled with those for the measure of worth to solve the tree and find the best alternatives stage by stage.

Staged funding over time can be approached using the evolving area of real options. Delaying an investment decision and considering the risks of the future can improve the overall E(PW) of a project, process, or system.

PROBLEMS

Sensitivity to Parameter Variation

18.1 Decker Scientific is considering an investment of $850,000 in a new product line. The company will make the investment only if it will result in a rate of return of 20% per year or higher. If the revenue is expected to be between $290,000 and $325,000 per year for each of 5 years, determine if the decision to invest is sensitive to the projected range of income using an annual worth analysis.

18.2 A company that manufactures high-speed submersible rotary-indexing spindles is considering an upgrade of production equipment to reduce costs over *the next 5 years*. The company can invest $80,000 now, 1 year from now, or 2 years from now. Depending on when the investment is made, the savings will vary. The saving estimates are $26,000, $31,000, or $37,000 per year if the investment is made now, 1 year from now, or 2 years from now, respectively. The company will only invest if the ROR is at least 20% per year.

Using a future worth analysis, determine if the timing of the investment will affect the return requirement and, if so, when the investment should be made. Solve by (*a*) *hand*, and (*b*) *spreadsheet*.

18.3 Home Automation is considering an investment of $500,000 in a new product line. The company will make the investment only if it will result in a rate of return of 15% per year or higher. The revenue is expected to be between $138,000 and $165,000 per year for 5 years. (*a*) Determine if the decision to invest is sensitive to the projected range of income using a PW analysis. (*b*) Use a *spreadsheet to* determine the annual revenue required to realize 15% per year.

18.4 The AW of the current process to make motion controllers is $–62,000 per year. A replacement process under consideration has estimates of $P =$ $64,000, AOC = $38,000, and $n =$ 3 years. Three engineers have given their opinions about what the salvage value of the new process may be in 3 years as $10,000, $14,500, and $18,000. (*a*) Is the decision to replace the process sensitive to the salvage value estimates at a MARR of 15% per year? (*b*) If sensitive, what is the lowest salvage value that justifies the replacement process?

18.5 Hemisphere Electric may purchase equipment to manufacture a new line of wireless devices for home appliance control. The first cost will be $80,000, and the life is estimated at 6 years with a salvage value of $10,000. Different people in marketing have provided revenue estimates that the devices will generate. The estimates range from a low of $10,000 to a high of $20,000, with an average of $16,000 per year. If the MARR is 8% per year, use PW to determine if these different estimates will change the decision to purchase the equipment.

18.6 MAG Industrial needs 1000 square meters of storage space. Purchasing land for $80,000 and erecting a temporary metal building at $70 per square meter is one option. The president expects to sell the land for $100,000 and the building for $20,000 after 3 years. Another option is to lease space for $30 per square meter per year payable at the beginning of each year. The MARR is 20% per year. (*a*) *By hand*, perform a present worth analysis of the building and leasing alternatives to determine the sensitivity of the decision if the construction cost decreases by 10% to $63 per square meter and the lease cost remains at $30 per square meter per year. (*b*) Write the three *spreadsheet* functions that display the correct PW values.

18.7 A major equipment purchase is being economically evaluated by a team of three design engineers and one production engineer at Raytheon Aerospace. The agreed-upon estimates are $P = \$560,000$, $n = 6$ years, and $S = \$100,000$. The group disagrees, however, on the estimated annual net savings the equipment will generate. Estimates are summarized below. (*a*) Use PW analysis and a *spreadsheet* to identify which estimates, if any, indicate that the equipment is economically justified at a MARR of 8% per year. (*b*) For any estimate(s) indicating that the purchase is *not* justified, determine the breakeven number of years for the purchase to be justified.

Team Member	Joseph	Janice	Carlos	Mehmet
Savings, $ per year	110,000	140,000	180,000	100,000

18.8 Consider two air-conditioning systems with the estimates below. (*a*) Use AW analysis to determine the sensitivity of the economic decision to MARR values of 4%, 6%, and 8% per year. (*b*) Develop the spreadsheet functions that will display the six AW values.

System	1	2
First cost, $	−10,000	−17,000
AOC, $ per year	−600	−150
Salvage value, $	−100	−300
New compressor and motor cost at midlife, $	−1,750	−3,000
Life, years	8	12

18.9 Determine if the selection of system GH or B3 is sensitive to variation in the return required by management. Depending on the type of project, the corporate MARR ranges from 8% to 16% per year on different projects. (*a*) Write the AW equations to find the values *by hand*. (*b*) Solve using a *spreadsheet* at 2% increments, develop the AW versus MARR graph, and find the breakeven MARR between GH and B3.

System	GH	B3
First cost, $	−50,000	−100,000
AOC, $ per year	−6,000	−1,500
Salvage value, $	30,000	0
Overhaul year 2, $	−17,000	−
Overhaul year 6, $	−	−30,000
Life, years	4	12

18.10 The production manager on the Ofon Phase 2 offshore platform operator by Total S.A. must purchase specialized environmental equipment or an equivalent service. The first cost is $250,000 with an AOC of $75,000. The manager has let it be known that he does not care about the salvage value because he thinks it will make no difference in the decision-making process. His supervisor estimates the salvage might be as high as $100,000 or as low as $10,000 in 3 years at which time the equipment will be unnecessary. Alternatively, a subcontractor can provide the service for $165,000 per year. Total's offshore project MARR is 15% per year. Determine if the decision to buy the equipment is sensitive to the salvage value. If the decision is sensitive, what should Total S.A. do before making the decision?

18.11 A biofuel subsidiary of Petrofac, Inc. is planning to borrow $12 million to acquire a small technology-based company. The rate on a 5-year loan is highly variable; it could be as low as 7%, as high as 15%, but is expected to be 10% per year. The company will only move forward with the acquisition offer if the AW is below $6.1 million. The M&O

cost is fixed at $3.1 million per year. The anticipated sales price of the company could be $2 million if the interest rate is 7% or as much as $2.5 million if the rate is 15%, but will most likely be about $2.3 million at the 10% per year rate. Is the decision to move forward with the acquisition sensitive to the loan interest rate and salvage value estimates?

18.12 A young couple planning ahead for their retirement has decided that $3 million is the amount they will need in order to retire comfortably 20 years from now. For the past 5 years, they have been able to invest one of their salaries ($50,000 per year, which includes employer contributions) while living off the other salary. They plan to start a family sometime in the next 10 years and when they will have their first child, one of the parents will quit working, causing the savings to decrease to $10,000 per year thereafter. If they have realized an average ROR of 10% per year on their investments, and expect to continue at this ROR, is reaching their goal of $3 million in 20 years sensitive to when they have their first child (i.e., between now and 10 years from now)? If so, how many years from now will they have to wait before they have their first child? Use a future worth analysis.

18.13 A company that manufactures clear PVC pipe is investigating the production options of batch and continuous processing. Estimated cash flows are:

Process	Batch	Continuous
First cost, $	−80,000	−140,000
Annual cost, $ per year	−52,000	−31,000
Salvage value, any year, $	10,000	25,000
Life, years	3–10	5

The chief operating officer (COO) has asked you to determine if the batch option would ever have a lower annual worth than the continuous flow system using an interest rate of 15% per year. The continuous flow process was previously determined to have its lowest cost over a 5-year life cycle, but the batch process can be used from 3 to 10 years. If selecting the batch process is sensitive to its useful life, what is the minimum life that makes it more attractive? Solve (a) by *hand*, and (b) by *spreadsheet*.

The following information is used in Problems 18.14 through 18.17.

An online patient diagnostics system for surgeons has the following point estimates: A first cost of $200,000

to install and $5000 annually to maintain over its expected life of 5 years. Added revenue is estimated to average $60,000 per year. Examine the sensitivity of present worth at a MARR of 10% per year to variation in selected parameter estimates, while others remain constant.

18.14 Sensitivity to first cost variation: $150,000 to $250,000 (−25% to + 25%).

18.15 Sensitivity to revenue variation: $45,000 to $75,000 (−25% to + 25%).

18.16 Sensitivity to life variation: 4 years to 7 years (−20% to + 40%).

18.17 Plot the results on a graph with % variation on the x-axis and PW of first cost, revenue, and life on the y-axis. Comment on the relative sensitivity of each parameter.

18.18 (a) George has been offered the opportunity to purchase a 9%, $10,000 bond due in 10 years. The bond dividend is paid semiannually. George expects an 8% per year, compounded semiannually return on his investments. Use a *spreadsheet* to graph the sensitivity in PW value if there is a ±30% change in (1) face value, (2) dividend rate, or (3) required nominal rate of return.

(b) If George purchases the $10,000 face-value bond at a premium of 5%, and all other estimates are correct, that is, 0% change, did he pay too much or too little? By how much?

Three-Estimate Sensitivity Analysis

18.19 An engineer must decide between two ways to pump concrete to the top of a seven-story building. Plan 1 requires the leasing of equipment for $60,000 initially and will cost between $0.40 and $0.95 per metric ton to operate, with a most likely cost of $0.50 per metric ton. The pumper is able to deliver 100 metric tons per 8-hour day. If leased, the asset will have a contract period of 5 years. Plan 2 is rental option that will cost $15,000 per year. In addition, an extra $15 per hour labor cost will be incurred for operating the rented equipment per 8-hour day. Which plan should the engineer recommend if the equipment will be needed for 50 days per year? The MARR is 12% per year.

18.20 A new anti-theft system incorporating MEMS technology is being economically evaluated separately by three engineers at Dragon Technologies. The first cost of the equipment will be $75,000, and the life is estimated at 6 years with a salvage value of $9000. The engineers made different estimates of the net savings that the equipment might generate. Jacob made an estimate of $10,000 per year. Susan states that this is too low and estimates $14,000, while Tyler estimates $18,000 per year. If the before-tax MARR is 8% per year, use PW to determine if these different estimates will change the decision to purchase the equipment. Perform the analysis (a) by *hand*, and (b) using a *spreadsheet*.

18.21 Jensen Systems purchases several parts for the instruments it makes via a fixed-price contract of $155,000 per year from a local supplier. The new president of Jensen wants to make the parts in-house through the purchase of equipment that will have a first cost of $240,000 with an estimated salvage value of $30,000 after 5 years. The AOC is difficult to estimate, but company engineers have made optimistic, most likely, and pessimistic estimates of $70,000, $85,000, and $120,000 per year, respectively. Determine if the company should purchase the equipment under any of the operating cost scenarios. The MARR is 20% per year.

18.22 An engineer has been offered an investment opportunity that will require a cash outlay of $40,000 now for a cash inflow of $3500 for each year of investment. However, she must state now the number of years she plans to retain the investment. Additionally, if the investment is retained for 6 years, a lump-sum amount of $36,000 will be returned to her; after 10 years, the lump-sum return is anticipated to be $49,000, and after 15 years, it is estimated to be $55,000. Money is currently worth 10% per year. (a) Is the decision sensitive to the retention period? If so, what investment period is best? (b) Write the format of the spreadsheet function that will display the correct PW values.

18.23 An expansion of the current BIM (Building Information Model) software has been proposed to First Financial, the building's owner. A total installed cost of $120,000 is expected to generate additional savings of $40,000 per year for 10 years, after which time the software will be replaced with no salvage value. The annual M&O cost is expected to be $10,000 the first year and increase by an arithmetic gradient G between $1000 and $5000 per year thereafter. Determine if the expanded BIM is sensitive to gradient increases of $1000 (optimistic), $3000 (most likely), and $5000 (pessimistic) per year. Use AW analysis and a MARR of 10% per year.

18.24 When the country's economy is expanding, AB Investment Company is optimistic and expects a MARR of 15% on new investments. However, in a receding economy the expected return is 8%. Normally a 10% per year return is anticipated. An expanding economy causes the estimates of project life to go down about 20%, and a receding economy causes the life values to increase about 10%. Additionally, the market value after the project's life varies widely; optimistically, it may be as high as 50% more or as low as 75% of the initial investment.
 (a) Summarize the optimistic, most likely, and pessimistic estimates in tabular form.
 (b) Use a *spreadsheet* to calculate and plot the sensitivity of PW versus (1) MARR, (2) life values, and (3) market value for possible investment in one of two locations. Use the most likely estimates for the other parameters.
 (c) Considering all the analyses, under which scenario, if any, should location Miami or Houston be rejected?
 All monetary values are in $1000 units.

Location	Miami	Houston
Initial investment, $	−100,000	−110,000
NCF, $ per year	15,000	19,000
Market value at project end, $	100,000	110,000
Life, years	20	20

Expected Value

18.25 There are four estimates made for the anticipated cycle time to produce a subcomponent. The estimates, in seconds, are 10, 20, 30, and 70. If the first two estimates have equal weights, which total 70%, and the last two are weighted equally with the remaining 30%, what is the expected time?

18.26 A-Z Technologies, a manufacturer of amplified pressure transducers, is trying to decide between a dual-speed and a variable-speed machine. The engineers are not sure about the salvage value of the variable speed machine, so they have asked several different used-equipment dealers for estimates. The results can be summarized as follows: there is a 35% chance of getting $18,000, a 41% chance of getting $24,000, and a 13% chance of getting $29,000. Also, there is an 11% chance that the company may have to pay $5000 to dispose of the equipment. Calculate the expected salvage value.

18.27 Calculate the expected flow rate (barrels per day) for each oil well using the estimated probabilities.

Estimated Flow, bbl/Day	100	200	300	400
	Probability of Flow			
North well	0.15	0.75	0.10	0.0
East well	0.35	0.15	0.45	0.05

18.28 Thermtech Science has performed an economic analysis of proposed service in a new region of the country. The three-estimate approach to sensitivity analysis has been applied. The optimistic and pessimistic values each have an estimated 15% chance of occurring. Use the FW values shown to determine the expected FW.

	Optimistic	Most Likely	Pessimistic
FW value, $	200,000	40,000	−25,000

18.29 The variable Y is defined by the relation 3^n for $n = 1$, 2, 3, 4 with probabilities of 0.4, 0.3, 0.233, and 0.067, respectively. Determine the expected value of Y.

18.30 Nationwide income from monthly sales data (rounded to the nearest $100,000) of Stay Flat vacuum hold-down tables for last year was collected. Estimate the expected value of monthly income if economic conditions remain the same.

Income, $/Month	Number of Months
500,000	4
600,000	2
700,000	1
800,000	2
900,000	3

18.31 (a) Determine the expected present worth of the following cash flow series if each series may be realized with the probability shown at the head of each column. Let $i = 20\%$ per year. (b) Determine the expected AW value for the same cash flow series.

	Annual Cash Flow, $ per Year		
Year	Prob. = 0.5	Prob. = 0.2	Prob. = 0.3
0	−5000	−6000	−4000
1	1000	500	3000
2	1000	1500	1200
3	1000	2000	−800

18.32 Jeremy has $6000 to invest. If he puts the money in a certificate of deposit (CD), he is assured of receiving an effective 2.35% per year for 5 years. If he invests the money in stocks, he has a 50-50 chance of one of the following cash flow sequences for the next 5 years.

	Annual Cash Flow, $ per Year	
	Prob. = 0.5	Prob. = 0.5
Year	Stock 1	Stock 2
0	−6000	−6000
1–4	250	600
5	6800	4000

Finally, Jeremy can invest his $6000 in real estate for the 5 years with the following cash flow and probability estimates.

	Annual Cash Flow, $ per Year		
Year	Prob. = 0.3	Prob. = 0.5	Prob. = 0.2
0	−6000	−6000	−6000
1	−425	0	500
2	−425	0	600
3	−425	0	700
4	−425	0	800
5	9500	7200	5200

(a) Write the spreadsheet function to display i^* for each opportunity.

(b) Which of the three investment opportunities offers the best expected rate of return?

Decision Trees

18.33 For the decision tree branch shown, determine the expected values of the two outcomes if decision D3 is already selected and the maximum outcome value is sought. (This decision branch is part of a larger tree.)

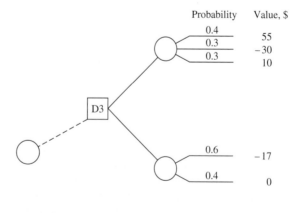

18.34 A large decision tree has an outcome branch detailed below. If decisions D1, D2, and D3 are all options in a 1-year time period, find the decision path that maximizes the outcome value. There are specific dollar investments necessary for decision nodes D1, D2, and D3, as indicated on each branch.

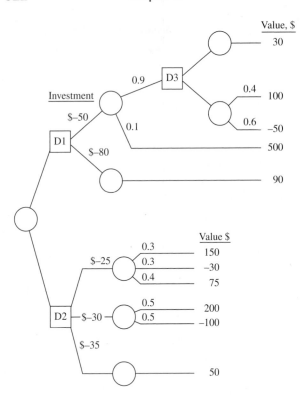

18.35 Decision D6, which has three possible choices (X, Y, or Z), must be made in year 3 of a 6-year study period in order to maximize E(PW). Using a MARR of 15% per year, the investment required in year 3, and the estimated cash flows for years 4 through 6, determine which decision should be made in year 3.

Investment, Cash Flow, $1000
Year 3

		Year			Outcome
	3	4	5	6	Prob.
High	$-200,000	$50	$50	$50	0.7
Low		40	30	20	0.3
High	$-75,000	30	40	50	
Low		30	30	30	0.55
High	$-250,000	190	170	150	0.7
Low		-30	-30	-30	0.3

18.36 Five thousand new smart camera systems are needed annually to expand the security surveillance of roads, buildings, airports, parks, etc. in a large metropolitan area. The system components can be obtained in one of three ways: (1) *Make* them in one of three plants partially owned by the government; (2) *Buy* them off the shelf from the one and only manufacturer; or (3) *Contract* to have them made to specifications by a vendor. The estimated annual cost for each alternative is dependent upon specifics of the plant, the producer, or the vendor. The

information below details the alternative, a probability of occurrence, and the estimated annual cost. (*a*) Construct and solve a decision tree to determine the least-cost alternative to provide the components. (*b*) Once you have chosen the alternative on the tree, use a *spreadsheet* to determine the change in probability that the *Buy* alternative requires to make it equally attractive with the chosen alternative. Assume that the probability change will be equally distributed to the other probability estimates. Write the new probability values.

Alternative	Outcomes		Probability	Annual Cost, $/Year
1. Make	Plant:			
		A	0.3	−250,000
		B	0.5	−400,000
		C	0.2	−350,000
2. Buy	Quantity:			
	<5000, pay premium		0.2	−550,000
	5000 available		0.7	−250,000
	>5000, forced to buy		0.1	−290,000
3. Contract	Delivery:			
	Timely delivery		0.5	−175,000
	Late delivery; then buy some off shelf		0.5	−450,000

18.37 The president of ChemTech is trying to decide whether to start a new product line or purchase a small company. It is not financially possible to do both. To make the product for a 3-year period will require an initial investment of $250,000. The expected annual cash flows with probabilities in parentheses are $75,000 (0.5), $90,000 (0.4), and $150,000 (0.1). To purchase the small company will cost $450,000 now. Market surveys indicate a 55% chance of increased sales for the company and a 45% chance of severe decreases with an annual cash flow of $25,000. If decreases are experienced in the first year, the company will be sold immediately (during year 1) at a price of $200,000. Increased sales could be $100,000 the first 2 years. If this occurs, a decision to expand after 2 years at an additional investment of $100,000 will be considered. This expansion could generate cash flows with indicated probabilities as follows: $120,000 (0.3), $140,000 (0.3), and $175,000 (0.4). If expansion is not chosen, the current size will be maintained with anticipated sales to continue. Assume there are no salvage values on any investments. Use the description given and a 15% per year return to do the following:

(*a*) Construct a decision tree including all values and probabilities.

(b) Determine the expected PW values at the "expansion/no expansion" decision node after 2 years provided sales are up.

(c) Determine what decision should be made now to offer the greatest return possible for ChemTech.

(d) Explain in words what would happen to the expected values at each decision node if the planning horizon were extended beyond 3 years and all cash flow values continued as forecasted in the description.

Real Options

18.38 Global Foundries (GF), a privately held company that invests in companies producing essential components for high-volume data storage, is valued at $4.0 billion dollars. A computer company that wants to get into cloud computing is considering the purchase of GF, but because of the uncertain economy, it would prefer to purchase an option that will allow it to buy the company for up to 1 year from now at a cost of $4.3 billion. What is the maximum amount the company should be willing to pay for the option if its MARR is 9% per year?

18.39 A nationwide real estate corporation is considering adding a new service line to sell the house of a client that buys another house through one of their agencies. It has determined that the first cost would be $80 million cash available for full implementation. However, John, the corporate vice president of sales, is not sure how well the service will be received, so he has projected added revenues using optimistic, most likely, and pessimistic estimates of $35 million, $25 million, and $10 million, respectively, with equal probability for each.

Instead of expanding now, the company could implement a pilot program for 1 year in a limited geographic area that will cost $4 million now. (The full-scale service will still cost $80 million if implemented later.) This will provide the company with the option to move forward or cancel the project. The criterion identified to move ahead with full-scale implementation is that added revenues during the pilot program must exceed $900,000. In this case, the pessimistic estimate will be eliminated, and equal probability placed on the remaining revenue projections. If the company uses a 5-year planning horizon and a MARR of 12% per year, should the company go ahead with the full-scale service now or take the option to implement the pilot program for 1 year?

18.40 Dupont is considering licensing a low liquid discharge (LLD) water treatment system from a small company that developed the process and owns the license. Dupont can purchase a 1-year option for $100,000 that will provide time to pilot test the LLD process or Dupont can acquire the license now at a cost of $1.8 million plus 25% of sales paid annually to the license owner. If they wait 1 year, the cost will increase to $1.9 million plus 30% of sales paid annually. If sales are estimated to be $1,000,000 per year over the 5-year license period, should Dupont purchase the license now or purchase the option now and possibly license it after the 1-year test period? Assume the MARR is 15% per year.

18.41 Abby has just negotiated a $15,000 price on a 2-year-old car and is with the salesman closing the deal. There is a 1-year sales warranty with the purchase; however, an extended warranty is available for $2500 that will cover the same repairs and component failures as the 1-year warranty for three additional years. Abby understands this to be a real options situation with the price of the option ($2500) paid to avoid future, unknown costs. To help with her decision, the salesman provided three typical sets of historical data (A, B, C) on estimated repair costs for used cars. The first year is shown as zero because it will be covered by the sales warranty.

Year	1	2	3	4
Repair cost, $/Year:				
A	0	−500	−1200	−850
B	0	−1000	−1400	−400
C	0	0	−500	−2000

The salesman said case C is the base case since it shows that the extended warranty is not needed because the repairs equal the warranty cost. Abby immediately recognized this to be the case only when $i = 0\%$.

(a) If Abby assumes that each repair cost scenario has an equal probability of occurring with her car, and money is worth a market interest rate of 8% per year to her, how much should she be willing to pay for the extended warranty that is offered at $2500?

(b) If the base case C actually occurs for her car and she does not purchase the warranty, what is the PW value of the expected future costs at $i = 8\%$ per year?

(c) At what market interest rate is case C economically equivalent to the extended warranty cost of $2500 now?

(d) Given all of this analysis, from a purely economic viewpoint, should Abby pay the $2500 for the extended warranty option? Why?

ADDITIONAL PROBLEMS AND FE EXAM REVIEW QUESTIONS

18.42 In evaluating the sensitivity of an alternative to its first cost, its AW was calculated for changes in the estimated first cost by –10%, +5%, and +15%. The resulting AW values were $+21,000, $–2410, and $–34,000, respectively. On the basis of these values, one could conclude that:
  (a) The attractiveness of the alternative is not sensitive to its first cost
  (b) The attractiveness of the alternative is slightly sensitive to its first cost
  (c) The attractiveness of the alternative is highly sensitive to its first cost
  (d) Cannot tell whether the attractiveness of the alternative is sensitive to its first cost or not

18.43 All of the following are steps in the procedure for conducting a sensitivity analysis, except:
  (a) Determine which parameters might vary from the most likely estimated value
  (b) Change the parameters in the range of -100% to $+100\%$
  (c) Select a measure of worth
  (d) Compute the results for each parameter using the measure of worth as a basis.

18.44 When the measure of worth is plotted versus percentage change for several parameters, the parameter that is the most sensitive in the economic analysis is the one:
  (a) That has the steepest curve
  (b) That has the flattest curve
  (c) With the largest present worth
  (d) With the shortest life

18.45 In conducting a sensitivity analysis of a proposed project, the present worth values of $–10,000, $40,000, and $50,000 were believed to have chances of 25%, 40%, and 35%, respectively. The expected PW is closest to:
  (a) $19,000   (b) $26,000
  (c) $28,500   (d) $31,000

18.46 When the sensitivity of several parameters is considered for one alternative using a single measure of worth and the measure of worth is plotted against percentage change for each parameter, the resulting graph is called a:
  (a) Spider graph
  (b) Distribution graph
  (c) Trend graph
  (d) Cash flow graph

18.47 When conducting a sensitivity analysis using three estimates for each parameter, all of the following are estimates that should be made, except:

  (a) Pessimistic
  (b) Improbable
  (c) Optimistic
  (d) Most likely

18.48 Revenue into the general fund of the state of Texas for any biennium is highly dependent on the price of oil. At a price average of $50 per barrel, general revenue will be $95 billion. At $68 and $75 per barrel, the revenue will be $118 billion and $125 billion, respectively. If the chances are estimated at 10%, 35%, and 55% for oil prices of $50, $68, and $75 per barrel for the next biennium, respectively, the expected revenue (in $ billion) is closest to:
  (a) $117.38   (b) $118.02
  (c) $118.92   (d) $119.50

18.49 A recent sensitivity analysis of a public works project indicates that the expected present worth is $83,000. If there is a 20% chance that the PW will be the pessimistic one of $45,000 and 50% chance that it will be the most likely one of $72,000, the optimistic PW is closest to:
  (a) $89,520   (b) $118,380
  (c) $126,670   (d) $138,540

18.50 A decision tree includes all of the following except:
  (a) Probability estimates for each outcome
  (b) Measure of worth as the selection criterion
  (c) Expected results from a decision at each stage
  (d) The MARR

18.51 A real options analysis is most valuable when:
  (a) The risk is low and stakes are high
  (b) The stakes are low and risk is high
  (c) The stakes are high and risk is high
  (d) The stakes are low and risk is low

The following information is used in Problems 18.52 and 18.53.

Four mutually exclusive alternatives are evaluated using three estimates or strategies (pessimistic, most likely, and optimistic) for several parameters. The resulting PW values over the LCM are determined as shown.

| | PW of Alternative, $ | | | |
Strategy	1	2	3	4
Pessimistic (P)	4,500	–6,000	3,700	–1,900
Most likely (ML)	6,000	–500	5,000	–100
Optimistic (O)	9,500	2,000	10,000	3,500

18.52 The best alternative to select under the pessimistic strategy is:
 (a) Alternative 1
 (b) Alternative 2
 (c) Alternative 3
 (d) Alternative 4

18.53 If none of the strategies is more likely than any other strategy, the alternative to select is:
 (a) 1
 (b) 1 and 2 are equally acceptable
 (c) 2
 (d) 3

CASE STUDY 1

SENSITIVITY TO THE ECONOMIC ENVIRONMENT

Background and Information

Berkshire Controllers usually finances its engineering projects with a combination of debt and equity capital. The resulting MARR ranges from a low of 4% per year if business is slow, to a high of 10% per year. Normally, a 7% per year return is expected. Also the life estimates for assets tend to go down about 20% from normal in a vigorous business environment and up about 10% in a receding economy. The following estimates are the most likely values for two expansion plans currently being evaluated. Plan A will be executed at one location; Plan B will require two locations. All monetary estimates are in $1000 units.

	Plan A	Plan B Location 1	Plan B Location 2
First cost, $	−10,000	−30,000	−5,000
AOC, $ per year	−500	−100	−200
Salvage value, $	1,000	5,000	−200
Estimated life, years	40	40	20

Case Study Questions

At the weekly meeting, you were asked to examine the following questions from Berkshire's president:

1. Are the PW values for Plans A and B sensitive to changes in the MARR?
2. Are the PW values sensitive to varying life estimates?
3. Is the breakeven point for the first cost of Plan A sensitive to the changes in MARR as business goes from vigorous to receding?

CASE STUDY 2

SENSITIVITY ANALYSIS OF PUBLIC SECTOR PROJECTS—WATER SUPPLY PLANS

Background

One of the most basic services provided by municipal governments is the delivery of a safe, reliable water supply. As cities grow and extend their boundaries to outlying areas, they often inherit water systems that were not constructed according to city codes. The upgrading of these systems is sometimes more expensive than installing one correctly in the first place. To avoid these problems, city officials sometimes install water systems beyond the existing city limits in anticipation of future growth. This case study was extracted from such a countywide water and wastewater management plan and is limited to only some of the water supply alternatives.

From about a dozen suggested plans, five methods were developed by an executive committee as alternative ways of providing water to the study area. These methods were then subjected to a preliminary evaluation to identify the most promising alternatives. Six attributes or factors were used in the initial rating: ability to serve the area, relative cost, engineering feasibility, institutional issues, environmental considerations, and lead time requirement. Each factor carried the same weighting and had values ranging from 1 to 5, with 5 being best. After the top three alternatives were identified, each was subjected to a detailed economic evaluation for selection of the best alternative. These detailed evaluations included an estimate of the capital investment of each alternative amortized over 20 years at 8% per year interest and the annual maintenance and operation (M&O) costs. The annual cost (an AW value) was then divided by the population served to arrive at a monthly cost per household.

Information

Table 18–5 presents the results of the screening using the six factors rated on a scale of 1 to 5. Alternatives 1A, 3, and 4 were determined to be the three best and were chosen for further evaluation.

Detailed Cost Estimates

All amounts are cost estimates.

Alternative 1A

Capital cost

Land with water rights: 1720 hectares @ $5000 per hectare	$8,600,000
Primary treatment plant	2,560,000
Booster station at plant	221,425
Reservoir at booster station	50,325
Site cost	40,260
Transmission line from river	3,020,000
Transmission line right-of-way	23,350
Percolation beds	2,093,500
Percolation bed piping	60,400
Production wells	510,000
Well field gathering system	77,000
Distribution system	1,450,000
Additional distribution system	3,784,800
Reservoirs	250,000
Reservoir site, land, and development	17,000
Subtotal	22,758,060
Engineering and contingencies	5,641,940
Total capital investment	$28,400,000

Maintenance and operation costs (annual)

Pumping 9,812,610 kWh per year @ $0.08 per kWh	$ 785,009
Fixed operating cost	180,520
Variable operating cost	46,730
Taxes for water rights	48,160
Total annual M&O cost	$1,060,419

Total annual cost = equivalent capital investment + M&O cost

$$= 28,400,000(A/P,8\%,20)$$
$$+ 1,060,419$$
$$= 2,892,540 + 1,060,419$$
$$= \$3,952,959$$

Average monthly household cost to serve 95% of 4980 households is

$$\text{Household cost} = (3,952,959)\left(\frac{1}{12}\right)\left(\frac{1}{4980}\right)\left(\frac{1}{0.95}\right)$$
$$= \$69.63 \text{ per month}$$

Alternative 3

Total capital investment = $29,600,000

Total annual M&O cost = $867,119

$$\text{Total annual cost} = 29,600,000(A/P,8\%,20)$$
$$+ 867,119$$
$$= 3,014,760 + 867,119$$
$$= \$3,881,879$$

Household cost = $68.38 per month

Alternative 4

Total capital investment = $29,000,000

Total annual M&O cost = $1,063,449

$$\text{Total annual cost} = 29,000,000(A/P,8\%,20)$$
$$+ 1,063,449$$
$$= 2,953,650 + 1,063,449$$
$$= \$4,017,099$$

Household cost = $70.76 per month

TABLE 18–5 Results of Rating Six Factors for Each Alternative, Case Study

		Factors						
Alternative	Description	Ability to Supply Area	Relative Cost	Engineering Feasibility	Institutional Issues	Environmental Considerations	Lead Time Requirement	Total
1A	Receive city water and recharge wells	5	4	3	4	5	3	24
3	Joint city and county plant	5	4	4	3	4	3	23
4	County treatment plant	4	4	3	3	4	3	21
8	Desalt groundwater	1	2	1	1	3	4	12
12	Develop military water	5	5	4	1	3	1	19

On the basis of the lowest monthly household cost, alternative 3 (joint city and county plant) is the most economically attractive.

Case Study Exercises

1. If the environmental-considerations factor is to have a weighting of twice as much as any of the other five factors, what is its percentage weighting?

2. If the ability-to-supply-area and relative-cost factors were each weighted 20% and the other four factors 15% each, which alternatives would be ranked in the top three?

3. By how much would the capital investment of alternative 4 have to decrease to make it more attractive than alternative 3?

4. If alternative 1A served 100% of the households instead of 95%, by how much would the monthly household cost decrease?

5. (a) Perform a sensitivity analysis on the two parameters of M&O costs and number of households to determine if alternative 3 remains the best economic choice. Three estimates are made for each parameter in Table 18–6. M&O costs may vary up (pessimistic) or down (optimistic) from the most likely estimates presented in the case statement. The estimated number of households (4980) is determined to be the pessimistic estimate. Growth of 2% up to 5% (optimistic) will tend to lower the monthly cost per household.

(b) Consider the monthly cost per household for alternative 4, the optimistic estimate. The number of households is 5% above 4980, or 5230. What is the number of households that would have to be available in order for this option to have exactly the same monthly household cost as that for alternative 3 at the optimistic estimate of 5230 households?

TABLE 18–6	Pessimistic, Most Likely, and Optimistic Estimates for Two Parameters	
	Annual M&O Costs	**Number of Households**
Alternative 1A		
Pessimistic	+1%	4980
Most likely	$1,060,419	+2%
Optimistic	−1%	+5%
Alternative 3		
Pessimistic	+5%	4980
Most likely	$867,119	+2%
Optimistic	0%	+5%
Alternative 4		
Pessimistic	+2%	4980
Most likely	$1,063,449	+2%
Optimistic	−10%	+5%

CHAPTER 19

More on Variation and Decision Making under Risk

© Brand X/JupiterImages

LEARNING OUTCOMES

Purpose: Incorporate decision making under risk into an engineering economy evaluation using probability, sampling, and simulation.

SECTION	TOPIC	LEARNING OUTCOME
19.1	Risk versus certainty	• Understand the approaches to decision making under risk and certainty.
19.2	Probability and distributions	• Construct a probability distribution and cumulative distribution for one variable.
19.3	Random sample	• Obtain a random sample from a cumulative distribution using a random number table.
19.4	μ, σ, and σ^2	• Estimate the population expected value, standard deviation, and variance from a random sample.
19.5	Simulation	• Use Monte Carlo sampling and spreadsheet-based simulation for alternative evaluation.

his chapter further expands our ability to analyze variation in estimates, to consider probability, and to **make decisions under risk.** Fundamentals discussed include variables; probability distributions, especially their graphs and properties of expected value and dispersion; random sampling; and the use of simulation to account for estimate variation in engineering economy studies.

Through coverage of variation and probability, this chapter complements topics in the first sections of Chapter 1, that is, the role of engineering economy in decision making and economic analysis in the problem-solving process. These techniques are more time consuming than using estimates made with certainty, so they should be used primarily for critical parameters.

19.1 Interpretation of Certainty, Risk, and Uncertainty ● ● ●

All things in the world vary—one from another, over time, and with different environments. We are guaranteed that variation will occur in engineering economy due to its emphasis on decision making for the future. Except for the use of breakeven analysis, sensitivity analysis, and a very brief introduction to expected values, virtually all our estimates have been **certain; that is, no variation** in the amount has entered into the computations of PW, AW, ROR, or any relations used. For example, the estimate that cash flow next year will be $+4500 is one of certainty. Decision making under certainty is, of course, not present in the real world now and surely not in the future. We can observe outcomes with a high degree of certainty, but even this depends upon the accuracy and precision of the scale or measuring instrument.

To allow a parameter of an engineering economy study to vary implies that *risk*, and possibly *uncertainty,* is introduced.

When there may be two or more observable values for a parameter *and* it is possible to estimate the chance that each value may occur, **risk** is present. Virtually all decision making is performed *under risk.*

Risk

As an illustration, decision making under risk is introduced when an annual cash flow estimate has a 50-50 chance of being either $-1000 or $+500.

Decision making under **uncertainty** means there are two or more values observable, but the chances of their occurring cannot be estimated or no one is willing to assign the chances. The observable values in uncertainty analysis are often referred to as *states of nature.*

For example, consider the states of nature to be the rate of national inflation in a particular country during the next 2 to 4 years: remain low, increase 2% to 6% annually, or increase 6% to 8% annually. If there is absolutely no indication that the three values are equally likely, or that one is more likely than the others, this is a statement that indicates decision making under uncertainty.

Example 19.1 explains how a parameter can be described and graphed to prepare for decision making under risk.

EXAMPLE 19.1

CMS in Fairfield, Virginia, received three bids each from vendors for two different pieces of large equipment, A and B. One of each piece of equipment must be purchased. Tom, an engineer at CMS, performed an evaluation of each bid and assigned it a rating between 0 and 100, with 100 points being the best of the three. The total for each piece of equipment is 100%. The bid amounts and ratings are shown at the top of Figure 19–1.

(*a*) Consider the ratings as the chance out of 100 that the bid will be chosen, and plot cost versus chance for each vendor.

(*b*) Since one each of A and B must be purchased, the total cost will vary somewhere between the sum of the lowest bids ($11 million) and the sum of the highest bids ($25 million). Plot this range with an equal chance of 1 in 14 that any amount in between these limits is possible.

(*c*) Discuss the significant difference between the values of the cost (*x* axis values) in the graphs in (*a*) and (*b*) above and how the chances are stated (*y* axis values).

Figure 19–1
Plot of cost estimates versus chance for (*a*) each piece of equipment, and (*b*) total cost range, Example 19.1.

Equipment A		Equipment B	
Bid, $1000	**Rating, %**	**Bid, $1000**	**Rating, %**
3,000	65	8,000	33.3
5,000	25	10,000	33.3
10,000	10	15,000	33.3

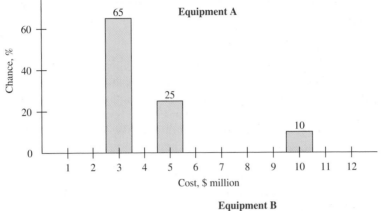

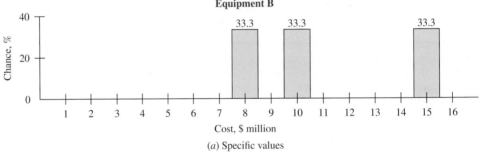

(*a*) Specific values

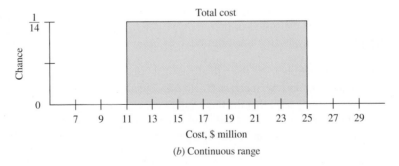

(*b*) Continuous range

Solution

(*a*) Figure 19–1*a* plots the specific bids for equipment A and B. The chances (ratings) for A and for B add to 100%. No values between the specific bids have any chance of occurring, according to the single-estimate bids from the three vendors.

(*b*) The range of total cost is between $11 million and $25 million, as shown in Figure 19–1*b*. Tom decided to make his estimate of total cost continuous between these two extremes. This means that the discrete sums of bids ($11 million, $15 million, and $25 million) are no longer used. Rather the entire range from $11 million to $25 million with a chance for every total cost in between is included. Every value has a chance of 1 in 14 of being observed. Now, the sum is a continuous value.

(*c*) In the graph for bid values (Figure 19–1*a*), only specific or discrete estimates are included on the *x* axis. In the graph for the sum of the cost for equipment A and B (Figure 19–1*b*), the *y* axis values are continuous over a specific range.

In the next section, the term *variable* is defined and two types of variables are explained—*discrete* and *continuous*—as illustrated here in an elementary form.

Before initiating an engineering economy study, it is important to decide if the analysis will be conducted with certainty for all parameters or if risk will be introduced. A summary of the meaning and use for each type of analysis follows.

Decision Making under Certainty   This is what we have done in most analyses thus far. Deterministic estimates are made and entered into measure of worth relations—PW, AW, FW, ROR, B/C—and decision making is based on the results. The values estimated can be considered the most likely to occur with all chance placed on the single-value estimate. A typical example is an asset's first cost estimate made with certainty, say, $P = \$50,000$. A plot of P versus chance has the general form of Figure 19–1*a* with one vertical bar at \$50,000 and 100% chance placed on it. The term *deterministic,* in lieu of *certainty*, is often used when **single-value** or **single-point estimates** are used exclusively.

In fact, sensitivity analysis using different values of an estimate is simply another form of analysis with certainty, except that the analysis is repeated with different values, *each estimated with certainty.* The resulting measure of worth values are calculated and graphically portrayed to determine the decision's sensitivity to different estimates for one or more parameters.

Decision Making under Risk   Now the element of chance is formally taken into account. However, it is more difficult to make a clear decision because the analysis attempts to accommodate **variation.** One or more parameters in an alternative will be allowed to vary. The estimates will be expressed as in Example 19.1 or in slightly more complex forms. Fundamentally, there are two ways to consider risk in an analysis:

Expected value analysis. Use the chance and parameter estimates to calculate expected values E(parameter) via formulas such as Equation [18.2]. Analysis results in E(cash flow), E(AOC), and the like; and the final result is the expected value for a measure of worth, such as E(PW), E(AW), E(ROR), or E(B/C). To select the alternative, choose the most favorable expected value of the measure of worth. In an elementary form, this is what we learned about expected values in Chapter 18. The computations may become more elaborate, but the principle is fundamentally the same.

Simulation analysis. Use the chance and parameter estimates to generate repeated computations of the measure of worth relation by randomly sampling from a plot for each varying parameter similar to those in Figure 19–1. When a representative and random sample is complete, an alternative is selected utilizing a table or plot of the results. Usually, graphics are an important part of decision making via simulation analysis. Basically, this is the approach discussed in the rest of this chapter.

Decision Making under Uncertainty   When chances are not known for the identified states of nature (or values) of the uncertain parameters, the use of expected value-based decision making under risk as outlined above is *not an option.* In fact, it is difficult to determine what criterion to use to even make the decision. If it is possible to agree that each state is equally likely, then all states have the same chance, and the situation reduces to one of decision making under risk, because expected values can be determined. Because of the relatively inconclusive approaches necessary to incorporate decision making under uncertainty into an engineering economy study, the techniques can be quite useful but are beyond the intended scope of this text.

In an engineering economy study, observed parameter values will vary from the value estimated at the time of the study. However, when performing the analysis, not all parameters should be considered as probabilistic (or at risk). Those that are estimable with a relatively high degree of certainty should be fixed for the study. Accordingly, the methods of sampling, simulation, and statistical data analysis are selectively used on parameters deemed important to the decision-making process. Parameters such as P, AOC, material and unit costs, sales price, revenues, etc., are the targets of decision making under risk. Anticipated variation in interest rates is more commonly addressed by sensitivity analysis.

The remainder of this chapter concentrates on decision making under risk as applied in an engineering economy study. Sections 19.2 to 19.4 provide foundation material necessary to design and correctly conduct a simulation analysis (Section 19.5).

Figure 19–2
(*a*) Discrete and continuous variable scales and (*b*) scales for a variable versus its probability.

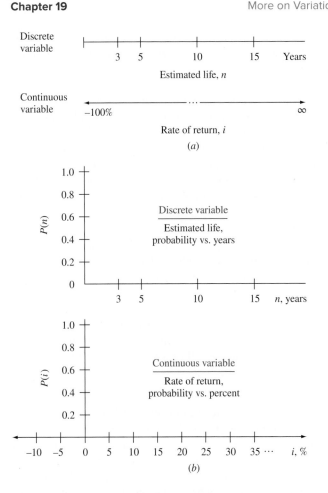

19.2 Elements Important to Decision Making under Risk ● ● ●

Some basics of probability and statistics are essential to correctly perform decision making under risk via expected value or simulation analysis. They are the *random variable, probability, probability distribution*, and *cumulative distribution*, as defined here. (If you are already familiar with them, this section will provide a review.)

A **random variable** or **variable** is a characteristic or parameter that can take on any one of several values. Variables are classified as *discrete* or *continuous*. Discrete variables have several specific, isolated values, while continuous variables can assume any value between two stated limits, called the *range* of the variable.

The estimated life of an asset is a discrete variable. For example, n may be expected to have values of $n = 3$, 5, 10, or 15 years, and no others. The rate of return is an example of a continuous variable; i can vary from -100% to ∞, that is, $-100\% \leq i < \infty$. The ranges of possible values for n (discrete) and i (continuous) are shown as the x axes in Figure 19–2*a*. (In probability texts, capital letters symbolize a variable, say X, and small letters x identify a specific value of the variable. Though correct, this level of rigor in terminology is not applied in this chapter.)

Probability is a number between 0 and 1.0 that expresses the chance in decimal form that a random variable (discrete or continuous) will take on any value from those identified for it. Probability is simply the amount of chance, divided by 100.

Probabilities are commonly identified by $P(X_i)$ or $P(X = X_i)$, which is read as the probability that the variable X takes on the value X_i. (Actually, for a continuous variable, the probability at a single value is zero, as shown in a later example.) The sum of all $P(X_i)$ for a variable must be 1.0, a requirement already discussed. The probability scale, like the percentage scale for chance in

Figure 19–1, is indicated on the ordinate (*y* axis) of a graph. Figure 19–2*b* shows the 0 to 1.0 range of probability for the variables *n* and *i*.

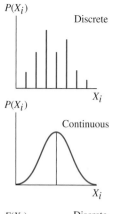

A **probability distribution** describes how probability is distributed over the different values of a variable. Discrete variable distributions look significantly different from continuous variable distributions, as indicated by the inset at the right.

The individual probability values are stated as

$$P(X_i) = \textbf{probability that } X \textbf{ equals } X_i \qquad \textbf{[19.1]}$$

The distribution may be developed in one of two ways: by listing each probability value for each possible variable value (see Example 19.2) or by a mathematical description or expression that states probability in terms of the possible variable values (Example 19.3).

Cumulative distribution, also called the **cumulative probability distribution,** is the accumulation of probability over all values of a variable up to and including a specified value.

Identified by $F(X_i)$, each cumulative value is calculated as

$$F(X_i) = \textbf{sum of all probabilities through the value } X_i$$
$$= P(X \leq X_i) \qquad \textbf{[19.2]}$$

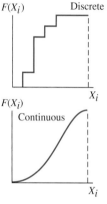

As with a probability distribution, cumulative distributions appear differently for discrete (stair-stepped) and continuous variables (smooth curve). Examples 19.2 and 19.3 illustrate cumulative distributions that correspond to specific probability distributions. These fundamentals about $F(X_i)$ are applied in the next section to develop a random sample.

EXAMPLE 19.2

Alvin is a medical doctor and biomedical engineering graduate who practices at Medical Center Hospital. He is planning to start prescribing an antibiotic that may reduce infection in patients with flesh wounds. Tests indicate the drug has been applied up to six times per day without harmful side effects. If no drug is used, there is always a positive probability that the infection will be reduced by a person's own immune system.

Published drug test results provide good probability estimates of positive reaction (i.e., reduction in the infection count) within 48 hours for increased treatments per day. Use the probabilities listed below to construct a probability distribution and a cumulative distribution for the total number of treatments per day.

Number of Added Treatments per Day	Probability of Infection Reduction for Each Added Treatment
0	0.07
1	0.08
2	0.10
3	0.12
4	0.13
5	0.25
6	0.25

Solution

Define the random variable *T* as the number of added treatments per day. Since *T* can take on only seven different values, it is a **discrete variable.** The probability of infection reduction is listed for each value in column 2 of Table 19–1. The cumulative probability $F(T_i)$ is determined using Equation [19.2] by adding all $P(T_i)$ values through T_i, as indicated in column 3.

Figure 19–3*a* and *b* shows plots of the probability distribution and cumulative distribution, respectively. The summing of probabilities to obtain $F(T_i)$ gives the cumulative distribution the stair-stepped appearance, and in all cases the final $F(T_i) = 1.0$, since the total of all $P(T_i)$ values must equal 1.0.

TABLE 19–1	Probability Distribution and Cumulative Distribution for Example 19.2	
(1)	(2)	(3)
Number per Day T_i	Probability $P(T_i)$	Cumulative Probability $F(T_i)$
0	0.07	0.07
1	0.08	0.15
2	0.10	0.25
3	0.12	0.37
4	0.13	0.50
5	0.25	0.75
6	0.25	1.00

Figure 19–3
(a) Probability distribution $P(T_i)$, and (b) cumulative distribution $F(T_i)$ for Example 19.2.

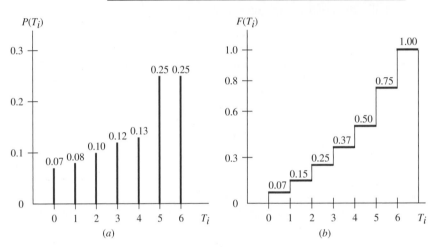

Comment

Rather than use a tabular form as in Table 19–1 to state $P(T_i)$ and $F(T_i)$ values, it is possible to express them for each value of the variable.

$$P(T_i) = \begin{cases} 0.07 & T_1 = 0 \\ 0.08 & T_2 = 1 \\ 0.10 & T_3 = 2 \\ 0.12 & T_4 = 3 \\ 0.13 & T_5 = 4 \\ 0.25 & T_6 = 5 \\ 0.25 & T_7 = 6 \end{cases} \qquad F(T_i) = \begin{cases} 0.07 & T_1 = 0 \\ 0.15 & T_2 = 1 \\ 0.25 & T_3 = 2 \\ 0.37 & T_4 = 3 \\ 0.50 & T_5 = 4 \\ 0.75 & T_6 = 5 \\ 1.00 & T_7 = 6 \end{cases}$$

In basic engineering economy situations, the probability distribution for a **continuous variable** is commonly expressed as a mathematical function, such as a *uniform distribution,* a *triangular distribution* (both discussed in Example 19.3 in terms of cash flow), or the more complex, but commonly used, *normal distribution.* For continuous variable distributions, the symbol $f(X)$ is routinely used instead of $P(X_i)$, and $F(X)$ is used instead of $F(X_i)$, simply because the point probability for a continuous variable is zero. Thus, $f(X)$ and $F(X)$ are continuous lines and curves.

EXAMPLE 19.3

As president of a manufacturing systems consultancy, Sallie has observed the monthly cash flows that have occurred over the last 3 years into company accounts from two longstanding clients. Sallie has concluded the following about the distribution of these monthly cash flows:

Client 1

Estimated low cash flow: $10,000
Estimated high cash flow: $15,000
Most likely cash flow: same for all values
Distribution of probability: uniform

Client 2

Estimated low cash flow: $20,000
Estimated high cash flow: $30,000
Most likely cash flow: $28,000
Distribution of probability: mode at $28,000

The *mode* is the most frequently observed value for a variable. Sallie assumes cash flow to be a continuous variable referred to as C. (*a*) Write and graph the two probability distributions and cumulative distributions for monthly cash flow, and (*b*) determine the probability that monthly cash flow is no more than $12,000 for client 1 and at least $25,000 for client 2.

Solution

All cash flow values are expressed in $1000 units.

Client 1: monthly cash flow distribution

(*a*) The distribution of cash flows for client 1, identified by the variable C_1, follows the *uniform distribution*. Probability and cumulative probability take the following general forms.

$$f(C_1) = \frac{1}{\text{high} - \text{low}} \qquad \text{low value} \leq C_1 \leq \text{high value}$$

$$f(C_1) = \frac{1}{H - L} \qquad L \leq C_1 \leq H \qquad\qquad [19.3]$$

$$F(C_1) = \frac{\text{value} - \text{low}}{\text{high} - \text{low}} \qquad \text{low value} \leq C_1 \leq \text{high value}$$

$$F(C_1) = \frac{C_1 - L}{H - L} \qquad L \leq C_1 \leq H \qquad\qquad [19.4]$$

For client 1, monthly cash flow is uniformly distributed with $L = \$10$, $H = \$15$, and $\$10 \leq C_1 \leq \$15$. Figure 19–4 is a plot of $f(C_1)$ and $F(C_1)$ from Equations [19.3] and [19.4].

$$f(C_1) = \frac{1}{5} = 0.2 \qquad \$10 \leq C_1 \leq \$15$$

$$F(C_1) = \frac{C_1 - 10}{5} \qquad \$10 \leq C_1 \leq \$15$$

(*b*) The probability that client 1 has a monthly cash flow of no more than $12 is easily determined from the $F(C_1)$ plot as 0.4, or a 40% chance. If the $F(C_1)$ relation is used directly, the computation is

$$F(\$12) = P(C_1 \leq \$12) = \frac{12 - 10}{5} = 0.4$$

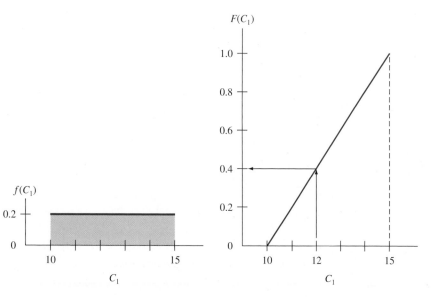

Figure 19–4

Uniform distribution for client 1 monthly cash flow, Example 19.3.

Client 2: monthly cash flow distribution

(a) The distribution of cash flows for client 2, identified by the variable C_2, follows the *triangular distribution*. This probability distribution has the shape of an upward-pointing triangle with the peak at the mode M, and downward-sloping lines joining the x axis on either side at the low (L) and high (H) values. The mode of the triangular distribution has the maximum probability value.

$$f(\text{mode}) = f(M) = \frac{2}{H - L} \tag{19.5}$$

The cumulative distribution is comprised of two curved line segments from 0 to 1 with a break point at the mode, where

$$F(\text{mode}) = F(M) = \frac{M - L}{H - L} \tag{19.6}$$

For C_2, the low value is $L = \$20$, the high is $H = \$30$, and the most likely cash flow is the mode $M = \$28$. The probability at M from Equation [19.5] is

$$f(28) = \frac{2}{30 - 20} = \frac{2}{10} = 0.2$$

The break point in the cumulative distribution occurs at $C_2 = 28$. Using Equation [19.6],

$$F(28) = \frac{28 - 20}{30 - 20} = 0.8$$

Figure 19–5 presents the plots for $f(C_2)$ and $F(C_2)$. Note that $f(C_2)$ is skewed since the mode is not at the midpoint of the range $H - L$, and $F(C_2)$ is a smooth S-shaped curve with an inflection point at the mode.

(b) From the cumulative distribution in Figure 19–5, there is an estimated 31.25% chance that cash flow is $25 or less. Therefore,

$$F(\$30) - F(\$25) = P(C_2 \geq \$25) = 1 - 0.3125 = 0.6875$$

Comment

The general relations $f(C_2)$ and $F(C_2)$ are not developed here. The variable C_2 is *not* a uniform distribution; it is triangular. Therefore, it requires the use of an integral to find cumulative probability values from the probability distribution $f(C_2)$.

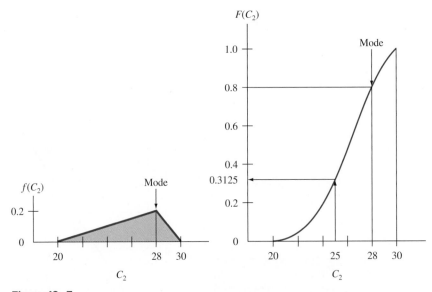

Figure 19–5
Triangular distribution for client 2 monthly cash flow, Example 19.3.

19.3 Random Samples ● ● ●

Estimating a parameter with a single value in previous chapters is the equivalent of taking a *random sample of size 1 from an entire population* of possible values. As an illustration, assume that estimates of first cost, annual operating cost, interest rate, and other parameters are used to compute one PW value in order to accept or reject an alternative. Each estimate is a sample of size 1 from an entire population of possible values for each parameter. Now, if a second estimate is made for each parameter and a second PW value is determined, a sample of size 2 has been taken. If all values in the population were known, the probability distribution and cumulative distribution would be known. Then a sample would not be necessary.

When we perform an engineering economy study and utilize decision making under certainty, we use one estimate for each parameter to calculate a measure of worth (i.e., a sample of size 1 for each parameter). The estimate is the most likely value, that is, one estimate of the expected value. We know that all parameters will vary somewhat; yet some are important enough, or will vary enough, that a probability distribution should be determined or assumed for it and the parameter treated as a random variable. This is using risk, and a sample from the parameter's probability distribution—$P(X)$ for discrete or $f(X)$ for continuous—helps formulate probability statements about the estimates. This approach complicates the analysis somewhat; however, it also provides a sense of confidence (or possibly a lack of confidence in some cases) about the decision made concerning the economic viability of the alternative based on the varying parameter. (We will further discuss this aspect later, after we learn how to correctly take a random sample from any probability distribution.)

> A **random sample** of size n is the selection in a random fashion of n values from a population with an assumed or known probability distribution, such that the values of the variable have the **same chance of occurring** in the sample as they are expected to occur in the population.

Suppose Yvon is an engineer with 20 years of experience working for the Aircraft Safety Commission. For a two-crew aircraft, there are three parachutes on board. The safety standard states that 99% of the time, all three chutes must be "fully ready for emergency deployment." Yvon is relatively sure that nationwide the probability distribution of N, the specific number of chutes fully ready, may be described by the probability distribution

$$P(N = N_i) = \begin{cases} 0.005 & N = 0 \text{ chutes ready} \\ 0.015 & N = 1 \text{ chute ready} \\ 0.060 & N = 2 \text{ chutes ready} \\ 0.920 & N = 3 \text{ chutes ready} \end{cases}$$

This means that the safety standard is clearly not met nationwide. Yvon is in the process of sampling 200 (randomly selected) corporate and private aircraft across the nation to determine how many chutes are classified as fully ready. If the sample is truly random and Yvon's probability distribution is a correct representation of actual parachute readiness, the observed N values in the 200 aircraft will approximate the same proportions as the population probabilities, that is, 1 aircraft with 0 chutes ready, etc. Since this is a sample, it is likely that the results won't track the population exactly. However, if the results are relatively close, the study indicates that the sample results may be useful in predicting parachute safety across the nation.

To develop a random sample, use **random numbers (RN)** generated from a uniform probability distribution for the discrete numbers 0 through 9, that is,

$$P(X_i) = 0.1 \qquad \text{for } X_i = 0, 1, 2, \ldots, 9$$

In tabular form, the random digits so generated are commonly clustered in groups of two digits, three digits, or more. Table 19–2 is a sample of 264 random digits clustered into two-digit numbers. This format is very useful because the numbers 00 to 99 conveniently relate to the cumulative distribution values 0.01 to 1.00. This makes it easy to select a two-digit RN and enter $F(X)$ to determine a value of the variable with the same proportions as it occurs in the probability distribution. To apply this logic manually and develop a random sample of size n from a known

TABLE 19–2	Random Digits Clustered into Two-Digit Numbers																				
51	82	88	18	19	81	03	88	91	46	39	19	28	94	70	76	33	15	64	20	14	52
73	48	28	59	78	38	54	54	93	32	70	60	78	64	92	40	72	71	77	56	39	27
10	42	18	31	23	80	80	26	74	71	03	90	55	61	61	28	41	49	00	79	96	78
45	44	79	29	81	58	66	70	24	82	91	94	42	10	61	60	79	30	01	26	31	42
68	65	26	71	44	37	93	94	93	72	84	39	77	01	97	74	17	19	46	61	49	67
75	52	14	99	67	74	06	50	97	46	27	88	10	10	70	66	22	56	18	32	06	24

discrete probability distribution $P(X)$ or a continuous variable distribution $f(X)$, the following procedure may be used:

1. Develop the cumulative distribution $F(X)$ from the probability distribution. Plot $F(X)$.
2. Assign the RN values from 00 to 99 to the $F(X)$ scale (the y axis) in the same proportion as the probabilities. For the parachute safety example, the probabilities from 0.0 to 0.15 are represented by the random numbers 00 to 14. Indicate the RNs on the graph.
3. To use a table of random numbers, determine the scheme or sequence of selecting RN values—down, up, across, diagonally. Any direction and pattern is acceptable, but the scheme should be used consistently for one entire sample.
4. Select the first number from the RN table, enter the $F(X)$ scale, and observe and record the corresponding variable value. Repeat this step until there are n values of the variable that constitute the random sample.
5. Use the n sample values for analysis and decision making under risk. These may include

   - Plotting the sample probability distribution
   - Developing probability statements about the parameter
   - Comparing sample results with the assumed population distribution
   - Determining sample statistics (Section 19.4)
   - Performing a simulation analysis (Section 19.5)

EXAMPLE 19.4

Develop a random sample of size 10 for the variable N, number of months, as described by the probability distribution

$$P(N = N_i) = \begin{cases} 0.20 & N = 24 \\ 0.50 & N = 30 \\ 0.30 & N = 36 \end{cases} \qquad [19.7]$$

Solution

Apply the procedure above, using the $P(N = N_i)$ values in Equation [19.7].

1. The cumulative distribution, Figure 19–6, is for the discrete variable N, which can assume three different values.
2. Assign 20 numbers (00 through 19) to $N_1 = 24$ months, where $P(N = 24) = 0.2$; 50 numbers to $N_2 = 30$; and 30 numbers to $N_3 = 36$.
3. Initially select any position in Table 19–2, and go across the row to the right and onto the row below toward the left. (Any routine can be developed, and a different sequence for each random sample may be used.)
4. Select the initial number 45 (4th row, 1st column), and enter Figure 19–6 in the RN range of 20 to 69 to obtain $N = 30$ months.
5. Select and record the remaining nine values from Table 19–2 as shown below.

RN	45	44	79	29	81	58	66	70	24	82
N	30	30	36	30	36	30	30	36	30	36

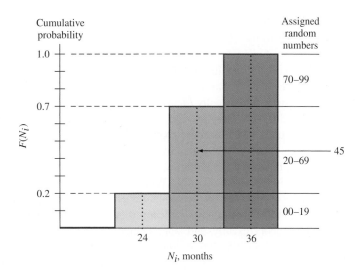

Figure 19–6
Cumulative distribution with random number values assigned in proportion to probabilities, Example 19.4.

Now, using the 10 values, develop the sample probabilities.

Months N	Times in Sample	Sample Probability	Equation [19.7] Probability
24	0	0.00	0.2
30	6	0.60	0.5
36	4	0.40	0.3

With only 10 values, we can expect the sample probability estimates to be different from the values in Equation [19.7]. Only the value $N = 24$ months is significantly different since no RN of 19 or less occurred. A larger sample will definitely make the probabilities closer to the original data.

To take a *random sample of size n for a continuous variable,* the procedure above is applied, except the random number values are assigned to the cumulative distribution on a continuous scale of 00 to 99 corresponding to the $F(X)$ values. As an illustration, consider Figure 19–4, where C_1 is the *uniformly distributed* cash flow variable for client 1 in Example 19.3. Here $L = \$10$, $H = \$15$, and $f(C_1) = 0.2$ for all values between L and H (all values are divided by \$1000). The $F(C_1)$ is repeated as Figure 19–7 with the assigned random number values shown on the right scale. If the two-digit RN of 45 is chosen, the corresponding C_1 is graphically estimated to be \$12.25. It can also be linearly interpolated as $\$12.25 = 10 + (45/100)(15 - 10)$.

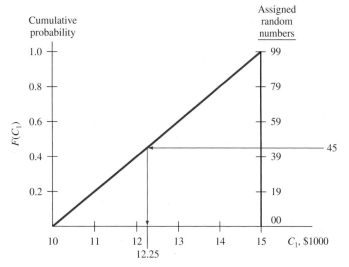

Figure 19–7
Random numbers assigned to the continuous variable of client 1 cash flows in Example 19.3.

For greater accuracy when developing a random sample, especially for a continuous variable, it is possible to use three-, four-, or five-digit RNs. These can be developed from Table 19–2 simply by combining digits in the columns and rows or obtained from tables with RNs printed in larger clusters of digits. In computer-based sampling, most simulation software packages have an RN generator built in that will generate values in the range of 0 to 1 from a continuous variable uniform distribution, usually identified by the symbol $U(0, 1)$. The RN values, usually between 0.00000 and 0.99999, are used to sample directly from the cumulative distribution employing essentially the same procedure explained here. The Excel functions RAND and RANDBE-TWEEN are described in Appendix A, Section A.3.

An initial question in random sampling usually concerns the **minimum size of n** required to ensure confidence in the results. Without detailing the mathematical logic, sampling theory, which is based upon the law of large numbers and the central limit theorem (check a basic statistics book to learn about these), indicates that an n of 30 is sufficient. However, since reality does not follow theory exactly, and since engineering economy often deals with sketchy estimates, samples in the *range of 100 to 200* are the common practice. But samples as small as 10 to 25 provide a much better foundation for decision making under risk than the single-point estimate for a parameter that is known to vary widely.

19.4 Sample Estimates: Mean and Standard Deviation ● ● ●

Two very important measures or properties of a random variable are the expected value and standard deviation. If the entire population for a variable were known, these properties would be calculated directly. Since they are usually not known, random samples are commonly used to estimate them via the sample mean and the sample standard deviation, respectively. The following is a brief introduction to the interpretation and calculation of these properties using a random sample of size n from the population.

The usual symbols are Greek letters for the true population measures and English letters for the sample estimates.

	True Population Measure		Sample Estimate	
	Symbol	Name	Symbol	Name
Expected value	μ or $E(X)$	Mu or true mean	$\overline{X}$	Sample mean
Standard deviation	σ or $\sqrt{\text{Var}(X)}$ or $\sqrt{\sigma^2}$	Sigma or true standard deviation	s or $\sqrt{s^2}$	Sample standard deviation

The **expected value** $E(X)$ is the long-run expected average if the variable is sampled many times.

The population expected value is not known exactly since the population itself is not known completely, so μ is estimated either by $E(X)$ from a distribution or by $\overline{X}$, the sample mean. Equation [18.2], repeated here as Equation [19.8], is used to compute the $E(X)$ of a probability distribution, and Equation [19.9] is the sample mean, also called the *sample average*.

Population: μ

Probability distribution: $E(X) = \Sigma X_i P(X_i)$ [19.8]

Sample: $\overline{X} = \dfrac{\text{sum of sample values}}{\text{sample size}}$

$$= \frac{\Sigma X_i}{n} = \frac{\Sigma f_i X_i}{n}$$ [19.9]

The f_i in the second form of Equation [19.9] is the frequency of X_i, that is, the number of times each value occurs in the sample. The resulting $\overline{X}$ is not necessarily an observed value of the variable; it is the long-run average value and can take on any value within the range of the variable. (We omit the subscript i on X and f when there is no confusion introduced.)

EXAMPLE 19.5

Jin Lee, an electrical engineer with Pacific Telecommunications, is planning to test several hypotheses about international roaming bills in North American and Asian countries. The variable of interest is X, the monthly bill in U.S. dollars (rounded to the nearest dollar). Two small and preliminary samples have been collected from different countries in North America and Asia. Estimate the population expected value. Do the samples (from a nonstatistical viewpoint) appear to be drawn from one population of bills or from two different populations?

North America, Sample 1, $	40	66	75	92	107	159	275
Asia, Sample 2, $	84	90	104	187	190		

Solution

Use Equation [19.9] for the sample mean.

Sample 1: $n = 7$ $\Sigma X_i = 814$ $\overline{X} = \$116.29$
Sample 2: $n = 5$ $\Sigma X_i = 655$ $\overline{X} = \$131.00$

Based solely on the preliminary sample averages, the approximate $15 difference, which is only 11% of the larger average bill, does not seem sufficiently large to conclude that the two populations are different. There are several statistical tests available to determine if samples come from the same or different populations. (Check a basic statistics text to learn about them.)

Comment

There are three commonly used measures of central tendency for data. The sample average is the most popular, but the *mode* and the *median* are also good measures. The mode, which is the most frequently observed value, was utilized in Example 19.3 for a triangular distribution. There is no specific mode in Jin Lee's two samples since all values are different. The *median is the middle value* of the sample. It is not biased by extreme sample values, as is the mean. The two medians in the samples are $92 and $104. Based solely on the medians, the conclusion is still that the samples do not necessarily come from two different populations of monthly roaming charges.

The **standard deviation** s or $s(X)$ is the dispersion or spread of values **about the expected value** $E(X)$ or sample average $\overline{X}$.

The sample standard deviation s estimates the property σ, which is the population measure of dispersion about the expected value of the variable. A probability distribution for data with strong central tendency is more closely clustered about the center of the data, and has a smaller s, than a wider, more dispersed distribution. In Figure 19–8, the samples with larger s values—s_1 and s_4—have a flatter, wider probability distribution.

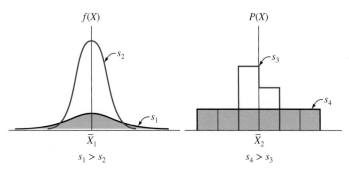

Figure 19–8
Sketches of continuous and discrete variable probability distributions with different standard deviation values; s_1 and s_4 are significantly larger than s_2 and s_3, respectively.

Actually, the variance s^2 is often quoted as the measure of dispersion. The standard deviation is simply the square root of the variance, so either measure can be used. The s value is what we use routinely in making computations about risk and probability. Mathematically, the formulas and symbols for variance and standard deviation of a discrete variable and a random sample of size n are as follows:

$$\text{Population:} \quad \sigma^2 = \text{Var}(X) \quad \text{and} \quad \sigma = \sqrt{\sigma^2} = \sqrt{\text{Var}(X)}$$

$$\text{Probability distribution:} \quad \text{Var}(X) = \Sigma[X_i - E(X)]^2 P(X_i) \qquad \text{[19.10]}$$

$$\text{Sample:} \quad s^2 = \frac{\text{sum of (sample value} - \text{sample average)}^2}{\text{sample size} - 1}$$

$$= \frac{\Sigma(X_i - \overline{X})^2}{n - 1} \qquad \text{[19.11]}$$

$$s = \sqrt{s^2}$$

Equation [19.11] for sample variance is usually applied in one of two more computationally convenient forms.

$$s^2 = \frac{\Sigma X_i^2}{n - 1} - \frac{n}{n - 1}\overline{X}^2 = \frac{\Sigma f_i X_i^2}{n - 1} - \frac{n}{n - 1}\overline{X}^2 \qquad \text{[19.12]}$$

The standard deviation uses the sample average as a basis about which to measure the spread or dispersion of data via the calculation $(X - \overline{X})$, which can have a minus or plus sign. To accurately measure the dispersion in both directions from the average, the quantity $(X - \overline{X})$ is squared. To return to the dimension of the variable itself, the square root of Equation [19.11] is extracted. The term $(X - \overline{X})^2$ is called the *mean-squared deviation,* and s has historically also been referred to as the *root-mean-square deviation.* The f_i in the second form of Equation [19.12] uses the frequency of each X_i value to calculate s^2.

One simple way to combine the average and standard deviation is to determine the percentage or fraction of the sample that is within ± 1, ± 2, or ± 3 standard deviations of the average, that is,

$$\overline{X} \pm ts \qquad \text{for } t = 1, 2, \text{ or } 3 \qquad \text{[19.13]}$$

In probability terms, this is stated as

$$P(\overline{X} - ts \leq X \leq \overline{X} + ts) \qquad \text{[19.14]}$$

Virtually all the sample values will always be within the $\pm 3s$ range of $\overline{X}$, but the percentage within $\pm 1s$ will vary depending on how the data points are distributed about $\overline{X}$. Example 19.6 illustrates the calculation of s to estimate σ and incorporates s with the sample average using $\overline{X} \pm ts$.

EXAMPLE 19.6

(*a*) Use the two samples of Example 19.5 to estimate the population variance and standard deviation for international roaming charges. (*b*) Determine the percentages of each sample that are inside the ranges of 1 and 2 standard deviations from the mean.

Solution

(*a*) For illustration purposes only, apply the two different relations to calculate s for the two samples. For sample 1 (North America) with $n = 7$, use X to identify the values. Table 19–3 presents the computation of $\Sigma(X - \overline{X})^2$ for Equation [19.11], with $\overline{X} = \$116.29$. The resulting s^2 and s values are

$$s^2 = \frac{37,743.40}{6} = 6290.57$$

$$s = \$79.31$$

TABLE 19–3	Computation of Standard Deviation Using Equation [19.11] with $\overline{X}$ = \$116.29, Example 19.6	
X, \$	$X - \overline{X}$	$(X - \overline{X})^2$
40	−76.29	5,820.16
66	−50.29	2,529.08
75	−41.29	1,704.86
92	−24.29	590.00
107	−9.29	86.30
159	+42.71	1,824.14
275	+158.71	25,188.86
814		37,743.40

TABLE 19–4	Computation of Standard Deviation Using Equation [19.12] with $\overline{Y}$ = \$131, Example 19.6
Y, \$	Y^2
84	7,056
90	8,100
104	10,816
187	34,969
190	36,100
655	97,041

For sample 2 (Asia), use Y to identify the values. With $n = 5$ and $\overline{Y} = 131$, Table 19–4 shows ΣY^2 for Equation [19.12]. Then

$$s^2 = \frac{97,041}{4} - \frac{5}{4}(131)^2 = 42,260.25 - 1.25(17,161) = 2809$$

$$s = \$53$$

The dispersion is smaller for the Asia sample (\$53) than for the North America sample (\$79.31).

(b) Equation [19.13] determines the ranges of $\overline{X} \pm 1s$ and $\overline{X} \pm 2s$. Count the number of sample data points between the limits, and calculate the corresponding percentage. See Figure 19–9 for a plot of the data and the standard deviation ranges.

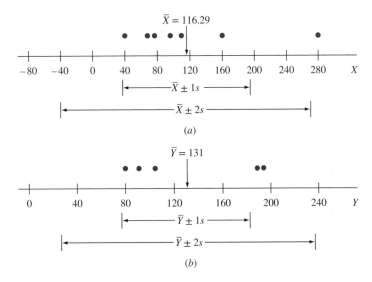

(a)

(b)

Figure 19–9
Values, averages, and standard deviation ranges for (a) North America, and (b) Asia samples, Example 19.6.

North America sample

$$\bar{X} \pm 1s = 116.29 \pm 79.31 \qquad \text{for a range of \$36.98 to \$195.60}$$

Six out of seven values are within this range, so the percentage is 85.7%.

$$\bar{X} \pm 2s = 116.29 \pm 158.62 \qquad \text{for a range of \$$-$42.33 to \$274.91}$$

There are still six of the seven values within the $\bar{X} \pm 2s$ range. The limit \$$-$42.33 is meaningful only from the probabilistic perspective; from the practical viewpoint, use zero, that is, no amount billed.

Asia sample

$$\bar{Y} \pm 1s = 131 \pm 53 \qquad \text{for a range of \$78 to \$184}$$

There are three of five values, or 60%, within the range.

$$\bar{Y} \pm 2s = 131 \pm 106 \qquad \text{for a range of \$25 to \$237}$$

All five of the values are within the $\bar{Y} \pm 2s$ range.

Comment

A second common measure of dispersion is the *range,* which is simply the largest minus the smallest sample values. In the two samples here, the range estimates are \$235 and \$106.

Only the hand computations for $E(X)$, s and s^2 have been demonstrated here. Calculators and spreadsheets all have functions to determine these values by simply entering the data.

Before we perform simulation analysis in engineering economy, it may be of use to summarize the expected value and standard deviation relations for a continuous variable, since Equations [19.8] through [19.12] address only discrete variables. The primary differences are that the summation symbol is replaced by the integral over the defined range of the variable, which we identify as R, and that $P(X)$ is replaced by the differential element $f(X)\,dX$. For a stated continuous probability distribution $f(X)$, the formulas are

Expected value: $\qquad\qquad E(X) = \int_R Xf(X)\,dX$ $\qquad\qquad$ [19.15]

Variance: $\qquad\qquad \text{Var}(X) = \int_R X^2 f(X)\,dX - [E(X)]^2$ $\qquad\qquad$ [19.16]

For a numerical example, again use the uniform distribution in Example 19.3 (Figure 19–4) over the range R from \$10 to \$15. If we identify the variable as X, rather than C_1, the following are correct:

$$f(X) = \frac{1}{5} = 0.2 \qquad \$10 \leq X \leq \$15$$

$$E(X) = \int_R X(0.2)\,dX = 0.1X^2 \Big|_{10}^{15} = 0.1(225 - 100) = \$12.5$$

$$\text{Var}(X) = \int_R X^2(0.2)\,dX - (12.5)^2 = \frac{0.2}{3}X^3 \Big|_{10}^{15} - (12.5)^2$$

$$= 0.06667(3375 - 1000) - 156.25 = 2.08$$

$$\sigma = \sqrt{2.08} = \$1.44$$

Therefore, the uniform distribution between $L = \$10$ and $H = \$15$ has an expected value of \$12.5 (the midpoint of the range, as expected) and a standard deviation of \$1.44.

EXAMPLE 19.7

Christy is the regional safety engineer for a chain of franchise-based gasoline and food stores. The home office has had many complaints and several legal actions from employees and customers about slips and falls due to liquids (water, oil, gas, soda, etc.) on concrete surfaces.

Corporate management has authorized each regional engineer to contract locally to apply to all exterior concrete surfaces a newly marketed product that absorbs up to 100 times its own weight in liquid and to charge a home office account for the installation. The authorizing letter to Christy states that, based upon their simulation and random samples that assume a normal population, the cost of the locally arranged installation should be about $10,000 and almost always is within the range of $8000 to $12,000.

Since Christy has a BS in applied math and you have a BS in engineering, you have been asked to write a brief but thorough summary about the normal distribution, explain the $8000 to $12,000 range statement, and explain the phrase "random samples that assume a normal population."

Solution

The following summaries about the normal distribution and sampling will help explain the authorization letter.

Normal distribution, probabilities, and random samples

The normal distribution is also referred to as the *bell-shaped curve,* the *Gaussian distribution*, or the *error distribution*. It is, by far, the most commonly used probability distribution in all applications. It places exactly one-half of the probability on either side of the mean or expected value. It is used for continuous variables over the entire range of numbers. The normal distribution is found to accurately predict many types of outcomes, such as IQ values; manufacturing errors about a specified size, volume, weight, etc.; and the distribution of sales revenues, costs, and many other business parameters around a specified mean, which is why it may apply in this situation.

The normal distribution, identified by the symbol $N(\mu, \sigma^2)$, where μ is the expected value or mean and σ^2 is the variance, or measure of spread, can be described as follows:

- The mean μ locates the probability distribution (Figure 19–10a), and the spread of the distribution varies with variance (Figure 19–10b), growing wider and flatter for larger variance values.
- When a sample is taken, the estimates are identified as sample mean $\overline{X}$ for μ and sample standard deviation s for σ.
- The normal probability distribution $f(X)$ for a variable X is quite complicated because its formula is

$$f(X) = \frac{1}{\sigma\sqrt{2\pi}} \exp\left\{-\left[\frac{(X - \mu)^2}{2\sigma^2}\right]\right\}$$

where exp represents the number $e = 2.71828$.

Since $f(X)$ is so unwieldy, random samples and probability statements are developed using a transformation, called the *standard normal distribution (SND),* which uses μ and σ (population) or $\overline{X}$ and s (sample) to compute values of the variable Z.

Population : $Z = \dfrac{\text{deviation from mean}}{\text{standard deviation}} = \dfrac{X - \mu}{\sigma}$ [19.17]

Sample: $Z = \dfrac{X - \overline{X}}{s}$ [19.18]

The SND for Z (Figure 19–10c) is the same as for X, except that it always has a mean of 0 and a standard deviation of 1, and it is identified by the symbol $N(0, 1)$. Therefore, the probability values under the SND curve can be stated exactly. It is always possible to transfer back to the original values from sample data by solving Equation [19.17] for X:

$$X = Z\sigma + \mu \qquad\qquad [19.19]$$

Figure 19–10
Normal distribution
showing (a) different
mean values μ,
(b) different standard
deviation values σ,
and (c) relation of
normal X to standard
normal Z.

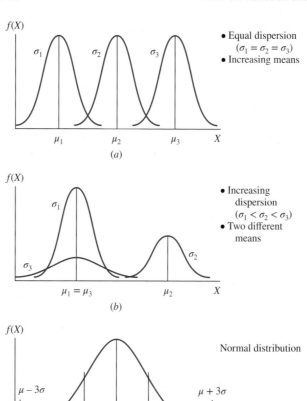

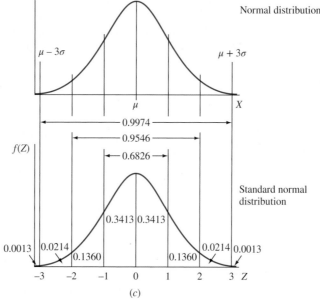

Several probability statements for Z and X are summarized in the following table and are shown on the distribution curve for Z in Figure 19–10c.

Variable X Range	Probability	Variable Z Range
$\mu + 1\sigma$	0.3413	0 to +1
$\mu \pm 1\sigma$	0.6826	−1 to +1
$\mu + 2\sigma$	0.4773	0 to +2
$\mu \pm 2\sigma$	0.9546	−2 to +2
$\mu + 3\sigma$	0.4987	0 to +3
$\mu \pm 3\sigma$	0.9974	−3 to +3

As an illustration, probability statements from this tabulation and Figure 19–10c for X and Z are as follows:

The probability that X is within 2σ of its mean is 0.9546.

The probability that Z is within 2σ of its mean, which is the same as between the values −2 and +2, is also 0.9546.

In order to take a random sample from a normal $N(\mu, \sigma^2)$ population, a specially prepared table of SND random numbers is used. (Tables of SND values are available in many statistics books.) The numbers are actually values from the Z or $N(0,1)$ distribution and have values such as -2.10, $+1.24$, etc. Translation from the Z value back to the sample values for X is via Equation [19.19].

Interpretation of the home office memo

The statement that virtually all the local contract amounts should be between \$8000 and \$12,000 may be interpreted as follows: A normal distribution is assumed with a mean of $\mu = $10,000$ and a standard deviation for $\sigma = \$667$, or a variance of $\sigma^2 = (\$667)^2$; that is, an $N[\$10,000, (\$667)^2]$ distribution is assumed. The value $\sigma = \$667$ is calculated using the fact that virtually all the probability (99.74%) is within 3σ of the mean, as stated above. Therefore,

$$3\sigma = \$2000 \quad \text{and} \quad \sigma = \$667 \quad \text{(rounded off)}$$

As an illustration, if six SND random numbers are selected and used to take a sample of size 6 from the normal distribution $N[\$10,000, (\$667)^2]$, the results are as follows:

SND Random Number Z	X Using Equation [19.19] $X = Z\sigma + \mu$
-2.10	$X = (-2.10)(667) + 10,000 = \$8599$
$+3.12$	$X = (+3.12)(667) + 10,000 = \$12,081$
-0.23	$X = (-0.23)(667) + 10,000 = \$9847$
$+1.24$	$X = (+1.24)(667) + 10,000 = \$10,827$
-2.61	$X = (-2.61)(667) + 10,000 = \$8259$
-0.99	$X = (-0.99)(667) + 10,000 = \$9340$

In this sample of six typical concrete surfacing contract amounts for sites in our region, the average is \$9825 and five of six values are within the range of \$8000 to \$12,000, with the sixth being only \$81 above the upper limit.

19.5 Monte Carlo Sampling and Simulation Analysis ●●◉

Up to this point, all alternative selections have been made using estimates with certainty, possibly followed by some testing of the decision via sensitivity analysis or expected values. In this section, we will use a simulation approach that incorporates the material of the previous sections to facilitate the engineering economy decision about one alternative or between two or more alternatives.

The random sampling technique discussed in Section 19.3 is called **Monte Carlo sampling.** The general procedure outlined below uses Monte Carlo sampling to obtain samples of size n for selected parameters of formulated alternatives. These parameters, expected to vary according to a stated probability distribution, require decision making under risk. All other parameters in an alternative are considered certain; that is, they are known, or they can be estimated with enough precision to consider them certain. An important assumption is made, usually without realizing it.

All parameters are **independent;** that is, one variable's distribution does not affect the value of any other variable of the alternative. This is referred to as the property of **independent random variables.**

The simulation approach to engineering economy analysis is summarized in the following basic steps:

Step 1. Formulation of alternative(s). Set up each alternative in the form to be considered using engineering economic analysis, and select the measure of worth upon which to base the decision. Determine the form of the relation(s) to calculate the measure of worth.

Step 2. Parameters with variation. Select the parameters in each alternative to be treated as random variables. Estimate values for all other (certain) parameters for the analysis.

Step 3. **Probability distributions.** Determine whether each variable is discrete or continuous, and describe a probability distribution for each variable in each alternative. Use standard distributions, where possible, to simplify the sampling process and to prepare for spreadsheet-based simulation.

Step 4. **Random sampling.** Incorporate the random sampling procedure of Section 19.3 (the first four steps) into this procedure. This results in the cumulative distribution, assignment of RNs, selection of the RNs, and a sample of size n for each variable.

Step 5. **Measure of worth calculation.** Compute n values of the selected measure of worth from the relation(s) determined in step 1. Use the estimates made with certainty and the n sample values for the varying parameters. (This is when the property of independent random variables is actually applied.)

Step 6. **Measure of worth description.** Construct the probability distribution of the measure of worth, using between 10 and 20 cells of data, and calculate measures such as $\overline{X}$, s, $\overline{X} \pm ts$, and relevant probabilities.

Step 7. **Conclusions.** Draw conclusions about each alternative, and decide which is to be selected. If the alternative(s) has(have) been previously evaluated under the assumption of certainty for all parameters, comparison of results may help with the final decision.

Example 19.8 illustrates this procedure using an abbreviated manual simulation analysis, and Example 19.9 utilizes spreadsheet simulation for the same estimates.

EXAMPLE 19.8

Yvonne Ramos is the CEO of a chain of 50 fitness centers in the United States and Canada. An equipment salesperson has offered Yvonne two long-term opportunities on new aerobic exercise systems, for which the usage is charged to customers on a per-use basis on top of the monthly fees paid by customers. As an enticement, the offer includes a guarantee of annual revenue for one of the systems for the first 5 years.

Since this is an entirely new and risky concept of revenue generation, Yvonne wants to do a careful analysis of each alternative. Details for the two systems follow:

System 1. First cost is $P = \$12,000$ for a set period of $n = 7$ years with no salvage value. No guarantee for annual net revenue is offered.

System 2. First cost is $P = \$8000$, there is no salvage value, and there is a guaranteed annual net revenue of $1000 for each of the first 5 years, but after this period, there is no guarantee. The equipment with updates may be useful up to 15 years, but the exact number is not known. Cancellation anytime after the initial 5 years is allowed, with no penalty.

For either system, new versions of the equipment will be installed with no added costs. If the MARR is 15% per year, use PW analysis to determine if neither, one, or both of the systems should be installed.

Solution by Hand

Estimates that Yvonne makes to use the simulation analysis procedure are included in the following steps.

Step 1. **Formulation of alternatives.** Using PW analysis, the relations for system 1 and system 2 are developed. The symbol NCF identifies the net cash flows (revenues), and NCF_G is the guaranteed NCF of $1000 for system 2.

$$PW_1 = -P_1 + NCF_1(P/A,15\%,n_1) \quad\quad [19.20]$$

$$PW_2 = -P_2 + NCF_G(P/A,15\%,5) \\ + NCF_2(P/A,15\%,n_2-5)(P/F,15\%,5) \quad\quad [19.21]$$

Step 2. **Parameters with variation.** Yvonne summarizes the parameters estimated with certainty and makes distribution assumptions about three parameters treated as random variables.

System 1

Certainty. $P_1 = \$12,000$; $n_1 = 7$ years.

Variable. NCF_1 is a continuous variable, uniformly distributed between $L = \$-4000$ and $H = \$6000$ per year, because this is considered a high-risk venture.

System 2

Certainty. $P_2 = \$8000$; $NCF_G = \$1000$ for first 5 years.

Variable. NCF_2 is a discrete variable, uniformly distributed over the values $L = \$1000$ to $H = \$6000$ only in $\$1000$ increments, that is, $\$1000$, $\$2000$, etc.

Variable. n_2 is a continuous variable that is uniformly distributed between $L = 6$ and $H = 15$ years.

Now, rewrite Equations [19.20] and [19.21] to reflect the estimates made with certainty.

$$\begin{aligned} PW_1 &= -12,000 + NCF_1(P/A,15\%,7) \\ &= -12,000 + NCF_1(4.1604) \end{aligned} \qquad [19.22]$$

$$\begin{aligned} PW_2 &= -8000 + 1000(P/A,15\%,5) \\ &\quad + NCF_2(P/A,15\%,n_2{-}5)(P/F,15\%,5) \\ &= -4648 + NCF_2(P/A,15\%,n_2{-}5)(0.4972) \end{aligned} \qquad [19.23]$$

Step 3.   Probability distributions.   Figure 19–11 (left side) shows the assumed probability distributions for NCF_1, NCF_2, and n_2.

Step 4.   Random sampling.   Yvonne decides on a sample of size 30 and applies the first four of the random sample steps in Section 19.3. Figure 19–11 (right side) shows the cumulative distributions (step 1) and assigns RNs to each variable (step 2). The RNs for NCF_2 identify the *x*-axis values so that all net cash flows will be in even $\$1000$ amounts. For the continuous variable n_2, three-digit RN values are used to make the numbers come out evenly, and they are shown in cells only as "indexers" for easy reference when a RN is used to find a variable value. However, we round the number to the next higher value of n_2 because it is likely the contract may be canceled on an anniversary date. Also, now the tabulated compound interest factors for $(n_2 - 5)$ years can be used directly (see Table 19–5).

   Once the first RN is selected randomly from Table 19–2, the sequence (step 3) used will be to proceed down the RN table column and then up the column to the left. Table 19–5 shows only the first five RN values selected for each sample and the corresponding variable values taken from the cumulative distributions in Figure 19–11 (step 4).

Step 5.   Measure of worth calculation.   With the five sample values in Table 19–5, calculate the PW values using Equations [19.22] and [19.23].

1. $PW_1 = -12,000 + (-2200)(4.1604)$ $= \$-21,153$
2. $PW_1 = -12,000 + 2000(4.1604)$ $= \$-3679$
3. $PW_1 = -12,000 + (-1100)(4.1604)$ $= \$-16,576$
4. $PW_1 = -12,000 + (-900)(4.1604)$ $= \$-15,744$
5. $PW_1 = -12,000 + 3100(4.1604)$ $= \$+897$

1. $PW_2 = -4648 + 1000(P/A,15\%,7)(0.4972)$ $= \$-2579$
2. $PW_2 = -4648 + 1000(P/A,15\%,5)(0.4972)$ $= \$-2981$
3. $PW_2 = -4648 + 5000(P/A,15\%,8)(0.4972)$ $= \$+6507$
4. $PW_2 = -4648 + 3000(P/A,15\%,10)(0.4972)$ $= \$+2838$
5. $PW_2 = -4648 + 4000(P/A,15\%,3)(0.4972)$ $= \$-107$

Now, 25 more RNs are selected for each variable from Table 19–2, and the PW values are calculated.

Step 6.   Measure of worth description.   Figure 19–12*a* and *b* presents the PW_1 and PW_2 probability distributions for the 30 samples with 14 and 15 cells, respectively, as well as the range of individual PW values and the $\overline{X}$ and s values.

Figure 19-11
Distributions used for random samples, Example 19.8.

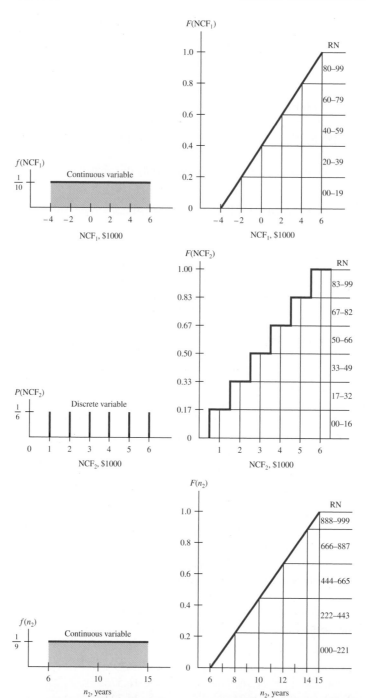

TABLE 19–5	Random Numbers and Variable Values for NCF$_1$, NCF$_2$, and n_2, Example 19.8					
	NCF$_1$		**NCF$_2$**		**n_2**	
RN*	**Value, $**	**RN†**	**Value, $**	**RN‡**	**Value, Year**	**Rounded§**
18	−2200	10	1000	586	11.3	12
59	+2000	10	1000	379	9.4	10
31	−1100	77	5000	740	12.7	13
29	−900	42	3000	967	14.4	15
71	+3100	55	4000	144	7.3	8

*Randomly start with row 1, column 4 in Table 19–2.
†Start with row 6, column 14.
‡Start with row 4, column 6.
§The n_2 value is rounded up.

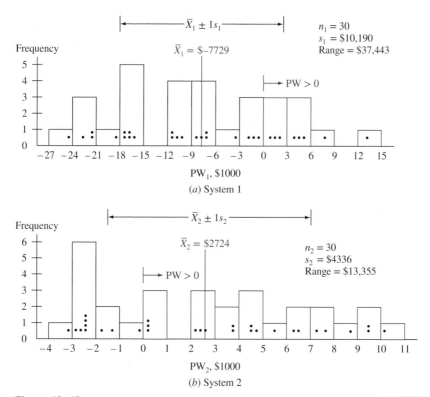

Figure 19–12
Probability distributions of simulated PW values for a sample of size 30, Example 19.8.

PW$_1$. Sample values range from $\$-24,481$ to $\$+12,962$. The calculated measures of the 30 values are

$$\overline{X}_1 = \$-7729$$
$$s_1 = \$10,190$$

PW$_2$. Sample values range from $\$-3031$ to $\$+10,324$. The sample measures are

$$\overline{X}_2 = \$2724$$
$$s_2 = \$4336$$

Step 7. **Conclusions.** Additional sample values will surely make the central tendency of the PW distributions more evident and may reduce the s values, which are quite large. Of course, many conclusions are possible once the PW distributions are known, but the following seem clear at this point.

> **System 1.** Based on this small sample of 30 observations, *do not accept* this alternative. The likelihood of making the MARR = 15% is relatively small, since the sample indicates a probability of 0.27 (8 out of 30 values) that the PW will be positive, and $\overline{X}_1$ is a large negative. Though appearing large, the standard deviation may be used to determine that about 20 of the 30 sample PW values (two-thirds) are within the limits $\overline{X} \pm 1s$, which are $\$-17,919$ and $\$2461$. A larger sample may alter this analysis somewhat.

> **System 2.** If Yvonne is willing to accept the longer-term commitment that may increase the NCF some years out, the sample of 30 observations indicates to *accept* this alternative. At a MARR of 15%, the simulation approximates the chance for a positive PW as 67% (20 of the 30 PW values in Figure 19–12b are positive). However, the probability of observing PW within the $\overline{X} \pm 1s$ limits ($\$-1612$ and $\$7060$) is 0.53 (16 of 30 sample values).

> **Conclusion at this point.** Reject system 1; accept system 2; and carefully watch net cash flow, especially after the initial 5-year period.

Comment

The estimates in Example 13.5 are very similar to those here, except all estimates were made with certainty ($NCF_1 = \$3000$, $NCF_2 = \$3000$, and $n_2 = 14$ years). The alternatives were evaluated by the payback period method at MARR = 15%, and the first alternative was selected. However, the subsequent PW analysis in Example 13.5 selected alternative 2 based, in part, upon the anticipated larger cash flow in the later years.

EXAMPLE 19.9

Help Yvonne Ramos set up a spreadsheet simulation for the three random variables and PW analysis in Example 19.8. Does the PW distribution vary appreciably from that developed using manual simulation? Do the decisions to reject the system 1 proposal and accept the system 2 proposal still seem reasonable?

Solution by Spreadsheet

Figures 19–13 and 19–14 are spreadsheet screen shots that accomplish the simulation portion of the analysis described above in steps 3 (determine probability distribution) through 6 (measure of worth description). Most spreadsheet systems are limited in the variety of distributions they can accept for sampling, but common ones such as uniform and normal are available.

Figure 19–13 shows the results of a small sample of 30 values from the three distributions using the RAND and IF functions. (See Section A.3 in Appendix A.)

NCF_1: Continuous uniform from -4000 to $\$6000$. The spreadsheet relation in column B translates RN1 values (column A) into NCF1 amounts.

NCF_2: Discrete uniform in $\$1000$ increments from $\$1000$ to $\$6000$. Column D cells display NCF2 in the $\$1000$ increments using the logical IF function to translate from the RN2 values.

n_2: Continuous uniform from 6 to 15 years. The results in column F are integer values obtained using the INT function operating on the RN3 values.

	A	B	C	D	E	F
1			Sample of size 30 simulated values			
2	RN1	NCF1, $	RN2	NCF2, $	RN3	N2, years
3	12.5625	-2,800	83.6176	6,000	556.2768	12
4	25.0262	-1,500	99.5425	6,000	8.7883	7
5	9.3856	-3,100	26.4693	2,000	507.3598	11
6	38.0199	-200	36.8475	3,000	681.5397	13
7	71.5088	3,100	83.4610	6,000	369.0917	10
8	66.7820	2,600	77.8699	5,000	91.3044	7
9	48.3324	800	8.4308	1,000	457.7487	11
10	39.3886	-100	52.8630	4,000	914.5432	15
11	21.5429	-1,900	57.4819	4,000	698.7624	13
12	44.4996	400	1.9322	1,000	744.2622	13
13	32.9911	-800	70.6307	5,000	190.8139	8
14	96.0249	5,600	61.0023	4,000	714.6685	13
15	99.6675	5,900	55.7741	4,000	648.2268	12
16	13.9560	-2,700	98.9107	6,000	199.9491	8
17	99.8535	5,900	10.7429	1,000	716.5830	13
18	63.2953	2,300	4.6540	1,000	133.4986	8
19	93.0860	5,300	56.7425	4,000	553.2489	11
20	52.6539	1,200	17.1873	2,000	809.5778	14
21	34.1609	-600	46.3758	3,000	810.4792	14
22	86.0288	4,600	99.6569	6,000	950.9657	15
23	46.9626	600	24.9754	2,000	0.9088	7
24	77.4690	3,700	52.2862	4,000	339.4470	10
25	28.5200	-1,200	52.3113	4,000	514.9377	11
26	82.9615	4,200	99.0275	6,000	912.6720	15
27	17.8793	-2,300	50.9493	4,000	800.4352	14
28	89.2411	4,900	6.2988	1,000	118.2531	8
29	9.0495	-3,100	54.6552	4,000	56.4377	7
30	1.4597	-3,900	24.6463	2,000	716.2222	13
31	66.2177	2,600	6.2064	1,000	91.4505	7
32	98.4803	5,800	94.8061	6,000	120.0419	8
33	`= RAND()*100`	`= INT((100*A32 -4000)/100)*100`	`= RAND()*100`	`= IF(C32<=16,1000, IF(C32<=32,2000, IF(C32<=49,3000, IF(C32<=66,4000, IF(C32<=82,5000, IF(C32<=100,6000,6000))))))`	`= RAND()*1000`	`= INT(0.009* E32+1)+6`

H ◄ ► H   Fig 19-13 Random Numbers   Fig 19-14 PW values   Sheet6   Sheet8   Sheet9   Sheet10   Sheet11   Sheet

Figure 19–13
Random sample of 30 values generated for spreadsheet simulation, Example 19.9.

	A	B	C	D	E	F	G	H	I
1				Information about Alternatives					
2	System 1, P1	$12,000		System 2, P2		$8,000			
3	n	7	Years		NCF	$1,000	5	Years	
4	MARR	15%			MARR	15%			
5								=SUM(G13:G42)	
6			Results of Analysis		=AVERAGE(F13:F42)				
7	# PW ≥ MARR		10				19		
8	# PW < MARR		20					11	
9	PW Average, $	-7,105				1,649			
10	PW Std Dev, $	13,199				3,871	=STDEV(F13:F42)		
11									
12			Present Worth Computations						
13	PW1	-23,649	0	1	PW2	7,763	1	0	
14	PW1	-18,241	0	1	PW2	202	1	0	
15	PW1	-24,897	0	1	PW2	-885	0	1	
16	PW1	-12,832	0	1	PW2	2,045	1	0	
17	PW1	897	1	0	PW2	5,352	1	0	
18	PW1	-1,183	0	1	PW2	-607	0	1	
19	PW1	-8,672	0	1	PW2	-2,766	0	1	
20	PW1	-12,416	0	1	PW2	5,333	1	0	
21	PW1	-19,905	0	1	PW2	4,276	1	0	
22	PW1	-10,336	0	1	PW2	-2,417	0	1	
23	PW1	-15,328	0	1	PW2	1,028	1	0	
24	PW1	11,298	1	0	PW2	4,276	1	0	
25	PW1	12,546	1	0	PW2	3,626	1	0	
26	PW1	-23,233	0	1	PW2	2,163	1	0	
27	PW1	12,546	1	0	PW2	-2,417	0	1	
28	PW1	-2,431	0	1	PW2	-3,513	0	1	
29	PW1	10,050	1	0	PW2	2,878	1	0	
30	PW1	-7,007	0	1	PW2	97	1	0	
31	PW1	-14,496	0	1	PW2	2,469	1	0	
32	PW1	7,138	1	0	PW2	10,323	1	0	
33	PW1	-9,504	0	1	PW2	-3,031	0	1	
34	PW1	3,394	1	0	PW2	2,019	1	0	
35	PW1	-16,993	0	1	PW2	2,878	1	0	
36	PW1	5,474	1	0	PW2	10,323	1	0	
37	PW1	-21,569	0	1	PW2	4,841	1	0	
38	PW1	8,386	1	0	PW2	-3,513	0	1	
39	PW1	-24,897	0	1	PW2	-1,415	0	1	
40	PW1	-28,226	0	1	PW2	-186	0	1	
41	PW1	-1,183	0	1	PW2	-3,840	0	1	
42	PW1	12,130	1	0	PW2	2,163	1	0	
43	Functions for last row	'= (PV($B$4,$B$3, -'Random Numbers'!B32))-$B$2	'= IF(B42 >= 0,1,0)	'= IF(B42 < 0,1,0)	'= -$F$2-PV($F$4,$G$3,$F$3) +PV($F$4,$G$3,,(PV($F$4, 'Random Numbers'!F32-$G$3, 'Random Numbers'!D32)))		'= IF(F42 >= 0,1,0)	'= IF(F42 < 0,1,0)	

H ◀ ▶ H   Fig 19-13 Random Numbers   **Fig 19-14 PW values**   Sheet6   Sheet8   Sheet9   Sheet10   Sheet11   Sheet

Figure 19–14
Simulation results for 30 PW values, Example 19.9.

Figure 19–14 presents the two alternatives' estimates in the top section. The PW1 and PW2 computations for the 30 repetitions of NCF1, NCF2, and N2 are the spreadsheet equivalent of Equations [19.22] and [19.23]. The tabular approach used here tallies the number of PW values below zero ($0) and equal to or exceeding zero using the IF operator. For example, cell C17 contains a 1, indicating PW1 > 0 when NCF1 = $3100 (in cell B7 of Figure 19–13), which was used to calculate PW1 = $897 by Equation [19.22]. Cells in rows 7 and 8 show the number of times in the 30 samples that system 1 and system 2 may return at least the MARR = 15% because the corresponding PW ≥ 0. Sample averages and standard deviations are also indicated.

Comparison between the hand and spreadsheet simulations is presented below.

	System 1 PW			System 2 PW		
	$\bar{X}$, $	s, $	No. of PW ≥ 0	$\bar{X}$, $	s, $	No. of PW ≥ 0
Hand	−7,729	10,190	8	2,724	4,336	20
Spreadsheet	−7,105	13,199	10	1,649	3,871	19

For the spreadsheet simulation, 10 (33%) of the PW1 values exceed zero, while the manual simulation included 8 (27%) positive values. These comparative results will change every time this spreadsheet is activated since the RAND function is set up (in this case) to produce a new RN each time. (It is possible to define RAND to keep the same RN values. See the Excel User's Guide.)

The conclusion to reject the system 1 proposal and accept system 2 is still appropriate for the spreadsheet simulation as it was for the hand solution, since there are comparable chances that PW ≥ 0.

CHAPTER SUMMARY

To perform decision making under risk implies that some parameters of an engineering alternative are treated as random variables. Assumptions about the shape of the variable's probability distribution are used to explain how the estimates of parameter values may vary. Additionally, measures such as the expected value and standard deviation describe the characteristic shape of the distribution. In this chapter, we learned several of the simple, but useful, discrete and continuous population distributions used in engineering economy—uniform and triangular—as well as specifying our own distribution or assuming the normal distribution.

Since the population's probability distribution for a parameter is not fully known, a random sample of size n is usually taken, and its sample average and standard deviation are determined. The results are used to make probability statements about the parameter, which help make the final decision with risk considered.

The Monte Carlo sampling method is combined with engineering economy relations for a measure of worth such as PW to implement a simulation approach to risk analysis. The results of such an analysis can then be compared with decisions when parameter estimates are made with certainty.

PROBLEMS

Certainty, Risk, and Uncertainty

19.1 Identify the following variables as either discrete or continuous:
(a) The number of times heads comes up in 100 tosses of a coin
(b) The number of accidents occurring in a specified section of a freeway
(c) The weight of boxes shipped from an Amazon warehouse
(d) The concentration of carbon dioxide in the air of San Diego versus time
(e) Optimistic, most likely, and pessimistic estimates of salvage value

19.2 For each situation below, determine (1) if the variable(s) is(are) discrete or continuous, and (2) if the information involves certainty, risk, and/or uncertainty.
(a) A friend in real estate tells you the price per square foot for new houses will go up slowly or rapidly during the next 6 months.
(b) Your manager informs the staff that there is an equal chance that sales will be between 50 and 55 units next month.
(c) Jane got paid yesterday and $800 was taken out in income taxes. The amount withheld next month will be larger because of a pay raise between 3% and 5%.
(d) There is a 20% chance of rain and a 30% chance of snow today.
(e) The first cost of a new front-end loader is $34,000 or $38,000 depending on the size purchased.

19.3 In the recent past, the production output has been between 10,000 and 20,000 units per week 75% of the time; however, it may fall below 10,000 or go above 20,000 more frequently in the future. The production manager wants to use E(output) to make a decision about upgrading automation software. Identify at least two additional pieces of information that must be obtained or assumed to finalize the output information for this use.

Probability and Distributions

19.4 The cost of flood damage from significant storms varies as a function of the severity of the storm. Estimate the expected flood damage due to the next significant storm.

Flood Flow, Million Acre-ft	Flood Damage, $ Millions	Probability
0.5	19	0.35
1.0	41	0.36
1.5	97	0.20
>2.0	210	0.09

19.5 AAA car-buying service surveyed 1000 households to determine the number of operating vehicles owned by residents at the same address. Use the results below to estimate the percentage of households that own (a) one or less vehicles, (b) one or two vehicles, and (c) more than three vehicles.

Number of Cars, C	Number of Households
0	120
1	560
2	260
3	32
4	22
5 or more	6

19.6 Royalties received by an investor in an oil well vary according to the price of oil. Data collected from stripper wells in a West Texas oilfield were used to develop the royalty probability relationships shown.
  (a) Is the variable discrete or continuous as shown?
  (b) What is the expected value of royalty income, $E(RI)$, per year
  (c) What are the chances that royalty income will be at least $10,500 per year?
  (d) Use a *spreadsheet* to plot the probability distribution of RI with the information shown.

Royalties, $/Year	6000	8500	9500	10,500	12,500	15,500
Probability	0.10	0.21	0.32	0.24	0.09	0.04

19.7 An engineer collected monthly operating expenses for the past 3 years for a micro-finishing department. Use the midpoints of each range in preparing your answers.
  (a) Develop a plot of the probability distribution by *hand* or using a *spreadsheet*, as instructed.
  (b) What is the probability that a month's expense may be above $40,000?
  (c) What is the probability that a month's expense may be $35,000?
  (d) What is the mode of the probability distribution?

Expense Range, $1000	1–10	10–20	20–30	30–40	40–50	50–60	60–70
Midpoint, $1000	5	15	25	35	45	55	65
Number of Months	2	5	8	7	6	5	3

19.8 A total of 100 maintenance costs, C, for ground-mounted transformers in high-density housing areas were collected by the City Electricity Authority. The costs are clustered into $200 cells with midpoints ranging from $600 to $2000. The number of times (frequency) each cell value was observed and its probability are shown. Find the expected value of the costs, $E(C)$, for use in a PW analysis.

Cell Midpoint, C_T	Frequency	Probability
600	6	0.06
800	10	0.10
1000	9	0.09
1200	15	0.15
1400	28	0.28
1600	15	0.15
1800	7	0.07
2000	10	0.10

19.9 Over a 1-week period, an officer of the state lottery commission sampled ticket purchasers at a single high-traffic location. The amounts labeled W for winnings, distributed back to the purchasers/winners, and the associated probabilities for 5000 tickets are shown below.
  (a) Is this variable discrete or continuous? Plot the cumulative distribution of W by *hand* or *spreadsheet*, as instructed.
  (b) Calculate the expected value of W per ticket.
  (c) If tickets cost $2, what is the expected long-term income to the state per ticket, based upon this sample?

Winnings W, $	0	2	5	10	100
Probability	0.95	0.025	0.015	0.0093	0.0007

19.10 Bob is working on two separate probability-related projects. The first involves the variable N, which is the number of consecutively filled bottles of an anticancer drug that weigh in above the weight specification limit. The variable N is described by the formula $(0.5)^N$ because each unit has a 50-50 chance of being below or above the limit. The second involves the battery life L, which varies between 2 and 5 months. The probability distribution is triangular with the mode at 5 months, which is the design life. Some batteries fail early, but 2 months is the smallest life experienced thus far.
  (a) Write out and plot the probability distributions and cumulative distributions for N and L for Bob.
  (b) Determine $P(N = 1, 2,$ or $3)$ consecutive units above the weight limit.

19.11 Carla is a statistician with a bank. She has collected debt-to-equity mix data on mature (M) and young (Y) companies. The debt percentages vary from 20% to 80%. Carla has defined D_M as a variable for the mature companies from 0 to 1, with $D_M = 0$ interpreted as the low of 20% debt and $D_M = 1.0$ as the high of 80% debt. The variable for young corporation debt percentages D_Y is similarly defined. The probability distributions used to describe D_M and D_Y are

$$f(D_M) = 3(1 - D_M)^2 \qquad 0 \le D_M \le 1$$
$$f(D_Y) = 2D_Y \qquad 0 = D_Y \le 1$$

  (a) Use different values of the debt percentage between 20% and 80% to calculate values for the probability distributions and then plot them.
  (b) Comment on the probability that a mature company or a young company will have a low debt percentage? A high debt percentage?

19.12 A discrete variable X can take on integer values of 1 to 10. A sample of size 50 results in the following probability estimates:

X_i	1	2	3	6	9	10
$P(X_i)$	0.2	0.2	0.2	0.1	0.2	0.1

(a) Write out and graph the cumulative distribution.

(b) Calculate the following probabilities using the cumulative distribution: (1) X is between 6 and 10, and (2) X takes the value 4, 5, or 6.

(c) Use the cumulative distribution to show that $P(X = 7 \text{ or } 8) = 0.0$. Even though this probability is zero, the statement about X is that it can take on integer values of 1 to 10. How do you explain the apparent contradiction in these two statements?

Random Samples

19.13 The net revenue R for a newly developed antifungal spray for citrus fruit is expected to be $3.1 million per year for the next 5 years. This expected amount is based on estimates that R could be $2.6, $2.8, $3.0, $3.2, $3.4 million, all with the same probability, or $3.6 million with half the probability of the other values. (a) Write the probability statements for each of the estimates. (b) Determine the expected value of R.

19.14 CINESA, a government-owned power company that normally uses natural gas for electricity generation, is purchasing fuels other than natural gas and power from a commercially built wind farm, often at extra costs, which are transferred to the customer. Total monthly fuel and wind-power costs are now averaging $6,800,000. An engineer with the utility has calculated the average revenue for the past 24 months using three fuel-mix and wind-power situations: all gas, < 30% other or wind, and ≥ 30% other or wind. The table below shows the number of months for each situation and the associated revenue. If the same situation persists for the next 2 years, determine whether the utility's revenue will be greater or less than the costs and by how much.

Fuel/Wind Situation	Months in Past 24	Average Revenue, $/Month
All gas	12	$5,270,000
< 30% other/wind	9	$7,850,000
≥ 30% other/wind	3	$12,130,000

19.15 A discrete variable T can take on integer values of 1 to 5. A sample of size 100 results in the following probability estimates:

T	1	2	3	4	5
$P(T)$	0.2	0.3	0.1	0.3	0.1

(a) Use the 40 random numbers (RN) below to estimate the probabilities for each value of T, and (b) determine the sample probabilities for the $T = 2$ and $T = 5$. Compare the sample probabilities with the probabilities in the problem statement.

RN: 10, 42, 18, 31, 23, 80, 80, 26, 74, 71, 03, 90, 55, 61, 61, 28, 41, 49, 00, 79, 96, 78, 42, 31, 26, 01, 30, 79, 60, 61, 10, 42, 94, 91, 82, 24, 70, 66, 58, 81

19.16 Gillian has developed estimates for strengthening the undercarriage on a medium-use, 70-year-old bridge for which there is no budget to replace the bridge for the next 5 years. Estimates are $P = \$800,000$, $S = 0$, maintenance cost, M&O = $10,000 per year, $n = 5$ years. Use a *spreadsheet* and MARR of 2% per year to do the following:

(a) Determine what the project will cost in PW terms.

(b) During the analysis, Gillian learned that some 25 similar projects had been completed over the last 10 years. With the help of her supervisor, Gillian obtained the information below on the first cost, P, and maintenance cost, M&O, for these projects. Repeat the PW evaluation using expected values for these two estimates. Is PW higher or lower than the value in part (a)?

First Cost, $			M&O Cost, $ per Year		
Range, $1000	Midpoint, $1000	Number of Projects	Range, $1000	Midpoint, $1000	Number of Projects
500–600	550	1	0–4	2	3
600–700	650	6	4–8	6	15
700–800	750	3	8–12	10	3
800–900	850	4	12–16	14	4
1000–1100	1050	8			
1100–1200	1150	2			
1300–1400	1350	1			

(c) Assume the bridge is strengthened and administratively moved to the toll division, which promises a net revenue of $180,000 per year above the M&O cost. Is the project economically justified at the first cost of $800,000? What is the projected response from the driving public to the toll on this 70-year-old bridge?

19.17 The percent price increase p on a variety of retail food prices over a 1-year period varied from 5% to 10% in all cases. Because of the distribution of *p* values, the assumed probability distribution for the next year is

$$f(X) = 2X \qquad 0 \le X \le 1$$

where: $X = 0$ when $p = 5\%$
$X = 1$ when $p = 10\%$

For a continuous variable, the cumulative distribution $F(X)$ is the integral of $f(X)$ over the same range of the variable. In this case

$$F(X) = X^2 \qquad 0 \le X \le 1$$

(*a*) Graphically assign RNs to the cumulative distribution, take a sample of size 30 for the variable, and transform the *X* values into interest rates *p*.

(*b*) Calculate the average *p* value for your sample.

19.18 (*a*) Use the RAND function in Excel© to generate 100 values from a *U*(0,1) distribution. Calculate the average and compare it to 0.5, the expected value for a random sample between 0 and 1.

(*b*) For the RAND function sample, cluster the results into cells of 0.1 width, that is, 0.0 to 0.1, 0.1 to 0.2, etc., where the upper-limit value is excluded from each cell. Determine the probability for each grouping from the results. Does your sample come close to having approximately 10% in each cell?

Sample Estimates—Average and Standard Deviation

19.19 Wastewater samples taken from an industrial discharge had the chemical oxygen demand (COD) concentrations below in mg/L. Determine the (*a*) arithmetic mean (by *hand* and *spreadsheet*), (*b*) median, (*c*) mode, and (*d*) standard deviation (by *hand* and *spreadsheet*).

COD values: 452, 364, 415, 395, 404, 470, 391, 395, 425, 430, 380

19.20 In order to determine if air quality readings in a certain assembly plant were within OSHA guidelines, an engineer collected the following data: 108, 99, 84, 93, 80, 90, 83, 83, 96, 85, and 89 ppb (parts per billion).

(*a*) Determine the arithmetic mean (by *hand* and *spreadsheet*).

(*b*) Calculate the standard deviation (by *hand* and *spreadsheet*).

(*c*) Determine the number of values and percentage of values that fall within ± 1 standard deviation of the mean.

19.21 Monthly maintenance costs, MC, for conveyor belts in an airport's centralized baggage collection and distribution facility are clustered into cell midpoints with associated frequencies as shown.

(*a*) Estimate the expected value and standard deviation of MC based on this sample.

(*b*) What is the best estimate of the percentage of MC that will fall within ± 1 standard deviation of the mean?

Cost MC, $	Frequency
200	4
400	8
600	9
800	14
1000	18
1200	25
1400	12
1600	10

19.22 A discrete variable *Q*, extra inventory units, has integer values from 1 to 10. When sampled 100 times, the following observed probability distribution was obtained:

Q	1	2	3	6	9	10
P(Q)	0.2	0.2	0.2	0.1	0.2	0.1

(*a*) Determine the sample average and standard deviation by *hand*.

(*b*) Determine the sample average and standard deviation by *spreadsheet*.

(*c*) Determine the values 1 and 2 standard deviations from the mean. Of the 100 sample points, how many fall within these two ranges?

19.23 (*a*) Use the relations in Section 19.4 for continuous variables to determine the expected value and standard deviation for the distribution of $f(D_Y)$ in Problem 19.11. (*b*) It is possible to calculate the probability of a continuous variable *Y* between two points (*a, b*) using the following integral:

$$P(a \le Y \le b) = \int_a^b f(Y)\, dy$$

Determine the probability that D_Y is within 2 standard deviations of the expected value.

19.24 Calculate the expected value of the variable *N* in Problem 19.10.

19.25 A newsstand manager is tracking *Y*, the number of weekly magazines left on the shelf when the new edition is delivered. Data collected over a 30-week period are summarized by the following probability distribution. Use a *spreadsheet* to plot the distribution and the estimates for expected value and 1 standard deviation on either side of *E*(*Y*) on the plot.

Y, Copies	3	7	10	12
P(Y)	1/3	1/4	1/3	1/12

Simulation

19.26 Jennifer, a senior project manager, estimated net cash flow after taxes (CFAT) for a large project she is working on. The additional CFAT of $2,800,000 in year 10 is the salvage value, estimated at 9% of the $31 million first cost for all capital assets.

Year	CFAT, $1000
0	−31,000
1–6	5,400
7–10	2,040
10	2,800

The PW value at the current MARR of 7% per year is

$$PW = -31{,}000 + 5400(P/A{,}7\%{,}6)$$
$$+ 2040(P/A{,}7\%{,}4)(P/F{,}7\%{,}6)$$
$$+ 2800(P/F{,}7\%{,}10)$$
$$= \$767$$

Jennifer believes the MARR will vary over a relatively narrow range, as will the CFAT, especially during the out years of 7 through 10. She is willing to accept the other estimates as certain. Use the following probability distribution assumptions for MARR and CFAT to perform a simulation—hand- or spreadsheet-based.

MARR. Uniform distribution over the range 6% to 10%.
CFAT, years 7 through 10. Uniform distribution over the range $1,600,000 to $2,400,000 for each year.

Plot the resulting PW distribution. Should the plan be accepted using decision making under certainty? Under risk?

19.27 Repeat Problem 19.26, except use the normal distribution for the CFAT in years 7 through 10 with an expected value of $2 million and a standard deviation of $500,000.

ADDITIONAL PROBLEMS AND FE EXAM REVIEW QUESTIONS

19.28 When single-value or single-point estimates are used exclusively in an economic analysis, the decision-making process is said to be:
(a) Probabilistic   (b) Uncertain
(c) Deterministic   (d) Monte Carlo

19.29 When there are at least two values for a parameter and it is not possible to estimate the chance that each may occur, this situation is known as:
(a) Uncertain   (b) Risk
(c) Deterministic   (d) Cost estimating

19.30 The shipping costs for fresh fruit items have been estimated and assigned the probabilities shown below. The expected value of the shipping costs is closest to:
(a) $36.33   (b) $39.21
(c) $41.28   (d) $45.11

Shipping Cost, $	34	38	55
Probability	0.22	0.31	0.47

19.31 In a table of two-digit random numbers, the probabilities of 0.16 to 0.20 can be represented by the numbers:
(a) 16 to 20   (b) 0 to 4
(c) 1 to 5   (d) 15 to 19

19.32 The following temperatures (in °F) were recorded inside the building of a remote pumping station: 99, 87, 93, 90, and 96. The variance of these readings is closest to:
(a) 11.7   (b) 22.5
(c) 31.1   (d) 38.6

19.33 For the income and probability values shown, the probability that the income in any year will be greater than $8500 is closest to:
(a) 0.38   (b) 0.49
(c) 0.70   (d) 0.92

Income, $/Year	6,200	8,500	9,600	10,300	12,600	15,500
Probability	0.15	0.23	0.32	0.24	0.09	0.04

19.34 The net revenue from a candy product called Mummies-Lite has averaged $15,000 per month for the past 12 months. If the value of $\sum(X_i - \overline{X})^2$ is $4,680,000, the standard deviation value is closest to:
(a) $381   (b) $652
(c) $958   (d) $1265

19.35 A survey of the types of cars parked at an NFL football stadium revealed that there were equal probabilities of finding cars identified as Type A, B, C, and D. If Type A cars were assigned random numbers 0 through 24, Type B numbers 25 through 49, Type C numbers 50 through 74, and Type D numbers 75 through 99, the sample probability of a Type C car from the 12 random numbers shown below is closest to:

RN: 75, 52, 14, 99, 67, 74, 06, 50, 97, 46, 27, 88

(a) 0.17   (b) 0.25
(c) 0.33   (d) 0.42

19.36 The dispersion of values about the sample average is represented by the:
(a) Variance
(b) Expected value
(c) Probability distribution
(d) Standard deviation

CASE STUDY

USING SIMULATION AND THREE-ESTIMATE SENSITIVITY ANALYSIS

Background

The Knox Brewing company makes specialty-named sodas and flavored drinks for retail grocery chains throughout North and Central America. During the past year, it has become obvious that a new bottle-capping machine is needed to replace the current 10-year-old system. Dr. Knox, the owner and president, knows the business quite well. You just handed him the first cost bids from three vendors for the machine. He looked carefully at the numbers and asked you to sit down. You were quite surprised, as this was the first time you had been in his office, and most other engineers at Knox have a great fear of "the Old Man."

Information

As he examined the three bids on first cost of the machine, he started to write some numbers, which, he explained, were his estimates of the annual operating cost, useful life, and possible salvage value for each of the machines sold by the three vendors.

After a few minutes, he told you to take these numbers and use some of that "new engineering knowledge" you acquired in college to determine which, if any, of these three bids made the best economic sense. He also told you to be innovative and use a computer and some probability to come up with a robust recommendation by tomorrow at 2 P.M.

You have used the estimates from the president to develop Table 19–6 of pessimistic (P), most likely (ML), and optimistic (O) estimates for each vendor's machine. In addition, you developed some possible distributions for the parameters that Dr. Knox estimated, namely, AOC, life, and salvage value. These are summarized in Table 19–7. You plan to use a simple Monte Carlo simulation to help formulate your recommendation for tomorrow.

Case Study Exercises

First, learn to use the RNG (random number generator) in Excel, if you have not already done so. It is necessary to sample from the normal distributions that you have specified in Table 19–7. RNG is part of the Analysis Tool-Pak accessed through the Office Button, Excel Options, Add-Ins path.

1. Prepare the simulation using a spreadsheet; determine which of the vendors offers the best machine from an economic perspective, and take into account the estimates made by Dr. Knox. Use a sample size of at least 50, and base your conclusions on the AW measure of worth.

2. Prepare a short presentation for Dr. Knox (and class) using your analysis.

TABLE 19–6	Parameter Estimates for Bottle-Capping Machine			
	First Cost, $	AOC, $ per Year	Salvage, $	Life, Years
Vendor 1				
P	−200,000	−11,000	0	3
ML	−200,000	−10,000	0	5
O	−200,000	−6,000	0	8
Vendor 2				
P	−150,000	−5,000	0	2
ML	−150,000	−3,500	5,000	4
O	−150,000	−2,000	8,000	7
Vendor 3				
P	−300,000	−8,000	5,000	5
ML	−300,000	−6,000	5,000	7
O	−300,000	−4,500	8,000	9

TABLE 19–7	Distribution Assumptions about AOC, Life, and Salvage		
Parameter	Vendor 1	Vendor 2	Vendor 3
AOC, $ per year	Normal Mean: 10,000 Std. dev.: 500	Normal Mean: 3500 Std. dev.: 500	Normal Mean: 6000 Std. dev.: 500
Salvage, $	Uniform 0 to 1000	Uniform 0 to 8000	Uniform 5000 to 8000
Life, years	Discrete uniform 3 to 8, equal probability	Discrete uniform 2 to 7, equal probability	Discrete uniform 5 to 9, equal probability

USING SPREADSHEETS AND MICROSOFT EXCEL®

This appendix explains the layout of a spreadsheet and the use of Microsoft Excel® (hereafter called Excel) functions in engineering economy. Refer to the Excel help system for your particular computer and version of Excel. Some specific commands and entries refer to Excel 2013 and may differ slightly from your version.

A.1 Introduction to Using Excel ● ● ●

Enter a Formula or Use an Excel Function

The **= sign is required** to perform any formula or function computation in a cell. The formulas and functions on the worksheet can be displayed by simultaneously pressing Ctrl and `. The symbol ` is usually in the upper left of the keyboard with the ~ (tilde) symbol. Pressing Ctrl+` a second time hides the formulas and functions.

1. Open Excel and a blank spreadsheet.
2. Move to cell C3. (Move the pointer to C3 and left-click.)
3. Type = –PV(5%,12,8) and <Enter>. This function will calculate the present value of 12 payments of $8 at a 5% per year interest rate. The display of $70.91 maintains the same plus sign as the entry of 8 due to the placement of the minus sign before the operator PV.

Another example: To calculate the future value of 12 payments of $8 at 6% per year interest, do the following:

1. Move to cell B3, and type INTEREST.
2. Move to cell C3, and type 6%.
3. Move to cell B4, and type PAYMENT.
4. Move to cell C4, and type 8 (to represent the size of each payment).
5. Move to cell B5, and type NUMBER OF PAYMENTS.
6. Move to cell C5, and type 12 (to represent the number of payments).
7. Move to cell B7, and type FUTURE VALUE.
8. Move to cell C7, and type = –FV(C3,C5,C4) and hit <Enter>. The answer will appear in cell C7 with the sign maintained as a plus.

To edit the values in cells

1. Move to cell C3 and type 5% (the previous value will be replaced).
2. The value in cell C7 will update.

Cell References in Formulas and Functions

If a cell reference is used in lieu of a specific number, it is possible to change the number once and perform sensitivity analysis on any variable that is referenced by the cell number, such as C5. This approach defines the referenced cell as a **global variable** for the worksheet. There are two types of cell references—relative and absolute.

Relative References   If a cell reference is entered, for example, A1, into a formula or function that is copied or dragged into another cell, the reference is changed relative to the movement of the original cell. If the formula in C5 is = A1 and it is copied into cell C6, the formula is changed to = A2. This feature is used when dragging a function through several cells, and the source entries must change with the column or row.

Absolute References   If adjusting cell references is not desired, place a **$ sign** in front of the part of the cell reference that is not to be adjusted—the column, row, or both.

For example, = $A$1 will retain the formula when it is moved anywhere on the worksheet. Similarly, = $A1 will retain the column A, but the relative reference on 1 will adjust the row number upon movement around the worksheet.

Absolute references are used in engineering economy for sensitivity analysis of parameters such as MARR, first cost, and annual cash flows. In these cases, a change in the absolute-reference cell entry can help determine the sensitivity of a result, such as ROR, PW, or AW.

Print the Spreadsheet

First define the portion (or all) of the spreadsheet to be printed.

1. Move the pointer to the top left cell of your spreadsheet.
2. Hold down the left-click button. (Do not release the left-click button.)
3. Drag the mouse to the lower right corner of your spreadsheet or to wherever you want to stop printing.
4. Release the left-click button. (It is ready to print.)
5. Left-click the FILE button (see Figure A–1, upper left).
6. Move the pointer down to select Print and left-click on Print when the screen appears. Note the number of copies can be adjusted to the right of Print.

Depending on your computer environment, you may have to select a network printer and queue your printout through a server. The Settings commands allow a variety of printing options. Additional options are available by left-clicking on the Page Setup tab at the bottom of the Settings table.

Save the Spreadsheet

You can save your spreadsheet at any time during or after completing your work. It is recommended that you save your work regularly.

1. Left-click the FILE button.
2. To save the spreadsheet the first time, left-click the Save As option. Decide where to save your file, for example, in Documents, or on a USB flash, etc.
3. Type the file name, for example, Prob 7.9, and left-click the Save button.

To save the spreadsheet after it has been saved the first time, that is, a file name has been assigned to it, left-click the Office button, move the pointer down, and left-click on Save.

Create a Scatter Chart

This chart is one of the most commonly used in scientific analysis, including engineering economy. It plots pairs of data and can place multiple series of entries on the Y axis. This chart is especially useful for results such as the PW versus i graph, where i is on the X axis and the Y axis displays the results of the NPV function for the alternatives.

1. Open Excel.
2. Enter the following headings and numbers in columns A, B, and C, respectively. See Figure A–1 for details.
   Column A, cell A1 through A8: Rate $i\%$; 4, 5, 6, 7, 8, 9, 10.
   Column B, cell B1 through B8: CF, A, $; 40, 50, 65, 72, 81, 80, 100.
   Column C, cell C1 through C8: CF, B, $; 100, 70, 65, 50, 20, 15, –15.
3. Move the mouse to A1, left-click, and hold while dragging to cell C8. All cells will be highlighted, including the title cell for each column.
4. If not all the columns for the chart are adjacent to one another, first press and hold the Control key on the keyboard during the entirety of step 3. After dragging over one column of data, momentarily release the left-click, then move to the top of the next (nonadjacent) column of the chart. Do not release the Control key until all columns to be plotted have been highlighted.
5. Left-click on the Insert button on the toolbar.
6. Select the Scatter option and choose a subtype of scatter chart. The graph appears with a legend (Figure A–1).

Now a large number of styling effects can be introduced for axis titles, legend, data series, etc. Note that only the bottom row of the title can be highlighted. If titles are not highlighted, the data sets are generically identified as series 1, series 2, etc. on the legend.

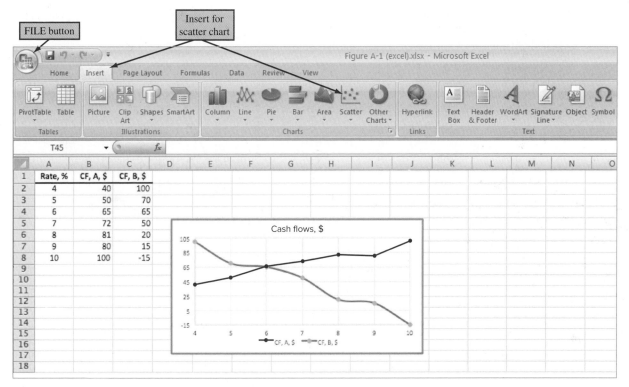

Figure A–1
Scatter chart for data entries and location of commonly used buttons.

Obtain Help While Using Excel

1. To get general help information, while Excel is open, left-click on the "?" (upper right).
2. Enter a topic or phrase. For example, if you want to know more about how to save a file, type the word Save. Click the Search Online Help button (could be a magnifying glass).
3. Select the appropriate matching words. You can browse through the options by left-clicking on any item.

A.2 Organization (Layout) of the Spreadsheet ●●●

A spreadsheet can be used in several ways to obtain answers to numerical questions. The first is as a rapid solution tool, often with the entry of only a few numbers or one predefined function. For example, to find the future worth in a single-cell operation, move the pointer to any cell and enter = FV(8%,5,−2500). The display of $14,666.50 is the 8% future worth at the end of year 5 of five equal payments of $2500 each.

A second use is more formal; it presents data, solutions, graphs, and tables developed on the spreadsheet and ready for presentation to others. Some fundamental guidelines in spreadsheet organization are presented here. A sample layout is presented in Figure A–2. As solutions become more complex, organization of the spreadsheet becomes increasingly important, especially for presentation to an audience via PowerPoint or similar software.

Cluster the data and the answers. It is advisable to organize the given or estimated data in the top left of the spreadsheet. A very brief label should be used to identify the data, for example, MARR = in cell A1 and the value, 12%, in cell B1. Then B1 can be the referenced cell for all entries requiring the MARR. Additionally, it may be worthwhile to cluster the answers into one area and frame it. Often, the answers are best placed at the bottom or top of the column of entries used in the formula or predefined function.

Enter titles for columns and rows. Each column or row should be labeled so its entries are clear to the reader. It is very easy to select from the wrong column or row when no brief title is present at the head of the data.

Enter revenue (income) and cost cash flows separately. When there are both revenue and cost cash flows involved, it is strongly recommended that the cash flow estimates for revenue (usually

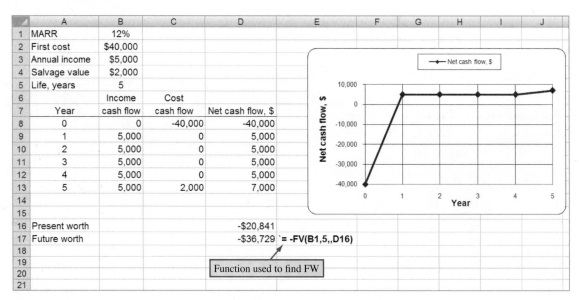

Figure A–2
Spreadsheet layout with cash flow estimates, results of functions, function formula detailed, and a scatter chart.

positive) and first cost, salvage value, and annual costs (usually negative, with salvage a positive number) be entered into two adjacent columns. Then a formula combining them in a third column displays the net cash flow. There are two immediate advantages to this practice: fewer errors are made when performing the summation and subtraction mentally, and changes for sensitivity analysis are more easily made.

Use cell references. The use of absolute and relative cell references is a must when any changes in entries are expected. For example, suppose the MARR is entered in cell B1 and three separate references are made to the MARR in functions on the spreadsheet. The absolute cell reference entry $B$1 in the three functions allows the MARR to be changed one time, not three.

Obtain a final answer through summing and embedding. When the formulas and functions are kept relatively simple, the final answer can be obtained using the SUM function. For example, if the present worth (PW) values of two columns of cash flows are determined separately, then the total PW is the SUM of the subtotals. This practice is especially useful when the cash flow series are complex.

Prepare for a chart. If a chart (graph) will be developed, plan ahead by leaving sufficient room on the right of the data and answers. Charts can be placed on the same worksheet or on a separate worksheet. Placement on the same worksheet is recommended, especially when the results of sensitivity analysis are plotted.

A.3 Spreadsheet Functions Important to Engineering Economy (Alphabetical Order) ● ● ●

DB (Declining Balance)

Calculates the depreciation amount for an asset for a specified period n using the declining balance method. The depreciation rate d used in the computation is determined from asset values S (salvage value) and B (basis or first cost) as $d = 1 - (S/B)^{1/n}$. This is Equation [16.12]. Three-decimal-place accuracy is used for d.

$$= DB(\text{cost, salvage, life, period, month})$$

cost	First cost or basis of the asset.
salvage	Salvage value.
life	Depreciation life (recovery period).
period	The period, year, for which the depreciation is to be calculated.
month	(optional entry) If this entry is omitted, a full year is assumed for the first year.

Example   A new machine costs $100,000 and is expected to last 10 years. At the end of 10 years, the salvage value of the machine is $50,000. What is the depreciation of the machine in the first year and the fifth year?

  Depreciation for the first year: = DB(100000,50000,10,1)

  Depreciation for the fifth year: = DB(100000,50000,10,5)

Note that a comma is not inserted in numbers with four or more digits (> 999) as this will display an error message or an incorrect answer. Because of the manner in which the DB function determines the fixed percentage d and the accuracy of the computations, it is **recommended that the DDB function (below) be used** for all declining balance depreciation rates. Simply use the optional factor entry for rates other than $d = 2/n$.

DDB (Double Declining Balance)

Calculates the depreciation of an asset for a specified period n using the double declining balance method. A factor can also be entered for some other declining balance depreciation method by specifying a factor in the function.

$$= \text{DDB(cost, salvage, life, period, factor)}$$

cost	First cost or basis of the asset.
salvage	Salvage value of the asset.
life	Depreciation life.
period	The period, a year, for which the depreciation is to be calculated.
factor	(optional entry) If this entry is omitted, the function will use a double declining method with two times the straight line rate. If, for example, the entry is 1.5, the 150% declining balance method will be used.

Example   A new machine costs $200,000 and is expected to last 10 years. The salvage value is $10,000. Calculate the depreciation of the machine for the first and the eighth years. Finally, calculate the depreciation for the fifth year using the 175% declining balance method.

  Depreciation for the first year: = DDB(200000,10000,10,1)

  Depreciation for the eighth year: = DDB(200000,10000,10,8)

  Depreciation for the fifth year using 175% DB: = DDB(200000,10000,10,5,1.75)

EFFECT (Effective Interest Rate)

Calculates the effective annual interest rate for a stated nominal annual rate and a given number of compounding periods per year. Excel uses Equation [4.7] to calculate the effective rate.

$$= \text{EFFECT(nominal, npery)}$$

nominal	Nominal interest rate for the year.
npery	Number of times interest is compounded per year.

Example   Claude has applied for a $10,000 loan. The bank officer told him that the interest rate is 8% per year and that interest is compounded monthly to conveniently match his monthly payments. What effective annual rate will Claude pay?

  Effective annual rate: = EFFECT(8%,12)

EFFECT can also be used to find **effective rates other than annually.** Enter the nominal rate for the time period of the required effective rate; npery is the number of times compounding occurs during the time period of the effective rate.

Example   Interest is stated as 3.5% per quarter with quarterly compounding. Find the effective semiannual rate.

  The 6-month nominal rate is 7%, and compounding is two times per 6 months.

  Effective semiannual rate: = EFFECT(7%,2)

FV (Future Value)

Calculates the future value (worth) based on periodic payments at a specific interest rate.

$$= \text{FV(rate, nper, pmt, pv, type)}$$

rate	Interest rate per compounding period.
nper	Number of compounding periods.
pmt	Constant payment amount.
pv	The present value amount. If pv is not specified, the function will assume it to be 0.
type	(optional entry) Either 0 or 1. A 0 represents payments made at the end of the period, and 1 represents payments at the beginning of the period. If omitted, 0 is assumed.

Example   Jack wants to start a savings account that can be increased as desired. He will deposit $12,000 to start the account and plans to add $500 to the account at the beginning of each month for the next 24 months. The bank pays 0.25% per month. How much will be in Jack's account at the end of 24 months? (Note: Enter a minus sign to retain a positive sign on the answer.)

Future value in 24 months: $= -\text{FV}(0.25\%,24,500,12000,1)$

IF (IF Logical Function)

Determines which of two entries is entered into a cell based on the outcome of a logical check on the outcome of another cell. The logical test can be a function or a simple value check, but it must use an equality or inequality sense. If the response is a text string, place it between quote marks (" "). The responses can themselves be IF functions. Up to seven IF functions can be nested for very complex logical tests.

$$= \text{IF(logical\_test,value\_if\_true,value\_if\_false)}$$

logical_test	Any worksheet function can be used here, including a mathematical operation.
value_if_true	Result if the logical_test argument is true.
value_if_false	Result if the logical_test argument is false.

Example   The entry in cell B4 should be "selected" if the PW value in cell B3 is greater than or equal to zero and "rejected" if PW < 0.

Entry in cell B4:   = IF(B3>=0,"selected","rejected")

Example   The entry in cell C5 should be "selected" if the PW value in cell C4 is greater than or equal to zero, "rejected" if PW < 0, and "fantastic" if PW ≥ 200.

Entry in cell C5:   = IF(C4<0,"rejected", IF(C4>=200,"fantastic","selected"))

IPMT (Interest Payment)

Calculates the interest accrued for a given period n based on constant periodic payments and interest rate.

$$= \text{IPMT(rate, per, nper, pv, fv, type)}$$

rate	Interest rate per compounding period.
per	Period for which interest is to be calculated.
nper	Number of compounding periods.
pv	Present value. If pv is not specified, the function will assume it to be 0.
fv	Future value. If fv is omitted, the function will assume it to be 0. The fv can also be considered a cash balance after the last payment is made.
type	(optional entry) Either 0 or 1. A 0 represents payments made at the end of the period, and 1 represents payments made at the beginning of the period. If omitted, 0 is assumed.

Example   Calculate the interest due in the 10th month for a 48-month, $20,000 loan. The interest rate is 0.25% per month.

Interest due: = IPMT(0.25%,10,48,20000)

IRR (Internal Rate of Return)

Calculates the internal rate of return between –100% and infinity for a series of cash flows at regular periods.

$$= \textbf{IRR(values, guess)}$$

values   A set of numbers in a spreadsheet column (or row) for which the rate of return will be calculated. The set of numbers must consist of at least *one* positive and *one* negative number. Negative numbers denote a payment made or cash outflow, and positive numbers denote income or cash inflow.

guess   (optional entry) To reduce the number of iterations, a *guessed rate of return* can be entered. In most cases, a guess is not required, and a 10% rate of return is initially assumed. If the #NUM! error appears, try using different values for guess. Inputting different guess values makes it possible to determine the multiple roots for the rate of return equation of a nonconventional (nonsimple) cash flow series.

Example   John wants to start a printing business. He will need $25,000 in capital and anticipates that the business will generate the following incomes during the first 5 years. Calculate his rate of return after 3 years and after 5 years.

Year 1 $5,000
Year 2 $7,500
Year 3 $8,000
Year 4 $10,000
Year 5 $15,000

Set up an array in the spreadsheet.

In cell A1, type −25000 (negative for payment).
In cell A2, type 5000 (positive for income).
In cell A3, type 7500.
In cell A4, type 8000.
In cell A5, type 10000.
In cell A6, type 15000.

Therefore, cells A1 through A6 contain the array of cash flows for the first 5 years, including the capital outlay. *Note that any years with a zero cash flow must have a zero entered* to ensure that the year value is correctly maintained for computation purposes.

To calculate the internal rate of return after 3 years, move to cell A7, and type = IRR(A1:A4).

To calculate the internal rate of return after 5 years and specify a guess value of 5%, move to cell A8, and type = IRR(A1:A6,5%).

MIRR (Modified Internal Rate of Return)

Calculates the modified internal rate of return for a series of cash flows and reinvestment of income and interest at a stated rate.

$$= \textbf{MIRR(values, finance\_rate, reinvest\_rate)}$$

values   Refers to an array of cells in the spreadsheet. Negative numbers represent payments, and positive numbers represent income. The series of payments and income must occur at regular periods and must contain at least *one* positive number and *one* negative number.

finance_rate   Interest rate on funds borrowed from external sources (i_b in Equation [7.9]).

reinvest_rate   Interest rate for reinvestment on positive cash flows (i_i in Equation [7.9]). (This is not the same reinvestment rate on the net investments when the cash flow series is nonconventional. See Section 7.5 for comments.)

Example   Jane opened a hobby store 4 years ago. When she started the business, Jane borrowed $50,000 from a bank at 12% per year. Since then, the business has yielded $10,000 the first year, $15,000 the second year, $18,000 the third year, and $21,000 the fourth year. Jane reinvests her profits, earning 8% per year. What is the modified rate of return after 3 years and after 4 years?

In cell A1, type −50000.

In cell A2, type 10000.

In cell A3, type 15000.

In cell A4, type 18000.

In cell A5, type 21000.

To calculate the modified rate of return after 3 years, move to cell A6, and type
= MIRR(A1:A4,12%,8%).

To calculate the modified rate of return after 4 years, move to cell A7, and type
= MIRR(A1:A5,12%,8%).

NOMINAL (Nominal Interest Rate)

Calculates the nominal **annual** interest rate for a stated effective **annual** rate and a given number of compounding periods per year. *This function is designed to display only nominal annual rates.*

$$= \text{NOMINAL(effective, npery)}$$

effective   Effective interest rate for the year.

npery   Number of times that interest is compounded per year.

Example   Last year, a corporate stock earned an effective return of 12.55% per year. Calculate the nominal annual rate, if interest is compounded quarterly and compounded continuously.

Nominal annual rate, quarterly compounding: = NOMINAL(12.55%,4)

Nominal annual rate, continuous compounding: = NOMINAL(12.55%,100000)

NPER (Number of Periods)

Calculates the number of periods for the present worth of an investment to equal the future value specified, based on uniform regular payments and a stated interest rate.

$$= \text{NPER(rate, pmt, pv, fv, type)}$$

rate   Interest rate per compounding period.

pmt   Amount paid during each compounding period.

pv   Present value (lump-sum amount).

fv   (optional entry) Future value or cash balance after the last payment. If fv is omitted, the function will assume a value of 0.

type   (optional entry) Enter 0 if payments are due at the end of the compounding period and 1 if payments are due at the beginning of the period. If omitted, 0 is assumed.

Example   Sally plans to open a savings account that pays 0.25% per month. Her initial deposit is $3000, and she plans to deposit $250 at the beginning of every month. How many payments does she have to make to accumulate $25,000 to buy a new car?

Number of payments: $= \text{NPER}(0.25\%, -250, -3000, 25000, 1)$

NPV (Net Present Value)

Calculates the net present value of a series of future cash flows at a stated interest rate.

$$= \text{NPV}(\textbf{rate, series})$$

rate   Interest rate per compounding period.

series   Series of costs and incomes set up in a range of cells in the spreadsheet.

Example   Mark is considering buying a sports store for $100,000 and expects to receive the following income during the next 6 years of business: $25,000, $40,000, $42,000, $44,000, $48,000, $50,000. The interest rate is 8% per year.

In cells A1 through A7, enter -100000, followed by the six annual incomes.

Present value: $= \text{NPV}(8\%, \text{A2:A7}) + \text{A1}$

The cell A1 value is already a present value. *Any year with a zero cash flow must have a 0 entered to ensure a correct result.*

PMT (Payments)

Calculates equivalent periodic amounts based on present value and/or future value at a constant interest rate.

$$= \text{PMT}(\textbf{rate, nper, pv, fv, type})$$

rate   Interest rate per compounding period.

nper   Total number of periods.

pv   Present value.

fv   Future value.

type   (optional entry) Enter 0 for payments due at the end of the compounding period and 1 if payment is due at the start of the compounding period. If omitted, 0 is assumed.

Example   Jim plans to take a $15,000 loan to buy a new car. The interest rate is 7% per year. He wants to pay the loan off in 5 years (60 months). What are his monthly payments?

Monthly payments: $= \text{PMT}(7\%/12, 60, 15000)$

PPMT (Principal Payment)

Calculates the payment on the principal based on uniform payments at a specified interest rate.

$$= \text{PPMT}(\textbf{rate, per, nper, pv, fv, type})$$

rate   Interest rate per compounding period.

per   Period for which the payment on the principal is required.

nper   Total number of periods.

pv   Present value.

fv   Future value.

type   (optional entry) Enter 0 for payments that are due at the end of the compounding period and 1 if payments are due at the start of the compounding period. If omitted, 0 is assumed.

Example   Jovita is planning to invest $10,000 in equipment, which is expected to last 10 years with no salvage value. The interest rate is 5%. What is the principal payment at the end of year 4 and year 8?

At the end of year 4: = PPMT(5%,4,10,−10000)

At the end of year 8: = PPMT(5%,8,10,−10000)

PV (Present Value)

Calculates the present value of a future series of equal cash flows and a single lump sum in the last period at a constant interest rate.

$$= PV(\text{rate, nper, pmt, fv, type})$$

rate   Interest rate per compounding period.

nper   Total number of periods.

pmt   Cash flow at regular intervals. Negative numbers represent payments (cash outflows), and positive numbers represent income.

fv   Future value or cash balance at the end of the last period.

type   (optional entry) Enter 0 if payments are due at the end of the compounding period and 1 if payments are due at the start of each compounding period. If omitted, 0 is assumed.

There are two primary differences between the PV function and the NPV function: PV allows for end or beginning of period cash flows, and PV requires that all amounts have the same value, whereas they may vary for the NPV function.

Example   Jose is considering leasing a car for $300 a month for 3 years (36 months). After the 36-month lease, he can purchase the car for $12,000. Using an interest rate of 8% per year, find the present value of this option.

Present value: = PV(8%/12,36,−300,−12000)

Note the minus signs on the pmt and fv amounts.

RAND (Random Number)

Returns an evenly distributed number that is (1) ≥ 0 and < 1; (2) ≥ 0 and < 100; or (3) between two specified numbers.

= RAND()	**for range 0 to 1**
= RAND()*100	**for range 0 to 100**
= RAND()*(b−a)+a	**for range a to b**

a = minimum integer to be generated
b = maximum integer to be generated

The Excel function RANDBETWEEN(a,b) may also be used to obtain a random number between two values.

Example   Grace needs random numbers between 5 and 10 with 3 digits after the decimal. What is the Excel function? Here a = 5 and b = 10.

Random number: = RAND()*5 + 5

Example   Randi wants to generate random numbers between the limits of −10 and 25. What is the Excel function? The minimum and maximum values are $a = -10$ and $b = 25$, so $b - a = 25 - (-10) = 35$.

Random number: = RAND()*35 − 10

RATE (Interest Rate)

Calculates the interest rate per compounding period for a series of payments or incomes.

$$= \text{RATE(nper, pmt, pv, fv, type, guess)}$$

nper   Total number of periods.

pmt   Payment amount made each compounding period.

pv   Present value.

fv   Future value (not including the pmt amount).

type   (optional entry) Enter 0 for payments due at the end of the compounding period and 1 if payments are due at the start of each compounding period. If omitted, 0 is assumed.

guess   (optional entry) To minimize computing time, include a guessed interest rate. If a value of guess is not specified, the function will assume a rate of 10%. This function usually converges to a solution if the rate is between 0% and 100%.

Example   Alysha wants to start a savings account at a bank. She will make an initial deposit of $1000 to open the account and plans to deposit $100 at the beginning of each month. She plans to do this for the next 3 years (36 months). At the end of 3 years, she wants to have at least $5000. What is the minimum interest required to achieve this result?

Interest rate: $= \text{RATE}(36,-100,-1000,5000,1)$

SLN (Straight Line Depreciation)

Calculates the straight line depreciation of an asset for a given year.

$$= \text{SLN(cost, salvage, life)}$$

cost   First cost or basis of the asset.

salvage   Salvage value.

life   Depreciation life.

Example   Maria purchased a printing machine for $100,000. The machine has an allowed depreciation life of 8 years and an estimated salvage value of $15,000. What is the depreciation each year?

Depreciation: $= \text{SLN}(100000,15000,8)$

SYD (Sum-of-Years-Digits Depreciation)

Calculates the sum-of-years-digits depreciation of an asset for a given year.

$$= \text{SYD(cost, salvage, life, period)}$$

cost   First cost or basis of the asset.

salvage   Salvage value.

life   Depreciation life.

period   The year for which the depreciation is sought.

Example   Jack bought equipment for $100,000 that has a depreciation life of 10 years. The salvage value is $10,000. What is the depreciation for year 1 and year 9?

Depreciation for year 1: $= \text{SYD}(100000,10000,10,1)$

Depreciation for year 9: $= \text{SYD}(100000,10000,10,9)$

VDB (Variable Declining Balance)

Calculates the depreciation using the declining balance method with a switch to straight line depreciation in the year in which straight line has a larger depreciation amount. This function automatically implements the switch from DB to SL depreciation, unless specifically instructed to not switch.

= **VDB (cost, salvage, life, start_period, end_period, factor, no_switch)**

cost	First cost of the asset.
salvage	Salvage value.
life	Depreciation life.
start_period	First period for depreciation to be calculated.
end_period	Last period for depreciation to be calculated.
factor	(optional entry) If omitted, the function will use the double declining rate of $2/n$, or twice the straight line rate. Other entries define the declining balance method, for example, 1.5 for 150% declining balance.
no_switch	(optional entry) If omitted or entered as FALSE, the function will switch from declining balance to straight line depreciation when the latter is greater than DB depreciation. If entered as TRUE, the function will not switch to SL depreciation at any time during the depreciation life.

Example Newly purchased equipment with a first cost of $300,000 has a depreciable life of 10 years with no salvage value. Calculate the 175% declining balance depreciation for the first year and the ninth year if switching to SL depreciation is acceptable and if switching is not permitted.

Depreciation for first year, with switching: = VDB(300000,0,10,0,1,1.75)

Depreciation for ninth year, with switching: = VDB(300000,0,10,8,9,1.75)

Depreciation for first year, no switching: = VDB(300000,0,10,0,1,1.75,TRUE)

Depreciation for ninth year, no switching: = VDB(300000,0,10,8,9,1.75,TRUE)

VDB (for MACRS Depreciation)

The VDB function can be adapted to generate the MACRS annual depreciation amount, when the start_period and end_period are replaced with the MAX and MIN functions, respectively. As above, the factor option should be entered if other than DDB rates start the MACRS depreciation. The VDB format is

= **VDB(cost,0,life,MAX(0,t−1.5),MIN(life,t−0.5),factor)**

Example Determine the MACRS depreciation for year 4 for a $350,000 asset that has a 20% salvage value and a MACRS recovery period of 3 years. $D_4 = \$25,926$ is the display.

Depreciation for year 4: = VDB(350000,0,3,MAX(0,4−1.5),MIN(3,4−0.5),2)

Example Find the MACRS depreciation in year 16 for a $350,000 asset with a recovery period of $n = 15$ years. The optional factor 1.5 is required here, since MACRS starts with 150% DB for $n = 15$-year and 20-year recovery periods. $D_{16} = \$10,334$.

Depreciation for year 16: = VDB(350000,0,15,MAX(0,16−1.5),MIN(15,16−0.5),1.5)

Other Functions

There are numerous additional financial functions available on Excel, as well as engineering, mathematics, trigonometry, statistics, data and time, logical, and information functions. These can be viewed by clicking the Formulas tab on the toolbar.

A.4 Goal Seek—A Tool for Breakeven and Sensitivity Analysis ●●●

Goal Seek is found on the Excel toolbar labeled Data, followed by What-if Analysis. This tool changes the value in a specific cell based on a numerical value in another (changing) cell as input by the user. It is a good tool for **sensitivity analysis, breakeven analysis,** and **"what if?" questions**

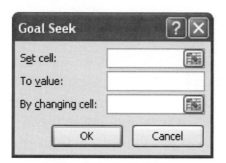

Figure A–3
Goal Seek template used to specify a cell, a value, and the changing cell.

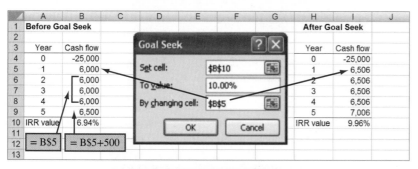

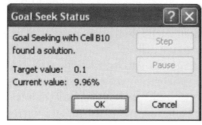

Figure A–4
Use of Goal Seek to determine an annual cash flow to increase the rate of return.

when no constraint relations or inequalities are needed. The initial Goal Seek template is pictured in Figure A–3. One of the cells (set or changing cell) must contain an equation or spreadsheet function that uses the other cell to determine a numeric value. Only a single cell can be identified as the changing cell; however, this limitation can be avoided by using equations rather than specific numerical inputs in any additional cells also to be changed. This is demonstrated below.

Example   A new asset will cost $25,000, generate an annual cash flow of $6000 over its 5-year life, and have an estimated $500 salvage value. The rate of return using the IRR function is 6.94%. Determine the annual cash flow necessary to raise the return to 10% per year.

Figure A–4 (top left) shows the cash flows and return displayed using the function = IRR(B4:B9) prior to the use of Goal Seek. Note that the initial $6000 is input in cell B5, but other years' cash flows are input as equations that refer to B5. The $500 salvage is added for the last year. This format allows Goal Seek to change only cell B5 while making the other cash flows have the same value. The tool finds the required cash flow of $6506 to approximate the 10% per year return. The Goal Seek Status inset indicates that a solution is found. Clicking OK saves all changed cells; clicking Cancel returns to the original values.

A.5 Solver—An Optimizing Tool for Capital Budgeting, Breakeven, and Sensitivity Analysis ● ● ●

Solver is a powerful spreadsheet tool to change the value in multiple (one or more) cells based on the value in a specific (target) cell. It is excellent when solving a **capital budgeting problem** to select from independent projects where budget constraints are present. (Section 12.4 details this application.) The initial Solver template is shown in Figure A–5.

Figure A–5
Solver template used to
specify optimization in a
target cell, multiple
changing cells, and
constraint relations.

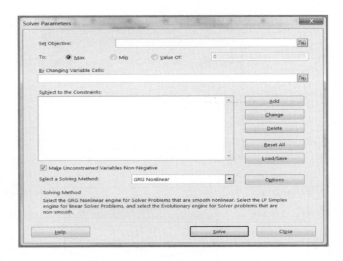

Set Target Cell box. Enter a cell reference or name. The target cell itself must contain a formula or function. The value in the cell can be maximized (Max), minimized (Min), or restricted to a specified value (Value Of).

By Changing Cells box. Enter the cell reference for each cell to be adjusted, using commas between nonadjacent cells. Each cell must be directly or indirectly related to the target cell. Solver proposes a value for the changing cell based on input provided about the target cell. The Guess button will list all possible changing cells related to the target cell.

Subject to the Constraints box. Enter any constraints that may apply, for example, $C$1 < $50,000. Integer and binary variables are determined in this box.

Options box. Choices here allow the user to specify various parameters of the solution: maximum time and number of iterations allowed, the precision and tolerance of the values determined, and the convergence requirements as the final solution is determined. Also, linear and nonlinear model assumptions can be set here. *If integer or binary variables are involved, the tolerance option must be set to a small number, say, 0.0001.* This is especially important for the binary variables when selecting from independent projects (Chapter 12). If tolerance remains at the default value of 5%, a project may be incorrectly included in the solution set at a very low level.

Solver Results box. This appears after Solve is clicked and a solution appears. It is possible, of course, that no solution can be found for the scenario described. It is possible to update the spreadsheet by clicking Keep Solver Solution, or return to the original entries using Restore Original Values.

A.6 Error Messages ● ● ●

If Excel is unable to complete a formula or function computation, an error message is displayed. Some of the common messages are:

#DIV/0!	Requires division by zero.
#N/A	Refers to a value that is not available.
#NAME?	Uses a name that Excel doesn't recognize.
#NULL!	Specifies an invalid intersection of two areas.
#NUM!	Uses a number incorrectly.
#REF!	Refers to a cell that is not valid.
#VALUE!	Uses an invalid argument or operand.
#####	Produces a result, or includes a constant numeric value, that is too long to fit in the cell. (Widen the column.)

APPENDIX B

BASICS OF ACCOUNTING REPORTS AND BUSINESS RATIOS

This appendix provides a fundamental description of financial statements. The documents discussed here will assist in reviewing or understanding basic financial statements and in gathering information useful in an engineering economy study.

B.1 The Balance Sheet ● ● ●

The fiscal year and the tax year are defined identically for a corporation or an individual—12 months in length. The fiscal year (FY) is commonly not the calendar year (CY) for a corporation. The U.S. government uses October through September as its FY. For example, October 2018 through September 2019 is FY2019. The fiscal or tax year is always the calendar year for an individual citizen.

At the end of each fiscal year, a company publishes a **balance sheet.** A sample balance sheet for S&B, Inc. is presented in Table B–1. This is an yearly presentation of the state of the firm at a particular time, for example, May 31, 2017; however, a balance sheet is also usually prepared quarterly and monthly. Three main categories are used.

Assets. This section is a summary of all resources owned by or owed to the company. There are two main classes of assets. *Current assets* represent shorter-lived working capital (cash, accounts receivable, etc.), which is more easily converted to cash, usually within 1 year. Longer-lived assets are referred to as *fixed assets* (land, buildings, etc.). Conversion of these holdings to cash in a short time would require a substantial corporate reorientation.

Liabilities. This section is a summary of all *financial obligations* (debts, mortgages, loans, etc.) of a corporation. Bond indebtedness is included here.

Net worth. Also called *owner's equity,* this section provides a summary of the financial value of ownership, including stocks issued and earnings retained by the corporation.

TABLE B–1 Sample Balance Sheet

S&B Incorporated Balance Sheet May 31, 2017			
Assets		**Liabilities**	
Current			
Cash	$10,500	Accounts payable	$19,700
Accounts receivable	18,700	Dividends payable	7,000
Interest accrued receivable	500	Long-term notes payable	16,000
Inventories	52,000	Bonds payable	20,000
Total current assets	$81,700	Total liabilities	$62,700
		Net Worth	
Fixed			
Land	$25,000	Common stock	$275,000
Building and equipment	438,000	Preferred stock	100,000
Less: Depreciation		Retained earnings	25,000
allowance $82,000	356,000		
Total fixed assets	381,000	Total net worth	400,000
Total assets	$462,700	Total liabilities and net worth	$462,700

The balance sheet is constructed using the relation

$$\text{Assets} = \text{liabilities} + \text{net worth}$$

In Table B–1, each major category is further divided into standard subcategories. For example, current assets is comprised of cash, accounts receivable, etc. Each subdivision has a specific interpretation, such as accounts receivable, which represents all money owed to the company by its customers.

B.2 Income Statement and Cost of Goods Sold Statement ● ● ●

A second important financial statement is the **income statement** (Table B–2). The income statement summarizes the profits or losses of the corporation for a stated period of time. Income statements always accompany balance sheets. The major categories of an income statement are

Revenues. This includes all *sales and interest revenue* that the company has received in the immediate past accounting period.

Expenses. This is a summary of *all expenses* (operating and others, including taxes) for the period. Some expense amounts are itemized in other statements, for example, cost of goods sold.

The final result of an income statement is the net operating profit after taxes (NOPAT), the amount used in Chapter 17, Sections 17.1 and 17.7. The income statement, published at the same time as the balance sheet, uses the basic equation

$$\text{Revenues} - \text{expenses} = \text{profit (or loss)}$$

The **cost of goods sold** is an important accounting term. It represents the net cost of producing the product marketed by the firm. Cost of goods sold may also be called *factory cost*. A statement of the cost of goods sold, such as that shown in Table B–3, is useful in determining exactly how much it costs to make a particular product over a stated time period, usually a year. The total of the cost of goods sold statement is entered as an expense item on the income statement. This total is determined using the relations

$$\text{Cost of goods sold} = \text{prime cost} + \text{indirect cost}$$
$$\text{Prime cost} = \text{direct materials} + \text{direct labor} \qquad \text{[B.1]}$$

TABLE B–2 Sample Income Statement

S&B Incorporated
Income Statement
Year Ended May 31, 2017

Revenues		
Sales	$505,000	
Interest revenue	3,500	
Total revenues		$508,500
Expenses		
Cost of goods sold (from Table B–3)	$290,000	
Selling	28,000	
Administrative	35,000	
Other	12,000	
Total expenses		365,000
Income before taxes		143,500
Taxes for year		64,575
Net operating profit after taxes (NOPAT)		$78,925

TABLE B–3 Sample Cost of Goods Sold Statement

S&B Incorporated
Statement of Cost of Goods Sold
Year Ended May 31, 2017

Materials		
Inventory June 1, 2016	$ 54,000	
Purchases during year	174,500	
Total	$228,500	
Less: Inventory May 31, 2017	50,000	
Cost of materials		$178,500
Direct labor		110,000
Prime cost		288,500
Indirect costs		7,000
Factory cost		295,500
Less: Increase in finished goods inventory during year		5,500
Cost of goods sold (into Table B–2)		$290,000

Indirect costs include all indirect costs (IDC) charged to a product, process, or cost center. Indirect cost allocation methods are discussed in Chapter 15.

B.3 Business Ratios ● ● ●

Accountants, financial analysts, and engineering economists frequently utilize business ratio analysis to evaluate the financial health (status) of a company over time and in relation to industry norms. Because the engineering economist must continually communicate with others, she or he should have a basic understanding of several ratios. For comparison purposes, it is necessary to compute the ratios for several companies in the same industry. Industrywide median ratio values are published annually by firms such as Dun and Bradstreet in *Industry Norms and Key Business Ratios.* The ratios are classified according to their role in measuring the corporation.

Solvency ratios. Assess ability to meet short-term and long-term financial obligations.

Efficiency ratios. Measure management's ability to use and control assets.

Profitability ratios. Evaluate the ability to earn a return for the owners of the corporation.

Numerical data for several important ratios are discussed here and are extracted from the S&B balance sheet and income statement, Tables B–1 and B–2.

Current Ratio This ratio is utilized to analyze the company's working capital condition. It is defined as

$$\text{Current ratio} = \frac{\text{current assets}}{\text{current liabilities}}$$

Current liabilities include all short-term debts, such as accounts and dividends payable. Note that only balance sheet data are utilized in the current ratio; that is, no association with revenues or expenses is made. For the balance sheet of Table B–1, current liabilities amount to $19,700 + $7000 = $26,700 and

$$\text{Current ratio} = \frac{81{,}700}{26{,}700} = 3.06$$

Since current liabilities are those debts payable in the next year, the current ratio value of 3.06 means that the current assets would cover short-term debts approximately three times. Current ratio values of 2 to 3 are common.

The current ratio assumes that the working capital invested in inventory can be converted to cash quite rapidly. Often, however, a better idea of a company's *immediate* financial position can be obtained by using the acid test ratio.

Acid Test Ratio (Quick Ratio or Liquidity Ratio)   This ratio is

$$\text{Acid test ratio} = \frac{\text{quick assets}}{\text{current liabilities}}$$

$$= \frac{\text{current assets} - \text{inventories}}{\text{current liabilities}}$$

It is meaningful for the emergency situation when the firm must cover short-term debts using its readily convertible assets. For S&B,

$$\text{Acid test ratio} = \frac{81{,}700 - 52{,}000}{26{,}700} = 1.11$$

Comparison of this and the current ratio shows that approximately two times the current debt of the company is invested in inventories. However, an acid test ratio of approximately 1.0 is generally regarded as a strong current position, regardless of the amount of assets in inventories.

Debt Ratio   This ratio is a measure of financial strength since it is defined as

$$\text{Debt ratio} = \frac{\text{total liabilities}}{\text{total assets}}$$

For S&B,

$$\text{Debt ratio} = \frac{62{,}700}{462{,}700} = 0.136$$

S&B is 13.6% creditor-owned and 86.4% stockholder-owned. A debt ratio in the range of 20% or less usually indicates a sound financial condition, with little fear of forced reorganization because of unpaid liabilities. However, a company with virtually no debts, that is, one with a very low debt ratio, may not have a promising future, because of its inexperience in dealing with short-term and long-term debt financing. The debt-equity (D-E) mix is another measure of financial strength.

Return on Sales Ratio   This often quoted ratio indicates the profit margin for the company. It is defined as

$$\text{Return on sales} = \frac{\text{net profit}}{\text{net sales}} (100\%)$$

Net profit is the after-tax value from the income statement. This ratio measures profit earned per sales dollar and indicates how well the corporation can sustain adverse conditions over time, such as falling prices, rising costs, and declining sales. For S&B,

$$\text{Return on sales} = \frac{78{,}925}{505{,}000} (100\%) = 15.6\%$$

Corporations may point to small return on sales ratios, say, 2.5% to 4.0%, as indications of sagging economic conditions. In truth, for a relatively large-volume, high-turnover business, an income ratio of 3% is quite healthy. Of course, a steadily decreasing ratio indicates rising company expenses, which absorb net profit after taxes.

Return on Assets Ratio   This is the key indicator of profitability since it evaluates the ability of the corporation to transfer assets into operating profit. The definition and value for S&B are

$$\text{Return on assets} = \frac{\text{net profit}}{\text{total assets}} (100\%)$$

$$= \frac{78{,}925}{462{,}700} (100\%) = 17.1\%$$

Efficient use of assets indicates that the company should earn a high return, while low returns usually accompany lower values of this ratio compared to the industry group ratios.

Inventory Turnover Ratio   Two different ratios are used here. They both indicate the number of times the average inventory value passes through the operations of the company. If turnover of inventory to *net sales* is desired, the formula is

$$\text{Net sales to inventory} = \frac{\text{net sales}}{\text{average inventory}}$$

where average inventory is the figure recorded in the balance sheet. For S&B this ratio is

$$\text{Net sales to inventory} = \frac{505,000}{52,000} = 9.71$$

This means that the average value of the inventory has been sold 9.71 times during the year. Values of this ratio vary greatly from one industry to another.

If inventory turnover is related to *cost of goods sold,* the ratio to use is

$$\text{Cost of goods sold to inventory} = \frac{\text{cost of goods sold}}{\text{average inventory}}$$

Now, average inventory is computed as the average of the beginning and ending inventory values in the statement of cost of goods sold. This ratio is commonly used as a measure of the inventory turnover rate in manufacturing companies. It varies with industries, but management likes to see it remain relatively constant as business increases. For S&B, using the values in Table B–3,

$$\text{Cost of goods sold to inventory} = \frac{290,000}{\frac{1}{2}(54,000 + 50,000)} = 5.58$$

There are, of course, many other ratios to use in various circumstances; however, the ones presented here are commonly used by both accountants and economic analysts.

EXAMPLE B.1

Sample values for financial ratios or percentages of four industry sectors are presented below. Compare the corresponding S&B, Inc. values with these norms, and comment on differences and similarities.

Ratio or Percentage	Motor Vehicles and Parts Manufacturing 336,105*	Air Transportation (Medium-Sized) 481,000*	Industrial Machinery Manufacturing 333,200*	Home Furnishings 442,000*
Current ratio	2.4	0.4	2.2	2.6
Quick ratio	1.6	0.3	1.5	1.2
Debt ratio	59.3%	96.8%	49.1%	52.4%
Return on assets	40.9%	8.1%	8.0%	5.1%

*North American Industry Classification System (NAICS) code for this industry sector.
SOURCE: L. Troy, *Almanac of Business and Industrial Financial Ratios,* CCH, Wolters Kluwer, USA.

Solution

It is not correct to compare ratios for one company with indexes in different industries, that is, with indexes for different NAICS codes. So the comparison below is for illustration purposes only. The corresponding values for S&B are

$$\text{Current ratio} = 3.06$$
$$\text{Quick ratio} = 1.11$$
$$\text{Debt ratio} = 13.5\%$$
$$\text{Return on assets} = 17.1\%$$

S&B has a current ratio larger than all four of these industries, since 3.06 indicates it can cover current liabilities three times compared with 2.6 and much less in the case of the "average" air transportation corporation. S&B has a significantly lower debt ratio than that of any of the sample industries, so it is likely more financially sound. Return on assets, which is a measure of ability to turn assets into profitability, is not as high at S&B as motor vehicles, but S&B competes well with the other industry sectors.

To make a fair comparison of S&B ratios with other values, it is necessary to have norm values *for its industry type* as well as ratio values for other corporations *in the same NAICS category* and about the same size in total assets. Corporate assets are classified in categories by $100,000 units, such as 100 to 250, 1001 to 5000, over 250,000, etc.

APPENDIX C

CODE OF ETHICS FOR ENGINEERS

Source: National Society of Professional Engineers (www.nspe.org).

National Society of Professional Engineers®

Code of Ethics for Engineers

Preamble

Engineering is an important and learned profession. As members of this profession, engineers are expected to exhibit the highest standards of honesty and integrity. Engineering has a direct and vital impact on the quality of life for all people. Accordingly, the services provided by engineers require honesty, impartiality, fairness, and equity, and must be dedicated to the protection of the public health, safety, and welfare. Engineers must perform under a standard of professional behavior that requires adherence to the highest principles of ethical conduct.

I. Fundamental Canons

Engineers, in the fulfillment of their professional duties, shall:

1. Hold paramount the safety, health, and welfare of the public.
2. Perform services only in areas of their competence.
3. Issue public statements only in an objective and truthful manner.
4. Act for each employer or client as faithful agents or trustees.
5. Avoid deceptive acts.
6. Conduct themselves honorably, responsibly, ethically, and lawfully so as to enhance the honor, reputation, and usefulness of the profession.

II. Rules of Practice

1. Engineers shall hold paramount the safety, health, and welfare of the public.
   a. If engineers' judgment is overruled under circumstances that endanger life or property, they shall notify their employer or client and such other authority as may be appropriate.
   b. Engineers shall approve only those engineering documents that are in conformity with applicable standards.
   c. Engineers shall not reveal facts, data, or information without the prior consent of the client or employer except as authorized or required by law or this Code.
   d. Engineers shall not permit the use of their name or associate in business ventures with any person or firm that they believe is engaged in fraudulent or dishonest enterprise.
   e. Engineers shall not aid or abet the unlawful practice of engineering by a person or firm.
   f. Engineers having knowledge of any alleged violation of this Code shall report thereon to appropriate professional bodies and, when relevant, also to public authorities, and cooperate with the proper authorities in furnishing such information or assistance as may be required.
2. Engineers shall perform services only in the areas of their competence.
   a. Engineers shall undertake assignments only when qualified by education or experience in the specific technical fields involved.
   b. Engineers shall not affix their signatures to any plans or documents dealing with subject matter in which they lack competence, nor to any plan or document not prepared under their direction and control.
   c. Engineers may accept assignments and assume responsibility for coordination of an entire project and sign and seal the engineering documents for the entire project, provided that each technical segment is signed and sealed only by the qualified engineers who prepared the segment.
3. Engineers shall issue public statements only in an objective and truthful manner.
   a. Engineers shall be objective and truthful in professional reports, statements, or testimony. They shall include all relevant and pertinent information in such reports, statements, or testimony, which should bear the date indicating when it was current.
   b. Engineers may express publicly technical opinions that are founded upon knowledge of the facts and competence in the subject matter.
   c. Engineers shall issue no statements, criticisms, or arguments on technical matters that are inspired or paid for by interested parties, unless they have prefaced their comments by explicitly identifying the interested parties on whose behalf they are speaking, and by revealing the existence of any interest the engineers may have in the matters.

4. Engineers shall act for each employer or client as faithful agents or trustees.
   a. Engineers shall disclose all known or potential conflicts of interest that could influence or appear to influence their judgment or the quality of their services.
   b. Engineers shall not accept compensation, financial or otherwise, from more than one party for services on the same project, or for services pertaining to the same project, unless the circumstances are fully disclosed and agreed to by all interested parties.
   c. Engineers shall not solicit or accept financial or other valuable consideration, directly or indirectly, from outside agents in connection with the work for which they are responsible.
   d. Engineers in public service as members, advisors, or employees of a governmental or quasi-governmental body or department shall not participate in decisions with respect to services solicited or provided by them or their organizations in private or public engineering practice.
   e. Engineers shall not solicit or accept a contract from a governmental body on which a principal or officer of their organization serves as a member.
5. Engineers shall avoid deceptive acts.
   a. Engineers shall not falsify their qualifications or permit misrepresentation of their or their associates' qualifications. They shall not misrepresent or exaggerate their responsibility in or for the subject matter of prior assignments. Brochures or other presentations incident to the solicitation of employment shall not misrepresent pertinent facts concerning employers, employees, associates, joint venturers, or past accomplishments.
   b. Engineers shall not offer, give, solicit, or receive, either directly or indirectly, any contribution to influence the award of a contract by public authority, or which may be reasonably construed by the public as having the effect or intent of influencing the awarding of a contract. They shall not offer any gift or other valuable consideration in order to secure work. They shall not pay a commission, percentage, or brokerage fee in order to secure work, except to a bona fide employee or bona fide established commercial or marketing agencies retained by them.

III. Professional Obligations

1. Engineers shall be guided in all their relations by the highest standards of honesty and integrity.
   a. Engineers shall acknowledge their errors and shall not distort or alter the facts.
   b. Engineers shall advise their clients or employers when they believe a project will not be successful.
   c. Engineers shall not accept outside employment to the detriment of their regular work or interest. Before accepting any outside engineering employment, they will notify their employers.
   d. Engineers shall not attempt to attract an engineer from another employer by false or misleading pretenses.
   e. Engineers shall not promote their own interest at the expense of the dignity and integrity of the profession.
2. Engineers shall at all times strive to serve the public interest.
   a. Engineers are encouraged to participate in civic affairs; career guidance for youths; and work for the advancement of the safety, health, and well-being of their community.
   b. Engineers shall not complete, sign, or seal plans and/or specifications that are not in conformity with applicable engineering standards. If the client or employer insists on such unprofessional conduct, they shall notify the proper authorities and withdraw from further service on the project.
   c. Engineers are encouraged to extend public knowledge and appreciation of engineering and its achievements.
   d. Engineers are encouraged to adhere to the principles of sustainable development[1] in order to protect the environment for future generations.

3. Engineers shall avoid all conduct or practice that deceives the public.
   a. Engineers shall avoid the use of statements containing a material misrepresentation of fact or omitting a material fact.
   b. Consistent with the foregoing, engineers may advertise for recruitment of personnel.
   c. Consistent with the foregoing, engineers may prepare articles for the lay or technical press, but such articles shall not imply credit to the author for work performed by others.
4. Engineers shall not disclose, without consent, confidential information concerning the business affairs or technical processes of any present or former client or employer, or public body on which they serve.
   a. Engineers shall not, without the consent of all interested parties, promote or arrange for new employment or practice in connection with a specific project for which the engineer has gained particular and specialized knowledge.
   b. Engineers shall not, without the consent of all interested parties, participate in or represent an adversary interest in connection with a specific project or proceeding in which the engineer has gained particular specialized knowledge on behalf of a former client or employer.
5. Engineers shall not be influenced in their professional duties by conflicting interests.
   a. Engineers shall not accept financial or other considerations, including free engineering designs, from material or equipment suppliers for specifying their product.
   b. Engineers shall not accept commissions or allowances, directly or indirectly, from contractors or other parties dealing with clients or employers of the engineer in connection with work for which the engineer is responsible.
6. Engineers shall not attempt to obtain employment or advancement or professional engagements by untruthfully criticizing other engineers, or by other improper or questionable methods.
   a. Engineers shall not request, propose, or accept a commission on a contingent basis under circumstances in which their judgment may be compromised.
   b. Engineers in salaried positions shall accept part-time engineering work only to the extent consistent with policies of the employer and in accordance with ethical considerations.
   c. Engineers shall not, without consent, use equipment, supplies, laboratory, or office facilities of an employer to carry on outside private practice.
7. Engineers shall not attempt to injure, maliciously or falsely, directly or indirectly, the professional reputation, prospects, practice, or employment of other engineers. Engineers who believe others are guilty of unethical or illegal practice shall present such information to the proper authority for action.
   a. Engineers in private practice shall not review the work of another engineer for the same client, except with the knowledge of such engineer, or unless the connection of such engineer with the work has been terminated.
   b. Engineers in governmental, industrial, or educational employ are entitled to review and evaluate the work of other engineers when so required by their employment duties.
   c. Engineers in sales or industrial employ are entitled to make engineering comparisons of represented products with products of other suppliers.
8. Engineers shall accept personal responsibility for their professional activities, provided, however, that engineers may seek indemnification for services arising out of their practice for other than gross negligence, where the engineer's interests cannot otherwise be protected.
   a. Engineers shall conform with state registration laws in the practice of engineering.
   b. Engineers shall not use association with a nonengineer, a corporation, or partnership as a "cloak" for unethical acts.

9. Engineers shall give credit for engineering work to those to whom credit is due, and will recognize the proprietary interests of others.
   a. Engineers shall, whenever possible, name the person or persons who may be individually responsible for designs, inventions, writings, or other accomplishments.
   b. Engineers using designs supplied by a client recognize that the designs remain the property of the client and may not be duplicated by the engineer for others without express permission.
   c. Engineers, before undertaking work for others in connection with which the engineer may make improvements, plans, designs, inventions, or other records that may justify copyrights or patents, should enter into a positive agreement regarding ownership.
   d. Engineers' designs, data, records, and notes referring exclusively to an employer's work are the employer's property. The employer should indemnify the engineer for use of the information for any purpose other than the original purpose.
   e. Engineers shall continue their professional development throughout their careers and should keep current in their specialty fields by engaging in professional practice, participating in continuing education courses, reading in the technical literature, and attending professional meetings and seminars.

Footnote 1   "Sustainable development" is the challenge of meeting human needs for natural resources, industrial products, energy, food, transportation, shelter, and effective waste management while conserving and protecting environmental quality and the natural resource base essential for future development.

As Revised July 2007

"By order of the United States District Court for the District of Columbia, former Section 11(c) of the NSPE Code of Ethics prohibiting competitive bidding, and all policy statements, opinions, rulings, or other guidelines interpreting its scope, have been rescinded as unlawfully interfering with the legal right of engineers, protected under the antitrust laws, to provide price information to prospective clients; accordingly, nothing contained in the NSPE Code of Ethics, policy statements, opinions, rulings, or other guidelines prohibits the submission of price quotations or competitive bids for engineering services at any time or in any amount."

Statement by NSPE Executive Committee

In order to correct misunderstandings which have been indicated in some instances since the issuance of the Supreme Court decision and the entry of the Final Judgment, it is noted that in its decision of April 25, 1978, the Supreme Court of the United States declared: "The Sherman Act does not require competitive bidding."

It is further noted that as made clear in the Supreme Court decision:
1. Engineers and firms may individually refuse to bid for engineering services.
2. Clients are not required to seek bids for engineering services.
3. Federal, state, and local laws governing procedures to procure engineering services are not affected, and remain in full force and effect.
4. State societies and local chapters are free to actively and aggressively seek legislation for professional selection and negotiation procedures by public agencies.
5. State registration board rules of professional conduct, including rules prohibiting competitive bidding for engineering services, are not affected and remain in full force and effect. State registration boards with authority to adopt rules of professional conduct may adopt rules governing procedures to obtain engineering services.
6. As noted by the Supreme Court, "nothing in the judgment prevents NSPE and its members from attempting to influence governmental action . . ."

Note: In regard to the question of application of the Code to corporations vis-a-vis real persons, business form or type should neither negate nor influence conformance of individuals to the Code. The Code deals with professional services, which services must be performed by real persons. Real persons in turn establish and implement policies within business structures. The Code is clearly written to apply to the Engineer, and it is incumbent on members of NSPE to endeavor to live up to its provisions. This applies to all pertinent sections of the Code.

**National Society of
Professional Engineers**®

1420 King Street
Alexandria, Virginia 22314-2794
703/684-2800 • Fax:703/836-4875
www.nspe.org
Publication date as revised: July 2007 • Publication #1102

ALTERNATE METHODS FOR EQUIVALENCE CALCULATIONS

Throughout the text, engineering economy factor formulas, tabulated factor values, or built-in spreadsheet functions have been used to obtain a value of P, F, A, i, or n. Because of advances in programmable and scientific calculators, many of the equivalence computations can be performed without the use of tables or spreadsheets, but rather with a handheld calculator. An overview of the possibilities is presented here.

Alternatively, the recognition that all equivalence calculations involve geometric series can likewise remove the need for tabulated values or spreadsheet functions. From the summation of the series, it is possible to perform calculator-based computations to obtain P, F, or A values. A brief introduction to this technique is presented in Section D.2.

D.1 Using Programmable Calculators ● ● ●

A basic way to calculate one parameter, given the other four, is to use a calculator that allows a present worth relation to be encoded. The software can then solve for any one of the parameters, when the remaining four are entered. For example, consider a PW relation in which all five parameters are included.

$$A(P/A,i,n) + F(P/F,i,n) + P = 0$$

The A, P, and F values can be positive (cash inflow) or negative (cash outflow) or zero, as long as there is at least one value with each sign. The interest rate i can be coded for entry as a percent or decimal. When the unknown variable is identified, the calculator's software can solve the equation for zero, thus providing the answer.

This is the approach taken by relatively simple scientific calculators, such as the Hewlett-Packard (HP) 33 and 35 series, which are acceptable for use when the FE exam is taken. By substituting the formulas for the factors, the actual relation entered into the calculator is

$$A\left[\frac{1 - (1 + i/100)^{-n}}{i/100}\right] + F[1 + (i/100)]^{-n} + P = 0$$

The HP calculator uses a slightly different symbol set than we have used thus far. The initial investment is called B rather than P, and the equal uniform amount is termed P rather than A. Once entered, the relation can be solved for any one variable, given values for the other four.

Another example that offers freedom from the spreadsheet and tables is an engineering calculator that has the same functions built in as those on a spreadsheet to determine P, F, A, i, or n. An example of this higher level is Texas Instrument's TI-Nspire series. The functions are basically the same as those on a spreadsheet. They are clustered under the heading of tvm (time value of money) functions. For example, the tvmPV function format is

$$\text{tvmPV(n,i,Pmt,FV,PpY,CpY,PmtAt)}$$

where   n = number of periods
           i = annual interest rate as a percent
       Pmt = equal uniform periodic amount A
        FV = future amount F
      PpY = payments per year (optional; default is 1)
      CpY = compounding periods per year (optional; default is 1)
   PmtAt = beginning- or end-of-period payments (optional; default is 0 = end)

It is easy to understand why it is possible to do a lot on a calculator with relatively well-behaved cash flow series. As the series become more complex, it is necessary to move to a spreadsheet for speed and versatility. Still, the use of tables or factor formulas is not necessary.

D.2 Using the Summation of a Geometric Series ● ● ●

A geometric progression is a series of n terms with a common ratio or base r. If c is a constant for each term, the series is written in the form

$$cr^a + cr^{a+1} + \cdots + cr^n = c\sum_{j=a}^{j=n} r^j$$

The sum S of a geometric series adds the terms using the closed-end form

$$S = \frac{r^{n+1} - r^a}{r - 1} \qquad \text{[D.1]}$$

Ristroph[1] and others have explained how the recognition that equivalence computations are simple applications of geometric series can be used to determine F, P, and A using Equation [D.1] and a simple hand-held calculator with exponentiation capability.

Before explaining how to apply this approach, we define the base r as follows when a future worth F or present worth P is sought.

To find F: $r = 1 + i$

To find P: $r = (1 + i)^{-1}$

A very familiar application of geometric series is the determination of the equivalent future worth F in year n for a single present worth amount P in year 0. This is the same as using a geometric series of only one term. As shown in Figure D–1, if $P = \$100$, $n = 10$ years, and $i = 10\%$ per year, when $r = 1 + i$,

$$F = P(1 + i)^n = P(r)^n = 100(1.1)^{10}$$

$$= 100(2.5937)$$

$$= \$259.37$$

This is identical to using the tabulated value (or formula) for the F/P factor.

$$F = P(F/P,i,n) = 100(F/P,10\%,10) = 100(2.5937)$$

$$= \$259.37$$

Figure D–2 shows a uniform annual series $A = \$100$ for 10 years. To calculate F, we can move each A value forward to year 10. Place a subscript j on each A value to indicate the year of occurrence, and determine F for each A_j value.

$$F = A_1(1 + i)^9 + A_2(1 + i)^8 + \cdots + A_9(1 + i)^1 + A_{10}(1 + i)^0$$

Figure D–1
Future worth of a single amount in year 0.

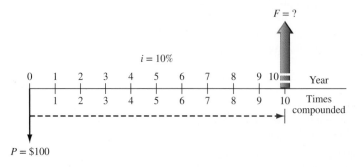

[1]J. H. Ristroph, "Engineering Economics: Time for New Directions?" *Proceedings*, ASEE Annual Conference, Austin, TX, June 2009.

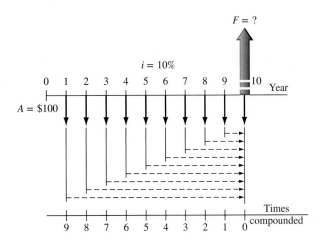

Figure D–2
Future worth of an A series from year 1 through year 10.

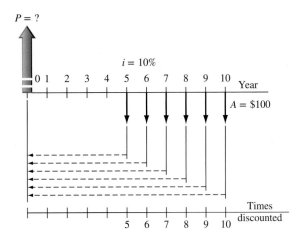

Figure D–3
Present worth of a shifted uniform series.

Note that the first value A_1 is compounded 9 times, not 10. With $r = 1 + i$, the geometric series and its sum S (from Equation [D.1]) can be developed. Removing the subscript on A,

$$F = A[r^9 + r^8 + \cdots + r^1 + r^0] = A\sum_{j=0}^{j=9} r^j$$

$$S = \frac{r^{10} - r^0}{r - 1} = \frac{(1.1)^{10} - 1}{0.1} = 15.9374$$

The F value is the same whether using the geometric series sum or the tabulated F/A factor.

$$F = 100(S) = 100(15.9374) = \$1593.74$$

$$F = 100(F/A,10\%,10) = 100(15.9374) = \$1593.74$$

As a final demonstration of the geometric series approach to equivalence computations, consider the shifted cash flow series in Figure D–3. The A series is present from years 5 through 10, and the P value is sought. If the tables are used, the solution is

$$P = A(P/A,10\%,6)(P/F,10\%,4) = 100(4.3553)(0.6830)$$

$$= \$297.47$$

As shown in the figure, this is a geometric series with the first A value discounted 5 years, therefore, $a = 5$. The last A term is discounted 10 years, making $n = 10$. Since P is sought, the geometric series base is $r = (1 + i)^{-1}$. Again using Equation [D.1] for the summation, we have

$$P = A\left[\sum_{j=5}^{j=10} r^j\right] = 100\left[\frac{r^{11} - r^5}{r - 1}\right] = 100\left[\frac{(1/1.1)^{11} - (1/1.1)^5}{(1/1.1) - 1}\right]$$

$$= 100\left[\frac{(0.9091)^{11} - (0.9091)^5}{0.9091 - 1}\right] = 100(2.9747)$$

$$= \$297.47$$

It is possible to develop similar relations to handle arithmetic and geometric series, conversions from P to A, and vice versa. Proponents of this approach point out the use of standard mathematical notation; the removal of a need to derive any factors; no need to remember the placement of P, F, and A values based on factor formula development; and the easy use of a calculator to determine the equivalence relations. As with the use of programmable calculators, the technique is excellent for well-behaved and reasonably complex series. When the series become quite involved, or when sensitivity analysis is required to reach an economic decision, it may be beneficial to use a spreadsheet. This often helps in performing sidebar and ancillary calculations that assist in the understanding of the problem, not just the math computations necessary to obtain an answer. However, once again, the use of tables and factors is not necessary.

APPENDIX E

GLOSSARY OF CONCEPTS AND TERMS

E.1 Important Concepts and Guidelines ● ● ●

Title

The following elements of engineering economy are identified throughout the text in the margin by this checkmark and a title below it. The numbers in parentheses indicate chapters where the concept or guideline is introduced or essential to obtaining a correct solution.

Time Value of Money   It is a fact that money *makes* money. This concept explains the change in the amount of money *over time* for both owned and borrowed funds. (1)

Economic Equivalence   A combination of time value of money and interest rate that makes different sums of money at different times have *equal economic value*. (1)

Cash Flow   The flow of money into and out of a company, project, or activity. *Revenues are cash inflows* and carry a positive (+) sign; *expenses are outflows* and carry a negative (−) sign. If only costs are involved, the − sign may be omitted, for example, benefit/cost (B/C) analysis. (1, 9)

End-of-Period Convention   To simplify calculations, cash flows (revenues and costs) are assumed to occur at the *end of a time period*. An interest period or fiscal period is commonly 1 *year*. A half-year convention is often used in depreciation calculations. (1, 16)

Cost of Capital   The interest rate incurred to obtain capital investment funds. COC is usually a *weighted average* that involves the cost of debt capital (loans, bonds, and mortgages) and equity capital (stocks and retained earnings). (1, 10)

Minimum Attractive Rate of Return (MARR)   A reasonable rate of return established for the evaluation of an economic alternative. Also called the *hurdle rate,* MARR is based on cost of capital, market trend, risk, etc. The inequality ROR ≥ MARR > COC is correct for an economically viable project. (1, 10)

Opportunity Cost   A forgone opportunity caused by the inability to pursue a project. Numerically, it is the *largest rate of return* of all the projects not funded due to the lack of capital funds. Stated differently, it is the ROR of the first project rejected because of unavailability of funds. (1, 10)

Nominal or Effective Interest Rate (r or i)   A nominal interest rate *does not include any compounding;* for example, 1% per month is the same as nominal 12% per year. Effective interest rate is the actual rate over a period of time because *compounding is imputed;* for example, 1% per month, compounded monthly, is an effective 12.683% per year. Inflation or deflation is not considered. (4)

Placement of Present Worth (P; PW)   In applying the $(P/A,i\%,n)$ factor, P or PW is always located *one interest period (year) prior to the first A amount*. The A or AW is a series of equal, end-of-period cash flows for n consecutive periods, expressed as money per time (say, $/year; €/year). (2, 3)

Placement of Future Worth (F; FW)   In applying the $(F/A,i\%,n)$ factor, F or FW is always located at the *end of the last interest period (year) of the A series*. (2, 3)

Placement of Arithmetic or Geometric Gradient Present Worth (P_G; P_g)   The ($P/G,i\%,n$) factor for an *arithmetic gradient* finds the P_G of only the gradient series *2 years prior* to the first appearance of the constant gradient G. The base amount A is treated separately from the gradient series.

The ($P/A,g,i,n$) factor for a *geometric gradient* determines P_g for the gradient and initial amount A_1 *two years prior* to the appearance of the first gradient amount. The initial amount A_1 *is* included in the value of P_g. (2, 3)

Equal-Service Requirement   Identical capacity of all alternatives operating over the *same amount of time* is mandated by the equal-service requirement. Estimated costs and revenues for equal service must be evaluated. PW analysis requires evaluation over the same number of years (periods) using the LCM (least common multiple) of lives; AW analysis is performed over one life cycle. Further, equal service assumes that all costs and revenues rise and fall in accordance with the overall rate of inflation or deflation over the total time period of the evaluation. (5, 6, 8)

LCM or Study Period   To select from mutually exclusive alternatives under the equal-service requirement for PW computations, the *LCM of lives with repurchase(s)* as necessary defines the study period. For a stated study period (planning horizon), evaluate cash flows *only over this period*, neglecting any beyond this time; estimated market values at termination of the study period are the salvage values. (5, 6, 11)

Salvage/Market Value   Expected trade-in, market, or scrap value at the *end of the estimated life* or the *study period*. In a replacement study, the defender's estimated market value at the end of a year is considered its "first cost" at the beginning of the next year. MACRS depreciation always reduces the book value to a salvage of zero. (5, 6, 11, 16)

Do Nothing   The DN alternative is always an option, unless one of the defined alternatives *must* be selected. DN is status quo; it generates *no new costs, revenues, or savings*. (5)

Revenue or Cost Alternative   Revenue alternatives have *costs and revenues* estimated; savings are considered negative costs and carry a + sign. Incremental evaluation requires comparison with DN for revenue alternatives. Cost alternatives have *only costs* estimated; revenues and savings are assumed equal between alternatives. (5, 8)

Rate of Return   An interest rate that equates a PW or AW relation to *zero*. Also defined as the rate on the unpaid balance of borrowed money, or rate earned on the unrecovered balance of an investment such that the *last cash flow brings the balance exactly to zero*. (7, 8)

Project Evaluation   *For a specified* MARR, determine a measure of worth for net cash flow series over the life or study period. Guidelines for a *single project* to be economically justified at the MARR (or discount rate) follow. (5, 6, 7, 9, 17)

Present worth: If PW $\geq$ 0   **Annual worth:** If AW $\geq$ 0

Future worth: If FW $\geq$ 0   **Rate of return:** If $i^* \geq$ MARR

Benefit/cost: If B/C $\geq$ 1.0   **Profitability index:** If PI $\geq$ 1.0

ME Alternative Selection   For mutually exclusive (select only one) alternatives, compare *two alternatives* at a time by determining a measure of worth for the incremental (Δ) cash flow series over the life or study period, adhering to the equal-service requirement. (5, 6, 8, 9, 10, 17)

Present worth or annual worth: Find PW or AW values at MARR; *select numerically largest* (least negative or most positive).

Rate of return: Order by *initial cost,* perform pairwise Δi^* comparison; if $\Delta i^* \geq$ MARR, select *larger cost* alternative; continue until one remains.

Benefit/cost: Order by *total equivalent cost,* perform pairwise $\Delta B/C$ comparison; if $\Delta B/C \geq$ 1.0, select *larger cost* alternative; continue until one remains.

Cost-effectiveness ratio: For service sector alternatives; order by *effectiveness measure;* perform pairwise $\Delta C/E$ comparison using *dominance;* select from nondominated alternatives without exceeding budget.

Independent Project Selection   No comparison between projects; only against DN. Calculate a measure of worth and select using the guidelines below. (5, 6, 8, 9, 12)

Present worth or annual worth: Find PW or AW at MARR; select all projects with PW or AW ≥ 0.

Rate of return: No incremental comparison; select all projects with overall $i^* \geq$ MARR.

Benefit/cost: No incremental comparison; select all projects with overall $B/C \geq 1.0$.

Cost-effectiveness ratio: For service sector projects; no incremental comparison; order by CER and select projects to not exceed budget.

When a capital budget limit is defined, independent projects are selected using the *capital budgeting process* based on PW values. The Solver spreadsheet tool is useful here.

Capital Recovery   CR is the equivalent annual amount an asset or system must earn to *recover the initial investment plus a stated rate of return.* Numerically, it is the AW value of the initial investment at a stated rate of return. The salvage value is considered in CR calculations. (6)

Economic Service Life   The ESL is the number of years *n* at which the *total AW of costs,* including salvage and AOC, is at its *minimum,* considering all the years the asset may provide service. (11)

Sunk Cost   Capital (money) that is lost and cannot be recovered. Sunk costs are not included when making decisions about the future. They should be handled using tax laws and write-off allowances, not the economic study. (11)

Inflation   Expressed as a percentage per time (% per year), it is an *increase* in the amount of money required to purchase the *same amount* of goods or services *over time.* Inflation occurs when the value of a currency decreases. Economic evaluations are performed using either a market (inflation-adjusted) interest rate or an inflation-free rate (constant-value terms). (1, 14)

Breakeven   For a single project, the value of a parameter that makes *two elements equal,* for example, sales necessary to equate revenues and costs. For two alternatives, breakeven is the value of a common variable at which the two are equally acceptable. Breakeven analysis is fundamental to make-buy decisions, replacement studies, payback analysis, sensitivity analysis, breakeven ROR analysis, and many others. The Goal Seek spreadsheet tool is useful in breakeven analysis. (8, 13)

Payback Period   Amount of time *n* before *recovery of the initial capital investment* is expected. Payback with $i > 0$ or simple payback at $i = 0$ is useful for preliminary or screening analysis to determine if a full PW, AW, or ROR analysis is needed. (13)

Direct/Indirect Costs   Direct costs are primarily human labor, machines, and materials associated with a product, process, system, or service. Indirect costs, which include support functions, utilities, management, legal, taxes, and the like, are more difficult to associate with a specific product or process. (15)

Value Added   Activities have added worth to a product or service from the perspective of a consumer, owner, or investor who is willing to pay more for an enhanced value. (17)

Sensitivity Analysis   Determination of how a measure of worth is affected by changes in estimated values of a parameter over a stated range. Parameters may be any cost factor, revenue, life, salvage value, inflation rate, etc. (18)

Risk Variation from an expected, desirable, or predicted value that may be detrimental to the product, process, or system. Risk represents an *absence of or deviation from certainty*. Probability estimates of variation (values) help evaluate risk and uncertainty using statistics and simulation. (10, 18, 19)

Decision Tree Constructed in the shape of a tree with branches that represent alternatives; each branch has a probability associated with it. Each final branch has an outcome with an estimated economic value. The tree is solved using a staged evaluation technique called the *rollback* process involving the outcome value and probabilities at each branch. The route through the branches with the best measure of worth, for example, PW, AW, B/C, determines the selected solution path. (18)

Real Options Analysis An analysis that evaluates the economic consequences of *delaying the funding of a decision* until later by purchasing an option now. Future costs must be estimated now, and risk is an important part of the analysis. The evaluation can usually be performed using a decision tree. (19)

E.2 Symbols and Terms ● ● ●

This section identifies and defines the common terms and their symbols used throughout the text. The numbers in parentheses indicate sections where the term is introduced and used in various applications.

Term	Symbol	Description
Annual amount or worth	A or AW	Equivalent uniform annual worth of all cash inflows and outflows over estimated life (1.5, 6.1).
Annual operating cost	AOC	Estimated annual costs to maintain and support an alternative (1.3).
Benefit/cost ratio	B/C	Ratio of a project's benefits to costs expressed in PW, AW, or FW terms (9.2).
Book value	BV	Remaining capital investment in an asset after depreciation is accounted for (16.1).
Breakeven point	Q_{BE}	Quantity at which revenues and costs are equal, or two alternatives are equivalent (13.1).
Capital budget	b	Amount of money available for capital investment projects (12.1).
Capital recovery	CR or A	Equivalent annual cost of owning an asset plus the required return on the initial investment (6.2).
Capitalized cost	CC or P	Present worth of an alternative that will last forever (or a long time) (5.5).
Cash flow	CF	Actual cash amounts that are receipts (inflow) and disbursements (outflow) (1.6).
Cash flow before or after taxes	CFBT or CFAT	Cash flow amount before relevant taxes or after taxes are applied (17.2).
Compounding frequency	m	Number of times interest is compounded per period (year) (4.1).
Cost-effectiveness ratio	CER	Ratio of equivalent cost to effectiveness measure to evaluate service sector projects (9.5).
Cost estimating relationships	C_2 or C_T	Relations that use design variables and changing costs over time to estimate current and future costs (15.3–4).
Cost of capital	WACC	Interest rate paid for the use of capital funds; includes both debt and equity funds. For debt and equity considered, it is weighted average cost of capital (1.9, 10.2).

Term	Symbol	Description
Debt-equity mix	D-E	Percentages of debt and equity investment capital used by a corporation (10.2).
Depreciation	D	Reduction in the value of assets using specific models and rules; there are book and tax depreciation methods (16.1).
Depreciation rate	d_t	Annual rate for reducing the value of assets using different depreciation methods (16.1).
Economic service life	ESL or n	Number of years at which the AW of costs is a minimum (11.2).
Effectiveness measure	E	A nonmonetary measure used in the cost-effectiveness ratio for service sector projects (9.5).
Expected value (average)	$\bar{X}$, μ, or $E(X)$	Long-run expected average if a random variable is sampled many times (18.3, 19.4).
Expenses, operating	OE	All corporate costs incurred in transacting business (17.1).
External rate of return	EROR	Unique ROR value determined using additional information when multiple IROR values are present for a cash flow series (7.5).
First cost	P	Total initial cost—purchase, construction, setup, etc. (1.3, 16.1).
Future amount or worth	F or FW	Amount at some future date considering time value of money (1.5, 5.4).
Gradient, arithmetic	G	Uniform change (+ or −) in cash flow each time period (2.5).
Gradient, geometric	g	Constant rate of change (+ or −) each time period (2.6).
Gross income	GI	Income from all sources for corporations or individuals (17.1).
Inflation rate	f	Rate that reflects changes in the value of a currency over time (14.1).
Interest	I	Amount earned or paid over time based on an initial amount and interest rate (1.4).
Interest rate	i or r	Interest expressed as a percentage of the original amount per time period; nominal (r) and effective (i) rates (1.4, 4.1).
Interest rate, inflation-adjusted	i_f	Interest rate adjusted to take inflation into account (14.1).
Internal rate of return	IROR	ROR value based on the cash flows themselves. Mathematically, there may be more than one IROR (7.2).
Life (estimated)	n	Number of years or periods over which an alternative or asset will be used; the evaluation time (1.5).
Life-cycle cost	LCC	Evaluation of costs for a system over all stages: feasibility to design to phase-out (6.5).
Measure of worth	Varies	Value, such as PW, AW, i^*, used to judge economic viability (1.1).
Minimum attractive rate of return	MARR	Minimum value of the rate of return for an alternative to be financially viable (1.9, 10.1).
Modified ROR	i' or MIRR	Unique ROR when a reinvestment rate i_i and external borrowing rate i_b are applied to multiple-rate cash flows (7.5).
Net cash flow	NCF	Resulting, actual amount of cash that flows in or out during a time period (1.6).
Net operating income	NOI	Difference between gross income and operating expenses (17.1).

Term	Symbol	Description
Net operating profit after taxes	NOPAT	Amount remaining after taxes are removed from taxable income (17.1).
Net present value	NPV	Another name for the present worth, PW.
Payback period	n_p	Number of years to recover the initial investment and a stated rate of return (13.3).
Present amount or worth	P or PW	Amount of money at the current time or a time denoted as *present* (1.5, 5.2).
Probability distribution	$P(X)$	Distribution of probability over different values of a variable (19.2).
Profitability index	PI	Ratio of PW of net cash flows to initial investment used for revenue projects; rewritten modified *B/C* ratio (9.2, 12.5).
Random variable	X	Parameter or characteristic that can take on any one of several values; discrete and continuous (19.2).
Rate of return	i^* or ROR	Compound interest rate on unpaid or unrecovered balances such that the final amount results in a zero balance (7.1).
Recovery period	n	Number of years to completely depreciate an asset (16.1).
Return on invested capital	i'' or ROIC	Unique ROR when a reinvestment rate i_i is applied to multiple-rate cash flows (7.5).
Salvage/market value	S or MV	Expected trade-in or market value when an asset is traded or disposed of (6.2, 11.1, 16.1).
Standard deviation	s or σ	Measure of dispersion or spread about the expected value or average (19.4).
Study period	n	Specified number of years over which an evaluation takes place (5.3, 11.5).
Taxable income	TI	Amount upon which income taxes are based (17.1).
Tax rate	T	Decimal rate, usually graduated, used to calculate corporate or individual taxes (17.1).
Tax rate, effective	T_e	Single-figure tax rate incorporating several rates and bases (17.1).
Time	t	Indicator for a time period (1.7).
Unadjusted basis	B	Depreciable amount of first cost, delivery, and installation costs of an asset (18.1).
Value added	EVA	Economic value added reflects net operating profit after taxes (NOPAT) after removing cost of invested capital during the year (17.7).
Value-added tax	VAT	An indirect consumption tax collected at each stage of production/distribution process; different from a sales tax paid by end user at purchase time (17.9).

REFERENCE MATERIALS

Textbooks on Engineering Economy and Related Topics ● ● ●

Blank, L. T., and A. Tarquin: *Basics of Engineering Economy,* 2nd ed., McGraw-Hill, New York, 2014.

Canada, J. R., W. G. Sullivan, D. J. Kulonda, and J. A. White: *Capital Investment Analysis for Engineering and Management,* 3rd ed., Pearson Prentice Hall, Upper Saddle River, NJ, 2005.

Collier, C. A., and C. R. Glagola: *Engineering Economic and Cost Analysis,* 3rd ed., Pearson Prentice Hall, Upper Saddle River, NJ, 1999.

Eschenbach, T. G.: *Engineering Economy: Applying Theory to Practice,* 3rd ed., Oxford University Press, New York, 2010.

Eschenbach, T. G., N. A. Lewis, J. C. Hartman, and L. E. Bussey: *The Economic Analysis of Industrial Projects,* 3rd ed., Oxford University Press, New York, 2016.

Fraser, N. M., E. M. Jewkes, I. Bernhardt, and M. Tajima: *Engineering Economics in Canada,* 3rd ed., Pearson Prentice Hall, Upper Saddle River, NJ, 2008.

Hartman, J. C.: *Engineering Economy and the Decision Making Process,* Pearson Prentice Hall, Upper Saddle River, NJ, 2007.

Newnan, D. G., J. P. Lavelle, and T. G. Eschenbach: *Engineering Economic Analysis,* 12th ed., Oxford University Press, New York, 2015.

Ostwald, P. F., and T. S. McLaren: *Cost Analysis and Estimating for Engineering and Management,* Pearson Prentice Hall, Upper Saddle River, NJ, 2004.

Park, C. S.: *Contemporary Engineering Economics,* 6th ed., Pearson Prentice Hall, Upper Saddle River, NJ, 2016.

Park, C. S.: *Fundamentals of Engineering Economics,* 3rd ed., Pearson Prentice Hall, Upper Saddle River, NJ, 2012.

Peurifoy, R. L., and G. D. Oberlender: *Estimating Construction Costs,* 6th ed., McGraw-Hill, New York, 2014.

Sullivan, W. G., E. M. Wicks, and C. P. Koelling: *Engineering Economy,* 16th ed., Pearson Prentice Hall, Upper Saddle River, NJ, 2014.

Thuesen, G. J., and W. J. Fabrycky: *Engineering Economy,* 9th ed., Pearson Prentice Hall, Upper Saddle River, NJ, 2001.

White, J. A., K. E. Case, D. B. Pratt, and M. H. Agee: *Principles of Engineering Economic Analysis,* 6th ed., John Wiley & Sons, New York, 2012.

White, J. A., K. S. Grasman, K. E. Case, K. L. Needy, and D. B Pratt: *Fundamentals of Engineering Economic Analysis,* 1st ed., John Wiley and Sons, New York, 2014.

Materials on Engineering Ethics ● ● ●

Harris, C. E., M. S. Pritchard, M. J. Rabins, R. James, and E. Englehardt: *Engineering Ethics: Concepts and Cases,* 5th ed., Wadsworth Cengage Learning, Boston, MA, 2014.

Martin, M. W., and R. Schinzinger: *Introduction to Engineering Ethics,* 2nd ed., McGraw-Hill, New York, 2010.

Selected Websites ● ● ●

Construction cost estimation index: www.construction.com
The Economist: www.economist.com
For this textbook: www.mhhe.com/blank
Plant cost estimation index: www.chemengonline.com
U.S. Internal Revenue Service: www.irs.gov
Wall Street Journal: www.wsj.com

U.S. Government Publications (Available at www.irs.gov) ● ● ●

Corporations, Publication 544, Internal Revenue Service, GPO, Washington, DC, annually.

Sales and Other Dispositions of Assets, Publication 542, Internal Revenue Service, GPO, Washington, DC, annually.

Your Federal Income Tax, Publication 17, Internal Revenue Service, GPO, Washington, DC, annually.

0.25%			TABLE 1	Discrete Cash Flow: Compound Interest Factors				0.25%
	Single Payments		**Uniform Series Payments**				**Arithmetic Gradients**	
	Compound Amount	Present Worth	Sinking Fund	Compound Amount	Capital Recovery	Present Worth	Gradient Present Worth	Gradient Uniform Series
n	F/P	P/F	A/F	F/A	A/P	P/A	P/G	A/G
1	1.0025	0.9975	1.00000	1.0000	1.00250	0.9975		
2	1.0050	0.9950	0.49938	2.0025	0.50188	1.9925	0.9950	0.4994
3	1.0075	0.9925	0.33250	3.0075	0.33500	2.9851	2.9801	0.9983
4	1.0100	0.9901	0.24906	4.0150	0.25156	3.9751	5.9503	1.4969
5	1.0126	0.9876	0.19900	5.0251	0.20150	4.9627	9.9007	1.9950
6	1.0151	0.9851	0.16563	6.0376	0.16813	5.9478	14.8263	2.4927
7	1.0176	0.9827	0.14179	7.0527	0.14429	6.9305	20.7223	2.9900
8	1.0202	0.9802	0.12391	8.0704	0.12641	7.9107	27.5839	3.4869
9	1.0227	0.9778	0.11000	9.0905	0.11250	8.8885	35.4061	3.9834
10	1.0253	0.9753	0.09888	10.1133	0.10138	9.8639	44.1842	4.4794
11	1.0278	0.9729	0.08978	11.1385	0.09228	10.8368	53.9133	4.9750
12	1.0304	0.9705	0.08219	12.1664	0.08469	11.8073	64.5886	5.4702
13	1.0330	0.9681	0.07578	13.1968	0.07828	12.7753	76.2053	5.9650
14	1.0356	0.9656	0.07028	14.2298	0.07278	13.7410	88.7587	6.4594
15	1.0382	0.9632	0.06551	15.2654	0.06801	14.7042	102.2441	6.9534
16	1.0408	0.9608	0.06134	16.3035	0.06384	15.6650	116.6567	7.4469
17	1.0434	0.9584	0.05766	17.3443	0.06016	16.6235	131.9917	7.9401
18	1.0460	0.9561	0.05438	18.3876	0.05688	17.5795	148.2446	8.4328
19	1.0486	0.9537	0.05146	19.4336	0.05396	18.5332	165.4106	8.9251
20	1.0512	0.9513	0.04882	20.4822	0.05132	19.4845	183.4851	9.4170
21	1.0538	0.9489	0.04644	21.5334	0.04894	20.4334	202.4634	9.9085
22	1.0565	0.9466	0.04427	22.5872	0.04677	21.3800	222.3410	10.3995
23	1.0591	0.9442	0.04229	23.6437	0.04479	22.3241	243.1131	10.8901
24	1.0618	0.9418	0.04048	24.7028	0.04298	23.2660	264.7753	11.3804
25	1.0644	0.9395	0.03881	25.7646	0.04131	24.2055	287.3230	11.8702
26	1.0671	0.9371	0.03727	26.8290	0.03977	25.1426	310.7516	12.3596
27	1.0697	0.9348	0.03585	27.8961	0.03835	26.0774	335.0566	12.8485
28	1.0724	0.9325	0.03452	28.9658	0.03702	27.0099	360.2334	13.3371
29	1.0751	0.9301	0.03329	30.0382	0.03579	27.9400	386.2776	13.8252
30	1.0778	0.9278	0.03214	31.1133	0.03464	28.8679	413.1847	14.3130
36	1.0941	0.9140	0.02658	37.6206	0.02908	34.3865	592.4988	17.2306
40	1.1050	0.9050	0.02380	42.0132	0.02630	38.0199	728.7399	19.1673
48	1.1273	0.8871	0.01963	50.9312	0.02213	45.1787	1040.06	23.0209
50	1.1330	0.8826	0.01880	53.1887	0.02130	46.9462	1125.78	23.9802
52	1.1386	0.8782	0.01803	55.4575	0.02053	48.7048	1214.59	24.9377
55	1.1472	0.8717	0.01698	58.8819	0.01948	51.3264	1353.53	26.3710
60	1.1616	0.8609	0.01547	64.6467	0.01797	55.6524	1600.08	28.7514
72	1.1969	0.8355	0.01269	78.7794	0.01519	65.8169	2265.56	34.4221
75	1.2059	0.8292	0.01214	82.3792	0.01464	68.3108	2447.61	35.8305
84	1.2334	0.8108	0.01071	93.3419	0.01321	75.6813	3029.76	40.0331
90	1.2520	0.7987	0.00992	100.7885	0.01242	80.5038	3446.87	42.8162
96	1.2709	0.7869	0.00923	108.3474	0.01173	85.2546	3886.28	45.5844
100	1.2836	0.7790	0.00881	113.4500	0.01131	88.3825	4191.24	47.4216
108	1.3095	0.7636	0.00808	123.8093	0.01058	94.5453	4829.01	51.0762
120	1.3494	0.7411	0.00716	139.7414	0.00966	103.5618	5852.11	56.5084
132	1.3904	0.7192	0.00640	156.1582	0.00890	112.3121	6950.01	61.8813
144	1.4327	0.6980	0.00578	173.0743	0.00828	120.8041	8117.41	67.1949
240	1.8208	0.5492	0.00305	328.3020	0.00555	180.3109	19,399	107.5863
360	2.4568	0.4070	0.00172	582.7369	0.00422	237.1894	36,264	152.8902
480	3.3151	0.3016	0.00108	926.0595	0.00358	279.3418	53,821	192.6699

| 0.5% | | | **TABLE 2** | Discrete Cash Flow: Compound Interest Factors | | | | 0.5% |

	Single Payments		Uniform Series Payments				Arithmetic Gradients	
	Compound Amount	Present Worth	Sinking Fund	Compound Amount	Capital Recovery	Present Worth	Gradient Present Worth	Gradient Uniform Series
n	F/P	P/F	A/F	F/A	A/P	P/A	P/G	A/G
1	1.0050	0.9950	1.00000	1.0000	1.00500	0.9950		
2	1.0100	0.9901	0.49875	2.0050	0.50375	1.9851	0.9901	0.4988
3	1.0151	0.9851	0.33167	3.0150	0.33667	2.9702	2.9604	0.9967
4	1.0202	0.9802	0.24813	4.0301	0.25313	3.9505	5.9011	1.4938
5	1.0253	0.9754	0.19801	5.0503	0.20301	4.9259	9.8026	1.9900
6	1.0304	0.9705	0.16460	6.0755	0.16960	5.8964	14.6552	2.4855
7	1.0355	0.9657	0.14073	7.1059	0.14573	6.8621	20.4493	2.9801
8	1.0407	0.9609	0.12283	8.1414	0.12783	7.8230	27.1755	3.4738
9	1.0459	0.9561	0.10891	9.1821	0.11391	8.7791	34.8244	3.9668
10	1.0511	0.9513	0.09777	10.2280	0.10277	9.7304	43.3865	4.4589
11	1.0564	0.9466	0.08866	11.2792	0.09366	10.6770	52.8526	4.9501
12	1.0617	0.9419	0.08107	12.3356	0.08607	11.6189	63.2136	5.4406
13	1.0670	0.9372	0.07464	13.3972	0.07964	12.5562	74.4602	5.9302
14	1.0723	0.9326	0.06914	14.4642	0.07414	13.4887	86.5835	6.4190
15	1.0777	0.9279	0.06436	15.5365	0.06936	14.4166	99.5743	6.9069
16	1.0831	0.9233	0.06019	16.6142	0.06519	15.3399	113.4238	7.3940
17	1.0885	0.9187	0.05651	17.6973	0.06151	16.2586	128.1231	7.8803
18	1.0939	0.9141	0.05323	18.7858	0.05823	17.1728	143.6634	8.3658
19	1.0994	0.9096	0.05030	19.8797	0.05530	18.0824	160.0360	8.8504
20	1.1049	0.9051	0.04767	20.9791	0.05267	18.9874	177.2322	9.3342
21	1.1104	0.9006	0.04528	22.0840	0.05028	19.8880	195.2434	9.8172
22	1.1160	0.8961	0.04311	23.1944	0.04811	20.7841	214.0611	10.2993
23	1.1216	0.8916	0.04113	24.3104	0.04613	21.6757	233.6768	10.7806
24	1.1272	0.8872	0.03932	25.4320	0.04432	22.5629	254.0820	11.2611
25	1.1328	0.8828	0.03765	26.5591	0.04265	23.4456	275.2686	11.7407
26	1.1385	0.8784	0.03611	27.6919	0.04111	24.3240	297.2281	12.2195
27	1.1442	0.8740	0.03469	28.8304	0.03969	25.1980	319.9523	12.6975
28	1.1499	0.8697	0.03336	29.9745	0.03836	26.0677	343.4332	13.1747
29	1.1556	0.8653	0.03213	31.1244	0.03713	26.9330	367.6625	13.6510
30	1.1614	0.8610	0.03098	32.2800	0.03598	27.7941	392.6324	14.1265
36	1.1967	0.8356	0.02542	39.3361	0.03042	32.8710	557.5598	16.9621
40	1.2208	0.8191	0.02265	44.1588	0.02765	36.1722	681.3347	18.8359
48	1.2705	0.7871	0.01849	54.0978	0.02349	42.5803	959.9188	22.5437
50	1.2832	0.7793	0.01765	56.6452	0.02265	44.1428	1035.70	23.4624
52	1.2961	0.7716	0.01689	59.2180	0.02189	45.6897	1113.82	24.3778
55	1.3156	0.7601	0.01584	63.1258	0.02084	47.9814	1235.27	25.7447
60	1.3489	0.7414	0.01433	69.7700	0.01933	51.7256	1448.65	28.0064
72	1.4320	0.6983	0.01157	86.4089	0.01657	60.3395	2012.35	33.3504
75	1.4536	0.6879	0.01102	90.7265	0.01602	62.4136	2163.75	34.6679
84	1.5204	0.6577	0.00961	104.0739	0.01461	68.4530	2640.66	38.5763
90	1.5666	0.6383	0.00883	113.3109	0.01383	72.3313	2976.08	41.1451
96	1.6141	0.6195	0.00814	122.8285	0.01314	76.0952	3324.18	43.6845
100	1.6467	0.6073	0.00773	129.3337	0.01273	78.5426	3562.79	45.3613
108	1.7137	0.5835	0.00701	142.7399	0.01201	83.2934	4054.37	48.6758
120	1.8194	0.5496	0.00610	163.8793	0.01110	90.0735	4823.51	53.5508
132	1.9316	0.5177	0.00537	186.3226	0.01037	96.4596	5624.59	58.3103
144	2.0508	0.4876	0.00476	210.1502	0.00976	102.4747	6451.31	62.9551
240	3.3102	0.3021	0.00216	462.0409	0.00716	139.5808	13,416	96.1131
360	6.0226	0.1660	0.00100	1004.52	0.00600	166.7916	21,403	128.3236
480	10.9575	0.0913	0.00050	1991.49	0.00550	181.7476	27,588	151.7949

| 0.75% | | TABLE 3 | Discrete Cash Flow: Compound Interest Factors | | | | | 0.75% |

	Single Payments		Uniform Series Payments				Arithmetic Gradients	
	Compound Amount	Present Worth	Sinking Fund	Compound Amount	Capital Recovery	Present Worth	Gradient Present Worth	Gradient Uniform Series
n	F/P	P/F	A/F	F/A	A/P	P/A	P/G	A/G
1	1.0075	0.9926	1.00000	1.0000	1.00750	0.9926		
2	1.0151	0.9852	0.49813	2.0075	0.50563	1.9777	0.9852	0.4981
3	1.0227	0.9778	0.33085	3.0226	0.33835	2.9556	2.9408	0.9950
4	1.0303	0.9706	0.24721	4.0452	0.25471	3.9261	5.8525	1.4907
5	1.0381	0.9633	0.19702	5.0756	0.20452	4.8894	9.7058	1.9851
6	1.0459	0.9562	0.16357	6.1136	0.17107	5.8456	14.4866	2.4782
7	1.0537	0.9490	0.13967	7.1595	0.14717	6.7946	20.1808	2.9701
8	1.0616	0.9420	0.12176	8.2132	0.12926	7.7366	26.7747	3.4608
9	1.0696	0.9350	0.10782	9.2748	0.11532	8.6716	34.2544	3.9502
10	1.0776	0.9280	0.09667	10.3443	0.10417	9.5996	42.6064	4.4384
11	1.0857	0.9211	0.08755	11.4219	0.09505	10.5207	51.8174	4.9253
12	1.0938	0.9142	0.07995	12.5076	0.08745	11.4349	61.8740	5.4110
13	1.1020	0.9074	0.07352	13.6014	0.08102	12.3423	72.7632	5.8954
14	1.1103	0.9007	0.06801	14.7034	0.07551	13.2430	84.4720	6.3786
15	1.1186	0.8940	0.06324	15.8137	0.07074	14.1370	96.9876	6.8606
16	1.1270	0.8873	0.05906	16.9323	0.06656	15.0243	110.2973	7.3413
17	1.1354	0.8807	0.05537	18.0593	0.06287	15.9050	124.3887	7.8207
18	1.1440	0.8742	0.05210	19.1947	0.05960	16.7792	139.2494	8.2989
19	1.1525	0.8676	0.04917	20.3387	0.05667	17.6468	154.8671	8.7759
20	1.1612	0.8612	0.04653	21.4912	0.05403	18.5080	171.2297	9.2516
21	1.1699	0.8548	0.04415	22.6524	0.05165	19.3628	188.3253	9.7261
22	1.1787	0.8484	0.04198	23.8223	0.04948	20.2112	206.1420	10.1994
23	1.1875	0.8421	0.04000	25.0010	0.04750	21.0533	224.6682	10.6714
24	1.1964	0.8358	0.03818	26.1885	0.04568	21.8891	243.8923	11.1422
25	1.2054	0.8296	0.03652	27.3849	0.04402	22.7188	263.8029	11.6117
26	1.2144	0.8234	0.03498	28.5903	0.04248	23.5422	284.3888	12.0800
27	1.2235	0.8173	0.03355	29.8047	0.04105	24.3595	305.6387	12.5470
28	1.2327	0.8112	0.03223	31.0282	0.03973	25.1707	327.5416	13.0128
29	1.2420	0.8052	0.03100	32.2609	0.03850	25.9759	350.0867	13.4774
30	1.2513	0.7992	0.02985	33.5029	0.03735	26.7751	373.2631	13.9407
36	1.3086	0.7641	0.02430	41.1527	0.03180	31.4468	524.9924	16.6946
40	1.3483	0.7416	0.02153	46.4465	0.02903	34.4469	637.4693	18.5058
48	1.4314	0.6986	0.01739	57.5207	0.02489	40.1848	886.8404	22.0691
50	1.4530	0.6883	0.01656	60.3943	0.02406	41.5664	953.8486	22.9476
52	1.4748	0.6780	0.01580	63.3111	0.02330	42.9276	1022.59	23.8211
55	1.5083	0.6630	0.01476	67.7688	0.02226	44.9316	1128.79	25.1223
60	1.5657	0.6387	0.01326	75.4241	0.02076	48.1734	1313.52	27.2665
72	1.7126	0.5839	0.01053	95.0070	0.01803	55.4768	1791.25	32.2882
75	1.7514	0.5710	0.00998	100.1833	0.01748	57.2027	1917.22	33.5163
84	1.8732	0.5338	0.00859	116.4269	0.01609	62.1540	2308.13	37.1357
90	1.9591	0.5104	0.00782	127.8790	0.01532	65.2746	2578.00	39.4946
96	2.0489	0.4881	0.00715	139.8562	0.01465	68.2584	2853.94	41.8107
100	2.1111	0.4737	0.00675	148.1445	0.01425	70.1746	3040.75	43.3311
108	2.2411	0.4462	0.00604	165.4832	0.01354	73.8394	3419.90	46.3154
120	2.4514	0.4079	0.00517	193.5143	0.01267	78.9417	3998.56	50.6521
132	2.6813	0.3730	0.00446	224.1748	0.01196	83.6064	4583.57	54.8232
144	2.9328	0.3410	0.00388	257.7116	0.01138	87.8711	5169.58	58.8314
240	6.0092	0.1664	0.00150	667.8869	0.00900	111.1450	9494.12	85.4210
360	14.7306	0.0679	0.00055	1830.74	0.00805	124.2819	13,312	107.1145
480	36.1099	0.0277	0.00021	4681.32	0.00771	129.6409	15,513	119.6620

	Single Payments		Uniform Series Payments				Arithmetic Gradients	
	Compound Amount	Present Worth	Sinking Fund	Compound Amount	Capital Recovery	Present Worth	Gradient Present Worth	Gradient Uniform Series
n	*F/P*	*P/F*	*A/F*	*F/A*	*A/P*	*P/A*	*P/G*	*A/G*
1	1.0100	0.9901	1.00000	1.0000	1.01000	0.9901		
2	1.0201	0.9803	0.49751	2.0100	0.50751	1.9704	0.9803	0.4975
3	1.0303	0.9706	0.33002	3.0301	0.34002	2.9410	2.9215	0.9934
4	1.0406	0.9610	0.24628	4.0604	0.25628	3.9020	5.8044	1.4876
5	1.0510	0.9515	0.19604	5.1010	0.20604	4.8534	9.6103	1.9801
6	1.0615	0.9420	0.16255	6.1520	0.17255	5.7955	14.3205	2.4710
7	1.0721	0.9327	0.13863	7.2135	0.14863	6.7282	19.9168	2.9602
8	1.0829	0.9235	0.12069	8.2857	0.13069	7.6517	26.3812	3.4478
9	1.0937	0.9143	0.10674	9.3685	0.11674	8.5660	33.6959	3.9337
10	1.1046	0.9053	0.09558	10.4622	0.10558	9.4713	41.8435	4.4179
11	1.1157	0.8963	0.08645	11.5668	0.09645	10.3676	50.8067	4.9005
12	1.1268	0.8874	0.07885	12.6825	0.08885	11.2551	60.5687	5.3815
13	1.1381	0.8787	0.07241	13.8093	0.08241	12.1337	71.1126	5.8607
14	1.1495	0.8700	0.06690	14.9474	0.07690	13.0037	82.4221	6.3384
15	1.1610	0.8613	0.06212	16.0969	0.07212	13.8651	94.4810	6.8143
16	1.1726	0.8528	0.05794	17.2579	0.06794	14.7179	107.2734	7.2886
17	1.1843	0.8444	0.05426	18.4304	0.06426	15.5623	120.7834	7.7613
18	1.1961	0.8360	0.05098	19.6147	0.06098	16.3983	134.9957	8.2323
19	1.2081	0.8277	0.04805	20.8109	0.05805	17.2260	149.8950	8.7017
20	1.2202	0.8195	0.04542	22.0190	0.05542	18.0456	165.4664	9.1694
21	1.2324	0.8114	0.04303	23.2392	0.05303	18.8570	181.6950	9.6354
22	1.2447	0.8034	0.04086	24.4716	0.05086	19.6604	198.5663	10.0998
23	1.2572	0.7954	0.03889	25.7163	0.04889	20.4558	216.0660	10.5626
24	1.2697	0.7876	0.03707	26.9735	0.04707	21.2434	234.1800	11.0237
25	1.2824	0.7798	0.03541	28.2432	0.04541	22.0232	252.8945	11.4831
26	1.2953	0.7720	0.03387	29.5256	0.04387	22.7952	272.1957	11.9409
27	1.3082	0.7644	0.03245	30.8209	0.04245	23.5596	292.0702	12.3971
28	1.3213	0.7568	0.03112	32.1291	0.04112	24.3164	312.5047	12.8516
29	1.3345	0.7493	0.02990	33.4504	0.03990	25.0658	333.4863	13.3044
30	1.3478	0.7419	0.02875	34.7849	0.03875	25.8077	355.0021	13.7557
36	1.4308	0.6989	0.02321	43.0769	0.03321	30.1075	494.6207	16.4285
40	1.4889	0.6717	0.02046	48.8864	0.03046	32.8347	596.8561	18.1776
48	1.6122	0.6203	0.01633	61.2226	0.02633	37.9740	820.1460	21.5976
50	1.6446	0.6080	0.01551	64.4632	0.02551	39.1961	879.4176	22.4363
52	1.6777	0.5961	0.01476	67.7689	0.02476	40.3942	939.9175	23.2686
55	1.7285	0.5785	0.01373	72.8525	0.02373	42.1472	1032.81	24.5049
60	1.8167	0.5504	0.01224	81.6697	0.02224	44.9550	1192.81	26.5333
72	2.0471	0.4885	0.00955	104.7099	0.01955	51.1504	1597.87	31.2386
75	2.1091	0.4741	0.00902	110.9128	0.01902	52.5871	1702.73	32.3793
84	2.3067	0.4335	0.00765	130.6723	0.01765	56.6485	2023.32	35.7170
90	2.4486	0.4084	0.00690	144.8633	0.01690	59.1609	2240.57	37.8724
96	2.5993	0.3847	0.00625	159.9273	0.01625	61.5277	2459.43	39.9727
100	2.7048	0.3697	0.00587	170.4814	0.01587	63.0289	2605.78	41.3426
108	2.9289	0.3414	0.00518	192.8926	0.01518	65.8578	2898.42	44.0103
120	3.3004	0.3030	0.00435	230.0387	0.01435	69.7005	3334.11	47.8349
132	3.7190	0.2689	0.00368	271.8959	0.01368	73.1108	3761.69	51.4520
144	4.1906	0.2386	0.00313	319.0616	0.01313	76.1372	4177.47	54.8676
240	10.8926	0.0918	0.00101	989.2554	0.01101	90.8194	6878.60	75.7393
360	35.9496	0.0278	0.00029	3494.96	0.01029	97.2183	8720.43	89.6995
480	118.6477	0.0084	0.00008	11,765	0.01008	99.1572	9511.16	95.9200

1.25%			TABLE 5	Discrete Cash Flow: Compound Interest Factors				1.25%
	Single Payments		Uniform Series Payments				Arithmetic Gradients	
	Compound Amount F/P	Present Worth P/F	Sinking Fund A/F	Compound Amount F/A	Capital Recovery A/P	Present Worth P/A	Gradient Present Worth P/G	Gradient Uniform Series A/G
n								
1	1.0125	0.9877	1.00000	1.0000	1.01250	0.9877		
2	1.0252	0.9755	0.49680	2.0125	0.50939	1.9631	0.9755	0.4969
3	1.0380	0.9634	0.32920	3.0377	0.34170	2.9265	2.9023	0.9917
4	1.0509	0.9515	0.24536	4.0756	0.25786	3.8781	5.7569	1.4845
5	1.0641	0.9398	0.19506	5.1266	0.20756	4.8178	9.5160	1.9752
6	1.0774	0.9282	0.16153	6.1907	0.17403	5.7460	14.1569	2.4638
7	1.0909	0.9167	0.13759	7.2680	0.15009	6.6627	19.6571	2.9503
8	1.1045	0.9054	0.11963	8.3589	0.13213	7.5681	25.9949	3.4348
9	1.1183	0.8942	0.10567	9.4634	0.11817	8.4623	33.1487	3.9172
10	1.1323	0.8832	0.09450	10.5817	0.10700	9.3455	41.0973	4.3975
11	1.1464	0.8723	0.08537	11.7139	0.09787	10.2178	49.8201	4.8758
12	1.1608	0.8615	0.07776	12.8604	0.09026	11.0793	59.2967	5.3520
13	1.1753	0.8509	0.07132	14.0211	0.08382	11.9302	69.5072	5.8262
14	1.1900	0.8404	0.06581	15.1964	0.07831	12.7706	80.4320	6.2982
15	1.2048	0.8300	0.06103	16.3863	0.07353	13.6005	92.0519	6.7682
16	1.2199	0.8197	0.05685	17.5912	0.06935	14.4203	104.3481	7.2362
17	1.2351	0.8096	0.05316	18.8111	0.06566	15.2299	117.3021	7.7021
18	1.2506	0.7996	0.04988	20.0462	0.06238	16.0295	130.8958	8.1659
19	1.2662	0.7898	0.04696	21.2968	0.05946	16.8193	145.1115	8.6277
20	1.2820	0.7800	0.04432	22.5630	0.05682	17.5993	159.9316	9.0874
21	1.2981	0.7704	0.04194	23.8450	0.05444	18.3697	175.3392	9.5450
22	1.3143	0.7609	0.03977	25.1431	0.05227	19.1306	191.3174	10.0006
23	1.3307	0.7515	0.03780	26.4574	0.05030	19.8820	207.8499	10.4542
24	1.3474	0.7422	0.03599	27.7881	0.04849	20.6242	224.9204	10.9056
25	1.3642	0.7330	0.03432	29.1354	0.04682	21.3573	242.5132	11.3551
26	1.3812	0.7240	0.03279	30.4996	0.04529	22.0813	260.6128	11.8024
27	1.3985	0.7150	0.03137	31.8809	0.04387	22.7963	279.2040	12.2478
28	1.4160	0.7062	0.03005	33.2794	0.04255	23.5025	298.2719	12.6911
29	1.4337	0.6975	0.02882	34.6954	0.04132	24.2000	317.8019	13.1323
30	1.4516	0.6889	0.02768	36.1291	0.04018	24.8889	337.7797	13.5715
36	1.5639	0.6394	0.02217	45.1155	0.03467	28.8473	466.2830	16.1639
40	1.6436	0.6084	0.01942	51.4896	0.03192	31.3269	559.2320	17.8515
48	1.8154	0.5509	0.01533	65.2284	0.02783	35.9315	759.2296	21.1299
50	1.8610	0.5373	0.01452	68.8818	0.02702	37.0129	811.6738	21.9295
52	1.9078	0.5242	0.01377	72.6271	0.02627	38.0677	864.9409	22.7211
55	1.9803	0.5050	0.01275	78.4225	0.02525	39.6017	946.2277	23.8936
60	2.1072	0.4746	0.01129	88.5745	0.02379	42.0346	1084.84	25.8083
72	2.4459	0.4088	0.00865	115.6736	0.02115	47.2925	1428.46	30.2047
75	2.5388	0.3939	0.00812	123.1035	0.02062	48.4890	1515.79	31.2605
84	2.8391	0.3522	0.00680	147.1290	0.01930	51.8222	1778.84	34.3258
90	3.0588	0.3269	0.00607	164.7050	0.01857	53.8461	1953.83	36.2855
96	3.2955	0.3034	0.00545	183.6411	0.01795	55.7246	2127.52	38.1793
100	3.4634	0.2887	0.00507	197.0723	0.01757	56.9013	2242.24	39.4058
108	3.8253	0.2614	0.00442	226.0226	0.01692	59.0865	2468.26	41.7737
120	4.4402	0.2252	0.00363	275.2171	0.01613	61.9828	2796.57	45.1184
132	5.1540	0.1940	0.00301	332.3198	0.01551	64.4781	3109.35	48.2234
144	5.9825	0.1672	0.00251	398.6021	0.01501	66.6277	3404.61	51.0990
240	19.7155	0.0507	0.00067	1497.24	0.01317	75.9423	5101.53	67.1764
360	87.5410	0.0114	0.00014	6923.28	0.01264	79.0861	5997.90	75.8401
480	388.7007	0.0026	0.00003	31,016	0.01253	79.7942	6284.74	78.7619

	Single Payments		Uniform Series Payments				Arithmetic Gradients	
1.5%			**TABLE 6**	**Discrete Cash Flow: Compound Interest Factors**				**1.5%**
	Compound Amount	Present Worth	Sinking Fund	Compound Amount	Capital Recovery	Present Worth	Gradient Present Worth	Gradient Uniform Series
n	*F/P*	*P/F*	*A/F*	*F/A*	*A/P*	*P/A*	*P/G*	*A/G*
1	1.0150	0.9852	1.00000	1.0000	1.01500	0.9852		
2	1.0302	0.9707	0.49628	2.0150	0.51128	1.9559	0.9707	0.4963
3	1.0457	0.9563	0.32838	3.0452	0.34338	2.9122	2.8833	0.9901
4	1.0614	0.9422	0.24444	4.0909	0.25944	3.8544	5.7098	1.4814
5	1.0773	0.9283	0.19409	5.1523	0.20909	4.7826	9.4229	1.9702
6	1.0934	0.9145	0.16053	6.2296	0.17553	5.6972	13.9956	2.4566
7	1.1098	0.9010	0.13656	7.3230	0.15156	6.5982	19.4018	2.9405
8	1.1265	0.8877	0.11858	8.4328	0.13358	7.4859	25.6157	3.4219
9	1.1434	0.8746	0.10461	9.5593	0.11961	8.3605	32.6125	3.9008
10	1.1605	0.8617	0.09343	10.7027	0.10843	9.2222	40.3675	4.3772
11	1.1779	0.8489	0.08429	11.8633	0.09929	10.0711	48.8568	4.8512
12	1.1956	0.8364	0.07668	13.0412	0.09168	10.9075	58.0571	5.3227
13	1.2136	0.8240	0.07024	14.2368	0.08524	11.7315	67.9454	5.7917
14	1.2318	0.8118	0.06472	15.4504	0.07972	12.5434	78.4994	6.2582
15	1.2502	0.7999	0.05994	16.6821	0.07494	13.3432	89.6974	6.7223
16	1.2690	0.7880	0.05577	17.9324	0.07077	14.1313	101.5178	7.1839
17	1.2880	0.7764	0.05208	19.2014	0.06708	14.9076	113.9400	7.6431
18	1.3073	0.7649	0.04881	20.4894	0.06381	15.6726	126.9435	8.0997
19	1.3270	0.7536	0.04588	21.7967	0.06088	16.4262	140.5084	8.5539
20	1.3469	0.7425	0.04325	23.1237	0.05825	17.1686	154.6154	9.0057
21	1.3671	0.7315	0.04087	24.4705	0.05587	17.9001	169.2453	9.4550
22	1.3876	0.7207	0.03870	25.8376	0.05370	18.6208	184.3798	9.9018
23	1.4084	0.7100	0.03673	27.2251	0.05173	19.3309	200.0006	10.3462
24	1.4295	0.6995	0.03492	28.6335	0.04992	20.0304	216.0901	10.7881
25	1.4509	0.6892	0.03326	30.0630	0.04826	20.7196	232.6310	11.2276
26	1.4727	0.6790	0.03173	31.5140	0.04673	21.3986	249.6065	11.6646
27	1.4948	0.6690	0.03032	32.9867	0.04532	22.0676	267.0002	12.0992
28	1.5172	0.6591	0.02900	34.4815	0.04400	22.7267	284.7958	12.5313
29	1.5400	0.6494	0.02778	35.9987	0.04278	23.3761	302.9779	12.9610
30	1.5631	0.6398	0.02664	37.5387	0.04164	24.0158	321.5310	13.3883
36	1.7091	0.5851	0.02115	47.2760	0.03615	27.6607	439.8303	15.9009
40	1.8140	0.5513	0.01843	54.2679	0.03343	29.9158	524.3568	17.5277
48	2.0435	0.4894	0.01437	69.5652	0.02937	34.0426	703.5462	20.6667
50	2.1052	0.4750	0.01357	73.6828	0.02857	34.9997	749.9636	21.4277
52	2.1689	0.4611	0.01283	77.9249	0.02783	35.9287	796.8774	22.1794
55	2.2679	0.4409	0.01183	84.5296	0.02683	37.2715	868.0285	23.2894
60	2.4432	0.4093	0.01039	96.2147	0.02539	39.3803	988.1674	25.0930
72	2.9212	0.3423	0.00781	128.0772	0.02281	43.8447	1279.79	29.1893
75	3.0546	0.3274	0.00730	136.9728	0.02230	44.8416	1352.56	30.1631
84	3.4926	0.2863	0.00602	166.1726	0.02102	47.5786	1568.51	32.9668
90	3.8189	0.2619	0.00532	187.9299	0.02032	49.2099	1709.54	34.7399
96	4.1758	0.2395	0.00472	211.7202	0.01972	50.7017	1847.47	36.4381
100	4.4320	0.2256	0.00437	228.8030	0.01937	51.6247	1937.45	37.5295
108	4.9927	0.2003	0.00376	266.1778	0.01876	53.3137	2112.13	39.6171
120	5.9693	0.1675	0.00302	331.2882	0.01802	55.4985	2359.71	42.5185
132	7.1370	0.1401	0.00244	409.1354	0.01744	57.3257	2588.71	45.1579
144	8.5332	0.1172	0.00199	502.2109	0.01699	58.8540	2798.58	47.5512
240	35.6328	0.0281	0.00043	2308.85	0.01543	64.7957	3870.69	59.7368
360	212.7038	0.0047	0.00007	14,114	0.01507	66.3532	4310.72	64.9662
480	1269.70	0.0008	0.00001	84,580	0.01501	66.6142	4415.74	66.2883

2%			**TABLE 7**	**Discrete Cash Flow: Compound Interest Factors**				**2%**
	Single Payments		**Uniform Series Payments**				**Arithmetic Gradients**	
	Compound Amount	Present Worth	Sinking Fund	Compound Amount	Capital Recovery	Present Worth	Gradient Present Worth	Gradient Uniform Series
n	*F/P*	*P/F*	*A/F*	*F/A*	*A/P*	*P/A*	*P/G*	*A/G*
1	1.0200	0.9804	1.00000	1.0000	1.02000	0.9804		
2	1.0404	0.9612	0.49505	2.0200	0.51505	1.9416	0.9612	0.4950
3	1.0612	0.9423	0.32675	3.0604	0.34675	2.8839	2.8458	0.9868
4	1.0824	0.9238	0.24262	4.1216	0.26262	3.8077	5.6173	1.4752
5	1.1041	0.9057	0.19216	5.2040	0.21216	4.7135	9.2403	1.9604
6	1.1262	0.8880	0.15853	6.3081	0.17853	5.6014	13.6801	2.4423
7	1.1487	0.8706	0.13451	7.4343	0.15451	6.4720	18.9035	2.9208
8	1.1717	0.8535	0.11651	8.5830	0.13651	7.3255	24.8779	3.3961
9	1.1951	0.8368	0.10252	9.7546	0.12252	8.1622	31.5720	3.8681
10	1.2190	0.8203	0.09133	10.9497	0.11133	8.9826	38.9551	4.3367
11	1.2434	0.8043	0.08218	12.1687	0.10218	9.7868	46.9977	4.8021
12	1.2682	0.7885	0.07456	13.4121	0.09456	10.5753	55.6712	5.2642
13	1.2936	0.7730	0.06812	14.6803	0.08812	11.3484	64.9475	5.7231
14	1.3195	0.7579	0.06260	15.9739	0.08260	12.1062	74.7999	6.1786
15	1.3459	0.7430	0.05783	17.2934	0.07783	12.8493	85.2021	6.6309
16	1.3728	0.7284	0.05365	18.6393	0.07365	13.5777	96.1288	7.0799
17	1.4002	0.7142	0.04997	20.0121	0.06997	14.2919	107.5554	7.5256
18	1.4282	0.7002	0.04670	21.4123	0.06670	14.9920	119.4581	7.9681
19	1.4568	0.6864	0.04378	22.8406	0.06378	15.6785	131.8139	8.4073
20	1.4859	0.6730	0.04116	24.2974	0.06116	16.3514	144.6003	8.8433
21	1.5157	0.6598	0.03878	25.7833	0.05878	17.0112	157.7959	9.2760
22	1.5460	0.6468	0.03663	27.2990	0.05663	17.6580	171.3795	9.7055
23	1.5769	0.6342	0.03467	28.8450	0.05467	18.2922	185.3309	10.1317
24	1.6084	0.6217	0.03287	30.4219	0.05287	18.9139	199.6305	10.5547
25	1.6406	0.6095	0.03122	32.0303	0.05122	19.5235	214.2592	10.9745
26	1.6734	0.5976	0.02970	33.6709	0.04970	20.1210	229.1987	11.3910
27	1.7069	0.5859	0.02829	35.3443	0.04829	20.7069	244.4311	11.8043
28	1.7410	0.5744	0.02699	37.0512	0.04699	21.2813	259.9392	12.2145
29	1.7758	0.5631	0.02578	38.7922	0.04578	21.8444	275.7064	12.6214
30	1.8114	0.5521	0.02465	40.5681	0.04465	22.3965	291.7164	13.0251
36	2.0399	0.4902	0.01923	51.9944	0.03923	25.4888	392.0405	15.3809
40	2.2080	0.4529	0.01656	60.4020	0.03656	27.3555	461.9931	16.8885
48	2.5871	0.3865	0.01260	79.3535	0.03260	30.6731	605.9657	19.7556
50	2.6916	0.3715	0.01182	84.5794	0.03182	31.4236	642.3606	20.4420
52	2.8003	0.3571	0.01111	90.0164	0.03111	32.1449	678.7849	21.1164
55	2.9717	0.3365	0.01014	98.5865	0.03014	33.1748	733.3527	22.1057
60	3.2810	0.3048	0.00877	114.0515	0.02877	34.7609	823.6975	23.6961
72	4.1611	0.2403	0.00633	158.0570	0.02633	37.9841	1034.06	27.2234
75	4.4158	0.2265	0.00586	170.7918	0.02586	38.6771	1084.64	28.0434
84	5.2773	0.1895	0.00468	213.8666	0.02468	40.5255	1230.42	30.3616
90	5.9431	0.1683	0.00405	247.1567	0.02405	41.5869	1322.17	31.7929
96	6.6929	0.1494	0.00351	284.6467	0.02351	42.5294	1409.30	33.1370
100	7.2446	0.1380	0.00320	312.2323	0.02320	43.0984	1464.75	33.9863
108	8.4883	0.1178	0.00267	374.4129	0.02267	44.1095	1569.30	35.5774
120	10.7652	0.0929	0.00205	488.2582	0.02205	45.3554	1710.42	37.7114
132	13.6528	0.0732	0.00158	632.6415	0.02158	46.3378	1833.47	39.5676
144	17.3151	0.0578	0.00123	815.7545	0.02123	47.1123	1939.79	41.1738
240	115.8887	0.0086	0.00017	5744.44	0.02017	49.5686	2374.88	47.9110
360	1247.56	0.0008	0.00002	62,328	0.02002	49.9599	2482.57	49.7112
480	13,430	0.0001			0.02000	49.9963	2498.03	49.9643

	Single Payments		Uniform Series Payments				Arithmetic Gradients	
	Compound Amount	Present Worth	Sinking Fund	Compound Amount	Capital Recovery	Present Worth	Gradient Present Worth	Gradient Uniform Series
n	F/P	P/F	A/F	F/A	A/P	P/A	P/G	A/G
1	1.0300	0.9709	1.00000	1.0000	1.03000	0.9709		
2	1.0609	0.9426	0.49261	2.0300	0.52261	1.9135	0.9426	0.4926
3	1.0927	0.9151	0.32353	3.0909	0.35353	2.8286	2.7729	0.9803
4	1.1255	0.8885	0.23903	4.1836	0.26903	3.7171	5.4383	1.4631
5	1.1593	0.8626	0.18835	5.3091	0.21835	4.5797	8.8888	1.9409
6	1.1941	0.8375	0.15460	6.4684	0.18460	5.4172	13.0762	2.4138
7	1.2299	0.8131	0.13051	7.6625	0.16051	6.2303	17.9547	2.8819
8	1.2668	0.7894	0.11246	8.8923	0.14246	7.0197	23.4806	3.3450
9	1.3048	0.7664	0.09843	10.1591	0.12843	7.7861	29.6119	3.8032
10	1.3439	0.7441	0.08723	11.4639	0.11723	8.5302	36.3088	4.2565
11	1.3842	0.7224	0.07808	12.8078	0.10808	9.2526	43.5330	4.7049
12	1.4258	0.7014	0.07046	14.1920	0.10046	9.9540	51.2482	5.1485
13	1.4685	0.6810	0.06403	15.6178	0.09403	10.6350	59.4196	5.5872
14	1.5126	0.6611	0.05853	17.0863	0.08853	11.2961	68.0141	6.0210
15	1.5580	0.6419	0.05377	18.5989	0.08377	11.9379	77.0002	6.4500
16	1.6047	0.6232	0.04961	20.1569	0.07961	12.5611	86.3477	6.8742
17	1.6528	0.6050	0.04595	21.7616	0.07595	13.1661	96.0280	7.2936
18	1.7024	0.5874	0.04271	23.4144	0.07271	13.7535	106.0137	7.7081
19	1.7535	0.5703	0.03981	25.1169	0.06981	14.3238	116.2788	8.1179
20	1.8061	0.5537	0.03722	26.8704	0.06722	14.8775	126.7987	8.5229
21	1.8603	0.5375	0.03487	28.6765	0.06487	15.4150	137.5496	8.9231
22	1.9161	0.5219	0.03275	30.5368	0.06275	15.9369	148.5094	9.3186
23	1.9736	0.5067	0.03081	32.4529	0.06081	16.4436	159.6566	9.7093
24	2.0328	0.4919	0.02905	34.4265	0.05905	16.9355	170.9711	10.0954
25	2.0938	0.4776	0.02743	36.4593	0.05743	17.4131	182.4336	10.4768
26	2.1566	0.4637	0.02594	38.5530	0.05594	17.8768	194.0260	10.8535
27	2.2213	0.4502	0.02456	40.7096	0.05456	18.3270	205.7309	11.2255
28	2.2879	0.4371	0.02329	42.9309	0.05329	18.7641	217.5320	11.5930
29	2.3566	0.4243	0.02211	45.2189	0.05211	19.1885	229.4137	11.9558
30	2.4273	0.4120	0.02102	47.5754	0.05102	19.6004	241.3613	12.3141
31	2.5001	0.4000	0.02000	50.0027	0.05000	20.0004	253.3609	12.6678
32	2.5751	0.3883	0.01905	52.5028	0.04905	20.3888	265.3993	13.0169
33	2.6523	0.3770	0.01816	55.0778	0.04816	20.7658	277.4642	13.3616
34	2.7319	0.3660	0.01732	57.7302	0.04732	21.1318	289.5437	13.7018
35	2.8139	0.3554	0.01654	60.4621	0.04654	21.4872	301.6267	14.0375
40	3.2620	0.3066	0.01326	75.4013	0.04326	23.1148	361.7499	15.6502
45	3.7816	0.2644	0.01079	92.7199	0.04079	24.5187	420.6325	17.1556
50	4.3839	0.2281	0.00887	112.7969	0.03887	25.7298	477.4803	18.5575
55	5.0821	0.1968	0.00735	136.0716	0.03735	26.7744	531.7411	19.8600
60	5.8916	0.1697	0.00613	163.0534	0.03613	27.6756	583.0526	21.0674
65	6.8300	0.1464	0.00515	194.3328	0.03515	28.4529	631.2010	22.1841
70	7.9178	0.1263	0.00434	230.5941	0.03434	29.1234	676.0869	23.2145
75	9.1789	0.1089	0.00367	272.6309	0.03367	29.7018	717.6978	24.1634
80	10.6409	0.0940	0.00311	321.3630	0.03311	30.2008	756.0865	25.0353
84	11.9764	0.0835	0.00273	365.8805	0.03273	30.5501	784.5434	25.6806
85	12.3357	0.0811	0.00265	377.8570	0.03265	30.6312	791.3529	25.8349
90	14.3005	0.0699	0.00226	443.3489	0.03226	31.0024	823.6302	26.5667
96	17.0755	0.0586	0.00187	535.8502	0.03187	31.3812	858.6377	27.3615
108	24.3456	0.0411	0.00129	778.1863	0.03129	31.9642	917.6013	28.7072
120	34.7110	0.0288	0.00089	1123.70	0.03089	32.3730	963.8635	29.7737

4%		TABLE 9	Discrete Cash Flow: Compound Interest Factors					4%
	Single Payments		Uniform Series Payments				Arithmetic Gradients	
	Compound Amount F/P	Present Worth P/F	Sinking Fund A/F	Compound Amount F/A	Capital Recovery A/P	Present Worth P/A	Gradient Present Worth P/G	Gradient Uniform Series A/G
n								
1	1.0400	0.9615	1.00000	1.0000	1.04000	0.9615		
2	1.0816	0.9246	0.49020	2.0400	0.53020	1.8861	0.9246	0.4902
3	1.1249	0.8890	0.32035	3.1216	0.36035	2.7751	2.7025	0.9739
4	1.1699	0.8548	0.23549	4.2465	0.27549	3.6299	5.2670	1.4510
5	1.2167	0.8219	0.18463	5.4163	0.22463	4.4518	8.5547	1.9216
6	1.2653	0.7903	0.15076	6.6330	0.19076	5.2421	12.5062	2.3857
7	1.3159	0.7599	0.12661	7.8983	0.16661	6.0021	17.0657	2.8433
8	1.3686	0.7307	0.10853	9.2142	0.14853	6.7327	22.1806	3.2944
9	1.4233	0.7026	0.09449	10.5828	0.13449	7.4353	27.8013	3.7391
10	1.4802	0.6756	0.08329	12.0061	0.12329	8.1109	33.8814	4.1773
11	1.5395	0.6496	0.07415	13.4864	0.11415	8.7605	40.3772	4.6090
12	1.6010	0.6246	0.06655	15.0258	0.10655	9.3851	47.2477	5.0343
13	1.6651	0.6006	0.06014	16.6268	0.10014	9.9856	54.4546	5.4533
14	1.7317	0.5775	0.05467	18.2919	0.09467	10.5631	61.9618	5.8659
15	1.8009	0.5553	0.04994	20.0236	0.08994	11.1184	69.7355	6.2721
16	1.8730	0.5339	0.04582	21.8245	0.08582	11.6523	77.7441	6.6720
17	1.9479	0.5134	0.04220	23.6975	0.08220	12.1657	85.9581	7.0656
18	2.0258	0.4936	0.03899	25.6454	0.07899	12.6593	94.3498	7.4530
19	2.1068	0.4746	0.03614	27.6712	0.07614	13.1339	102.8933	7.8342
20	2.1911	0.4564	0.03358	29.7781	0.07358	13.5903	111.5647	8.2091
21	2.2788	0.4388	0.03128	31.9692	0.07128	14.0292	120.3414	8.5779
22	2.3699	0.4220	0.02920	34.2480	0.06920	14.4511	129.2024	8.9407
23	2.4647	0.4057	0.02731	36.6179	0.06731	14.8568	138.1284	9.2973
24	2.5633	0.3901	0.02559	39.0826	0.06559	15.2470	147.1012	9.6479
25	2.6658	0.3751	0.02401	41.6459	0.06401	15.6221	156.1040	9.9925
26	2.7725	0.3607	0.02257	44.3117	0.06257	15.9828	165.1212	10.3312
27	2.8834	0.3468	0.02124	47.0842	0.06124	16.3296	174.1385	10.6640
28	2.9987	0.3335	0.02001	49.9676	0.06001	16.6631	183.1424	10.9909
29	3.1187	0.3207	0.01888	52.9663	0.05888	16.9837	192.1206	11.3120
30	3.2434	0.3083	0.01783	56.0849	0.05783	17.2920	201.0618	11.6274
31	3.3731	0.2965	0.01686	59.3283	0.05686	17.5885	209.9556	11.9371
32	3.5081	0.2851	0.01595	62.7015	0.05595	17.8736	218.7924	12.2411
33	3.6484	0.2741	0.01510	66.2095	0.05510	18.1476	227.5634	12.5396
34	3.7943	0.2636	0.01431	69.8579	0.05431	18.4112	236.2607	12.8324
35	3.9461	0.2534	0.01358	73.6522	0.05358	18.6646	244.8768	13.1198
40	4.8010	0.2083	0.01052	95.0255	0.05052	19.7928	286.5303	14.4765
45	5.8412	0.1712	0.00826	121.0294	0.04826	20.7200	325.4028	15.7047
50	7.1067	0.1407	0.00655	152.6671	0.04655	21.4822	361.1638	16.8122
55	8.6464	0.1157	0.00523	191.1592	0.04523	22.1086	393.6890	17.8070
60	10.5196	0.0951	0.00420	237.9907	0.04420	22.6235	422.9966	18.6972
65	12.7987	0.0781	0.00339	294.9684	0.04339	23.0467	449.2014	19.4909
70	15.5716	0.0642	0.00275	364.2905	0.04275	23.3945	472.4789	20.1961
75	18.9453	0.0528	0.00223	448.6314	0.04223	23.6804	493.0408	20.8206
80	23.0498	0.0434	0.00181	551.2450	0.04181	23.9154	511.1161	21.3718
85	28.0436	0.0357	0.00148	676.0901	0.04148	24.1085	526.9384	21.8569
90	34.1193	0.0293	0.00121	827.9833	0.04121	24.2673	540.7369	22.2826
96	43.1718	0.0232	0.00095	1054.30	0.04095	24.4209	554.9312	22.7236
108	69.1195	0.0145	0.00059	1702.99	0.04059	24.6383	576.8949	23.4146
120	110.6626	0.0090	0.00036	2741.56	0.04036	24.7741	592.2428	23.9057
144	283.6618	0.0035	0.00014	7066.55	0.04014	24.9119	610.1055	24.4906

5%			TABLE 10	Discrete Cash Flow: Compound Interest Factors				5%
	Single Payments		Uniform Series Payments				Arithmetic Gradients	
	Compound Amount F/P	Present Worth P/F	Sinking Fund A/F	Compound Amount F/A	Capital Recovery A/P	Present Worth P/A	Gradient Present Worth P/G	Gradient Uniform Series A/G
n								
1	1.0500	0.9524	1.00000	1.0000	1.05000	0.9524		
2	1.1025	0.9070	0.48780	2.0500	0.53780	1.8594	0.9070	0.4878
3	1.1576	0.8638	0.31721	3.1525	0.36721	2.7232	2.6347	0.9675
4	1.2155	0.8227	0.23201	4.3101	0.28201	3.5460	5.1028	1.4391
5	1.2763	0.7835	0.18097	5.5256	0.23097	4.3295	8.2369	1.9025
6	1.3401	0.7462	0.14702	6.8019	0.19702	5.0757	11.9680	2.3579
7	1.4071	0.7107	0.12282	8.1420	0.17282	5.7864	16.2321	2.8052
8	1.4775	0.6768	0.10472	9.5491	0.15472	6.4632	20.9700	3.2445
9	1.5513	0.6446	0.09069	11.0266	0.14069	7.1078	26.1268	3.6758
10	1.6289	0.6139	0.07950	12.5779	0.12950	7.7217	31.6520	4.0991
11	1.7103	0.5847	0.07039	14.2068	0.12039	8.3064	37.4988	4.5144
12	1.7959	0.5568	0.06283	15.9171	0.11283	8.8633	43.6241	4.9219
13	1.8856	0.5303	0.05646	17.7130	0.10646	9.3936	49.9879	5.3215
14	1.9799	0.5051	0.05102	19.5986	0.10102	9.8986	56.5538	5.7133
15	2.0789	0.4810	0.04634	21.5786	0.09634	10.3797	63.2880	6.0973
16	2.1829	0.4581	0.04227	23.6575	0.09227	10.8378	70.1597	6.4736
17	2.2920	0.4363	0.03870	25.8404	0.08870	11.2741	77.1405	6.8423
18	2.4066	0.4155	0.03555	28.1324	0.08555	11.6896	84.2043	7.2034
19	2.5270	0.3957	0.03275	30.5390	0.08275	12.0853	91.3275	7.5569
20	2.6533	0.3769	0.03024	33.0660	0.08024	12.4622	98.4884	7.9030
21	2.7860	0.3589	0.02800	35.7193	0.07800	12.8212	105.6673	8.2416
22	2.9253	0.3418	0.02597	38.5052	0.07597	13.1630	112.8461	8.5730
23	3.0715	0.3256	0.02414	41.4305	0.07414	13.4886	120.0087	8.8971
24	3.2251	0.3101	0.02247	44.5020	0.07247	13.7986	127.1402	9.2140
25	3.3864	0.2953	0.02095	47.7271	0.07095	14.0939	134.2275	9.5238
26	3.5557	0.2812	0.01956	51.1135	0.06956	14.3752	141.2585	9.8266
27	3.7335	0.2678	0.01829	54.6691	0.06829	14.6430	148.2226	10.1224
28	3.9201	0.2551	0.01712	58.4026	0.06712	14.8981	155.1101	10.4114
29	4.1161	0.2429	0.01605	62.3227	0.06605	15.1411	161.9126	10.6936
30	4.3219	0.2314	0.01505	66.4388	0.06505	15.3725	168.6226	10.9691
31	4.5380	0.2204	0.01413	70.7608	0.06413	15.5928	175.2333	11.2381
32	4.7649	0.2099	0.01328	75.2988	0.06328	15.8027	181.7392	11.5005
33	5.0032	0.1999	0.01249	80.0638	0.06249	16.0025	188.1351	11.7566
34	5.2533	0.1904	0.01176	85.0670	0.06176	16.1929	194.4168	12.0063
35	5.5160	0.1813	0.01107	90.3203	0.06107	16.3742	200.5807	12.2498
40	7.0400	0.1420	0.00828	120.7998	0.05828	17.1591	229.5452	13.3775
45	8.9850	0.1113	0.00626	159.7002	0.05626	17.7741	255.3145	14.3644
50	11.4674	0.0872	0.00478	209.3480	0.05478	18.2559	277.9148	15.2233
55	14.6356	0.0683	0.00367	272.7126	0.05367	18.6335	297.5104	15.9664
60	18.6792	0.0535	0.00283	353.5837	0.05283	18.9293	314.3432	16.6062
65	23.8399	0.0419	0.00219	456.7980	0.05219	19.1611	328.6910	17.1541
70	30.4264	0.0329	0.00170	588.5285	0.05170	19.3427	340.8409	17.6212
75	38.8327	0.0258	0.00132	756.6537	0.05132	19.4850	351.0721	18.0176
80	49.5614	0.0202	0.00103	971.2288	0.05103	19.5965	359.6460	18.3526
85	63.2544	0.0158	0.00080	1245.09	0.05080	19.6838	366.8007	18.6346
90	80.7304	0.0124	0.00063	1594.61	0.05063	19.7523	372.7488	18.8712
95	103.0347	0.0097	0.00049	2040.69	0.05049	19.8059	377.6774	19.0689
96	108.1864	0.0092	0.00047	2143.73	0.05047	19.8151	378.5555	19.1044
98	119.2755	0.0084	0.00042	2365.51	0.05042	19.8323	380.2139	19.1714
100	131.5013	0.0076	0.00038	2610.03	0.05038	19.8479	381.7492	19.2337

6%			TABLE 11	Discrete Cash Flow: Compound Interest Factors				6%
	Single Payments		Uniform Series Payments				Arithmetic Gradients	
	Compound Amount F/P	Present Worth P/F	Sinking Fund A/F	Compound Amount F/A	Capital Recovery A/P	Present Worth P/A	Gradient Present Worth P/G	Gradient Uniform Series A/G
n								
1	1.0600	0.9434	1.00000	1.0000	1.06000	0.9434		
2	1.1236	0.8900	0.48544	2.0600	0.54544	1.8334	0.8900	0.4854
3	1.1910	0.8396	0.31411	3.1836	0.37411	2.6730	2.5692	0.9612
4	1.2625	0.7921	0.22859	4.3746	0.28859	3.4651	4.9455	1.4272
5	1.3382	0.7473	0.17740	5.6371	0.23740	4.2124	7.9345	1.8836
6	1.4185	0.7050	0.14336	6.9753	0.20336	4.9173	11.4594	2.3304
7	1.5036	0.6651	0.11914	8.3938	0.17914	5.5824	15.4497	2.7676
8	1.5938	0.6274	0.10104	9.8975	0.16104	6.2098	19.8416	3.1952
9	1.6895	0.5919	0.08702	11.4913	0.14702	6.8017	24.5768	3.6133
10	1.7908	0.5584	0.07587	13.1808	0.13587	7.3601	29.6023	4.0220
11	1.8983	0.5268	0.06679	14.9716	0.12679	7.8869	34.8702	4.4213
12	2.0122	0.4970	0.05928	16.8699	0.11928	8.3838	40.3369	4.8113
13	2.1329	0.4688	0.05296	18.8821	0.11296	8.8527	45.9629	5.1920
14	2.2609	0.4423	0.04758	21.0151	0.10758	9.2950	51.7128	5.5635
15	2.3966	0.4173	0.04296	23.2760	0.10296	9.7122	57.5546	5.9260
16	2.5404	0.3936	0.03895	25.6725	0.09895	10.1059	63.4592	6.2794
17	2.6928	0.3714	0.03544	28.2129	0.09544	10.4773	69.4011	6.6240
18	2.8543	0.3503	0.03236	30.9057	0.09236	10.8276	75.3569	6.9597
19	3.0256	0.3305	0.02962	33.7600	0.08962	11.1581	81.3062	7.2867
20	3.2071	0.3118	0.02718	36.7856	0.08718	11.4699	87.2304	7.6051
21	3.3996	0.2942	0.02500	39.9927	0.08500	11.7641	93.1136	7.9151
22	3.6035	0.2775	0.02305	43.3923	0.08305	12.0416	98.9412	8.2166
23	3.8197	0.2618	0.02128	46.9958	0.08128	12.3034	104.7007	8.5099
24	4.0489	0.2470	0.01968	50.8156	0.07968	12.5504	110.3812	8.7951
25	4.2919	0.2330	0.01823	54.8645	0.07823	12.7834	115.9732	9.0722
26	4.5494	0.2198	0.01690	59.1564	0.07690	13.0032	121.4684	9.3414
27	4.8223	0.2074	0.01570	63.7058	0.07570	13.2105	126.8600	9.6029
28	5.1117	0.1956	0.01459	68.5281	0.07459	13.4062	132.1420	9.8568
29	5.4184	0.1846	0.01358	73.6398	0.07358	13.5907	137.3096	10.1032
30	5.7435	0.1741	0.01265	79.0582	0.07265	13.7648	142.3588	10.3422
31	6.0881	0.1643	0.01179	84.8017	0.07179	13.9291	147.2864	10.5740
32	6.4534	0.1550	0.01100	90.8898	0.07100	14.0840	152.0901	10.7988
33	6.8406	0.1462	0.01027	97.3432	0.07027	14.2302	156.7681	11.0166
34	7.2510	0.1379	0.00960	104.1838	0.06960	14.3681	161.3192	11.2276
35	7.6861	0.1301	0.00897	111.4348	0.06897	14.4982	165.7427	11.4319
40	10.2857	0.0972	0.00646	154.7620	0.06646	15.0463	185.9568	12.3590
45	13.7646	0.0727	0.00470	212.7435	0.06470	15.4558	203.1096	13.1413
50	18.4202	0.0543	0.00344	290.3359	0.06344	15.7619	217.4574	13.7964
55	24.6503	0.0406	0.00254	394.1720	0.06254	15.9905	229.3222	14.3411
60	32.9877	0.0303	0.00188	533.1282	0.06188	16.1614	239.0428	14.7909
65	44.1450	0.0227	0.00139	719.0829	0.06139	16.2891	246.9450	15.1601
70	59.0759	0.0169	0.00103	967.9322	0.06103	16.3845	253.3271	15.4613
75	79.0569	0.0126	0.00077	1300.95	0.06077	16.4558	258.4527	15.7058
80	105.7960	0.0095	0.00057	1746.60	0.06057	16.5091	262.5493	15.9033
85	141.5789	0.0071	0.00043	2342.98	0.06043	16.5489	265.8096	16.0620
90	189.4645	0.0053	0.00032	3141.08	0.06032	16.5787	268.3946	16.1891
95	253.5463	0.0039	0.00024	4209.10	0.06024	16.6009	270.4375	16.2905
96	268.7590	0.0037	0.00022	4462.65	0.06022	16.6047	270.7909	16.3081
98	301.9776	0.0033	0.00020	5016.29	0.06020	16.6115	271.4491	16.3411
100	339.3021	0.0029	0.00018	5638.37	0.06018	16.6175	272.0471	16.3711

| 7% | | | | TABLE 12 | Discrete Cash Flow: Compound Interest Factors | | | | 7% |

	Single Payments		Uniform Series Payments				Arithmetic Gradients	
	Compound Amount F/P	Present Worth P/F	Sinking Fund A/F	Compound Amount F/A	Capital Recovery A/P	Present Worth P/A	Gradient Present Worth P/G	Gradient Uniform Series A/G
n								
1	1.0700	0.9346	1.00000	1.0000	1.07000	0.9346		
2	1.1449	0.8734	0.48309	2.0700	0.55309	1.8080	0.8734	0.4831
3	1.2250	0.8163	0.31105	3.2149	0.38105	2.6243	2.5060	0.9549
4	1.3108	0.7629	0.22523	4.4399	0.29523	3.3872	4.7947	1.4155
5	1.4026	0.7130	0.17389	5.7507	0.24389	4.1002	7.6467	1.8650
6	1.5007	0.6663	0.13980	7.1533	0.20980	4.7665	10.9784	2.3032
7	1.6058	0.6227	0.11555	8.6540	0.18555	5.3893	14.7149	2.7304
8	1.7182	0.5820	0.09747	10.2598	0.16747	5.9713	18.7889	3.1465
9	1.8385	0.5439	0.08349	11.9780	0.15349	6.5152	23.1404	3.5517
10	1.9672	0.5083	0.07238	13.8164	0.14238	7.0236	27.7156	3.9461
11	2.1049	0.4751	0.06336	15.7836	0.13336	7.4987	32.4665	4.3296
12	2.2522	0.4440	0.05590	17.8885	0.12590	7.9427	37.3506	4.7025
13	2.4098	0.4150	0.04965	20.1406	0.11965	8.3577	42.3302	5.0648
14	2.5785	0.3878	0.04434	22.5505	0.11434	8.7455	47.3718	5.4167
15	2.7590	0.3624	0.03979	25.1290	0.10979	9.1079	52.4461	5.7583
16	2.9522	0.3387	0.03586	27.8881	0.10586	9.4466	57.5271	6.0897
17	3.1588	0.3166	0.03243	30.8402	0.10243	9.7632	62.5923	6.4110
18	3.3799	0.2959	0.02941	33.9990	0.09941	10.0591	67.6219	6.7225
19	3.6165	0.2765	0.02675	37.3790	0.09675	10.3356	72.5991	7.0242
20	3.8697	0.2584	0.02439	40.9955	0.09439	10.5940	77.5091	7.3163
21	4.1406	0.2415	0.02229	44.8652	0.09229	10.8355	82.3393	7.5990
22	4.4304	0.2257	0.02041	49.0057	0.09041	11.0612	87.0793	7.8725
23	4.7405	0.2109	0.01871	53.4361	0.08871	11.2722	91.7201	8.1369
24	5.0724	0.1971	0.01719	58.1767	0.08719	11.4693	96.2545	8.3923
25	5.4274	0.1842	0.01581	63.2490	0.08581	11.6536	100.6765	8.6391
26	5.8074	0.1722	0.01456	68.6765	0.08456	11.8258	104.9814	8.8773
27	6.2139	0.1609	0.01343	74.4838	0.08343	11.9867	109.1656	9.1072
28	6.6488	0.1504	0.01239	80.6977	0.08239	12.1371	113.2264	9.3289
29	7.1143	0.1406	0.01145	87.3465	0.08145	12.2777	117.1622	9.5427
30	7.6123	0.1314	0.01059	94.4608	0.08059	12.4090	120.9718	9.7487
31	8.1451	0.1228	0.00980	102.0730	0.07980	12.5318	124.6550	9.9471
32	8.7153	0.1147	0.00907	110.2182	0.07907	12.6466	128.2120	10.1381
33	9.3253	0.1072	0.00841	118.9334	0.07841	12.7538	131.6435	10.3219
34	9.9781	0.1002	0.00780	128.2588	0.07780	12.8540	134.9507	10.4987
35	10.6766	0.0937	0.00723	138.2369	0.07723	12.9477	138.1353	10.6687
40	14.9745	0.0668	0.00501	199.6351	0.07501	13.3317	152.2928	11.4233
45	21.0025	0.0476	0.00350	285.7493	0.07350	13.6055	163.7559	12.0360
50	29.4570	0.0339	0.00246	406.5289	0.07246	13.8007	172.9051	12.5287
55	41.3150	0.0242	0.00174	575.9286	0.07174	13.9399	180.1243	12.9215
60	57.9464	0.0173	0.00123	813.5204	0.07123	14.0392	185.7677	13.2321
65	81.2729	0.0123	0.00087	1146.76	0.07087	14.1099	190.1452	13.4760
70	113.9894	0.0088	0.00062	1614.13	0.07062	14.1604	193.5185	13.6662
75	159.8760	0.0063	0.00044	2269.66	0.07044	14.1964	196.1035	13.8136
80	224.2344	0.0045	0.00031	3189.06	0.07031	14.2220	198.0748	13.9273
85	314.5003	0.0032	0.00022	4478.58	0.07022	14.2403	199.5717	14.0146
90	441.1030	0.0023	0.00016	6287.19	0.07016	14.2533	200.7042	14.0812
95	618.6697	0.0016	0.00011	8823.85	0.07011	14.2626	201.5581	14.1319
96	661.9766	0.0015	0.00011	9442.52	0.07011	14.2641	201.7016	14.1405
98	757.8970	0.0013	0.00009	10,813	0.07009	14.2669	201.9651	14.1562
100	867.7163	0.0012	0.00008	12,382	0.07008	14.2693	202.2001	14.1703

8%			**TABLE 13**	**Discrete Cash Flow: Compound Interest Factors**				**8%**
	Single Payments		Uniform Series Payments				Arithmetic Gradients	
n	Compound Amount **F/P**	Present Worth **P/F**	Sinking Fund **A/F**	Compound Amount **F/A**	Capital Recovery **A/P**	Present Worth **P/A**	Gradient Present Worth **P/G**	Gradient Uniform Series **A/G**
1	1.0800	0.9259	1.00000	1.0000	1.08000	0.9259		
2	1.1664	0.8573	0.48077	2.0800	0.56077	1.7833	0.8573	0.4808
3	1.2597	0.7938	0.30803	3.2464	0.38803	2.5771	2.4450	0.9487
4	1.3605	0.7350	0.22192	4.5061	0.30192	3.3121	4.6501	1.4040
5	1.4693	0.6806	0.17046	5.8666	0.25046	3.9927	7.3724	1.8465
6	1.5869	0.6302	0.13632	7.3359	0.21632	4.6229	10.5233	2.2763
7	1.7138	0.5835	0.11207	8.9228	0.19207	5.2064	14.0242	2.6937
8	1.8509	0.5403	0.09401	10.6366	0.17401	5.7466	17.8061	3.0985
9	1.9990	0.5002	0.08008	12.4876	0.16008	6.2469	21.8081	3.4910
10	2.1589	0.4632	0.06903	14.4866	0.14903	6.7101	25.9768	3.8713
11	2.3316	0.4289	0.06008	16.6455	0.14008	7.1390	30.2657	4.2395
12	2.5182	0.3971	0.05270	18.9771	0.13270	7.5361	34.6339	4.5957
13	2.7196	0.3677	0.04652	21.4953	0.12652	7.9038	39.0463	4.9402
14	2.9372	0.3405	0.04130	24.2149	0.12130	8.2442	43.4723	5.2731
15	3.1722	0.3152	0.03683	27.1521	0.11683	8.5595	47.8857	5.5945
16	3.4259	0.2919	0.03298	30.3243	0.11298	8.8514	52.2640	5.9046
17	3.7000	0.2703	0.02963	33.7502	0.10963	9.1216	56.5883	6.2037
18	3.9960	0.2502	0.02670	37.4502	0.10670	9.3719	60.8426	6.4920
19	4.3157	0.2317	0.02413	41.4463	0.10413	9.6036	65.0134	6.7697
20	4.6610	0.2145	0.02185	45.7620	0.10185	9.8181	69.0898	7.0369
21	5.0338	0.1987	0.01983	50.4229	0.09983	10.0168	73.0629	7.2940
22	5.4365	0.1839	0.01803	55.4568	0.09803	10.2007	76.9257	7.5412
23	5.8715	0.1703	0.01642	60.8933	0.09642	10.3711	80.6726	7.7786
24	6.3412	0.1577	0.01498	66.7648	0.09498	10.5288	84.2997	8.0066
25	6.8485	0.1460	0.01368	73.1059	0.09368	10.6748	87.8041	8.2254
26	7.3964	0.1352	0.01251	79.9544	0.09251	10.8100	91.1842	8.4352
27	7.9881	0.1252	0.01145	87.3508	0.09145	10.9352	94.4390	8.6363
28	8.6271	0.1159	0.01049	95.3388	0.09049	11.0511	97.5687	8.8289
29	9.3173	0.1073	0.00962	103.9659	0.08962	11.1584	100.5738	9.0133
30	10.0627	0.0994	0.00883	113.2832	0.08883	11.2578	103.4558	9.1897
31	10.8677	0.0920	0.00811	123.3459	0.08811	11.3498	106.2163	9.3584
32	11.7371	0.0852	0.00745	134.2135	0.08745	11.4350	108.8575	9.5197
33	12.6760	0.0789	0.00685	145.9506	0.08685	11.5139	111.3819	9.6737
34	13.6901	0.0730	0.00630	158.6267	0.08630	11.5869	113.7924	9.8208
35	14.7853	0.0676	0.00580	172.3168	0.08580	11.6546	116.0920	9.9611
40	21.7245	0.0460	0.00386	259.0565	0.08386	11.9246	126.0422	10.5699
45	31.9204	0.0313	0.00259	386.5056	0.08259	12.1084	133.7331	11.0447
50	46.9016	0.0213	0.00174	573.7702	0.08174	12.2335	139.5928	11.4107
55	68.9139	0.0145	0.00118	848.9232	0.08118	12.3186	144.0065	11.6902
60	101.2571	0.0099	0.00080	1253.21	0.08080	12.3766	147.3000	11.9015
65	148.7798	0.0067	0.00054	1847.25	0.08054	12.4160	149.7387	12.0602
70	218.6064	0.0046	0.00037	2720.08	0.08037	12.4428	151.5326	12.1783
75	321.2045	0.0031	0.00025	4002.56	0.08025	12.4611	152.8448	12.2658
80	471.9548	0.0021	0.00017	5886.94	0.08017	12.4735	153.8001	12.3301
85	693.4565	0.0014	0.00012	8655.71	0.08012	12.4820	154.4925	12.3772
90	1018.92	0.0010	0.00008	12,724	0.08008	12.4877	154.9925	12.4116
95	1497.12	0.0007	0.00005	18,702	0.08005	12.4917	155.3524	12.4365
96	1616.89	0.0006	0.00005	20,199	0.08005	12.4923	155.4112	12.4406
98	1885.94	0.0005	0.00004	23,562	0.08004	12.4934	155.5176	12.4480
100	2199.76	0.0005	0.00004	27,485	0.08004	12.4943	155.6107	12.4545

9%			TABLE 14	Discrete Cash Flow: Compound Interest Factors				9%
	Single Payments		Uniform Series Payments				Arithmetic Gradients	
	Compound Amount F/P	Present Worth P/F	Sinking Fund A/F	Compound Amount F/A	Capital Recovery A/P	Present Worth P/A	Gradient Present Worth P/G	Gradient Uniform Series A/G
n								
1	1.0900	0.9174	1.00000	1.0000	1.09000	0.9174		
2	1.1881	0.8417	0.47847	2.0900	0.56847	1.7591	0.8417	0.4785
3	1.2950	0.7722	0.30505	3.2781	0.39505	2.5313	2.3860	0.9426
4	1.4116	0.7084	0.21867	4.5731	0.30867	3.2397	4.5113	1.3925
5	1.5386	0.6499	0.16709	5.9847	0.25709	3.8897	7.1110	1.8282
6	1.6771	0.5963	0.13292	7.5233	0.22292	4.4859	10.0924	2.2498
7	1.8280	0.5470	0.10869	9.2004	0.19869	5.0330	13.3746	2.6574
8	1.9926	0.5019	0.09067	11.0285	0.18067	5.5348	16.8877	3.0512
9	2.1719	0.4604	0.07680	13.0210	0.16680	5.9952	20.5711	3.4312
10	2.3674	0.4224	0.06582	15.1929	0.15582	6.4177	24.3728	3.7978
11	2.5804	0.3875	0.05695	17.5603	0.14695	6.8052	28.2481	4.1510
12	2.8127	0.3555	0.04965	20.1407	0.13965	7.1607	32.1590	4.4910
13	3.0658	0.3262	0.04357	22.9534	0.13357	7.4869	36.0731	4.8182
14	3.3417	0.2992	0.03843	26.0192	0.12843	7.7862	39.9633	5.1326
15	3.6425	0.2745	0.03406	29.3609	0.12406	8.0607	43.8069	5.4346
16	3.9703	0.2519	0.03030	33.0034	0.12030	8.3126	47.5849	5.7245
17	4.3276	0.2311	0.02705	36.9737	0.11705	8.5436	51.2821	6.0024
18	4.7171	0.2120	0.02421	41.3013	0.11421	8.7556	54.8860	6.2687
19	5.1417	0.1945	0.02173	46.0185	0.11173	8.9501	58.3868	6.5236
20	5.6044	0.1784	0.01955	51.1601	0.10955	9.1285	61.7770	6.7674
21	6.1088	0.1637	0.01762	56.7645	0.10762	9.2922	65.0509	7.0006
22	6.6586	0.1502	0.01590	62.8733	0.10590	9.4424	68.2048	7.2232
23	7.2579	0.1378	0.01438	69.5319	0.10438	9.5802	71.2359	7.4357
24	7.9111	0.1264	0.01302	76.7898	0.10302	9.7066	74.1433	7.6384
25	8.6231	0.1160	0.01181	84.7009	0.10181	9.8226	76.9265	7.8316
26	9.3992	0.1064	0.01072	93.3240	0.10072	9.9290	79.5863	8.0156
27	10.2451	0.0976	0.00973	102.7231	0.09973	10.0266	82.1241	8.1906
28	11.1671	0.0895	0.00885	112.9682	0.09885	10.1161	84.5419	8.3571
29	12.1722	0.0822	0.00806	124.1354	0.09806	10.1983	86.8422	8.5154
30	13.2677	0.0754	0.00734	136.3075	0.09734	10.2737	89.0280	8.6657
31	14.4618	0.0691	0.00669	149.5752	0.09669	10.3428	91.1024	8.8083
32	15.7633	0.0634	0.00610	164.0370	0.09610	10.4062	93.0690	8.9436
33	17.1820	0.0582	0.00556	179.8003	0.09556	10.4644	94.9314	9.0718
34	18.7284	0.0534	0.00508	196.9823	0.09508	10.5178	96.6935	9.1933
35	20.4140	0.0490	0.00464	215.7108	0.09464	10.5668	98.3590	9.3083
40	31.4094	0.0318	0.00296	337.8824	0.09296	10.7574	105.3762	9.7957
45	48.3273	0.0207	0.00190	525.8587	0.09190	10.8812	110.5561	10.1603
50	74.3575	0.0134	0.00123	815.0836	0.09123	10.9617	114.3251	10.4295
55	114.4083	0.0087	0.00079	1260.09	0.09079	11.0140	117.0362	10.6261
60	176.0313	0.0057	0.00051	1944.79	0.09051	11.0480	118.9683	10.7683
65	270.8460	0.0037	0.00033	2998.29	0.09033	11.0701	120.3344	10.8702
70	416.7301	0.0024	0.00022	4619.22	0.09022	11.0844	121.2942	10.9427
75	641.1909	0.0016	0.00014	7113.23	0.09014	11.0938	121.9646	10.9940
80	986.5517	0.0010	0.00009	10,951	0.09009	11.0998	122.4306	11.0299
85	1517.93	0.0007	0.00006	16,855	0.09006	11.1038	122.7533	11.0551
90	2335.53	0.0004	0.00004	25,939	0.09004	11.1064	122.9758	11.0726
95	3593.50	0.0003	0.00003	39,917	0.09003	11.1080	123.1287	11.0847
96	3916.91	0.0003	0.00002	43,510	0.09002	11.1083	123.1529	11.0866
98	4653.68	0.0002	0.00002	51,696	0.09002	11.1087	123.1963	11.0900
100	5529.04	0.0002	0.00002	61,423	0.09002	11.1091	123.2335	11.0930

10%			TABLE 15	Discrete Cash Flow: Compound Interest Factors				10%
	Single Payments		Uniform Series Payments				Arithmetic Gradients	
	Compound Amount	Present Worth	Sinking Fund	Compound Amount	Capital Recovery	Present Worth	Gradient Present Worth	Gradient Uniform Series
n	F/P	P/F	A/F	F/A	A/P	P/A	P/G	A/G
1	1.1000	0.9091	1.00000	1.0000	1.10000	0.9091		
2	1.2100	0.8264	0.47619	2.1000	0.57619	1.7355	0.8264	0.4762
3	1.3310	0.7513	0.30211	3.3100	0.40211	2.4869	2.3291	0.9366
4	1.4641	0.6830	0.21547	4.6410	0.31547	3.1699	4.3781	1.3812
5	1.6105	0.6209	0.16380	6.1051	0.26380	3.7908	6.8618	1.8101
6	1.7716	0.5645	0.12961	7.7156	0.22961	4.3553	9.6842	2.2236
7	1.9487	0.5132	0.10541	9.4872	0.20541	4.8684	12.7631	2.6216
8	2.1436	0.4665	0.08744	11.4359	0.18744	5.3349	16.0287	3.0045
9	2.3579	0.4241	0.07364	13.5795	0.17364	5.7590	19.4215	3.3724
10	2.5937	0.3855	0.06275	15.9374	0.16275	6.1446	22.8913	3.7255
11	2.8531	0.3505	0.05396	18.5312	0.15396	6.4951	26.3963	4.0641
12	3.1384	0.3186	0.04676	21.3843	0.14676	6.8137	29.9012	4.3884
13	3.4523	0.2897	0.04078	24.5227	0.14078	7.1034	33.3772	4.6988
14	3.7975	0.2633	0.03575	27.9750	0.13575	7.3667	36.8005	4.9955
15	4.1772	0.2394	0.03147	31.7725	0.13147	7.6061	40.1520	5.2789
16	4.5950	0.2176	0.02782	35.9497	0.12782	7.8237	43.4164	5.5493
17	5.0545	0.1978	0.02466	40.5447	0.12466	8.0216	46.5819	5.8071
18	5.5599	0.1799	0.02193	45.5992	0.12193	8.2014	49.6395	6.0526
19	6.1159	0.1635	0.01955	51.1591	0.11955	8.3649	52.5827	6.2861
20	6.7275	0.1486	0.01746	57.2750	0.11746	8.5136	55.4069	6.5081
21	7.4002	0.1351	0.01562	64.0025	0.11562	8.6487	58.1095	6.7189
22	8.1403	0.1228	0.01401	71.4027	0.11401	8.7715	60.6893	6.9189
23	8.9543	0.1117	0.01257	79.5430	0.11257	8.8832	63.1462	7.1085
24	9.8497	0.1015	0.01130	88.4973	0.11130	8.9847	65.4813	7.2881
25	10.8347	0.0923	0.01017	98.3471	0.11017	9.0770	67.6964	7.4580
26	11.9182	0.0839	0.00916	109.1818	0.10916	9.1609	69.7940	7.6186
27	13.1100	0.0763	0.00826	121.0999	0.10826	9.2372	71.7773	7.7704
28	14.4210	0.0693	0.00745	134.2099	0.10745	9.3066	73.6495	7.9137
29	15.8631	0.0630	0.00673	148.6309	0.10673	9.3696	75.4146	8.0489
30	17.4494	0.0573	0.00608	164.4940	0.10608	9.4269	77.0766	8.1762
31	19.1943	0.0521	0.00550	181.9434	0.10550	9.4790	78.6395	8.2962
32	21.1138	0.0474	0.00497	201.1378	0.10497	9.5264	80.1078	8.4091
33	23.2252	0.0431	0.00450	222.2515	0.10450	9.5694	81.4856	8.5152
34	25.5477	0.0391	0.00407	245.4767	0.10407	9.6086	82.7773	8.6149
35	28.1024	0.0356	0.00369	271.0244	0.10369	9.6442	83.9872	8.7086
40	45.2593	0.0221	0.00226	442.5926	0.10226	9.7791	88.9525	9.0962
45	72.8905	0.0137	0.00139	718.9048	0.10139	9.8628	92.4544	9.3740
50	117.3909	0.0085	0.00086	1163.91	0.10086	9.9148	94.8889	9.5704
55	189.0591	0.0053	0.00053	1880.59	0.10053	9.9471	96.5619	9.7075
60	304.4816	0.0033	0.00033	3034.82	0.10033	9.9672	97.7010	9.8023
65	490.3707	0.0020	0.00020	4893.71	0.10020	9.9796	98.4705	9.8672
70	789.7470	0.0013	0.00013	7887.47	0.10013	9.9873	98.9870	9.9113
75	1271.90	0.0008	0.00008	12,709	0.10008	9.9921	99.3317	9.9410
80	2048.40	0.0005	0.00005	20,474	0.10005	9.9951	99.5606	9.9609
85	3298.97	0.0003	0.00003	32,980	0.10003	9.9970	99.7120	9.9742
90	5313.02	0.0002	0.00002	53,120	0.10002	9.9981	99.8118	9.9831
95	8556.68	0.0001	0.00001	85,557	0.10001	9.9988	99.8773	9.9889
96	9412.34	0.0001	0.00001	94,113	0.10001	9.9989	99.8874	9.9898
98	11,389	0.0001	0.00001		0.10001	9.9991	99.9052	9.9914
100	13,781	0.0001	0.00001		0.10001	9.9993	99.9202	9.9927

| 11% | | | TABLE 16 | Discrete Cash Flow: Compound Interest Factors | | | | 11% |

	Single Payments		Uniform Series Payments				Arithmetic Gradients	
	Compound Amount F/P	Present Worth P/F	Sinking Fund A/F	Compound Amount F/A	Capital Recovery A/P	Present Worth P/A	Gradient Present Worth P/G	Gradient Uniform Series A/G
n								
1	1.1100	0.9009	1.00000	1.0000	1.11000	0.9009		
2	1.2321	0.8116	0.47393	2.1100	0.58393	1.7125	0.8116	0.4739
3	1.3676	0.7312	0.29921	3.3421	0.40921	2.4437	2.2740	0.9306
4	1.5181	0.6587	0.21233	4.7097	0.32233	3.1024	4.2502	1.3700
5	1.6851	0.5935	0.16057	6.2278	0.27057	3.6959	6.6240	1.7923
6	1.8704	0.5346	0.12638	7.9129	0.23638	4.2305	9.2972	2.1976
7	2.0762	0.4817	0.10222	9.7833	0.21222	4.7122	12.1872	2.5863
8	2.3045	0.4339	0.08432	11.8594	0.19432	5.1461	15.2246	2.9585
9	2.5580	0.3909	0.07060	14.1640	0.18060	5.5370	18.3520	3.3144
10	2.8394	0.3522	0.05980	16.7220	0.16980	5.8892	21.5217	3.6544
11	3.1518	0.3173	0.05112	19.5614	0.16112	6.2065	24.6945	3.9788
12	3.4985	0.2858	0.04403	22.7132	0.15403	6.4924	27.8388	4.2879
13	3.8833	0.2575	0.03815	26.2116	0.14815	6.7499	30.9290	4.5822
14	4.3104	0.2320	0.03323	30.0949	0.14323	6.9819	33.9449	4.8619
15	4.7846	0.2090	0.02907	34.4054	0.13907	7.1909	36.8709	5.1275
16	5.3109	0.1883	0.02552	39.1899	0.13552	7.3792	39.6953	5.3794
17	5.8951	0.1696	0.02247	44.5008	0.13247	7.5488	42.4095	5.6180
18	6.5436	0.1528	0.01984	50.3959	0.12984	7.7016	45.0074	5.8439
19	7.2633	0.1377	0.01756	56.9395	0.12756	7.8393	47.4856	6.0574
20	8.0623	0.1240	0.01558	64.2028	0.12558	7.9633	49.8423	6.2590
21	8.9492	0.1117	0.01384	72.2651	0.12384	8.0751	52.0771	6.4491
22	9.9336	0.1007	0.01231	81.2143	0.12231	8.1757	54.1912	6.6283
23	11.0263	0.0907	0.01097	91.1479	0.12097	8.2664	56.1864	6.7969
24	12.2392	0.0817	0.00979	102.1742	0.11979	8.3481	58.0656	6.9555
25	13.5855	0.0736	0.00874	114.4133	0.11874	8.4217	59.8322	7.1045
26	15.0799	0.0663	0.00781	127.9988	0.11781	8.4881	61.4900	7.2443
27	16.7386	0.0597	0.00699	143.0786	0.11699	8.5478	63.0433	7.3754
28	18.5799	0.0538	0.00626	159.8173	0.11626	8.6016	64.4965	7.4982
29	20.6237	0.0485	0.00561	178.3972	0.11561	8.6501	65.8542	7.6131
30	22.8923	0.0437	0.00502	199.0209	0.11502	8.6938	67.1210	7.7206
31	25.4104	0.0394	0.00451	221.9132	0.11451	8.7331	68.3016	7.8210
32	28.2056	0.0355	0.00404	247.3236	0.11404	8.7686	69.4007	7.9147
33	31.3082	0.0319	0.00363	275.5292	0.11363	8.8005	70.4228	8.0021
34	34.7521	0.0288	0.00326	306.8374	0.11326	8.8293	71.3724	8.0836
35	38.5749	0.0259	0.00293	341.5896	0.11293	8.8552	72.2538	8.1594
40	65.0009	0.0154	0.00172	581.8261	0.11172	8.9511	75.7789	8.4659
45	109.5302	0.0091	0.00101	986.6386	0.11101	9.0079	78.1551	8.6763
50	184.5648	0.0054	0.00060	1668.77	0.11060	9.0417	79.7341	8.8185
55	311.0025	0.0032	0.00035	2818.20	0.11035	9.0617	80.7712	8.9135
60	524.0572	0.0019	0.00021	4755.07	0.11021	9.0736	81.4461	8.9762
65	883.0669	0.0011	0.00012	8018.79	0.11012	9.0806	81.8819	9.0172
70	1488.02	0.0007	0.00007	13,518	0.11007	9.0848	82.1614	9.0438
75	2507.40	0.0004	0.00004	22,785	0.11004	9.0873	82.3397	9.0610
80	4225.11	0.0002	0.00003	38,401	0.11003	9.0888	82.4529	9.0720
85	7119.56	0.0001	0.00002	64,714	0.11002	9.0896	82.5245	9.0790

12%			TABLE 17	Discrete Cash Flow: Compound Interest Factors				12%
	Single Payments		Uniform Series Payments				Arithmetic Gradients	
	Compound Amount	Present Worth	Sinking Fund	Compound Amount	Capital Recovery	Present Worth	Gradient Present Worth	Gradient Uniform Series
n	F/P	P/F	A/F	F/A	A/P	P/A	P/G	A/G
1	1.1200	0.8929	1.00000	1.0000	1.12000	0.8929		
2	1.2544	0.7972	0.47170	2.1200	0.59170	1.6901	0.7972	0.4717
3	1.4049	0.7118	0.29635	3.3744	0.41635	2.4018	2.2208	0.9246
4	1.5735	0.6355	0.20923	4.7793	0.32923	3.0373	4.1273	1.3589
5	1.7623	0.5674	0.15741	6.3528	0.27741	3.6048	6.3970	1.7746
6	1.9738	0.5066	0.12323	8.1152	0.24323	4.1114	8.9302	2.1720
7	2.2107	0.4523	0.09912	10.0890	0.21912	4.5638	11.6443	2.5512
8	2.4760	0.4039	0.08130	12.2997	0.20130	4.9676	14.4714	2.9131
9	2.7731	0.3606	0.06768	14.7757	0.18768	5.3282	17.3563	3.2574
10	3.1058	0.3220	0.05698	17.5487	0.17698	5.6502	20.2541	3.5847
11	3.4785	0.2875	0.04842	20.6546	0.16842	5.9377	23.1288	3.8953
12	3.8960	0.2567	0.04144	24.1331	0.16144	6.1944	25.9523	4.1897
13	4.3635	0.2292	0.03568	28.0291	0.15568	6.4235	28.7024	4.4683
14	4.8871	0.2046	0.03087	32.3926	0.15087	6.6282	31.3624	4.7317
15	5.4736	0.1827	0.02682	37.2797	0.14682	6.8109	33.9202	4.9803
16	6.1304	0.1631	0.02339	42.7533	0.14339	6.9740	36.3670	5.2147
17	6.8660	0.1456	0.02046	48.8837	0.14046	7.1196	38.6973	5.4353
18	7.6900	0.1300	0.01794	55.7497	0.13794	7.2497	40.9080	5.6427
19	8.6128	0.1161	0.01576	63.4397	0.13576	7.3658	42.9979	5.8375
20	9.6463	0.1037	0.01388	72.0524	0.13388	7.4694	44.9676	6.0202
21	10.8038	0.0926	0.01224	81.6987	0.13224	7.5620	46.8188	6.1913
22	12.1003	0.0826	0.01081	92.5026	0.13081	7.6446	48.5543	6.3514
23	13.5523	0.0738	0.00956	104.6029	0.12956	7.7184	50.1776	6.5010
24	15.1786	0.0659	0.00846	118.1552	0.12846	7.7843	51.6929	6.6406
25	17.0001	0.0588	0.00750	133.3339	0.12750	7.8431	53.1046	6.7708
26	19.0401	0.0525	0.00665	150.3339	0.12665	7.8957	54.4177	6.8921
27	21.3249	0.0469	0.00590	169.3740	0.12590	7.9426	55.6369	7.0049
28	23.8839	0.0419	0.00524	190.6989	0.12524	7.9844	56.7674	7.1098
29	26.7499	0.0374	0.00466	214.5828	0.12466	8.0218	57.8141	7.2071
30	29.9599	0.0334	0.00414	241.3327	0.12414	8.0552	58.7821	7.2974
31	33.5551	0.0298	0.00369	271.2926	0.12369	8.0850	59.6761	7.3811
32	37.5817	0.0266	0.00328	304.8477	0.12328	8.1116	60.5010	7.4586
33	42.0915	0.0238	0.00292	342.4294	0.12292	8.1354	61.2612	7.5302
34	47.1425	0.0212	0.00260	384.5210	0.12260	8.1566	61.9612	7.5965
35	52.7996	0.0189	0.00232	431.6635	0.12232	8.1755	62.6052	7.6577
40	93.0510	0.0107	0.00130	767.0914	0.12130	8.2438	65.1159	7.8988
45	163.9876	0.0061	0.0074	1358.23	0.12074	8.2825	66.7342	8.0572
50	289.0022	0.0035	0.00042	2400.02	0.12042	8.3045	67.7624	8.1597
55	509.3206	0.0020	0.00024	4236.01	0.12024	8.3170	68.4082	8.2251
60	897.5969	0.0011	0.00013	7471.64	0.12013	8.3240	68.8100	8.2664
65	1581.87	0.0006	0.00008	13,174	0.12008	8.3281	69.0581	8.2922
70	2787.80	0.0004	0.00004	23,223	0.12004	8.3303	69.2103	8.3082
75	4913.06	0.0002	0.00002	40,934	0.12002	8.3316	69.3031	8.3181
80	8658.48	0.0001	0.00001	72,146	0.12001	8.3324	69.3594	8.3241
85	15,259	0.0001	0.00001		0.12001	8.3328	69.3935	8.3278

14%			TABLE 18	Discrete Cash Flow: Compound Interest Factors				14%
	Single Payments		Uniform Series Payments				Arithmetic Gradients	
	Compound Amount	Present Worth	Sinking Fund	Compound Amount	Capital Recovery	Present Worth	Gradient Present Worth	Gradient Uniform Series
n	F/P	P/F	A/F	F/A	A/P	P/A	P/G	A/G
1	1.1400	0.8772	1.00000	1.0000	1.14000	0.8772		
2	1.2996	0.7695	0.46729	2.1400	0.60729	1.6467	0.7695	0.4673
3	1.4815	0.6750	0.29073	3.4396	0.43073	2.3216	2.1194	0.9129
4	1.6890	0.5921	0.20320	4.9211	0.34320	2.9137	3.8957	1.3370
5	1.9254	0.5194	0.15128	6.6101	0.29128	3.4331	5.9731	1.7399
6	2.1950	0.4556	0.11716	8.5355	0.25716	3.8887	8.2511	2.1218
7	2.5023	0.3996	0.09319	10.7305	0.23319	4.2883	10.6489	2.4832
8	2.8526	0.3506	0.07557	13.2328	0.21557	4.6389	13.1028	2.8246
9	3.2519	0.3075	0.06217	16.0853	0.20217	4.9464	15.5629	3.1463
10	3.7072	0.2697	0.05171	19.3373	0.19171	5.2161	17.9906	3.4490
11	4.2262	0.2366	0.04339	23.0445	0.18339	5.4527	20.3567	3.7333
12	4.8179	0.2076	0.03667	27.2707	0.17667	5.6603	22.6399	3.9998
13	5.4924	0.1821	0.03116	32.0887	0.17116	5.8424	24.8247	4.2491
14	6.2613	0.1597	0.02661	37.5811	0.16661	6.0021	26.9009	4.4819
15	7.1379	0.1401	0.02281	43.8424	0.16281	6.1422	28.8623	4.6990
16	8.1372	0.1229	0.01962	50.9804	0.15962	6.2651	30.7057	4.9011
17	9.2765	0.1078	0.01692	59.1176	0.15692	6.3729	32.4305	5.0888
18	10.5752	0.0946	0.01462	68.3941	0.15462	6.4674	34.0380	5.2630
19	12.0557	0.0829	0.01266	78.9692	0.15266	6.5504	35.5311	5.4243
20	13.7435	0.0728	0.01099	91.0249	0.15099	6.6231	36.9135	5.5734
21	15.6676	0.0638	0.00954	104.7684	0.14954	6.6870	38.1901	5.7111
22	17.8610	0.0560	0.00830	120.4360	0.14830	6.7429	39.3658	5.8381
23	20.3616	0.0491	0.00723	138.2970	0.14723	6.7921	40.4463	5.9549
24	23.2122	0.0431	0.00630	158.6586	0.14630	6.8351	41.4371	6.0624
25	26.4619	0.0378	0.00550	181.8708	0.14550	6.8729	42.3441	6.1610
26	30.1666	0.0331	0.00480	208.3327	0.14480	6.9061	43.1728	6.2514
27	34.3899	0.0291	0.00419	238.4993	0.14419	6.9352	43.9289	6.3342
28	39.2045	0.0255	0.00366	272.8892	0.14366	6.9607	44.6176	6.4100
29	44.6931	0.0224	0.00320	312.0937	0.14320	6.9830	45.2441	6.4791
30	50.9502	0.0196	0.00280	356.7868	0.14280	7.0027	45.8132	6.5423
31	58.0832	0.0172	0.00245	407.7370	0.14245	7.0199	46.3297	6.5998
32	66.2148	0.0151	0.00215	465.8202	0.14215	7.0350	46.7979	6.6522
33	75.4849	0.0132	0.00188	532.0350	0.14188	7.0482	47.2218	6.6998
34	86.0528	0.0116	0.00165	607.5199	0.14165	7.0599	47.6053	6.7431
35	98.1002	0.0102	0.00144	693.5727	0.14144	7.0700	47.9519	6.7824
40	188.8835	0.0053	0.00075	1342.03	0.14075	7.1050	49.2376	6.9300
45	363.6791	0.0027	0.00039	2590.56	0.14039	7.1232	49.9963	7.0188
50	700.2330	0.0014	0.00020	4994.52	0.14020	7.1327	50.4375	7.0714
55	1348.24	0.0007	0.00010	9623.13	0.14010	7.1376	50.6912	7.1020
60	2595.92	0.0004	0.00005	18,535	0.14005	7.1401	50.8357	7.1197
65	4998.22	0.0002	0.00003	35,694	0.14003	7.1414	50.9173	7.1298
70	9623.64	0.0001	0.00001	68,733	0.14001	7.1421	50.9632	7.1356
75	18,530	0.0001	0.00001		0.14001	7.1425	50.9887	7.1388
80	35,677				0.14000	7.1427	51.0030	7.1406
85	68,693				0.14000	7.1428	51.0108	7.1416

15%			TABLE 19	Discrete Cash Flow: Compound Interest Factors				15%
	Single Payments		Uniform Series Payments				Arithmetic Gradients	
	Compound Amount F/P	Present Worth P/F	Sinking Fund A/F	Compound Amount F/A	Capital Recovery A/P	Present Worth P/A	Gradient Present Worth P/G	Gradient Uniform Series A/G
n								
1	1.1500	0.8696	1.00000	1.0000	1.15000	0.8696		
2	1.3225	0.7561	0.46512	2.1500	0.61512	1.6257	0.7561	0.4651
3	1.5209	0.6575	0.28798	3.4725	0.43798	2.2832	2.0712	0.9071
4	1.7490	0.5718	0.20027	4.9934	0.35027	2.8550	3.7864	1.3263
5	2.0114	0.4972	0.14832	6.7424	0.29832	3.3522	5.7751	1.7228
6	2.3131	0.4323	0.11424	8.7537	0.26424	3.7845	7.9368	2.0972
7	2.6600	0.3759	0.09036	11.0668	0.24036	4.1604	10.1924	2.4498
8	3.0590	0.3269	0.07285	13.7268	0.22285	4.4873	12.4807	2.7813
9	3.5179	0.2843	0.05957	16.7858	0.20957	4.7716	14.7548	3.0922
10	4.0456	0.2472	0.04925	20.3037	0.19925	5.0188	16.9795	3.3832
11	4.6524	0.2149	0.04107	24.3493	0.19107	5.2337	19.1289	3.6549
12	5.3503	0.1869	0.03448	29.0017	0.18448	5.4206	21.1849	3.9082
13	6.1528	0.1625	0.02911	34.3519	0.17911	5.5831	23.1352	4.1438
14	7.0757	0.1413	0.02469	40.5047	0.17469	5.7245	24.9725	4.3624
15	8.1371	0.1229	0.02102	47.5804	0.17102	5.8474	26.6930	4.5650
16	9.3576	0.1069	0.01795	55.7175	0.16795	5.9542	28.2960	4.7522
17	10.7613	0.0929	0.01537	65.0751	0.16537	6.0472	29.7828	4.9251
18	12.3755	0.0808	0.01319	75.8364	0.16319	6.1280	31.1565	5.0843
19	14.2318	0.0703	0.01134	88.2118	0.16134	6.1982	32.4213	5.2307
20	16.3665	0.0611	0.00976	102.4436	0.15976	6.2593	33.5822	5.3651
21	18.8215	0.0531	0.00842	118.8101	0.15842	6.3125	34.6448	5.4883
22	21.6447	0.0462	0.00727	137.6316	0.15727	6.3587	35.6150	5.6010
23	24.8915	0.0402	0.00628	159.2764	0.15628	6.3988	36.4988	5.7040
24	28.6252	0.0349	0.00543	184.1678	0.15543	6.4338	37.3023	5.7979
25	32.9190	0.0304	0.00470	212.7930	0.15470	6.4641	38.0314	5.8834
26	37.8568	0.0264	0.00407	245.7120	0.15407	6.4906	38.6918	5.9612
27	43.5353	0.0230	0.00353	283.5688	0.15353	6.5135	39.2890	6.0319
28	50.0656	0.0200	0.00306	327.1041	0.15306	6.5335	39.8283	6.0960
29	57.5755	0.0174	0.00265	377.1697	0.15265	6.5509	40.3146	6.1541
30	66.2118	0.0151	0.00230	434.7451	0.15230	6.5660	40.7526	6.2066
31	76.1435	0.0131	0.00200	500.9569	0.15200	6.5791	41.1466	6.2541
32	87.5651	0.0114	0.00173	577.1005	0.15173	6.5905	41.5006	6.2970
33	100.6998	0.0099	0.00150	664.6655	0.15150	6.6005	41.8184	6.3357
34	115.8048	0.0086	0.00131	765.3654	0.15131	6.6091	42.1033	6.3705
35	133.1755	0.0075	0.00113	881.1702	0.15113	6.6166	42.3586	6.4019
40	267.8635	0.0037	0.00056	1779.09	0.15056	6.6418	43.2830	6.5168
45	538.7693	0.0019	0.00028	3585.13	0.15028	6.6543	43.8051	6.5830
50	1083.66	0.0009	0.00014	7217.72	0.15014	6.6605	44.0958	6.6205
55	2179.62	0.0005	0.00007	14,524	0.15007	6.6636	44.2558	6.6414
60	4384.00	0.0002	0.00003	29,220	0.15003	6.6651	44.3431	6.6530
65	8817.79	0.0001	0.00002	58,779	0.15002	6.6659	44.3903	6.6593
70	17,736	0.0001	0.00001		0.15001	6.6663	44.4156	6.6627
75	35,673				0.15000	6.6665	44.4292	6.6646
80	71,751				0.15000	6.6666	44.4364	6.6656
85					0.15000	6.6666	44.4402	6.6661

16%			**TABLE 20**	**Discrete Cash Flow: Compound Interest Factors**				16%
	Single Payments		**Uniform Series Payments**				**Arithmetic Gradients**	
	Compound Amount	Present Worth	Sinking Fund	Compound Amount	Capital Recovery	Present Worth	Gradient Present Worth	Gradient Uniform Series
n	*F/P*	*P/F*	*A/F*	*F/A*	*A/P*	*P/A*	*P/G*	*A/G*
1	1.1600	0.8621	1.00000	1.0000	1.16000	0.8621		
2	1.3456	0.7432	0.46296	2.1600	0.62296	1.6052	0.7432	0.4630
3	1.5609	0.6407	0.28526	3.5056	0.44526	2.2459	2.0245	0.9014
4	1.8106	0.5523	0.19738	5.0665	0.35738	2.7982	3.6814	1.3156
5	2.1003	0.4761	0.14541	6.8771	0.30541	3.2743	5.5858	1.7060
6	2.4364	0.4104	0.11139	8.9775	0.27139	3.6847	7.6380	2.0729
7	2.8262	0.3538	0.08761	11.4139	0.24761	4.0386	9.7610	2.4169
8	3.2784	0.3050	0.07022	14.2401	0.23022	4.3436	11.8962	2.7388
9	3.8030	0.2630	0.05708	17.5185	0.21708	4.6065	13.9998	3.0391
10	4.4114	0.2267	0.04690	21.3215	0.20690	4.8332	16.0399	3.3187
11	5.1173	0.1954	0.03886	25.7329	0.19886	5.0286	17.9941	3.5783
12	5.9360	0.1685	0.03241	30.8502	0.19241	5.1971	19.8472	3.8189
13	6.8858	0.1452	0.02718	36.7862	0.18718	5.3423	21.5899	4.0413
14	7.9875	0.1252	0.02290	43.6720	0.18290	5.4675	23.2175	4.2464
15	9.2655	0.1079	0.01936	51.6595	0.17936	5.5755	24.7284	4.4352
16	10.7480	0.0930	0.01641	60.9250	0.17641	5.6685	26.1241	4.6086
17	12.4677	0.0802	0.01395	71.6730	0.17395	5.7487	27.4074	4.7676
18	14.4625	0.0691	0.01188	84.1407	0.17188	5.8178	28.5828	4.9130
19	16.7765	0.0596	0.01014	98.6032	0.17014	5.8775	29.6557	5.0457
20	19.4608	0.0514	0.00867	115.3797	0.16867	5.9288	30.6321	5.1666
22	26.1864	0.0382	0.00635	157.4150	0.16635	6.0113	32.3200	5.3765
24	35.2364	0.0284	0.00467	213.9776	0.16467	6.0726	33.6970	5.5490
26	47.4141	0.0211	0.00345	290.0883	0.16345	6.1182	34.8114	5.6898
28	63.8004	0.0157	0.00255	392.5028	0.16255	6.1520	35.7073	5.8041
30	85.8499	0.0116	0.00189	530.3117	0.16189	6.1772	36.4234	5.8964
32	115.5196	0.0087	0.00140	715.7475	0.16140	6.1959	36.9930	5.9706
34	155.4432	0.0064	0.00104	965.2698	0.16104	6.2098	37.4441	6.0299
35	180.3141	0.0055	0.00089	1120.71	0.16089	6.2153	37.6327	6.0548
36	209.1643	0.0048	0.00077	1301.03	0.16077	6.2201	37.8000	6.0771
38	281.4515	0.0036	0.00057	1752.82	0.16057	6.2278	38.0799	6.1145
40	378.7212	0.0026	0.00042	2360.76	0.16042	6.2335	38.2992	6.1441
45	795.4438	0.0013	0.00020	4965.27	0.16020	6.2421	38.6598	6.1934
50	1670.70	0.0006	0.00010	10,436	0.16010	6.2463	38.8521	6.2201
55	3509.05	0.0003	0.00005	21,925	0.16005	6.2482	38.9534	6.2343
60	7370.20	0.0001	0.00002	46,058	0.16002	6.2492	39.0063	6.2419

18%			**TABLE 21**	**Discrete Cash Flow: Compound Interest Factors**				*18%*
	Single Payments		**Uniform Series Payments**				**Arithmetic Gradients**	
	Compound Amount	Present Worth	Sinking Fund	Compound Amount	Capital Recovery	Present Worth	Gradient Present Worth	Gradient Uniform Series
n	*F/P*	*P/F*	*A/F*	*F/A*	*A/P*	*P/A*	*P/G*	*A/G*
1	1.1800	0.8475	1.00000	1.0000	1.18000	0.8475		
2	1.3924	0.7182	0.45872	2.1800	0.63872	1.5656	0.7182	0.4587
3	1.6430	0.6086	0.27992	3.5724	0.45992	2.1743	1.9354	0.8902
4	1.9388	0.5158	0.19174	5.2154	0.37174	2.6901	3.4828	1.2947
5	2.2878	0.4371	0.13978	7.1542	0.31978	3.1272	5.2312	1.6728
6	2.6996	0.3704	0.10591	9.4420	0.28591	3.4976	7.0834	2.0252
7	3.1855	0.3139	0.08236	12.1415	0.26236	3.8115	8.9670	2.3526
8	3.7589	0.2660	0.06524	15.3270	0.24524	4.0776	10.8292	2.6558
9	4.4355	0.2255	0.05239	19.0859	0.23239	4.3030	12.6329	2.9358
10	5.2338	0.1911	0.04251	23.5213	0.22251	4.4941	14.3525	3.1936
11	6.1759	0.1619	0.03478	28.7551	0.21478	4.6560	15.9716	3.4303
12	7.2876	0.1372	0.02863	34.9311	0.20863	4.7932	17.4811	3.6470
13	8.5994	0.1163	0.02369	42.2187	0.20369	4.9095	18.8765	3.8449
14	10.1472	0.0985	0.01968	50.8180	0.19968	5.0081	20.1576	4.0250
15	11.9737	0.0835	0.01640	60.9653	0.19640	5.0916	21.3269	4.1887
16	14.1290	0.0708	0.01371	72.9390	0.19371	5.1624	22.3885	4.3369
17	16.6722	0.0600	0.01149	87.0680	0.19149	5.2223	23.3482	4.4708
18	19.6733	0.0508	0.00964	103.7403	0.18964	5.2732	24.2123	4.5916
19	23.2144	0.0431	0.00810	123.4135	0.18810	5.3162	24.9877	4.7003
20	27.3930	0.0365	0.00682	146.6280	0.18682	5.3527	25.6813	4.7978
22	38.1421	0.0262	0.00485	206.3448	0.18485	5.4099	26.8506	4.9632
24	53.1090	0.0188	0.00345	289.4945	0.18345	5.4509	27.7725	5.0950
26	73.9490	0.0135	0.00247	405.2721	0.18247	5.4804	28.4935	5.1991
28	102.9666	0.0097	0.00177	566.4809	0.18177	5.5016	29.0537	5.2810
30	143.3706	0.0070	0.00126	790.9480	0.18126	5.5168	29.4864	5.3448
32	199.6293	0.0050	0.00091	1103.50	0.18091	5.5277	29.8191	5.3945
34	277.9638	0.0036	0.00065	1538.69	0.18065	5.5356	30.0736	5.4328
35	327.9973	0.0030	0.00055	1816.65	0.18055	5.5386	30.1773	5.4485
36	387.0368	0.0026	0.00047	2144.65	0.18047	5.5412	30.2677	5.4623
38	538.9100	0.0019	0.00033	2988.39	0.18033	5.5452	30.4152	5.4849
40	750.3783	0.0013	0.00024	4163.21	0.18024	5.5482	30.5269	5.5022
45	1716.68	0.0006	0.00010	9531.58	0.18010	5.5523	30.7006	5.5293
50	3927.36	0.0003	0.00005	21,813	0.18005	5.5541	30.7856	5.5428
55	8984.84	0.0001	0.00002	49,910	0.18002	5.5549	30.8268	5.5494
60	20,555			114,190	0.18001	5.5553	30.8465	5.5526

20%			TABLE 22	Discrete Cash Flow: Compound Interest Factors				20%
	Single Payments		Uniform Series Payments				Arithmetic Gradients	
	Compound Amount F/P	Present Worth P/F	Sinking Fund A/F	Compound Amount F/A	Capital Recovery A/P	Present Worth P/A	Gradient Present Worth P/G	Gradient Uniform Series A/G
n								
1	1.2000	0.8333	1.00000	1.0000	1.20000	0.8333		
2	1.4400	0.6944	0.45455	2.2000	0.65455	1.5278	0.6944	0.4545
3	1.7280	0.5787	0.27473	3.6400	0.47473	2.1065	1.8519	0.8791
4	2.0736	0.4823	0.18629	5.3680	0.38629	2.5887	3.2986	1.2742
5	2.4883	0.4019	0.13438	7.4416	0.33438	2.9906	4.9061	1.6405
6	2.9860	0.3349	0.10071	9.9299	0.30071	3.3255	6.5806	1.9788
7	3.5832	0.2791	0.07742	12.9159	0.27742	3.6046	8.2551	2.2902
8	4.2998	0.2326	0.06061	16.4991	0.26061	3.8372	9.8831	2.5756
9	5.1598	0.1938	0.04808	20.7989	0.24808	4.0310	11.4335	2.8364
10	6.1917	0.1615	0.03852	25.9587	0.23852	4.1925	12.8871	3.0739
11	7.4301	0.1346	0.03110	32.1504	0.23110	4.3271	14.2330	3.2893
12	8.9161	0.1122	0.02526	39.5805	0.22526	4.4392	15.4667	3.4841
13	10.6993	0.0935	0.02062	48.4966	0.22062	4.5327	16.5883	3.6597
14	12.8392	0.0779	0.01689	59.1959	0.21689	4.6106	17.6008	3.8175
15	15.4070	0.0649	0.01388	72.0351	0.21388	4.6755	18.5095	3.9588
16	18.4884	0.0541	0.01144	87.4421	0.21144	4.7296	19.3208	4.0851
17	22.1861	0.0451	0.00944	105.9306	0.20944	4.7746	20.0419	4.1976
18	26.6233	0.0376	0.00781	128.1167	0.20781	4.8122	20.6805	4.2975
19	31.9480	0.0313	0.00646	154.7400	0.20646	4.8435	21.2439	4.3861
20	38.3376	0.0261	0.00536	186.6880	0.20536	4.8696	21.7395	4.4643
22	55.2061	0.0181	0.00369	271.0307	0.20369	4.9094	22.5546	4.5941
24	79.4968	0.0126	0.00255	392.4842	0.20255	4.9371	23.1760	4.6943
26	114.4755	0.0087	0.00176	567.3773	0.20176	4.9563	23.6460	4.7709
28	164.8447	0.0061	0.00122	819.2233	0.20122	4.9697	23.9991	4.8291
30	237.3763	0.0042	0.00085	1181.88	0.20085	4.9789	24.2628	4.8731
32	341.8219	0.0029	0.00059	1704.11	0.20059	4.9854	24.4588	4.9061
34	492.2235	0.0020	0.00041	2456.12	0.20041	4.9898	24.6038	4.9308
35	590.6682	0.0017	0.00034	2948.34	0.20034	4.9915	24.6614	4.9406
36	708.8019	0.0014	0.00028	3539.01	0.20028	4.9929	24.7108	4.9491
38	1020.67	0.0010	0.00020	5098.37	0.20020	4.9951	24.7894	4.9627
40	1469.77	0.0007	0.00014	7343.86	0.20014	4.9966	24.8469	4.9728
45	3657.26	0.0003	0.00005	18,281	0.20005	4.9986	24.9316	4.9877
50	9100.44	0.0001	0.00002	45,497	0.20002	4.9995	24.9698	4.9945
55	22,645		0.00001		0.20001	4.9998	24.9868	4.9976

22%			TABLE 23	Discrete Cash Flow: Compound Interest Factors				22%
	Single Payments		Uniform Series Payments				Arithmetic Gradients	
	Compound Amount	Present Worth	Sinking Fund	Compound Amount	Capital Recovery	Present Worth	Gradient Present Worth	Gradient Uniform Series
n	F/P	P/F	A/F	F/A	A/P	P/A	P/G	A/G
1	1.2200	0.8197	1.00000	1.0000	1.22000	0.8197		
2	1.4884	0.6719	0.45045	2.2200	0.67045	1.4915	0.6719	0.4505
3	1.8158	0.5507	0.26966	3.7084	0.48966	2.0422	1.7733	0.8683
4	2.2153	0.4514	0.18102	5.5242	0.40102	2.4936	3.1275	1.2542
5	2.7027	0.3700	0.12921	7.7396	0.34921	2.8636	4.6075	1.6090
6	3.2973	0.3033	0.09576	10.4423	0.31576	3.1669	6.1239	1.9337
7	4.0227	0.2486	0.07278	13.7396	0.29278	3.4155	7.6154	2.2297
8	4.9077	0.2038	0.05630	17.7623	0.27630	3.6193	9.0417	2.4982
9	5.9874	0.1670	0.04411	22.6700	0.26411	3.7863	10.3779	2.7409
10	7.3046	0.1369	0.03489	28.6574	0.25489	3.9232	11.6100	2.9593
11	8.9117	0.1122	0.02781	35.9620	0.24781	4.0354	12.7321	3.1551
12	10.8722	0.0920	0.02228	44.8737	0.24228	4.1274	13.7438	3.3299
13	13.2641	0.0754	0.01794	55.7459	0.23794	4.2028	14.6485	3.4855
14	16.1822	0.0618	0.01449	69.0100	0.23449	4.2646	15.4519	3.6233
15	19.7423	0.0507	0.01174	85.1922	0.23174	4.3152	16.1610	3.7451
16	24.0856	0.0415	0.00953	104.9345	0.22953	4.3567	16.7838	3.8524
17	29.3844	0.0340	0.00775	129.0201	0.22775	4.3908	17.3283	3.9465
18	35.8490	0.0279	0.00631	158.4045	0.22631	4.4187	17.8025	4.0289
19	43.7358	0.0229	0.00515	194.2535	0.22515	4.4415	18.2141	4.1009
20	53.3576	0.0187	0.00420	237.9893	0.22420	4.4603	18.5702	4.1635
22	79.4175	0.0126	0.00281	356.4432	0.22281	4.4882	19.1418	4.2649
24	118.2050	0.0085	0.00188	532.7501	0.22188	4.5070	19.5635	4.3407
26	175.9364	0.0057	0.00126	795.1653	0.22126	4.5196	19.8720	4.3968
28	261.8637	0.0038	0.00084	1185.74	0.22084	4.5281	20.0962	4.4381
30	389.7579	0.0026	0.00057	1767.08	0.22057	4.5338	20.2583	4.4683
32	580.1156	0.0017	0.00038	2632.34	0.22038	4.5376	20.3748	4.4902
34	863.4441	0.0012	0.00026	3920.20	0.22026	4.5402	20.4582	4.5060
35	1053.40	0.0009	0.00021	4783.64	0.22021	4.5411	20.4905	4.5122
36	1285.15	0.0008	0.00017	5837.05	0.22017	4.5419	20.5178	4.5174
38	1912.82	0.0005	0.00012	8690.08	0.22012	4.5431	20.5601	4.5256
40	2847.04	0.0004	0.00008	12,937	0.22008	4.5439	20.5900	4.5314
45	7694.71	0.0001	0.00003	34,971	0.22003	4.5449	20.6319	4.5396
50	20,797		0.00001	94,525	0.22001	4.5452	20.6492	4.5431
55	56,207				0.22000	4.5454	20.6563	4.5445

24%			TABLE 24	Discrete Cash Flow: Compound Interest Factors			24%
	Single Payments		Uniform Series Payments				Arithmetic Gradients
	Compound Amount	Present Worth	Sinking Fund	Compound Amount	Capital Recovery	Present Worth	Gradient Present Worth	Gradient Uniform Series
n	F/P	P/F	A/F	F/A	A/P	P/A	P/G	A/G
1	1.2400	0.8065	1.00000	1.0000	1.24000	0.8065		
2	1.5376	0.6504	0.44643	2.2400	0.68643	1.4568	0.6504	0.4464
3	1.9066	0.5245	0.26472	3.7776	0.50472	1.9813	1.6993	0.8577
4	2.3642	0.4230	0.17593	5.6842	0.41593	2.4043	2.9683	1.2346
5	2.9316	0.3411	0.12425	8.0484	0.36425	2.7454	4.3327	1.5782
6	3.6352	0.2751	0.09107	10.9801	0.33107	3.0205	5.7081	1.8898
7	4.5077	0.2218	0.06842	14.6153	0.30842	3.2423	7.0392	2.1710
8	5.5895	0.1789	0.05229	19.1229	0.29229	3.4212	8.2915	2.4236
9	6.9310	0.1443	0.04047	24.7125	0.28047	3.5655	9.4458	2.6492
10	8.5944	0.1164	0.03160	31.6434	0.27160	3.6819	10.4930	2.8499
11	10.6571	0.0938	0.02485	40.2379	0.26485	3.7757	11.4313	3.0276
12	13.2148	0.0757	0.01965	50.8950	0.25965	3.8514	12.2637	3.1843
13	16.3863	0.0610	0.01560	64.1097	0.25560	3.9124	12.9960	3.3218
14	20.3191	0.0492	0.01242	80.4961	0.25242	3.9616	13.6358	3.4420
15	25.1956	0.0397	0.00992	100.8151	0.24992	4.0013	14.1915	3.5467
16	31.2426	0.0320	0.00794	126.0108	0.24794	4.0333	14.6716	3.6376
17	38.7408	0.0258	0.00636	157.2534	0.24636	4.0591	15.0846	3.7162
18	48.0386	0.0208	0.00510	195.9942	0.24510	4.0799	15.4385	3.7840
19	59.5679	0.0168	0.00410	244.0328	0.24410	4.0967	15.7406	3.8423
20	73.8641	0.0135	0.00329	303.6006	0.24329	4.1103	15.9979	3.8922
22	113.5735	0.0088	0.00213	469.0563	0.24213	4.1300	16.4011	3.9712
24	174.6306	0.0057	0.00138	723.4610	0.24138	4.1428	16.6891	4.0284
26	268.5121	0.0037	0.00090	1114.63	0.24090	4.1511	16.8930	4.0695
28	412.8642	0.0024	0.00058	1716.10	0.24058	4.1566	17.0365	4.0987
30	634.8199	0.0016	0.00038	2640.92	0.24038	4.1601	17.1369	4.1193
32	976.0991	0.0010	0.00025	4062.91	0.24025	4.1624	17.2067	4.1338
34	1500.85	0.0007	0.00016	6249.38	0.24016	4.1639	17.2552	4.1440
35	1861.05	0.0005	0.00013	7750.23	0.24013	4.1664	17.2734	4.1479
36	2307.71	0.0004	0.00010	9611.28	0.24010	4.1649	17.2886	4.1511
38	3548.33	0.0003	0.00007	14,781	0.24007	4.1655	17.3116	4.1560
40	5455.91	0.0002	0.00004	22,729	0.24004	4.1659	17.3274	4.1593
45	15,995	0.0001	0.00002	66,640	0.24002	4.1664	17.3483	4.1639
50	46,890		0.00001		0.24001	4.1666	17.3563	4.1653
55					0.24000	4.1666	17.3593	4.1663

25%		TABLE 25	Discrete Cash Flow: Compound Interest Factors				25%
	Single Payments		**Uniform Series Payments**				**Arithmetic Gradients**	
	Compound Amount	Present Worth	Sinking Fund	Compound Amount	Capital Recovery	Present Worth	Gradient Present Worth	Gradient Uniform Series
n	*F/P*	*P/F*	*A/F*	*F/A*	*A/P*	*P/A*	*P/G*	*A/G*
1	1.2500	0.8000	1.00000	1.0000	1.25000	0.8000		
2	1.5625	0.6400	0.44444	2.2500	0.69444	1.4400	0.6400	0.4444
3	1.9531	0.5120	0.26230	3.8125	0.51230	1.9520	1.6640	0.8525
4	2.4414	0.4096	0.17344	5.7656	0.42344	2.3616	2.8928	1.2249
5	3.0518	0.3277	0.12185	8.2070	0.37185	2.6893	4.2035	1.5631
6	3.8147	0.2621	0.08882	11.2588	0.33882	2.9514	5.5142	1.8683
7	4.7684	0.2097	0.06634	15.0735	0.31634	3.1611	6.7725	2.1424
8	5.9605	0.1678	0.05040	19.8419	0.30040	3.3289	7.9469	2.3872
9	7.4506	0.1342	0.03876	25.8023	0.28876	3.4631	9.0207	2.6048
10	9.3132	0.1074	0.03007	33.2529	0.28007	3.5705	9.9870	2.7971
11	11.6415	0.0859	0.02349	42.5661	0.27349	3.6564	10.8460	2.9663
12	14.5519	0.0687	0.01845	54.2077	0.26845	3.7251	11.6020	3.1145
13	18.1899	0.0550	0.01454	68.7596	0.26454	3.7801	12.2617	3.2437
14	22.7374	0.0440	0.01150	86.9495	0.26150	3.8241	12.8334	3.3559
15	28.4217	0.0352	0.00912	109.6868	0.25912	3.8593	13.3260	3.4530
16	35.5271	0.0281	0.00724	138.1085	0.25724	3.8874	13.7482	3.5366
17	44.4089	0.0225	0.00576	173.6357	0.25576	3.9099	14.1085	3.6084
18	55.5112	0.0180	0.00459	218.0446	0.25459	3.9279	14.4147	3.6698
19	69.3889	0.0144	0.00366	273.5558	0.25366	3.9424	14.6741	3.7222
20	86.7362	0.0115	0.00292	342.9447	0.25292	3.9539	14.8932	3.7667
22	135.5253	0.0074	0.00186	538.1011	0.25186	3.9705	15.2326	3.8365
24	211.7582	0.0047	0.00119	843.0329	0.25119	3.9811	15.4711	3.8861
26	330.8722	0.0030	0.00076	1319.49	0.25076	3.9879	15.6373	3.9212
28	516.9879	0.0019	0.00048	2063.95	0.25048	3.9923	15.7524	3.9457
30	807.7936	0.0012	0.00031	3227.17	0.25031	3.9950	15.8316	3.9628
32	1262.18	0.0008	0.00020	5044.71	0.25020	3.9968	15.8859	3.9746
34	1972.15	0.0005	0.00013	7884.61	0.25013	3.9980	15.9229	3.9828
35	2465.19	0.0004	0.00010	9856.76	.025010	3.9984	15.9367	3.9858
36	3081.49	0.0003	0.00008	12,322	0.25008	3.9987	15.9481	3.9883
38	4814.82	0.0002	0.00005	19,255	0.25005	3.9992	15.9651	3.9921
40	7523.16	0.0001	0.00003	30,089	0.25003	3.9995	15.9766	3.9947
45	22,959		0.00001	91,831	0.25001	3.9998	15.9915	3.9980
50	70,065				0.25000	3.9999	15.9969	3.9993
55					0.25000	4.0000	15.9989	3.9997

| 30% | | TABLE 26 | Discrete Cash Flow: Compound Interest Factors | | | | 30% |

	Single Payments		Uniform Series Payments				Arithmetic Gradients	
	Compound Amount	Present Worth	Sinking Fund	Compound Amount	Capital Recovery	Present Worth	Gradient Present Worth	Gradient Uniform Series
n	F/P	P/F	A/F	F/A	A/P	P/A	P/G	A/G
1	1.3000	0.7692	1.00000	1.0000	1.30000	0.7692		
2	1.6900	0.5917	0.43478	2.3000	0.73478	1.3609	0.5917	0.4348
3	2.1970	0.4552	0.25063	3.9900	0.55063	1.8161	1.5020	0.8271
4	2.8561	0.3501	0.16163	6.1870	0.46163	2.1662	2.5524	1.1783
5	3.7129	0.2693	0.11058	9.0431	0.41058	2.4356	3.6297	1.4903
6	4.8268	0.2072	0.07839	12.7560	0.37839	2.6427	4.6656	1.7654
7	6.2749	0.1594	0.05687	17.5828	0.35687	2.8021	5.6218	2.0063
8	8.1573	0.1226	0.04192	23.8577	0.34192	2.9247	6.4800	2.2156
9	10.6045	0.0943	0.03124	32.0150	0.33124	3.0190	7.2343	2.3963
10	13.7858	0.0725	0.02346	42.6195	0.32346	3.0915	7.8872	2.5512
11	17.9216	0.0558	0.01773	56.4053	0.31773	3.1473	8.4452	2.6833
12	23.2981	0.0429	0.01345	74.3270	0.31345	3.1903	8.9173	2.7952
13	30.2875	0.0330	0.01024	97.6250	0.31024	3.2233	9.3135	2.8895
14	39.3738	0.0254	0.00782	127.9125	0.30782	3.2487	9.6437	2.9685
15	51.1859	0.0195	0.00598	167.2863	0.30598	3.2682	9.9172	3.0344
16	66.5417	0.0150	0.00458	218.4722	0.30458	3.2832	10.1426	3.0892
17	86.5042	0.0116	0.00351	285.0139	0.30351	3.2948	10.3276	3.1345
18	112.4554	0.0089	0.00269	371.5180	0.30269	3.3037	10.4788	3.1718
19	146.1920	0.0068	0.00207	483.9734	0.30207	3.3105	10.6019	3.2025
20	190.0496	0.0053	0.00159	630.1655	0.30159	3.3158	10.7019	3.2275
22	321.1839	0.0031	0.00094	1067.28	0.30094	3.3230	10.8482	3.2646
24	542.8008	0.0018	0.00055	1806.00	0.30055	3.3272	10.9433	3.2890
25	705.6410	0.0014	0.00043	2348.80	0.30043	3.3286	10.9773	3.2979
26	917.3333	0.0011	0.00033	3054.44	0.30033	3.3297	11.0045	3.3050
28	1550.29	0.0006	0.00019	5164.31	0.30019	3.3312	11.0437	3.3153
30	2620.00	0.0004	0.00011	8729.99	0.30011	3.3321	11.0687	3.3219
32	4427.79	0.0002	0.00007	14,756	0.30007	3.3326	11.0845	3.3261
34	7482.97	0.0001	0.00004	24,940	0.30004	3.3329	11.0945	3.3288
35	9727.86	0.0001	0.00003	32,423	0.30003	3.3330	11.0980	3.3297

35%		**TABLE 27**	**Discrete Cash Flow: Compound Interest Factors**					*35%*
	Single Payments		**Uniform Series Payments**				**Arithmetic Gradients**	
	Compound Amount	Present Worth	Sinking Fund	Compound Amount	Capital Recovery	Present Worth	Gradient Present Worth	Gradient Uniform Series
n	*F/P*	*P/F*	*A/F*	*F/A*	*A/P*	*P/A*	*P/G*	*A/G*
1	1.3500	0.7407	1.00000	1.0000	1.35000	0.7407		
2	1.8225	0.5487	0.42553	2.3500	0.77553	1.2894	0.5487	0.4255
3	2.4604	0.4064	0.23966	4.1725	0.58966	1.6959	1.3616	0.8029
4	3.3215	0.3011	0.15076	6.6329	0.50076	1.9969	2.2648	1.1341
5	4.4840	0.2230	0.10046	9.9544	0.45046	2.2200	3.1568	1.4220
6	6.0534	0.1652	0.06926	14.4384	0.41926	2.3852	3.9828	1.6698
7	8.1722	0.1224	0.04880	20.4919	0.39880	2.5075	4.7170	1.8811
8	11.0324	0.0906	0.03489	28.6640	0.38489	2.5982	5.3515	2.0597
9	14.8937	0.0671	0.02519	39.6964	0.37519	2.6653	5.8886	2.2094
10	20.1066	0.0497	0.01832	54.5902	0.36832	2.7150	6.3363	2.3338
11	27.1439	0.0368	0.01339	74.6967	0.36339	2.7519	6.7047	2.4364
12	36.6442	0.0273	0.00982	101.8406	0.35982	2.7792	7.0049	2.5205
13	49.4697	0.0202	0.00722	138.4848	0.35722	2.7994	7.2474	2.5889
14	66.7841	0.0150	0.00532	187.9544	0.35532	2.8144	7.4421	2.6443
15	90.1585	0.0111	0.00393	254.7385	0.35393	2.8255	7.5974	2.6889
16	121.7139	0.0082	0.00290	344.8970	0.35290	2.8337	7.7206	2.7246
17	164.3138	0.0061	0.00214	466.6109	0.35214	2.8398	7.8180	2.7530
18	221.8236	0.0045	0.00158	630.9247	0.35158	2.8443	7.8946	2.7756
19	299.4619	0.0033	0.00117	852.7483	0.35117	2.8476	7.9547	2.7935
20	404.2736	0.0025	0.00087	1152.21	0.35087	2.8501	8.0017	2.8075
22	736.7886	0.0014	0.00048	2102.25	0.35048	2.8533	8.0669	2.8272
24	1342.80	0.0007	0.00026	3833.71	0.35026	2.8550	8.1061	2.8393
25	1812.78	0.0006	0.00019	5176.50	0.35019	2.8556	8.1194	2.8433
26	2447.25	0.0004	0.00014	6989.28	0.35014	2.8560	8.1296	2.8465
28	4460.11	0.0002	0.00008	12,740	0.35008	2.8565	8.1435	2.8509
30	8128.55	0.0001	0.00004	23,222	0.35004	2.8568	8.1517	2.8535
32	14,814	0.0001	0.00002	42,324	0.35002	2.8569	8.1565	2.8550
34	26,999		0.00001	77,137	0.35001	2.8570	8.1594	2.8559
35	36,449		0.00001		0.35001	2.8571	8.1603	2.8562

40%				TABLE 28	Discrete Cash Flow: Compound Interest Factors			40%
	Single Payments		Uniform Series Payments				Arithmetic Gradients	
	Compound Amount	Present Worth	Sinking Fund	Compound Amount	Capital Recovery	Present Worth	Gradient Present Worth	Gradient Uniform Series
n	F/P	P/F	A/F	F/A	A/P	P/A	P/G	A/G
1	1.4000	0.7143	1.00000	1.0000	1.40000	0.7143		
2	1.9600	0.5102	0.41667	2.4000	0.81667	1.2245	0.5102	0.4167
3	2.7440	0.3644	0.22936	4.3600	0.62936	1.5889	1.2391	0.7798
4	3.8416	0.2603	0.14077	7.1040	0.54077	1.8492	2.0200	1.0923
5	5.3782	0.1859	0.09136	10.9456	0.49136	2.0352	2.7637	1.3580
6	7.5295	0.1328	0.06126	16.3238	0.46126	2.1680	3.4278	1.5811
7	10.5414	0.0949	0.04192	23.8534	0.44192	2.2628	3.9970	1.7664
8	14.7579	0.0678	0.02907	34.3947	0.42907	2.3306	4.4713	1.9185
9	20.6610	0.0484	0.02034	49.1526	0.42034	2.3790	4.8585	2.0422
10	28.9255	0.0346	0.01432	69.8137	0.41432	2.4136	5.1696	2.1419
11	40.4957	0.0247	0.01013	98.7391	0.41013	2.4383	5.4166	2.2215
12	56.6939	0.0176	0.00718	139.2348	0.40718	2.4559	5.6106	2.2845
13	79.3715	0.0126	0.00510	195.9287	0.40510	2.4685	5.7618	2.3341
14	111.1201	0.0090	0.00363	275.3002	0.40363	2.4775	5.8788	2.3729
15	155.5681	0.0064	0.00259	386.4202	0.40259	2.4839	5.9688	2.4030
16	217.7953	0.0046	0.00185	541.9883	0.40185	2.4885	6.0376	2.4262
17	304.9135	0.0033	0.00132	759.7837	0.40132	2.4918	6.0901	2.4441
18	426.8789	0.0023	0.00094	1064.70	0.40094	2.4941	6.1299	2.4577
19	597.6304	0.0017	0.00067	1491.58	0.40067	2.4958	6.1601	2.4682
20	836.6826	0.0012	0.00048	2089.21	0.40048	2.4970	6.1828	2.4761
22	1639.90	0.0006	0.00024	4097.24	0.40024	2.4985	6.2127	2.4866
24	3214.20	0.0003	0.00012	8033.00	0.40012	2.4992	6.2294	2.4925
25	4499.88	0.0002	0.00009	11,247	0.40009	2.4994	6.2347	2.4944
26	6299.83	0.0002	0.00006	15,747	0.40006	2.4996	6.2387	2.4959
28	12,348	0.0001	0.00003	30,867	0.40003	2.4998	6.2438	2.4977
30	24,201		0.00002	60,501	0.40002	2.4999	6.2466	2.4988
32	47,435		0.00001		0.40001	2.4999	6.2482	2.4993
34	92,972				0.40000	2.5000	6.2490	2.4996
35					0.40000	2.5000	6.2493	2.4997

	Single Payments		Uniform Series Payments				Arithmetic Gradients	
	Compound Amount F/P	Present Worth P/F	Sinking Fund A/F	Compound Amount F/A	Capital Recovery A/P	Present Worth P/A	Gradient Present Worth P/G	Gradient Uniform Series A/G
n								
1	1.5000	0.6667	1.00000	1.0000	1.50000	0.6667		
2	2.2500	0.4444	0.40000	2.5000	0.90000	1.1111	0.4444	0.4000
3	3.3750	0.2963	0.21053	4.7500	0.71053	1.4074	1.0370	0.7368
4	5.0625	0.1975	0.12308	8.1250	0.62308	1.6049	1.6296	1.0154
5	7.5938	0.1317	0.07583	13.1875	0.57583	1.7366	2.1564	1.2417
6	11.3906	0.0878	0.04812	20.7813	0.54812	1.8244	2.5953	1.4226
7	17.0859	0.0585	0.03108	32.1719	0.53108	1.8829	2.9465	1.5648
8	25.6289	0.0390	0.02030	49.2578	0.52030	1.9220	3.2196	1.6752
9	38.4434	0.0260	0.01335	74.8867	0.51335	1.9480	3.4277	1.7596
10	57.6650	0.0173	0.00882	113.3301	0.50882	1.9653	3.5838	1.8235
11	86.4976	0.0116	0.00585	170.9951	0.50585	1.9769	3.6994	1.8713
12	129.7463	0.0077	0.00388	257.4927	0.50388	1.9846	3.7842	1.9068
13	194.6195	0.0051	0.00258	387.2390	0.50258	1.9897	3.8459	1.9329
14	291.9293	0.0034	0.00172	581.8585	0.50172	1.9931	3.8904	1.9519
15	437.8939	0.0023	0.00114	873.7878	0.50114	1.9954	3.9224	1.9657
16	656.8408	0.0015	0.00076	1311.68	0.50076	1.9970	3.9452	1.9756
17	985.2613	0.0010	0.00051	1968.52	0.50051	1.9980	3.9614	1.9827
18	1477.89	0.0007	0.00034	2953.78	0.50034	1.9986	3.9729	1.9878
19	2216.84	0.0005	0.00023	4431.68	0.50023	1.9991	3.9811	1.9914
20	3325.26	0.0003	0.00015	6648.51	0.50015	1.9994	3.9868	1.9940
22	7481.83	0.0001	0.00007	14,962	0.50007	1.9997	3.9936	1.9971
24	16,834	0.0001	0.00003	33,666	0.50003	1.9999	3.9969	1.9986
25	25,251		0.00002	50,500	0.50002	1.9999	3.9979	1.9990
26	37,877		0.00001	75,752	0.50001	1.9999	3.9985	1.9993
28	85,223		0.00001		0.50001	2.0000	3.9993	1.9997
30					0.50000	2.0000	3.9997	1.9998
32					0.50000	2.0000	3.9998	1.9999
34					0.50000	2.0000	3.9999	2.0000
35					0.50000	2.0000	3.9999	2.0000

50%   **TABLE 29**   **Discrete Cash Flow: Compound Interest Factors**   **50%**

INDEX